Drummer / Osswald / Rudolph / Baur
Plastics Handbook

Dietmar Drummer
Tim A. Osswald
Natalie Rudolph
Erwin Baur

Plastics Handbook

The Resource for Plastics Engineers

6th Edition

HANSER

Print-ISBN: 978-1-56990-176-2
E-Book-ISBN: 978-1-56990-373-5

Bibliographic information of the German National Library:
The German National Library lists this publication in the German National Bibliography; detailed bibliographic data are available on the Internet at http://dnb.d-nb.de.

Vilshofener Straße 10 | 81679 Munich | info@hanser.de
www.hanserpublications.com
www.hanser-fachbuch.de
Editor: Dr. Mark Smith
Production Management: Der Buchmacher – Arthur Lenner, Windach
Cover concept: Marc Müller-Bremer, www.rebranding.de, Munich
Cover design: Thomas West
Cover picture: © Tim A. Osswald
Typesetting: Eberl & Koesel Studio, Kempten

Preface

This 6th edition of the Plastics Handbook is a completely reworked version of a reference that traces its origins to the German Saechtling Kunststoff Taschenbuch, first published in 1936 and now in its 32nd edition. Since the previous edition, the German and English versions have been fully synchronized, while continuing to address the needs of their respective markets. In this edition, we have added a new chapter on Material- and Processing-Oriented Design and Simulation to reflect the updates introduced in the latest German edition. The plastics industry continues to evolve rapidly, shaped by major factors such as energy efficiency, the development of new materials, and emerging technologies like additive manufacturing. For that reason, we view the Plastics Handbook as a living project—one that will continue to adapt and grow alongside the industry. We look forward to the innovations the coming years will bring and to the opportunity of incorporating them into future editions. This edition would not have been possible without the invaluable contributions of Nicole Janke, who reworked some of the figures with dedication and precision. We are also grateful to Dr. Mark Smith for his careful review of the manuscript and to Rebecca Wehrmann for her translation and production work, which ensured a result both accurate and visually pleasing. Finally, our deepest thanks go to our families for their unwavering love and support.

Fall 2025

Dietmar Drummer, Tim A. Osswald, and Natalie Rudolph

Contents

1 Common Abbreviations and Acronyms in Plastics Technology

The plastics industry typically categorizes plastic materials by their chemical family and assigns material abbreviations (or acronyms) with respect to this family. While this is common practice, it does not reflect reality in the plastics industry, because the materials are sold under their trade names, each with a very specific property spectrum. In fact, the trade name is the only criterion for identification (similar to an order number). Standards defining the properties of specific material classes, such as are common practice for metals, have been introduced for thermosetting materials only. The materials within one family typically exhibit a wide range of properties.

However, in order to structure our knowledge of these materials, it is necessary to categorize them in a logical and comprehensible way.

We will introduce the common method of assigning abbreviations in this book. Here, a compromise needs to be struck between an unambiguous classification that follows strict rules and the popular notations commonly used. Although there are several standards regarding plastic material notations, they are inconsistent and contradictory, sometimes even within the same standard.

In Section 1.1 the abbreviations of the plastic materials covered in this book are summarized, preferably considering the chemical composition of the polymers and avoiding additional specifications that relate to physical properties or processing technologies. However, in light of the fact that notations such as “A” for amorphous or “B” for block copolymers are often used in the literature, they are used here at times as well. We discourage using them though because they lack general validity and often cause contradictions.

Table 1.1 contains bold listings, indicating that they are standardized. They are based on suggestions made in connection with the plastics database CAMPUS. Here, the current ISO-standards are reflected; however, exceptions are permitted as long as they are widely used. This approach allowed for CAMPUS to define a list of so-called base polymers that covers almost the entire plastic materials market. This list is routinely reviewed and expanded when necessary.

Table 1.1 provides the abbreviations for plastics and rubbers, whereas Table 1.2 provides abbreviations for *plasticizers*.

1.1 Alphabetical List of Plastics Name Abbreviations, Chemical Notation

Table 1.1 Plastics and Rubbers

Abbreviation	Chemical notation	Section
*	Pyrrone	6.16
*	Polycyclone	6.16
*	Polyphenylene (polyarylene)	6.16
*	Polytriazine	6.14.1.4
ABS	Acrylonitrile-butadiene-styrene copolymer	6.3.2
ACM	Acrylate rubber (AEM, ANM)	6.23.4
ACS	Acrylonitrile-chlorinated polyethylene-styrene	6.3.2
AECM	Acrylic ester-ethylene rubber	6.23.4
AEM	Acrylate ethylene polymethylene rubber	6.23.4
AES	Acrylonitrile ethylene propylene diene styrene copolymer	6.3.2
AFMU	Nitroso rubber	6.23.8
AMMA	Acrylonitrile methyl methacrylate	6.6.2.2
APE-CS	see ACS	
ASA	**Acrylonitrile styrene acrylic ester copolymer**	6.3.2
AU	Polyester urethane rubber	6.23.8
BIIR	Bromobutyl rubber	6.23.3
BR	Butadiene rubber	6.23.3
CA	Cellulose acetate	6.18.1
CAB	Cellulose acetobutyrate	6.18.1
CAP	Cellulose acetopropionate	6.18.1

Abbreviation	Chemical notation	Section
CF	Cresol formaldehyde	6.21
CH	Hydrated cellulose, cellulose film	6.18.1
CIIR	Chlorobutyl rubber	6.23.3
CM	Chlorinated polyethylene rubber	6.23.4
CMC	Carboxymethyl cellulose	6.18.1
CN	Cellulose nitrate, celluloid	6.18.1
CO	Epichlorohydrin rubber	6.23.5
COC	**Cyclic polyolefin copolymers**	6.2.4
COP	COC copolymer	6.2.4.6
CP	Cellulose propionate	6.18.1
CR	Chloroprene rubber	6.23.3
CSF	Casein formaldehyde, artificial horn	6.18.3
CSM	Chlorosulfonated polyethylene rubber	6.23.4
CTA	Cellulose triacetate	6.18.1
E/P	**Ethylene propylene copolymer**	6.2.4
EAM	Ethylene vinyl acetate rubber	6.23.4
EAMA	Ethylene acrylic acid ester maleic acid anhydride copolymer	6.2.4
EB	Ethylene butene	6.2.4
EBA	Ethylene butyl acrylate	6.2.4
EC	Ethyl cellulose	6.18.1
ECB	Ethylene copolymer bitumen blend	6.2.4
ECO	Epichlorohydrin rubber	6.23.5
ECTFE	Ethylene chlorotrifluoroethylene	6.2.4
EEA	Ethylene ethyl acrylate copolymer	6.2.4
EIM	Ionomer copolymer	6.2.4
EMA	Ethylene methacrylic acid ester copolymer	6.2.4
EP	**Epoxy resin**	6.21
EP(D)M	see EPDM	
EPDM	Ethylene propylene diene rubber	6.23.4

Table 1.1 Plastics and Rubbers *(continued)*

Abbreviation	Chemical notation	Section
EPM	Ethylene propylene rubber	6.23.4
ET	Polyethylene oxide tetrasulfide rubber	6.23.7
ETER	Epichlorohydrin ethylene oxide rubber (terpolymer)	6.23.5
ETFE	Ethylene tetrafluoroethylene copolymer	6.5.2
EU	Polyether urethane rubber	6.23.8
EVAC	**Ethylene vinyl acetate copolymer**	6.2.4
EVAL	Ethylene vinyl alcohol, old abbreviation EVOH	6.2.4
FEP	Polyfluoroethylene propylene	6.5.2
FF	Furan formaldehyde	6.21
FFKM	Perfluoro rubber	6.23.4
FKM	Fluoro rubber	6.23.4
FPM	Propylene tetrafluoroethylene rubber	6.23.4
FVMQ	Fluorosilicone rubber	6.23.6
HCR	High consistency rubber (high temperature vulcanizing silicone rubber)	6.23.6
HNBR	Hydrated NBR rubber	6.23.3
HTV	see HCR	
ICP	Intrinsically conductive polymers	6.19
IIR	Butyl rubber (CIIR, BIIR)	6.23.3
IR	Isoprene rubber	6.23.3
LCP	**Liquid crystal polymer**	6.15
LSR	Liquid silicone rubber	6.23.6
MABS	**Methyl methacrylate acrylonitrile butadiene styrene**	6.6.2
MBS	**Methacrylate butadiene styrene**	6.6.2
MC	Methylcellulose (cellulose derivate)	6.18.1
MF	**Melamine formaldehyde**	6.21
MFA	Tetrafluoroethylene perfluoromethyl vinyl ether copolymer	6.5.2.7
MMA-EML	Methyl methacrylate-exo-methylene lactone	6.6.3.3

Abbreviation	Chemical notation	Section
MPF	**Melamine phenolic formaldehyde**	6.21
MQ	Polydimethyl silicone rubber	6.23.6
MS	see PMS	
MUF	Melamine urea formaldehyde	6.21
MUPF	Melamine urea phenolic formaldehyde	6.21
NBR	Acrylonitrile butadiene rubber	6.23.3
NCR	Acrylonitrile chloroprene rubber	6.23.3
NR	Natural rubber	6.23.3
PA	Polyamide	6.8
PA 11	**Polyamide from aminoundecanoic acid**	6.8.1
PA 12	**Polyamide from dodecanoic acid**	6.8.1
PA 46	**Polyamide from polytetramethylene adipic acid**	6.8.1
PA 6	**Polyamide from ε-caprolactam**	6.8.1
PA 610	**Polyamide from hexamethylene diamine sebacic acid**	6.8.1
PA 612	**Polyamide from hexamethylene diamine dodecanoic acid**	6.8.1
PA 66	**Polyamide from hexamethylene diamine adipic acid**	6.8.1
PA 69	**Polyamide from hexamethylene diamine azelaic acid**	6.8.1
PAA	Polyacrylic acid ester	6.6.1
PAC	Polyacetylene	6.19
PAEK	**Polyarylether ketone**	6.13.1
PAI	Polyamide imide	6.14.2
PAN	Polyacrylonitrile	6.6.1
PANI	Polyaniline, polyphenylene amine	6.19
PAR	Polyarylate	6.9.3.1
PARI	Polyarylimide	6.14.2
PB	Polybutene	6.2.8
PBA	Polybutyl acrylate	6.6.1
PBI	Polybenzimidazole	6.14.1.3
PBMI	Polybismaleinimide	6.14.1.2

Table 1.1 Plastics and Rubbers *(continued)*

Abbreviation	Chemical notation	Section
PBN	Polybutylene naphthalate	6.9.3.2
PBO	Poly(p-phenylene-2,6-benzobisoxazole)	6.14
PBT	**Polybutylene terephthalate**	6.9.2.2
PC	**Polycarbonate (from bisphenol-A)**	6.9.1.1
PCTFE	Polychlorotrifluoroethylene	6.5.1
PDAP	Polydiallylphthalate resin	6.21.4.5
PDCPD	Polydicyclopentadiene	6.2.9.2
PE	**Polyethylene**	6.2.1
PE-HD	Polyethylene-high density	6.2.1
PE-HMW	Polyethylene-high molecular weight	6.2.1
PE-LD	Polyethylene-low density	6.2.1
PE-LLD	Polyethylene-linear low density	6.2.1
PE-MD	Polyethylene-medium density	6.2.1
PE-UHMW	Polyethylene-ultra high molecular weight	6.2.1
PE-ULD	Polyethylene-ultra low density	6.2.1
PE-VLD	Polyethylene-very low density	6.2.1
PE-X	Polyethylene, cross-linked	6.2.2
PEA	Polyester amide	6.15
PEDT	Polyethylenedioxythiophene	6.19
PEEEK	Polyether ether ether ketone	6.13.1
PEEK	Polyether ether ketone	6.13.1
PEEKEK	Polyether ether ketone ether ketone	6.13.1
PEEKK	Polyether ether ketone ketone	6.13.1
PEI	**Polyetherimide**	6.14.2.2
PEK	Polyether ketone	6.13.1
PEKEKK	Polyether ketone ether ketone ketone	6.13.1
PEKK	Polyether ketone ketone	6.13.1

Abbreviation	Chemical notation	Section
PEN	**Polyethylene naphthalate**	6.9.3
PEOX	Polyethylene oxide	6.12
PESI	Polyester imide	6.15
PES	Polyether sulfone	6.10.2
PESU	see PES	
PET	**Polyethylene terephthalate**	6.9.2.1
PET-G	**Polyethylene terephthalate, glycol modified**	6.9.2.1
PF	**Phenolic formaldehyde resin**	6.21.3.2
PFA	Perfluoroalkoxy	6.5.2.4
PFMT	Polyperfluorotrimethyl triazine rubber	6.23.10
PFU	Polyfuran	6.19
PHA	Polyhydroxyalkanoate	6.18.2
PHB	Polyhydroxybutyrate	6.18.2
PHV	Polyhydroxy valeric acid	6.18.2
PI	**Polyimide**	6.14
PIB	Polyisobutene	6.2.8
PISO	Polyimide sulfone	6.14.2.3
PK	**Polyketone**	6.13.2
PLA	Polylactide	6.18.4
PMA	Polymethylacrylate	6.6.1
PMI	Polymethacrylimide	6.14.2
PMMA	**Polymethylmethacrylate**	6.6.2
PMMI	Polymethacrylmethylimide	6.6.3.1, 6.14.2
PMP	Poly-4-methylpentene-1	6.2.9.1
PMS	Poly-α-methylstyrene	6.3.1.1
PNF	Fluoro-phosphazene rubber	6.23.9
PNR	Polynorbornene rubber	6.23.3
PO	Polypropylene oxide rubber	6.23.5

Table 1.1 Plastics and Rubbers *(continued)*

Abbreviation	Chemical notation	Section
PO	General notation for polyolefins, polyolefin derivatives and copolymers	6.2
POM	**Polyoxymethylene (polyacetal resin, polyformaldehyde)**	6.7
PP	**Polypropylene**	6.2.5
PPA	Polyphthalamide	6.8.3.1
PPB	Polyphenylene butadiene	6.19
PPE	**Polyphenylene ether, old notation PPO**	6.11
PPMS	Poly-para-methylstyrene	6.3.1.1
PPOX	Polypropylene oxide	6.12
PPP	Poly-para-phenylene	6.19
PPS	**Polyphenylene sulfide**	6.10.1
PPSU	**Polyphenylene sulfone**	6.10.2
PPTA	Poly-p-phenylene terephthalamide	6.8
PPV	Polyphenylene vinylene	6.19
PPY	Polypyrrole	6.19
PPYR	Polyparapyridine	6.19
PPYV	Polyparapyridine vinylene	6.19
PS	**Polystyrene**	6.3.1.1
PSAC	Polysaccharide, starch	6.18.1
PSU	**Polysulfone**	6.10.2
PT	Polythiophene	6.19
PTFE	**Polytetrafluoroethylene**	6.5.1
PTHF	Polytetrahydrofuran	6.12
PTT	Polytrimethylene terephthalate	6.9.2.4
PUR	**Polyurethane**	6.17
PVAC	Polyvinyl acetate	6.4.5
PVAL	Polyvinyl alcohol	6.4.5
PVB	Polyvinyl butyral	6.4.5

Abbreviation	Chemical notation	Section
PVC	**Polyvinyl chloride**	6.4.1, 6.4.2
PVDC	Polyvinylidene chloride	6.4.5
PVDF	**Polyvinylidene fluoride**	6.5.1
PVF	Polyvinyl fluoride	6.5.1
PVFM	Polyvinyl formal	6.4.5
PVK	Polyvinyl carbazole	6.4.5
PVME	Polyvinyl methyl ether	6.4.5
PVMQ	Polymethylsiloxane phenyl vinyl rubber	6.22.6
PVP	Polyvinyl pyrrolidone	6.4.5
PZ	Phosphazene rubber with phenoxy groups	6.23.9
RF	Resorcinol formaldehyde resin	6.21, 6.22.1
RTV	Room-temperature-vulcanizing silicone rubber	6.23.6
SAN	**Styrene acrylonitrile copolymer**	6.3.2
SB	**Styrene butadiene**	6.3.2
SBMMA	Styrene butadiene methyl methacrylate	6.3.2
SBR	Styrene butadiene rubber	6.23.3
SBS	Styrene butadiene styrene	6.3.2
SCR	Styrene chloroprene rubber	6.23.3
SEBS	Styrene ethene butene styrene	6.3.2
SEPS	Styrene ethene propene styrene	6.3.2
SI	Silicone, silicone resin	6.21
SIMA	Styrene isoprene maleic acid anhydride	6.3.2
SIR	Styrene isoprene rubber	6.23.3
SIS	Styrene isoprene styrene block copolymer	6.3.2.1
SMAB	Styrene maleic acid anhydride butadiene	6.3.2
SMAH	**Styrene maleic acid anhydride**	6.3.2
TCF	Thiocarbonyldifluoride copolymer rubber	6.23.7
TFEHFPVDF	Tetrafluoroethylene hexafluoropropylene vinylidene fluoride	6.5.2
TFEP	Tetrafluoroethylene hexafluoropropylene	6.5.2

Table 1.1 Plastics and Rubbers *(continued)*

Abbreviation	Chemical notation	Section
TOR	Polyoctenamer	6.23.3
TPA	**Thermoplastic elastomers based on polyamide**	6.20
TPC	**Thermoplastic elastomers based on copolyester**	6.20
TPE	Thermoplastic elastomers	6.20
TPE-A	see TPA	6.20
TPE-C	see TPC	6.20
TPE-O	see TPO	6.20
TPE-S	see TPS	6.20
TPE-U	see TPU	6.20
TPE-V	see TPV	6.20
TPO	**Thermoplastic elastomers based on olefins**	6.20
TPS	**Thermoplastic elastomers based on styrene**	6.20
TPU	**Thermoplastic elastomers based on polyurethane**	6.20
TPV	**Thermoplastic elastomers based on cross-linked rubber**	6.20
TPZ	**Other thermoplastic elastomers**	6.20
UF	Urea formaldehyde resin	6.21
UP	**Unsaturated polyester resin**	6.21
VCE	Vinyl chloride ethylene	6.4.3
VCEMAK	Vinyl chloride ethylene methyl methacrylate	6.4.3
VCEVAC	Vinyl chloride ethylene vinyl acetate	6.4.3
VCMAAN	Vinyl chloride maleic acid anhydride acrylonitrile	6.4.3
VCMAH	Vinyl chloride maleic acid anhydride	6.4.3
VCMAI	Vinyl chloride maleic imide	6.4.3
VCMAK	Vinyl chloride methacrylate	6.4.3
VCMMA	Vinyl chloride methyl methacrylate	6.4.3
VCOAK	Vinyl chloride octylacrylate	6.4.3
VCPAEAN	Vinyl chloride acrylate rubber – acrylonitrile copolymer	6.4.3
VCPE-C	Vinyl chloride-chlorinated ethylene	6.4.3

Abbreviation	Chemical notation	Section
VCVAC	Vinyl chloride vinyl acetate	6.4.3, 6.4.4
VCVDC	Vinyl chloride vinylidene chloride copolymer	6.4.3
VCVDCAN	Vinyl chloride vinylidene chloride acrylonitrile copolymer	6.4.3
VDFHFP	Vinylidene chloride hexafluoropropylene copolymer	6.4.3
VE	Vinyl ester resin	6.21
VF	Vulcanized fiber	6.18.1
VMQ	Polymethylsiloxane vinyl rubber	6.23.6
VU	Vinyl ester urethane	6.21.1.3
XBR	Butadiene rubber, containing carboxylic groups	6.23.3
XCR	Chloroprene rubber, containing carboxylic groups	6.23.3
XF	Xylene formaldehyde resin	6.21.1.1
XNBR	Acrylonitrile butadiene rubber, containing carboxylic groups	6.23.3
XSBR	Styrene butadiene rubber, containing carboxylic groups	6.23.3

* There are no known acronyms for these plastic materials.

Table 1.2 Common Plasticizers

Abbreviation	Chemical notation	Abbreviation	Chemical notation
ASE	Alkylsulfone acid ester	DODP	Dioctyldecyl phthalate
BBP	Benzyl butyl phthalate	DOP, DEHP	Dioctyl phthalate
DBA	Dibutyl adipate	DOS	Dioctyl sebacate
DBP	Dibutyl phthalate	DOZ	Dioctyl azelate
DBS	Dibutyl sebacate	DPCF	Diphenyl cresyl phosphate
DCHP	Dicyclohexyl phthalate	DPOF	Diphenyl octyl phosphate
DEP	Diethyl phthalate	DPP	Dipropyl phthalate
DHXP	Dihexyl phthalate	ELO	Epoxidized linseed oil
DIBP	Diisobutyl phthalate	ESO	Epoxidized soybean oil
DIDP	Diisodecyl phthalate	ODA	Octyl decyl adipate
DINA	Diisononyl adipate	ODP	Octyl decyl phthalate

Table 1.2 Common Plasticizers *(continued)*

Abbreviation	Chemical notation
DMP	Dimethyl phthalate
DMS	Dimethyl sebacate
DNA	Dinonyl adipate
DNODP	Di-n-octyl-n-decyl phthalate
DNOP	Di-n-octyl phthalate
DNP	Dinonyl phthalate
DOA (DEHA)	Dioctyl adipate, also diethylhexyl adipate, DEHA no longer used
PO	Paraffin oil
TBP	Tributyl phosphate
TCEF	Trichloroethyl phosphate
TCF	Tricresyl phosphate
TIOTM	Tri-iso-octyl trimellitate
TOF	Trioctyl phosphate
TPP	Triphenyl phosphate

1.2 Common Units, ISO- and US-Units

Length

	mm	m	km	in	ft	yd	mile
1 mm =	1	10^{-3}	10^{-6}	0.0394	0.0033	–	–
1 m =	10^{3}	1	10^{-3}	39.37	3.281	1.094	–
1 km =	10^{6}	10^{-3}	1	39,370	3281	1094	0.6214
1 inch =	25.40	0.0254	–	1	0.0833	0.0278	–
1 foot =	304.8	0.3048	–	12	1	0.3333	–
1 yard =	914.4	0.9144	–	36	3	1	–
1 statute mile =		1609	1609	–	5280	1760	1

Area

	cm^2	m^2	a	ha	km^2	in^2	ft^2	yd^2
1 cm^2 =	1	10^{-4}	–	–	–	0.155	–	–
1 m^2 =	10^4	1	0.01	10^{-4}	10^{-6}	1550	10.76	1.196
1 a =	–	100	1	0.01	10^{-4}	–	1076	119.6
1 ha =	–	10^4	100	1	0.01	–	–	–
1 km^2 =	–	10^6	10^4	100	1	–	–	–
1 square inch =	6.452	–	–	–	–	1	–	–
1 square foot =	929	0.0929	–	–	–	144	1	0.1111
1 square yard =	8361	0.8361	–	–	–	1256	9	1

Volume

	cm^3	dm^3	m^3	in^3	ft^3	yd^3	gal (US)
1 cm^3 =	1	10^{-3}	10^{-6}	0.061	–	–	–
1 dm^3 [1] =	10^3	1	10^{-3}	61.02	0.0353	–	0.2642
1 m^3 =	10^6	10^3	1	61023	35.31	1.308	264.2
1 cubic inch =	16.39	0.0164	–	1	–	–	–
1 cubic foot =	–	28.32	0.0283	1728	1	0.037	7.841
1 cubic yard =	–	764.6	0.7646	46656	27	1	202
1 gallon (US) =	3785	3.785	–	281	0.1337	–	1

[1] 1 Liter (1) = 1.0 dm^3

Force

	N	dyn^2	kp	
Newton 1 N =	1	10^5	0.101972	
Kilopond[1] 1 kp =	9.80665	980 665	1	
Pound-force 1 lbf =	4.44822		0.4536	kilogram-force (kgf) = kp

[1] Unit no longer approved

Mass

	g	kg	t	oz	lb
1 g =	1	10^{-3}	10^{-5}	0.0353	
1 kg =	10^{3}	1	10^{-3}	35.27	2.205
1 t =	10^{6}	10^{3}	1	–	2205
1 ounce (oz) =	28.35	0.0284	–	1	0.0625
1 pound (lb) =	453.6	0.4536	–	16	1
Metric carat 1 ct = 0.200 g for gemstones. For gold alloys, 1 carat represents $^{1}/_{24}$ mass fraction gold; grain 1 gr = ($^{1}/_{700}$) lb = 0.0648 g					

Pressure

Rounded values. The unit for pressure is Pascal (Pa)

	Pa = N/m^2	bar	kp/cm^2 = at	atm	Torr	lbf/in^2	
1 Pa = 1 N/m^2 =	1	10^{-5}	$10.2 \cdot 10^{-6}$	$9.869 \cdot 10^{-6}$	$7.5 \cdot 10^{-3}$	$145.05 \cdot 10^{-6}$	–
1 bar =	10^{5}	1	1.02	0.987	750	14.505	1 µbar = 1 dyn/cm^2 [1)]
1 kp/cm^2 = 1 at = (techn. atmosph.)[1)]	98 100	0.981	1	0.968	735.5	14.224	1 at = 10 m WS (4 °C)
1 atm = (phys. atmosph.)[1)]	101 325	1.013	1.033	1	760	14.7	1 Torr = 1/760 atm
1 Torr [1)] =	133.32	$1.333 \cdot 10^{-3}$	$1.36 \cdot 10^{-3}$	$1.316 \cdot 10^{-3}$	1	0.01934	1 Torr = 1 mm Hg (0 °C)
1 lbf/in^2 =	6894.8	0.6895	0.0703	0.06804	51.715	1	–

[1)] Unit no longer approved

Energy (work, heat)

	J	kWh	kpm	PSh	kcal	Btu	Details
1 J = 1 Ws = 1 Nm = 1 kg m^2/s^2	1	$277.8 \cdot 10^{-9}$	0.101972	$377.5 \cdot 10^{-9}$	$238.8 \cdot 10^{-6}$	$984 \cdot 10^{-6}$	1 J = 1 Nm = 10^7 erg 1 eV = $1.602 \cdot 10^{-19}$ J
1 kWh =	$3.6 \cdot 10^6$	1	$367 \cdot 10^3$	1.360	859.8	3412	
1 kpm [1)] =	9.80665	$2.724 \cdot 10^{-6}$	1	$3.704 \cdot 10^{-3}$	$2.342 \cdot 10^{-3}$	$9.294 \cdot 10^{-3}$	
1 PSh [1)] =	$2.648 \cdot 10^6$	0.7355	$270 \cdot 10^3$	1	632.4	2509	
1 kcal [1)] =	4186.8	$1.163 \cdot 10^{-3}$	426.9	$1.581 \cdot 10^{-3}$	1	3.968	
1 Btu = (British Thermal Unit)	1055	$293 \cdot 10^{-6}$	107.6	$398.5 \cdot 10^{-6}$	0.252	1	

[1)] Unit no longer approved

Power

	W	kW	kpm/s	PS	hp	
1 W =	1	10^{-3}	0.101972	$1.36 \cdot 10^{-3}$	$1.341 \cdot 10^{-3}$	1 W = 1 J/s = 10^7 erg/s
1 kW =	10^3	1	101.975	1.36	1.341	
1 kpm/s [1)] =	9.80665	$9.80665 \cdot 10^{-3}$	1	0.0133	0.0131	
1 PS [1)] =	735.5	0.7355	75	1	0.986	
1 hp (horsepower) =	745.7	0.746	76.04	1.014	1	1 hp = 550 ft lbf/s

[1)] Unit no longer approved

2 Introduction

2.1 Economic Significance of Plastics

As an independent group of materials alongside metals and ceramic materials, plastics have achieved a very high level of technological and economic importance. The considerable growth of this group of materials has meant that plastics now have a larger production volume worldwide production volume than, for example, steel and aluminum (Figure 2.1).

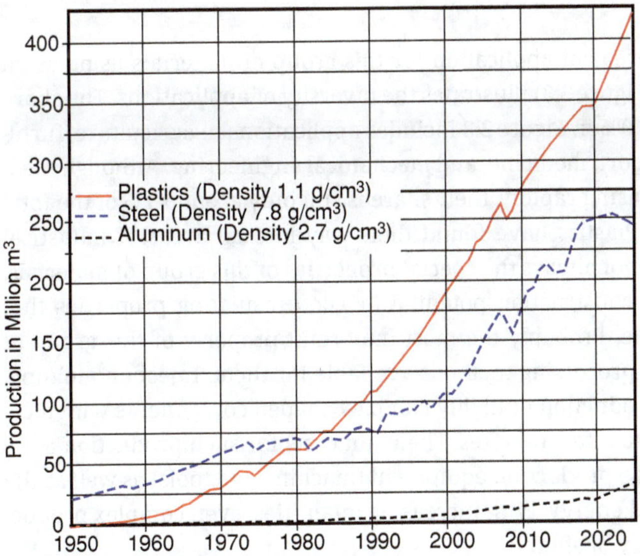

Figure 2.1 Production volume of different materials
[Source: Plastics Europe, World Steel Association, The International Aluminum Institute]

Figure 2.2 shows the rapid growth in plastics production in the various regions of the world between 1990 and 2023. We have indisputably arrived in the "plastics era" in the 21st century. This group of materials is an essential basis for technological development and an indicator of the economic growth of an industrialized society. There is a clear correlation between the production of this material and the economic and technological performance of a region.

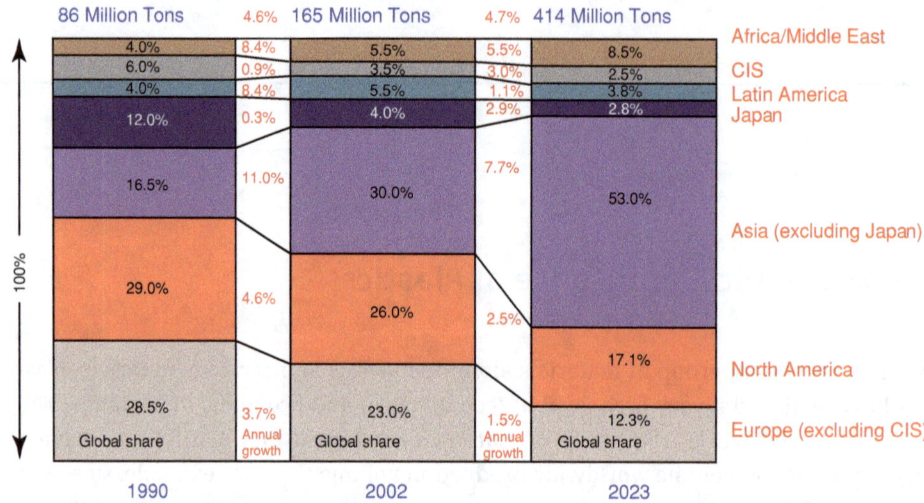

Figure 2.2 World plastics production since 1990
[Source: Plastics Europe]

An analysis of the main areas of application for this group of materials using as an example Europe in 2022 (Figure 2.3) illustrates the diversity of applications. The share of other applications at 24.0% in Figure 2.3 includes applications in agriculture, furniture, household, leisure, sport, medicine, and mechanical engineering. Although medical applications are increasing rapidly, their share is still only around 1% of the total amount of plastics used. Plastics have found their way into all areas of industrial production. However, it is not always the special properties of this group of materials, such as their lightweight construction potential or good insulating properties that make their use worthwhile. Probably the most important property of this group of materials is the economic processing methods available for them. Injection molding, for example, allows the production of highly complex shaped components with a cycle time of a few seconds to a few minutes. The main costs for such production arise from the depreciation of the production equipment (machine and tools) as well as the costs for raw materials and energy. With increasing quantities, even complex production facilities become highly profitable.

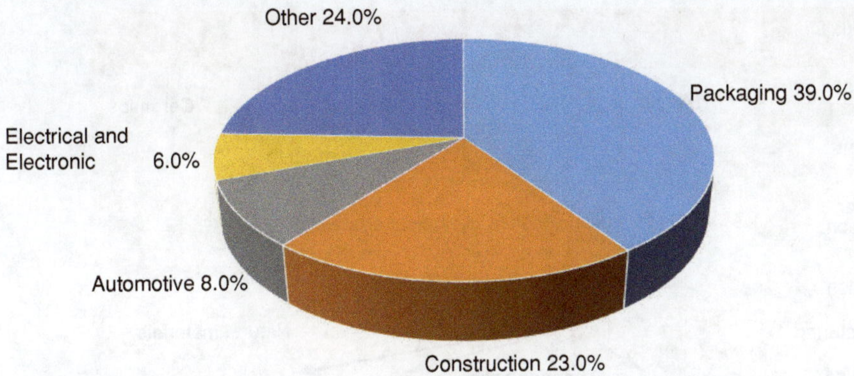

Figure 2.3 Main areas of application for plastics in Europe (2022)
[Source: Plastics Europe]

If we look at the industry working with plastics materials, it can be divided into the following three areas:

- Plastics production,
- Plastics processing, and
- Manufacture of machines for the above.

The plastics industry makes a significant contribution to the well-being of the European economy. More than 1.5 million people work in almost 52,000 companies that generate an estimated annual turnover of 365 billion euros in the EU27+3 countries.

2.2 Classification of Plastics

Polymers are organic or semi-organic materials with high molecular mass (molecular weight), *i. e.*, they are composed of very large molecules (macromolecules), which significantly determine the distinct characteristics of these materials. Figure 2.4 reflects the classification of plastic materials in the general field of material science. Here, distinguishing characteristics are chemical structure, type of polymerization, and the processing and service properties.

In this book, the focus will be on thermoplastic materials, reflecting their economic importance. However, many statements made for thermoplastics are valid for all polymeric materials. A clear definition of the terminology used is missing in many texts, *e. g.*, the terms plastic and polymer are often used synonymously. In our view, the term *polymer* is applicable to all materials with a macromolecular structure, whereas the term *plastics* only describes polymers that are modified with additives to meet the requirements of industrial processing technologies, such as processing aids, stabilizers, pigments, fillers, and others.

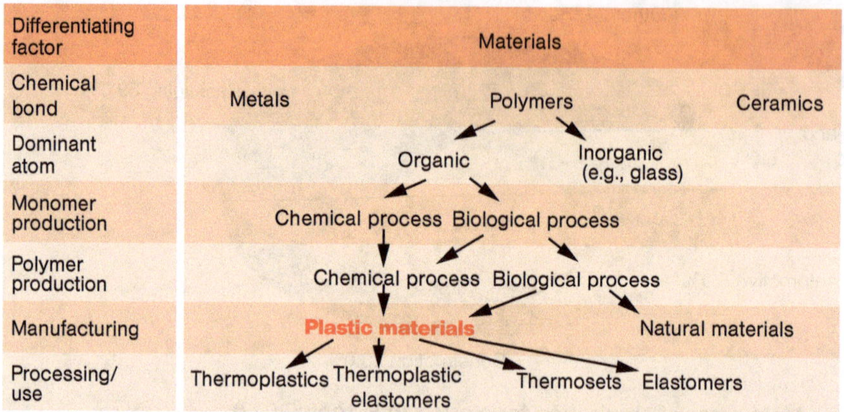

Figure 2.4 Classification of plastic materials in the general field of material science

Despite this clear definition, we were not always able to strictly adhere to this terminology. One reason for this shortcoming is the fact that even in the language of science and in economic statistics the differentiation is not consistent. Even in material standards, the term polymer is often used when in fact they cover plastic materials. In these cases, it is not an option to change the term *polymer* to *plastics*, because that would make it impossible to retrieve the standard.

The following general statements can be made for plastic materials:

- The variety of plastic material classes and types is unparalleled by any other material class. Plastics represent an extremely large span of property profiles, and the slogan of the early days of plastic material development that euphorically declared them "**tailored materials**" has become reality. In almost every area of modern life plastic materials have established themselves as manufacturing materials of choice or as materials with specific functionalities.
- Their complex chemical and morphological structures together with their wide variety in terms of composition and modifiability result in highly complex **material behaviors** that strongly determine service and processing conditions. Examples of properties to be considered include their visco- and entropy elasticity, non-Newtonian flow, complex aging characteristics, semi-crystallinity, liquid crystallinity, orientation- and modification-dependent anisotropy, stress cracking, and many others. A variety of test procedures is necessary to comprehensively describe plastic properties and to provide meaningful characteristic values (single-point data) or property functions (multiple-point data) (see Chapter 3).
- Plastics technology provides a large number of different **processes** for the compounding, processing, and post-processing of plastic materials. The molding and shaping processes play a dominant role because they offer highly productive and energy efficient material utilization within a minimal number of process steps. In

some cases, material shaping and conversion (*e. g.*, tempering, vulcanization) happen simultaneously during processing. The properties of the final product are significantly influenced by the processing conditions. Therefore, process optimization and quality control have to be emphasized appropriately in plastics manufacturing operations (see Chapter 4).

- It is not possible to efficiently **design** plastic components without considering the constraints introduced by material properties and manufacturing processes. This fact, together with the high degree of material and geometry specialization of many manufacturing technologies, requires the consideration of the close interdependence of design, material, and process decisions.
- The manufacturing costs of plastic parts are significantly determined by **material costs**. In addition to economic goals, ecological requirements are increasingly considered during material selection (see Chapter 6). Appropriate material selection and material-saving designs are effective approaches to achieve economic production and ecologically expedient plastic applications. Integration of functionalities also facilitates the reduction of processing steps and thus processing costs. Component design that allows for easy recycling offers additional benefits.

The majority of plastic materials are used in systems in which they perform structural, protective, or separating roles. Here, the mechanical properties of plastics materials play a dominant role.

Plastic materials are increasingly used as functional materials, *e. g.*, they are selected for specific applications because of their optical, electrical, or specific mechanical properties. These materials include UV-curing resins, photo-addressable macromolecules, and electrically switchable polymers. Generally, these materials are referred to as functional polymers (more precisely: functional plastics). They are locally placed in the component, *e. g.*, in layers or punctiform arrangements, and assume specific functions (*e. g.*, data storage). Growing importance is predicted for this group of materials in the future. Because they are currently used only in rather specific applications, these materials will not be covered in this book.

Historically, textiles are not covered in the field of plastics, because textile manufacturing traditionally utilized natural materials only. However, over the course of time, textile technology adapted manufacturing technologies geared towards plastic materials.

Also historically based is the differentiation between thermoplastic and thermoset materials on the one hand and elastomers on the other. However, the development of thermoplastic elastomers revealed that there was no technical rationale for this differentiation

In plastic processing we distinguished between three characteristic phases:

- Compounding,
- Processing,
- Post-processing and finishing.

During the compounding phase, the plastic raw materials or precursors are transformed into processable molding compounds. If a chemical reaction occurs during the processing phase, the molding material is a reactive precursor and fundamentally different from the final molding material. A second processing step is used, if necessary, to shape, machine, or weld semi-finished plastic parts into final products or to assemble components and/or semi-finished parts into finished products. During post-processing the parts may be finished (*e.g.*, by deflashing), surface treated (*e.g.*, by printing, flocking, metal coating), or their material properties may be changed (conditioning, tempering), depending on application requirements. Molding compounds are available in different types and conditions: powder, chips, dry blends, agglomerates, pellets, prepregs, premix, paste (plastisol), solution, suspension, emulsion, and latex, among others.

2.3 Composition of Plastics

A material's properties are defined by its structure. Plastic materials offer the most structural variety of all materials used by man. Both synthetically produced plastics, plastics resulting from material conversion, as well as bio-plastics, such as proteins and carbohydrates, follow the same structural principles. Elementary knowledge of polymeric structure is necessary in order to navigate the field of plastics and rubbers and to assess material properties. In the following, we will present this knowledge in a concise but simple way that does not claim to be complete or go into academic depth. In the subsequent chapters, more detailed information on structure-property relationships will be provided.

2.3.1 Chemical Structure (Constitution and Configuration of Macromolecules)

Polymers are organic macromolecules consisting of many repeating units. They are synthesized from low-molecular units *(monomers)* by different polymerization reactions. The predominant types of chemical bonds in polymers are atomic bonds (covalent bonds). In very few cases we also see ionic bonds (electrovalence). If the concentration of ions results in water-soluble polymers, the ionic polymers are called *polyelectrolytes;* they are called ionomers if the concentration of ionic dissociated bonds leads to non-water-soluble polymers.

The atomic building blocks of polymers are primarily the non-metallic elements carbon (C), hydrogen (H), and oxygen (O). Nitrogen (N), chlorine (Cl), fluorine (F), and sulfur (S) *(heteroatoms)* are also recurrent elements in polymers. The so-called semi-organic polymers contain the metalloid elements silicon (Si) (in silicone and polysiloxane) and boron (B). Although other elementary compositions can sometimes be found in polymers, we will not discuss them in more detail here because of their very specific nature. The structural characteristics of the mentioned elemental building blocks define the properties common to all organic polymers:

- Low electrical conductance (electrical insulators),
- Low thermal conductance (thermal insulators),
- Low specific weight (density 0.8 to 2.2 g/cm^3),
- Limited thermal resistance, because the splitting of covalent bonds is irreversible.

In addition, the properties of final plastic parts and components are significantly influenced by the conformation configuration of the macromolecules.

2.3.1.1 Conformation

The conformation describes the preferential spatial positions of the atoms in a molecule, which is described by the polarity, flexibility, and regularity of the macromolecule. Typically, carbon atoms are tetravalent, which means that they are surrounded by four substituents in a symmetric, tetrahedral geometry.

The arguably best known example of a molecule with tetrahedral geometry is methane, CH_4, schematically shown in Figure 2.5. The figure shows that the tetrahedral arrangement of the substituents (here hydrogen) creates a bond angle of 109.5°. This angle is maintained between the carbon atoms in the macromolecule, as illustrated in Figure 2.5, right. The figure also shows that each axis of the carbon chain is able to freely rotate.

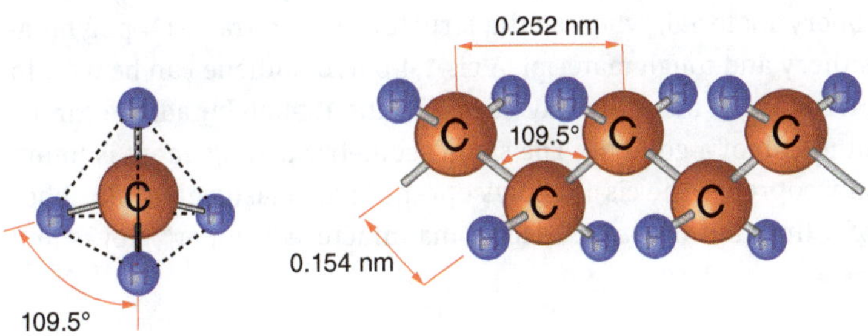

Figure 2.5 Left: schematic illustration of methane (CH_4); right: detail of a polyethylene molecule

2.3.1.2 Configuration

The configuration provides information about the distribution and spatial organization of the molecule. For polymers composed of asymmetric monomers it is possible to place the side groups on the carbon-carbon backbone in different directions during polymerization. The order in which they are arranged is called the tacticity. The polymers whose side groups are all on the same side are called isotactic and are assigned the suffix "I" as in polypropylene: PP-I. The polymers whose side groups are regularly alternating on both sides are called syndiotactic (*e. g.*, PP-S). Irregular arrangement of the side groups is called atactic or random configurations (*e. g.*, PP-R).

Figure 2.6 illustrates the three different tacticity cases for polypropylene. A polymer's tacticity determines the degree of crystallinity that it can reach. For example, polypropylene with a high isotactic content will reach a high degree of crystallinity, and as a result will be stiff, strong, and hard.

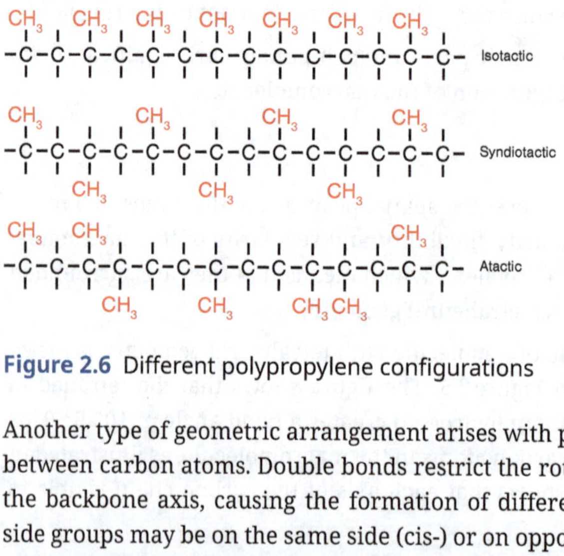

Figure 2.6 Different polypropylene configurations

Another type of geometric arrangement arises with polymers that have double bonds between carbon atoms. Double bonds restrict the rotation of the carbon atoms about the backbone axis, causing the formation of different, so-called stereoisomers. The side groups may be on the same side (cis-) or on opposite sides (trans-) of the chain, as schematically shown in Figure 2.7. The arrangement in a cis-1,4-polybutadiene results in a very elastic rubbery material, whereas the structure of the trans-1,4-polybutadiene results in a leathery and tough material. A cis-1,4-polybutadiene can be used to manufacture the outer tread of an automotive tire. A trans-1,4-polybutadiene can be used to make the outer skin of a golf ball. The same geometric arrangement is found in natural rubber, polyisoprene. The cis-1,4-polyisoprene is the elastic natural rubber used for the body of a tire, and the latex used to manufacture "rubber" gloves and condoms.

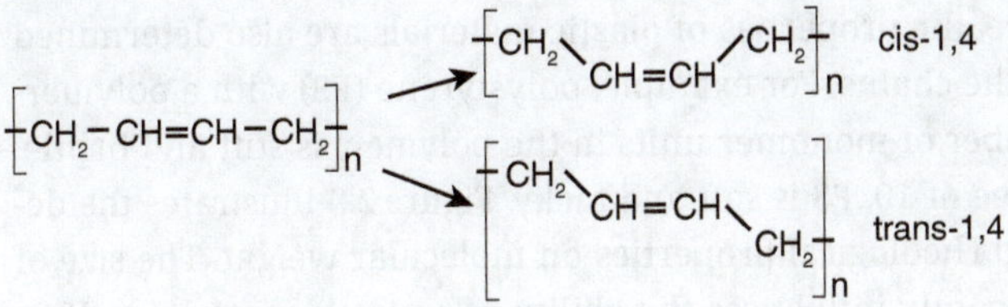

Figure 2.7 Chemical structure of cis-1,4- and trans-1,4-polybutadiene molecules

2.3.1.3 Constitution

Chemical reactions cause chain branching, and the type and length of the branches are determined by the type of polymerization reaction chosen. The chain branches influence the final structure, crystallinity, and thus the properties of the plastic material. Figure 2.8 schematically compares different molecular structures of polyethylene (PE).

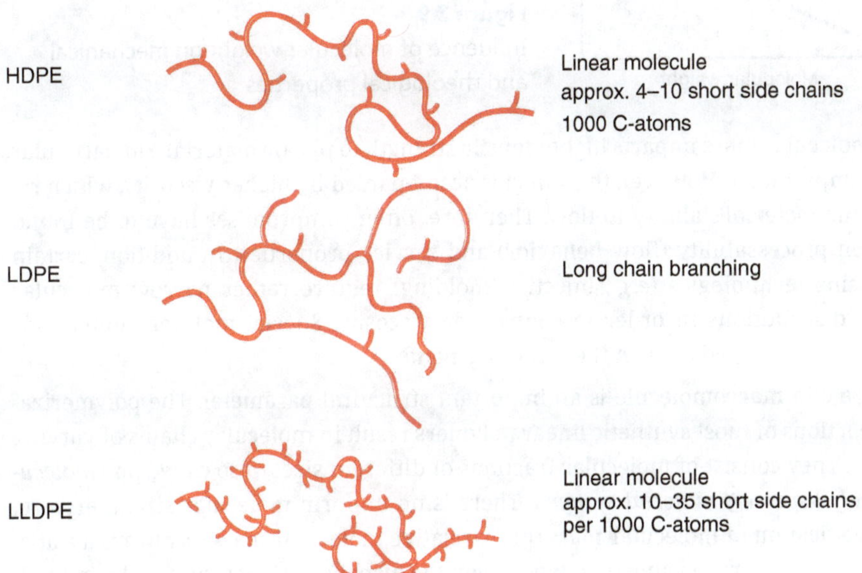

Figure 2.8 Molecular structure of various polyethylenes (schematic)

High density polyethylene (PE-HD) features between 4 and 10 short side chains per 1000 carbon atoms. Low density polyethylene (PE-LD) has the same number of side chains; however, they are significantly longer and often branched themselves. Linear low density polyethylene (PE-LLD) features between 10 and 35 short side chains per 1000 carbon atoms. Molecular chains with less and shorter side chains will crystallize more easily and therefore exhibit higher densities.

Other than by their structure, the properties of plastic materials are also determined by the molecular weight of the chains. For example, polystyrene (PS) with a polymerization degree of 1000 (number of monomer units in the polymer) is stiff and brittle. With a polymerization degree of 10, PS is soft and tacky. Figure 2.9 illustrates the dependence of mechanical and rheological properties on molecular weight. The size of the molecular chains significantly influences the ability of a plastic material to flow in the molten state or in solution. Therefore, there are material characteristic values and functions describing the flow behavior of polymers that are directly correlated to molecular weight.

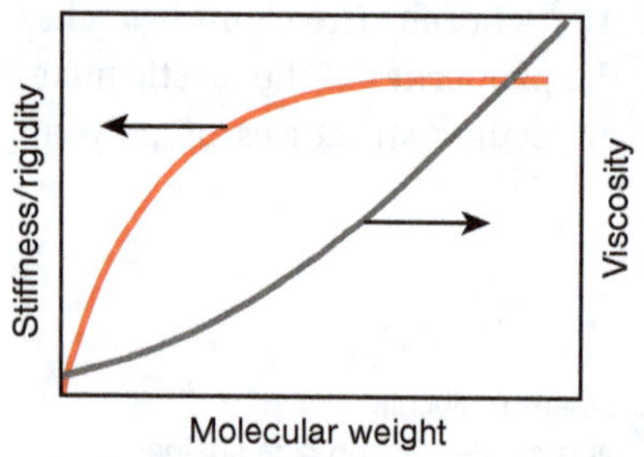

Figure 2.9 Influence of molecular weight on mechanical and rheological properties

High molecular mass imparts higher tensile strength to plastic materials, in particular under impact load. However, this effect is accompanied by higher viscosity, which reduces the material's ability to flow. Therefore, often compromises have to be found between processability (flow behavior) and service properties. In addition, certain processing technologies (*e.g.*, injection molding) require rather narrow molecular weight distributions in order to control the viscosity. Special molding compounds have been developed to meet these requirements.

The size of a macromolecule is an important structural parameter. The polymerization reactions of most synthetic linear polymers result in molecular chains of varying lengths. They consist of molecular fractions of different sizes, also called *polymolecular ("polydisperse") material systems*. There is no uniform molecular size; therefore, the specification of molecular mass (here: relative molecular mass) is always a statistical mean, or average value that depends on the distribution spectrum of the individual molecules and on the way the average value was determined. With the exception of some natural polymers, most polymers exhibit a molecular weight distribution similar to the one shown in Figure 2.10.

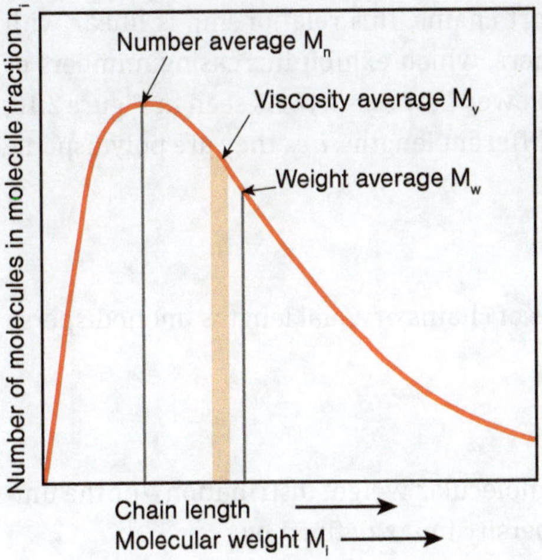

Figure 2.10
Molecular weight distribution of a typical thermoplastic material

For such a distribution function, we can define a number-, weight-, and viscosity average. In practical applications, the weight average (M_w) is more meaningful when investigating property correlations. The number average (M_n) is used when investigating the heterogeneous nature of polymers. It is defined as

$$\overline{M}_n = \frac{\sum m_i}{\sum n_i} = \frac{\sum n_i \cdot M_i}{\sum n_i}$$

Where:
m_i = the total mass of all molecules
M_i = molecular weight
n_i = number of molecules with i repeat units

The weight average (M_w) is defined as

$$\overline{M}_w = \frac{\sum n_i \cdot M_i^2}{\sum n_i \cdot M_i} = \frac{\sum m_i \cdot M_i}{\sum m_i}$$

The viscosity average (M_v) is calculated by

$$\overline{M}_v = \left(\frac{\sum m_i \cdot M_i^{\alpha+1}}{\sum m_i} \right)^{1/\alpha}$$

Where:
α = material constant that relates the *intrinsic viscosity* $[\eta]$ to the molecular weight

This correlation is sometimes identified as the *Mark-Houwink Equation*, written as

$$[\eta] = k\overline{M}_v^{\alpha}$$

For low molecular weight polymers or short chains, this relationship is linear with $\alpha = 1$, and for high molecular weight polymers, which exhibit increasing numbers of entanglements, this relationship follows a Power Law ($\alpha = 3.4$). As seen in Figure 2.10, most plastic materials consist of chains of different lengths, *i. e.*, they are polydisperse and can be described as follows:

$$\overline{M}_w > \overline{M}_v > \overline{M}_n$$

In the few cases in which a polymer consists of chains of equal lengths (monodisperse polymers), we can say

$$\overline{M}_w = \overline{M}_v = \overline{M}_n$$

A measure of the broadness of a polymer's molecular weight distribution – or the uniformity of its chain lengths – is the polydispersity index defined by

$$PI = \frac{\overline{M}_w}{\overline{M}_n} - 1$$

The broadness of the molecular weight distribution significantly influences the mechanical and rheological properties of plastic materials. Plastic material grades with low polydispersity index values typically exhibit higher strength and better flowability. Table 2.1 provides an overview of the polydispersity of various plastic materials.

Table 2.1 Molecular Dispersity as a Function of Polymerization Method

Polymer	$\frac{\overline{M}_w}{\overline{M}_n}$
Polymers with consistent molecular weight	1
Polycondensation and polyaddition products	2
Vinyl polymers	2-5
Highly branched polymers	20–50

What is the molecular mass that defines a polymer? Considering the minimum molecular size at which entanglements in melts and solutions can be detected, the lower limit for the molecular weight of a polymer can be determined as approx. 10^4 g/mol. Molecular entanglements represent temporary physical cross-links that create qualitatively new material properties (*e. g.*, entropy elasticity) not seen in lower molecular materials. An upper limit to the molecular mass of polymers cannot be determined. Plastics used in technical applications have molecular weights ranging from 10^4–10^7, and typically from $5 \cdot 10^4$–$5 \cdot 10^5$ g/mol. Macromolecules with weights in a transitional range of 10^2–10^4 g/mol are also called oligomers.

For a visual representation of the macromolecules of engineering plastics in a disentangled, stretched state, envision 1 mm thick threads with lengths ranging from 2 to 200 m. Specific biopolymers may contain macromolecules with lengths exceeding this by several orders of magnitude.

The different polymerization reactions will be covered in the following. The reactions forming macromolecules are controlled by suitable reaction conditions (temperature, pressure, monomer concentration, reaction medium) and by either activating or inhibiting and structure-controlling additives (catalysts, activators, accelerators, inhibitors). During *polymerization* the monomers connect through a chain reaction by either splitting of multiple bonds or by ring opening to form *polymers*. Here, *polycondensation* reactions are step reactions during which monomers are polymerized by the elimination of reaction products (*e. g.*, water). In *polyaddition* reactions, intermolecular migration of hydrogen atoms without the elimination of any reaction products creates *polyadducts*. An example of this reaction is the urethane reaction. There are a number of other polymerization reactions; however, they will not be discussed in detail. It is not possible to unambiguously relate polymers to specific reaction types, because polymer synthesis may occur in a number of different ways. Polyacetal (POM), for example, can be synthesized by each of the above mentioned reaction mechanisms. Other polymers, such as polyamide (PA), are synthesized in several steps using different reaction mechanisms, such as condensation and addition polymerization.

From a technical point of view, polymerization reactions can be categorized with respect to the reaction medium into bulk-, suspension-, emulsion-, and solution polymerizations. When the type of polymerization reaction is relevant for the material properties of the polymer, an appropriate designation will be added to the polymer acronym.

Appropriate reaction control (*e. g.*, stereo-catalysis; metallocene catalysts) allows tailoring of the structure of engineering polymers, which explains why it is possible for plastic materials with identical elemental compositions to exhibit a wide range of properties. For example, the soft, rubber-like, amorphous, atactic (random) PP-R is of very little practical interest, while the harder, semi-crystalline, isotactic PP-I has found wide-spread applications. The so-called impact-modified polypropylenes are a blend of these two structures.

The technical synthesis of 100% PP-I is not achievable. The atactic content can be determined as a heptane-soluble phase and is declared in form of the “isotacticity index” as a material characteristic.

2.3.1.4 Major Plastic Material Groups

In this section we will introduce the major groups of plastic materials currently in use. In Chapter 6 many of these materials are covered in more detail regarding their properties and processing options.

Linear homopolymers should also be mentioned here; they are either polymerized directly from unsaturated monomers or created by polymer-analogue conversions. The latter are represented by polyvinyl alcohol and by polyvinyl acetals, which are chemically derived from polyvinyl acetate. The many possible combinations of monomers result in the respective variety of copolymers.

Often, these plastic materials are categorized by different attributes. For example, all plastic materials based on olefinic precursors are categorized as polyolefins (PO): PE, PP, PB, PMB, and PIB.

Polyacetals and polyether result when heteroatoms, such as oxygen, are part of the backbone. When additional "bulky" aromatic units are incorporated (polyaryl ether, polyaryl ether ketone), the stiffer macromolecules cause a significant increase in heat distortion temperature of these plastic materials.

In the early days of plastic material development, natural products, such as cellulose, were subjected to conversion reactions that resulted in plastic materials suitable for technical applications, some of which are still (or again) in use today. The stiff cellulose molecule impedes thermoplastic processing so that these materials are blended with plasticizers or other polymers. The high polarity of these blends results in high levels of water absorption, and in some cases even in water solubility, such as in carboxy methyl(ated) cellulose (CMC) used as wallpaper paste.

Polyamides are linear polycondensation products that are widely used as structural materials, fibers, and films, among other applications. Some of these polymers are highly heat resistant because of their hydrogen bridges and aromatic groups; however, they are also hydrophilic and exhibit relatively high water absorption rates. Varying the non-polar methylene groups [—CH_2—] allows the creation of almost all transitional phases between polyolefins and polyamides with high levels of amide group concentration [—NHCO—]. As a result, a variety of copolyamides makes up the wide property palette.

Linear polyesters are also polycondensation products; however, they have gained technical importance only in the form of aromatic polyaryl esters. Specific copolyesters contain extremely stiff molecular segments (mesogens) that render them anisotropic even in the molten state; they represent liquid crystalline, self-reinforced polymers (LCP).

The chemical structure of polyimides provides the highest level of molecular stiffness. Some of these materials, such as PPI, PBI, PBMI, can no longer be processed like thermoplastic materials. They exhibit very high heat aging and heat distortion resistance and are typically used as conductive or semi-conductive polymers.

Polyaryl ether sulfones and polyaryl sulfides represent a group of technically important thermoplastic materials. Their well-balanced properties are based on their chemical structure, in particular the combination of aromatic and sulfuric chain units, which makes them uniquely suitable for structural applications.

All plastic materials described so far are linear polymers that can be processed in their molten state by molding or shaping methods; if they are infusible, they are processed as solutions, dispersions, or powders by coating or sintering technologies. Polymers with double bonds in their macromolecules can be chemically crosslinked to form elastomers (*e.g.*, by sulfur vulcanization), representing typical rubbers. Recently, polycyclic olefins have gained importance as particularly heat resistant thermoplastic materials.

The production of crosslinked plastics (thermosets) or elastomers (rubbers) requires soluble or melt-processable molding materials that are composed of largely linear or partially crosslinked macromolecules or of low-molecular compounds. The actual molding compound is generated only during the crosslinking reaction *(curing, vulcanization)*. The starting materials must contain tri- or higher functional molecules in order to facilitate the crosslinking reaction. Here, the plastics manufacturer needs to control not only the molding operation, but also the chemical synthesis. If it is not possible to initiate the reaction with low molecular, reactive compounds, the use of "prefabricated" macromolecules *(prepolymers)* is an option. These are typically oligomers in terms of their molecular mass. Potential crosslinking reactions include all types of polymerization reactions discussed so far. The most important crosslinked plastics are phenolic plastics (PF), aminoplastics (UF, MF), unsaturated polyester resins (UP), epoxy resins (EP), polyurethane thermosets and -elastomers (PUR), and silicone resins and -elastomers. With appropriate chemical modifications each represents a wide variety of types. A more detailed description of crosslinked plastics can be found in Chapter 6.

2.3.2 Morphological Structure (Conformation and Aggregation of Macromolecules)

Now that we have covered chemical composition and molecular architecture of macromolecules, we will investigate the state of motion and order in macromolecules under varying external conditions, *e. g.*, changes in temperature or deformation under load.

Amorphous structures lack a continuous, long-range order of their molecules. The molecules assume the lowest state of energy in a statistical intertwined coil. Therefore, a certain short-range order is only detectable in a range of approx. 1 nm, and it depends on intermolecular forces and the shape of the molecule. If the intermolecular forces are weakened, *e. g.*, by increasing temperatures or through adsorption of solvent molecules (solvation), micro-Brownian motion is released and all polymers assume this amorphous intertwined state, independently of their specific structure in the solid state.

The amorphous solid state resembles the state of solidified glass and is therefore often referred to as "*glassy state*". Because of their random structure, the characteristic size of the largest ordered region of amorphous materials is on the order of a carbon-carbon bond. This dimension is much smaller than the wavelength of visible light, and thus amorphous thermoplastics are generally transparent. Figure 2.11 shows the shear modulus G' (a measure for the linear elastic deformation of a plastic material) of polystyrene as a function of temperature. Polystyrene is one of the most commonly used amorphous plastic materials. We distinguish two major regions: for low temperatures or in the glassy state the modulus is relatively constant; with increasing temperatures the modulus decreases significantly. At high temperatures, the modulus is negligible – the material is soft enough to flow. Although amorphous materials do not exhibit a sharp transition between the "solid" and the "fluid" state, we identify the temperature range separating these two states as the *glass transition range or glass transition temperature* T_g. For the polystyrene in Figure 2.11 the glass transition temperature is approx. 110 °C. At temperatures below T_g the material behaves like a viscoelastic solid, identified as the *energyelastic zone* in the figure. Once the material is heated above its T_g, it enters the *entropy-elastic zone*, where it can be readily molded, *e. g.*, utilizing thermoforming processes. However, in order for the material to flow, the temperature has to be increased beyond the *softening temperature* T_s. Then the material behaves like a viscoelastic fluid. The softening temperature is also characterized as the point at which the viscous parts of the deformation (the loss modulus) outweigh the elastic parts of the deformation (the storage modulus). For amorphous materials, this zone is typically 50 K above the glass transition temperature. The exact determination of *glass transition* and *softening temperatures* relies on the testing method and also depends on the type and duration of the load during testing. There is a delay between the "load" exerted by the test method and the "molecular response". Therefore, dynamic-mechanical tests with high load cycle frequencies result in significantly higher glass transition temperatures (dynamic glass transition temperature $T_{g\ dyn}$) than calorimetric or dilatometric (elongation) tests with slow temperature changes (static glass transition temperature $T_{g\ stat}$). Depending on type of plastic and level of glass transition temperature, differences of up to 50 K are to be expected. A descriptive way to explain this phenomenon is the statement that plastic materials are the more rigid or stiff the faster the load cycle during testing.

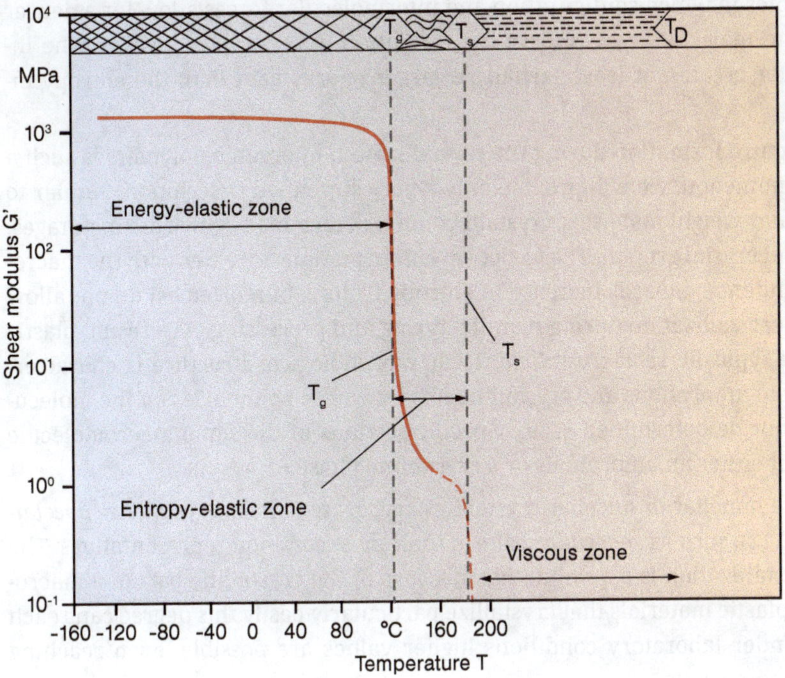

Figure 2.11 Shear modulus of PS as a function of temperature

When plastic materials are exposed to loads or deformations, we observe a time-dependent increase in strain at constant stress *(creep, retardation)* and a time-dependent decrease in stress after being deformed *(relaxation)*. This behavior, also called *viscoelasticity*, is determined by the simultaneous effect of three deformation mechanisms:

- Energy-elastic (also spontaneous-elastic) deformation,
- Entropy-elastic (also rubber-elastic) deformation, and
- Plastic (also viscous) deformation.

The difference in behavior between plastic materials and rubbers is determined only by the gradually different deformation mechanisms in the respective thermal state. The different mechanisms of molecular motion allow for widely different deformations, ranging from approx. 0.1% for energy-elastic deformations, to almost 100% for rubber-elastic deformations, to deformations of up to 10^4% for plastic melts. The *non-Newtonian flow* behavior (*e.g.*, shear thinning) of plastic melts and solutions is also attributed to these deformation mechanics, see Chapter 3.

The amorphous coil structure is a well-suited example to explain the main structure-property relationships encountered in plastic materials. In very uniformly structured macromolecules in which configuration and intermolecular forces allow for a denser packing of the molecules, we observe the formation of *crystallites*, because the increase in order creates, at least partially, a larger energy gain than the entropy increase.

Because structure formation during the crystallization of organic polymers is such a complex phenomenon, we will provide only a very simplified description. Similar to lower molecular weight materials, crystallization is determined by two distinct stages: nucleation and crystal growth. The variety of conformations together with the macromolecules' tendency towards increase in entropy (coiling of molecules) do not allow for complete crystallization during manufacturing and processing. As a result, plastic materials are typically *semi-crystalline*. Their morphological structure is characterized by a mix of amorphous and crystalline phases whose boundaries on the molecular level cannot be determined easily. Specific sections of the same macromolecule may be part of either an amorphous or a crystalline phase.

In general, the lamellar or micellar crystallites aggregate to form *crystalline superlattices* (Figure 2.12), such as *spherulites*, one of their most common representatives. The degree of crystallization is a quantitative measure of the crystalline part in a macromolecule. In plastic materials that crystallize particularly easily this degree can reach 60 to 80%. Under laboratory conditions higher values are possible, even reaching *polymeric mono-crystals*.

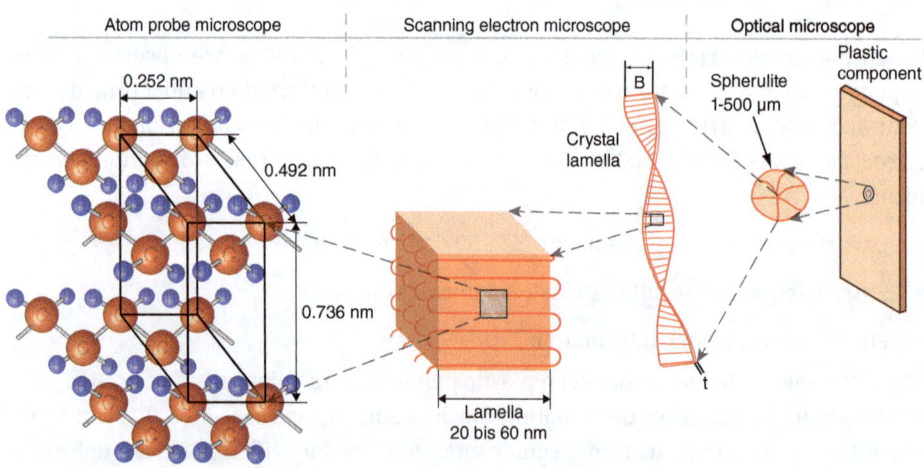

Figure 2.12 Morphology of a semi-crystalline thermoplastic material

Figure 2.13 shows the dynamic shear modulus as a function of temperature for a PE-HD, the most common semi-crystalline plastic material. This curve presents data measured at one test frequency. The figure clearly shows two distinct transitions: one at about −110 °C, the *glass transition temperature*, and one near 140 °C, the melting temperature. Above the *melting temperature*, the shear modulus is negligible and the material will flow. Crystalline arrangement begins to develop as the temperature decreases below the melting point. Between the melting and glass transition temperatures, the material behaves as a leathery solid. Once the temperature decreases below the glass transition temperature, the amorphous regions between the crystalline structures vitrify, resulting in stiff and often also brittle material behavior.

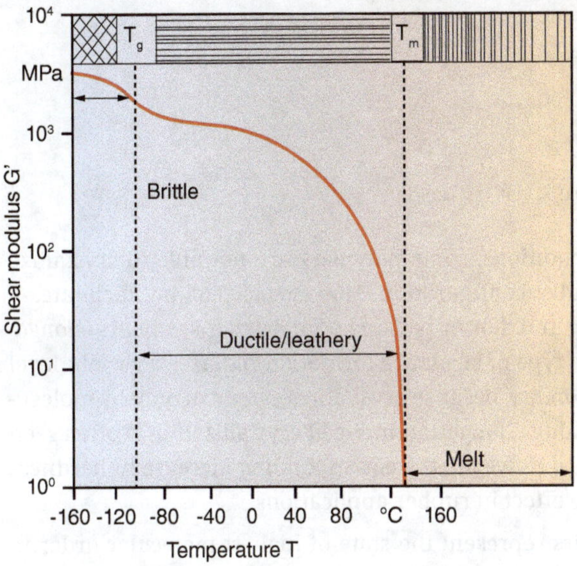

Figure 2.13 Shear modulus of PE-HD as a function of temperature

During processing of plastic materials, inhomogeneous crystal nucleation or differential cooling rates in the material may create locally different crystalline structures, *e. g.*, the fine spherulitic surface layers in injection molded components. This in turn results in anisotropic material properties. The particular characteristics of the molecular structure of semi-crystalline plastics have been tailored in various ways in order to create desired properties:

- Stretching or shearing of macromolecules are used to create molecular rearrangement and the formation of new oriented, crystalline structures, respectively. This results in a significant increase in stability and stiffness, in particular in a desired direction. Examples of such crystalline structures are the fibrils and shish-kebab structures shown in Figure 2.14. A number of processing variations are common in the chemical fiber industry and in plastic engineering, including fiber stretching, biaxial film stretching, and injection- and compression molding stretching. These manipulations result in anisotropic structures with unidirectional or orthogonal orientation.

- Highly homogeneous and fine-grained crystallite structures are obtained by controlling the cooling conditions and/or adding finely dispersed nucleating agents. These structures are advantageous in particular for dimensionally stable and wear resistant components.

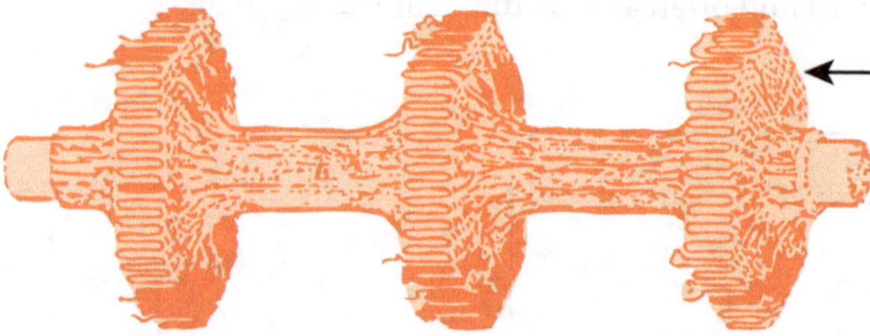

Figure 2.14 Shish-kebab structure

Under common solidification conditions, some polymers are not able to crystallize because of their specific molecular configuration. However, stretching facilitates a partial crystallization by aligning and denser packing of molecular segments *(elongation-induced crystallization)*. This type of structure formation requires a certain level of micro-Brownian molecular mobility, because crystallites can form only by molecular rearrangement during stretching. Elongation-induced crystallization is often seen with rubbers (*e.g.*, NR, IR, IIR, CR). However, the corresponding increase in hardness is not necessarily a welcome side effect in rubber applications.

Crystalline structures in polymers represent the state of highest molecular order in terms of dense molecular aggregation and the corresponding stronger intermolecular forces. Yet, they are still imperfect results of a "frozen" crystallization imbalance so that an increase in temperature will cause melting over a varying temperature interval rather than at a specific melting point. Melting requires additional latent thermal energy in the form of *melting heat*. During rapid solidification, crystallization sets in at temperatures that may be up to 20 to 40 K below the melt temperature range. The use of nucleating agents may decrease this difference. Crystalline regions increase hardness, stiffness, and wear resistance in plastic materials. They also decrease permeation, diffusion, solubility, and swelling. In many respects, the effects of crystallization on the properties of plastic materials are similar to those seen after chemical crosslinking. Therefore, crystalline structures can be described as thermally reversible *physical cross-linking points*.

Extremely stiff macromolecules, *e.g.*, ladder polymers (double-strand polymers with a ladder-like structure), are neither able to form an amorphous coil structure, nor can they be arranged in crystal lattices. The molecules can be compared to rather inflexible rods *(mesogens)* with limited micro-Brownian mobility. If macro-Brownian motion is possible, they form anisotropic parallel structures (model: tree trunks in a river) that are called *mesophases*. Although there are intermolecular forces between the mesogens, their strength is far smaller than those in crystal lattice structures. In other words, this is a quasi-crystal degree of order with a low energy level. In a liquid-like state, the mesophase structure is typically retained, and polymers exhibiting such structures are called *liquid crystal polymers (LCP)*. Because even minimal macro-Brownian molecular motion will cause a unidirectional orientation of the mesogens in the direction of stretching or flow, the result are anisotropic structures with excellent mechanical properties, also called *self-reinforcing polymers*. The mesophase structure is considered an intermediate between amorphous and crystalline structure and such materials are called *mesomorphous*.

Continuous ladder polymers (*e.g.*, aramids) cannot be plasticized by heating; a liquid-like state can only be achieved in solution *(lyotropic LCP)*. In order to obtain *thermotropic* LCPs, the ladder structures have to be combined with flexible molecular segments, so-called *spacers*. In these polymers the mesophase structure is limited to the mesogens. The mesogens in the solid phase are arranged differently than in the melt, which leads to a variety of mesophase structures (*e.g.*, nematic, smectic) that exhibit different orientation and packing symmetries. Mesophase transitions occur at certain temperatures without general changes in the mesomorphous (anisotropic) structure of the LCPs. Above a certain temperature (as long as there is no thermal degradation), the mesophase structure is transformed to an un-ordered, isotropic structure. All these transitions require only low levels of heat of transformation. The spacer segments in the macromolecules may exhibit any kind of polymeric morphology (amorphous, crystalline).

2.3.2.1 Different Classes of Plastic Materials

Figure 2.4 explained the classification of plastic materials in thermoplastic, thermosetting, elastomeric, and thermoplastic elastomeric materials based on a qualitative differentiation of their molecular mobility at room temperature (RT). Here, the identification as a thermoplastic material indicates only the potential, not the imperative for a thermoplastic state. Figure 2.15 relates the classes of plastic materials introduced in Figure 2.4 to their respective macro- and micro-Brownian molecular mobility.

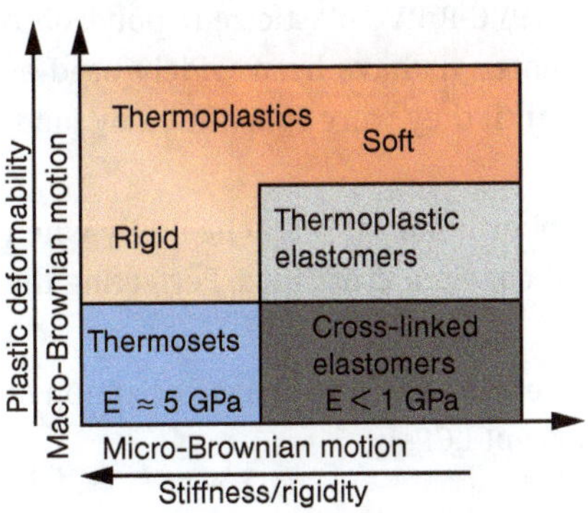

Figure 2.15
Classification of plastics according to their molecular mobility at room temperature (RT)

The materials classified here are end product materials, not resins or precursors. Compared to thermoplastics and thermoplastic elastomers, thermosets and crosslinking elastomers exhibit lower macro-Brownian mobility (free mobility of complete molecular chains: sliding of chains past each other, formation and releasing of entanglements) because of their covalent (chemical) crosslinks. Therefore, plastic deformation of thermosets and rubbers is typically not observed; only incomplete crosslinking allows for a limited degree of plastic deformation.

The differentiation between thermosets and rubbers is a question of convention and typically refers to differences in hardness and stiffness. This approach to a classification is directly related to micro-Brownian molecular mobility (rotational and translational movement around the bond axis of atoms or small chain segments) of the material. Materials covered by this include rigid rubber, flexible EP-, UP-, PUR-X-, SI-X-resins, and others.

Following this categorization, thermosets and thermoplastic elastomers are identified as non-crosslinked plastic materials, *i. e.,* they consist of either linear or branched macromolecules. Only this type of structure allows for a thermoplastic state with highest macro-Brownian molecular mobility. Plastic materials that combine rubber-like properties with the ability to be processed like thermoplastics have gained increasing importance since approx. 1990. This class of materials is identified by the term *thermoplastic elastomers.* The current efforts regarding the nomenclature of these plastic materials are yet another example of the long struggle with classification systems in the plastics technology community. It also sheds light on the fact that the physical composition of thermoplastic elastomers represents a complex problem. In essence, these materials are supposed to be ideal rubber materials that can be processed like thermoplastics. This demand means that the highest possible micro-Brownian molecular mobility has to be combined with sufficient rubber-elastic (entropy-elastic) deformability and low macro-Brownian mobility (little plastic deformation). This requires a loosely crosslinked plastic material with low intermolecular forces and thermally reversible crosslinks. Such *physical crosslinks* can be realized in a number of different ways.

The entanglements of the molecules in thermoplastic materials represent a certain "rubber potential". In fact, soft polyvinyl chloride (PVC-P, PVC plasticized), polyisobutene (PIB), and other soft-elastic thermoplastic materials have been widely used as rubber-like substitutes in a range of applications until they were substituted by modern thermoplastic elastomers.

However, Figure 2.15 also shows that the ideal rubber material is reached only when the physical crosslinks attain the bond strength of covalent crosslinks. Perfecting the ionic structures may offer a promising approach.

In the context of these structural considerations, we will provide a characterization of the most important classes of plastic materials (except LCPs):

- *Amorphous Thermoplastics.* Linear and branched macromolecules with amorphous structure. Thermo-plastically softening (*e.g.*, PMMA molding compounds) or for extremely high molecular mass materials only thermo-elastically softening (*e.g.*, PMMA cast semi-finished products). Generally soluble in suitable solvents. $T_g < RT$ for low intermolecular forces (*e.g.*, non-polar PIB) or when intermolecular forces are decreased by solvation (*e.g.*, addition of plasticizers, PVC-P) and flexible macromolecules. These plastic materials are soft-elastic at RT and increasingly rubber-like with higher molecular mass. Stronger intermolecular forces and stiffer molecules result in $T_g > RT$, *i.e.*, at RT they are rigid-elastic plastic materials. In extreme cases (*e.g.*, ladder polymers) thermal degradation will set in prior to softening.
- *Semi-crystalline Thermoplastics.* General characteristics of the molecules are similar to those of amorphous plastics. In addition, semi-crystallinity causes higher hardness and lower solubility. When $T_g < RT$, the crystalline phase results in horn-like, ductile-rigid behavior at RT. A thermoplastic state is attainable only above T_m. If T_m is much higher than T_g, materials reach a thermoplastic state immediately after the crystallites melt (*e.g.*, polyamides). When this is not the case or for materials with very high molecular mass (*e.g.*, high-molecular polyamides), the material reaches a thermo-elastic state. Extreme reduction of molecular mobility may prevent both thermo-elastic softening (*e.g.*, anhydrous cellulose with $T_g = 225\,°C$) and a thermoplastic state (*e.g.*, PE-UHMW, ultra-high molecular weight PE) prior to degradation. The solubility at RT of such materials may be severely limited, and for some there is no currently known solvent (*e.g.*, polytetrafluoroethylene, PTFE).
- *Thermoplastic Elastomers.* Thermally reversible crosslinked macromolecules with low crosslinking density created by amorphous and semi-crystalline rigid segments (domains) or by ion clusters. Also blends of thermoplastic materials with high contents of crosslinked and non-crosslinked elastomer phases. Thermo-plastically softening and soluble, elastomer-like plastic materials.
- *Crosslinked Elastomers (Rubber).* Thermally irreversibly (covalently) crosslinked macromolecules with predominantly amorphous structure and typically low crosslinking density (low hardness). Thermo-plastic softening not possible. Generally not soluble, but suitable media cause high levels of swelling. Pronounced rubber-elasticity (entropy elasticity) without plastic deformation in service temperature range. Stretching-induced crystallization relatively common.
- *Thermosets.* Thermally irreversibly (covalently) crosslinked macromolecules with amorphous structure and high crosslinking density. Thermo-elastic softening more or less pronounced depending on the strength of the intermolecular forces and crosslinking density. For very high crosslinking densities of no practical interest. Generally not soluble, but depending on the degree of crosslinking more or less swelling in suitable media.

2.4 Effects of Processing on Material Properties

The properties of a plastic material in a manufactured component are determined by the processing and its many variable parameters together with component geometry, material properties, and other factors (Figure 2.16). Because of these typically complex interdependencies, it has become common practice in plastics technology to define "intrinsic characteristics". These characteristics are attributes describing the state of a plastic material, which include:

- Morphology,
- Orientation,
- Residual stresses,
- Crystallization (crystallization parameter),
- Molecular composition (average molecular weight, molecular weight distribution),
- Change in additives/reinforcements (*e.g.*, stabilizer content, fiber length).

The intrinsic characteristics describe the change in properties during use and allow the determination of processing conditions on component properties.

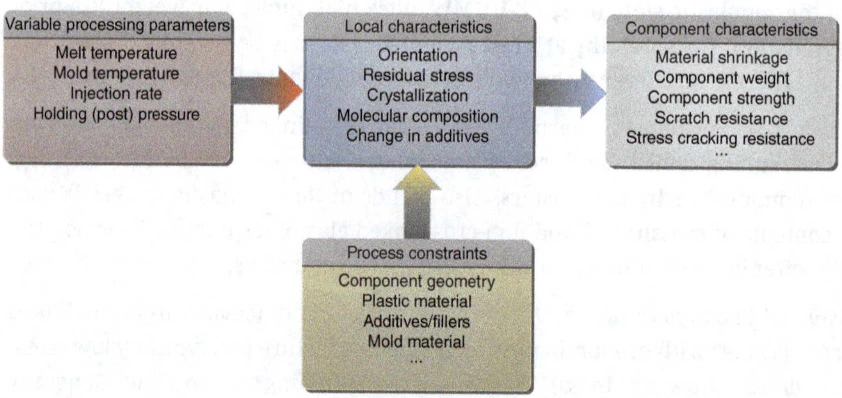

Figure 2.16 Factors influencing intrinsic properties

2.4.1 Residual Stress

Residual stresses are the direct result of pressure and temperature control during the solidification process (Figure 2.17). Internal stress will increase with:

- A heterogeneous temperature distribution in the component,
- High cooling rates,

- High coefficients of thermal expansion, and
- High Young's modulus of the material.

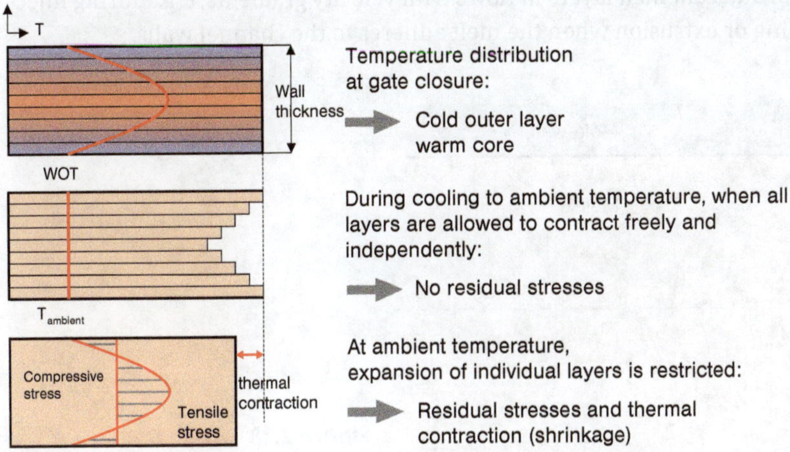

Figure 2.17 Layer model: formation of residual stresses during cooling

Residual stresses represent mechanical loads that act on the component just like external stresses and need to be considered during strength analysis (in a first approximation by reducing the strength parameter).

Differences in coefficients of thermal expansion in composite materials may also cause residual stresses subject to the temperature interval experienced (bi-metal effect). Tempering (heating over a long period of time) to just below the melting temperature and subsequent slow cooling allow for a release of these stresses. In components made from materials that are prone to stress cracking, even small external loads or exposure to crack-inducing media may cause crack formation.

The effects of processing, and in particular those of residual stresses, on the component properties are most pronounced in characteristics such as shrinkage and warpage. They will be covered in more detail in Sections 3.2.2.2 and 3.2.2.3.

2.4.2 Molecular Orientation

In a stationary plastic melt, the molecular chains are in an unordered, coiled state (LCPs are the only exception), while in a moving melt stream the molecules align in a preferred direction, an effect that is also called molecular orientation. Figure 2.18 schematically shows how a circular melt element turns into an oriented element during flow. Molecular orientation is based on two distinct phenomena:

- Local stretching of a melt element, *e. g.*, in a narrowing flow channel or when a melt front expands similar to a balloon being inflated; this effect is also seen during blow molding and thermoforming.
- Shearing of adjacent melt layers in flows with velocity gradients, *e. g.*, during injection molding or extrusion when the melt adheres to the channel walls.

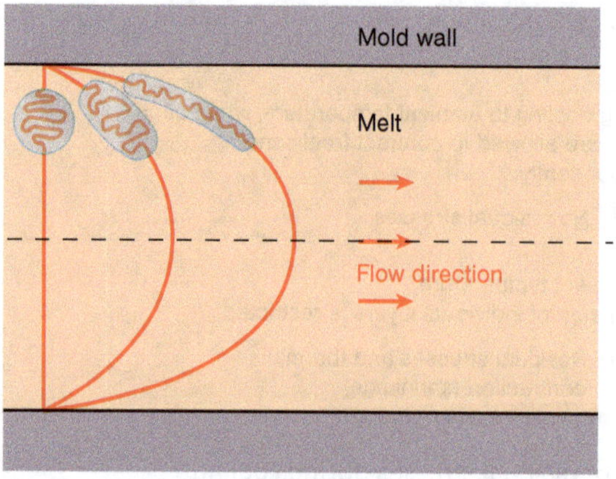

Figure 2.18 Formation of molecular orientation

Determining the level of orientation may be difficult, and in particular for complex components it takes a certain level of experience and the help of simulation programs. Stretching of the flow front and its sudden solidification on contact with the mold wall typically result in a thin, highly bi-axially oriented layer at the component surface. During filling, the high shear differential below the surface results in a layer that is primarily oriented in flow direction. The center of the component wall experiences low shear differentials and thus little orientation.

Because of molecular Brownian motion, orientations in the melt quickly decrease, the molecules relax. Under the conditions prevalent during injection molding, relaxation takes only a few seconds; however, because the mold also cools very fast, orientation is partially frozen in the plastic component. The degree to which orientation is frozen-in depends on the melt temperature, the level of stretching and shear flow, the solidification behavior of the melt, the temperature at the mold wall, and the thickness of the component wall. Although high injection rates during injection molding will cause high levels of orientation, they also result in longer residence times at higher temperatures, which in turn allow for higher levels of relaxation. In consequence, the level of frozen-in orientation may be lower than at slower injection rates. Low melt temperatures, cool molds, and thinner component walls result in higher levels of orientation.

Once at room temperature, the frozen-in orientations will no longer relax. However, if the component experiences elevated service temperatures for longer periods, these orientations may relax, causing shrinkage or warpage, unless this deformation is prevented by suitable constraints, similar to the mold constraint during injection molding, see Sections 3.2.2.2 and 3.2.2.3.

Orientation always results in anisotropic properties: parallel to the direction of orientation strength, stiffness, thermal expansion, and mold shrinkage will increase.

Orientations in transparent materials can be qualitatively evaluated by optical polarization methods, because oriented regions are birefringent, see Section 3.6. X-ray diffraction analysis is particularly suited for a quantitative evaluation of semi-crystalline plastic materials. Measuring shrinkage is another quantitative approach. Here, complete components, sections, or microtome cuts are stored at elevated temperatures so that orientations can relax and cause deformations. The degree of deformation is a measure for the frozen-in orientation.

2.4.3 Crystallization Behavior

The intrinsic properties of semi-crystalline thermoplastic materials are also determined by the level of crystallinity. The degree of crystallinity together with general structural properties (*e. g.*, the diameter of the spherulites) are crystallinity characteristics.

- The higher the degree of crystallinity, the stiffer, stronger, but also the more brittle the component. The density of the plastic material is directly correlated to its degree of crystallinity. Low degrees of crystallinity or very fine crystal structures result in opaque to transparent parts. The degree of crystallinity is determined by the chemical structure of the plastic material, the type of nucleation, the processing technique, and also potential heat treatment (tempering). Lower tool temperatures and high rates of cooling result in low degrees of crystallinity and fine crystalline structures. A typical example is bottles made from PET. They are cooled so rapidly that the normally white material solidifies into transparent bottles. The same effect is seen for components with thin walls and on the surfaces of components. Long melt residence times at high temperatures result in high degrees of crystallization and coarser crystalline structures, which will occur in particular in the interior of thick-walled components. In injection molded parts, heat is conducted through the component surface to the tool. The low thermal conductivity of plastic materials causes a temperature gradient in the component, which in turn results in a layer-specific degree of crystallinity, Figure 2.19.
- Post-crystallization is observed with extended use of molded components, in particular at elevated temperatures. It causes a decrease in volume that may cause warpage or clearance problems. Tempering or annealing, *i. e.*, exposure to elevated temperatures for a certain period, may preempt post shrinkage. During processing, the spherulites can be aligned by shearing – similar to the macromolecules –

and subsequently cause anisotropic behavior of the component. The higher the degree of order of the macromolecules in the melt (orientation), the easier they are able to crystallize (shear-induced crystallization).

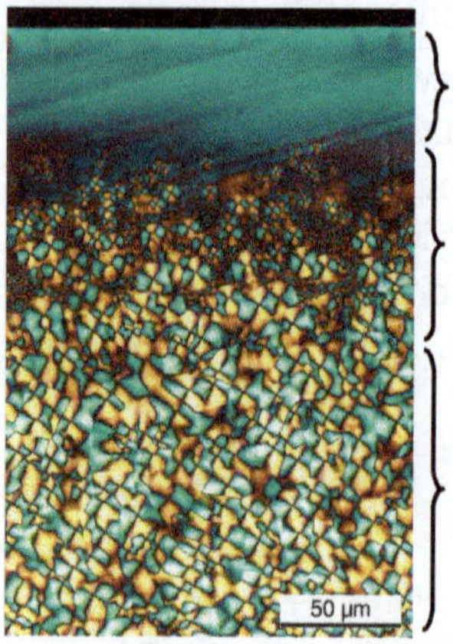

Figure 2.19 Structure of an injection molded component made of PA 66 [1)]

The crystalline portion in a given polymeric material can be identified by X-ray diffraction, differential scanning calorimetry (DSC), or by nuclear resonance spectroscopy. When a calibration curve is available, the direct relationship between density and the degree of crystallinity allows the inference of crystallinity from density.

2.5 Modifications of Plastic Materials

In the context covered here, material modifications will include blending of homopolymers and the addition of fillers and other additives for a variety of specific functions. This field is rather complex and we will give only a general overview here. More detailed information is provided in the following chapters.

2.5.1 Copolymers and Polymer Blends

Copolymers are plastic materials containing two or more types of monomers within their chain. A copolymer consisting of two different monomers is also called a *bi-polymer*. Copolymers consisting of three different monomer units are called *ter-polymers*.

---

1) C. M. Dallner, M. O. Kobes, R. W. Feulner, E. Schmachtenberg: Influence of the morphology on the tribological behaviour of assembly injection moulded Microdrives, PPS-23, May 27–31, 2007, Salvador, Brazil (2007)

Depending on the arrangement of the different monomer units within the chain, we distinguish between four different types of copolymers: *statistical*, alternating, *block-* and *graft-copolymers*. Ethylene-propylene is a typical example of a copolymer. Although both monomers are semi-crystalline thermoplastics, blends with ratios between 35/65 und 65/35 do not exhibit a melting temperature: the material is an elastomer. These so-called EDM-materials (ethylene propylene-diene rubbers) are increasingly utilized for industrial applications because of their good weather resistance. In contrast, all ethylene-propylene block copolymers exhibit a melting temperature, independent of their blend ratios.

Plastic mixtures, blends, and alloys are created by mixing or blending two or more plastic materials. The goal is to create a material with improved properties compared to those of its components. The most common polymer blends include PP-PC (polycarbonate), PVC-ABS (acrylonitrile butadiene styrene), and PE-PTFE. The miscibility of polymers is generally rather limited (disadvantageous energy states), and on a molecular level homogeneous mixtures (blends) are an exception.

The formation of a morphological multi-phase system may be advantageous, because many useful properties of the initial components can be retained. A prerequisite for high-value blends is a stable blend morphology. Dispersed phase and matrix must be uniformly distributed and exhibit good phase adhesion. With current state-of-the-art mixing and compounding technology "phase droplets" below the wavelength of visible light can be achieved (transparent blends). A variety of methods and technologies is used to improve miscibility and adhesion:

- Matching of initial compatibility,
- Matching of quantitative composition and compounding temperature control,
- Chemical modification by ionic bonds, dynamic crosslinking of dispersed phases, partial graft copolymerization and others. Here, the reactions may take place during blending,
- Addition of compatibilizers, and
- Formation of interpenetrating networks (IPN) for crosslinked polymer structures, which result in particularly good mechanical properties.

Originally, blending was used to improve processability and achieve impact modification, in particular for thermoforming applications, typically using rubbers as blend components (elastomer modified plastics). Today, polymer blends are used to improve any given property, including increasing the heat distortion temperature, decreasing stress crack formation, improving coated and plated surfaces, and reducing flammability. Utilizing blend technologies for thermoplastic elastomers should be particularly noted. The increasing trend for material recycling has proven to be an additional incentive for blend technology.

2.5.2 Plastic Composites

Plastic composites represent a form of property and/or material modification that can be classified by the type of bonding:

- Multi-layer composites and
- Reinforced composites.

Multi-layer composites consist of geometrically distinct layers of different materials. They include *multi-layer films, sheets, pipes, hoses, and sandwich designs* consisting of light plastic, honey comb, or homogenous cores with relatively thin-walled surface layers, *multi-component parts* manufactured by multi-component and/or multi-color injection molding, back-injection molding, and by *insert- and outsert methods*. Multi-layer structures allow the development of composite films with clearly defined permeation behavior and of light yet very rigid sandwich designs for aeronautical and automotive applications. They also facilitate a variety of other property combinations with regard to strength, stiffness, gas permeability, flammability, decorative appearance, thermal insulation, ease of assembly, among others. Multi-layer composites can be formed by the combination of different plastic materials as well as by combinations with other materials (metals, wood, etc.). A technological focus of this material modification is to achieve sufficient adhesion between the layers using chemical and/or mechanical bonding principles.

Reinforced composites consist of a polymer matrix into which fillers have been incorporated for reinforcement or to improve specific properties. The distribution of these fillers determines whether the composites exhibit isotropic or anisotropic behavior. Typical examples are *high-performance fiber composites* made from high-tensile and/or high-modulus fibers or textiles and a plastic matrix. In addition, this class of materials also includes plastic/concrete composites, laminated materials (fabric-, paper-, and glass mat based laminates), long-fiber reinforced thermoplastics, and the wide range of filled and reinforced thermosets and laminates. Chapter 6 will discuss these materials in more detail.

However, plastic components reflect the excellent mechanical properties of fiber-reinforced materials only if they contain high ratios of well-ordered fibers, which requires special, high-cost manufacturing methods (see Chapter 4).

The majority of plastic materials is shaped/molded using flow processes. To ensure the flowability of the plastic material, the reinforcing fibers must not exceed certain lengths. The type and ratio of reinforcing material determines the applicable manufacturing process (see Figure 2.20). Plastic manufacturing processes suitable for large-scale production typically allow for fiber lengths of approx. 10 mm. It should be noted that fiber lengths typically decrease during manufacturing so that the fiber length in the component is always smaller than in the original molding compound.

	Injection molding	Compression molding	Resin injection and transfer molding, autoclave processes
Manufacturing process	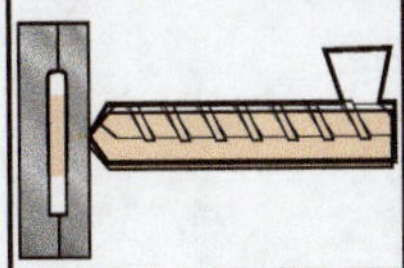	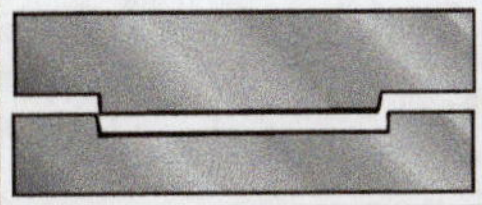	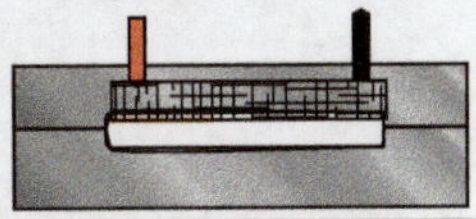
Pellets	Pellets	Sheet and bulk molding compounds, long fiber reinforced thermoplastics	Textile semi-finished products, preforms, prepregs
Characteristics	Short fibers: L < 4.5 mm Φ < 25%	Short fibers: 4.5 < L < 40 mm Φ < 25%	Continuous fibers: L ≈ component size Φ > 30%
Fiber orientation in component		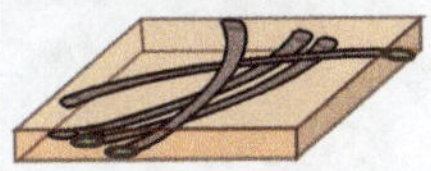	

← Suitable for mass production   Mechanical properties →

Figure 2.20 Correlation between manufacturing process and achievable fiber length

In general, a wide range of fillers can be incorporated into plastic materials in order to tailor the material's properties. Figure 2.21 shows matrix reinforcement (*i. e.*, the change in Young's modulus as a result of the filler) as a function of type, shape, and orientation of the filler particles at different filler contents.

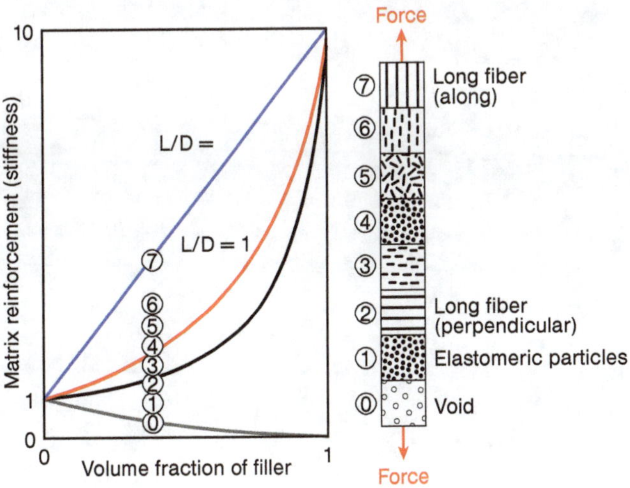

Figure 2.21 Relationship between stiffness and type, shape, and orientation of filler particles

3 Material Properties and Testing Methods

In this chapter we will introduce the characteristic properties of plastics and the testing methods required to determine them. It is our goal to develop and improve a general understanding of this group of materials. Our descriptions will be general and qualitative and the diagrams displaying material characteristics are meant to help gain insight into the behavior of this group of materials. Some tables and figures provide a comparison of selected plastics; however, these comparisons are also only means to foster a general understanding.

It should be noted that plastics are almost always modified by functional or other additives. The properties of commercially available plastic materials may vary from the "average values" provided here. Before selecting a specific plastic material, it is recommended to verify properties through the manufacturer. Another highly recommended source of material property information is CAMPUS, a free database provided by 30 plastic material producers. This database contains material data for materials under their respective tradenames. Both space and timeliness are prohibitive factors for a printed work to provide the plethora of information available. More detailed information regarding the CAMPUS database and free downloads can be found at *www.CAMPUSplastics.com*.

Figure 3.1 compares the stress/strain diagrams of unfilled PBT grades to exemplify the wide variation of plastic characteristics (data according to CAMPUS). Large differences in property profiles can be found even within one plastic class.

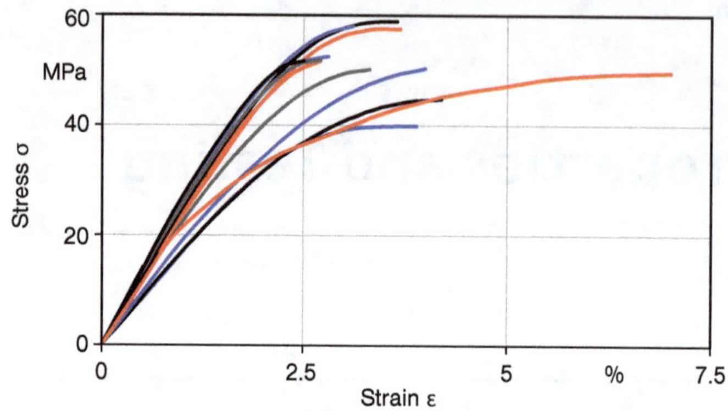

Figure 3.1 Range of stress-strain diagrams for PBT at 23 °C

3.1 Significance of Characteristic Values

Material characteristics are the determining factors during all phases of development and design. They affect the initial, conceptual design steps, material selection, and dimensioning of molded components. In order to meet these requirements, material characteristics have to be

- comparable,
- meaningful, and
- determined by a rational process.

Databases and material data sheets provide up to 200 different characteristics of a product. This large amount of information makes it difficult if not impossible to gain and maintain a comprehensive overview.

In order to select meaningful characteristics here, we refer to a globally respected source: ISO 10350: "Plastics – Acquisition and presentation of comparable single-point data, molding materials" (see Table 3.1). The reference to this standard will also determine the order in which we will describe specific characteristics in this chapter. Another important standard is ISO 11403: "Plastics – Acquisition and presentation of comparable multipoint data, mechanical properties" (see Table 3.2).

Table 3.1 CAMPUS Characteristic Data Catalog According to ISO 10350

Property	Symbol	ISO-(IEC)-standard	Specimen (in mm)	Unit	Notes
Rheological characteristics					
Melt volume rate	MVR	1133	Compound	cm^3/10 min	
Mold shrinkage *parallel*	S_{Mp}	294-4 (thermoplastics) 2577 (thermosets)	60 · 60 · 2	%	See ISO 294-3, thermoplastics and ISO 10724-2, thermosets
Mold shrinkage *normal*	S_{Mn}				
Mechanical characteristics, 23 °C					
Tensile modulus	E_t	527-1/2	ISO 3167 (multi-purpose specimen)	MPa	Elongation 0.05 to 0.25%
Yield stress	σ_y			MPa	
Yield strain	ε_y			%	
Nominal strain at break	ε_{tB}			%	After yield
Stress at 50% strain	σ_{50}			MPa	For specimens without σ_y up to $\varepsilon = 50\%$
Stress at break	σ_B			MPa	For specimen without yield
Strain at break	ε_B			%	For specimen without yield
Tensile creep modulus: 1 h	E_{tc} 1	899-1		MPa	Elongation 0.5%
Tensile creep modulus: 1000 h	E_{tc} 10^3				
Charpy impact strength, unnotched at +23 °C	a_{cU} +23	179/1eU	80 · 10 · 4 (general purpose specimen)	kJ/m^2	
at −30 °C	a_{cU} −30				
Charpy impact strength, notched at +23 °C	a_{cA} +23	179/1eA			
at −30 °C	a_{cA} −30				

Table 3.1 CAMPUS Characteristic Data Catalog According to ISO 10350 *(continued)*

Property	Symbol	ISO-(IEC)-standard	Specimen (in mm)	Unit	Notes
Tensile impact strength at +23 °C	a_t 1	8256/1	80 · 10 · 4 (multi-purpose specimen) with double V-notch	kJ/m^2	When Charpy impact strength cannot be determined
Puncture impact properties at +23 and −30 °C	F_M	6603-2	60 · 60 · 2	N	
Max. force at +23 and −30 °C	W_P			J	
Thermal characteristics					
Melting temperature	T_m	11357-1 and -3	Material	°C	Temperature increase 10 K/min
Glass transition temperature	T_g	11357-1 and -2			
Temperature of deflection under load at 1.8 MPa	T_f 1.8	75-1/2	80 · 10 · 4 (multi-purpose specimen)	°C	Select 1.8 MPa and a second load level
at 0.45 MPa	T_f 0.45				
at 8.0 MPa	T_f 8.0				
Vicat softening temperature	T_v 50/50	306	≥ 10 · 10 · 4 (multi-purpose specimen)	°C	Temperature increase 10 K/min, load increase 50 N
Coefficient of linear thermal expansion, parallel	α_p	11359-1/2	Center of 80 · 10 · 4 (multi-purpose specimen)	E^{-6}/K	Increase in elongation between 23 and 55 °C
Normal	α_n				
Burning behavior, 1.5 mm thickness	B50/1.5 B500/1.5	60695-11-10	125 · 13 · 1.5 ≥ 150 · 150 · 1.5	Class	Classes for B50: V-0, V-1, V-2, HB Classes for B500: 5VA, 5VB, N
-.- mm thickness	B50/-.- B500/-.-	60695-11-10/-20	125 · 13 · 1.5 ≥ 150 · 150 · 1.5		
Oxygen index	OI23	4589-1/2	80 · 10 · 4 (multi-purpose specimen)	%	Method A: Ignition at upper surface

Property	Symbol	ISO-(IEC)-standard	Specimen (in mm)	Unit	Notes
Electrical characteristics					
Relative permittivity 100 Hz 1 MHz	 $\varepsilon_r 100$ $\varepsilon_r 1M$	IEC 60250	60 · 60 · 2 (60 · 60 · 1)	E^{-4}	
Dissipation factor 100 Hz 1 MHz	 tan δ100 tan δ1M				
Specific volume resistivity	ρ_e	IEC 60093	60 · 60 · 2 (60 · 60 · 1)	Ω · m	1-minute values
Specific surface resistivity	σ_e			W	1-minute values, 500 V
Electric strength	$E_B 1$	IEC 60243-1	≥ 60 · ≥ 60 · 1	kV/mm	Voltage increase 2 kV/s
Comparative tracking index	CTI	IEC 60112	≥ 15 · ≥ 15 · 4 (multi-purpose specimen)		Testing medium A
Other characteristics, 23 °C					
Water absorption	W_W	62 and 15512	Thickness ≥ 1	%	Saturation value in water, 23 °C
Moisture absorption	W_H				Saturation value at 23 °C/50% r.h.
Density	ρ	ISO 1183	(Multi-purpose specimen)	kg/m^3	
Resin-specific characteristics					
Viscosity number	VN		Material	cm^3/g	
Characteristic density	ρ_I	1872-1		kg/m^3	Only for PE
Luminous transmission index	τ_t	13468-1/ -2	60 · 60 · 2	%	
Processing conditions for testing specimens					

Table 3.1 CAMPUS Characteristic Data Catalog According to ISO 10350 *(continued)*

Property	Symbol	ISO-(IEC)-standard	Specimen (in mm)	Unit	Notes
Injection molding of thermoplastics					Where: P = Thermoplastics S = Thermosets I = Injection molding C = Compression molding R = Cooling rate M = Molding $_c$ = Cure
Compound temperature	T_M IP	294-1 and 3		°C	
Mold temperature	T_C P				
Injection rate	v_I P			mm/s	
Holding pressure	P_H			MPa	
Injection molding of thermosets					
Temperature of the plasticized compound	T_M IS	10724		°C	
Mold temperature	T_C S				
Injection rate	v_I S			mm/s	
Curing time	t_C I			min	
Compression molding of thermoplastics					
Mold temperature	T_MCP	293		°C	
Cooling rate	R_C			K/min	
Compression time	t_M			min	
Demolding temperature	T_D			°C	
Compression molding of thermosets					
Mold temperature	T_MCS	295		°C	
Curing time	t_CC			min	

These two standards form the basis for the international database of plastic characteristics (CAMPUS). It provides the characteristic values for specific products according to Table 3.1 and Table 3.2.

Table 3.2 CAMPUS Characteristic Diagrams According to ISO 11403

Diagrams	Parameter	ISO Standard
Viscosity/Shear rate	Temperature	11443
Shear modulus/Temperature		6721-1,2,7
Stress/Strain	Temperature	527-1,2
Secant modulus/Strain	Temperature	
Stress/Strain (isochronous) T1 to T6	Temperature, time	899-1
Creep modulus/Time T1 to T6	Stress	
Specific volume/Temperature (p-v-T)	Pressure	17744
Specific enthalpy per mass/Temperature (DSC)		11357-1 and 4

Each characteristic value is determined using standardized testing methods that are defined by ISO standards. The manufacture of testing specimens also follows standardized rules, which is important to ensure that the values are comparable, in particular because the characteristic values depend strongly on the processing conditions during injection molding of the specimens. Although the CAMPUS database was designed predominantly for thermoplastics, the data can also be applied to elastomers, thermoplastic elastomers, and thermosets.

In the following we will provide more detail with regard to the characteristics used to evaluate the behavior of plastics during processing. This includes the molding of components, their intended use, and component overload. In general, it is important to note that characteristic material values do not represent universal material constants. Similar to characteristics of other materials they depend on the following influencing factors:

- Temperature,
- Duration and rate of load,
- Load frequency,
- Environmental influences,
- Material inhomogeneities, such as flow lines, notches, etc., and
- Manufacturing conditions of components and the resulting shrinkage and warpage.

The following sections will provide specific examples.

3.2 Processing Characteristics

Thermoplastic and thermoset components are manufactured by molding at elevated temperatures. For thermoplastics this process can be repeated several times. Thermosets, on the other hand, cure during molding and cannot be fused repeatedly. Cast resin components, also considered thermosetting materials, are manufactured using special processes, see Section 4.6.

The manufacturing process determines both properties and costs of a molded component. Therefore, knowledge of the following properties is important for material selection and preparation for production.

3.2.1 Rheological Behavior

The number and arrangement of gates in an injection mold is determined by the flowability of the plastic melt at processing temperature. Together with the material's solidification behavior it also determines the achievable wall thickness of the component. The minimal wall thickness is a measure for the required material input. **Viscosity** is the measure for the flowability of a material. The rheology of plastics is a complex field because of the numerous influencing factors (plastic melts are shear thinning, visco-elastic, and flow behavior is temperature and pressure dependent). Shear viscosity plays a determining role in the majority of manufacturing processes. Therefore, viscosity is generally measured by devices in which the melt flows under shear load.

3.2.1.1 Flow Behavior, Viscosity

The rheological behavior of thermoplastics is described by flow curves that are determined by capillary rheometers. They characterize the correlation between shear viscosity in Pa · s and the shear rate in s^{-1}.

The shear viscosity is a measure for the relation between shear stress and shear rate in a flowing substance and thus a measure for the resistance of the melt to flow. Figure 3.2 shows the change in viscosity with increasing shear rate for different thermoplastics. At low shear rates, plastics exhibit Newtonian behavior, *i. e.*, the viscosity is independent on the shear rate and exhibits a so-called Newtonian plateau. This is also referred to as zero-viscosity. It is possible to reach a second Newtonian plateau at very high shear rates. In other words, here the viscosity is again shear rate independent. This phenomenon can be explained by the fact that at high shear rates, all molecular chain entanglements are relaxed and the chains cannot slip any further.

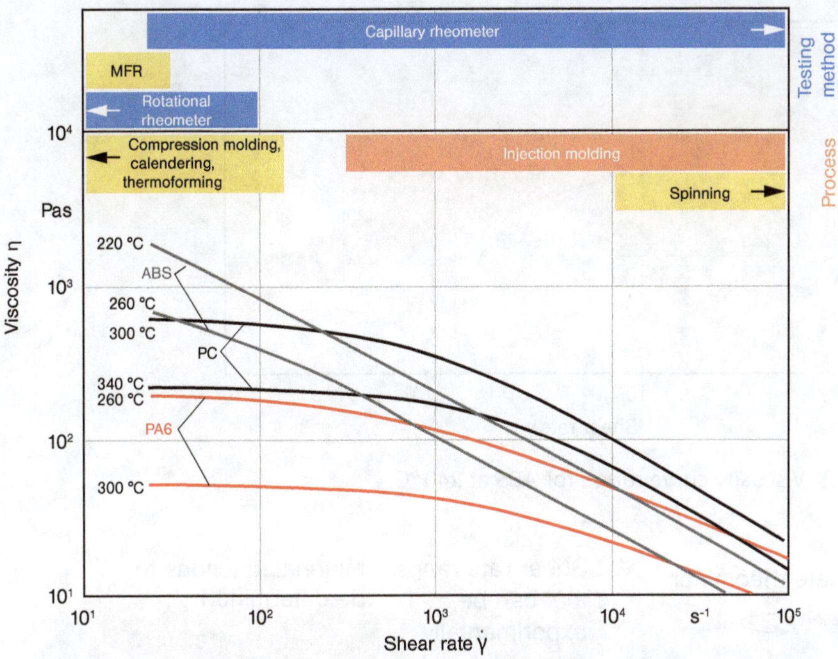

Figure 3.2 Melt viscosity of selected thermoplastics at normal upper and lower permissible melt temperature, respectively

Temperature also influences viscosity: with decreasing temperature the viscosity will increase. Figure 3.2 also shows the shear rate ranges encountered in different processing technologies and identifies the testing methods used for their determination. Figure 3.3 shows the order of magnitude of possible variations in the viscosity curves for ABS at constant temperature. Flow path calculations based on these viscosity curves resulted in variations of up to 30%. This example shows that caution is warranted when operating with averaged values for rheological data. They should be used exclusively for comparisons between material classes – never as the basis for design or dimensioning.

Figure 3.4 shows the two most commonly used rheometer designs and their areas of application. Rotational rheometers, such as the depicted cone-plate system, facilitate measurements at low shear rates. These typically complex measurements allow for the fundamental analysis of rheological phenomena, such as principal stress differences and the derivation of rheological models, because they detect changes on the molecular level. Shear viscosity, which is a processing relevant parameter, is typically measured using capillary rheometers. It is determined at constant temperature by measuring the material exiting a capillary at a specific pressure.

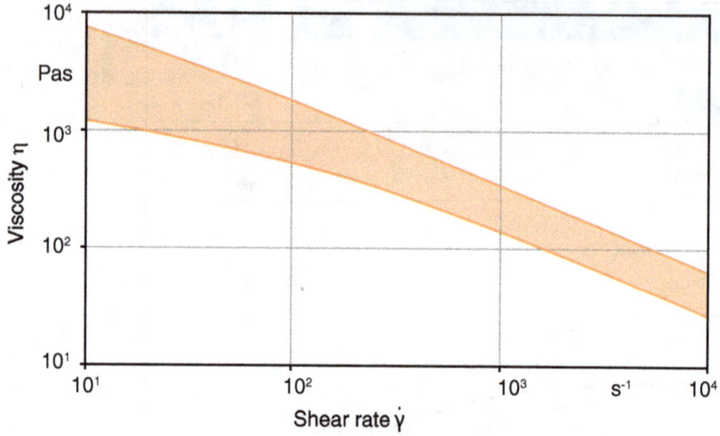

Figure 3.3 Viscosity curve range for ABS at 240 °C

	Shear rate range that can be experimentally measured	Material functions to be determined
Cone-plate rheometer 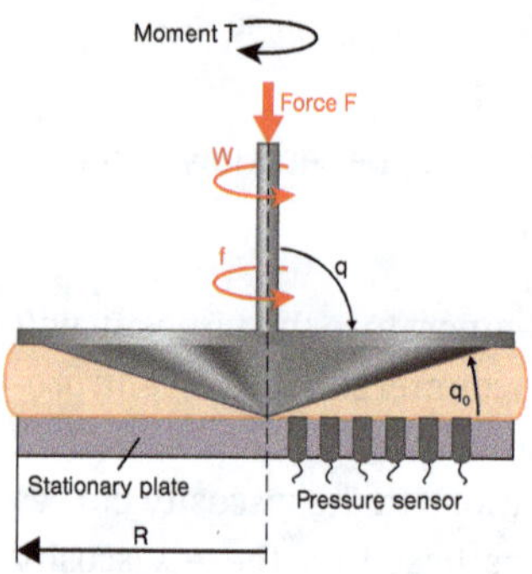	$(10^{-6}$ to $10^{2})s^{-1}$	Rotating operation: $\tau, \eta = f(T, p, \dot{\gamma}, t)$ Oscillating operation: $\eta', \eta''; G', G''$
Capillary rheometer 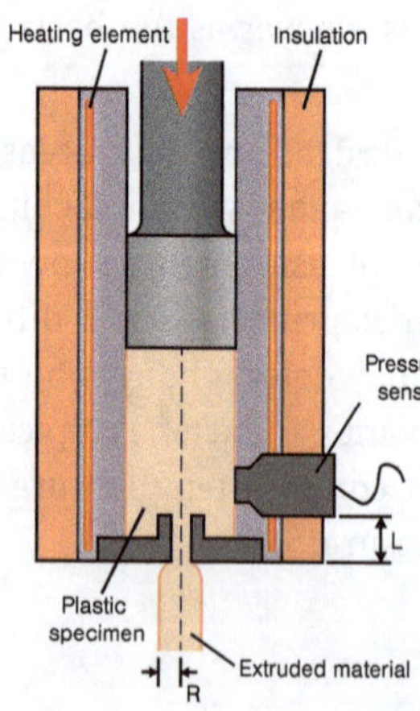	$(10^{1}$ to $10^{6})s^{-1}$	$\tau, \eta, \sigma = f(p, T, \gamma)$

Symbol	Meaning	Unit
T	= Temperature	°C
t	= Time	s
p	= Pressure	Pa
σ	= Tensile stress	Pa
τ	= Shear stress	Pa
γ	= Shearing	m
γ̇	= Shear rate	$m \cdot s^{-1}$
η	= Viscosity	Pa · s
G	= Shear modulus	Pa
η'; G'	= Elastic part	Pa · s
h''; G''	= Viscous part	Pa · s

Figure 3.4 Rheometer types

Online- and inline rheometers help monitor both injection molding and extrusion processes. While in the past partial flows were sampled from the volume flow (bypass principle), today viscosity is measured directly in the main volume flow. In order to do this, pressure and temperature sensors are placed along the flow length that allow to calculate the viscosity taking into consideration the time of contact with the flow front, the pressure drop, and the geometric dimensions of the cavity. However, this approach is possible only at the shear rate predetermined by the process.

Figure 3.5 shows the flow curves of different plastics with Figure 3.5 a) showing the viscosity curves of polyolefins with different flow behaviors. The easy flowing grades exhibit lower viscosities. Many plastics contain additives, such as flow promoters or color pigments, which affect processing behavior as well as product properties (*e. g.*, color). As seen in Figure 3.5 b), the addition of white pigments (TiO_2) increases the viscosity of the compound [1]. The data also indicate that this effect is more pronounced for particles with larger diameters. The increase in viscosity caused by the addition of fillers is also dependent on the viscosity of the plastic itself: the lower the viscosity, the higher the increase in viscosity. This is further illustrated in Figure 3.5 c) in the case of an easy flowing PE-LD [1]. The viscosity of plastic grades that are offered by the manufacturer already compounded with color pigments typically matches that of the base (uncolored) grade. For master batches however, the processing conditions have to be adjusted accordingly by the processor. Figure 3.5 d) illustrates the effect of fillers with even larger particles, such as fibers or the copper platelets shown here, as a function of temperature [2]. At high temperatures the filler content has less influence than at lower temperatures.

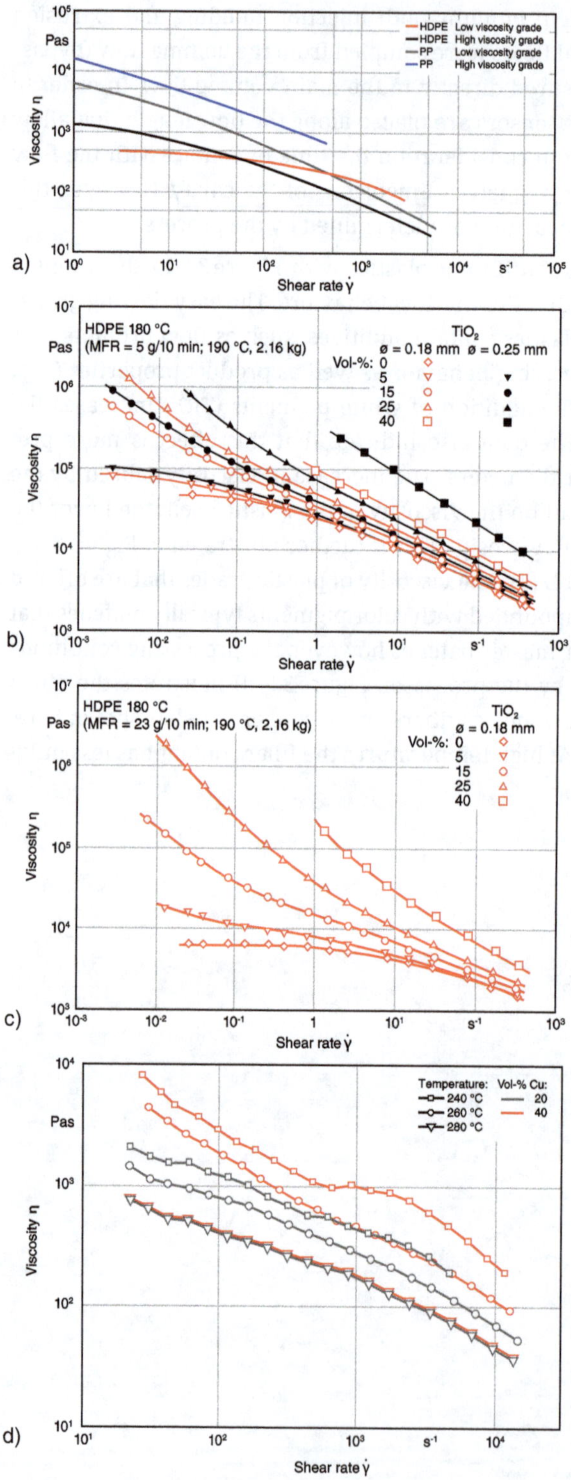

Figure 3.5
Flow curves of selected plastics

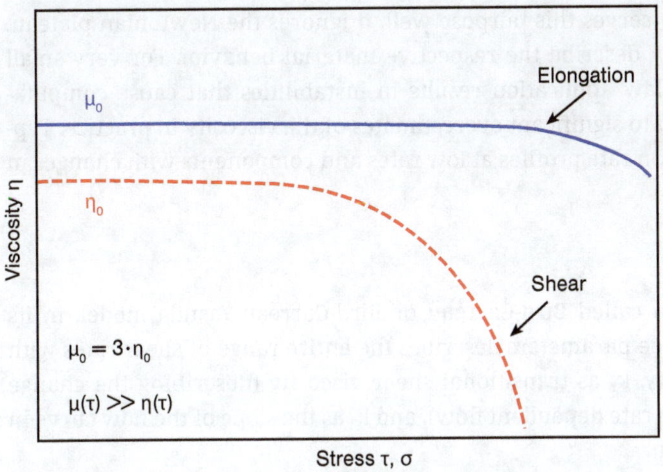

Figure 3.6 Schematic representation of the difference between shear and elongational viscosity

Elongational Viscosity

In plastic processes, such as blow molding, fiber spinning, thermoforming, foaming, compression molding, as well as in certain extrusion dies, the major mode of deformation is elongational. In addition, there are a number of processes in which mixed forms of shear and elongational flow can be detected. Figure 3.6 illustrates the differences between shear and elongational viscosities. At equal stress the elongational viscosity is significantly higher than the shear viscosity. Close to the Newtonian plateau it exceeds the shear viscosity by a factor of 3. The ratio of both viscosities is often used as a rule of thumb for the determination of the elongational viscosity, because the actual measurement is very difficult and thus only few data are available.

3.2.1.2 Viscosity Models

In order to take flow behavior into consideration during the design of plastic components and their respective injection molds or extrusion dies, the correlations between viscosity, shear rate, and temperature are described by constitutive equations. The most important such models will be detailed in the following.

Power Law

Ostwald and *de Waele* suggested to apply the power law (general: $y = a \cdot x^b$) to describe the viscosity drop in the shear thinning region:

$$\eta = m(T)\dot{\gamma}^{n-1} \tag{3.1}$$

Although the power law serves this purpose well, it ignores the Newtonian plateau. The parameters m and n describe the respective material behavior. For very small shear rates, the power law application results in instabilities that cause computational problems and lead to significant overestimates of the viscosity in practice. Typical examples are injection rate profiles at low rates and components with changes in wall thickness.

Carreau Model

The Carreau model, also called Bird-Carreau or Bird-Carreau-Yasuda model, in its simplified form with three parameters describes the entire range of shear rates with k_1 as zero shear viscosity, k_2 as transitional shear viscosity (describing the change from Newtonian to shear rate dependent flow), and k_3 as the slope of the flow curve in the shear thinning region.

$$\eta\,(\dot{\gamma}) = \frac{k_1}{[1 + k_2\dot{\gamma}]^{k_3}} \tag{3.2}$$

This model can be used over a wide range of shear rates because it also covers the shear rate-independent behavior at low shear rates. In its more complex variation, the model may also be used to represent the second Newtonian plateau. At high shear rates the completely stretched molecular chains also exhibit shear-independent behavior. Given the fact that this case is not relevant in industrial practice, simulation uses the model in the form presented in Equation 3.2.

Cross Model

Another model that is able to represent the transition from zero shear to shear thinning viscosity is the Cross model:

$$\eta\,(\dot{\gamma}) = \frac{\eta_0}{1 + \left[\frac{\eta_0}{\tau}\dot{\gamma}\right]^{1-n}} \tag{3.3}$$

Where:
η_0 = zero shear viscosity
τ = transition to the shear thinning region
n = flow exponent

Both models are used interchangeably in common simulation programs.

Temperature Dependency

The fact that viscosity is dependent on temperature is taken into consideration by two different approaches. For semi-crystalline plastics at high processing temperatures ($T > T_g + 100\,°C$), the Arrhenius approach is used:

$$a_T(T) = \exp\left(\frac{E_0}{R}\left(\frac{1}{T} - \frac{1}{T_0}\right)\right) \tag{3.4}$$

Where:
E_0 = activation energy
R = universal gas constant
T_0 = reference temperature

For amorphous thermoplastics and lower temperature ranges between processing temperature and T_g, the Williams-Landel-Ferry (WLF) equation is used:

$$a_T(T) = \exp\left(\frac{-C_1(T - T_0)}{C_2 + T - T_0}\right) \tag{3.5}$$

Where:
C_1 and C_2 =material constants
T_0 = reference temperature

In both models, the shift factors are multiplied with the respective values of the zero shear viscosity. Both models also recognize the pressure dependency of the viscosity, a factor that is the particular focus of scientific research. However, because there are only few experimental characteristic data available, the pressure shift factor is typically not considered.

3.2.1.3 Melt Volume Flow Rate (MVR) and Melt Mass Flow Rate (MFR)

The melt volume flow rate (MVR) is used as a comparative index for the flow behavior of a melt; however, it is not suitable for its calculation. It is defined as the volume of melt exiting a capillary with defined dimensions within 10 minutes (cm^3/10 min) under given temperature and pressure conditions (Figure 3.7). Also commonly used is the melt mass flow rate, also called melt flow index, which is determined under the same conditions, but measured in g/10 min. The respective values can be easily converted using the melt density. When comparing such data, it is important to make sure they were determined under the same conditions, *i. e.*, under the same temperature and pressure.

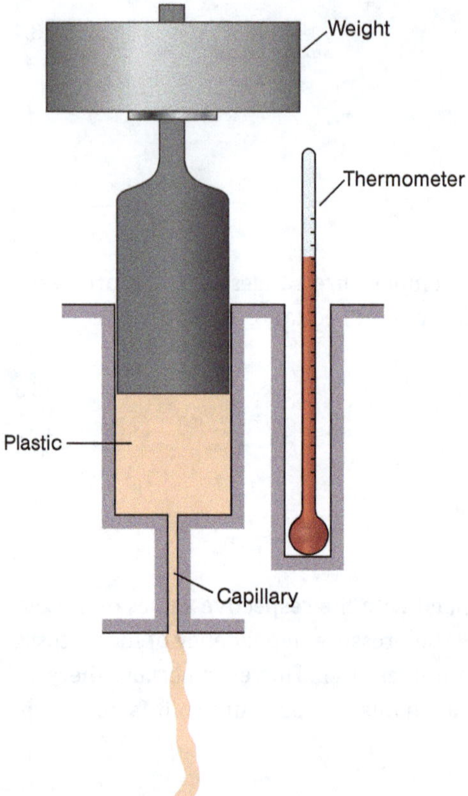

Figure 3.7
Schematic of melt volume flow measurement (flow indexer)

3.2.2 Solidification Behavior

3.2.2.1 Sealing Time

In industrial practice, the time until a molded component is solidified is determined by weight. Here, a series of tests is used to determine the duration of holding time after which the component weight no longer increases. It is not easy to determine the sealing time with the help of these tests because close to the sealing time the component weight increases only incrementally as a function of holding pressure.

3.2.2.2 Shrinkage

Molded plastic components are expected to have the required dimensions (within a certain tolerance range) and to be warpage free after completion of the processing cycle. In addition, these dimensions should be maintained through the component's service life and under the respective service conditions. Often, point values are used as reference points or in order to compare the warpage of different materials; however, sometimes a range of values is also provided, see Table 8.1.

The dimensions of molded plastic components are smaller than those of the molds used to manufacture them. This effect is caused by the differences in thermal and pressure dependent expansion and contraction behavior between plastics and steel molds. Pressure (p) – volume (v) – temperature (T) diagrams are used to describe this behavior of thermoplastic materials (see also Section 3.4.6 for a detailed description of pvT-behavior). These diagrams can also be used to monitor the effect of the thermodynamic process of shrinkage during the injection molding process. Figure 3.8 shows the change in specific volume as a function of pressure for an amorphous thermoplastic material. The plotted curve represents one potential process line at a specific point in an injection molded component.

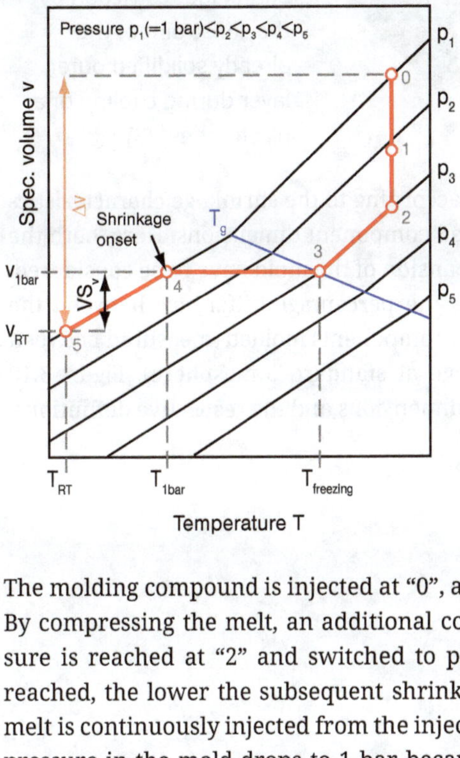

Figure 3.8
Schematic pvT diagram of an injection molding process for an amorphous thermoplastic material

The molding compound is injected at "0", and at "1" the cavity is volumetrically filled. By compressing the melt, an additional compaction is achieved until injection pressure is reached at "2" and switched to packing pressure. The higher the pressure reached, the lower the subsequent shrinkage. Isobaric cooling sets in until "3", *i. e.*, melt is continuously injected from the injection barrel. The gate freezes at "3", and the pressure in the mold drops to 1 bar because of the melt contraction during cooling ("4"). The temperature at which atmospheric pressure is reached is critical for the volume shrinkage VS_v. Once this temperature is reached, the molded component will begin to shrink away from the mold wall. The specific volume shrinkage VS_v can be calculated using the specific volume at "4" (= V_4) and at room temperature "5" (= V_5):

$$VS_V = \frac{V_4 - V_5}{V_4}$$

Assuming isotropic shrinkage, linear shrinkage is

$$VS_l = \frac{VS_v}{3}$$

Even when the mold does not restrain shrinkage, the geometric constraint imposed by the already solidified outer layer causes a much larger through-thickness (*d*) shrinkage in the component wall (in general 90–95% of volume shrinkage) than in length or width direction (Figure 3.9).

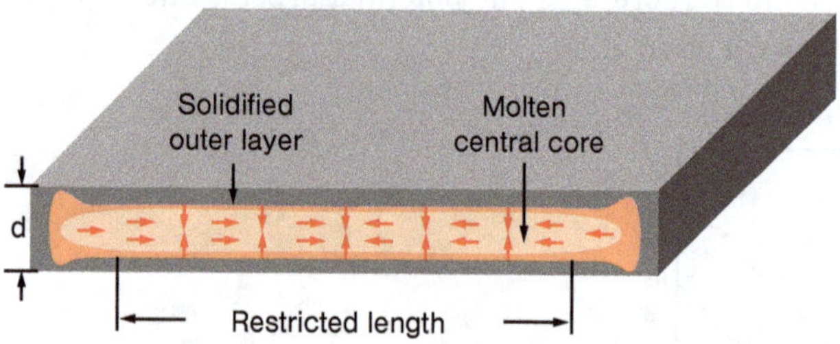

Figure 3.9 Shrinkage restriction by the constraint of the already solidified outer layer during cooling of a plastic sheet [3]

The cavity of the mold has to be increased according to the shrinkage characteristics of the plastic in order to produce the desired component dimensions. Here, both the behavior of the plastic material and the expansion of the mold have to be considered. The processing shrinkage VS is defined as the percentage difference between the dimensions of the cold mold and those of the components molded (measured between 24 h and 168 h after processing and storage at standard atmosphere). Figure 3.10 shows the potential changes in component dimensions and the respective definitions.

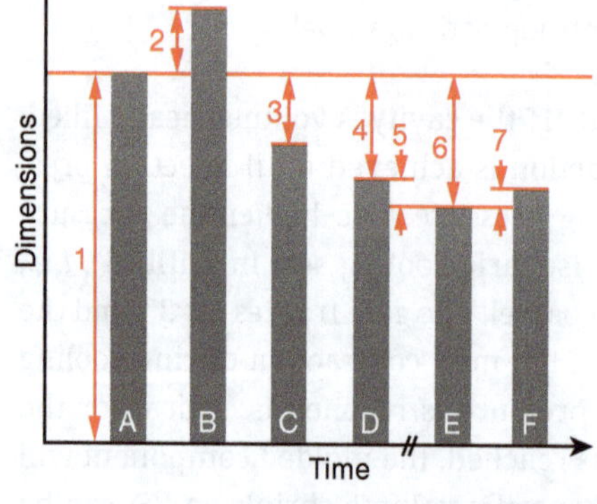

Figure 3.10 Dimensional changes in molds and molded components

1: Mold dimension; 2: Thermal expansion of mold; 3: Ejection shrinkage; 4: Mold shrinkage; 5: Post shrinkage; 6: Total shrinkage; 7: Potential change in length caused by conditioning.

A: Dimensions of cold mold; B: Dimensions of heated mold under pressure; C: Component after ejection; D: Component after 24 h in standard atmosphere; E: Component after long period of time or after heat aging; F: Component, *e. g.*, after water absorption for PA

Ejection shrinkage is caused by the already mentioned higher contraction of plastic materials during cooling as well as by volume reduction during crosslinking of thermosets or crystallization of semi-crystalline thermoplastics. Post-shrinkage (PS) is caused by chemical reactions, material loss, post-crystallization, or retardation. For amorphous thermoplastics, post-shrinkage is negligible; for semi-crystalline thermoplastics it ranges from 0.2 to 0.5%; and for mineral filled plastics it is ≤ 0.2%. Among thermosets, post-shrinkage is 0.1–0.6% for PF, 0.6–2.2% for UF, and 0–0.1% for UP. At room temperature, post-shrinkage is reached after several months. Heat aging at material specific temperatures accelerates post-shrinkage.

Additional factors influencing the shrinkage behavior include processing conditions, component geometry, and gating. Figure 3.11 details the effect of the processing parameters on shrinkage for selected materials. Because of the complex interactions between the factors, these rules are valid only for very simple geometries such as sheets. Amorphous thermoplastics and their shrinkage behavior is less processing dependent, and they exhibit less shrinkage than semi-crystalline plastics.

With increasing packing pressure the mold shrinkage decreases. However, increasing packing pressure with PP and PS also has a regressive effect on shrinkage reduction while for other materials, linear reduction may be assumed.

The packing time significantly influences shrinkage compensation because it determines how much molten material will be delivered into the cavity after the initial injection. This influence is more pronounced for semi-crystalline materials because of their higher contraction potential. However, the packing time cannot be arbitrarily extended, because once the gate is frozen, no more molten material can be injected into the cavity.

With regard to material temperature, there are two opposing mechanisms affecting shrinkage. On the one hand, in absolute terms higher melt temperatures cause higher volume contractions (larger Δv in Figure 3.8). On the other hand, the reduction in melt viscosity allows for better pressure transfer and thus shrinkage reduction. As a rule, the improved pressure transfer is the predominant effect. With increasing mold temperatures, the effects caused by the changed thermal conditions, which affect crystallization, dominate the effects caused by changed flow conditions. Therefore, increasing mold temperatures cause an increase in shrinkage in semi-crystalline thermoplastics. For amorphous thermoplastics, this effect is not significant. Lower ejection temperatures offer a third temperature variable that allows for longer cooling times and thus for a longer retention of mold constraints. Both semi-crystalline and amorphous plastics exhibit lower shrinkage here.

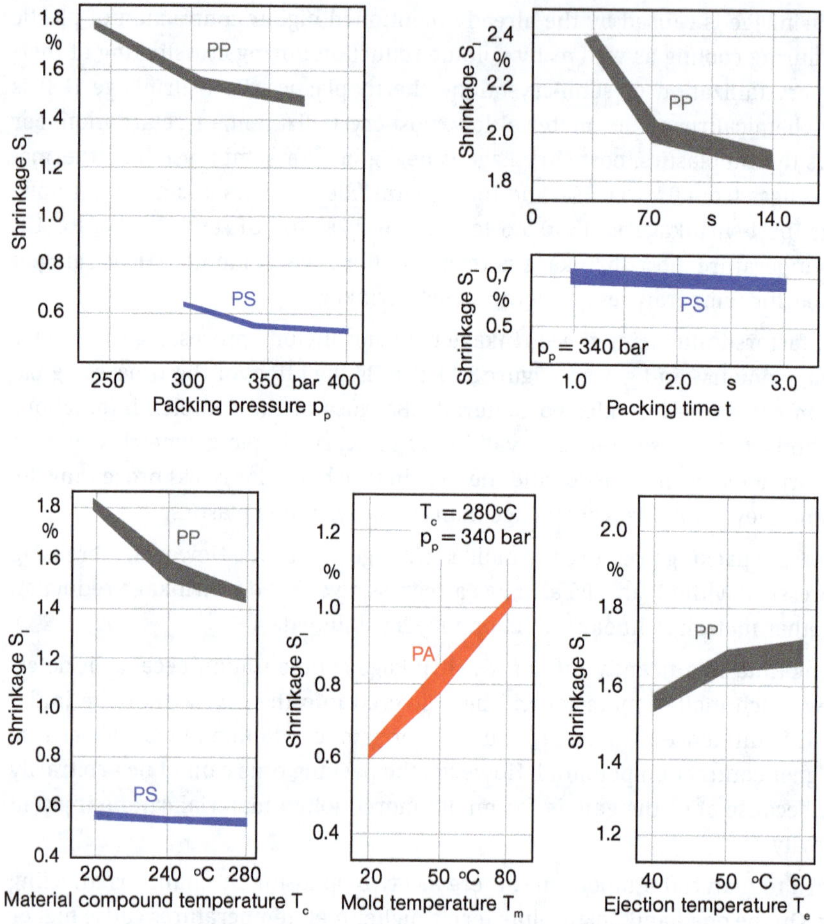

Figure 3.11 Effect of processing conditions on shrinkage of amorphous and semi-crystalline thermoplastics. Examples: PS, PP, and PA [4]
Parameters (unless defined otherwise):
T_c = 240 °C, T_m = 40 °C, t_F = 0.4 s, p_N = 310 bar, d = 2 mm

With regard to component geometry, two factors are of particular impact. So-called restraining geometries, such as undercuts, reduce or even prevent shrinkage. Wall thickness also affects shrinkage because it influences thermal conditions. At identical mold and compound temperatures, a thicker component will take longer to cool. In semi-crystalline thermoplastics this improves crystallization conditions, allowing for higher degrees of crystallization and thus higher levels of shrinkage. In amorphous thermoplastics this correlation is less pronounced. Shrinkage will vary in components that exhibit changes in wall thickness, causing warpage of the component. This effect is stronger in semi-crystalline than in amorphous thermoplastics.

The gating configuration is typically predetermined by component geometry. In order to ensure the least amount of shrinkage, it is critical to allow for packing pressure. In addition, the time for packing pressure can be increased by placing the sprue at a location with maximum wall thickness, thus increasing the time span during which shrinkage can be compensated by injection of additional melt.

Shrinkage in flow direction and perpendicular to flow may be different because of orientation effects on the macromolecules and/or elongational fillers during processing.

The difference between longitudinal and cross or transverse shrinkage is also called differential shrinkage. High values indicate semi-crystalline thermoplastics, in particular fiber reinforced grades, and thermosets with organic fillers such as wood flour. This difference is one of the reasons for warpage in injection molded components.

Other phenomena associated with shrinkage are residual stress, sink marks, and voids. Residual stresses are described in more detail in Section 2.4.1.

Sink marks are local depressions on the component's surface. They are created when the solidified outer layer cannot withstand the contraction forces exerted by the melt core layer of the component (Figure 3.12, left). Therefore, they occur primarily with large melt cores and simultaneous thin outer layers. The contraction forces increase when the ratio of melt core to the overall cross section increases. Sink marks often occur on the flat side of ribbed walls because there is poor heat transfer at the transition. When the solidified outer layer is able to withstand the contraction forces, voids will form in the thick-walled regions of the component (Figure 3.12, right).

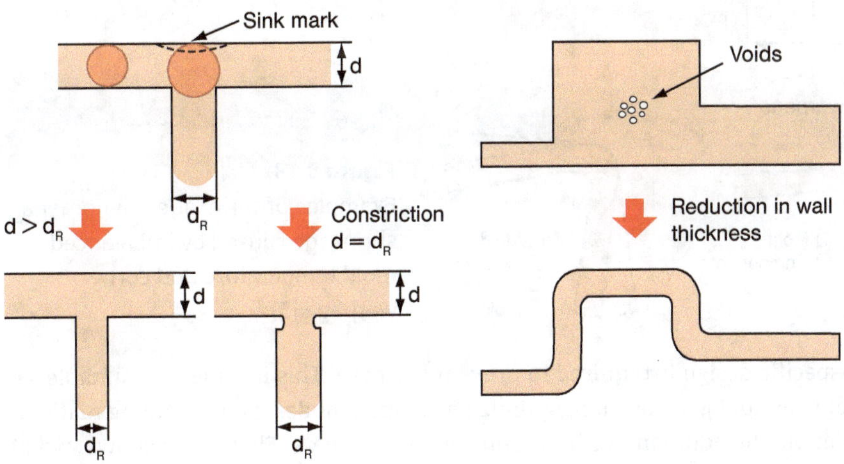

Figure 3.12 Formation of sink marks (left) and voids (right) and how to avoid them using appropriate designs

3.2.2.3 Warpage

In industrial practice, warpage of plastic components poses an even bigger problem than shrinkage, and it always is the result of shrinkage differentials. As described in the previous section, shrinkage is caused by molecular or fiber orientation, differences in wall thickness, locally different mold temperatures and packing pressure effects. Because the shrinkage of semi-crystalline thermoplastics is particularly dependent on these process parameters, the warpage exhibited is also typically higher than that of amorphous thermoplastics. In thermosets, warpage can be caused by differential shrinkage during curing, *e.g.*, by local differences in the degree of curing.

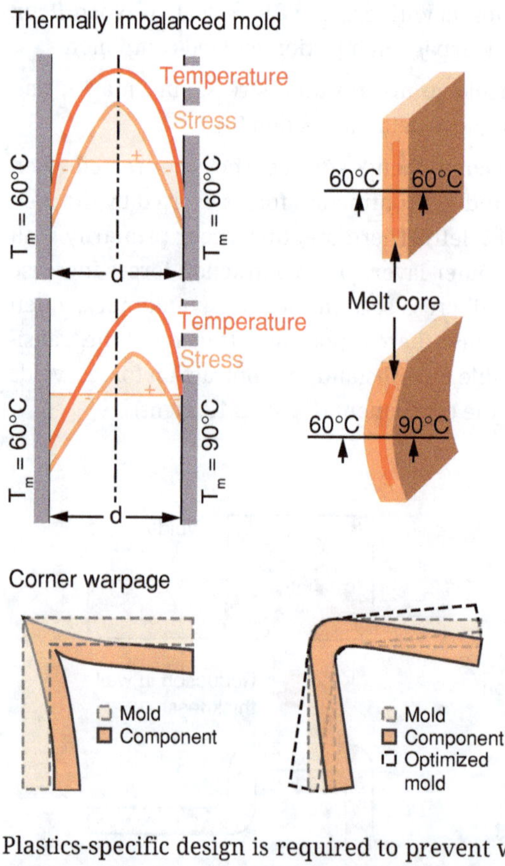

Figure 3.13
Examples of shrinkage phenomena: shrinkage caused by imbalanced mold temperature and corner warpage [5]

Plastics-specific design is required to prevent warpage. This includes the suitable selection of gates and gate location, optimizing component design by avoiding material accumulation, the addition of fillers, and rounded corners. Figure 3.13 shows two of these warpage phenomena. An inhomogeneous temperature distribution in different areas of the molded component causes differential shrinkage and thus leads to local variations in component properties. In the example shown in Figure 3.13 (top) the two halves of the component are at different temperatures. The component part at higher temperatures will experience higher shrinkage. During the imbalanced cooling the temperature profile will shift and cause imbalanced stresses. The compressive stresses are denoted by "–", the tensile stresses by "+". Material accumulation at corners and the associated poor heat transfer result in higher contraction at the inner side of the corner, also called spring-in effect. Here, warpage can be prevented by addressing wall thickness (rounding of corners) or by optimizing the mold. Fiber reinforced composite components (Figure 3.13 bottom right) with large fiber length/component thickness ratio exhibit higher coefficients of thermal expansion in direction of thickness than in length or width direction. The warpage observed here is also called spring-in effect. It depends not only on material accumulation, change in temperature, and mold angle, but also on the difference in coefficient of expansion in thickness and length direction. Here, warpage is mitigated by adaptation and/or optimization of the mold.

3.2.2.4 Tolerances

The tolerances achievable during the manufacture of molded plastic components depend on the shrinkage and crystallization behavior of the respective plastic material. Empirical data can be found in Chapter 8.

3.2.2.5 Filler Orientation

Longitudinal fillers, such as glass, carbon, graphite or textile and natural fibers, assume a state of orientation in a flowing melt similar to the molecules, see Section 2.4.2. The main direction of orientation is the flow direction. In the case of radial flows, *e. g.*, when a round disk is filled via a central gate, the fibers in the interior of the component are oriented tangentially by elongational flow, *i. e.*, they are oriented transversely to the flow direction. Because the orientation of the fillers cannot relax, the state of orientation is exclusively determined by the melt deformation processes during molding of the component. While a change in processing parameters offers little opportunity to influence the state of orientation in the component, the selection of gate and gate position significantly influences fiber orientation in the molded component. The orientation of fiber-like fillers creates higher anisotropies than molecular orientation. While shrinkage, thermal expansion, and strain at break decrease in flow direction, the mechanical characteristics show distinct increases. Filler orientation can be determined using X-ray or optical analysis of polished/thin sections. Figure 3.14 shows the orientation through the sheet thickness d when PA 6 with 30 vol.-% copper platelets is injection molded into thin sheets.

For 1 mm thick sheets, almost all fillers – both in the outer layers as well as in the center of the specimen – are oriented in flow direction. Sheets with a thickness of 2 mm exhibit the layer formation shown in Figure 3.14 (right) as a function of shear rate profile, temperature, and viscosity of the plastic material (see also material flow curves in Figure 3.5 d).

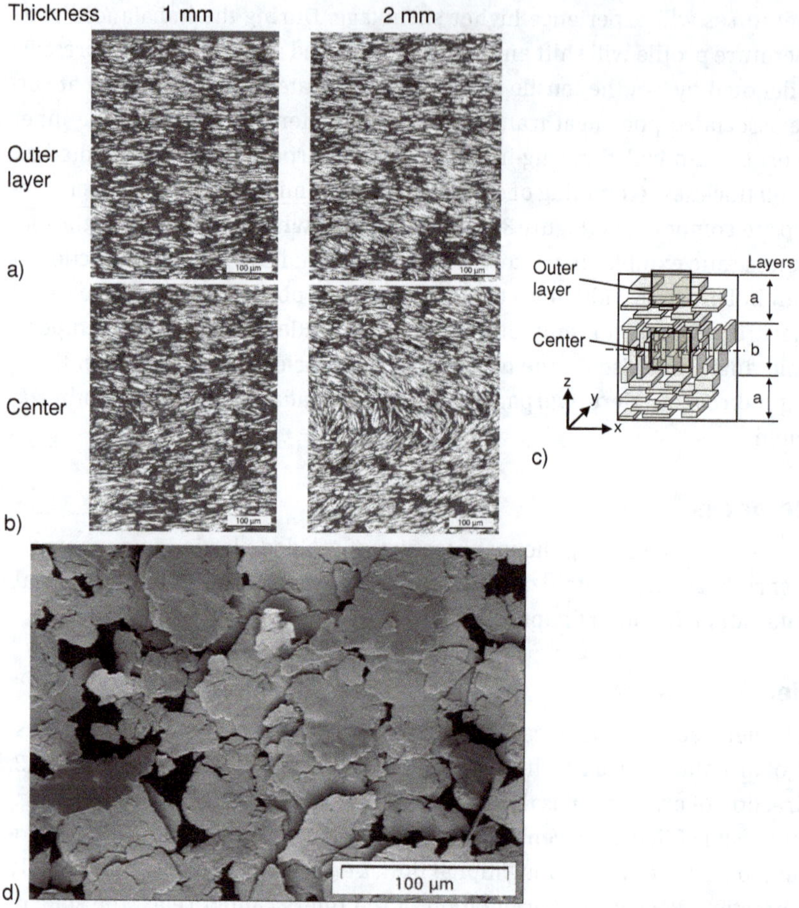

Figure 3.14 Change in filler orientation caused by variation in specimen thickness *d* for PA 6 + 30 vol.-% Cu compound; dimensions: 50 x 50 x *d* [mm^3]; flow direction from left to right [2]

a) Micrograph of outer layer

b) Micrograph of specimen center

c) Sketch of the positions of extracted polished sections

d) REM micrograph of the Cu filler

3.2.3 Testing of Particulate Solids

The size of the particles typically found in plastics processing can range from several millimeters, as in plastic granulates, to a few micrometers in fine powders. Understanding the movement and compaction of solid particles is important not only in processes such as rotational molding (see Section 4.12) and Selective Laser Sintering (see Section 4.10) but also in everyday polymer processes when it comes to controlling the flow and feed of granulate in the hopper and in the feed zone of extruders and injection molding machines.

3.2.3.1 Determination of Particle Size Distribution

There are several methods for determining particle size and size distribution. One of the most widely used methods is traditional sieve analysis. In this method, bulk materials are separated into individual fractions or portions in a sieve tower using different sieve mesh sizes. The size distribution can be determined from the mass fractions of the respective portions. In addition to classic sieve analysis, optical measurement methods have also become established in this field. Imaging measurement methods can be divided into static and dynamic methods. In static methods, a powder sample is atomized onto a slide and then each particle is automatically evaluated using microscopic images. In dynamic methods, the particles are separated and passed through a measuring chamber in free fall. Here, the particles are recorded and evaluated by a camera. In addition to particle size distribution, these measurement methods can also determine the shape factors of the particles. Another method for determining particle size distribution is laser diffraction. A dispersed particle sample is penetrated by a laser beam and the intensity of the scattered light from the laser beam is measured. This data can be used to calculate the size of the particles. In addition, particle shapes and surfaces can be evaluated using classical light microscopy or scanning electron microscopy.

3.2.3.2 Mechanics of Particulate Solids

Particulate solids, powders, or granulated materials are encountered everywhere in polymer processing and the size of the particles can be from several millimeters (plastic pellets) to a few micrometers (fine powders). Understanding the movement and compaction of these solids is important both in processes such as rotational molding and Selective Laser Sintering and in everyday polymer processes (when controlling the flow and feeding of pellets).

The mechanics of particulate materials is dominated by an inner friction that is a result of the friction between individual particles. A material element confined inside a volume and subjected to an axial stress σ_1, as depicted in Figure 3.15, feels a lateral stress σ_2 that varies between 0 and σ_1, depending on the inner friction parameter κ, also referred to as *stress ratio*. The relation between the axial and the lateral stresses can be written as

$$\sigma_2 = \kappa\sigma_1 \tag{3.6}$$

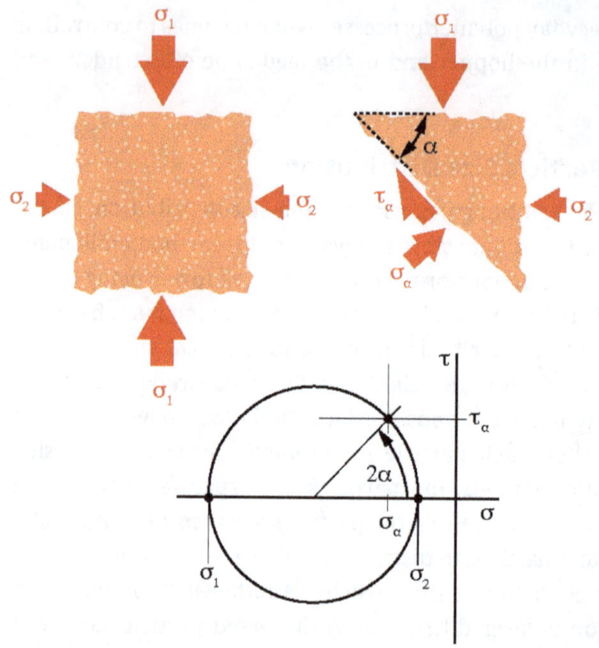

Figure 3.15 Schematic of stress transmission within particulate media

A common method of assessing the inner friction parameter is by measuring the angle of repose θ of a sprinkled powder heap as schematically depicted in Figure 3.16. The empirical relation between the angle of repose and the inner friction parameter is given by

$$\mu \approx \tan(0.75\theta) \tag{3.7}$$

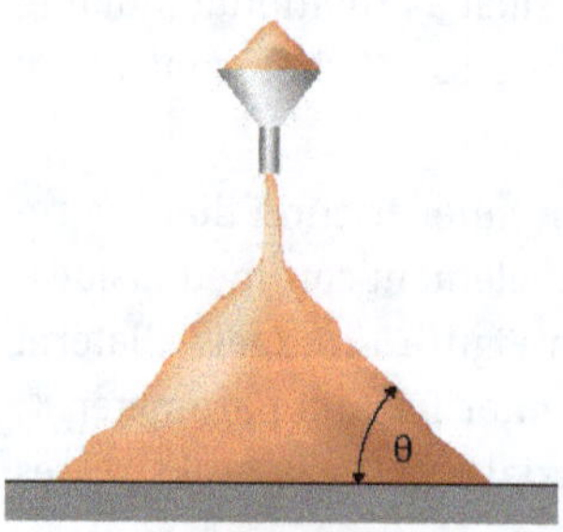

Figure 3.16 Schematic of the angle of repose measurement

3.2.3.3 Adhesive Forces and Flowability

The flow of powders or particulate solids is dominated by adhesive forces between the particles. These adhesive forces can depend on the chemical composition of the powder, moisture content, and even van der Waals forces in very fine powders. Furthermore, as a powder is compacted it becomes a more cohesive material with reduced flowability. This effect is taken advantage of during the manufacture of pills or tablets. As the compaction pressure increases, the flowability of the powder goes down until a perfectly cohesive solid behavior is reached.

It is possible to measure the effect of compaction on the flowability of a powder using the system schematically depicted in Figure 3.17, where a powder is compacted inside a cylinder. First, the powder is sprinkled into the cylinder filling a specified volume with the loose powder with a bulk density ρ_b, as shown in Figure 3.17(a). The cylinder is then closed and an axial stress σ_1 is applied until the density of the powder increases to ρ_c, as shown in Figure 3.17 (b). Since the compaction of a powder is caused by all normal stresses acting on the powder, one can use the trace of the stress tensor to compute the hydrostatic stress σ_H that leads to density changes, as the hydrostatic stress can be related to the trace of the stress tensor[1]. In the abovementioned case we can write

$$\sigma_H = \frac{1}{3}(\sigma_{11} + \sigma_{22} + \sigma_{33}) \tag{3.8}$$

We can now relate the density of the powder to the hydrostatic stress as

$$\rho_c = \rho_\infty + (\rho_b - \rho_\infty)e^{-\beta\sigma_H} \tag{3.9}$$

Where:

β = compressibility of the particulate solid

ρ_∞ = density of the solid material

A schematic of the density as a function of compaction stress is presented in Figure 3.18. As with the density, one can define the porosity of the compacted powder using

$$\phi_c = \phi_b e^{-\beta\sigma_H} = \frac{\text{void volume}}{\text{total volume}} \tag{3.10}$$

Where:

ϕ_b = density of the bulk or loose powder

Once fully compacted, as $\sigma_H \rightarrow \infty$, the porosity reduces to $\phi_c \rightarrow 0$. It is important to note that, in order to avoid density variations along the height of the compacted powder cylinder, the test must be done with a cylinder of relatively low height and with an inner wall that has a low coefficient of friction.

1 The trace of a stress tensor is often also referred to as the first invariant of the stress tensor. In a Cartesian coordinate system the first invariant of the stress tensor is written as $I_\sigma = \sigma_{xx} + \sigma_{yy} + \sigma_{zz}$.

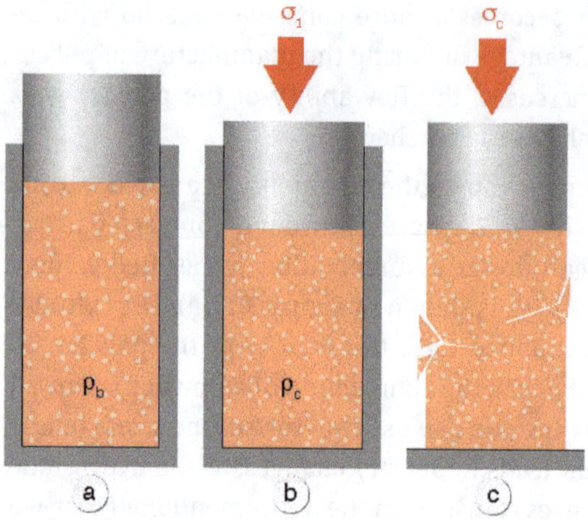

Figure 3.17
Schematic of a uniaxial compaction process

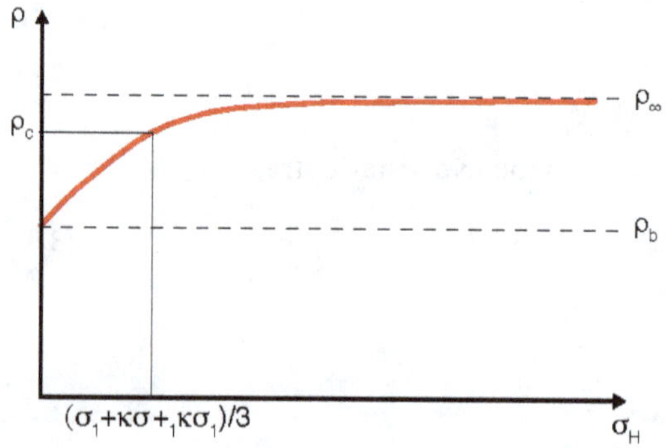

Figure 3.18
Schematic of the density distribution of a powder subjected to hydrostatic stresses

Once compacted, the outer cylinder walls are removed, freeing the compacted powder outer walls, as shown in Figure 3.17 (c). Finally, once the compacted powder body is free, an axial stress σ_1 is applied such that the powder cylinder crumbles at a stress σ_c. Figure 3.19 presents the strength of the compacted powder, σ_c, as a function of the axial compaction stress σ_1. The graph also presents the effect that the storage or consolidation time has on the strength of the powder. It is well known that even loose powders will have a lower flowability when stored for an extended period of time. This effect is often referred to as *caking*.

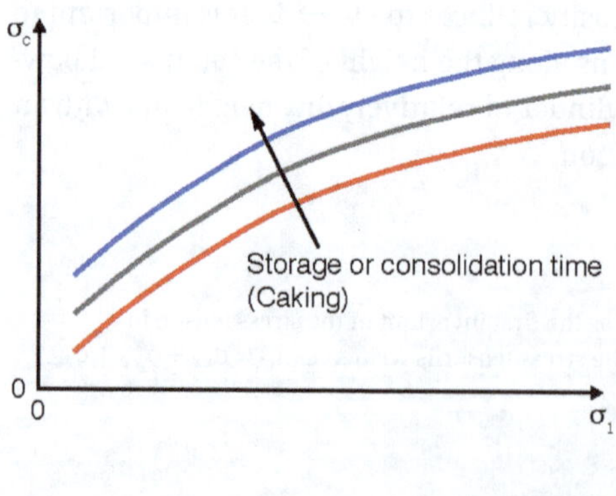

Figure 3.19
Flow functions for different storage times

The flowability can be computed using

$$f_c = \frac{\sigma_1}{\sigma_c} \tag{3.11}$$

Figure 3.20 presents constant flowability lines superposed on a curve of the strength of the compacted powder versus the compaction stress. Various qualitative descriptions are also presented. A powder with a flowability number larger than 10 ($f_c \geq 10$) is considered a free-flowing powder. On the other hand, a particulate with a flowability number smaller than 1 ($f_c \leq 1$) is a powder that cannot flow.

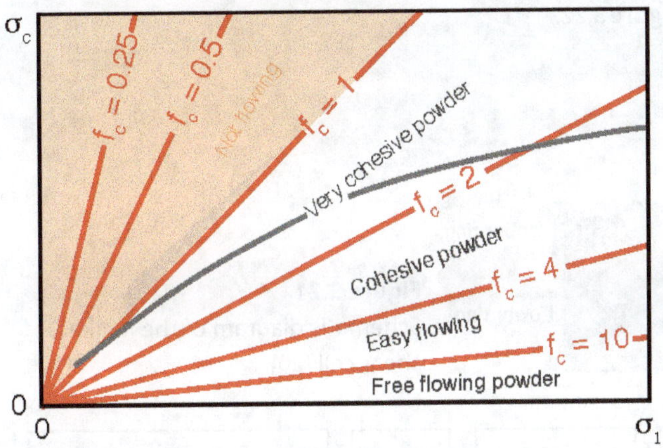

Figure 3.20 Constant flowability lines in relation to the flow function

3.2.3.4 Flowability and the Yield Locus

The flowability number gives us an idea of the ability of a powder to flow after compaction; however, it does not reveal the necessary stresses required to force the powder to flow under stressed conditions. Furthermore, the flowability test is difficult to perform since it is nearly impossible to make a cylinder with frictionless walls. As will be demonstrated later, wall friction has a profound effect on the stress distribution inside the cylinder and consequently leads to a final density and flowability distribution within the powder inside the cylinder.

In order to understand the flow characteristics of a powder, it is necessary to measure the required shear stress that will cause the material to flow under any compaction and stress conditions a powder may be subjected to. Such curves are referred to as *yield loci*. The measurements needed for a yield locus can be performed with shear testers such as the Jenike shear cell [20] or the Schulze ring shear tester [21]. The Jenike shear cell, schematically depicted in Figure 3.21, has the ability to stress the powder as it is being sheared. To generate one yield locus, a standard procedure must be followed. The upper and lower rings are aligned as shown in Figure 3.21, and the cell is closed with the lid. The powder is pre-stressed to a maximum consolidation stress of σ_p by applying a normal force F_N. While the maximum consolidation stress σ_p

is maintained, a side shear force F_S is applied and increased until the upper ring starts moving and the shear stress within the cross-sectional area reaches a steady-state value of τ_p, as depicted in Figure 3.22. The point (σ_p, τ_p) is recorded and plotted on a σ vs. τ graph as shown in Figure 3.22. At this point, the ring is returned to its initial position, aligning the upper and lower rings, and the normal force F_N is decreased, such that it generates a stress within the powder of σ_1. A shear force F_S is applied until the material flows at a corresponding shear stress τ_1. The new data point is recorded in the σ vs. τ graph. A new specimen is prepared, pre-stressed to σ_p, and sheared under that stress until the material flows at τ_p. A new stress σ_2, which is lower than σ_p and different to σ_1, is applied to the powder and a new point is generated on the yield locus curve, as shown in Figure 3.22.

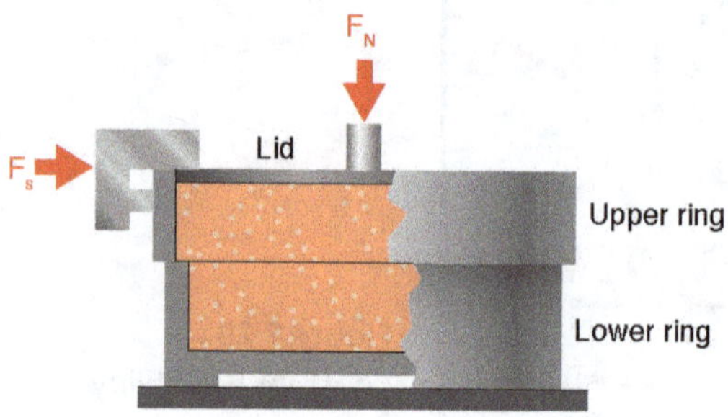

Figure 3.21 Schematic diagram of the Jenike shear cell [20]

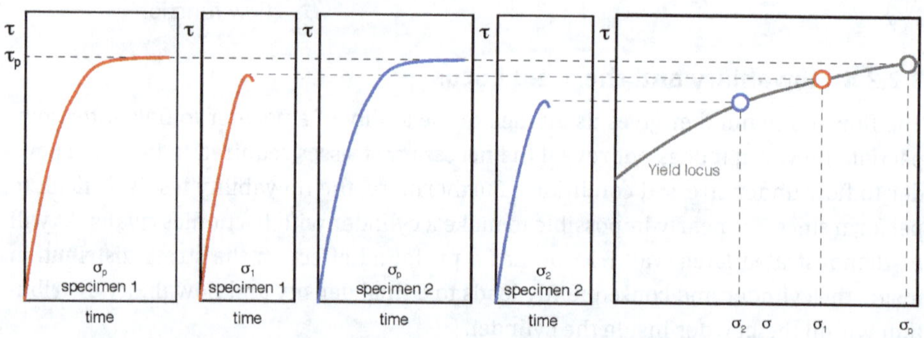

Figure 3.22 Procedure to extract the yield locus using a shear cell

3.3 Mechanical Properties

The mechanical behavior of plastic materials depends on a variety of factors:

- Rate of deformation,
- Duration and frequency of load,
- Specimen geometry,

- Processing parameters during specimen manufacture, and
- Ambient conditions, in particular temperature.

The tensile test is a suitable means to exemplify the effects of these factors on the behavior of plastic materials.

3.3.1 Short-Term Behavior

3.3.1.1 Accelerated Tensile Test

The mechanical behavior of plastics, *i. e.*, the correlation between stress σ and the resulting strain ε or between the load F acting on a component and the resulting deformation ΔL, is best characterized using tensile testing. During short-term tensile tests, tensile bars (dog-bone shaped specimens, see Figure 3.23, top) are elongated at a constant rate while the load F and the change in length ΔL of the measured length L_0 is recorded. The stress σ (in N/mm²) can be calculated by dividing the load by the original cross sectional area A_0 of the specimen, while the strain ε (in %) is derived from the change in length divided by the original specimen L_0. In general, the slope of the load/deformation curve is used to calculate the modulus of elasticity E as the ratio between stress and strain.

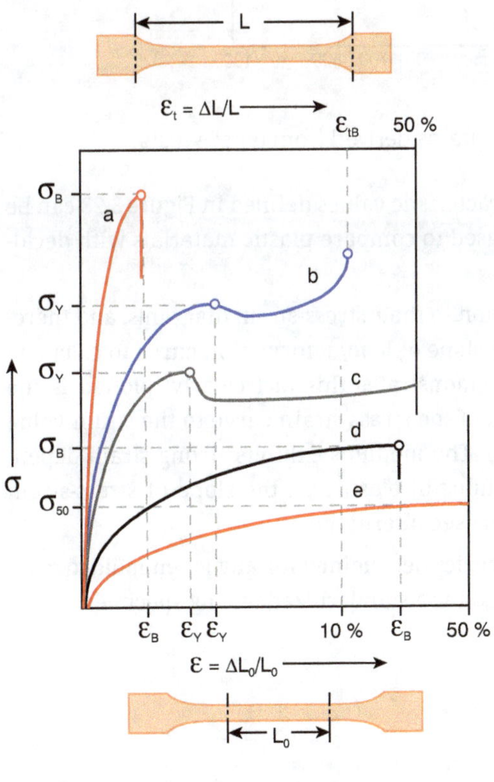

Figure 3.23
Schematic σ/ε-diagrams based on CAMPUS data for tensile tests;
a: brittle plastic, b: ductile plastic, c: stretched plastic, d and e: plasticized plastic
σ_B = stress at break ε_B = strain at break
σ_y = yield stress ε_y = yield strain
σ_{50} = stress at 50% strain ε_{tB} = nominal strain at break

Figure 3.23 uses the stress-strain diagram to show the characteristic data (according to CAMPUS) typically used to evaluate plastic materials. For plastics exhibiting yield stress or major deformation prior to break, nominal strain at break (in %) is determined using the change in gripping distance and the original clamping distance.

It is used as a comparative value, describing a plastic's elasticity under tensile load.

Figure 3.24 shows real stress-strain diagrams for several plastic classes. Note that the y-axis in the left part of the figure (strain up to 20%) is depicted at a 5 times larger scale than that of the y-axis on the right part of the figure to allow for a clearer depiction.

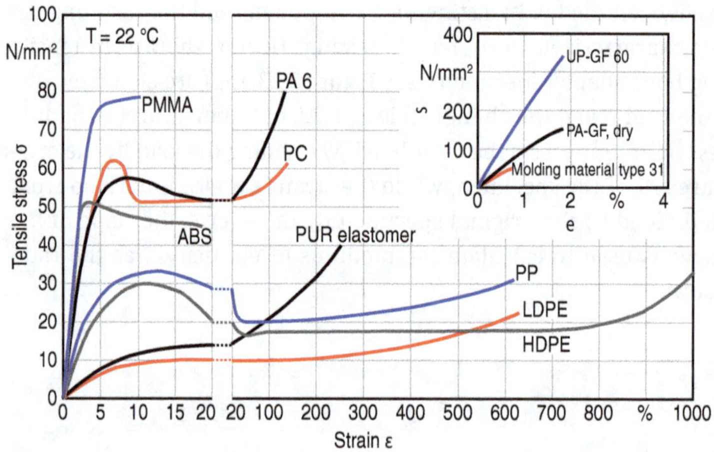

Figure 3.24 Examples of stress-strain diagrams derived from tensile tests

With the help of these diagrams the characteristic values defined in Figure 3.23 can be determined. The diagrams can also be used to compare plastic materials with decidedly different levels of flexibility.

Most plastics do not exhibit a linear section in their stress-strain diagrams, and therefore it is not permissible here to use the slope of load/deformation curve to calculate the modulus of elasticity. Figure 3.25 demonstrates this fact clearly: plotted is the slope of the straight line from the origin of the stress-strain curve to the strain value considered (also called secant modulus). The modulus exhibits strong strain dependence. In order to determine the "modulus of elasticity", the slope of stress-strain curve between 0.05% and 0.25% strain is used alternatively.

It is important to note that the characteristics determined for any given molded component may not reflect the data measured for a standardized testing specimen.

A special tensile test is used with notched film specimens to determine both edge tear and tear propagation resistance (force required to propagate tearing in proportion to specimen thickness) according to Graves.

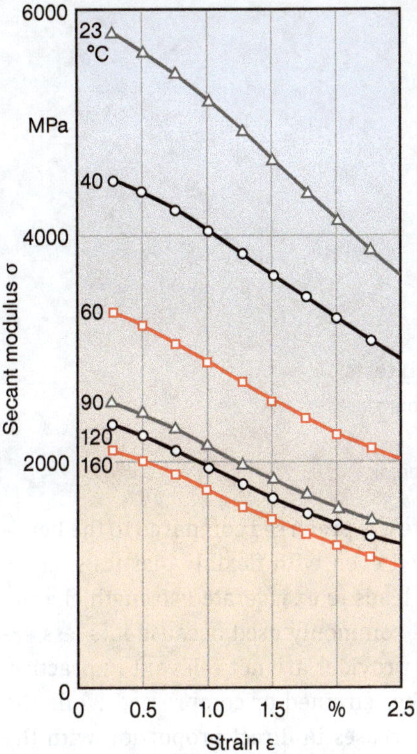

Figure 3.25
Secant modulus for glass fiber reinforced PA 6

3.3.1.2 Short-Term Bend Test

In short-term bend tests, the ends of a beam-like specimen (80 mm × 10 mm × 4 mm) are placed on two supports and then a load is applied to the middle of the specimen using a bending punch. The measured loads and bending deflections are used to calculate the characteristics defined in Figure 3.26:

$$\text{Flexural stress } \sigma_b = \frac{3 \cdot F \cdot l_v}{2 \cdot b \cdot h^2} \text{ in N/mm}^2 \quad (3.12)$$

$$\text{Outer fiber strain } \varepsilon_b = \frac{600 \cdot h \cdot f}{l_v^2} \text{ in \%} \quad (3.13)$$

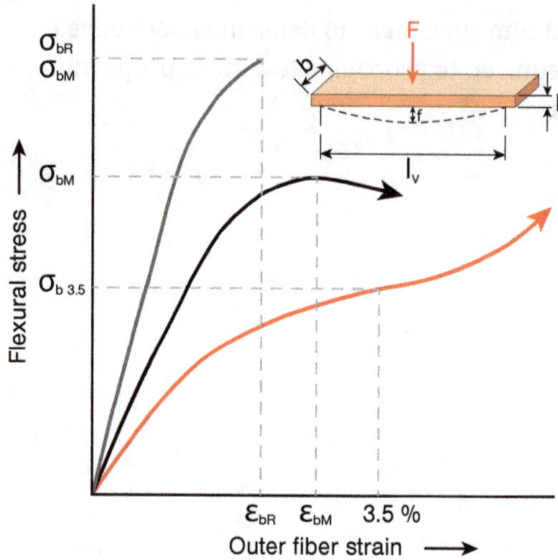

Figure 3.26 Schematic σ/ε diagrams for bending tests

σ_{bM} = Bending strength; ε_{bM} = Outer fiber strain at maximum stress

$\sigma_{b3.5}$ = 3.5% bending stress

σ_{bR} = Bending stress at break; ε_{bR} = Outer fiber strain at break

There are a number of reasons why the tensile test is preferred compared to the bending test. For one, tensile tests can be performed even with flexible plastics such as elastomers. In addition, the bending test often leads to exaggerated strength characteristics. Unfortunately, the bending test is still commonly used because it is less expensive, even though the characteristic data it provides are not relevant in practice. When a beam is bent, the outer fibers are either strained or compressed while the neutral fibers are not deformed. The strain increases in direct proportion with the distance to the neutral fiber. If the correlation between stress and strain is linear (as determined in the stress-strain diagram), the stress distribution through the beam thickness is also linear. However, plastics may also be under load outside the linear range (in particular under short-term loads) and then the stress profile through the beam cross section is no longer linear. In this case, the equation representing the elastic line that is employed to calculate the bending stress is no longer valid and the calculated stress values are too high. The further the stress-strain lines (determined with the help of the stress-strain diagram) deviate from linearity, the higher is the difference between the calculated bending strength values and the tensile strength and/or yield stress values that were determined by tensile testing. Then bending strength is only a notional stress that actually does not reflect the component, because failure will occur as soon as the tensile strength is exceeded. For these reasons alone, bending tests should be performed only with long fiber reinforced plastics to determine strength characteristics, *e. g.*, interlaminar shear strength.

3.3.1.3 Compression Test

Compression tests are used to determine material behavior under uniaxial compression load. However, in industrial practice it is relevant only for the evaluation of fiber reinforced and foamed plastics. Other than in high fiber reinforced plastics, compression stress does not cause failure. Therefore, compression testing is rarely used for the evaluation of plastics. Highly uniaxially reinforced plastics may delaminate under compression stress. These products are compression tested by laterally supporting a slim specimen, thus preventing buckling. Characteristics analogous to those derived from tensile tests are determined here.

3.3.1.4 Indentation Test, Hardness Measurements

Indentation tests determine the indentation resistance as well as the indentation depth of a defined body into a plastic surface. There is a certain correlation between hardness and modulus of elasticity when the elastic deformation predominates during testing; otherwise the characteristic values do not provide suitable comparative data for component design.

Plastics and elastomers can be evaluated with respect to their *Shore hardness*. The resistance against the penetration of a truncated cone (Shore A for flexible plastics and elastomers) or of a cone with rounded tip (Shore D for rigid elastomers and thermoplastics) is measured as the deformation of a spring 3 seconds (or 15 seconds for plastics with decidedly plastic behavior) after indentation of the tester into the surface of the specimen (Figure 3.27).

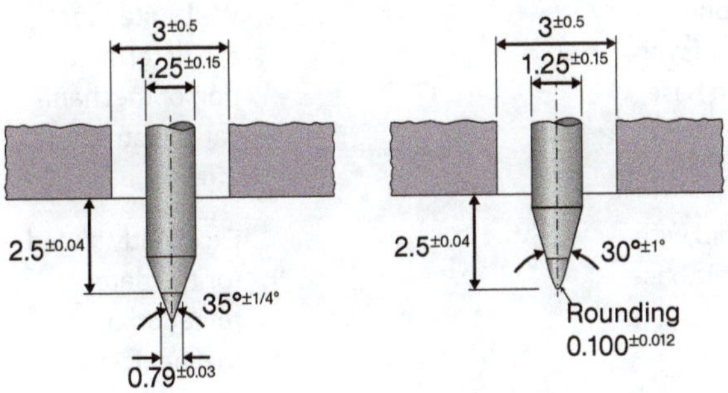

Figure 3.27 Indenter for the determination of Shore hardness

Table 3.3 provides an overview of the methods available for the determination of hardness, most of which were developed for metals and are of only limited suitability for applications in plastics technology.

Table 3.3 Comparison of Hardness Tests for Plastics

Test	Standard	Test			Application
		Indenter	Under load, depth of indentation	After load removal	
Shore A	ISO 868	Figure 3.27	X		Soft elastomers and thermoplastics
Shore D	ISO 868	Figure 3.27	X		Rigid elastomers and thermoplastics
Ball indentation hardness H	ISO 2039-1	Ball, Ø 5 mm	X		Thermoplastics and thermosets
Rockwell-α-hardness Rα	ISO 2039-2	Ball, Ø depending on hardness scale R, L, M, E	X		Suitable for the determination of mechanical anisotropy effects
Rockwell HR	ISO 2039-2	Ball, Ø depending on hardness scale		X	
Vickers hardness HV	ISO 6507	Diamond pyramid with square base		X	Well suited for the determination of mechanical anisotropy effects
Knoop-hardness HKn	ASTM D1474	Pyramid with rhombic base		X	Excellently suited for the determination of mechanical anisotropy effects

Anisotropy effects: *e. g.*, orientations and residual stresses, filler orientation, hardness profile in weld lines

3.3.2 Impact Behavior

3.3.2.1 Characteristics

The shear rate during short-term tensile tests is selected so that the characteristic strength values are reached within approx. 1 min. Plastics are viscoelastic materials and therefore their characteristics depend on the rate of the applied load. Figure 3.28 uses an ABS + PA 6 blend as an example to show that the yield stress increases with increasing shear rate while the strain at break decreases. The modulus of elasticity also increases with increasing loading rate. This indicates that data determined with the help of short-term tests have no relevance for processes taking place in extremely short time intervals, *e. g.*, crash simulation in automotive engineering.

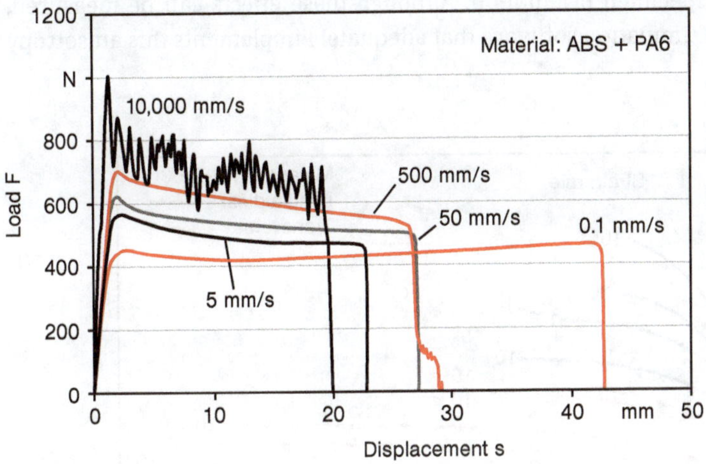

Figure 3.28 Typical tensile test results for different strain rates

In order to represent very high speeds in tensile tests, these tests are performed with suitable hydraulic systems or drop towers. However, these tests are very complex. When multi-axial load conditions have to be calculated, additional compression, shear, and torsional tests will be necessary – an effort only rarely made in industrial practice. The multiple superpositions of the occurring oscillations make it difficult to interpret the test results.

In particular, it should be noted that the analysis of the measured load-deformation diagrams (so-called technical stress-strain diagrams) usually performed (and standardized) for quasi-static tensile tests is not sufficient for tests performed at high speeds. In order to facilitate the use of simulation software, diagrams depicting "true stress vs. true strain", which include the change in geometry, in particular the change in specimen cross section, are required for the stress calculation. By taking the reduction in cross section and/or the constriction behavior into consideration, the in fact much higher stresses in the constricted specimen areas can be correctly predicted. Below the yield point, these effects can be approximated or determined by measuring transverse strain. At higher strains, optical measurements with subsequent complex image interpretation are conducted. This approach allows for the exact determination of local strains.

All high-speed tests are confronted with the issue that in order to reach high strain rates (beyond 1000 s^{-1}) they require short test specimens, which eliminates the universal testing specimen. Currently, no new standardized specimen geometries have been established.

The same effects are encountered in load tests at crash speeds that are seen in quasi-static tests, *i. e.*, temperature dependence and anisotropic behavior. With increasing temperature, the stress values decrease, significantly improving strain. Figure 3.29 shows the effect of specimen orientation. Although these effects can be measured, there is currently no simulation software that adequately implements this anisotropy in a material model.

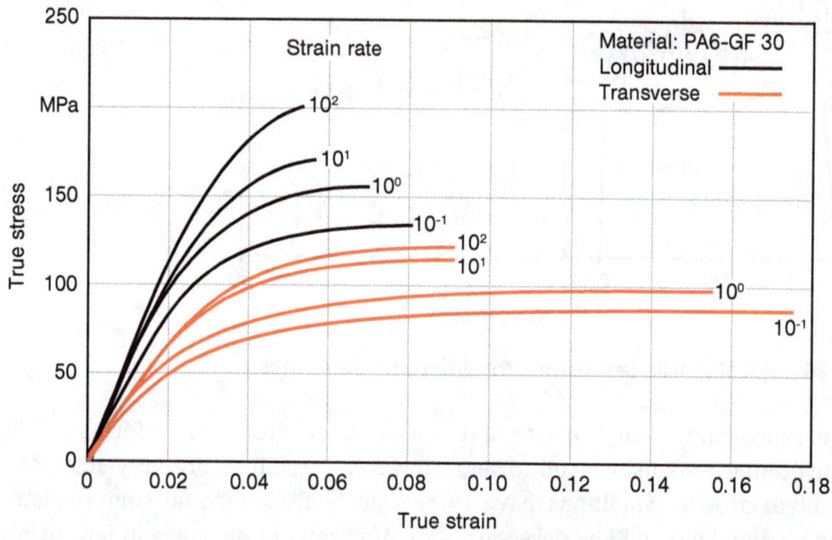

Figure 3.29 Effect of specimen orientation on stress profiles at high strain rates

3.3.2.2 Impact, Flexural, and Tensile Tests According to CAMPUS

The impact behavior (a measure for the toughness) of plastics is determined via standardized impact or impact penetration test procedures. Here, rod- or sheet-like specimens (which may be notched) are tested with the help of pendulum impact machines or drop towers until they fail, or in penetration tests often until the first specimen damage is detected. The most common testing arrangements are shown in Figure 3.30.

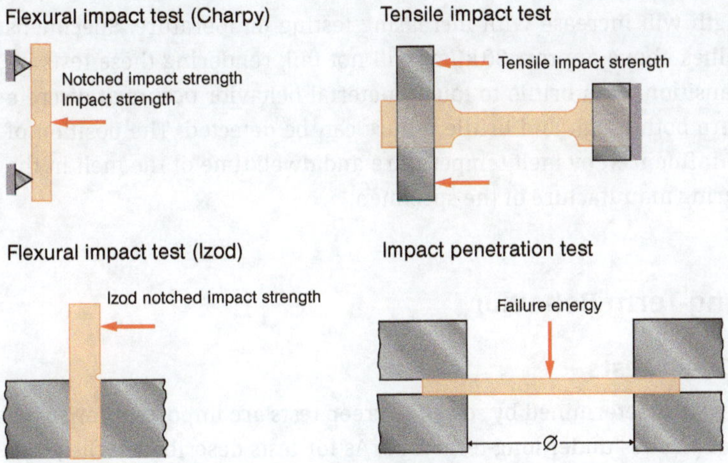

Figure 3.30 Test set-up for impact strength tests

Depending on testing set-up, specimen type and dimensions and type of notch, a variety of different impact strength characteristics can be determined. Values determined by different testing methods are typically not comparable because the testing parameters significantly affect the magnitude of the characteristic values. Even a list of plastics in the order of their impact strength values depends on the testing method, which indicates that these values provide only an estimate of the impact behavior of molded components. They are not suitable for design or dimensioning calculations. However, they are useful during the material selection phase when several grades of the same plastic material are under consideration. Here it is critical that all characteristic values were determined using the same testing procedure. The following procedures are available to determine the impact strength of different plastics:

- Charpy impact strength and Izod impact strength for brittle plastics.
- Charpy notched impact strength and Izod notched impact strength for tough plastics; both are performed with 80 mm × 10 mm × 4 mm specimens, impact on narrow side of notched specimen (clamped on both sides = Charpy, clamped on one side = Izod), V-shaped notch.
- Plastics that do not exhibit failure in these tests are evaluated by tensile-impact-strength tests, using rod-shaped specimens (80 mm × 10 mm × 4 mm) with double V-notches (radius 1 mm). All tests report the impact strength value as the impact energy (kJ/m^2) until failure at the smallest specimen cross section.

The impact penetration test also provides characteristic data with regard to specimen deformation and energy absorption and can therefore be used for the evaluation of crash behavior. Impact strength values are highly susceptible to the effects of both inadvertent and deliberate modifications of the plastic material, processing parameters during the manufacturing of specimens, notch geometry, and testing temperature. Impact strength will increase with increasing testing temperature. Specimens with toughness values above approx. 30 kJ/m² will not fail, rendering these tests inconclusive. The transition from brittle to tough material behavior occurs at discrete temperatures, where both tough and brittle failure can be detected. The position of these areas can be influenced by melt temperature and dwell time of the melt in the injection barrel during manufacture of the specimen.

3.3.3 Static Long-Term Behavior

3.3.3.1 Tensile Creep Tests

The characteristic values determined by so-called creep tests are important for structural applications of plastics under long-term load. As for tests described earlier, tensile tests are preferred over compression or flexural tests. Tensile creep tests can be performed as either retardation or relaxation tests; however, because the retardation test is easier to perform than the relaxation test, it is used almost exclusively today.

During creep tests, specimens are under constant uniaxial loads (typically weights) in a controlled testing environment. The elongation of the measured length and the time until failure of the specimen are recorded. Using the resulting creep-time lines (Figure 3.31), corresponding stress/strain values at specified load times can be determined, resulting in isochronous stress-strain diagrams (Figure 3.32) that are meaningful for design purposes (they are also provided by CAMPUS). When isochronous stress-strain diagrams are linear, the investigated plastic material is visco-elastic. In this case, the load level does not affect creep or relaxation behavior. In order to determine component behavior, it will suffice to consider temperature- and time-dependent characteristics. However, plastics generally exhibit non-linear visco-elastic behavior, *i. e.*, the isochronous lines are curved.

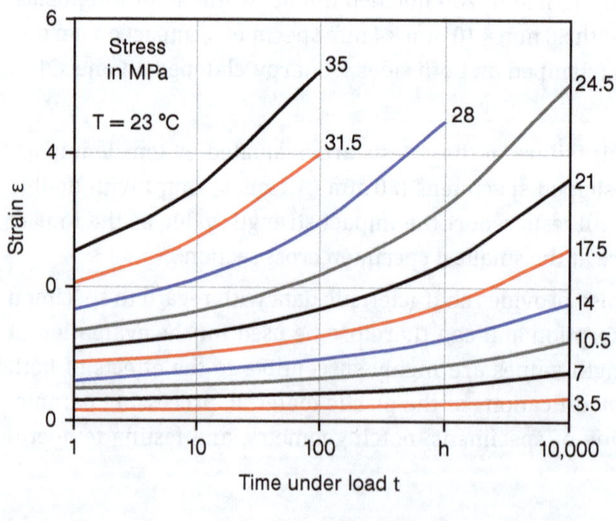

Figure 3.31 Creep-time diagram for PBT at 23 °C

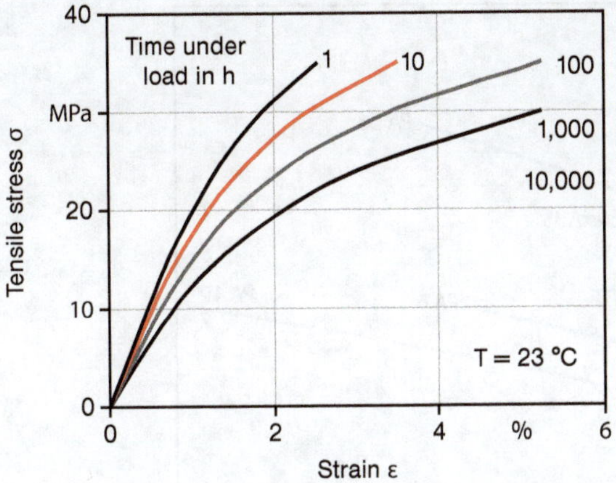

Figure 3.32
Isochronous stress-strain diagram for PBT at 23 °C

The deformation behavior of plastic materials is typically described with the help of the modulus of elasticity. Similar to the stress-strain lines resulting from tensile tests, the slope of the isochronous stress-strain diagrams can be used to calculate a modulus. It will decline with increasing load time, temperature, and stress and thus describes the creeping behavior of the material. This modulus is called the creep modulus E_c (Figure 3.33) when it is determined via creep tests, and relaxation modulus E_r when it is determined via relaxation tests.

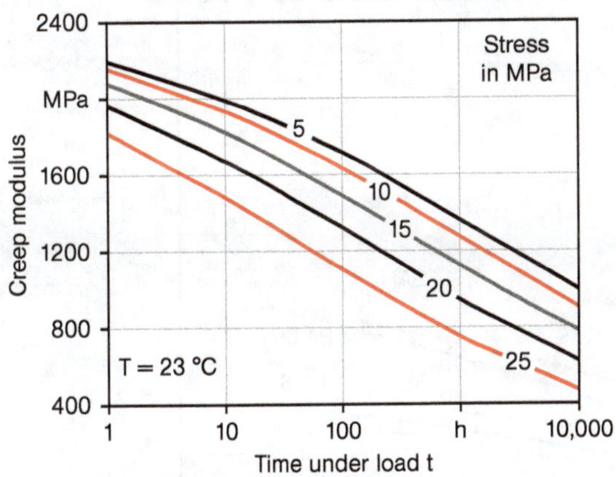

Figure 3.33
Secant creep modulus of PBT at 23 °C

Both modulus values coincide for identical times and load levels (either a stress or a strain value can be selected as a measure for the load level).

The time-creep lines have to be evaluated at the respective relevant service temperature because the mechanical behavior of plastic materials is strongly dependent on temperature in addition to time and load. Therefore, material data sheets often provide several sets of diagrams (see Figure 3.31 to Figure 3.33) for various service temperatures (typically at temperature intervals of 20 °C). Figure 3.34 to Figure 3.37 show such isochronous stress-strain diagrams for different plastic materials.

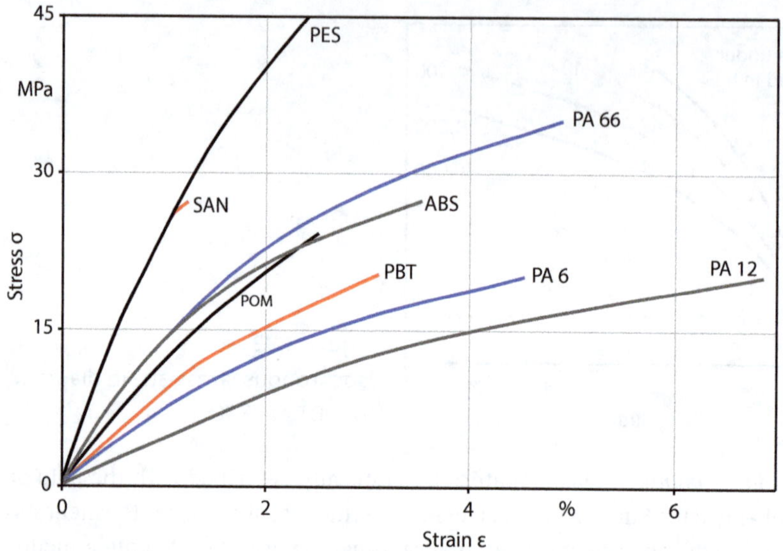

Figure 3.34 Isochronous stress-strain diagrams for unfilled thermoplastics at 23 °C and 1000 h

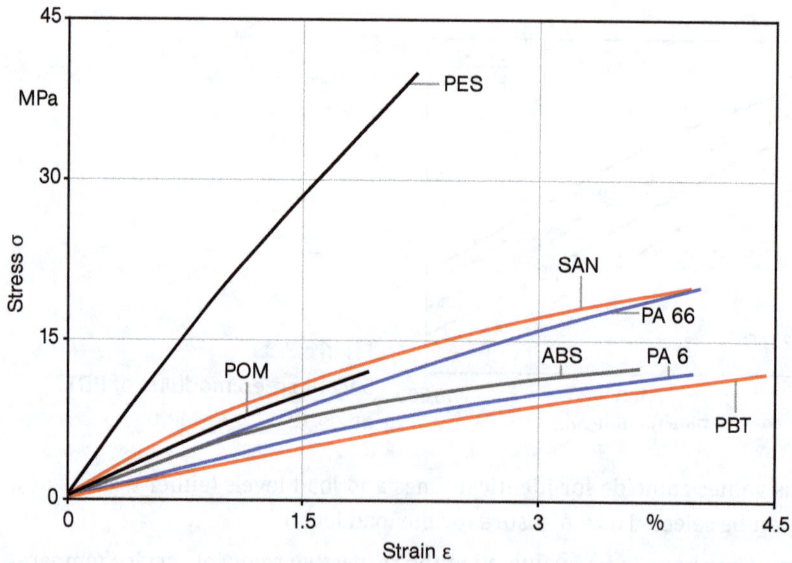

Figure 3.35 Isochronous stress-strain diagrams for unfilled thermoplastics at 60 °C and 1000 h

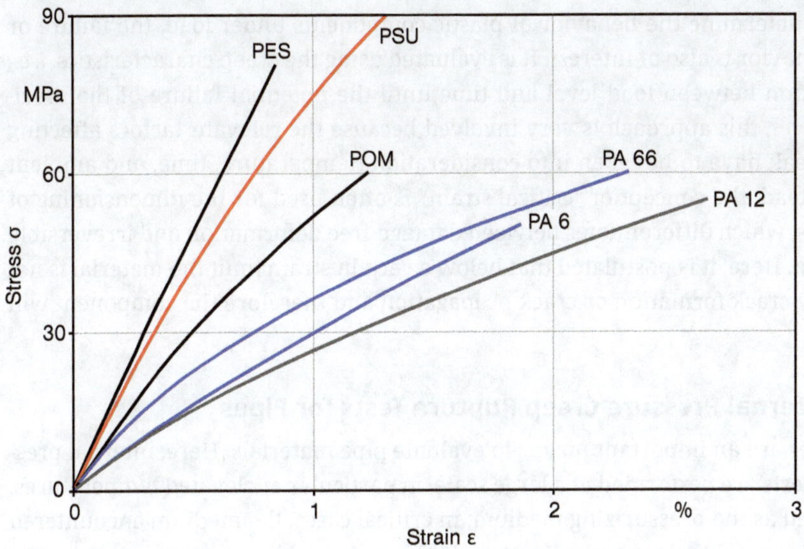

Figure 3.36 Isochronous stress-strain diagrams for filled thermoplastics (25 to 35% GF) at 23 °C and 1000 h

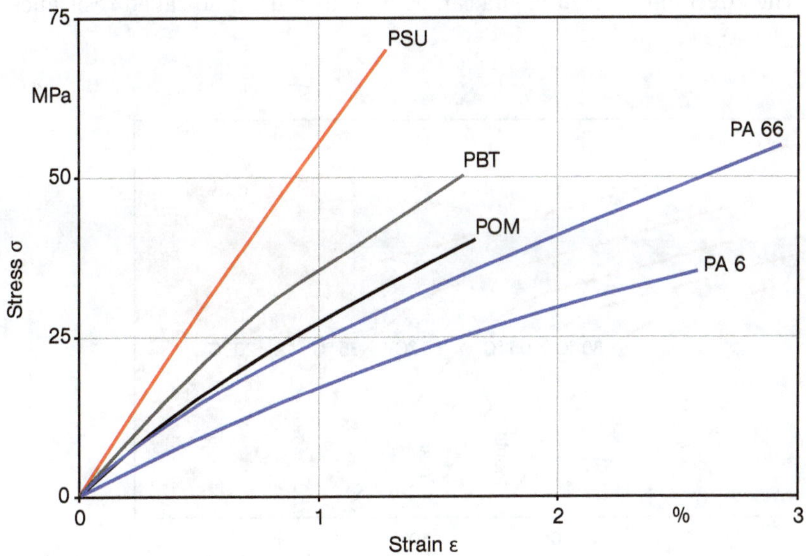

Figure 3.37 Isochronous stress-strain diagrams for filled thermoplastics (25 to 35% GF) at 60 °C and 1000 h

In order to determine the behavior of plastic components under load, the failure or fracture behavior is also of interest. It is evaluated using the creep characteristics, *i. e.*, the correlation between load level and time until the potential failure of the specimen. However, this approach is very involved because the relevant factors affecting stress at break have to be taken into consideration (temperature, time, and ambient media). Instead, the concept of "critical strain" is often used for the dimensioning of components, which differentiates between damage-free deformation and irreversible deformation. Here, it is postulated that below a certain strain limit the material is not damaged by crack formation or crack propagation and therefore the component will not fail.

3.3.3.2 Internal Pressure Creep Rupture Tests for Pipes

Creep curves are an important means to evaluate pipe materials. Here, internal pressure creep tests are performed at a large scale, in particular at elevated temperatures. Water is used as the pressurizing medium; in critical cases, the medium encountered under service conditions may be employed. The permissible long-term stresses (for water pipes 50 years) are calculated using these test results under consideration of a safety factor. The drops in the creep lines seen in Figure 3.38 indicate that the investigated PE exhibited stress crack formation under the influence of water and oxidative aging, see Section 3.7.3. For other media, this drop may appear after different times under load. The effect of different media can be described using so-called resistance factors F_{CR}.

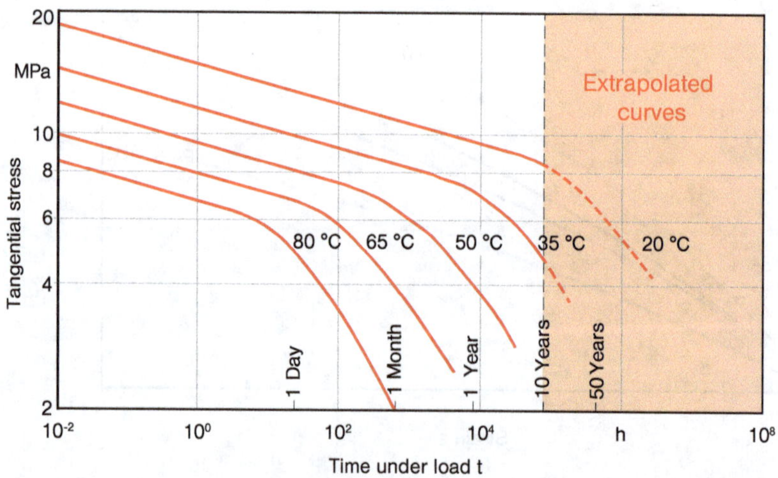

Figure 3.38 Creep resistance of PE-HD pipes under internal water pressure

These factors indicate by which time factor the steep part of the creep lines has to be shifted compared to water after specific times under load.

3.3.4 Dynamic Long-Term Behavior

The most common cause for the failure of plastic components under the conditions of their intended service is dynamic or repeated loading, often in the presence of notches, weld lines, or anisotropies (molecular or reinforcing fiber orientation) and the simultaneous exposure to chemicals (media initiated stress cracking) or other ambient loads. This section will therefore cover this type of load in more detail.

When designing plastic components that are exposed to repeated alternating loads, the strength characteristics determined under increasing uniform loading or deformation or under static long-term loading are not suitable. Exposure to repeated loads or oscillation typically causes component failure at lower stress or deformation.

Fatigue tests are used to determine the behavior of plastic specimens or components under oscillating loads. Fatigue testing of components is recommended for critical applications because, similar to the values determined by impact tests, the characteristics are highly dependent on processing parameters during manufacturing of the injection molded components and the load conditions. The load profile for fatigue tests (stress or strain) is typically almost sinusoidal.

Other than metals, plastics exhibit high mechanical damping, expressed as the mechanical loss factor tan δ (Table 3.4). Under load, this damping effect causes irreversible energy absorption, which will lead to heating of the component (in particular under repeated loads) because plastics are poor heat conductors, delaying heat transfer to the surroundings. This effect has to be considered during testing and result analysis, especially because even small increases in temperature may cause significant changes in material properties.

Table 3.4 Mechanical Loss Factor tan δ

Material	20 °C	60 °C
PE-LD	0.17	0.06
PP	0.07	0.07
PS	0.013	0.028
ABS	0.015	0.028
PVC rigid	0.018	0.025
PTFE	0.075	0.06
PMMA	0.08	0.10
POM	0.014	0.015
PC	0.008	0.010

Table 3.4 Mechanical Loss Factor tan δ *(continued)*

Material	20 °C	60 °C
PA 6 dry	0.01	0.16
PA 6 ambient moisture	0.15	0.06
UP	0.02	0.4
EP	0.02	0.02
PF Type 31	0.016	0.022
MF Type 151	0.016	0.022
Steel	0.00002	0.001
Copper	0.0002	0.001

In a test series, a variety of thermoplastics and thermosets were subjected to dynamic loads and the number of cycles to failure was recorded. Figure 3.39 shows the stress amplitude as a function of the number of cycles to failure. The example of epoxy (EP) shows that thermosets exhibit higher stress at break values than thermoplastics. This effect is explained by the fact that thermosets are more brittle and exhibit lower damping and/or friction values. Therefore, there is a lower temperature increase in the specimen. Higher load frequencies create higher temperature increases and thus lead to earlier failure.

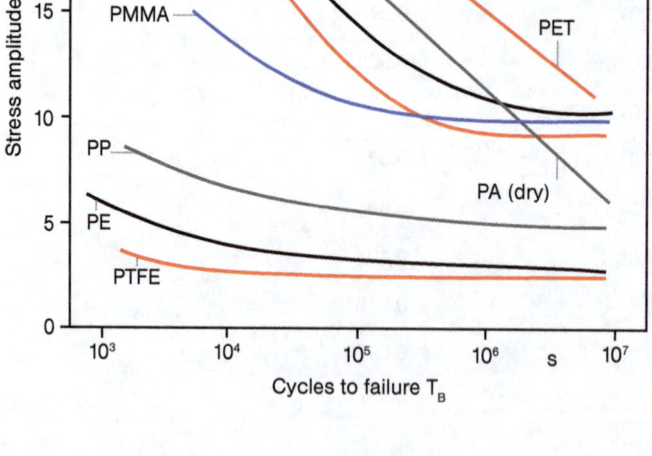

Figure 3.39 Fatigue curves for various thermoplastics and thermosets at a test frequency of 30 Hz about zero mean stress

A newer testing method uses a thermocouple attached to the specimen or non-contact temperature measurements to control the loading frequency so that the specimen temperature increases only by a defined amount above ambient temperature, *i. e.*, stays constant during the test. It is currently not clear how the resulting variation in frequency at different stress levels/material damping levels will affect the comparability of the results.

When plastic components are injection molded, weld lines, *i. e.*, areas in the mold where two melt streams meet, are not always avoidable. They are not necessarily detrimental to strength properties measured in static tests; however, they may be weak points under dynamic as well as under impact load. The same is valid for notches of any kind. Reduction factors have proven useful for the characterization of the susceptibility of plastics to such effects. They represent the ratio of strength characteristics (determined in short-term tensile tests such as yield stress and tensile strength) and fatigue strength determined by fatigue tests.

In order to perform fatigue tests of plastic materials, flat specimens are subjected to flexural deformations. Because of stress relaxation and temperature increase, the stress absorbed by the specimen decreases with increasing number of load cycles. In order to make the results comparable, all parameters describing loads have to be measured continuously and recorded together with the test results.

Servo-hydraulically controlled test machines allow for ideal plastic fatigue tests during which strains and deformations that can be kept constant; however, these are rather involved tests.

3.3.5 Moduli and Poisson's Ratio

During uniaxial tensile tests, a longitudinal elongation is imposed on the specimen. The ratio of stress and strain is called the modulus of elasticity (Young's modulus). Perpendicular to the stress direction, a transverse contraction is observed. The ratio of relative transverse contraction ε_t to longitudinal elongation ε is called Poisson's ratio υ. The values for Poisson's ratio range from 0.3 to 0.5 for homogeneous and isotropic materials. A Poisson's ratio of 0.5 means that the volume does not change under load. Thermoplastics with Young's moduli between 2000 and 3000 N/mm^2 exhibit Poisson's ratios between 0.3 and 0.35. The lower the Young's modulus, the closer the Poisson ratio approaches 0.5, which represents the value characteristic for the rubber-elastic state. This value is also reached by thermoplastics at high temperatures (Figure 3.40).

Gienke und *Meder* offer approaches to approximate Poisson's ratio using the Young's modulus [6]. The highest Young's modulus E_0 known for a specific material is related to a Poisson's ratio of 0.3, while a Poisson's ratio of 0.5 correlates to a Young's modulus approaching zero. Using the linear rule of mixtures, any time-, temperature-, and strain-dependent Young's modulus E^* can be assigned a corresponding Poisson's ratio:

$$\nu = 0.3 + 0.2 \cdot (1 - {E^*}/{E_0}) \tag{3.14}$$

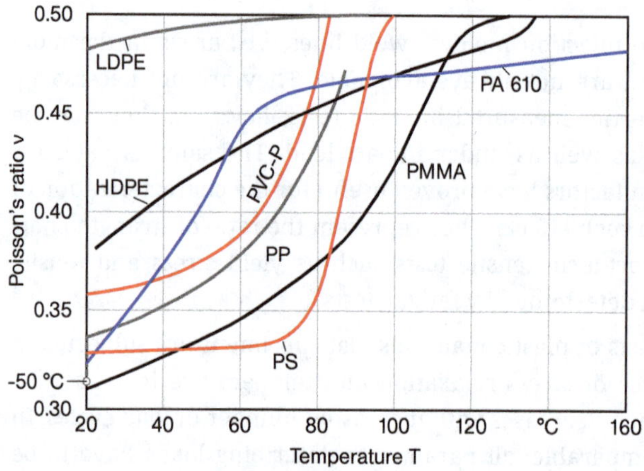

Figure 3.40 Poisson's ratio as a function of temperature

For anisotropic, *e. g.*, fiber reinforced, plastics, the Poisson's ratio – similar to the modulus of elasticity – is dependent on fiber content, fiber orientation, and load direction with respect to the reinforcement direction (Figure 3.41).

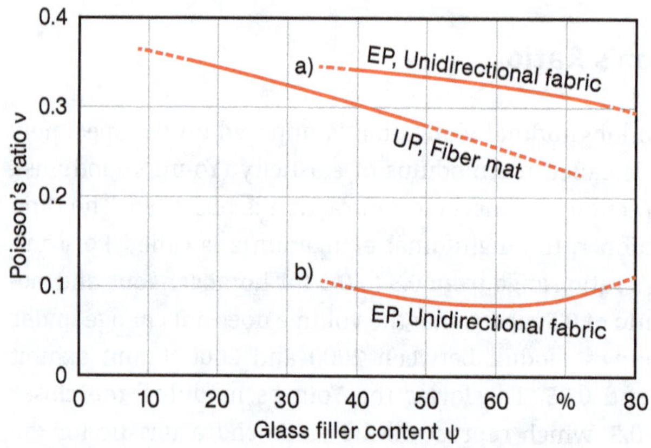

Figure 3.41 Poisson's ratio as a function of fiber content and fiber orientation

The Poisson's ratio is required to calculate multi-axial deformation states when determining the strength of materials. The following basic equations serve this purpose (isotropic materials):

Tension/compression load		
$\sigma = \varepsilon \cdot E$	σ	= Tension/compression stress
	ε	= Strain
	E	= Modulus of elasticity (Young's modulus)
Shear loads		
$\tau = \gamma \cdot G$	τ	Shear stress
	γ	Shear, angle of shear
	G	Shear modulus
Uniform, *e. g.*, hydrostatic, pressure		
$p = \Delta v/v \cdot B$	p	Pressure
	$\Delta v/v$	Relative change in volume
	B	Bulk modulus
Load on a laterally deformed sheet in direction of the surface normal		
$\sigma = \varepsilon \cdot L$	L	Longitudinal-wave modulus

Table 3.5 Correlation between Moduli

	Expressed by				
Parameter	E, ε	G, ε	K, ε	E, G	E, K
Poisson's ratio υ	–	–	–	$\frac{E}{2G} - 1$	$\frac{1}{2} - \left(1 - \frac{E}{3K}\right)$
Young's modulus E	–	$2G(1+\nu)$	$3K(1+\nu)$	–	–
Shear modulus G	$\frac{E}{2(1+\nu)}$	–	$\frac{3K(1-2\nu)}{2(1-\nu)}$	–	$\frac{E}{3-\frac{E}{3K}}$
Compression modulus B	$\frac{E}{3(1-2\nu)}$	$\frac{2G(1+\nu)}{3(1-2\nu)}$	–	$\frac{G}{3\left(3\frac{G}{E}-1\right)}$	–
Longitudinal-wave modulus L	$\frac{E(1-\nu)}{(1+\nu)(1-2\nu)}$	$\frac{2G(1-\nu)}{(1-2\nu)}$	$\frac{3K(1-\nu)}{(1-\nu)}$	$\frac{4G-E}{3-\frac{E}{G}}$	$\frac{3K+E}{3-\frac{E}{3K}}$

In general, only moduli of elasticity are available to designers for the calculations of a component. With the help of Poisson's ratio, the four moduli E, G, B, and L can be converted (Table 3.5). It is important to note that the correlations listed in the table are valid only when all values refer to the same temperature, load rate, and time under load.

3.3.6 Density

For solid plastics, the density d_B is determined by:

$$d_B = m/V$$

Where:

V = volume of the entire material, *i. e.*, including potential pores or voids

m = mass of the specimen

The bulk density is of interest not only when designing semi-finished products and molded components, it can also provide information on the correlation with technological properties, *e. g.*, for polyolefins. Standard physical methods are used for testing: buoyancy and floating equilibrium method, pycnometer, and density gradient column. The buoyancy method is the most commonly used method because it facilitates quick measurements. The pycnometer allows density measurements for powders and small molded components with undercuts or hollow sections where air pockets could form when using the buoyancy method, thus distorting the results. The floating equilibrium method also provides accurate results and is used when the density of the specimen can be approximated, *e. g.*, during receiving inspection tests or quality control. Density gradient columns are used only in R & D facilities.

Foamed materials constitute a special case for the determination of a material's density. The bulk density of a foamed material is calculated as the quotient of its mass and volume and it is reported in kg/m^3. Using geometrically simple specimens with a volume of at least 100 cm^3 it is determined by length and weight measurements. Structural and integral foams are manufactured with a cell distribution profile where the density continuously decreases from an almost solid outer skin to the center (Figure 3.42). By removing and measuring individual layers, the bulk density profile can be determined.

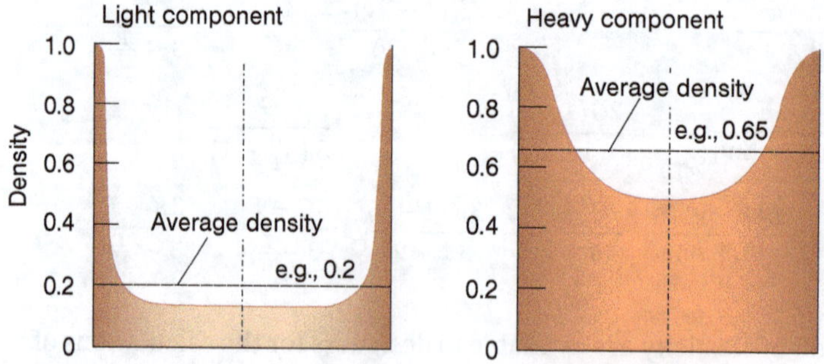

Figure 3.42 Density profile of structural foam components

3.4 Thermal Properties

Even at room temperature, the properties of plastics materials strongly depend on temperature. Therefore, service-relevant characteristics are generally determined in a wide, service-relevant temperature range.

3.4.1 Permissible Service Temperatures

The service temperature range, *i. e.*, the permissible upper and lower temperature limits for the application of plastics, depend on two factors:

- the reversible softening or embrittlement under short-term temperature loads, and
- the behavior under long-term temperature influence with or without additional loads.

3.4.1.1 Effect of Short-Term Temperature Influence

The best way to characterize material behavior under short-term temperature influence is to determine mechanical properties as a function of temperature. The respective diagrams reveal characteristic steps or maxima that indicate glass transitions (see Section 2.3.2). In general, all characteristics values may be used here, *e. g.*, the Young's modulus. However, this requires either complex measuring techniques and instrumentation or a large number of tests at different temperatures. A very simple method is the torsion pendulum, and an increasingly popular method is dynamic mechanical analysis (DMA, see Section 3.10.3), both of which can generally be used for all types of plastics and across a wide temperature range.

This range covers temperatures from −180 °C to melting temperature for semi-crystalline materials and to approx. 50–80 °C above glass transition temperature for amorphous plastics. Thermosets can be tested up to their respective degradation temperatures.

The test results, all functions of temperature, include modulus and damping values, the shear modulus G, and the loss factor d = tan δ. The latter is a measure for the amount of mechanical work that was expended during loading and could not be recovered after load relaxation, but was converted to heat. G und tan δ as function of temperature are basically the calling cards of plastic materials. Chapter 2 provides more detail on the schematic profile of these characteristics for both semi-crystalline and amorphous plastics (Figure 2.11 and Figure 2.13). Figure 3.43 to Figure 3.45 provide an overview of the shear modulus curves for different materials.

The glass transition temperature T_g of amorphous thermoplastics, such as PS and SAN, is characterized by a steep modulus decrease (Figure 3.43). Below this temperature the brittleness of the material increases (energy-elastic) and thus it is only used in this temperature range. Above T_g amorphous thermoplastics are flexible and not dimensionally stable (entropy-elastic).

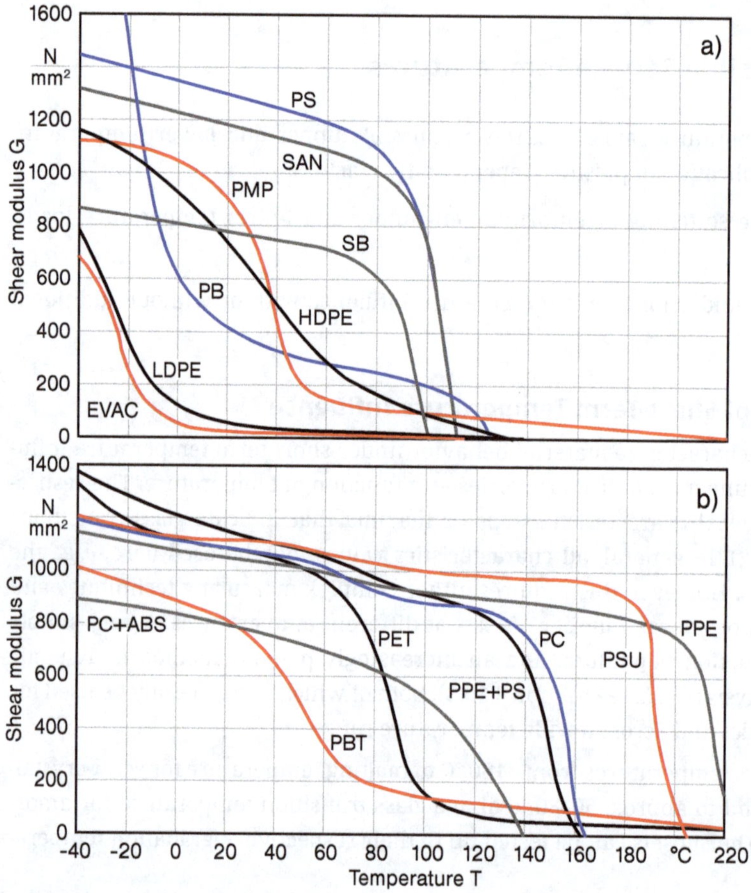

Figure 3.43 Shear modulus as a function of temperature for amorphous and semi-crystalline plastics

Semi-crystalline thermoplastics also exhibit a decrease in modulus at T_g; however, it is significantly smaller than that of amorphous plastics. The larger decrease occurs at the crystallite melting temperature T_s. Above T_s crystallites can no longer exist; the plastic material transitions into the molten state – it turns viscous.

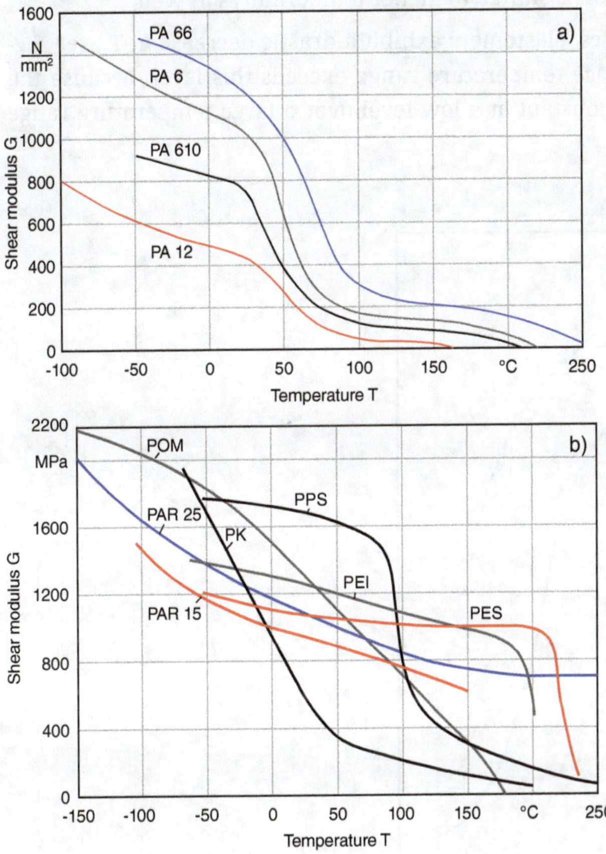

Figure 3.44 Shear modulus as a function of temperature for engineering and high-performing thermoplastics

Figure 3.45 uses UP resin as an example to show the shear modulus profile for thermosets. Because of the high level of molecular crosslinking they exhibit high modulus and low damping values across the entire service temperature range. The upper range of service temperatures for thermosets is limited by their chemical degradation, not because they turn viscous. Reinforcing fibers significantly increase the shear modulus and thus the transition is shifted to higher temperatures as well.

Like amorphous thermoplastics, elastomers exhibit a drastic decrease at T_g, see Figure 5.21. However, their service temperature range exceeds this level, because the moduli remain more or less constant at a low level over a large temperature range (entropy-elastic range).

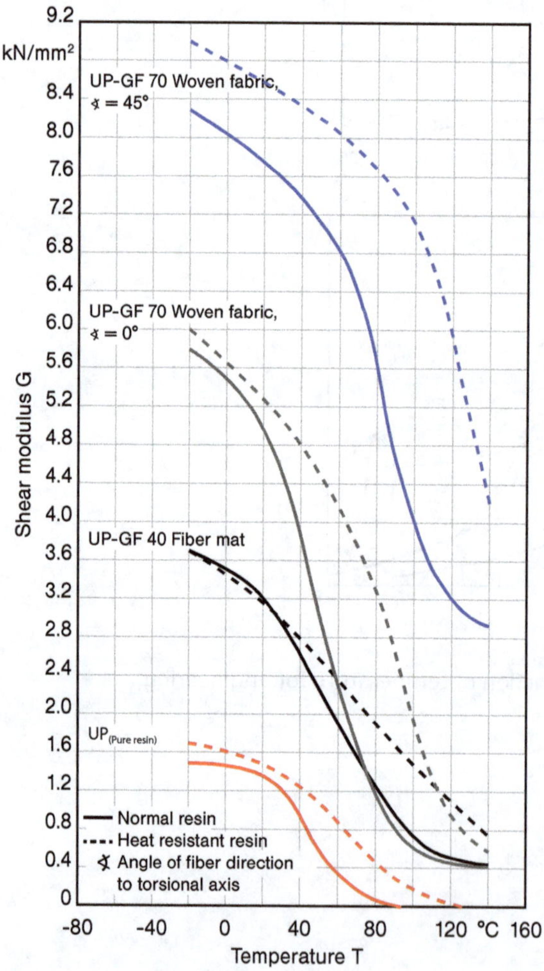

Figure 3.45 Shear modulus as a function of time for reinforced UP resin

Table 3.6 details two procedures used to determine dimensional stability (heat distortion resistance) under thermal load: *Vicat* softening and heat distortion temperature (HDT). Both methods subject the specimens to defined loads and heat them at a specified rate while measuring the increase in deformation. Vicat softening temperature and heat distortion temperature HDT are defined as the temperatures at which the deformation reaches a specified value (Table 3.6).

Table 3.6 Methods to Determine Heat Distortion Resistance

	HDT	Vicat
Test standard	ISO 75-1/2, ASTM D 648	ISO 306, ASTM D 1525
Test set-up (dimensions in mm)	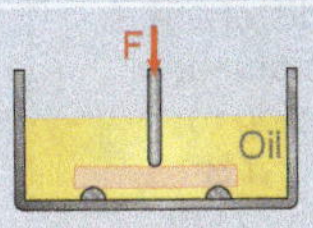	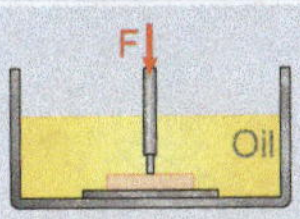
Type of load	Bending	Penetration of a needle A = 1 mm^2
Load magnitude	Bending stress σ_b A: 1.85 N/mm^2 *) B: 0.46 N/mm^2 C: 8 N/mm^2	Pressure A: 10 N B: 50 N*)
Specified deformation	Deflection depending on specimen height 0.21 to 0.33 mm**)	Penetration depth 1 mm
Temperature	Increase 2 K/min	Increase 50 K/h*) or 120 K/h
Heat transfer medium	Liquid bath	Liquid bath, in specific cases air
Specimen l × b × h in mm	l = 110 b = 3.0 . . . 4.2 (13) h = 9.8 . . . 15	10 × 10 × 3 . . . 6.4
Example of result (abbreviated notation)	HDT/A = 95 °C	VST/B 50 = 82 °C

*) Preferred method
**) Corresponding to an outer fiber strain of 0.2%

These characteristic values are suitable to differentiate the softening behavior within a class of materials. However, in order to determine permissible service temperatures, additional testing or empirical data are required. Figure 3.46 shows the shear modulus/temperature curves and Vicat and HDT values, and the shaded temperature regions represent empirical upper service temperatures for short-term heat exposure without additional loads. The values do not seem to correlate well, which may be caused by the fact that the determined Vicat and HDT values depend strongly on the testing conditions. Figure 3.47 shows this dependence using HDT values as a function of bending stress.

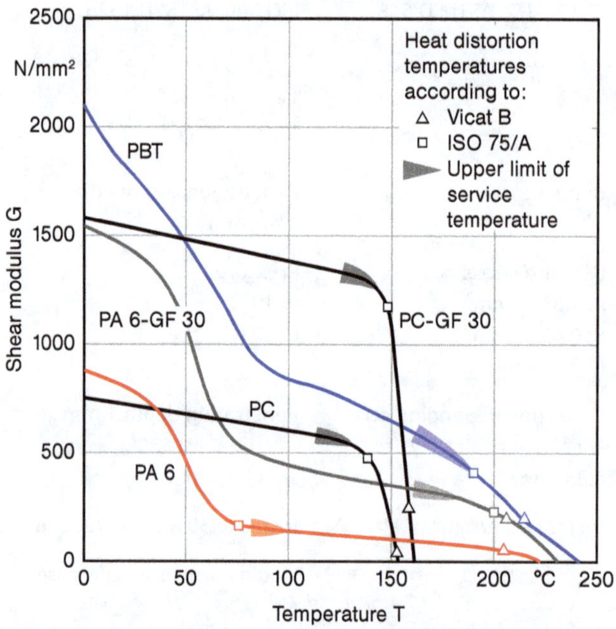

Figure 3.46 Comparison of "heat distortion resistance" values

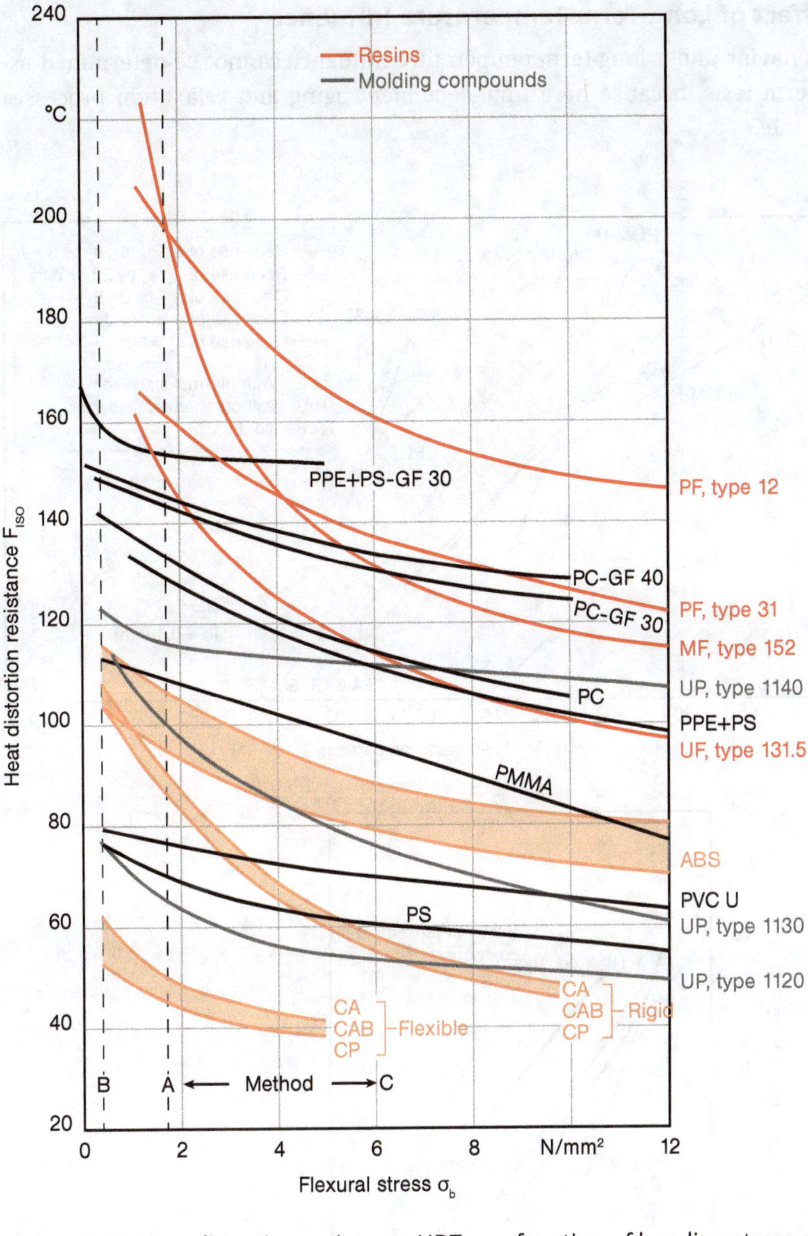

Figure 3.47 Heat distortion resistance HDT as a function of bending stress according to ISO 75

3.4.1.2 Effect of Long-Term Temperature Influence

Material behavior under long-term temperature influence cannot be determined using short-term tests, because here time-dependent aging and relaxation processes also play a role.

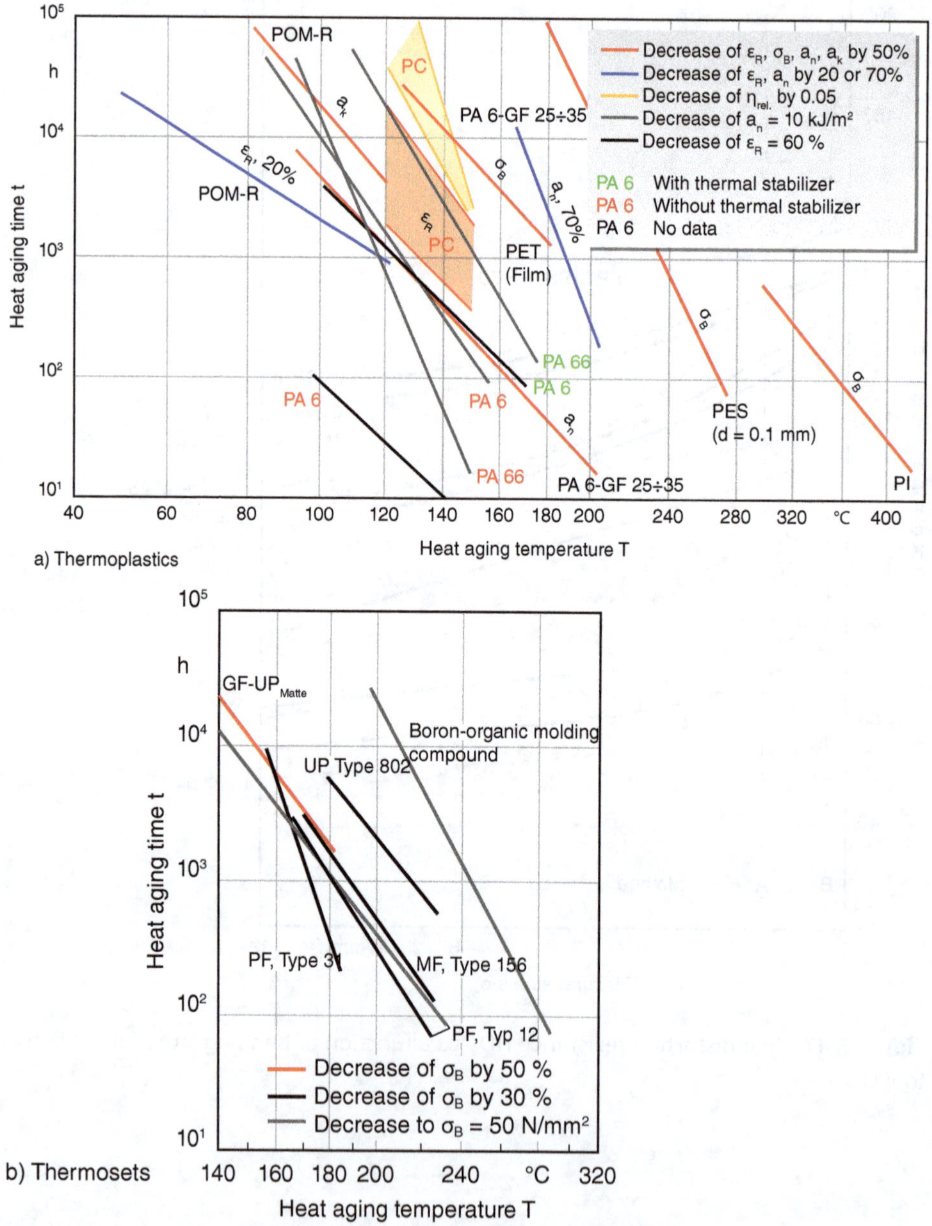

Figure 3.48 Temperature-time limits for selected thermoplastics (a) and thermosets (b)

The aging behavior of plastic not subjected to other loads is determined by temperature indices (TI). Prior and after long-term storage (5000 to 20,000 h) of specimens in ventilated heating chambers at elevated temperatures, selected properties are evaluated. Typically, there is a linear correlation between the logarithm of the time of heat aging that leads to a specific change in properties and the temperature (Figure 3.48). The temperature index is expressed as the temperature at which (after a defined time) a measured characteristic value (measured at room temperature) has decreased to a specified percentage of the original value. Note that these indices are valid only for one characteristic value and its change. The selection of limiting conditions depends on the requirements on the component. Certain mechanical or electrical characteristics may be subject to regulations.

If the plastic specimen is subject to mechanical loads or deformations in additions to temperature loads, the behavior can be evaluated only with the help of creep and relaxation tests at the respective temperatures – possibly also with respect to exposure to certain media encountered in service, see Section 3.3.3.

3.4.2 Specific Heat Capacity, Specific Enthalpy, Heat Value

The specific heat capacity c_p is defined as the heat Q required to increase the temperature of 1 kg of mass of a material by 1 K:

$$Q = m \cdot c_p \cdot \Delta T \tag{3.15}$$

Heat capacity is typically determined at constant pressure (measurement see Section 3.10.5). The specific heat determined at constant volume c_v is approx. 10% lower. Heat capacities determined at room temperature range from 1 to 2 kJ/kg·K and are thus significantly higher than those of metals. However, when considered with respect to mass rather than volume unit, the heat capacity values for both material classes are in the same order of magnitude.

Figure 3.49 shows that the specific heat capacity is temperature dependent. The specific enthalpy is the energy of a material with respect to mass at a specific temperature and can therefore be calculated by integration of the heat capacity curve from zero to this specific temperature. Thus, the energy required for heating a material during molding or shaping operations can be determined using the enthalpy-temperature curves shown in Figure 3.50 and Figure 3.51. This energy is calculated as the difference between the values at the initial and considered final temperature.

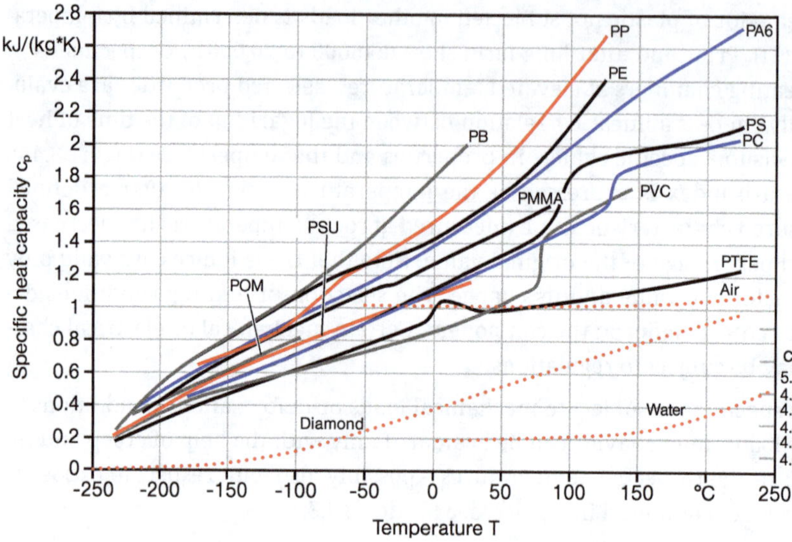

Figure 3.49 Specific heat capacity as a function of temperature

For filled or reinforced plastics, the specific heat capacity of the compound C_m can be calculated using the specific heat capacity values of the components (filler C_f and plastic C) and the weight proportion ψ of the filler:

$$C_m = \psi \cdot C_f + (1 - \psi) \cdot C \tag{3.16}$$

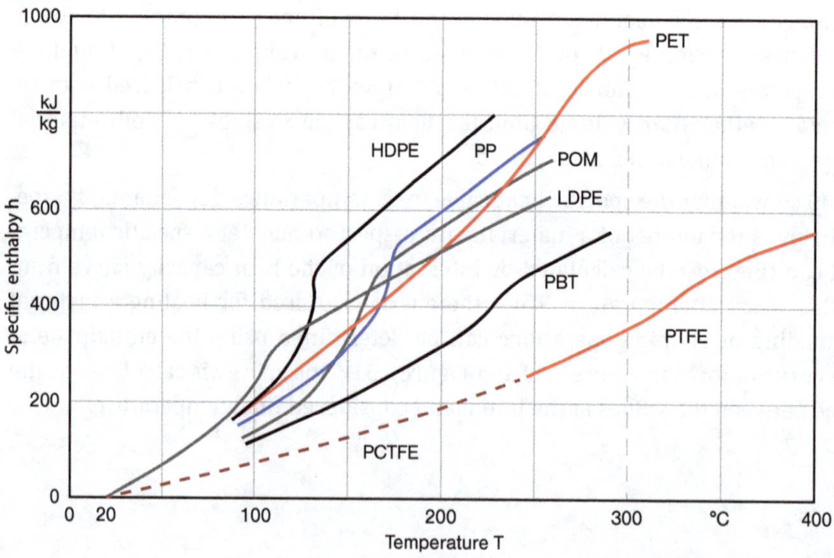

Figure 3.50 Specific enthalpy as a function of temperature for semi-crystalline plastics

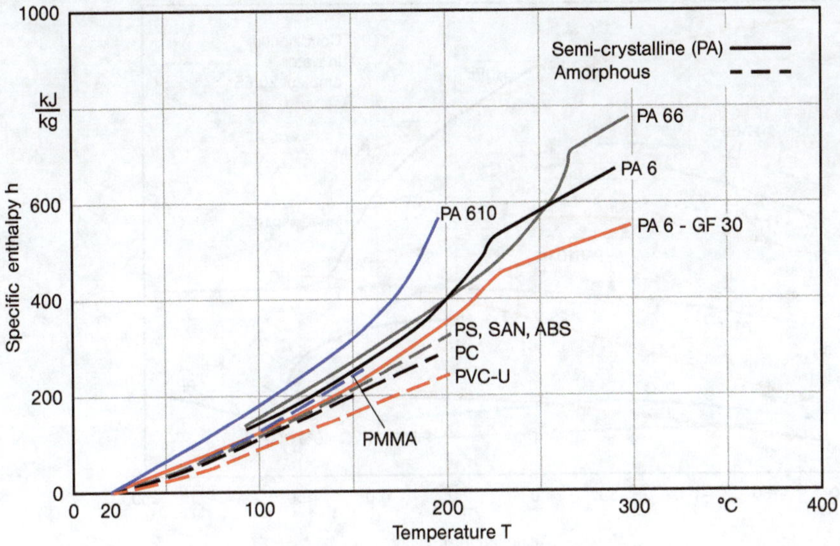

Figure 3.51 Specific enthalpy as a function of temperature for selected polyamides (semi-crystalline) and amorphous thermoplastics

The heat value (in kWh/kg) is an important characteristic value for the evaluation of thermal recyclability. For plastics it is in the same range as common fuels.

3.4.3 Thermal Conductivity

The thermal conductivity λ (in W/K · m) represents the heat Q (in J) that is conducted in 1 s through a layer with the area A = 1 m^2 and a thickness of s = 1 m when the temperature differential ΔT between both surfaces is 1 K. The heat per thickness is stated in watts: $Q = \lambda \cdot A \cdot \Delta T/s$. The corresponding measurements are detailed in Section 3.10.8.

Compared to metals, plastics exhibit very low thermal conductivity. The thermal conductivity values of different plastics vary widely in the temperature range of technical interest, however, they depend little on temperature (see Figure 3.52 and Figure 3.53). Drawing and the resulting orientation of macromolecules or fibers causes an increase in thermal conductivity in the preferred direction (| |) and a decrease perpendicular to it (⊥), see Figure 3.54):

$$3/\lambda = 1/\lambda_{||} + 2/\lambda_{\perp} \quad (3.17)$$

Where:

λ = thermal conductivity of isotropic material

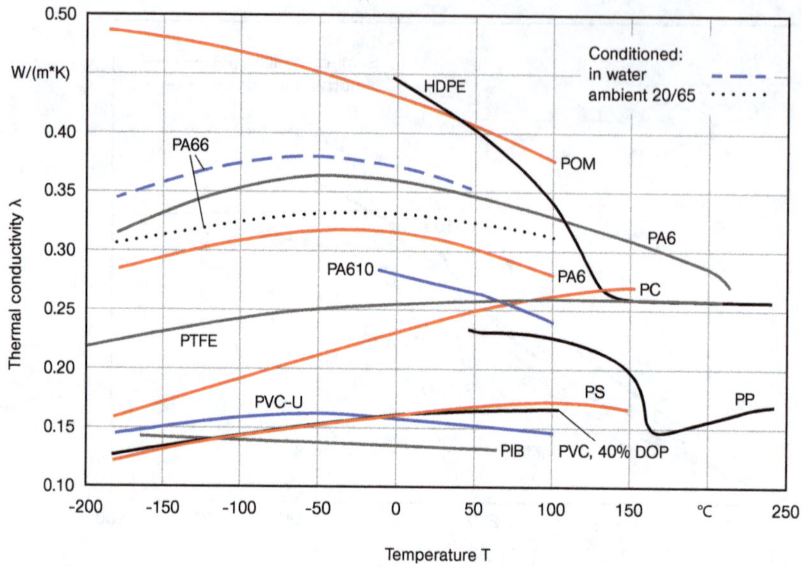

Figure 3.52 Thermal conductivity as a function of temperature for selected thermoplastics

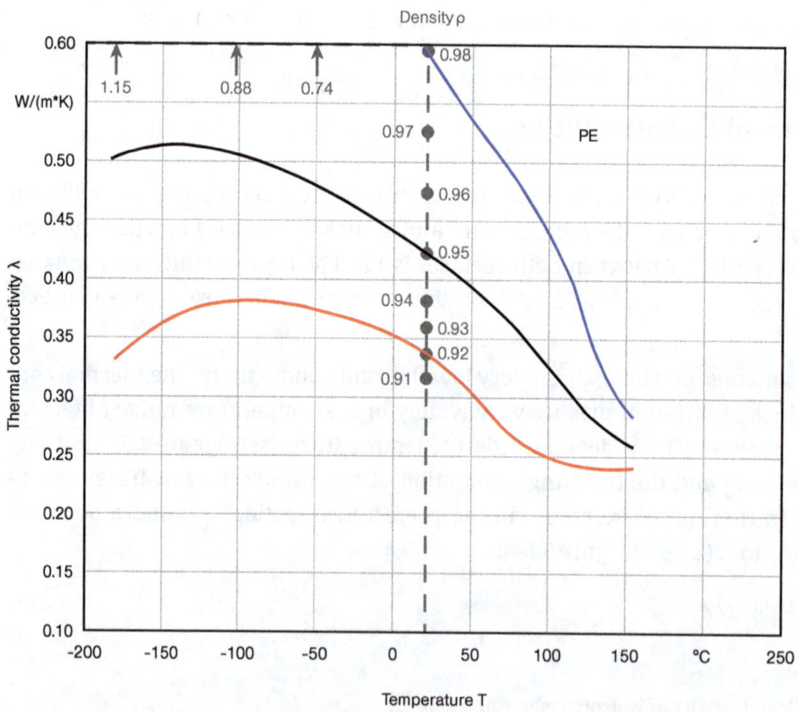

Figure 3.53 Thermal conductivity of PE materials with varying density as a function of temperature

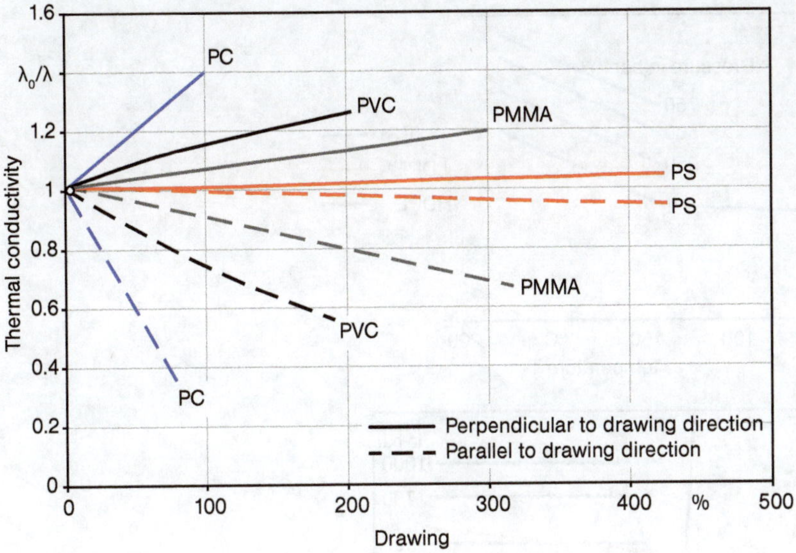

Figure 3.54 Thermal conductivity as a function of draw ratio

Thermal conductivity increases with increasing pressure (Figure 3.55). The addition of fillers affects thermal conductivity: silica powder, glass, metal, aluminum nitride powder, and carbon fibers increase thermal conductivity, while foaming with air, CO_2, or other blowing agents improves heat insulation. Both options are used with almost all plastic types.

The difficult measurements and anisotropy effects cause wide variations in the reported values.

The thermal conductivity of mixtures λ_m can be calculated from the thermal conductivity of the filler λ_F and that of the plastic material λ and the volume fraction of the filler ϕ using the following rule of mixtures:

$$\lambda_m = \frac{\lambda_F \cdot \lambda}{\varphi \cdot \lambda + (1 - \varphi) \cdot \lambda_F} \tag{3.18}$$

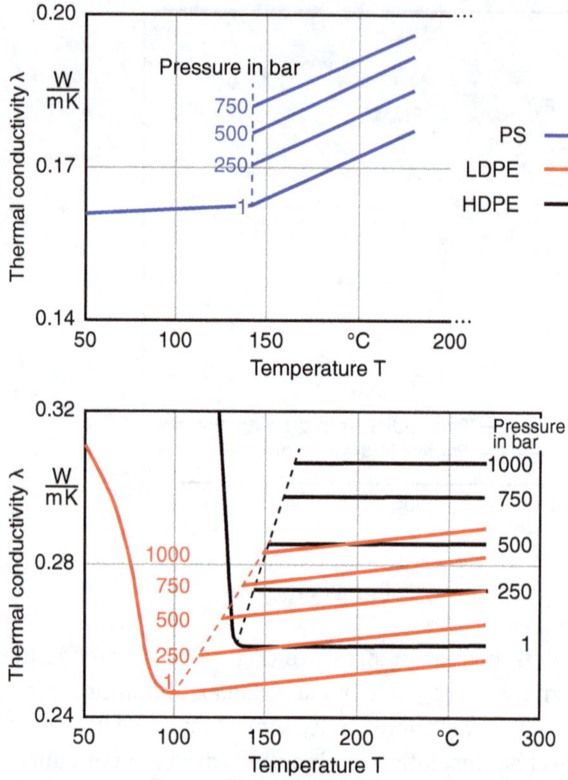

Figure 3.55 Thermal conductivity as a function of pressure for amorphous (top) and semi-crystalline thermoplastics (bottom)

3.4.4 Thermal Effusivity

When both specific heat capacity c_p and thermal conductivity λ are known, additional characteristic numbers can be derived. The thermal effusivity *b*, which depends also on the density ρ, is a measure for a material's ability to exchange thermal energy with its surroundings:

$$b = \sqrt{\lambda \cdot c_p \cdot \rho} \tag{3.19}$$

With the help of the thermal effusivity value, the contact temperature T_K can be calculated. It refers to the temperature encountered when two bodies A and B with different temperatures T come into contact:

$$T_K = \frac{b_A \cdot T_A + b_B \cdot T_B}{b_A + b_B} \tag{3.20}$$

Where:

T = absolute temperature

Because of plastics' comparatively low thermal effusivity, cold plastic components feel warmer than highly conductive metals or water, which exhibit markedly lower contact temperatures.

3.4.5 Thermal Diffusivity

Heat transfer, which is critical for all heating and cooling processes, is determined by a material's heat diffusivity a:

$$a = \frac{\lambda}{c_p \cdot \rho} \tag{3.21}$$

Where:
λ = thermal conductivity
c_p = specific heat capacity
ρ = density

It can be used to approximate the average cooling time t_K (time until the average ejection temperature is reached) for an injection molded component – an important factor for injection molding processes:

$$t_K = \frac{s^2}{\pi^2 \cdot a} \cdot \ln\left(\frac{8}{\pi^2} \cdot \frac{T_M - T_W}{T_E - T_W}\right) \tag{3.22}$$

Where:
s = wall thickness
T_M = average melt temperature
T_W = mold wall temperature
T_E = average ejection temperature

Equation 3.22 shows that the cooling time increases with the square of the wall thickness, *i. e.*, an increase in wall thickness by a factor 2 will lead to an increase in cooling time by a factor of 2^2.

The thermal diffusivity of mixtures a_m can be calculated from the thermal diffusivity of the filler a_F and the volume fraction of the filler ψ according to Equation 3.23:

$$a_m = (1 - \psi) \cdot a + 0.36 \cdot \psi \cdot a_F + 0.64 \cdot \psi^3 \cdot a_F \tag{3.23}$$

3.4.6 Coefficient of Thermal Expansion

The expansion of plastic materials with increasing temperature affects the manufacturing of both molded components and semi-finished products (see Section 3.2.2.2) and the service properties of plastic parts. For information regarding measurements, see Section 3.10.7. Because plastics exhibit rather high coefficients of thermal expansion compared to metals, joining of these two materials can lead to difficulties (bimetal effect).

We distinguish between linear (α) and cubic (β) coefficients of thermal expansion. The linear coefficient of thermal expansion states by how much a temperature change by 1 K will increase or decrease 1 m of a specific material. The cubic coefficient of thermal expansion states by how much volume 1 m^3 of a specific material will increase or decrease in response to a temperature change of 1 K. Both are stated in 1/K. Equation 3.24 is valid for homogeneous, solid bodies.

$$\beta = 3 \cdot \alpha \tag{3.24}$$

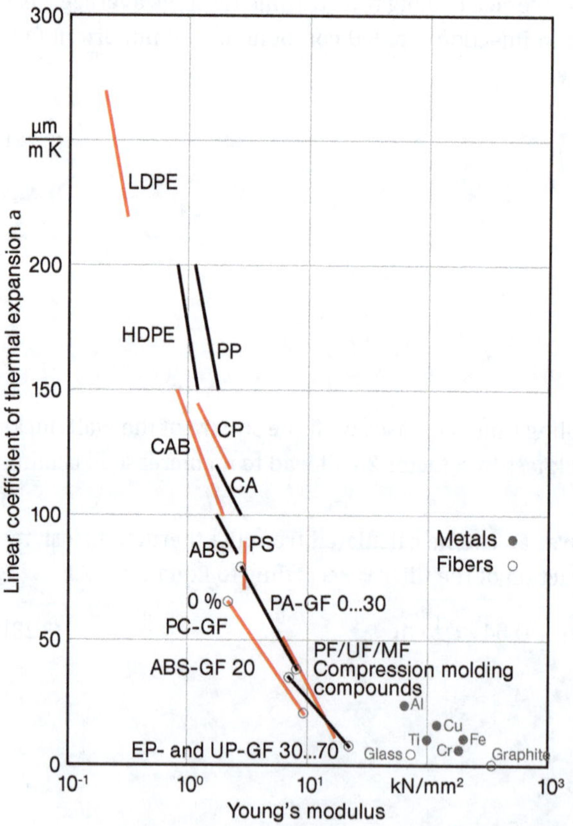

Figure 3.56 Relationship between linear coefficient of thermal expansion and modulus of elasticity

In general, the coefficient of thermal expansion decreases with increasing modulus of elasticity (Figure 3.56). Log-log plots are almost linear and Equation 3.25 holds:

$$\alpha = 1/E^m \tag{3.25}$$

The coefficient of thermal expansion is temperature dependent (Figure 3.57). Combining plastics with materials that exhibit lower coefficients of thermal expansion, such as glass or graphite, reduces the coefficient of expansion of these so-called composite materials, and brings it more in line with the coefficients of metals. Elongated fillers and reinforcing materials that are oriented in the direction of flow cause anisotropic coefficients of thermal expansion. The same effect is encountered with molecular orientations caused by drawing, see Figure 3.58.

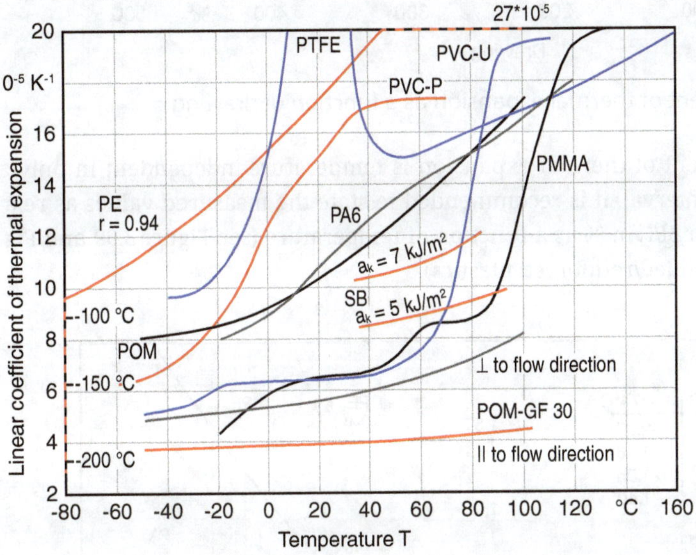

Figure 3.57 Coefficient of thermal expansion as a function of temperature

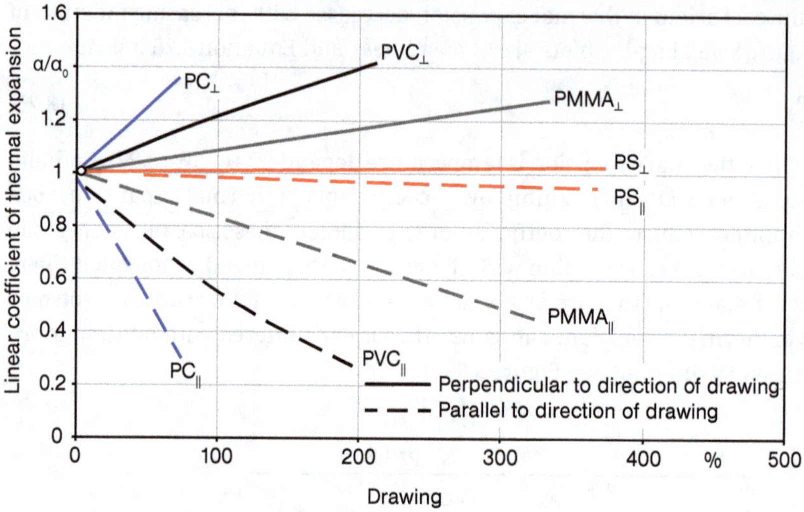

Figure 3.58 Coefficient of thermal expansion as a function of drawing

Because the coefficient of thermal expansion is temperature-independent in only a small temperature interval, it is recommended to state the measured values as relative change in length $\Delta l/l_0$ in % as a function of temperature (see Figure 3.59 and Figure 3.60 for selected filled/reinforced plastics).

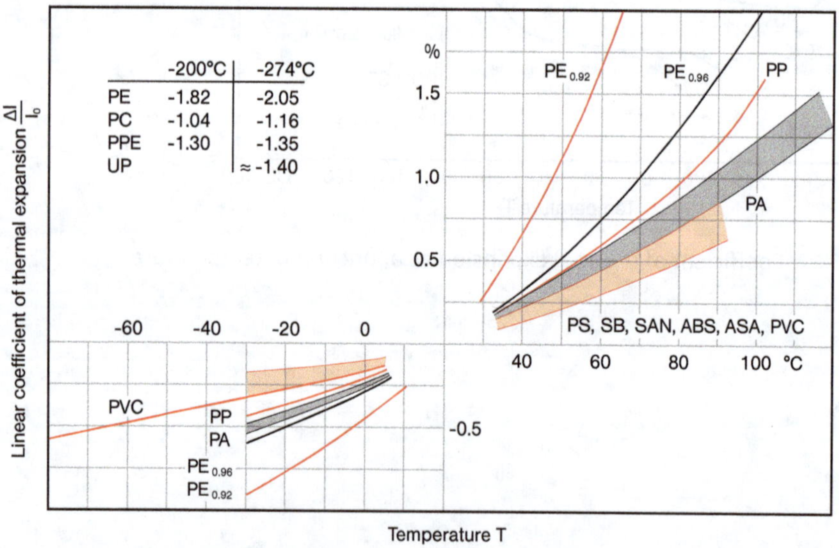

Figure 3.59 Correlation between relative change in length and temperature; index for PE = density

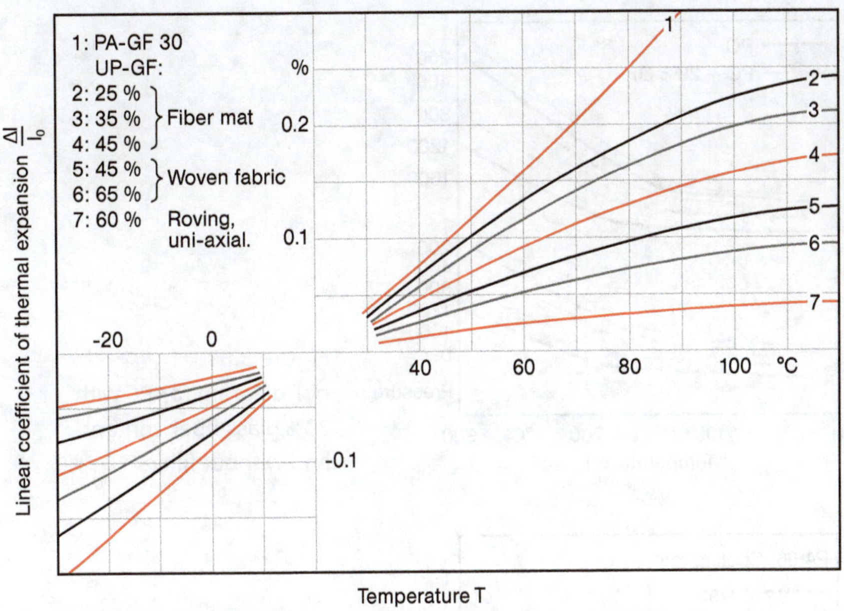

Figure 3.60 Relative change in length as a function of temperature

In order to determine the thermal expansion in the melt, so-called pvT-diagrams are necessary. They state the specific volume at different, injection molding relevant, temperatures as a function of pressure. These data facilitate the optimization of the injection molding process, in particular with respect to mold shrinkage, see Section 3.2.2. Figure 3.61 shows the specific volume profile of polycarbonate with and without fillers. The two different slopes of the individual curves represent the change in specific volume in the solid and melt state – separated by the glass transition temperature T_g. Depending on the type of plastic, the volume expands during the transition from the solid to the melt state by a factor of 2 to 4. With increasing pressure, the specific volume decreases, *i. e.*, the density increases. Increasing pressure also causes a shift of T_g to higher temperatures. The comparison of filled and unfilled plastics shows that T_g is not affected by filler. However, the compressibility is affected, *i. e.*, the change in specific volume with increasing pressure: it is lower for filled materials.

Figure 3.62 (above) shows the pvT-diagrams of a PA 66 as a representative of semi-crystalline thermoplastics. The crystallization temperature T_c can be identified as the inflection point in the graph. Above this temperature the material is in the molten state. Below T_c the material begins to crystallize. Crystallization causes high packing densities and thus a higher volume reduction than for amorphous thermoplastics, identified by the characteristic, hyperbolic graph.

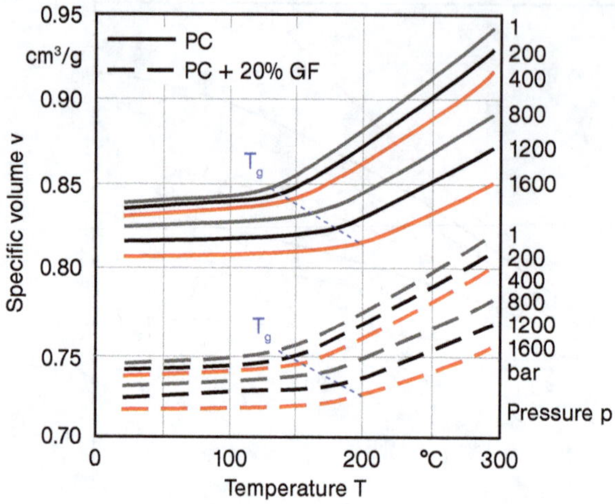

Figure 3.61
pvT-diagram for PC with 20% glass fiber content and without fillers

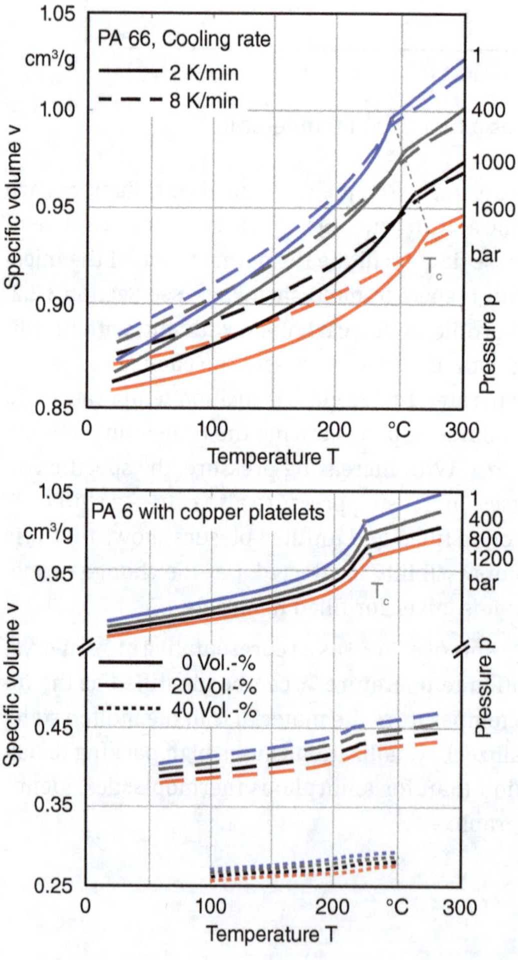

Figure 3.62
pvT-diagram of semi-crystalline thermoplastics; above: PA 66 at different cooling rates [7]; below: PA 6 with different filler contents [2]

Also shown in the figure is the effect of the cooling rate on the specific volume and thus on shrinkage. Higher cooling rates leave less time for rearrangement or crystallization of the macromolecular chains. Packing is less dense and the specific volume at room temperature is higher. At the very high cooling rates encountered during injection molding (up to 200 K/min) the effect is even more pronounced than shown here. Processors take advantage of this effect in special applications. For example, during bottle manufacture, PBT is quenched so rapidly that the originally semi-crystalline material solidifies amorphous and thus transparent. In the lower part of Figure 3.62, the pvT profile of filled and unfilled PA 6 is shown (the same material shown in Figure 3.5 d and Figure 3.14). The figure shows that with increasing filler content the characteristics of the filler material dominate (copper: melting point: 1085 °C, specific volume: 0.11 cm^3/g, coefficient of thermal expansion: $17 \cdot 10^{-6}$/K). Here, the specific volume, the slope of the graphs, and the change in slope from solid to melt state decrease.

Figure 3.63 shows typical specific volume profiles for different plastic materials at 1 bar. Distinct differences can be seen in the profiles, but also in absolute values between different plastics. Therefore, our description will be limited to some exemplary representatives. Additional information can be found in the database CAMPUS. Given that pvT data are not available for all commercially available plastic grades, in certain cases data obtained for other grades within the same plastics family may be used. Therefore, CAMPUS includes also generic data for pvT-diagrams. The availability of data is limited because the necessary measurements are time-consuming and instrumentation is not widely accessible.

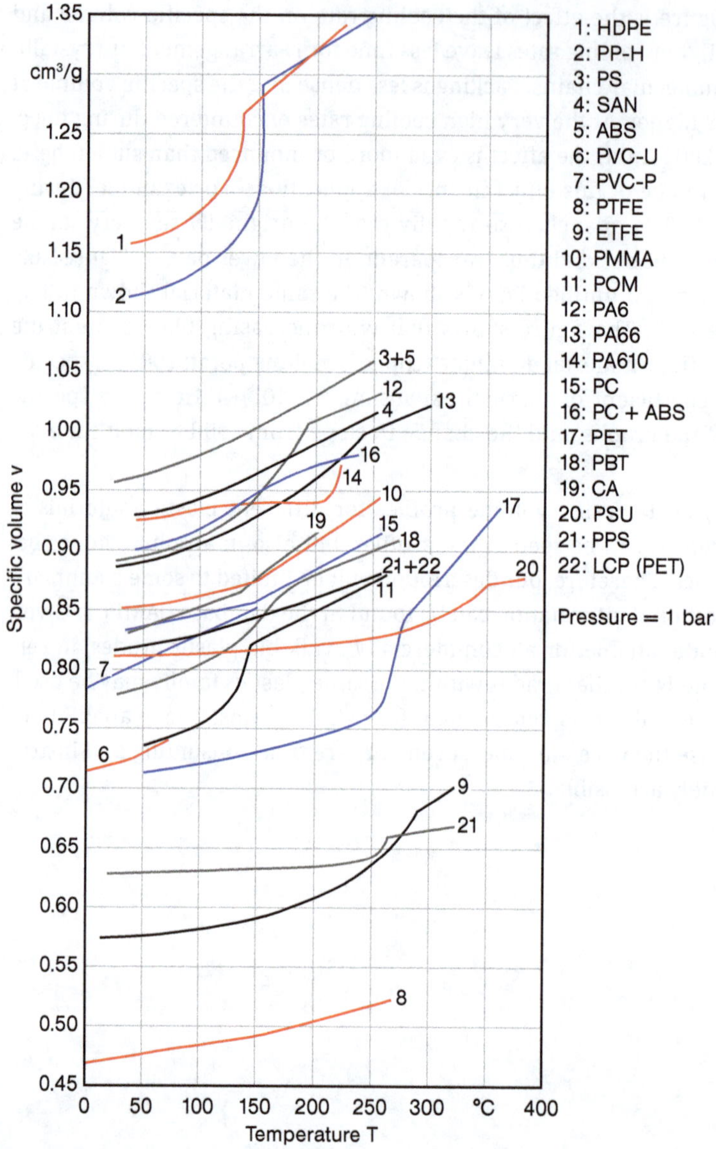

Figure 3.63 Specific volume at 1 bar as a function of temperature (1 bar curve derived from the respective pvT diagrams)

3.5 Electrical Properties

In general, plastics are electrical insulators because of the predominantly disordered amorphous or partially amorphous structure of these materials. Because the electrical charges cannot be dissipated, plastics are susceptible to electrostatic discharge. Individual plastic materials exhibit large differences in electrical and dielectric characteristics depending on their respective chain structure and morphology. Therefore, any structural changes caused by temperature or pressure influences, chemical or physical aging, contamination, or UV exposure can be tracked with the help of these characteristics. This is also the reason why these characteristics are the focus of research in polymer science. Processors use these characteristics to facilitate in-line process control, *e. g.*, during rubber crosslinking and thermoset curing processes.

In addition to mechanical characteristics, the designer may also take advantage of electrical characteristics in many areas of applications. Plastics differ as much with respect to their electrical properties as they differ in terms of mechanical characteristics, and both are strongly dependent on geometry, loads, and processing conditions during manufacturing of the components. In many cases, testing of the electrical properties of components is mandatory. Comparability of electrical characteristics (as of many others) can only be ensured by standardizing specimen manufacture and testing procedures, see Section 3.1 and Table 3.1.

The use of plastic materials in electrical and electronic applications is mainly due to their insulating properties. Examples include polyethylene for cable insulation and polystyrene covers for current-carrying components. Also, the electrical properties of plastics can be tailored by the addition of suitable fillers. There is furthermore research into to area of polymer semiconductors and intrinsically conductive polymers. However, these have not yet broken through into the mainstream and are mainly still in the research stage. When a dielectric, such as a plastic material, is under the influence of an electric field, two different effects take place, Figure 3.64:

a) Orientation polarization: permanent dipoles are oriented by the field, and

b) Displacement orientation: mobile charge carriers, *i. e.*, electrons and ions, undergo orientation by the field, which leads to electrical conductivity of the plastic.

Electrical and dielectrical properties generally depend on many factors, including temperature and pressure, so that they are ideally reported as multi-point rather than as single point values. The Maxwell equation is used to describe them independent of the nature of the charge carriers or the type of conduction mechanism. At low electrical field strength E (for many plastics $< 10^6\,\mathrm{Vm^{-1}}$), there is a linear relationship between E and the dielectric displacement D:

$$D = \varepsilon^* \varepsilon_0 E \tag{3.26}$$

Where:

ε_0 = dielectric permittivity of vacuum

ε^* = complex dielectric function or permittivity of the plastic

The latter is time- or frequency-dependent, because in plastics (other than in metals) charge transfer or conductivity is time-dependent.

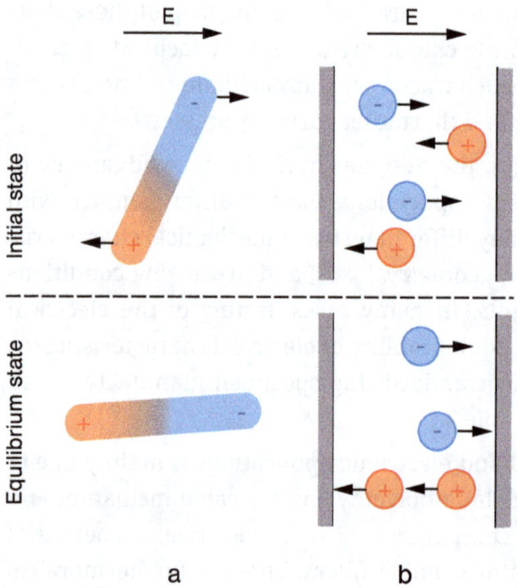

Figure 3.64
Mechanisms of charge transport in a dielectric such as a plastic [8];
a: orientation polarization, b: displacement polarization

The relationship between E and D is illustrated in Figure 3.65. The figure shows that the sinusoidal voltage excites the charge carriers, causing current flow with a shift in amplitude and phase (Δ amplitude and Δ phase) by the phase angle δ.

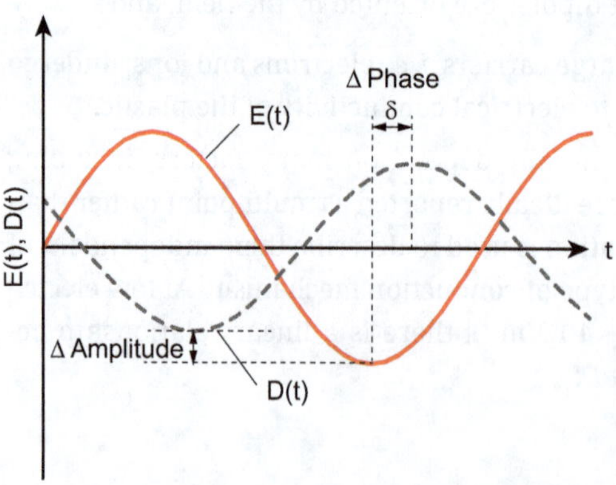

Figure 3.65
Correlation between voltage and resulting current

3.5.1 Electrical Insulation Properties

The electrical insulation behavior of a material is characterized by its resistance to the flow of electrical current. We distinguish between volume resistivity, which considers only the current flowing through the specimen, excluding the current flowing on the surface, and surface resistivity that is measured between electrodes on the specimen surface.

3.5.1.1 Volume Resistivity/Conductivity

The volume resistivity R_V of a specimen is the ratio of applied voltage U to the resulting current I: $R_D = U/I$ (in watts). For simple geometries, the volume resistivity can be used to calculate the specific volume resistivity ρ_D by multiplying R_D and the electrode area F with respect to the specimen thickness t in current direction: $\rho_D = R_D \cdot F/t$ (in $\Omega \cdot$ m or [$\Omega \cdot$ cm). Optimum contact during measurement is critical. The specific volume resistivity ρ_D of most plastics ranges from 10^8 to 10^{18} $\Omega \cdot$ cm. Non-polar plastics, such as polyolefins (PE, PP), PS, and PTFE exhibit the highest resistivity values $\rho_D > 10^{16}$ $\Omega \cdot$ cm. Assuming defined specimen geometries, this implies a volume resistivity R_D of 10^6 to 10^{16} Ω. Plastic materials with volume resistivities $\rho_D < 10^6$ $\Omega \cdot$ cm are used as electrical insulators.

Table 3.7 compiles the specific volume resistivities for a selection of common plastic materials. These values provide a guideline only, because here too, the properties are strongly dependent on temperature and other factors.

Table 3.7 Specific Volume Resistivity at 23 °C for Selected Plastics [9]

Thermoplastics	ρ_D [$\Omega \cdot$ cm]	Thermosets	ρ_D [$\Omega \cdot$ cm]
Unfilled		**Resins based on**	
Fluoropolymers	ca. 10^{18}	Phenolic resin	10^8–10^{13}
Polyolefins	ca. 10^{18}	Urethane	10^{11}–10^{12}
PS	> 10^{16}	MF	10^8–10^{11}
PA (dry)	10^{14}–10^{15}	UP	10^{11}–10^{13}
PC	ca. 10^{18}	EP	10^{13}–10^{15}
PMMA	> 10^{15}	Silicone resin	10^{14}
PVC-U	10^{15}–10^{16}		
PVC-P	10^{12}–10^{15}		

Table 3.7 Specific Volume Resistivity at 23 °C for Selected Plastics [9] *(continued)*

Thermoplastics	ρ_D [Ω · cm]	Thermosets	ρ_D [Ω · cm]
Filled		**Casting resins based on**	
ABS + 15% CF	10^4	UP	10^{16}–10^{17}
PA 6 + 30% GF	3×10^{14}	EP	10^{16}–10^{17}
PA 66 + 40% CF	70–80	PUR	10^{11}–10^{16}

Figure 3.66 shows the decrease in specific volume resistivity with increasing temperature. This effect can be explained by the increase in molecular mobility with increasing temperature. Some plastic materials, such as PBT and POM, exhibit less pronounced decreases than non-polar plastics such as PE or PS. Figure 3.67 illustrates the decrease in volume resistivity for polyamides with increasing water content.

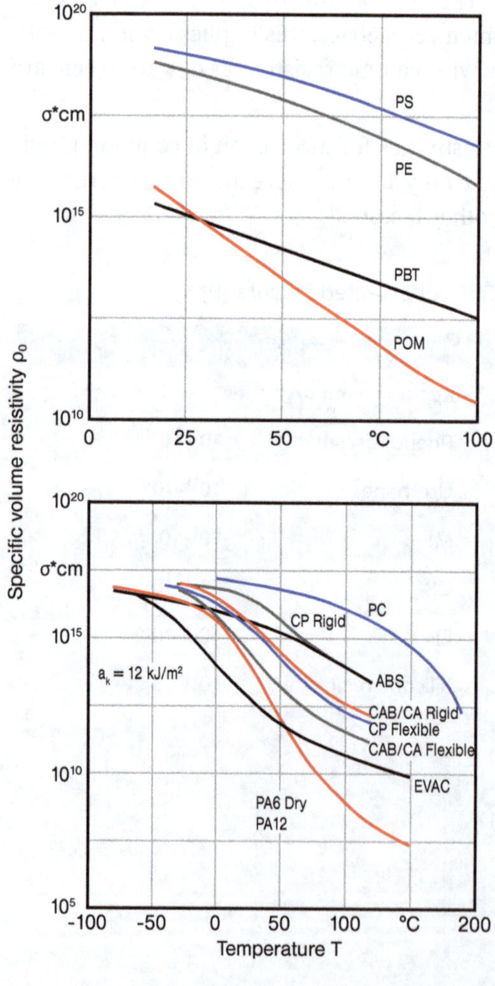

Figure 3.66 Specific volume resistance as a function of temperature

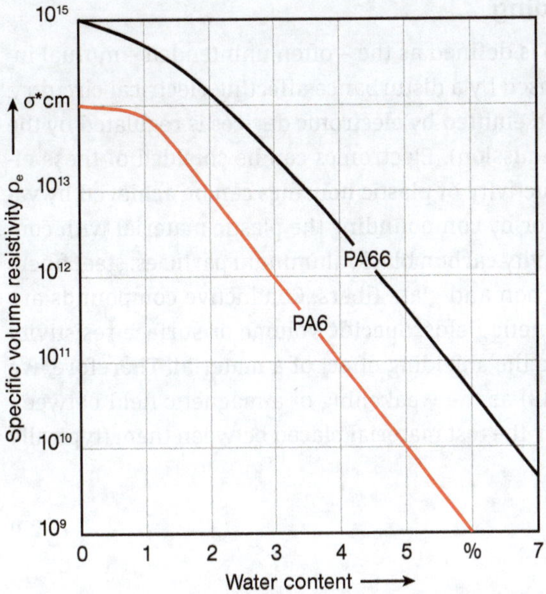

Figure 3.67 Specific volume resistance as a function of water content

The electrical conductivity S (in Siemens) is a measure for the capability of a material to conduct an electrical current. It is identical to the reciprocal electrical resistivity value:

$$S = \frac{1}{R_D}[(\Omega)^{-1}] \tag{3.27}$$

The specific electrical conductivity σ is:

$$\sigma = \frac{1}{\rho_D}[(\Omega \cdot m)^{-1}] \tag{3.28}$$

3.5.1.2 Surface Resistivity

The surface resistivity R_o provides information on the resistivity of a plastic specimen's surface, which may differ significantly from its volume resistivity. It depends, among other factors, on ambient conditions, such as humidity, and surface contaminations. It can be calculated as the ratio of a DC voltage U (between two electrodes at the specimen surface) to the resulting current between the electrodes: $R_o = U/I$ (in Ω). The specific surface resistivity ρ_o (Ω) is the surface resistivity with respect to a square area. Here, the volume resistivity is also included to a non-determinable degree so that the determined values cannot be transferred to other geometries.

3.5.1.3 Electromagnetic Shielding

Electromagnetic interference (EMI) is defined as the – often unintended – mutual interference of electronic devices, caused by a disturbance affecting electrical circuitry. The maximum permissible radiation emitted by electronic devices is regulated by the FCC (Federal Communications Commission). Electronics can be shielded of these effects by conductive housings. Conductivity of plastic housings can be achieved by vapor deposition of aluminum layers or by compounding the plastic material with conductive fillers such as high-conductivity carbon black, aluminum particles, steel fibers (0.7 to 1.4 vol.-%), or metallized carbon and glass fibers. Conductive compounds are less suitable as shields against magnetic fields. Specific volume or surface resistivity values are not sufficient to evaluate the shielding effect of a material. Therefore, we measure magnetic shielding S (in dB) as the weakening of a magnetic field between two coaxial-loop antennae caused by the test material placed between them (typically as a function of frequency):

$$S = -20 \log \frac{H1}{H0} \tag{3.29}$$

Where:

H1 = strength of the magnetic field with testing material

H0 = strength without testing material (Figure 3.68)

A coaxial measuring cell is used to determine the electromagnetic shielding attenuation ES (in dB). Analogous to shielding S, shielding attenuation is measured as the ratio of the measured electrical or magnetic field strength with and without shielding.

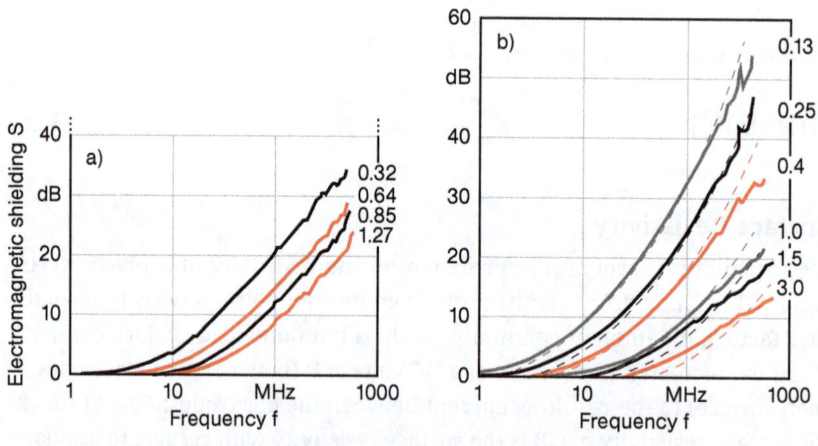

Figure 3.68 Electromagnetic shielding S as a function of frequency; Parameter: specific volume resistivity (ρ_D) of the plastic material

a: Plastic with vapor-depositioned aluminum coating

b: Plastic filled with steel fibers

3.5.2 Dielectric Properties

3.5.2.1 Dielectric Strength

High electric voltages can cause a breakdown of plastic materials. The electric breakdown resistance or dielectric strength E_d is the quotient of voltage U_d at which breakdown occurs and the distance between the electrodes, *i. e.*, for the specimen thickness d: $E_d = U_d/d$ (in kV/mm). It depends strongly on electrode shape and specimen thickness and is therefore not a material constant (Figure 3.69 and Figure 3.70). The database CAMPUS requires a specimen thickness of 1 mm in order to ensure comparability. Figure 3.71 and Figure 3.72 show dielectric strength as a function of temperature and frequency (dielectric heating!).

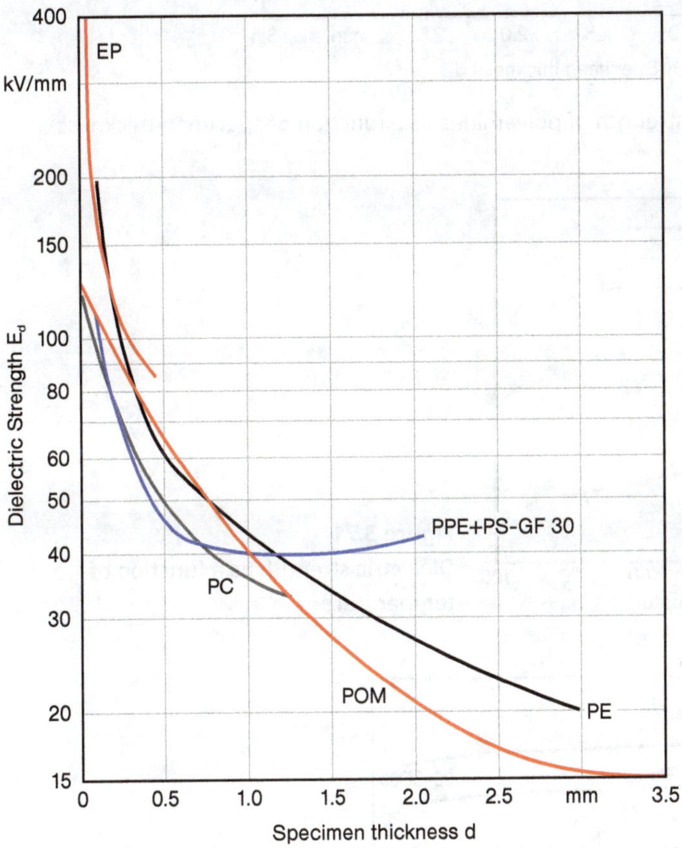

Figure 3.69 Dielectric strength as a function of specimen thickness

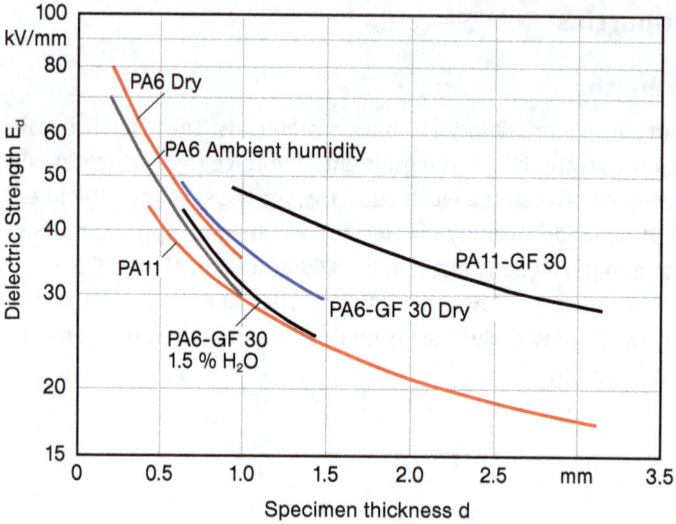

Figure 3.70 Dielectric strength of polyamides as a function of specimen thickness

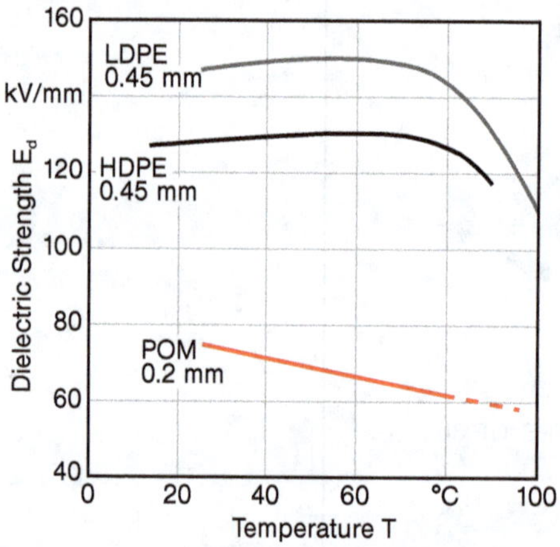

Figure 3.71 Dielectric strength as a function of temperature

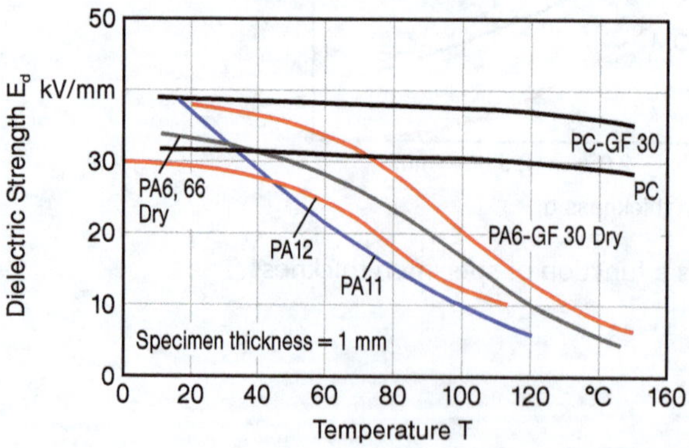

Figure 3.72 Dielectric strength of polyamides and PC as a function of temperature

3.5.2.2 Long-Term Dielectric Strength

The described short-term dielectric strength provides only reference values with respect to long-term voltage loads. Analogous to mechanical strength, the permissible electrical voltage depends on the time under load. Dielectrical strength is basically constant for load durations from 1 µs to 1 s. Then an electrical breakdown occurs, caused by electrons. When the time under load is longer or for plastics that heat up because of either sufficient electrical conductivity or because of dielectric losses in alternating electrical fields, thermal breakdown will occur because dielectric strength decreases with increasing temperature. Materials with low conductivity and low dielectric losses, such as PE, often do not experience thermal breakdown, see Figure 3.73. Here, electrical breakdown conditions directly transition to erosion breakdown conditions. The reason for this breakdown is glow discharges during which both mechanical and chemical influences affect the surface. Figure 3.74 and Figure 3.75 illustrate the time-dependence of dielectric strength.

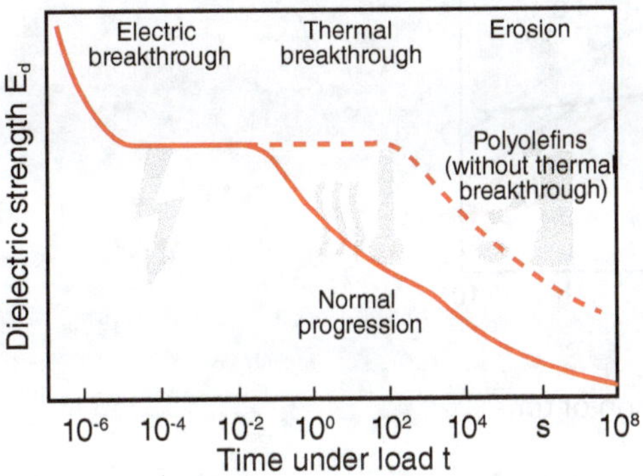

Figure 3.73
Dielectric strength as a function of time under load, schematic depiction

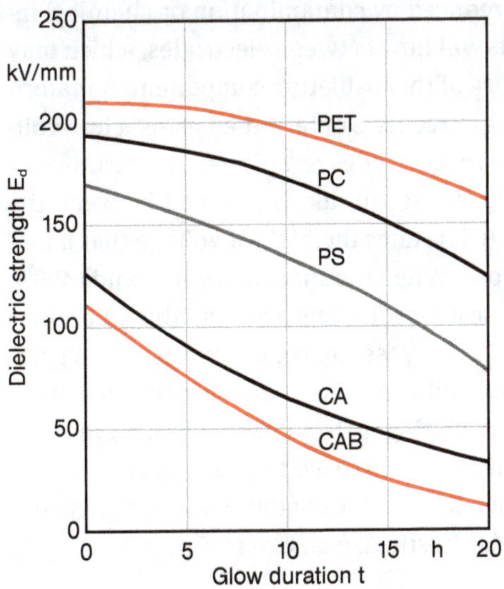

Figure 3.74
Dielectric strength as a function of time

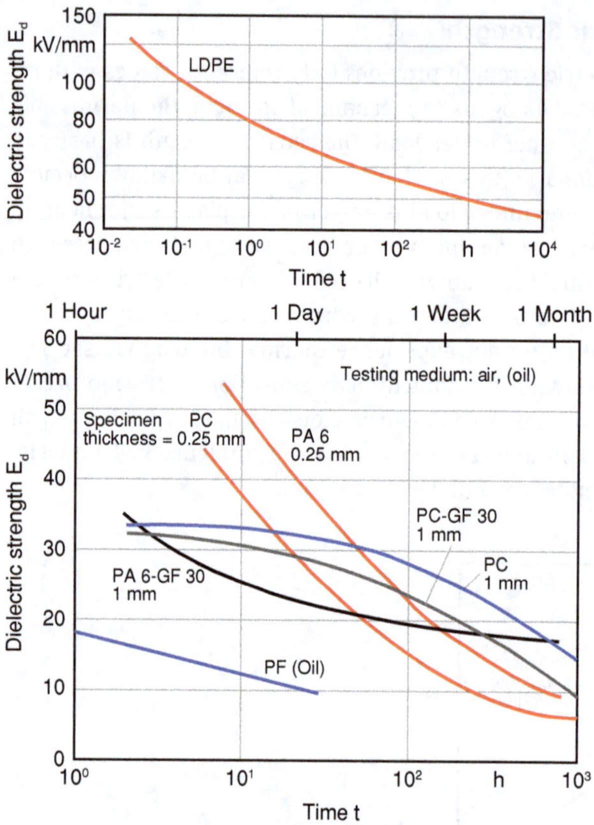

Figure 3.75 Dielectric strength as a function of time

3.5.2.3 Tracking Index

The surface resistivity of a material can be reduced by contamination or chemical influences to a point at which tracking currents will flow between electrodes, which may cause arcs and ultimately the complete failure of the insulating component. A material's susceptibility to the formation of tracking currents is determined using a low-voltage test (up to 600 V), see IEC 60112. Here, two electrodes are placed at a distance of 4 mm on the surface to be tested. A conductive test solution is dropped between the electrodes. The comparative tracking index (CTI) states the highest voltage that a material will resist without tracking after 50 drops. When the specimen withstands 425 V, the tracking index is CTI 425. Two different testing solutions are available (A and B); solution A, which is required according to CAMPUS testing conditions, is less aggressive. Insulating materials that are used at high voltages or outdoors are tested using a high-voltage testing method; see IEC 60587. Insulating materials are classified in three classes according the most commonly method A. Three voltage levels (2.5, 3.5, or 4.5 kV) are applied to the specimen under specified testing conditions for 6 h. Depending on the level the specimen withstands, it is classified, *e. g.*, Class 3.5 A.

3.5.2.4 Arc Resistance

Arc resistance is determined by attaching two carbon electrodes to the surface of the specimen and then pulling them apart at a rate of 1 mm/s after ignition of a low-voltage/high-current arc (200 V). The current is limited by a 50 Ω series resistor. The length of the arc, the conductivity of the arc gap during testing, and the specimen damage after cooling are evaluated. Typically, arc resistance is a characteristic value measured and provided by manufacturers for their devices.

3.5.2.5 Electrolytic Corrosion

Insulating materials that are in contact with electric conductors may cause signs of corrosion in the latter. The formation of corrosion depends on the combination of metal/insulator. For testing purposes, insulator films are brought into contact with brass testing films and exposed to 100 V direct current for 4 days under humid/warm conditions (40 °C/92% rel. hum.). Characteristic values for electrolytic corrosion effects are determined from measured changes at the poles.

3.5.3 Dielectric Behavior

When using plastics in capacitators or in high-frequency applications, the respective dielectric properties, *i. e.*, dielectric constant ε_r or ε' and dielectric dissipation factor tan δ have to be considered. The fundamentals have been explained at the beginning of the chapter. The dielectric constant of a plastic material states the increase in capacitance of a capacitor with plastic as the dielectric is compared to the capacitance of the same capacitor with air as the dielectric. The dielectric dissipation factor is a measure for the energy that is transformed to heat in an insulator in an alternating electrical field (and thus is no longer available as electrical energy). The dielectric properties are material constants that are largely independent of the methods used to determine them. However, temperature and frequency affect their values, and therefore they are typically recorded as a function of these parameters (Figure 3.76 and Figure 3.77).

The dissipation factor tan δ is a critical material characteristic to describe the heating of a plastic material in a dielectric. The heat capacity N, *i. e.*, the energy dissipated in the form of heat, is

$$N \sim U^2 \omega \varepsilon'' \quad (3.30)$$

or

$$N \sim U^2 \omega \varepsilon' \tan \delta \quad (3.31)$$

Where:

U = voltage

ω = frequency

Plastic materials with high dissipation factors can therefore be heated (*e.g.*, welded) in an alternating electrical field.

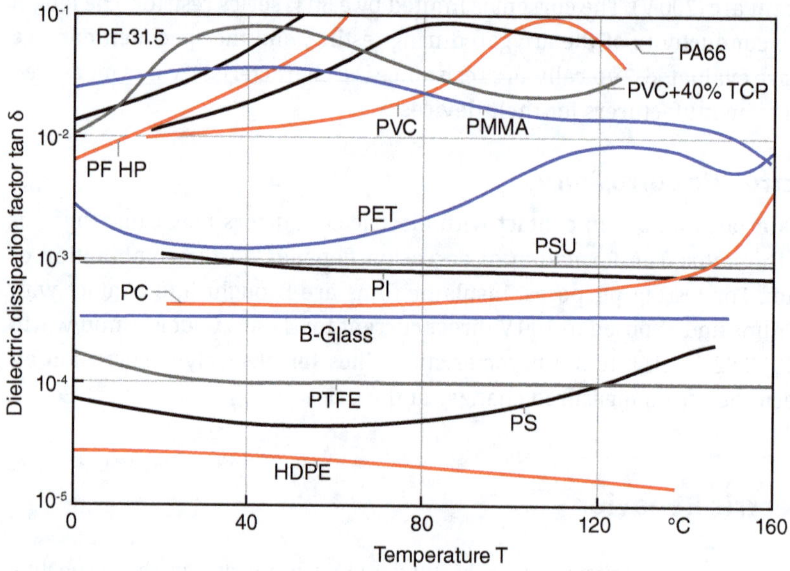

Figure 3.76 Dielectric dissipation factor (tan δ) as a function of temperature at 50 Hz

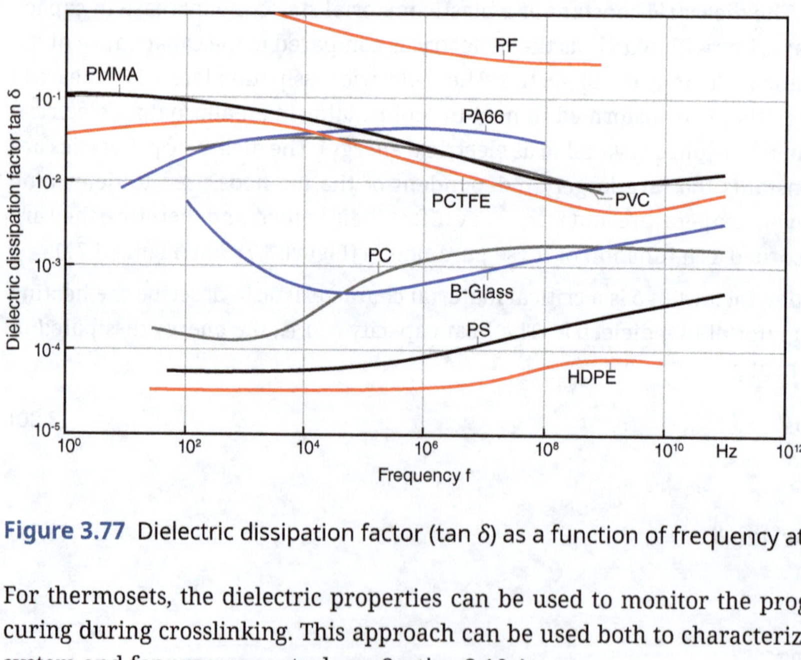

Figure 3.77 Dielectric dissipation factor (tan δ) as a function of frequency at 20 °C

For thermosets, the dielectric properties can be used to monitor the progression of curing during crosslinking. This approach can be used both to characterize the resin system and for process control, see Section 3.10.4.

3.5.4 Electrostatic Behavior

Because plastics are non-conductive materials, they can be electrostatically charged. The charge is caused by friction, or better by contact and separation of surfaces. Demolding of a molded component may already result in a charge. The level of charge depends on the plastic material and the friction partner. The charges may be positive or negative and they dissipate only very slowly because of the low conductivity of the plastic material (electrostatic discharge, ESD).

Electrostatic charges may attract dust from the air and cause arcing. The latter poses a danger when transferring combustible liquids from plastic containers or in carpeted areas exposed to explosion hazards.

The specific surface resistivity values indicate the electrostatic behavior of a plastic material, because these values determine the dissipation of charges, see Section 3.5.1.2. Surface resistivities above $10^{14}\,\Omega$ indicate high susceptibility to electrostatic charges. Materials with surface resistivities between $10^{9}\,\Omega$ and $10^{11}\,\Omega$ exhibit lower chargeability, and at $10^{9}\,\Omega$ or below a material can no longer be electrostatically charged, *i. e.*, the danger of ignition or explosion is eliminated.

Only thermoset molding compounds exhibit surface resistivities below $10^{11}\,\Omega$, while the values for dry, unfilled thermoplastics are above $10^{12}\,\Omega$. Hygroscopic plastics, such as PA, exhibit more favorable behavior.

A material's absolute chargeability is difficult to measure in a reproducible way because it depends strongly on the test method, friction partner, and ambient conditions during the test.

3.5.5 Conductive Plastics

Desired conductive properties in plastic materials can be tailored by the addition of electrically conductive fillers, as seen in the percolation profiles in Figure 3.78. The electrical properties are mainly determined by filler content. Above a certain filler content, the conductivity increases sharply because a conductive filler network is formed. This drop in surface resistivity is also called percolation threshold. The position of the percolation threshold depends on both the filler surface and the filler shape.

Most *conductive thermoplastic compounds* contain conductive carbon black fillers. The electric resistance of these compounds is affected by carbon black content and the specific carbon black surface. Below percolation concentration the electric resistance is barely lower than that of the unfilled thermoplastic.

Disadvantages of thermoplastic/conductive filler compounds include:

- High melt viscosity,
- Low mechanical strength and toughness,

- Increase of electrical resistance under alternating temperature loads,
- Inhomogeneity of electrical resistance,
- Low reproducibility for antistatic applications in the region of steep resistivity drop, and
- Low near-field shielding at low frequencies.

Intrinsically conductive plastic compounds exhibit these disadvantages to a lower degree or not at all. Here, the lower limit for volume resistivity values is 0.01 Ω · cm, while for thermoplastic/carbon black compounds these limits are at 1 Ω · cm.

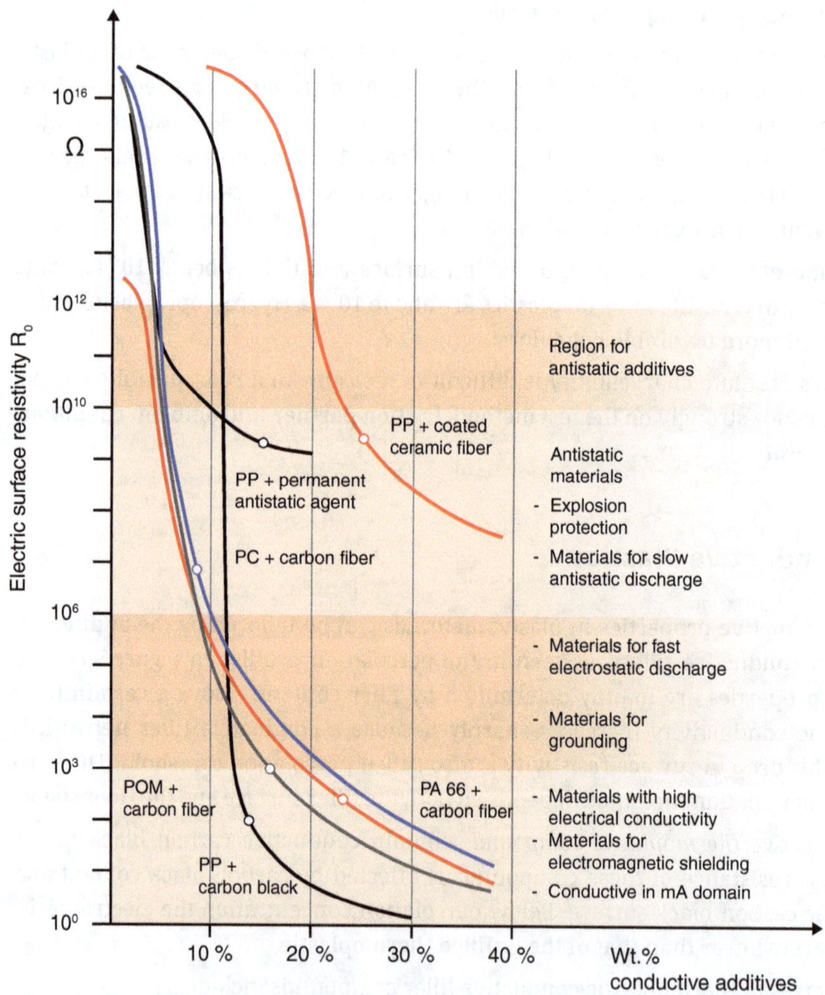

Figure 3.78 Electric surface resistivity as a function of filler type and content (percolation profile)

3.6 Optical Behavior

The fact that plastics are also referred to as "organic glasses" indicates that they can be used in optical applications. Today's processes and machines facilitate the manufacture of high-precision molded components and semi-finished products. For example, lenses, prisms, panes, optical fibers, and photoelastic models can all be manufactured using machining, casting, compression molding, embossing, injection molding, or extrusion. In order to evaluate the applicability of plastic materials for these products, it is critical to be aware of not only their optical characteristics, but also of the tolerances achievable when manufacturing these components. This is of particular concern when using plastic materials for the production of optical data carriers, which require highly accurate reproduction of minute surface details. Another example comprises components with changes in wall thickness, such as automotive headlights, where it is critical to consider the different shrinkage values during the design of the component.

3.6.1 Transparency

Compared to metals, wood products, and other inorganic materials, many plastic materials offer the advantage of transparency. The optical characteristics of transparent plastics show no significant differences compared to those of inorganic glasses. Often, they exhibit higher transparency in the visible spectrum than glass, and some are also more transparent in the ultraviolet and infrared region. Plastic materials can be roughly distinguished in three categories with respect to their transparency: crystal clear (transparent), translucent, and opaque.

The transmittance τ is used to measure the transparency of a material: it is defined as the quotient of transmitted light Φ_n and incident light Φ_v:

$$\tau = \frac{\Phi_n}{\Phi_v} \cdot 100[\%] \tag{3.32}$$

It considers absorption as well as scattering and the losses caused by reflection on the front and back of the specimen. Transmittance is typically determined in air and expressed as a function of wavelength, Figure 3.79.

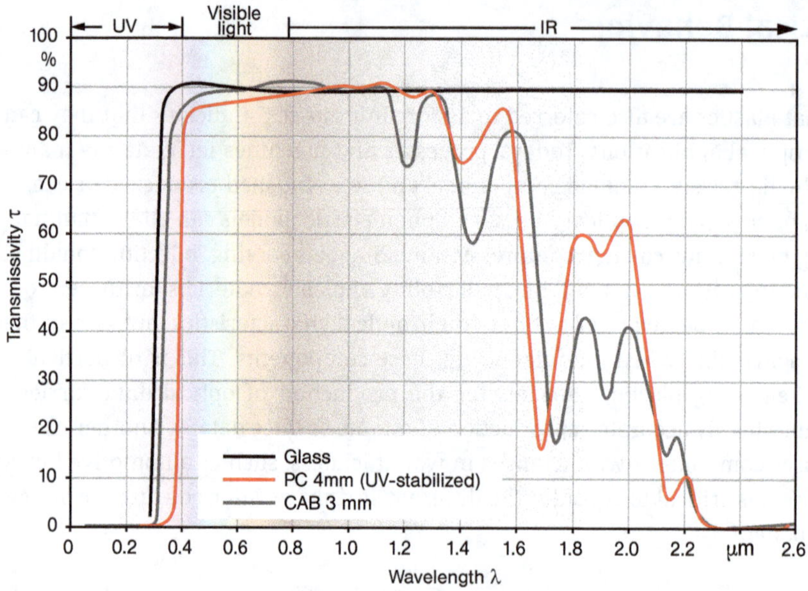

Figure 3.79 Transmittance as a function of wavelength

3.6.2 Gloss, Reflection, and Haze

Gloss is primarily created by the directed reflection of light. It is essentially a subjective property: even a slight shift of the body with respect to the observer will change the impression of gloss significantly; therefore, it is difficult to define "gloss". Figure 3.80 illustrates different methods of gloss measurement. The gloss level *g* is defined by the ratio of the intensity of the reflected light I_r and the intensity of a perfect standard surface $I_{r,s}$, such as a mirror, see part (1a) in Figure 3.80. Typically, the measurement is performed at an angle of 45°; however, other angles are also used. In Figure 3.80 (1b), *g* is measured as the ratio of intensity of light exiting at a reflection angle of 45° (I_r) to the light intensity of scattered light I_α (exiting perpendicular to the body). Results obtained by different methods are not comparable.

The gloss of a plastic part depends on its surface, which in turn is determined by the surface of the mold and its temperature. Highly polished molds and high mold temperatures result in glossy surfaces, see part (2a) in Figure 3.80. Highly filled plastics and some blends result in more matte surfaces, see parts (2b) und (2c). Gloss measurements are the preferred non-destructive approach to control surface changes during UV- and weathering tests.

Transmittance of directed light is negatively affected by crystallization, fillers, roughness, contamination, or scratching of the surface, which cause haze. The haze number is a measure for haze. It is the ratio of scattered light reflected by the specimen at an angle of 80° to the axis of the incident ray to the almost perpendicularly impinging primary light. According to another definition (ASTM D 1003) haze is the portion of light transmitted by the specimen that deviates from the direction of incident light by more than 2.5°.

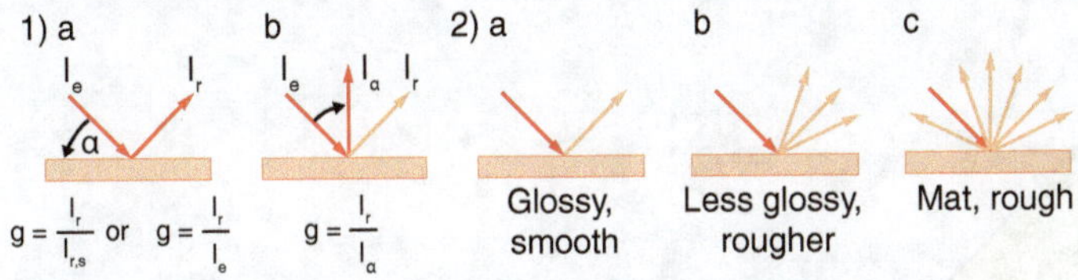

Figure 3.80 Left: Schematic illustration of different gloss measurement methods on plastic surfaces (1a and b); right: different types of surface reflection (2a: directed, 2b: dispersed, and 2c: diffused)

3.6.3 Color

3.6.3.1 Fundamentals

Color is a visual impression of a surface gained by an observer. In order to understand why a surface appears colored, the concepts of light, color, and complementary color have to be explained.

Visible light is comprised of electromagnetic waves with wavelengths ranging from 400 nm to 800 nm, and it is only a small region in the overall spectrum of electromagnetic waves. Every wavelength in this "visible region" corresponds to a specific color. "White light" is the composition of all wavelengths in the visible range.

Two colors are called complementary when their combination results in white light. A color circle can be used to determine the respective complimentary color. The color circle is filled with all possible color transitions.

When "white light" impinges on a colored surface, certain wavelengths are absorbed by the surface and free and/or bonding electrons in the pigments are excited so that the surface appears in its complementary color. The coloring effect of pigments is based on this phenomenon.

3.6.3.2 Colorimetry

Objective measuring methods are required to characterize color, *e.g.*, in quality control. For this purpose, the CIE-Lab system (Commission international de l'éclairage) is globally accepted. The intensity of the reflected light is recorded as a function of wavelength in so-called radiance curves. The colors are arranged in a horizontal plane similar to the color circle (with the only difference that we now know that the circle is in fact a quasi-elliptical), see Figure 3.81. Here, all colors are represented in their maximum saturation. The additional z-axis represents the color intensity, shown in Figure 3.82. The perceivable color is defined by the coordinates L, a, and b.

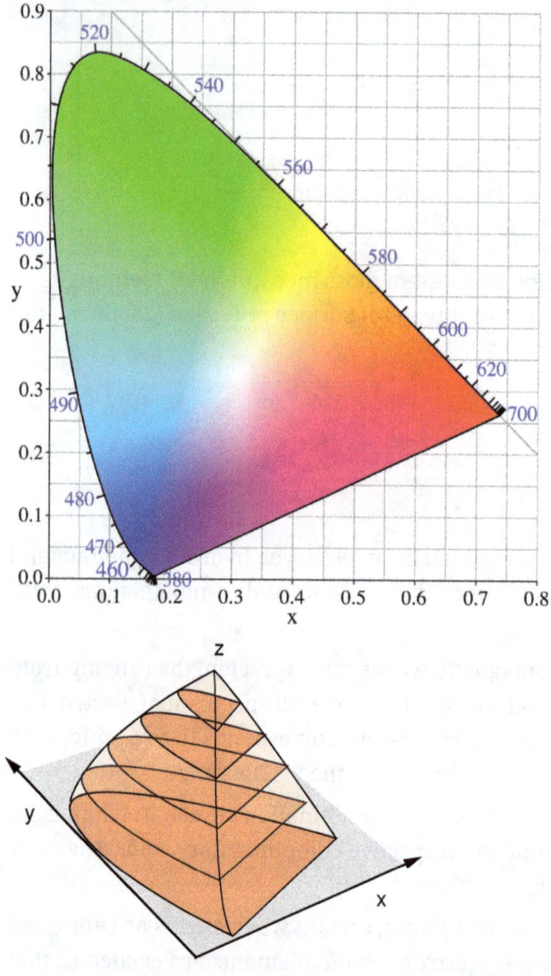

Figure 3.81
Above: CIE-standard chromaticity table; below: three-dimensional illustration of intensity

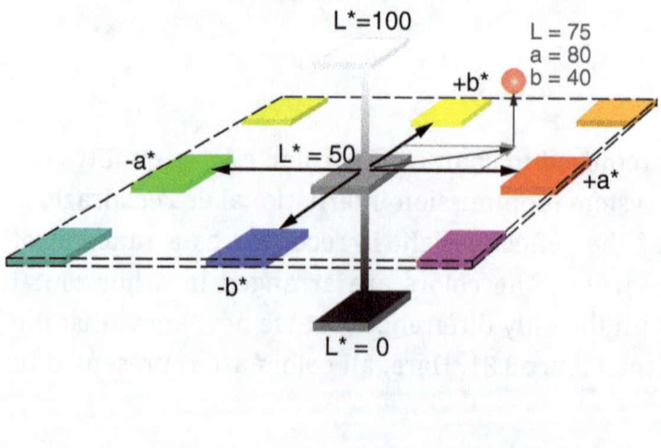

Figure 3.82
CIE-Lab system for color characterization

According to the theory of complimentary colors, green (a = −127) and red (a = +127) are opposite each other on the a*-axis. The b*-axis represents the complimentary colors blue (b = −127) and yellow (b = +127). The L*-axis is perpendicular to this plane and represents color intensity. Black (L = 0) and white (L = 100) are the terminal points, with shades of gray as intermediate values. If the a-component equals 0, only colors ranging from blue to yellow can be represented with intensities ranging from white to black. Analogously, colors from red to green are represented when the b-component equals 0.

3.6.4 Index of Refraction, Birefringence

When light passes from one medium into another, *e. g.*, from air into a plastic material, it is refracted. The degree of light refraction is characterized by the relative index of refraction of medium 2 with respect to medium 1.

$$n_{21} = \frac{\sin\alpha}{\sin\beta} = \frac{c_1}{c_2} \tag{3.33}$$

where α and β are the angles that a light ray will form in the first and second medium with the axis of incidence, and c_1 and c_2 are the respective rates of propagation of light in these media. In general, indices of refraction refer to vacuum as the first medium. The speed of light c_0 in a vacuum is higher than in any other medium. Assuming the speed of light in a medium is *c*, and the index of refraction with respect to vacuum is *n*, we can calculate for light passing from a vacuum into a medium:

$$n = \frac{\sin\alpha}{\sin\beta} = \frac{c_1}{c_2} \tag{3.34}$$

Because the speed of light differs only little between air and vacuum, *n* is generally valid also for the transition from air into a solid or liquid medium.

The refraction index depends on the wavelength of the incident light, Figure 3.83. Therefore, it is determined for defined wavelengths, typically with the help of the Sodium D-line n_D at 589 nm.

With increasing temperature the index of refraction will decrease. The differential quotient $\Delta n/\Delta T$ for selected plastic materials is provided in Table 3.8. It is significantly higher above the glass transition temperature T_g than below. Compared to glass, it is one order of magnitude higher.

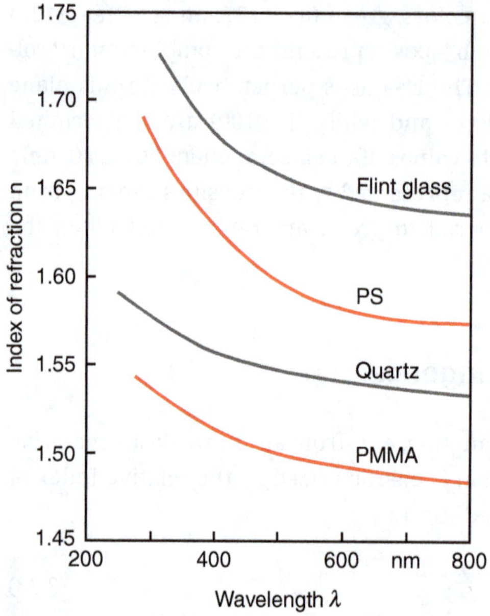

Figure 3.83 Index of refraction as a function of wavelength

Table 3.8 Temperature Coefficients for Indices of Refraction for Polymers

Material	T_g (°C)	Coefficient (-dn/dt) · 10^4	
		T < Tg	T > Tg
Polymethylacrylate	0	1.2	3.1
Polymethylmethacrylate	105	1.1	2.1
Polyethylmethacrylate	47	1.1	2.0
Polypropylmethacrylate	33	1.3	2.9
Polybutylmethacrylate	17	1.6	2.9
Polyvinyl acetate	24	1.0	3.1
Polyvinyl chloroacetate	23	1.1	3.0
PVC/PvAz = 95/5	71	1.0	2.6
PVC/PvAz = 88/12	63	1.2	3.2
PVDC/PVC (Geon 205)	55	1.0	2.8
Polystyrene	75 ± 4	1.7	4.6
Styrene/butadiene copolymer (85/15)	40	1.1	3.3

The previous statements are valid for optically isotropic materials. However, orientation of macromolecules or crystal structures, influence of mechanical loads, or residual stresses (see Section 2.4.1) render plastics optically anisotropic, *i. e.*, the index of refraction becomes directional. When a light ray passes through an anisotropic material, it is split into two rays with different speeds of light depending on their direction of propagation in the medium. One part is refracted normally (ordinary ray) while the other experiences a direction change (extraordinary ray) perpendicular to the ordinary ray. This phenomenon is called birefringence, and it is calculated from the orientation-dependent differences in refractive indices: $\Delta n = n_1 - n_2$. The longer path of the extraordinary ray compared to the ordinary ray creates an optical retardation. Birefringence is used to qualitatively confirm orientations and stresses. When an optically anisotropic specimen is evaluated in a polariscope or a polarization microscope between polarizer and analyzer under linearly polarized light, the resulting interference color can be attributed to a birefringence using a color table and the specimen thickness. Figure 3.84 shows an optimized chromaticity table ("Michel-Levy" color table), while Figure 3.85 shows a polarized optical image of a transparent, injection molded component.

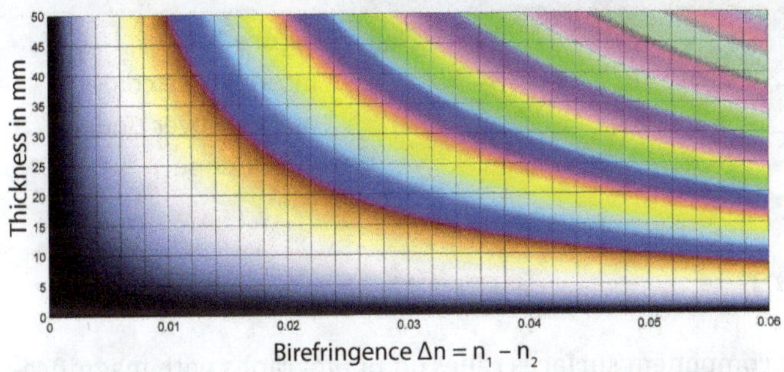

Figure 3.84 Optimized Michel-Levy interference chromaticity table for the determination of birefringence as a function of specimen thickness [10]

Optical polarization is typically used only for the qualitative evaluation of stress differences. It is not suitable for very thick or curved components. When plastics are used for optical data carriers, birefringence should be minimized.

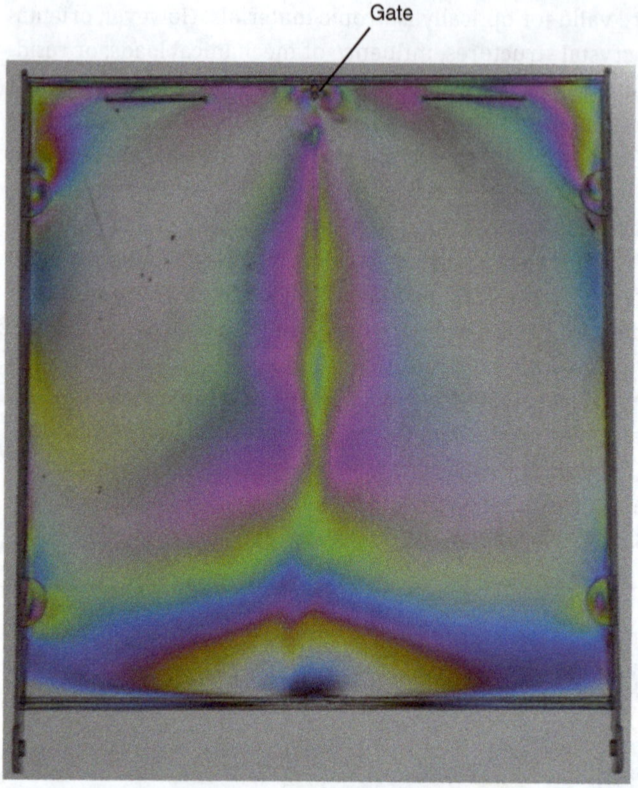

Figure 3.85 Interference colors of a CD cover under polariscope

3.6.5 Surface Texture

The evaluation of mold or component surfaces relies on photographs with magnifications ranging from 20–1000 times or on roughness characteristics "*R*". The latter are determined using electrical stylus instruments, which are also critical for the determination of *draft angles* in injection molds. Table 3.9 provides the minimal draft angle values for non-reinforced PA and PC as a function of mean roughness index R_a. For glass fiber reinforced components, draft angles should be selected one step higher.

The plastic material is supposed to replicate the surface texture of the mold. However, the replicability depends on a number of parameters. It increases with decreasing melt viscosity, increasing injection rate, mold pressure, mold temperature, and wall thickness. Therefore, the mold surface texture must take into consideration the type of plastic material and component to be molded.

Depending on the desired degree of gloss, grain sizes ranging from P240 to P500 or diamond pastes with a grain size of 30 µm are used for *polishing*. For optical components, special requirements have to be met in terms of polishing the mold steel.

Table 3.9 Minimum Draft Angle as a Function of Roughness

Mean roughness index R_a	Draft angle in °	
	PA	PC
< 0.4	0.5	1.0
1.6	0.5	1.5
2.24	1.5	2.0
3.15	1.5	2.0
4.5	2.0	3.0
6.3	2.5	4.0
9.0	3.0	5.0
12.5	4.0	6.0
18.0	5.0	7.0

Velvety to coarse grained surfaces can be achieved by *erosion*. Extremely fine textures should be avoided because they are susceptible to scratching.

Cavity surfaces are corrosively machined by photoetching. Typically, the etching procedure is repeated several times in order to create a uniform surface roughness. The degree of gloss can be corrected/adjusted by subsequent jet blasting with glass beads or plastic pellets.

3.7 Resistance to Environmental Influences

Under service conditions, plastics are exposed not only to mechanical loads but also to environmental influences, such as chemicals, weather, high-energy radiation, or fire, which may drastically affect their behavior. Here, it is relevant whether mechanical and environmental loads are affecting the material simultaneously.

When plastics are in contact with solids, gases, or liquids, these materials may penetrate the polymer structure until an equilibrium is reached and thus reduce the intermolecular bonds, increasing the mobility of the molecules (Figure 3.86 and Figure 3.87). This results in a decrease of the modulus of elasticity, hardness, strength, a lowering of the glass transition temperature, and a change in electrical and physical characteristics. These are physical processes and as such typically reversible, as long as no plastic constituents are removed or its structure is changed, *e. g.*, by secondary crystallization.

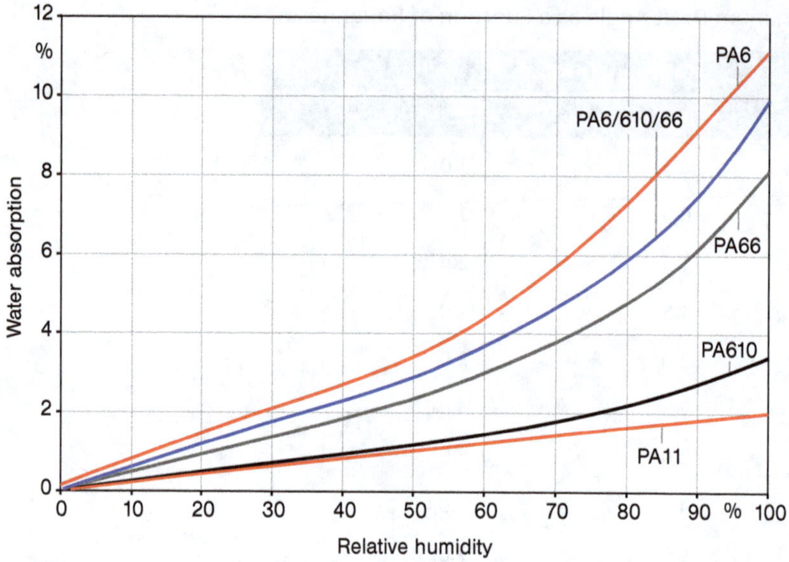

Figure 3.86 Equilibrium water content as a function of relative humidity for polyamides

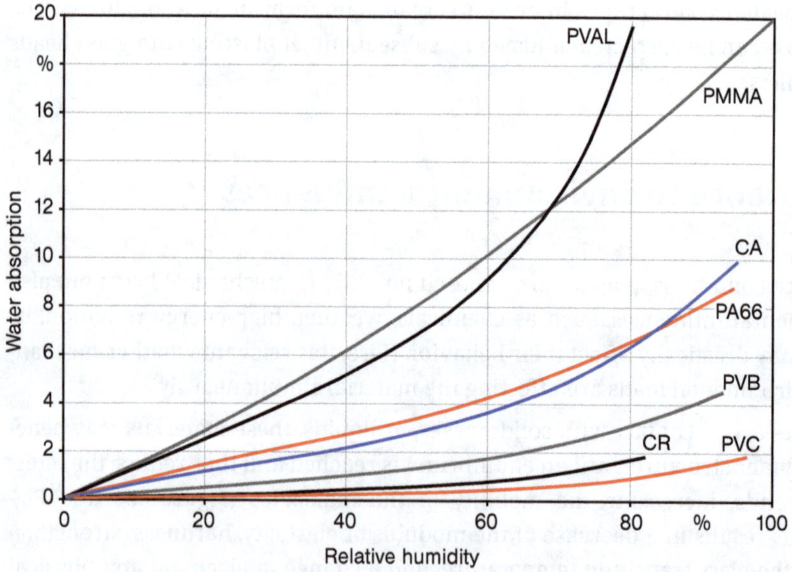

Figure 3.87 Equilibrium water content as a function of relative humidity for elected thermoplastics

3.7.1 Water, Moisture

Absorption of water or humidity is of particular importance for the behavior of polyamides. Absorption is evaluated after immersing specimens in water as a function of immersion time, or measured gravimetrically after exposure to specified temperatures and humidity after specified times. The values (in mg or %) determined using these conventional methods are not comparable quantitatively because the specimen dimensions and testing parameters affect the results. CAMPUS therefore requires only the saturation values for immersion in water or under standard conditions (23 °C and 50% rel. humidity).

Water absorption Φ can be calculated using Fick's 2nd Law, the saturation values j_S, and the coefficient of diffusion *D*. Figure 3.88 shows the solution of the respective differential equation in dimensionless form (left) with an example graphically displayed on the right. With this diagram, the water distribution in sheet-like specimens can be determined for any immersion time as long as the specimen is immersed on both sides and the coefficients of diffusion are known.

The diagram in Figure 3.89 provides a similar tool to evaluate the average moisture content. The coefficients of diffusion can be calculated using the results of immersion tests. However, the fact that they will increase significantly with increasing temperature has to be taken into consideration, Figure 3.90 and Figure 3.91. The saturation values for water and moisture absorption, on the other hand, depend only little on temperature, Figure 3.92.

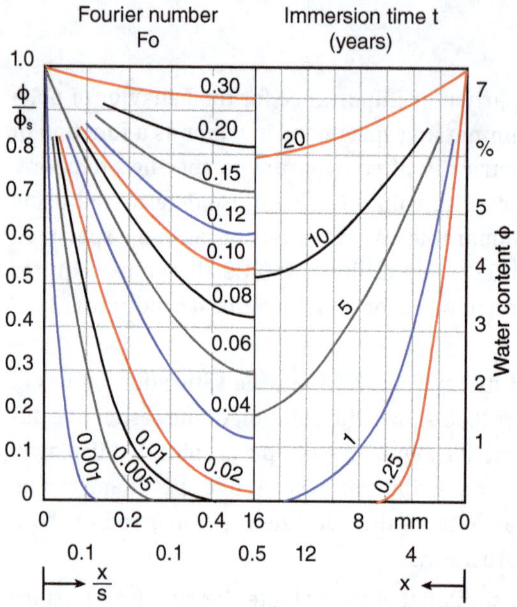

		Example PA 6 casting resin Sought: time after which φ = 4.5% is reached at x = 8 mm
y-axis:	φ = water content	= 4.5 % (given)
	φ_s = saturation value	= 7.5 % (given)
	φ/φ_s	= 0.6
x-axis:	x = edge distance	= 8 mm (given)
	s = wall thickness	= 32 mm (given)
	x/s	= 0.25
Parameter:	F_0 = Fourier number	
	$= \frac{D \cdot t}{s^2}$	= 0.08 (from diagram)
	D = coefficient of diffusion	= $0.32*10^{6}$ mm²/s
	t = time	$= \frac{F_0 \cdot s^2}{D} = 2.56*10^{8}$ s = 8.1 years
	$\bar{\varphi}$ = average water content	$= \varphi_s * \bar{\varphi}/\varphi_s = 7.5 * 0.63\ \% = 4.7\ \%$
	φ/φ_s	= 0.63

Figure 3.88 Diagram for the evaluation of moisture distribution in fully immersed flat plastic components

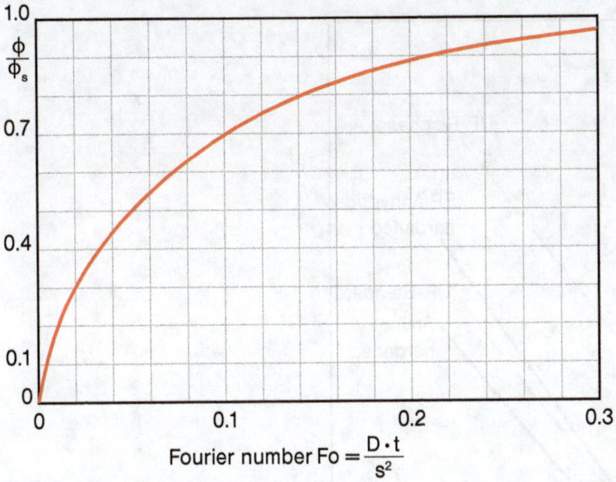

Figure 3.89 Diagram for the evaluation of average moisture content (see Figure 3.88)

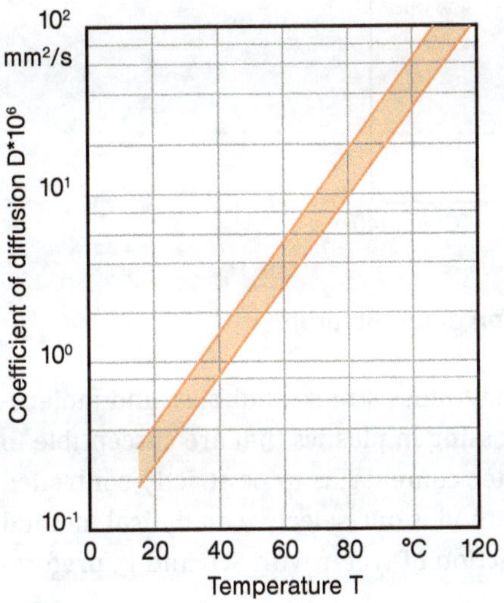

Figure 3.90 Coefficients of diffusion for PA 6, 12, and 66 as a function of temperature

Water absorption also causes an increase in volume. For isotropic materials it can be assumed as a general rule that the linear swelling ε_q is smaller by a factor of 0.3 than the water absorption. In glass fiber reinforced components, swelling is restrained in particular in fiber direction so that it is lower in the direction of orientation than in the thickness direction of a sheet.

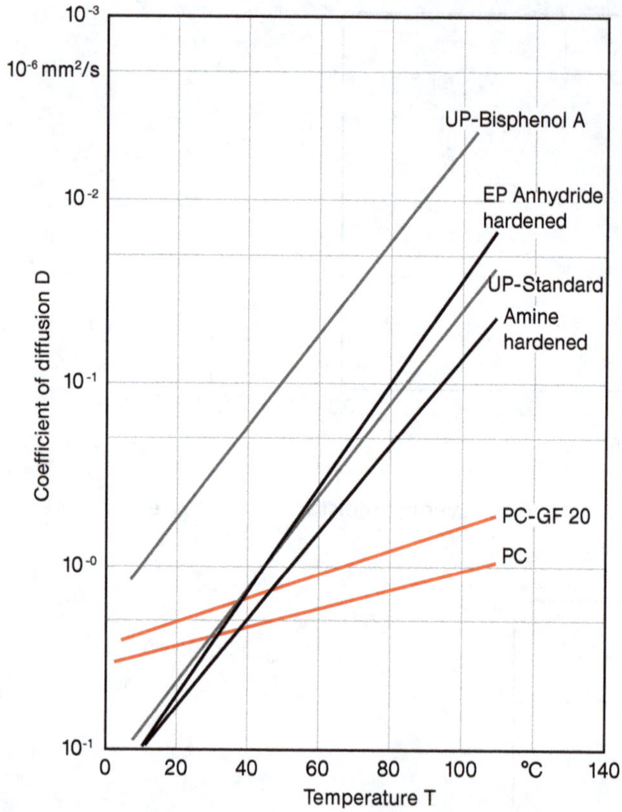

Figure 3.91 Coefficient of diffusion as a function of temperature

Even very low water contents in the pellets may cause streaks, bubbles, and molecular decomposition during thermoplastic processing in plastics that are susceptible to hydrolysis, such as polyesters. Therefore, water content has to be strictly controlled prior to processing. Karl Fischer's titration method is one selective analytical method to determine water content. It utilizes the reaction of water with SO_2 and J_2, preferably in methanolic solution.

Here, the water is volatilized in an inert gas atmosphere. Calibration of the measurement system is not necessary. It is possible to distinguish between absorptively bound surface water and chemically bound water and/or water formed during decomposition by employing an increasing temperature gradient.

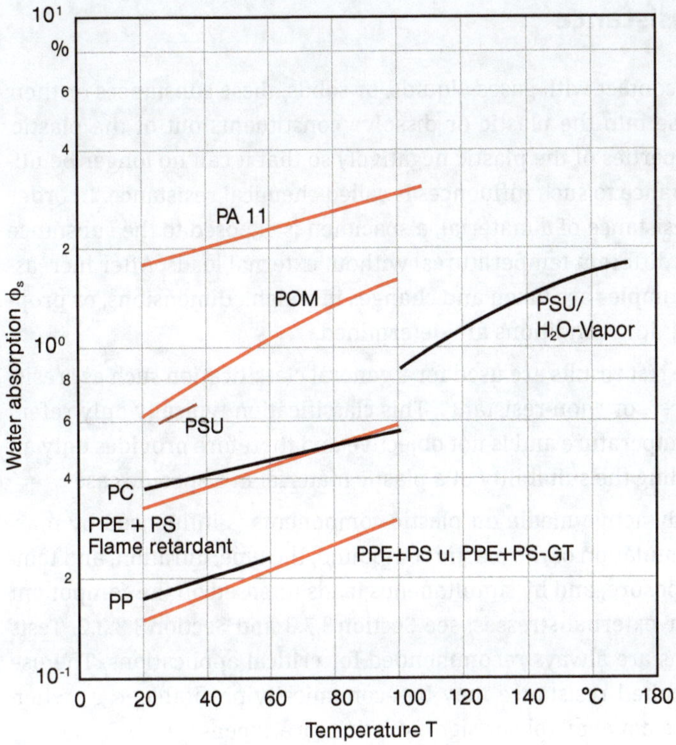

Figure 3.92 Saturation water absorption as a function of temperature

A more time-consuming method to determine the water content is drying in a vacuum desiccator. This method comes with the risk of measuring not only water, but also other volatile components. In industrial practice, so-called Tomasetti volatile indicator (TVI) tests have proven successful. Here, the pellets are heated between two glass plates on a heating table while monitoring the formation of bubbles. If no bubble formation is registered during melting, the water content is sufficiently low for thermoplastic processing, Figure 3.93.

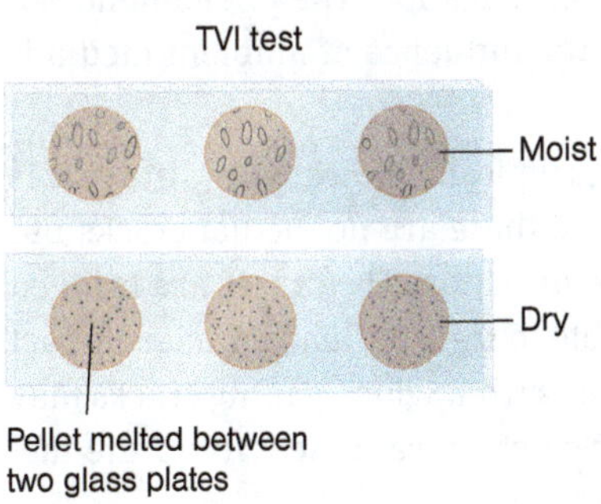

Figure 3.93
TVI test, moisture test prior to thermoplastic processing

3.7.2 Chemical Resistance

When plastics come in contact with gases, liquids, or solids, these substances or their constituents may diffuse into the plastic or dissolve constituents out of the plastic. This may affect the properties of the plastic negatively so that it can no longer be utilized. A material's resistance to such influences is called chemical resistance. In order to measure chemical resistance of a material, a specimen is exposed to the substance to be tested (possibly at different temperatures) without external loads. After increasingly longer intervals, samples are taken and changes in weight, dimensions, or properties compared to the initial conditions are determined.

In the simplest case, the test results are used for a general classification such as "resistant", "limited resistance", or "non-resistant". This classification typically only refers to resistance at room temperature and is not objective and therefore provides only an initial indication regarding the suitability of a plastic material in a specific case.

The effects of chemically active media on plastic components is influenced by marginal changes in the formulation of the plastic compound, the type, duration, and temperature of medium exposure, and by simultaneous loads imposed on the component by frozen-in residual or external stresses, see Section 3.7.3 and Section 3.3.3.2. Tests under service conditions are always recommended for critical applications. The use of plastics with only limited resistance may be economically profitable, *e.g.*, when alternative materials are not available or significantly more expensive.

3.7.3 Stress Cracking Resistance

The phenomenon of stress corrosion is well known for metals. Under the simultaneous influence of certain media and tensile stresses, it causes brittle fracture along the grain boundaries or across the grain. The cause is typically electrochemical in nature. This type of brittle cracking occurs in plastics as a result of residual cooling stresses or external tensile stresses and exposure to certain media. Here, we primarily observe purely physical processes, *e.g.*, wetting, diffusion, and swelling. However, oxidation or hydrolysis at the crack tip may also affect crack propagation. The phenomenon of brittle failure in otherwise tough materials under the influence of ambient media is also called environmental stress cracking (ESC).

The initial phase of damage in transparent plastics is typically indicated by fine crazes originating at the surface that reflect light. However, these are not actual cracks but rather regions in which the plastic is drawn in load direction. The crazes are bridged by stretched molecular chains and are therefore able to transfer loads. Under impact however, they act like notches and thus as weak points from which gaping cracks may form under long-term loads, leading to fracture. The chemical reactivity of the medium in which the plastic is tested has considerable influence on stress crack formation, Figure 3.94.

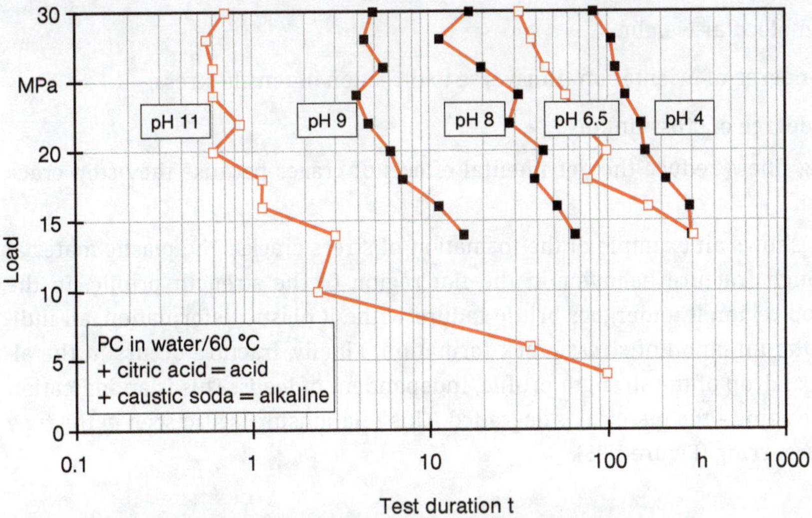

Figure 3.94 Effect of pH-value on service life of polycarbonate

Because crack formation is a function of time, stress crack resistance is evaluated using creep tests. Here, we distinguish between tests at constant load and tests at constant strain. The determination of creep resistance under tension in different media and at different temperatures is a 1-step testing method. There are several standard methods commonly used to impose constant deformation on specimens while they are simultaneously exposed to a medium for a specified time:

- The ball and pin indentation tests, where balls or pins with slightly larger dimensions are pressed into a hole in the specimen.
- The bent strip test, where the specimen is clamped on a forming plate with a constant radius of curvature.

Damage is evaluated by recording cracks or by determining the specimen's residual strength or strain at break after exposure to the medium. The analysis of the test results requires experience and familiarity with the testing method, in particular because stress crack formation is affected by a variety of factors. It increases with:

- Higher tensile loads, also residual and cooling stresses,
- Longer time of exposure,
- Higher temperature,

- Higher reactivity of the ambient medium (*e.g.*, oxygen concentration, surfactant, pH-value),
- Higher crystallinity,
- Lower molecular weight,
- Higher degree of orientation transverse to direction of tensile stress,
- Lower degree of cross-linking.

Reinforcing fibers reduce the detrimental effects of crazes because they stop crack propagation.

Figure 3.38 shows an example of the formation of stress cracks. The plastic material exhibits tough fracture behavior in the flat region of the strength profile, in the steeper drop region it undergoes brittle failure without plastic deformation, an indication for the initiation of stress crack formation. Finally, fracture occurs in the almost vertical drop of the strength profile, independent of loads. This is an indication of oxidative aging – the material is degraded. This relationship is also seen in the type of failure occurring (Figure 3.95).

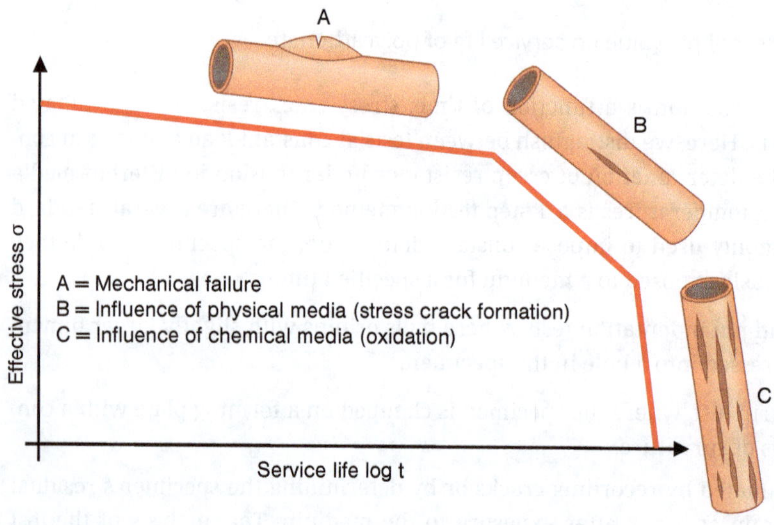

Figure 3.95 Schematic creep rupture diagram for polyethylene

There is no relationship between chemical resistance and stress cracking resistance. On the contrary: a plastic material that was determined to be chemically resistant after immersion in a specific medium may well be susceptible to stress crack formation. This can be explained by the fact that stress crack formation mechanisms set in only when the material is exposed to stresses (which could also be residual stresses created during processing or the effects of temperature changes).

3.7.4 Weathering

Weathering is defined as the influence of all factors encountered when plastic components are used outdoors: solar radiation, oxygen, ozone, other volatile components of air, temperature, moisture, precipitation, and dust deposits. The intensity and relative duration of these influences strongly depend on the geographic location and the season. Therefore, it is difficult to provide a general characterization of weather resistance. Depending on the intended service location of a plastic component, a variety of exposure scenarios are tested and subsequently the changes in color, gloss, dimensions, density, and physical and mechanical characteristics are determined. Tests under the following exposure conditions are common:

- Open air exposure to UV light under window panes, optionally in different climate zones,
- Open air weathering in different climate zones,
- Artificial (laboratory) weathering – with or without cyclic water spray,
- Reference temperatures/standard climate/constant testing climate/humid alternating climate /condensation water climate.

There are a number of approaches to interpret the results of exposure to artificial UV light or weathering chambers and predict behavior in service. The effects measured in artificial or accelerated weathering tests vary from product to product and from device to device, showing that these tests are suitable only for the pre-selection of plastic materials or for the evaluation of product development and modification. The wide variety of commercially available radiation devices complicates the comparability of results.

3.7.5 High-Energy Radiation

High-energy (ionizing) radiation includes all particle and wave radiation with energy levels much higher than their molecular bonding energy; they include: high-energy electrons, protons, α-particles and heavy nuclei, neutron-, X-, and γ-rays. The effect of radiation on a specific material is determined by its absorption behavior, and less by the type of radiation. Similar to other resistance investigations, a material is exposed to radiation in order to determine the energy dose (in J/kg) that causes a specified change in characteristics. The dose rate (the absorbed energy per mass and time unit in J/kg · h) plays a role when radiation triggers oxidation processes in the plastic material. At high dose rates, the respective radiation times until specified material damages occur are short so that the influence of oxygen is negligible. On the other hand, low dose rates and the respective long exposure times allow for oxygen to diffuse into the plastic and thus causing significantly higher damage (Figure 3.96).

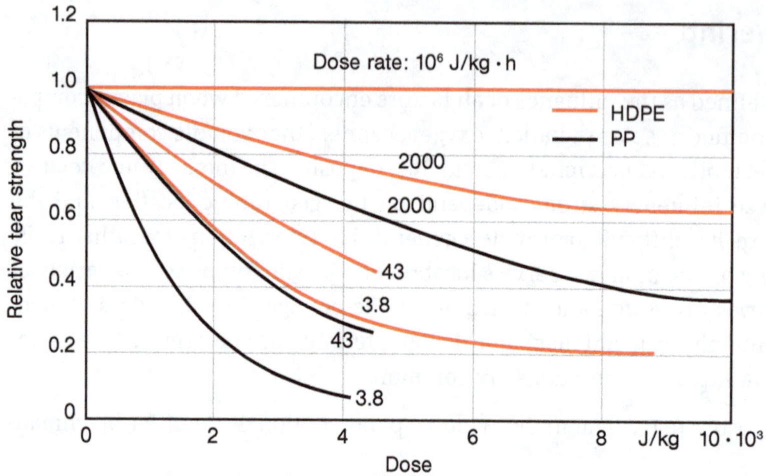

Figure 3.96 Effect of dose and dose rate on resistance to high-energy radiation

3.7.6 Resistance to Organisms

In order to determine a material's resistance to organisms such as mold, algae, and bacteria, plastic specimens are exposed to these biological media. Testing criteria include the effect of the plastic specimen on the growth of the medium and the damage detected on the specimens. Most plastics exhibit good microbial resistance; however, low-molecular additives may negatively affect this property. In the tropics and other hot and humid climates, termite resistance is of particular interest. Possible rodent attack has to be investigated by feeding tests. The harder and smoother a plastic material's surface, the less likely it is to be attacked.

3.7.7 Migration and Permeation

Absorption, solubility, and diffusion of liquids, volatile compounds, and gases in plastics and plastic interfaces are of technical importance even if they do not lead to significant changes in the overall characteristics of a plastic material. The same statement is valid for the migration of plastic components into their surroundings.

Testing methods applied in industrial practice measure the threat to adjoining materials (*e.g.*, by stress crack formation) caused by the migration of plasticizers and bleeding of colorants. Here, plastic films are pressed against an absorbing contact surface at elevated temperatures and changes in weight and surface properties are measured after a certain exposure time. The volatility of plasticizers and other additives is measured using active carbon adsorber methods. Extractable components are determined in order to evaluate the suitability of plastic materials for applications in food contact.

The following effects are observed when a medium permeates a plastic wall: "weeping" of contaminating or hazardous components from container or pipe walls, their penetration behind plastic linings, the evaporation of fuels through gas tank walls, and the absorption of oxygen in the material flowing through pipelines or into the content of plastic containers, which in turn may cause corrosion of metal parts in production lines. Specifically designed test procedures or simulation by long-term tests are required to sufficiently analyze these processes.

The gas and vapor permeation characteristics of plastics are of particular interest for the packaging industry. They are determined with the help of films.

Because the permeation P of gases and vapors is inversely proportional to the thickness of a film, the following equation holds:

$$P = P_0 \cdot d_0/d$$

Where:
P = Permeation value of a layer thickness d
P_0 = Permeation value of a reference layer thickness d_0

For multi-layer composites, it holds:

$$1/P = 1/P_1 + 1/P_2 + \cdots 1/P_i$$

Where:
P_1; P_2; P_i = Permeation value of each layer

3.7.7.1 Water Vapor Permeability

The water vapor permeability of a plastic material refers to the weight of water vapor that diffuses through 1 m² of the tested film within 24 h at a humidity difference of 85% (which equals a water vapor partial pressure differential of 19.68 mbar) and a temperature of 20 °C:

$$\text{Water vapor permeability in } \left[\frac{g}{m^2 \cdot 24\,h}\right] \quad \text{(ISO 1195)}$$

The thickness of the film has to be specified because it strongly influences the measurement (Figure 3.97). Multiplication with the specimen thickness in relation to the partial pressure differential provides the permeation coefficient.

$$\text{Permeation coefficient in } \left[\frac{\text{g}}{\text{cm} \cdot \text{h} \cdot \text{mbar}}\right]$$

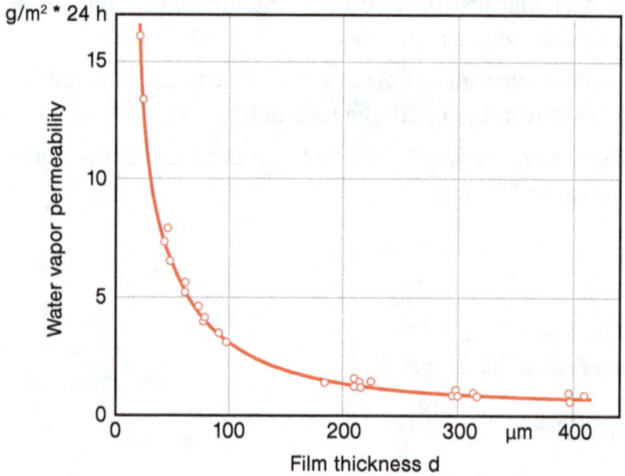

Figure 3.97 Water vapor permeability of PVC-U as a function of film thickness

Figure 3.98 shows the water vapor permeability of selected plastic films, while the effect of temperature is shown in Figure 3.99.

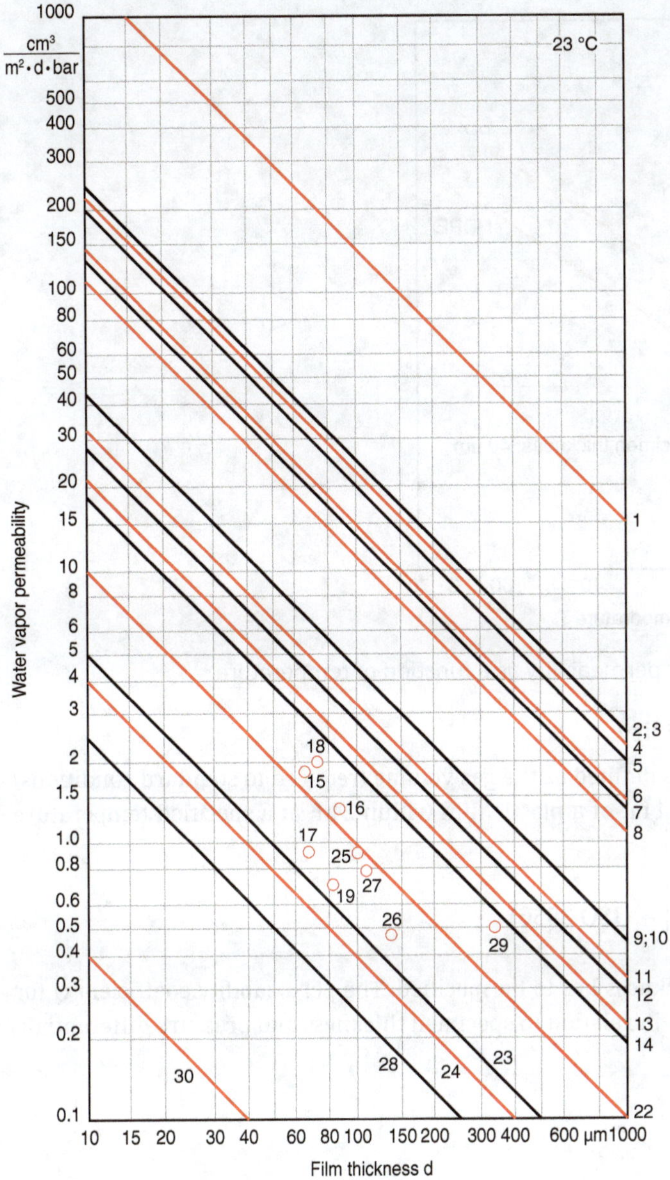

1: CTA
2: PVC-P (25 % WM)*
3: ABS
4: SAN
5: PA-Cop.
6: PA 6
7: PS
8: PS-HI
9: EVA
10: PVC-P (22 % WM)
11: EVA
12: PVC-U (E-PVC)
13: PVC-U (S-PVC)
14: PET
15: PET/PE (12/50)

Figure 3.98 Water vapor permeability as a function of film thickness

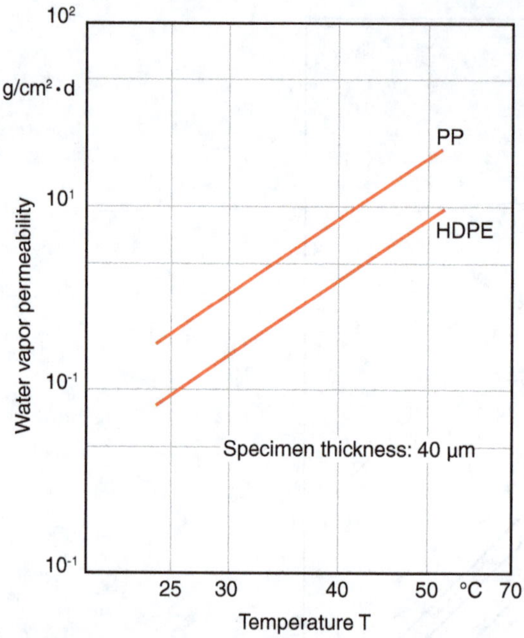

Figure 3.99 Water vapor permeability as a function of temperature

3.7.7.2 Gas Permeability

The gas permeability q is defined as the gas volume (reduced to standard conditions) that permeates through 1 m^2 of a plastic film within 24 h at a specified temperature and pressure differential:

$$q \text{ in } \left[\frac{\text{cm}^3}{\text{m}^2 \cdot 24\,\text{h} \cdot \text{bar}}\right] \quad \text{(ISO 2556)}$$

Again, the specimen thickness has to be specified. The permeability coefficient Q for gases is the permeability in relation to specimen thickness and pressure differential:

$$\text{Q in } \left[\frac{\text{cm}^3(\text{NTP})}{\text{cm} \cdot \text{s} \cdot \text{mbar}}\right]$$

Permeability characteristics increase sharply with increasing temperature (Figure 3.100) and depend strongly both on the measurement conditions and the kind and condition of the tested film. Published permeability values therefore offer only a qualitative standard of comparison. In critical applications, measurements of the actual product are required. Sometimes, different units are used for permeability values and permeability coefficients, see Table 3.10.

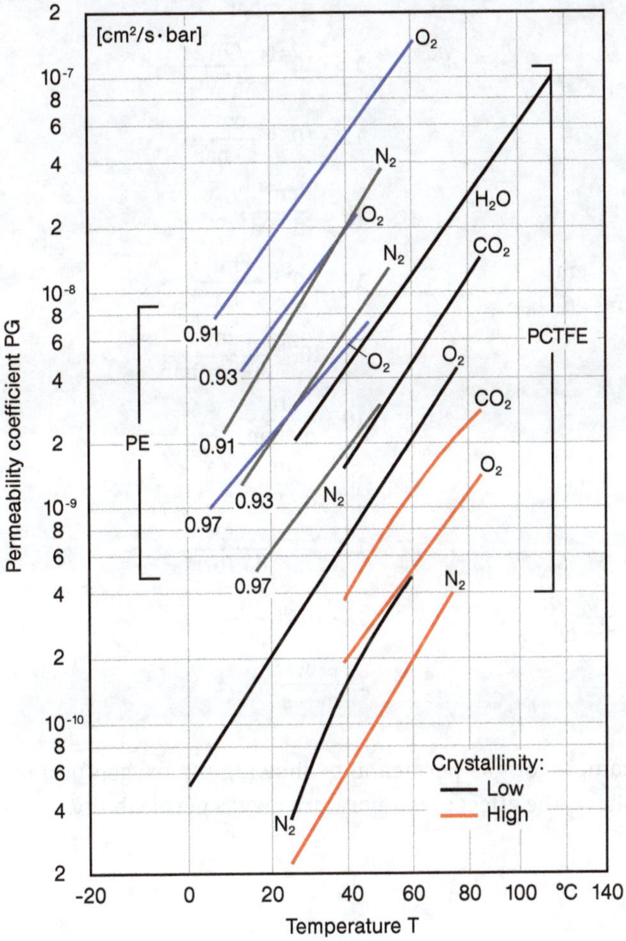

Figure 3.100 Permeability coefficients for selected PE and PCTFE as a function of temperature

Table 3.10 Conversion Table for the Permeability Characteristics of Gases and Water Vapor

Parameter	Recommended unit	Conversion factor
Permeability coefficient	$1 \frac{cm^3 \cdot 100\ \mu m}{m^2 \cdot d \cdot bar} \hat{=}$	$10^{-2} \frac{cm^3 \cdot 100\ \mu m}{dm^2 \cdot d \cdot bar}$ $10^{-9} \frac{cm^2}{d \cdot mbar}$ $1.5 \cdot 10^{-14} \frac{cm^2}{s \cdot Torr}$ $1.2 \cdot 10^{-11} \frac{cm^2}{s \cdot bar}$ $10^{-11} \frac{cm^2}{d \cdot Pa}$
Gas permeability	$1 \frac{cm^3}{m^2 \cdot d \cdot bar} \hat{=}$	$1.2 \cdot 10^{-9} \frac{cm}{s \cdot bar}$ $12 \cdot 10^{-6} \frac{cm^3}{m^2 \cdot s \cdot bar}$ $10^{-4} \frac{cm}{d \cdot bar}$ $10^{-9} \frac{cm}{d \cdot Pa}$ $1.3 \cdot 10^{-7} \frac{cm}{d \cdot Torr}$
Water vapor permeability	$1 \frac{g}{m^2 \cdot d} \hat{=}$	$12 \cdot 10^{-6} \frac{g}{m^2 \cdot s}$ $12 \frac{\mu g}{m^2 \cdot s}$

Figure 3.101 provides an example for gas permeability showing the oxygen permeation of films; Figure 3.102 shows the effect of temperature on gas permeability.

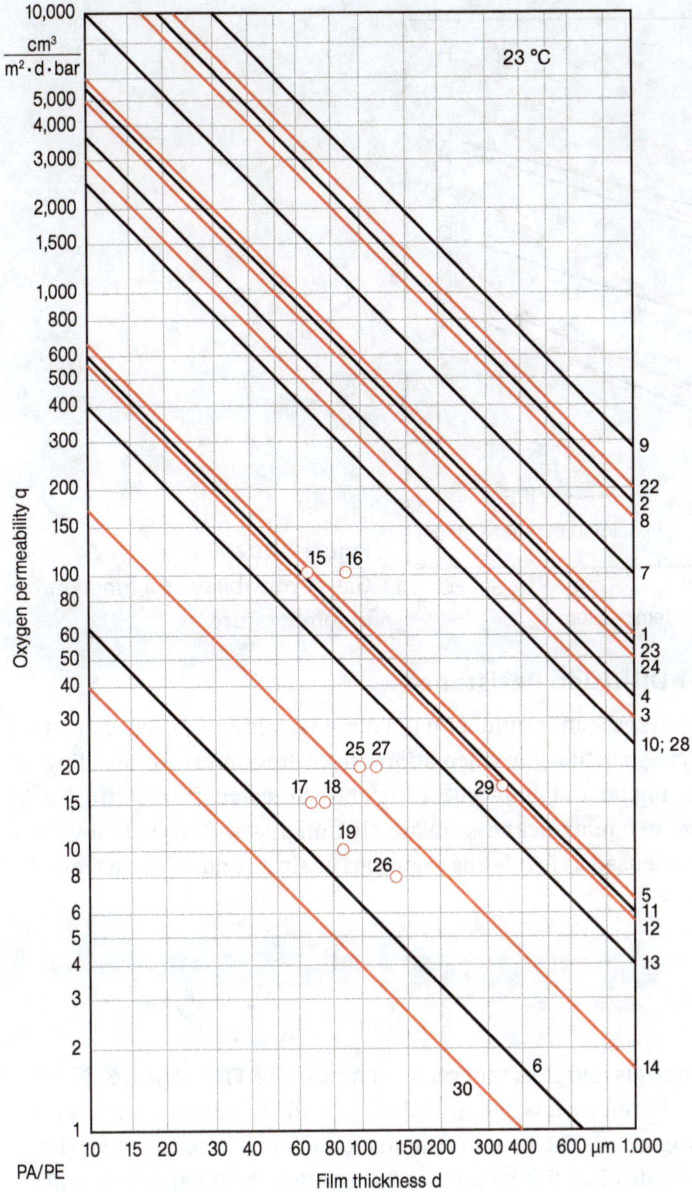

PA/PE

16: PET/PE (12/75)
17: PET/PE-X (15-50)
18: PA/PE (35/50)
19: PA/PE-X (35/60)
22: LDPE
23: PP
24: HDPE
25: PA/PE (40/60)
26: PA/PE-X (60-75)
27: PA/PP (40/75)
28: PP-O
29: PVC/PE (250/75)
30: PVDC

Figure 3.101 Oxygen permeability as a function of film thickness

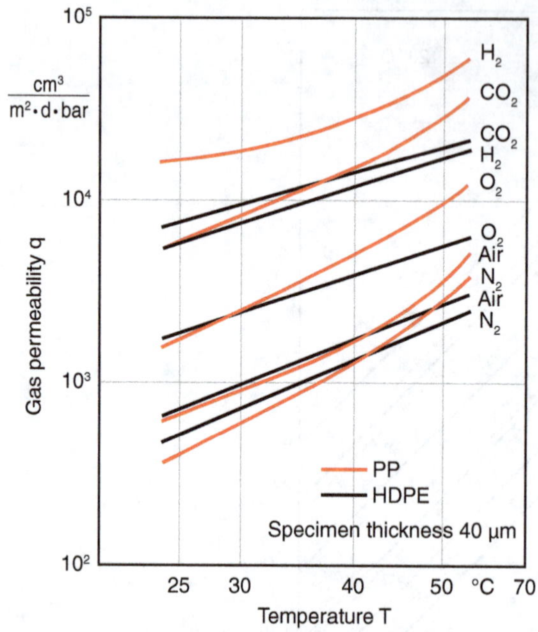

Figure 3.102 Gas permeability as a function of temperature

3.7.7.3 Water Vapor Diffusion Resistance

The water vapor diffusion resistance number µ is a dimensionless material characteristic used in building construction for calculations regarding moisture proofing. It states how many times higher the material's resistance to water vapor diffusion is compared to an air layer of equal thickness. Relatively thin plastic films with µ-values ranging from 10^4 to 10^5 are used as insulating materials against condensation water in walls and insulated roofs.

3.7.8 Fire Behavior

Plastics are organic materials and thus inherently combustible. Fire behavior is not a material characteristic; therefore, it is not possible to provide a general assessment of plastics during or after ignition. The heat of combustion value of a plastic material is a determining characteristic of its fire behavior after ignition (heat value in kWh/kg). The heat values for selected common plastics are summarized in Table 3.11. Other factors influencing fire behavior include the material's ignitability, its contribution to flame propagation, and parallel fire phenomena such as smoke formation, and toxicity and corrosivity of fire effluents. All these fire phenomena and their effects depend not only on the material, but also on ambient factors, such as the type, duration, and intensity of the ignition source, air supply at the location of the fire, quantity, shape, and arrangement of the plastic component in space, and others. The complexity of fire behavior resulted in the development of many testing procedures, and in certain applications plastic components have to pass all of these tests in order to gain approval. Passing several different flammability tests ensures that a material has been evaluated in all relevant fire risk scenarios.

Table 3.11 Heat Value of Selected Thermoplastics in kWh/kg

Material	kJ/kg	kWh/kg
POM	17,000	4.7
PESU	17,000	4.7
PVC	20,000	5.6
PMMA	26,000	7.2
PBT	31,000	8.6
PA	36,000	10.0
PSU	36,000	10.0
ABS	37,000	10.3
PS	42,000	11.7
PE	46,000	12.8
PP	46,000	12.8

The most commonly used method to evaluate flammability of a material was introduced by Underwriters' Laboratories Inc. (UL Subject 94), which is now also covered by ISO standard 60695. UL offers certification and monitoring to manufacturers of plastic resins or components. Specimens to be tested are fixed either horizontally or vertically before they are exposed to the flame of a Bunsen burner. The following parameters are assessed: burning velocity, flame spread, after burn, and dripping. Materials are designated a flammability classification: "no", "HB", "V-2", "V-1", "V0", "5VA", and "5VB". The listed sequence represents increasing rigor of the applied test.

In addition, there are numerous, sometimes industry sector-specific tests and characteristics that will be covered here only in limited detail.

The ignition and auto-ignition temperatures are defined as the temperatures at which the gases developing during heating will ignite with or without additional ignition flame, respectively.

The limiting oxygen index (LOI) is the minimum oxygen concentration (in %) in an oxygen/nitrogen atmosphere that allows a vertical specimen to burn in a candle-like manner. Of all fire characteristics, the LOI closest resembles a material characteristic because it does not depend on geometry or other conditions.

Incandescent wire tests are used to determine whether a specimen or a component poses a flammability risk when in contact with a glowing wire.

In construction engineering, flammability is a critical factor for material selection. There is a variety of different tests. In many cases, the evaluation is performed with components under actual service conditions and real fire load in specifically built test buildings and building front testing rigs.

Flammability is also of importance in transportation and shipping applications. In addition to the previously mentioned UL tests, automotive interiors have to meet the requirements of the Federal Motor Vehicle Safety Standards 302 (FMV SS 302) (see Figure 3.103). Here, the flame velocity of specimens exposed to Bunsen burner flames for 15 s is limited to 4 inch/min.

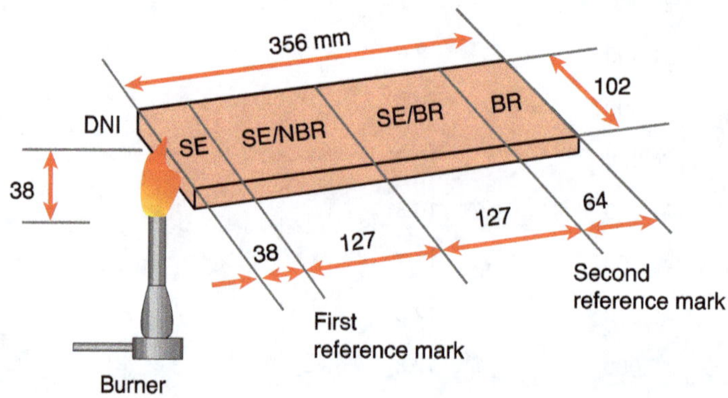

Test conditions
- Thickness of test specimen as in application
- Horizontal specimen
- Flame application using a Bunsen burner (15 sec)

DNI	does not ignite
SE	self-estinguishing
SE/NBR	self-extinguishing/ no burn rate
SE/BR	self-extinguishing/ with burn rate
BR	burn rate

Requirements: Maximum rate of flame speed: 102 mm/min

Figure 3.103 Test set up and explanation of results according to FMVSS 302 [Source: BASF AG]

There are additional test procedures for car body panels and in particular for gas tanks.

For obvious reasons, material flammability in aeronautical applications is one of the most critical selection criteria, thus this industry has to meet the most stringent regulations. The regulations of the country that grants certification for aviation application determine the respective requirements; however, most countries have adopted the Federal Aviation Regulations (FAR). Most important here is FAR 25. The practical tests include exposure to Bunsen burner flames at different angles. Depending on the application, the tests require different testing set-ups and define different criteria for passing.

3.8 Biodegradation

Günter Auernhammer, Dieter Fischer, Holger Ruckdäschel, Christian Laforsch, Martin Löder, Seema Agarwal, Kathrin Harre, Andreas Fery

Biodegradation is a specific type of plastic degradation that occurs through the biological activity of microorganisms such as bacteria and fungi. The primary processes involved in the biodegradation of plastics are enzyme-catalyzed hydrolysis and oxidation, which result in macromolecular chain scission. In addition, biodegradation can occur through biological activity that does not directly involve enzymes, such as the use of reactive metabolites, reactive oxygen species, etc. Biodegradation occurs in two main stages:

- In the first step, large macromolecular chains (high molecular weight) break down into low-molar-mass oligomers or monomers. This step occurs primarily outside the microorganisms, mainly catalyzed by extracellular enzymes.
- In the second step, these oligomers/monomers undergo bioassimilation and mineralization inside microorganisms to biomass and simpler compounds, such as carbon dioxide/methane, nitrates/ammonium, along with water. In the presence of oxygen (aerobic biodegradation), carbon dioxide, nitrates, and water are the main products, whereas methane and ammonia are mainly formed under anaerobic (absence of oxygen) conditions.

The outcome of biodegradation is a macromolecular chain scission with changes in chemical structure, composition, and physical and chemical properties, with ultimate conversion to small assimilated and mineralized molecules

Ultimate biodegradation: The biodegradation process can halt at any stage, depending on factors like the degradation products, environmental conditions, and material properties. If the process reaches completion, resulting in the breakdown to ultimate end products carbon dioxide/methane, nitrates/ammonia, and water, it is termed ultimate biodegradation.

Bulk biodegradation and surface erosion: Bulk biodegradation and biodegradation by surface erosion are two distinct mechanisms by which biodegradable plastics degrade. In bulk biodegradation, enzymes, and water penetrate into the bulk of the plastic, causing it to degrade uniformly throughout its entire structure. In contrast, in biodegradation by surface erosion, degradation occurs only on the surface of the plastic. The outer plastic layer erodes over time while the internal structure remains intact until the outer layers have been fully degraded. The material maintains its structural integrity for a longer period, gradually thinning as the outer layers erode. More hydrophobic plastic surfaces undergo biodegradation by surface erosion. Distinguishing between the two mechanisms typically involves observing how a material degrades over time. The material degrading by surface erosion shows a linear and steady mass loss, almost no change in the molecular weight and composition as observed by spectroscopic methods, uniform thinning of the material keeping its form, and morphological changes as observed by electron / optical microscopes only on the surface and not in the cross-section.

Facts:

- Biodegradation can occur in various environments such as water bodies, soil, compost, and even within human bodies.
- The rate of biodegradation for the same plastic varies based on factors such as the degradation environment, molecular weight, additives, form, and size of the material.
- The rate of biodegradation in the same environment is different for different plastics with varied chemical structures.
- The in vitro degradation of plastic by an isolated enzyme is not considered biodegradation since no biological activity is involved.
- Polymers with C–C backbones are generally non-biodegradable, though there are some exceptions.
- Aliphatic polyesters are biodegradable, whereas aromatic polyesters are not. However, aliphatic-aromatic polyesters can be biodegradable if the aromatic content does not exceed a certain threshold. Polylactide (PLA), polyglycolide (PGA), and polycaprolactone (PCL) are examples of biodegradable aliphatic polyesters, whereas polybutylene adipate terephthalate (PBAT), an aliphatic-aromatic polyester, is often used in compostable packaging and films.

Biodegradable polymers are not a comprehensive solution on their own for addressing the microplastic problem. While the three "Rs" (recycle, reuse, and reduce) remain the gold standards for managing increasing plastic waste, the use of biodegradable polymers is promising in applications where plastic waste collection is difficult or impossible, such as in agricultural fields (e.g., mulch films, dispensers) and in water bodies (e.g., fishing nets).

Composting plastics can be seen as a form of recycling (organic recycling), where the biodegradation process yields by-products like water and carbon dioxide. These by-products play a critical role in plant growth, and the resulting biomass can be harnessed as a renewable resource for the chemical industry to produce bio-based monomers and polymers, creating a sustainable, circular system where waste contributes directly to the creation of new materials. Keeping this in mind, the use of biodegradable polymers for food packaging is a promising application.

Due to the complexity and varying interpretations of biodegradability, numerous testing methods (US: ASTM standards, Europe: EN norms, global: ISO standards) have been established across different regions and applied to verify biodegradability under specific conditions for particular applications.

3.9 Friction and Wear Behavior

Robert Feulner

The friction and wear behavior of materials are the outputs of the so-called tribological system. It is subject to external factors such as temperature, ambient media (moisture, lubricants, contamination), and type of motion (sliding, rolling) among many others.

3.9.1 Friction

Friction is defined as the resistance to the relative movement of two touching surfaces. We distinguish between the resistance to the initiation of relative movement (static friction, prior to motion) and to maintaining this movement (kinetic – or dynamic – friction after motion has been initiated). Static friction is described by the static coefficient of friction and kinetic friction by the dynamic coefficient of friction.

3.9.1.1 Friction Mechanisms

Friction mechanisms are defined as the motion-inhibiting, *i. e.*, energy dissipating, basic processes occurring in the contact area of two bodies. The number of microcontacts increases approximately in linear proportion with the normal force F_N (in N). Each contact represents an elementary motion resistance event. Depending on the kind of contact between the friction partners, we encounter different levels of elementary resistance. The sum of all elementary resistance events constitutes the friction force F_R (in N). The friction behavior of plastic on metal surfaces can be described by two friction mechanisms: adhesion and deformation of the contact area [11, 12]. The adhesive effect decreases with increasing roughness of the metal surface because the contact area decreases. By contrast, the deformative effect increases as a result of the increased number of roughness peaks. These inverse effects create friction resistance minima, see Figure 3.104.

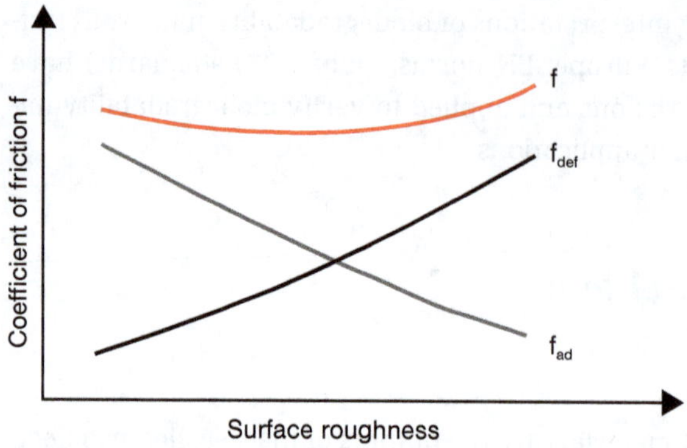

Figure 3.104 Effect of surface roughness of the metal friction partner on the adhesive and deformative component of the coefficient of friction [13]

f = Coefficient of friction [-]

f_{def} = Deformative coefficient of friction [-]

f_{ad} = Adhesive coefficient of friction [-]

Coulomb's friction model is used to determine the friction of a specific tribological system. Here, the coefficient of friction f is defined as the ratio of friction force F_R and normal force F_N:

$$f = \frac{F_R}{F_N} \tag{3.35}$$

The coefficient of friction f is the sum of an adhesive and a deformative component:

$$f = f_{def} + f_{ad} \tag{3.36}$$

3.9.1.2 Stick-Slip Effect

Large differences in the coefficients of static and kinetic friction cause auto-excited friction oscillations in systems with low stiffness, see Figure 3.105. A common example of this phenomenon is the stick-slip behavior of windshield wipers under certain conditions. The deformation and subsequent spring back of the component cause oscillations of the friction forces that generate heat and noise.

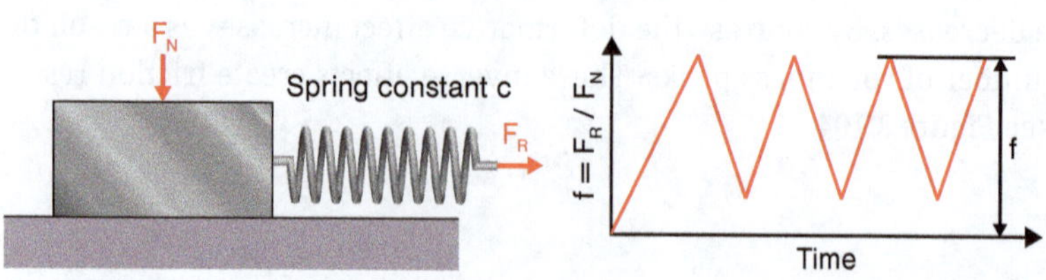

Figure 3.105 Stick-slip effect: left: physical illustration; right: oscillating frictional force

3.9.2 Wear

Wear is defined as progressive material abrasion. It is created by mechanical or chemical load on the surface.

There are several ways to specify wear as a metrological parameter, one of which is the "linear wear" w_l illustrated in Figure 3.106.

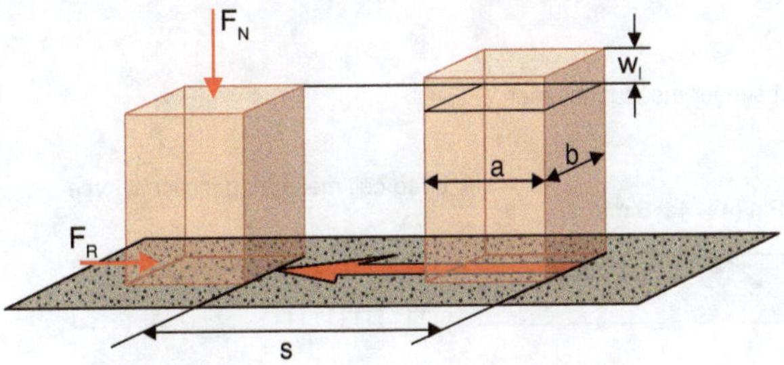

Figure 3.106 Schematic illustration of linear wear w_l

The linear wear rate $w_{l/s}$ is defined as the quotient of linear wear w_l and glide path s:

$$w_{l/s} = \frac{w_l}{s} \tag{3.37}$$

The volumetric wear rate $w_{v/s}$ is defined as the quotient of volumetric wear w_v (= $w_l \cdot a \cdot b$) and glide path s:

$$w_{v/s} = \frac{w_v}{s} \tag{3.38}$$

In general, the wear behavior of a material is described using the coefficient of wear k, which is defined as the quotient of volumetric wear rate and normal force F_N:

$$k = \frac{w_{v/s}}{F_N} \tag{3.39}$$

Model systems, such as the pin-on-disk test, measure the decrease in length of a pin-shaped specimen that has a defined contact area and contact pressure *p* with the respective friction partner. The equation used here for the coefficient of friction is:

$$k = \frac{w_{l/s}}{p} \tag{3.40}$$

For more complex component tests, *e.g.*, for gears, the coefficient of wear has to be adjusted for line contact and rolling motion. More detailed information can be found in [14, 15].

3.9.3 Specimens and Testing Systems

In order to determine tribological characteristics to facilitate material comparisons, the load collective always has to be reproducible. The pin-on-disk test offers a number of advantages compared to more complex component tests, such as transmission tests and others. They include specimen manufacturing and preparation and the option to adjust defined loads such as sliding speed and contact pressure. Figure 3.107 illustrates the schematic design of a pin-on-disk test [16].

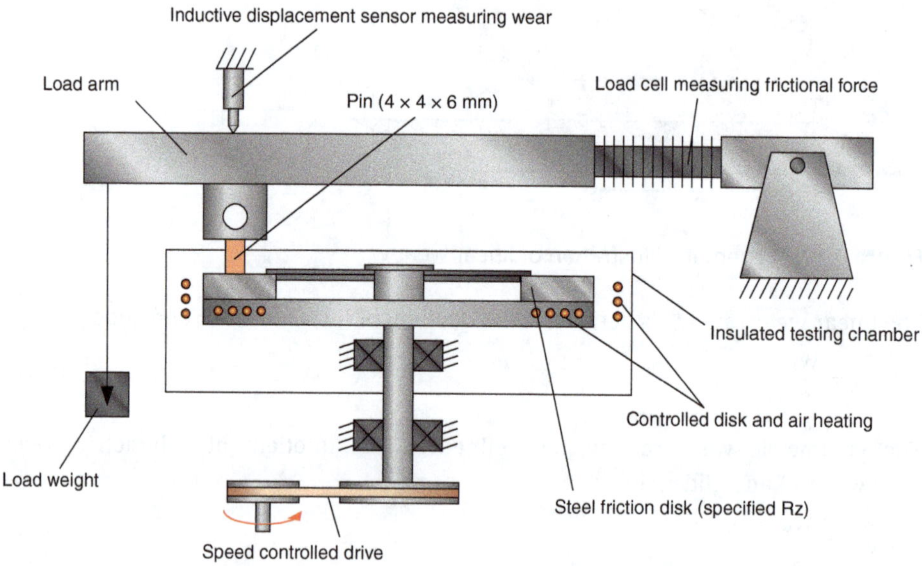

Figure 3.107 Schematic of pin-on-disk test

The plastic test pin is fixed in a holder and pressed (with a constant force) against a disk spinning at constant speed. The disk is made of 100Cr6 steel and polished to a defined roughness. An inductive displacement sensor records the pin's reduction in length and a load cell the deflection force (frictional force) caused by the friction. The tests determine the coefficient of wear k and the coefficient of friction f (Figure 3.108). For fiber-reinforced plastics, the coefficients of wear are determined in the state of the system when the reinforcing fibers are in frictional contact, *i. e.*, ensuring that the outer component layers, which often are low in fiber content, do not affect the test results. If the low-fiber content outer layers are friction partners, it is permissible to assume the tribological behavior of non-fiber reinforced plastics.

Fiber reinforced specimens exhibit fiber orientation that depends on material, processing, and component geometry, and that affects the measured system parameters. In order to ensure reproducible test results, specimens should be sampled from the center of an injection molded sheet (*e. g.*, 115 × 115 × 4 mm). The test pins should have a friction surface area of 4 × 4 mm². In order to ensure the required 6 mm length of the test pins, they are sampled in flow direction.

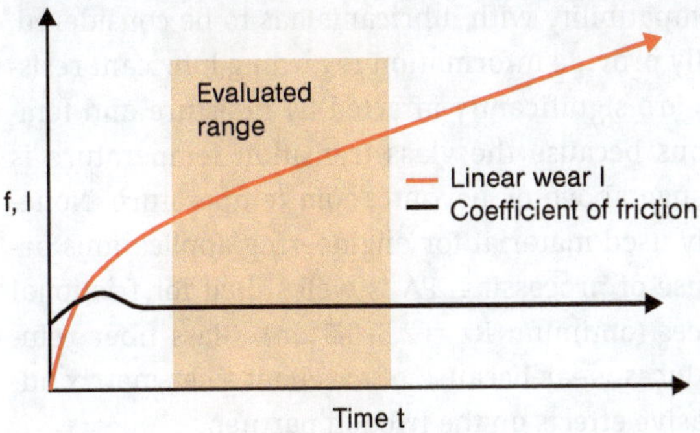

Figure 3.108
Analysis of tribological test results

In general, there is large number of different testing systems suitable for the determination of tribological material characteristics. For example, in cases when both friction partners are in permanent, complete overlap under service conditions, *i. e.*, the wear particles remain in the contact area, the disk-disk system is of interest. In addition, there are systems representing rolling (*i. e.*, with slip) or alternating loads among many other systems.

3.9.4 Tribological Properties of Thermoplastics

Engineering plastics, such as polyoxymethylene (POM), polyamide (PA), and polybutylene terephthalate (PBT), are thermoplastic materials of choice in industrial applications. Each provides specific advantages and disadvantages, some of which will be discussed in the following [13, 14, 16–18]. Amorphous thermoplastics, such as acrylonitrile butadiene styrene (ABS) and polycarbonate (PC), show significantly lower performance in tribological applications, see Table 8.22.

- Polyoxymethylene (POM):

  In contact with smooth (Rz = 1–2 µm) metallic surfaces, POM exhibits particularly low wear of the contact areas. The mold shrinkage of POM is high. Components made from POM are distinguished by high dimensional stability because its media absorption rate is low. The material is often used for gears, pulleys, and actuators. It is also well suited for applications in contact with fuels and gasoline. Fiber reinforcement negatively affects friction and wear characteristics because under friction, the glass fibers may be separated from the plastic matrix and turn into abrasive friction partners, thus increasing wear. The service temperature is typically above the glass transition temperature of −50 °C.

- Polyamide (PA):

  Given that polyamides absorb significant amounts of water, the respective dimensional change has to be considered when designing machine components (expansion ~50% of water absorption). In addition, water or other media, such as lubricants, may also be absorbed. Compatibility with lubricants has to be considered. Lubricant manufacturers typically provide information regarding lubricant resistance. Polyamide characteristics are significantly affected by moisture and temperature under service conditions because the glass transition temperature is moisture-dependent and may range above or beyond room temperature. Nonetheless, PA is the most commonly used material for engineering applications because of its high ductility and ease of processing. PA is well suited for frictional contact with rough metal surfaces (optimum Rz = 2.5–3.5 µm). Glass fiber reinforcement of PA significantly reduces wear because of excellent fiber-matrix adhesion. However, it has high abrasive effects on the friction partner.

- Polybutylene terephthalate (PBT):

  PBT is increasingly used for electrical applications and housings. Compared to other materials, processing of PBT is more demanding. The material exhibits high long-term service temperatures. The glass transition temperature (60 °C) is typically in the service temperature range. PBT is well suited as friction partner for POM and PA.

The above mentioned thermoplastics are the materials of choice for both tribological and economic reasons. For applications with high thermal loads (caused by high ambient temperature or by high friction-induced heat), PA 6, PA 66, partially aromatic polyamides, or PA 46 are possible alternatives. It is also possible to crosslink a polyamide by compounding a crosslink enhancer into the polymer and subsequently expose the component to β- or γ-radiation, thus effectively preventing a softening of the material [18].

Typical high-temperature thermoplastics are the more expensive polyphenylene sulfide (PPS) and polyetheretherketone (PEEK). PPS is a comparatively brittle material that is typically used with high levels of reinforcement. When reinforced by carbon fibers, PPS exhibits excellent wear characteristics in underwater applications. Reinforcement is often recommended for PEEK because the mechanical properties are considerably improved by, *e.g.*, carbon fiber reinforcement. The high material costs justify the high-value reinforcement.

Material Selection Guidelines

Despite the complexity of tribological systems, it is possible to design well performing systems as long as some basic rules are followed. They include [12, 17, 18]:

1. Amorphous thermoplastics, such as ABS and PC, exhibit considerably worse friction and wear characteristics compared to semi-crystalline thermoplastics such as POM, PA, PPS, or PEEK.

2. The pairing of similar plastic materials under dry running conditions should be avoided. Under these conditions, local friction welding occurs, *i. e.*, high friction, high wear, and often noise (stick-slip effect). This effect may also occur when metal and plastic are friction partners.
3. Compounding at least one friction partner with a lubricant, such as PTFE, PE, or silicone oil, is recommended when two similar plastic materials are friction partners.
4. POM is a suitable friction partner for smooth metal surfaces (Rz ~ 0.5–1.5 µm) in terms of friction and wear characteristics. For rougher surfaces PA is preferred (optimum Rz ~ 2.5 µm).
5. Glass fiber reinforcement has an abrasive effect on the friction partner as soon as the low fiber content surface is worn. With soft metals, such as brass, the use of aramid fibers (aromatic polyamide, also known as Kevlar®) is of interest.
6. Glass- and carbon fiber reinforcement of plastics has a positive effect in terms of their wear characteristics (*e. g.*, with PPS, PA 6, PA 66, PA 46, PEEK). An exception is POM, where the glass fibers are extracted from the polymer matrix, which causes increased wear. Without these reinforcing materials, POM has been successfully used in heavy-duty applications such as gears, pulleys, and others. Besides its wear resistance, another reason for the use of POM is its high dimensional stability.
7. For plastic/plastic gears, it is recommended to pair different thermoplastic materials. In order to maximize transferable loads, the material with higher thermal resistance (*e. g.*, PA) should be used for the smaller wheel, while the less thermally resistant material should be used for the larger wheel (*e. g.*, POM). In metal/plastic pairs, the smaller wheel should be made from metal. The same approach is recommended for other tribological pairings with similar properties.
8. PPS is a rather brittle material. For tribological applications, fiber reinforcement is recommended. This combination provides a material with excellent wear characteristics that is particularly suited for applications in humid/moist environments (hydrolysis resistant).
9. For component design purposes, the effect of temperature, pressure, and humidity (the latter is particularly relevant for PA) may be estimated by the following consideration: volumetric wear increases proportionally with contact pressure per unit area or with sliding travel. Temperature is less relevant for volumetric wear, unless it reaches (or surpasses) the glass transition or melting temperature of the material. This simplified approach may not be used when mechanical or thermal loads cause the material to creep.
10. Characteristics of radiation cross-linked polyamides: the thermal resistance significantly increases. Therefore, the resistance to small surface roughness, high friction rates (both induce frictional heat), and high ambient temperatures is significantly

higher. Here, the tribological characteristics are positively affected by the low mold temperatures experienced during injection molding, because radiation causes cross-linking primarily in the amorphous regions of the plastic (exception!).

11. The internal structure of the material (see Section 2.4) significantly influences its wear characteristics. When, for example, low injection molding temperatures are selected, the resulting low level of crystallinity will cause comparatively lower tribological characteristics of the amorphous outer layers (in industrial practice, reductions up to ~ 50% have been reported), see Figure 3.109 [12].

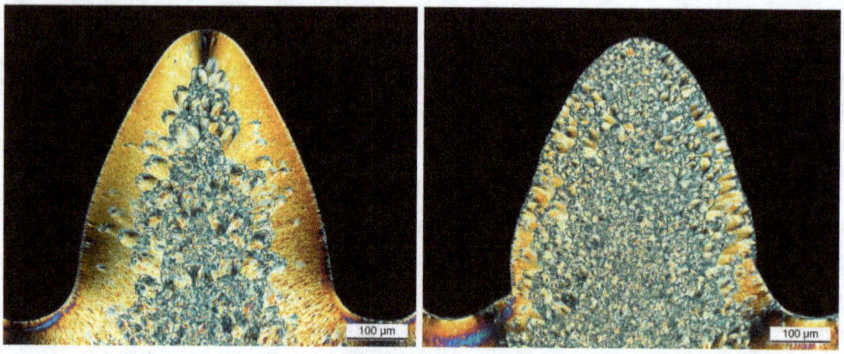

Figure 3.109 Structure formation in injection molded gear tooth made from POM [17];
left: low (T_{mold} = 50 °C); right: recommended mold temperature (T_{mold} = 90 °C, CAMPUS)

The following table summarizes the effects of reinforcing fibers and radiation cross-linking on tribological characteristics. Additional information can be found in Chapter 8.

Table 3.12 Changes in Tribological Characteristics Caused by Reinforcing Fibers/Fillers and Radiation Cross-linking [12, 14, 16–18]

	Wear polymer	Wear/friction partner	Friction	Stick-slip	Notes
Glass fibers GF Glass beads GB	↓ (PA)* ↑↑ (POM)*	↑↑	↑↑	–	* Depending on fiber-matrix adhesion
Carbon fibers CF	↓↓	↑	↓	–	
Aramid fibers	↓↓	–	↓	–	Well suited for soft metal friction partners (aluminum, brass, bronze)

	Wear polymer	Wear/friction partner	Friction	Stick-slip	Notes
Graphite	↓↓*	–	↓↓*	–	* In particular in humid environments
MoS_2	↓*	–	↓*	–	* In particular for low humidity/vacuum
PTFE (TF), PE	↓↓*	–	↓↓*	↓↓*	* For polar matrix polymers (POM, PA, PBT, etc.) these non-polar fillers are tribologically advantageous
Silicone oil Si	↓↓*	–	↓↓*	↓↓*	Well suited for high-speed bearings, migrates to the surface * For polar matrices
Radiation cross-linking X	↓(↓↓*)	–	↑	–	* Radiation cross-linking prevents melting of the material, *i. e.*, higher tribological and thermal loads can be sustained. Radiation also increases surface tension (friction)

3.10 Analytical Evaluations

In the following, we will briefly describe the most important physical methods used to analyze the structure and chemical composition of plastics.

3.10.1 Infrared (IR) and Raman Spectroscopy

Both methods evaluate the absorption spectra of translational and rotational modes in a system (of molecules and/or molecular parts or groups) in a wave range between 750 nm and 1 mm. They are typically used for analytical evaluations (detection or identification of constituents); however, they also provide information on the conformation and tacticity of macromolecules, on crystalline content, and molecular orientation.

Today, infrared spectroscopy is one of the most important methods for the identification of plastics. Common IR-spectrometers record absorption spectra in a range of wavelengths from 2 to 250 µm. They represent the fingerprint of every plastic material, as seen in Figure 3.110 using polycarbonate as an example.

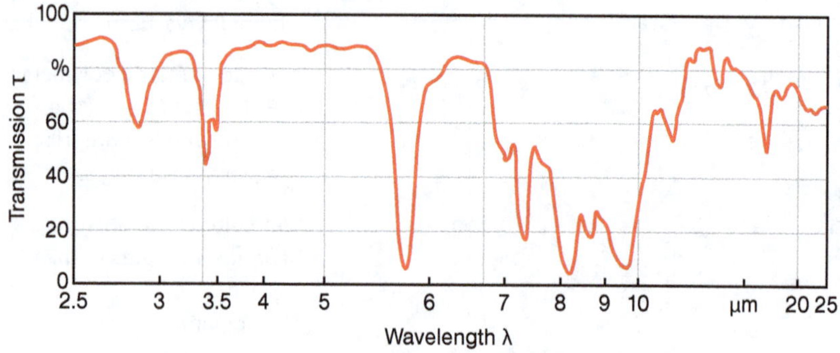

Figure 3.110 Infrared spectrum of a polycarbonate film

With appropriate calibration, the near (NIR) and medium (MIR) infrared range (800 to 2500 nm) allows for the sorting of plastic waste streams (PE, PP, PS, PET, PVC), even of complex mixed plastic streams (recycling). IR-spectroscopy is also used for in-line analysis of polymer reactions in the melt, *e. g.*, in extruders.

3.10.2 Nuclear Magnetic Resonance (NMR)

This method is based on the effect of a magnetic field on the magnetic moment of a nucleus. High-resolution NMR-spectroscopy is used to evaluate polymer solutions, while broadband-NMR analyzes solid polymer specimens. This method allows analytical evaluations as well as investigations regarding tacticity, molecular branching, composition of copolymers, or cross-linking density of rubber-elastic plastics.

Material defects and behavior under load can also be analyzed. Using an NMR-MOUSE (MObile Universal SurfacE Scanner) allows for the examination of thickness and homogeneity of tire layers.

3.10.3 Dynamic-Mechanical Spectroscopy (DMA)

Mechanical spectroscopy is used to determine Young's and shear moduli as well as mechanical damping coefficients as a function of temperature and/or frequency. The temperature typically ranges from −180 °C to 40–80 °C above T_g for amorphous and up to melting temperature for semi-crystalline thermoplastics. Specimens (typically strips cut from the center section of a tensile test bar or metal strips coated with the plastic material to be tested) are excited to torsional or flexural vibrations. The most basic test set-up is the torsion pendulum, Figure 3.111. Here, the disk is utilized as an inertial member and the subsequent oscillation period and amplitude are measured. The oscillation frequency is used to calculate the modulus, while the logarithmic decrement of the dampened oscillation is used to determine the material's mechanical loss factor. As a function of time, both values exhibit characteristic steps or maxima that indicate glass transitions, see also Section 3.4 and Figure 3.43–Figure 3.45. This method allows the identification specific polymer classes and to differentiate between polymers with similar chemical constitution, tacticity, degree of cross-linking, or crystallinity. This approach is also valid for blends and copolymers with identical or similar composition.

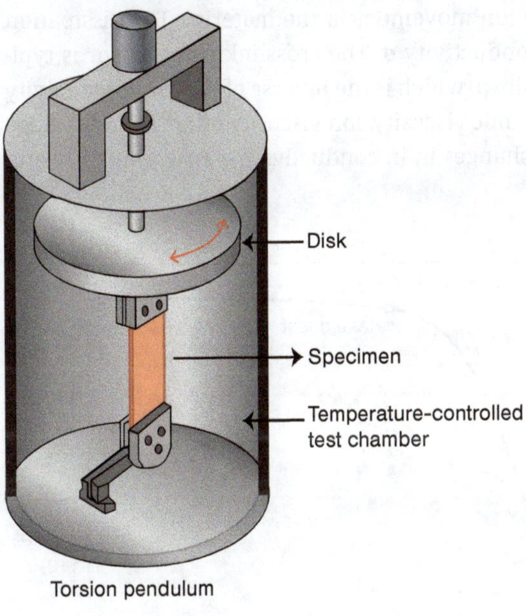

Figure 3.111
Schematic of torsion pendulum

3.10.4 Dielectric Spectroscopy (DEA)

Dielectric spectroscopy is comparable to mechanical spectroscopy. The dielectric constant ε_r and the dielectric dissipation factor tan δ are determined as functions of temperature and frequency. However, only polar molecular groups are excited here, limiting the informational value compared to mechanical spectroscopy. On the other hand, this method is significantly easier and less expensive and allows for measurements in a wide range of temperatures and frequencies, see also Section 3.5.3.

In thermosets, the dielectric properties can be used to analyze the progression of crosslinking during curing. Therefore, DEA is used for resin characterization as well as for process monitoring and control.

When an alternating electrical field is applied to a mixture, the dipoles and ions in the mixture orient themselves according to the field, see Figure 3.64. As long as the resin system is still in the liquid state, mobility is very high, allowing for dipoles and ions to follow the changing field and completely relax within the available time, even at high frequencies. The amplitude and phase shift of the return signal depends on the mobility of the ions and dipoles, see Figure 3.65. Once the curing reaction sets in, the material turns more viscous. The mobility of the charge carriers decreases and thus the amplitude of the resulting signal decreases while the phase shift increases. The measured amplitude is linked to the dielectric constant ε′.

The dielectric dissipation factor ε″ is calculated from the phase shift; it is a measure for the energy dissipation caused by ion movement in the material. The dissipation factor ε″ is proportional to the ion conductivity σ. The crosslinking behavior is typically investigated using the ion viscosity μ, which is the inverse of the ion conductivity and thus correlates well with the dynamic viscosity. Ion viscosity offers the advantage of being able to detect even minute changes in in conductivity, *e. g.*, during diffusion controlled crosslinking.

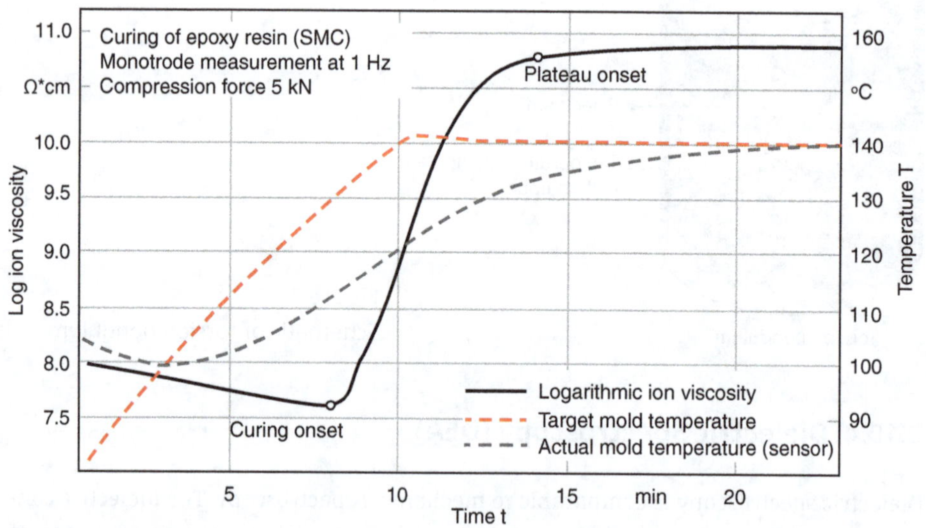

Figure 3.112 DEA-inline measurement in a compression mold to optimize EP curing [Source: Netzsch Gerätebau GmbH]

Figure 3.112 shows an example of using DMA to optimize the curing profile of an epoxy resin during an SMC process. The viscosity decreases with increasing temperature until it reaches a minimum at which curing begins. After approx. 13 min, the ion viscosity reaches a plateau. The process time can now be controlled to reach the desired degree of curing. Dielectric analysis can be used for in-line process monitoring, *e. g.*, to program the thresholds for actuating the compression mold.

3.10.5 Differential Thermal Analysis (DTA) and Differential Calorimetry (DSC)

Both methods rely on the same principle: the specimen to be analyzed and a reference specimen are heated and cooled together and the difference in temperature (DTA) or the difference in specific heat (DSC) between the two specimens is determined as a function of temperature. Today, DSC is basically the only method employed in industrial environments. Differential scanning calorimeters use two crucibles that are heated at the same rate in an oven: the sample to be analyzed (0.1 to 20 mg weight) and an empty reference crucible (Figure 3.113). The temperature difference between both crucibles is proportional to the specific heat capacity of the sample. The system can be calibrated with inert reference materials, *e. g.*, indium. In power compensating scanning calorimeters both crucibles are heated in separate heating units while ensuring that there is no temperature difference. The respective required heat output and the known specific heat capacity of the reference sample are used to determine the heat capacity of the analyzed sample as a function of temperature. The method allows the identification of endo- and exothermal processes caused by crystallization, glass transition (Figure 3.114), or de-orientation processes. The initial heating provides information on the material's thermal and mechanical history. For example, post-crystallization of a semi-crystalline material may be attributed to quenching during manufacturing, or a curing peak of a thermoset may be due to incomplete curing. The second heating period (after specified cooling) is used to characterize the material's behavior [19].

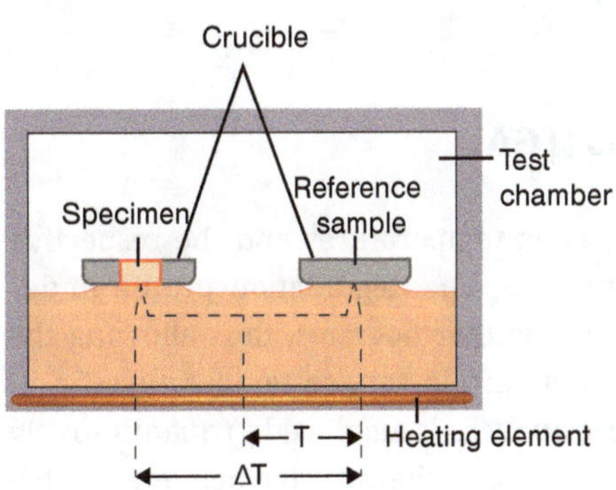

Figure 3.113
Schematic of DSC test set-up

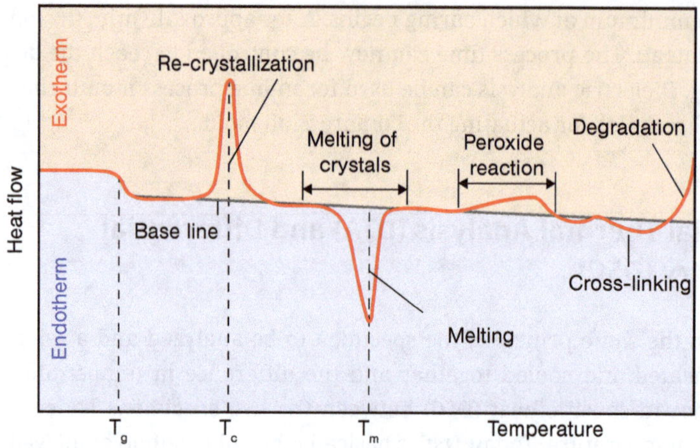

Figure 3.114 Typical DSC-profile of a semi-crystalline thermoplastic during heating with characteristic transition temperatures and effects

Differential scanning calorimetry may also be used to determine the state of stabilization of a plastic material. Here, DSC measures the time until the start of an exothermic reaction under oxygen atmosphere and constant temperature (oxidation induction time, OIT). The value of this characteristic number allows a qualitative assessment of the material's state of aging.

Superimposing thermal effects can be separated using temperature modulated DSC (TM-DSC). Here, the basic linear heating rate is superimposed with a sinusoidal temperature modulation, which makes it possible to distinguish between reversible (glass transition, melting, and crystallization) and irreversible processes (curing, degradation, evaporation of volatiles, relaxation, etc.).

High-pressure DSC (up to 15 MPa) suppresses the evaporation of volatiles so that it is possible to distinguish between evaporation and other reaction effects. Suitable alternatives are high-pressure crucibles that prevent volatilization up to approx. 5 to 10 MPa (depending on model).

3.10.6 Thermogravimetric Analysis (TGA)

A small material sample is exposed to increasing temperatures and the respective weight loss is continuously recorded, Figure 3.115. The degradation profile under short-term load can be derived from the resulting thermograms, thus allowing the determination of the suitability of a material at certain service temperatures, *e.g.*, serving as thermal shields. Thermogravimetric analysis is employed to quantitatively determine material composition and filler contents, to characterize degradation behavior, and to a lesser extent to determine the moisture content or to detect volatilization of low-molecular components.

The gases released during TGA can be detected with a high level of accuracy using Fourier-transform infrared spectroscopy (FTIR, see Section 3.10.1) or mass spectroscopy.

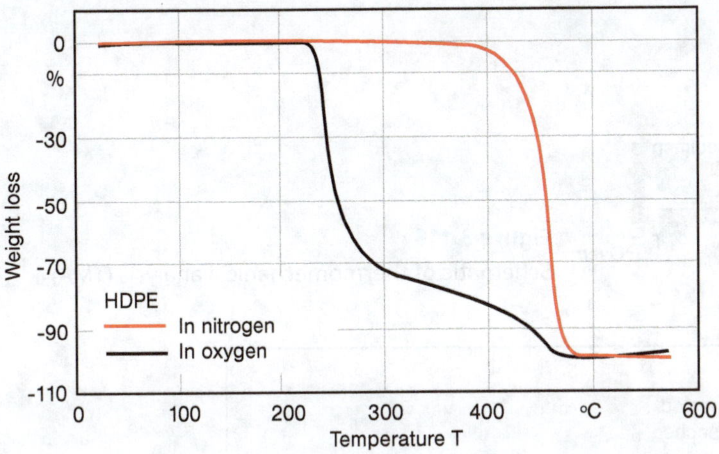

Figure 3.115 Weight loss of PE-HD for a specimen weighing 3.1 mg in a nitrogen atmosphere (curve 1) and for a specimen weighing 0.8 mg in an oxygen atmosphere (curve 2), TGA (heating rate: 20 K/min)

3.10.7 Dilatometry, Thermomechanical Analysis (TMA), Specific Volume (pvT)

These testing methods are used to determine the change in length (TMA) or volume (pvT) of solid specimens as a function of temperature. This provides the temperature-dependent coefficient of expansion or pvT-diagrams (relationship between pressure, volume, and temperature) that are of particular interest for the evaluation of plastic shrinkage during manufacturing, see Figure 3.8 and Figure 3.61.

The linear coefficient of expansion of solids can be measured using thermomechanical analysis until the softening point is reached. Here, a cylindrical or rectangular specimen is exposed to a stamp at low force and the change in length is measured at low heating rates (Figure 3.116).

Analogous to DSC measurements, the thermal and mechanical history of the material is evaluated during the initial heating phase. That includes post-crystallization and curing, but also residual stresses and orientations.

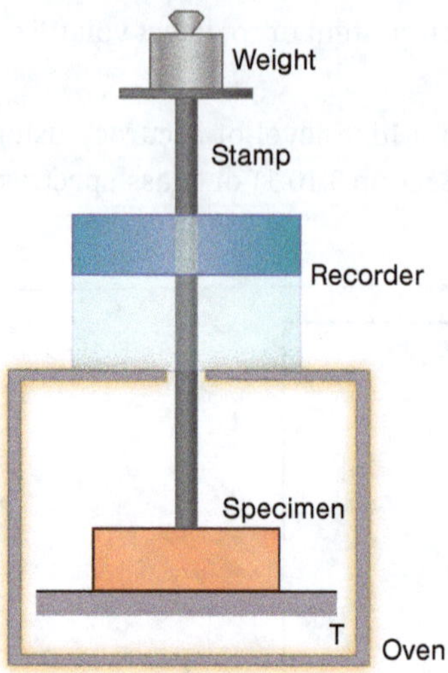

Figure 3.116
Schematic of thermomechanical analysis (TMA)

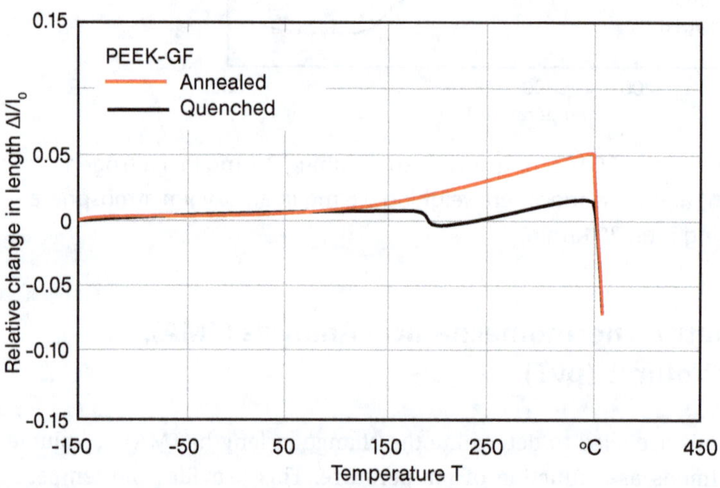

Figure 3.117 Relative change in length Dl/l_0 in curve for tempered (curve 1) and quenched (curve 2) PEEK-GF specimen (heating rate: 10 K/min)

Anisotropy of the coefficient of expansion indicates molecular orientation or the existence of a filler (see also Section 3.4.6). Figure 3.117 shows the effect of tempering: shrinkage of the quenched specimen (curve 2) above 175 °C is the result of post-crystallization. Therefore, it is expedient to determine thermal expansion in each spatial direction.

When such effects are encountered, it is recommended to determine characteristic transition temperatures, such as glass transition temperature, during the second heating phase.

Pressure-volume-temperature (pvT) devices are suitable for the measurement of the coefficient of volumetric expansion and/or for measurements in both melt and solid state. These devices can also be used to determine compressibility, changes in density or specific volume, glass transition temperature, and crystallization and melt temperature as a function of pressure and temperature according to ISO 17744. With the confining fluid method, Figure 3.118 top, the pelletized sample material is added to a temperature-controlled mercury bath (A1). The bellows apply pressure, and an inductive displacement sensor detects the position. The change in volume during cooling and heating (A2) and under pressure (A3) is determined with the help of the volume change in the complete system. With the given values for compressibility and expansion of mercury, the specific volume of the sample material can be determined. Figure 3.118 (bottom) shows the piston-die measurement method. In contrast to the confining fluid method, the pellets have to be melted via the cylinder walls after filling (B1). When measurement is initiated, the melt is compressed (B2) and the change in length of the sample is measured while it cools under constant pressure (B3). The weight of the sample and the known cylinder geometry are used together with these data to determine the specific volume.

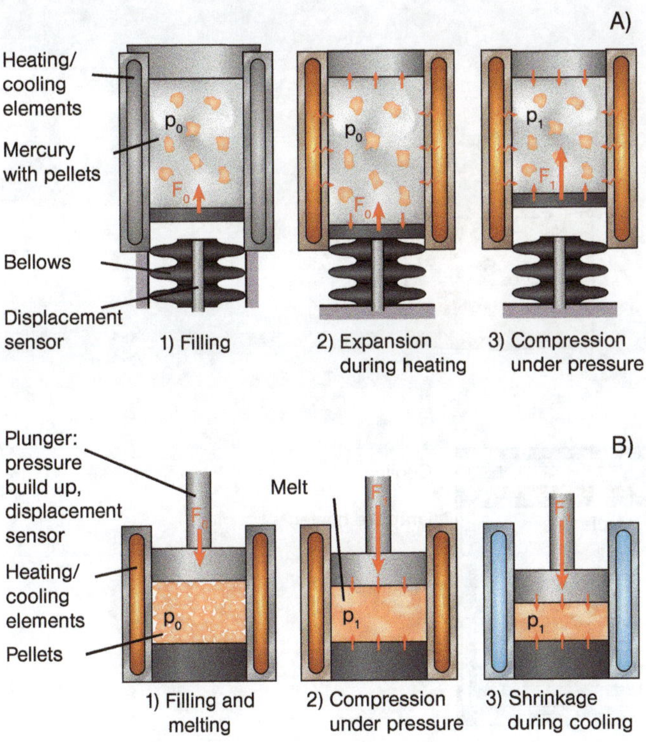

Figure 3.118 Comparison of the two common pvT-measurement methods;
A: confining fluid principle, B: piston-die principle

In order to determine injection molding data, isobaric cooling measurements are recommended because they closely resemble the manufacturing process, thus eliminating the risk of influencing data by the measurement procedure. Exemplary measurements are shown in Figure 3.61 and Figure 3.62.

3.10.8 Thermal Conductivity

Plastics modified to exhibit higher heat conductivity are gaining increasing importance, and therefore measuring this property also becomes a priority. We distinguish between steady-state methods, where heat flow through the sample is constant over time, and transient methods, where thermal conductivity is calculated from a time-dependent heat flow. Geometric requirements and the determination of component-specific characteristics led to an increasing predominance of transient methods for the determination of thermal conductivity (such as TPS and pulse methods). Specific test set-ups and additional analyses allow for the calculation of one-dimensional characteristic values and thus for the interpretation of orientation-dependent component properties, see Figure 3.119.

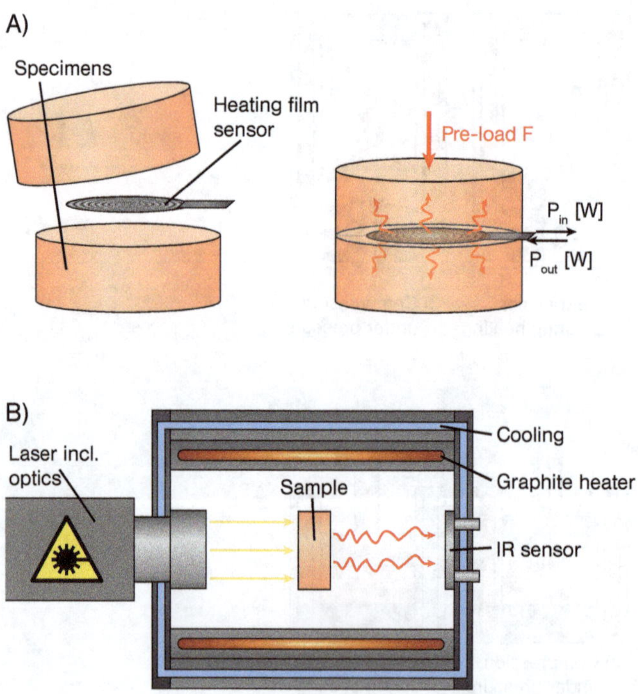

Figure 3.119 Measurement methods for the determination of thermal conductivity;
A: Hot disk method (TPS method)
B: Laser-flash analysis (pulse method)

Hot disk methods use a heating element (*e.g.*, a film or a wire) between two sample sheets. When the heating element is employed, the temperature in the sample sheets increases so that the change in temperature can be measured as a function of time. These data can be used directly to calculate the material's heat conductivity. Because here the heat flow is not oriented, the determined value is also called integral heat conductivity.

Impulse methods introduce a pulse of energy on one sample side via a Xenon-light flash while taking time-dependent temperature measurements on the other sample side using an IR detector. The thermal diffusivity a thus determined together with the density ρ (*e.g.*, obtained by pvT measurements) and the specific heat capacity c_p (obtained by DSC measurements) allow for the calculation of the thermal conductivity λ according to

$$\lambda = a \cdot \rho \cdot c_p \tag{3.41}$$

In order to measure thermal conductivity in the melt, solid state, or under pressure, a pvT device based on the heating probe method can be used (Figure 3.118 B). Here, a certain amount of heat (adapted to the heat conductivity of the sample) is introduced into the center of the cylindrical sample via an electric heating probe. The resulting increase in temperature at the sample walls is measured in parallel and both values are used to calculate the material's heat conductivity.

3.10.9 Solution Viscosimetry

The relationship between a polymer's viscosity in solution and its molecular weight is used to determine molecular weights. The viscosities of solvent and solution (0.2 to 1 g polymer in 100 ml solvent) are determined using a capillary viscometer (Ubbelohde) (η viscosity of the solution, h_s viscosity of the solvent, C concentration in g/cm^3). The following characteristics can be derived as measures of molecular weight:

The viscosity number (inherent viscosity):

$$J = \left[\frac{\eta}{\eta_s} - 1\right] \cdot \frac{1}{C} \tag{3.42}$$

The viscosity ratio:

$$\eta_{rel} = \frac{\eta}{\eta_s} \tag{3.43}$$

The relative change in viscosity:

$$\eta_{sp} = \left[\frac{\eta}{\eta_s} - 1\right] \tag{3.44}$$

These data are often used in research and development labs and for production control. They are well suited to detect degradation caused by light, weather, high-energy radiation, or the exposure to chemicals or heat.

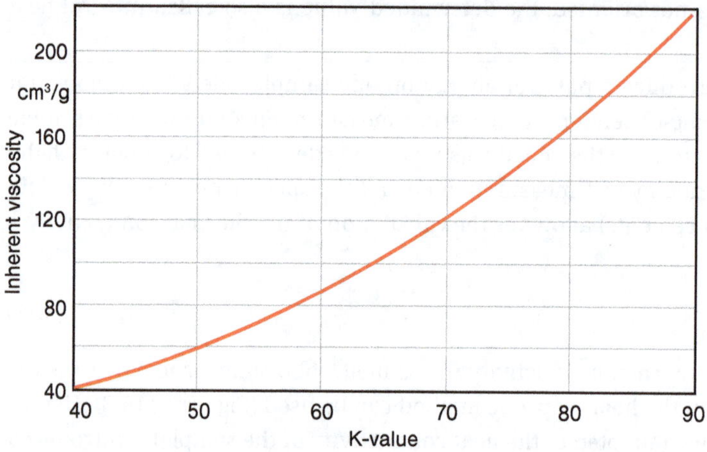

Figure 3.120 Correlation between heat transfer coefficient and inherent viscosity

The viscosity value depends on the solvent used, the concentration of the solution, the measuring temperature, and to a lesser extent also on the shear rate. The inherent viscosity and the derived heat transfer coefficient (Figure 3.120) are used to characterize plastic types within a material group.

3.10.10 Chromatography

A common element of all chromatographic processes is the separation of mixtures using the partitioning effect of a mobile phase (the mixture in solution) traveling through a stationary phase, see Figure 3.121. These technologies are used to separate and concentrate soluble or volatile materials, and in combination with detectors to identify the separated materials.

The following chromatography variations are available: LC = liquid chromatography, TLC = thin layer chromatography, GC = gas chromatography, GPC = gel permeation chromatography, SCF = supercritical fluid chromatography.

These technologies are mainly used in plastic analysis to evaluate low-molecular polymer degradation products and to identify additives. It is possible to determine molecular weight distribution with the help of GPC.

When GPC and IR-spectroscopy are coupled, the GPC is equipped with a flow-through cell that is positioned in the path of the IR-spectrometer (GPC-FTIR-spectrometer). This allows for the separation of polymer blends.

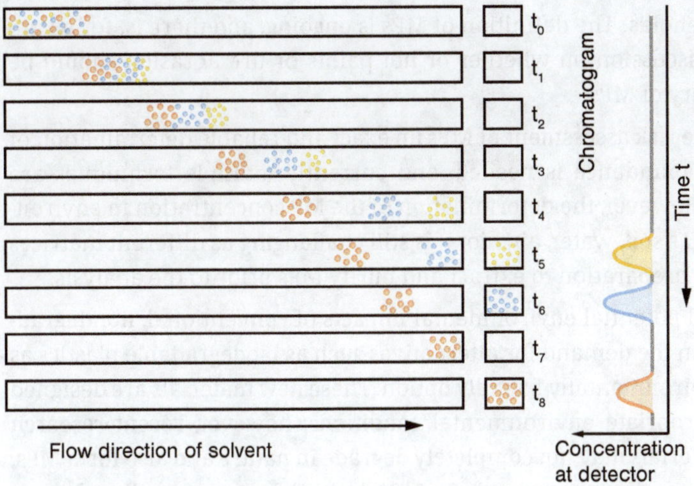

Figure 3.121 Schematic of the separation procedure used for chromatography. The three colors represent three different kinds of molecules; their positions are shown at different times t_0 to t_8 in the separatory column and the detector. The chromatogram represents the test result

3.11 Microplastics: Characterization and Separation Methods

Günter Auernhammer, Dieter Fischer, Holger Ruckdäschel, Christian Laforsch, Martin Löder, Seema Agarwal, Kathrin Harre, Andreas Fery

3.11.1 Introduction

Plastics are an essential part of our everyday life and, in tandem, the increasing plastics production results in an ever-increasing amount of plastic waste entering the environment due to, for example, littering or inadequate waste management. The environmental effects of larger plastic items have been in focus for decades; however, research on the occurrence and environmental risks of plastic particles smaller than 1 mm, so-called microplastics (MPs), kick-started only in the early 2000s. Although MP particles are a very diverse suite of particles that differ in size, shape, polymer type, density, added chemical compounds, surface properties, environmental coating and biofouling as well as other physicochemical properties, these particles can be differentiated according to their origin. Primary MPs are intentionally produced in that size range and used as, for example, a sandblasting medium, exfoliants in cosmetics, or thickening agents in paints. Secondary MPs originate from the fragmentation or the abrasion of particles from plastic items during their use in, for instance, industry, construction, or households. The definition of MPs is ongoing, and there is, for example, a controversial discussion on whether or not paints or tire abrasion should be included in the category of MPs.

As a basis for a reliable risk assessment of MPs an exact and reliable determination of their environmental abundance is needed, and currently multiple techniques are used to analyze MPs. However, the determination of the MP concentration in environmental samples (such as soil, water, air, biota) is still challenging as different matrices need a special sample preparation to extract and purify MPs prior to the analysis.

The accumulation and potential environmental impacts of conventional, nondegradable plastics resulted in the demand for alternatives such as biodegradable plastics as a potentially more environmentally friendly option. These new materials are designed to biodegrade in appropriate environmental conditions; however, recent research showed that these materials may not completely degrade in nature and also form MPs, eliciting similar effects as those reported from MPs derived from commodity plastics.

In the following, we will give a rough overview on methods to characterize and separate MPs.

3.11.2 Characterization of Microplastics

Relevant parameters for the characterization of MPs are the particle size and shape, the polymer type(s), the surface properties, and the effects of aging. In order to ascertain these parameters, a reliable determination with appropriate analytical methods is required. Basically, there are two different kinds of methods, mass-based methods and particle-based methods.

Identification and Quantification of Microplastics

Real-world samples from the environment (oceans, rivers, lakes, sediments, soils, atmosphere), from biota (plants, animals, humans), or in food and packaging contain also inorganic and organic components. These have to be removed without destroying the MP particles through different separation methods (as described in Section 3.11.3) and chemical or enzymatic sample purification treatments. As MPs are ubiquitous in our environment, it is essential to work in a microplastics-free laboratory and to measure daily blanks in all steps of the pretreatment and measurement. After pretreatment, fractional vacuum filtration of all particles on filters is the next step to separate the MP particles into size fractions for subsequent analysis.

Mass-based methods determine the total amount of MP in a sample. The most common methods are thermoanalytical methods such as pyrolysis–gas chromatography–mass spectroscopy (Py-GC/MS) and thermal extraction desorption GC/MS (TED-GC/MS). The results are MP mass per volume for sampled waters (µg/l) or MP per total dry mass for sampled solids (mg/kg). Mass-based methods involve thermal decomposition of the particles in a complex pyrolytic process and generate characteristic gaseous indicator products by separation with GC and analysis by MS.

Particle-based methods determine the number of particles, particle size, and particle size distribution, and which polymers are present. These include the most common spectroscopic methods such as infrared (FTIR, Fourier-transform infrared spectroscopy) and Raman microspectroscopy and fluorescence microscopy. The results are the number of MP particles per volume for sampled waters (number/l) or number of MP particles per total dry mass for sampled solids (number/kg). Typically, particle-based methods identify particles on a filter using FTIR and Raman spectroscopy and fluorescence microscopy. Out of up to 100,000 particles on a filter, often less than 1% are MP particles. An automated analysis of all or a statistically relevant number of MP particles is therefore essential for a quantitative analysis. An advanced method for this is a combination of optical particle analysis and FTIR/Raman spectroscopy. The optical particle identification software determines the shape and dimensions of all particles and stores their coordinates for subsequent FTIR/Raman measurements. After the measurements, all spectra are automatically identified using spectral databases, including polymers, additives, inorganic and organic substances, and pigments, or using machine-learning methods. The results are a combination of optical (size, shape, color) and chemical (spectral) information for all particles on a filter for the sample under investigation. For quantitative analysis of MP in the environment, there are public databases where all particle-related results should be uploaded.

Surface Properties

In most cases, the interaction of MP particles with organisms or cells involves a coating of the MP particles with a biofilm or ecocorona. Ecocorona refers to a coating of MP particles with organic and inorganic components after incubation in the environmental medium. In an aqueous environment these can be, for example, humic substances, FeO(OH), and biomolecules (such as polysaccharides, proteins, lipids). A biofilm is a surface coating of MP particles by an accumulating matrix that additionally contains extracellular polymeric material containing a community of microorganisms. These coatings significantly alter the surface properties of the MP particle. For the interaction with cells and organisms, the zeta potential, the surface mechanics, and the adhesiveness of the MP particles are dominant. The zeta potential (the electrical potential at a short distance from the particle surface, called the shear plane) is a good proxy for cell adhesion strength and the probability of cell uptake under otherwise identical conditions.

The zeta potential of MP particles is determined by electrokinetic methods based on measurements of either electrophoretic mobility (the movement of particles in an applied electric field) or streaming potential (the electric potential resulting from a flow of electrolyte solution through a porous plug of particles). Typical measurements examine an ensemble of MP particles. The sample preparation must ensure that the

MP particles are separated from other particles in the sample without disrupting the biofilm or ecocorona. These measurements result in pH-dependent values of the zeta potential, where the pH dependence is essential for the interpretation of the data. For all of these measurements, the control of the external conditions, including ionic strength and pH, is essential.

For the determination of mechanical surface properties and adhesion, atomic force microscopy (AFM)-based methods – for example, colloidal probe AFM (CP-AFM), nanoindentation, or microfluidic shear force measurements – are most suitable. These are all particle-based methods. The sample preparation requirements are similar to those for the zeta potential determination; see above. CP-AFM and nanoindentation work in the direction normal to the surface of interest. Typically, the MP particles are immobilized on a surface and a well-defined probe is moved towards them. These methods provide information about the attraction or repulsion between the MP particles and the probe (before contact), the softness of the MP particle surface (during contact), and the adhesion between the MP particles and the probe surface (during separation). Shear flow measurements provide a statistical quantification of how many MP particles adhere to a given surface under well-defined flow conditions.

3.11.3 Microplastic Separation Methods

Microplastics are usually found in low abundance in the environment. Therefore, separation methods are often used in MP analysis to concentrate MPs from environmental samples of different matrices, including air, water, sediments, soils, food, tissue, and plant and animal samples. This process ensures the required MP concentration and representativeness of the sample for analysis. In many cases, the removal of interfering substances that may be present in excess is also a critical aspect of the analysis. There are also methods to separate and remove MPs from, for example, surface runoff or rivers to reduce their impact on the environment, but these are not discussed in detail here. When selecting a separation method, different factors should be considered: the type of environmental sample, the desired level of enrichment, the applicability to the particular size class of MP particles, and the sample amount that can be processed with the particular method. For an optimum separation result, a combination of these methods might be needed to minimize interfering matrix effects in complex environmental samples.

Size-based separation methods typically use filtration or sieving techniques. It is important to note that the pore or mesh size determines the size class that is sampled quantitatively. MPs in aqueous media (ocean, river, lake, tap water) can be concentrated by using different mesh types, wet sieving, or vacuum-based filtration during sampling. Similarly, sediment and soil samples can be sieved dry or wet after suspension. MPs in air samples are usually concentrated on filters with active pumping systems. In other sample matrices, such as food, tissue samples, and animal or plant material, matrix removal may be required prior to the application of any size-based method. Dry sieving is often used as a preliminary step to classify samples prior to the implementation of density-based methods, while wet sieving is used after density separation. Depending on the matrix and mesh/pore size, several thousand liters can be concentrated for aqueous samples, while other matrices can typically be processed in the kilogram range at most. Visual sorting is another crude method in which the collected samples and debris are examined by either the naked eye or a microscope to distinguish plastics from organic matter and other materials.

Density-based separation methods are the most widely used methods for extracting MPs from the mineral fraction in sediment or soil samples. These methods rely on the difference in density between plastic particles and mineral grains. When the sample is suspended in a high-density salt solution, the sample fraction lighter than the solution, including MPs, floats and the fraction heavier than the solution, mainly minerals, settles to the bottom. It is important to note that depending on the density of the salt solution, heavier plastics may not be separated. Density-based separation approaches typically use equipment that allows recovery of the floating light fraction, including MPs. Density separation methods are efficient for MP particles in the low double-digit micrometer range and are widely implemented due to their relative ease of use and relatively low cost. Depending on the volume of the equipment chosen, density separation can be applied to samples up to the kilogram range.

Continuous flow centrifugation is a promising technique for sampling MPs (and nanoplastics) from water bodies. It allows the processing of several cubic meters of water. A major advantage of this method is that a high throughput without size discrimination can be achieved simply by choosing the right combination of centrifugal force and time.

Hydrophobicity-based separation methods (*i.e.*, "oil separation") have been used to separate MPs from sediments, suspended solids from rivers or the sea, and agricultural soils. Oil separation uses similar equipment to density separation and separates MP particles by exploiting the wetting properties of two immiscible liquid phases. Oil separation is particularly suitable for particle sizes below 250 µm. High enrichments can be achieved in this size range compared to similar sample sizes with density separation. A particularly practical variant works with oils that solidify at low temperatures – for example, in a refrigerator – and therefore can be separated easily from the aqueous phase. The applicability is limited by changes in the hydrophobicity of the MP particles due to aging processes, which can lead to insufficient or even false negative results.

Separation based on the electrostatic properties of MP is a method that exploits the differences in electrical chargeability – for example, triboelectricity – of MPs and can be particularly effective for separating them from other particles with significantly different electrical properties. It is mainly used to separate MPs from dry sediments and soils and is suitable for larger amounts of sediment, with a throughput measured in kilograms of dry matter per hour. The method is based on the application of an electric field that separates charged particles from each other.

Microplastic particles can be separated from each other and from inorganic and biological components of the sample by electroseparation based on their surface charge or dielectric methods based on their properties. This method is particularly advantageous for the separation of particles larger than 125 µm. Occasionally, magnetic seed filtration is used to separate MP from dilute aqueous suspensions. In this process, magnetic particles are agglomerated with the MP in a first step so that they can be separated in a second step in a magnetic field.

References

[1] Minagawa, N., White, J. L., The Influence of Titanium Dioxide on the Rheological and Extrusion Properties of Polymer Melts, *Journal of Applied Polymer Science*, 20 (1976), pp. 501–523

[2] Heinle, C., Simulationsgestützte Entwicklung von Bauteilen aus wärmeleitenden Kunststoffen, Dissertation, Universität Erlangen-Nürnberg, Erlangen, Germany (2012)

[3] Stitz, S., Analyse der Formteilbildung beim Spritzgießen von Plastomeren als Grundlage für die Prozesssteuerung, Dissertation, RWTH Aachen, Aachen, Germany (1973)

[4] Pötsch, H. G., Prozesssimulation zur Abschätzung von Schwindung und Verzug thermoplastischer Spritzgussteile, Disserartion, RWTH Aachen, Aachen, Germany (1990)

[5] Zöllner, O., Grundlagen zur Schwindung von thermoplastischen Kunststoffen, company brochure Bayer AG, Leverkusen, Germany

[6] Giencke, E., Meder, G., Ermittlung der Kriechfunktionen von zweiachsig beanspruchten Harzen und Laminaten aus den Meßwerten für einachsig beanspruchter Harze, Materialprüfung 23 (1981) 3, pp. 75–85

[7] N. N., *Kenndaten für die Verarbeitung thermoplastischer Kunststoffe*, VDMA, Ed., Verein Dt. Maschinebau-Anstalten e. V., p. 548 ff., Carl Hanser Verlag, Munich (1979)

[8] Beyer, M., *Elektrisches und Dielektrisches Verhalten von Epoxidformstoffen*, Schering-Institut für Hochspannungstechnik und Hochspannungsanlagen (1991)

[9] Grellmann, W., Seidler, S., *Polymer Testing*, 2nd ed., Carl Hanser Verlag, Munich (2013)

[10] Sorensen, B. E., A revised Michel-Lévy interference colour chart based on first-principles calculations, *European Journal of Mineralogy*, 25 (2013) 1, pp. 5–10

[11] Czichos, H., Habig, K.-H, *Tribologie-Handbuch*, 2nd ed., Vieweg Verlag, Wiesbaden, Germany (2003)

[12] Künkel, R., Auswahl und Optimierung von Kunststoffen für tribologische Systeme. Dissertation, Universität Erlangen-Nürnberg, Erlangen, Germany (2005)

[13] Ehrenstein, G. W., *Mit Kunststoffen konstruieren*, 3rd ed., Carl Hanser Verlag, Munich Vienna (2007)

[14] Feulner, R., Verschleiß trocken laufender Kunststoffgetriebe – Kennwertermittlung und Auslegung. Dissertation, Universität Erlangen-Nürnberg, Erlangen, Germany (2008)

[15] Feulner, R., Kobes, M. O., Merken, D., Drummer, D., Ermittlung und Anwendung von Verschleißkennwerten basierend auf Modell- und Bauteilversuchen für die Auslegung von Getrieben aus Thermoplastkunststoffen. *Tribologie + Schmierungstechnik*, 57 (2010) 4, pp. 9–20

[16] Feulner, R., Künkel, R., Kobes, M. O., Ehrenstein, G. W., Tribologisches Verhalten faserverstärkter Kunststoffe. *Tribologie + Schmierungstechnik*, 56 (2009) 3, pp. 11–17

[17] Dallner, C. M., Kobes, M. O., Feulner, R. W., Schmachtenberg, E., Influence of the morphology on the tribological behaviour of assembly injection moulded Microdrives, PPS-23, Salvador, Brazil (2007)

[18] Feulner, R., Brocka, Z., Seefried, A., Kobes, M., Hülder, G., Osswald, T. A., The effects of e-beam irradiation induced cross linking on the friction and wear of polyamide 66 in sliding contact. *Wear*, 268 (2010) pp. 905–910

[19] Ehrenstein, G. W., Riedel, G., Trawiel, P., *Thermal Analysis of Plastics*, Carl Hanser Verlag, Munich (2004)

[20] Jenike, A. W.: Storage and Flow of Solids, Bull. No. 123, Eng. Exp. Stat., University of Utah, 1964

[21] Schulze, D.: *Powders and Bulk Solids – Behavior, Characterization, Storage and Flow*, Springer, Berlin, 2008

4 Plastic Processing Technologies

The manufacturing of plastic parts includes all forming processes. In general, manufacturing engineering distinguishes between:

Shaping operations: This involves transforming a plastic pellet, powder, or resin into a final product or into a preform using extrusion or molding processes such as injection and compression molding.

Secondary shaping operation: A semi-finished product or a preform is transformed into a final product using heat and/or pressure.

Material removal: A semi-finished product or a preform is transformed into a final product by material removal.

Joining operations: Two or more semi-finished products or preforms are assembled by welding or bonding operations.

Most plastic parts are manufactured using shaping operations. Here, the material is deformed into its final shape at low processing temperatures (between room temperature and 350 °C), at which the materials used for tools, dies, and molds are durable and largely wear-resistant. Therefore, production equipment can be re-used many times, allowing for the injection molding of 10^6 to 10^7 parts using the same tool, thus distributing the high mold costs over a high production volume. Another advantage of polymer molding processes is the fact that the shape of the finished product is obtained without the need for secondary operations (trimming, turning, sanding or polishing), because the part is a precise replicate of the mold surface. The injection molding of CDs and DVDs illustrates the accuracy of plastic processing technologies. Here, the mold surface replicates digital information onto the surface of the data storage medium in the form of minute structures with lateral dimensions of 4 µm and depths of 1 µm. Currently, it is possible to realize five to six times smaller features, *e. g.*, for functional surfaces such as for self-cleaning ("Lotus" effect) or anti-reflection coatings ("moth eye" effect).

In the past years, we have seen trends to develop more complex manufacturing systems to shape and join intricate components that use various materials or semi-finished products. Examples include the injection molding of components with rigid and flexible segments or the co-extrusion of multilayer films with specific layer compositions. Here, the design of the manufacturing equipment has to meet the requirements set by component development. Therefore, in plastic processing we often find manufacturing equipment specifically tailored to a specific component or semi-finished product. The design of plastic products includes the design of the respective manufacturing systems as well as the consideration of production-relevant factors.

Prerequisite for all shaping operations is the pliability of the material. Thermoplastics and thermoplastic elastomers are shaped and formed by heating them above glass transition or melting temperatures and then freezing them into their final shape by lowering the temperature. The solidification of the material resets the crystallization, molecular or fiber orientation, and residual stress distribution of the final part. Therefore, the final properties of the component are always dependent on the manufacturing parameters.

Components made from elastomers or thermosets cannot be manufactured by solidification from the molten state because these materials do not melt. Instead, the processes that synthesize macromolecular materials from low-molecular molecules are used with pre-synthesized products (resins and rubbers) as basic materials. The initially molten basic materials (at higher temperatures if necessary) are further synthesized (crosslinked) during the shaping process. This chemical process initiates the solidification of the component in the mold (typically accompanied by an increase in temperature).

The solidification mechanism (by cooling for thermoplastics or thermoplastic elastomers and by crosslinking for elastomers and thermosets) significantly determines process selection and management during plastic processing operations. Crosslinking plastics are typically compression molded while thermoplastics are injection molded and extruded. However, closer examination shows that basically all processing methods are suitable for all plastic material classes.

4.1 Compounding

Compounding describes all necessary steps to be taken before the polymeric material is processed into the finished product. Such steps include the addition of additives such as dyes and pigments, fillers and reinforcements, plasticizers, lubricants, stabilizers, flame retardants, foaming agents, solvents, or other polymers, or the material's transformation into a suitable form for further processing (powder, paste, or pellets). The most important material preparation operations are mixing, dispersing, kneading, dissolving, granulating or pelletizing, and drying.

4.1.1 Mixing Processes

The properties of the finished product in almost all polymer processes depend in part on how well the material was mixed, and all processing methods include a mixing step of the base materials. For example, an integral part of a screw extruder is a mixing zone. In fact, most twin screw extruders are primarily used as mixing devices. Hence, a better understanding of the mixing process helps to optimize processing conditions and increase part quality.

The process of polymer blending or mixing is accomplished by distributing or dispersing a minor or secondary component within a major component serving as a matrix. The major component can be thought of as the continuous phase and the minor components as distributed or dispersed phases in the form of droplets, filaments, or agglomerates. When creating a polymer blend, one must always keep in mind that the blend will probably be re-melted in subsequent processing or shaping processes. Coalescence may compromise the finely dispersed structure created during compounding by causing phase separation. To avoid this problem, compatibilizers, which are macromolecules used to ensure compatibility in the boundary layers between the two phases, are common.

We distinguish between three general categories of polymer blends (for examples see Table 4.1):

- Homogeneous mixtures of compatible polymers,
- Single-phase mixtures of partly incompatible polymers, and
- Multi-phase mixtures of incompatible polymers.

The mixing process itself is defined by three fundamental mechanisms:

- Distributive mixing,
- Dispersive mixing, and
- Coalescence.

Table 4.1 Common Polymer Blends

Blends of compatible plastic materials	Natural rubber and polybutadiene Polyamides (*e. g.*, PA 6 and PA 66) Polyphenylene ether and polystyrene
Blends of partially incompatible plastic materials	Polyethylene and polyisobutene Polyethylene and polypropylene (< 5% PE in PP) Polycarbonate and polyethylene terephthalate
Blends of incompatible plastic materials	Polystyrene and polyethylene Polyamide and polystyrene Polypropylene and polystyrene

Distributive or laminar mixing of compatible materials describes the distribution of a component A in a matrix B. This distribution is achieved by imposing large strains on the material. This increases the interfacial area between the two or more phases and decreases the local dimensions, or striation thicknesses, of the secondary phases. Imposing large strains on the blend is not always sufficient to achieve a homogeneous mixture. The type of mixing device and the initial orientation and position of the two or more fluid components play a significant role in the quality of the mixture.

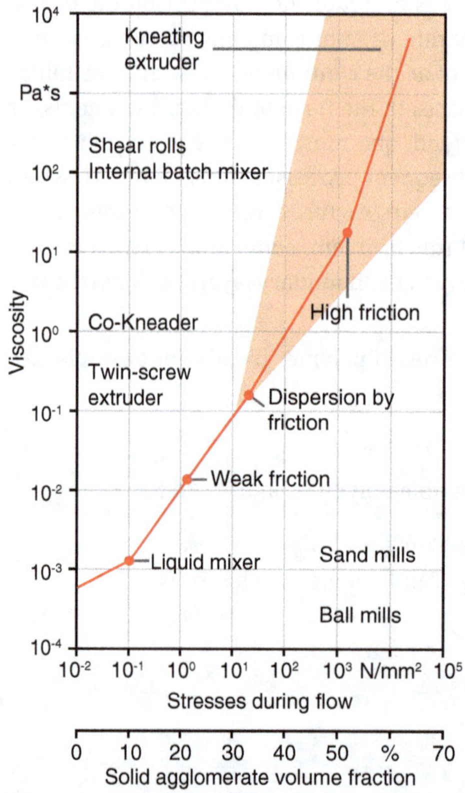

Figure 4.1 Use of dispersive mixing equipment depending on compound viscosity and volume fraction of solid agglomerates

Dispersive mixing in polymer processing involves breaking a secondary immiscible fluid or an agglomerate of solid particles and dispersing them throughout the matrix. Here, the imposed strain is not as important as the imposed stress which causes the system to break up. Hence, the type of flow inside a mixer plays a significant role on the break-up of solid particle clumps or fluid droplets when dispersing them throughout the matrix. The most common example of dispersive mixing of particulate solid agglomerates is the dispersion and mixing of carbon black into a rubber compound. Figure 4.1 relates the viscosity of the compound to the volume fraction of the solid agglomerates and the type of mixing device. When breaking up a polymer melt within a matrix, *i. e.*, when compounding a polymer blend, the droplets inside the incompatible matrix tend to stay or become spherical because of the natural tendency of the drop to maintain the lowest possible surface-to-volume ratio. However, the flow field within the mixer applies a stress on the droplets, causing them to deform. If this stress is high enough, it will eventually cause the drops to elongate and finally break up. This phenomenon of dispersion and distribution continues to repeat itself until the stresses caused by the flow field can no longer overcome the surface tension of the ever smaller droplets that are formed.

The dispersion of solid agglomerates or melt droplets is most effectively achieved by stretching or elongational flow. Hence, devices that stretch the melt, instead of shearing it, lead to more effective mixers while significantly lowering energy consumption.

Coalescence describes the coming together of single drops, *e. g.*, by collision in the flow field. In essence, it is the opposite of dispersion. Depending on filler content and surface tension, coalescence may predominate dispersion. In this case, mixing will cause a coarsening of the blend morphology.

4.1.1.1 Mixing Devices

We generally distinguish between two categories of mixers:

- Discontinuous mixers and
- Continuous mixers.

Rotating Solid Particulate Mixers

The different designs have varying types of mixing drums and thus rely on different kinds of mixing mechanism. There are simple roll drums, drum hoop mixers, tumbling mixers (for larger volumes), double conical mixers, and V-mixers. They are used as pre-mixers for free flowing materials and additives to be further processed on screw kneaders, extruders, or injection molding machines. They are also used to post-mix and homogenize charges of final compounds and for subsequent drying operations.

Solid Particulate Mixers with Rotating Elements

This mixer type comes in a wide variety of designs. The internal impeller speeds range from less than 2 m/s to speeds up to 50 m/s. As the speed increases, so does the energy input and dispersion of the mixed components. However, that also makes the mixing process less gentle. These types of mixers can be discontinuous (Figure 4.2) or continuous (Figure 4.3). In continuous systems, the mixing impellers also impose a conveying action on the particulates or mixture.

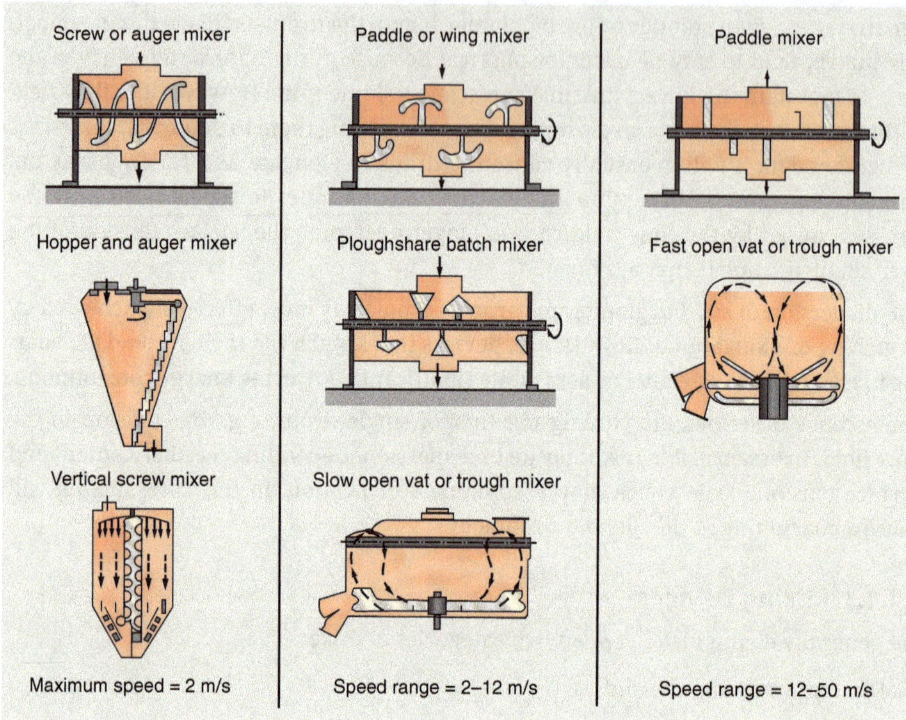

Figure 4.2 Solid particulate (discontinuous) batch mixing devices

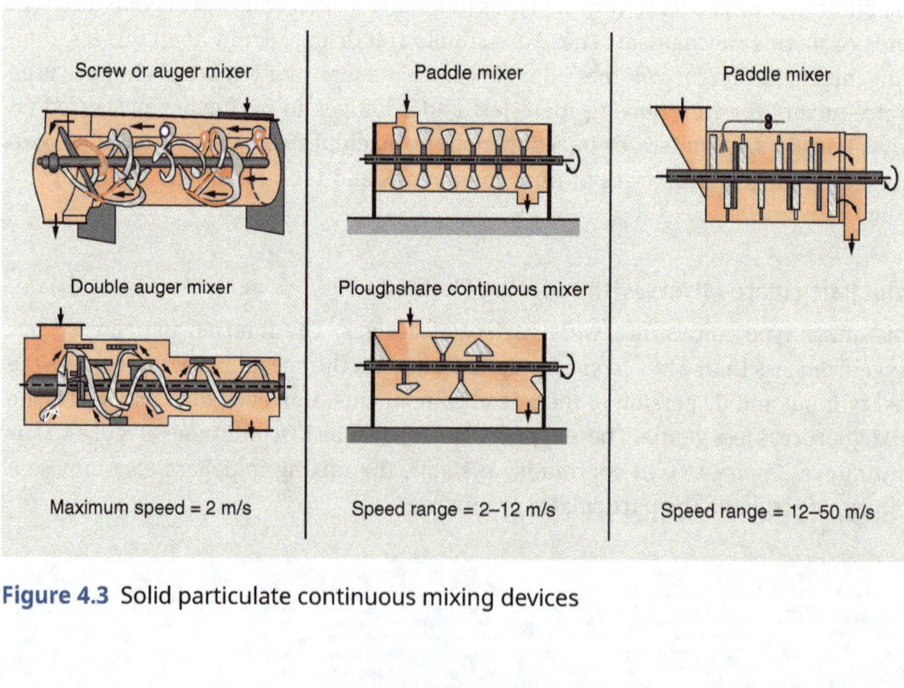

Figure 4.3 Solid particulate continuous mixing devices

Typically, these mixers are equipped with different types of mixing elements or impellers such as augers, paddles, or spirals. Conical hopper mixers with rotating auger have a capacity of up to 30 m^3, while silo mixers with an auger have a capacity of up to 100 m^3. Discontinuous mixers with horizontal impellers have a capacity of up to 30 m^3, and continuous systems have a throughput of up to 450 m^3/h.

Mixers with rotating elements are used to perform all kinds of mixing tasks. Heated/cooled combinations of through and paddle mixers are used to heat and plasticize PVC. Silo mixers and hopper/auger mixers are used to homogenize large pellet volumes.

4.1.1.2 Screw-Type Mixers for Viscous Materials

Screw-type mixing devices are used to continuously compound plastics into processable materials. The material is fed as premixed solid agglomerates or metered into the mixers by dosing systems, heated, and homogeneously mixed on the micro scale. Volatile components can be removed in de-gassing zones, and chemical reactions can take place. We distinguish between single-screw and twin-screw compounders.

Single-Screw Compounder

Single-screw compounders generally resemble plasticizing units in extruders, see Section 4.2. In a *plasticator*, a metering screw transfers flexible PVC to a conical mixing and shearing section, where it is plasticized by shearing and discharged by a subsequent screw.

The single-screw *co-kneader* performs both rotational and axially oscillating movements (Figure 4.4). The material is sheared in both axial and radial directions in the spaces between the screw flights and the fixed kneading pins. Calenders either discharge the material directly at the end of the kneader or it is transferred to a pelletizer via a single-screw, typically arranged at right angles to the kneader.

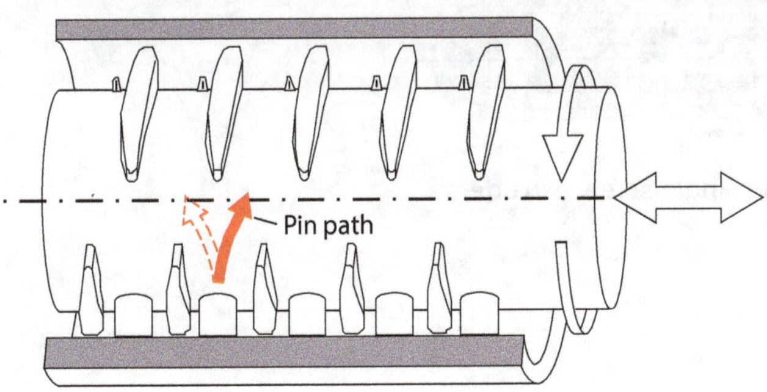

Figure 4.4 Schematic diagram of a co-kneader

Typical single-screw extruders for injection molding or extrusion provide only little mixing effect. The mixing effect can be enhanced by the introduction of mixing or shearing elements as seen in Figure 4.5 and Figure 4.6. In order to improve optical and mechanical properties, PE-LD is homogenized in water-cooled extruders with screw lengths of 24 to 40 D using two to four consecutive Maillefer zones.

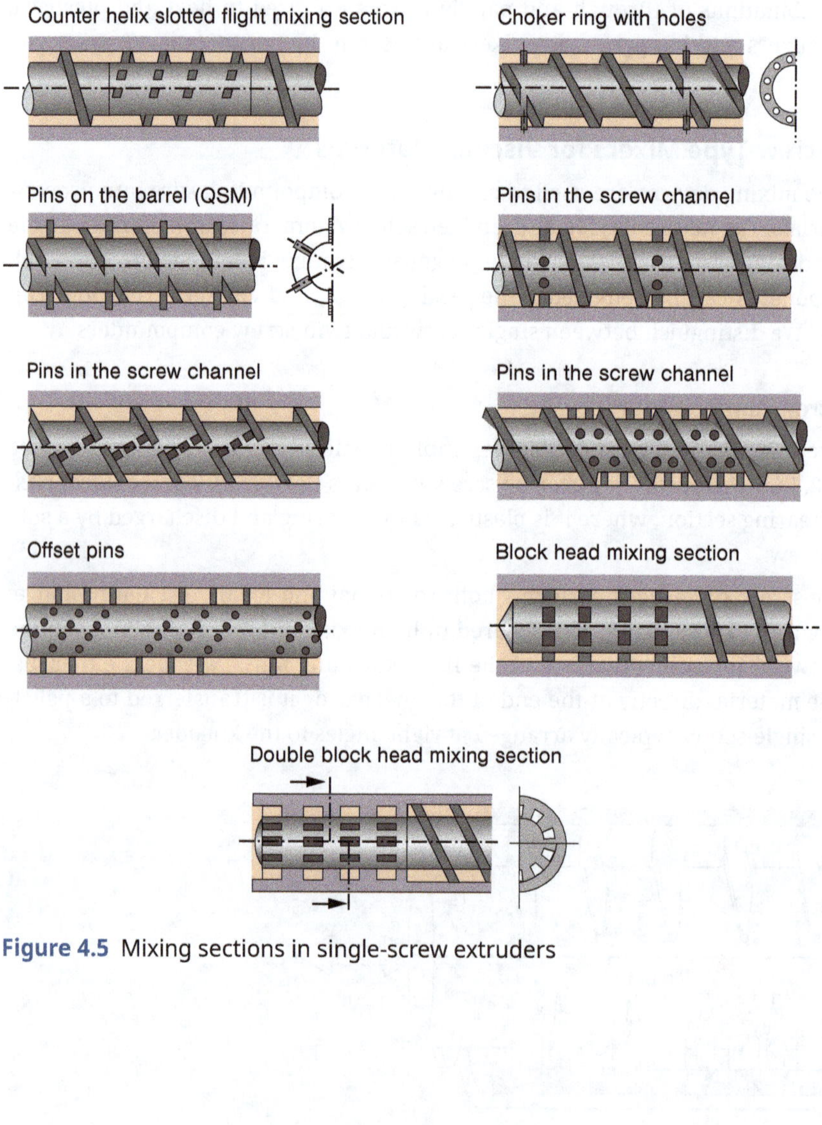

Figure 4.5 Mixing sections in single-screw extruders

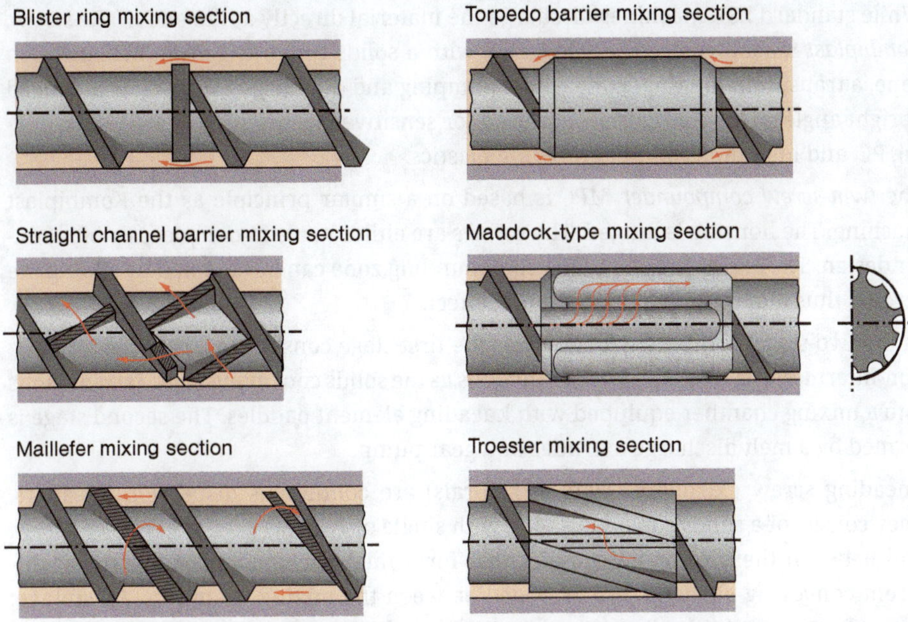

Figure 4.6 Dispersive mixing sections in single-screw extruders

Twin-Screw Compounders

Co-rotating screw kneaders with intermeshing screws are the most commonly used compounders. In the so-called *ZSK*- or *ZE-machines,* whose abbreviation originated from the German words for twin-screw extruder, the screws follow modular design principles (Figure 4.7). Each screw element is slipped on the screw shaft. The type and sequence of the conveying, kneading, and sealing elements are selected depending on the requirements of the task. In order to facilitate devolatization or addition of solids, melts, pastes, or fluids, specific sections can be added. ZSK equipment is self-wiping and can be heated electrically, using fluids, or steam.

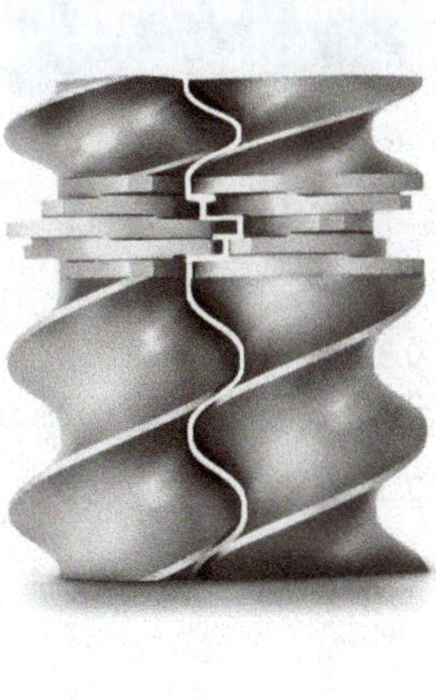

Figure 4.7
Sealing profile for screw elements and kneading disks in a twin screw compounder

While standard ZSK machines discharge the material directly at the end of the screw, *Kombiplast* machines use a single screw with a solids conveying and devolatization zone, a transition (melting) zone, and a pumping and discharge zone that is arranged at right angles. This configuration is used for sensitive plastics such as PVC, crosslinking PE, and high-molecular engineering plastics.

The *twin-screw compounder MPC* is based on a similar principle as the Kombiplast machine. The housing is hinged. The screws are either made in one piece or of modular design. The gap at the end of the compounding zone can be adjusted by sliding the screws, thus manipulating the kneading effect.

In the two-stage, continuous *FCM mixer* the first stage consists of a counter-rotating, non-intermeshing twin screw that functions as the solids conveying zone. It transitions into a mixing chamber equipped with kneading element paddles. The second stage is formed by a melt discharge extruder or a gear pump.

Kneading screw extruders (Type KEX, Drais) are continuous dispersion kneaders. They consist of a solids conveying zone with single or twin screws, a discharge screw, and between them rotor-stator pairs that form mixing chambers. A variety of different conveying elements are arranged between the mixing chambers. Advantage: large effective surface, adjustable shearing force, high homogenization effect.

Table 4.2 provides an overview of screw compounders. A special type of screw compounder is the *shear roll extruder*, Figure 4.8. It combines the advantages of a closed extruder with those of an open rolling mill. It is used to pack, melt, homogenize, disperse, dry, and pelletize materials of medium to high viscosity in the temperature range from 20 to 280 °C. The two very long, counter-rotating rolls are arranged horizontally next to each other. They are equipped with also counter-rotating, helical, and sharp-edged shearing grooves that convey the material from one end of the rolls the other.

Table 4.2 Typical Sizes, Power Consumption, and Throughput of Screw Compounders

Compounder type	Screw diameter mm	Power kW	Energy input kWh/kg	Throughput kg/h
Plasticizer	330–555	22–160	0.07–0.1	200–2300
Co-kneader	46–400	11–650	0.08–0.4	10–8000
Quantec	50–170	30–1150	0.02–0.4	400–16,000
ZSK MEGAcompounder	25–380	75–16,000	0.1–0.35	10–75,000
ZSK MEGAvolume	34–125	75–2600	0.02–0.15	10–18,000
Kombiplast	25–133	17–355	0.05–0.28	60–7500
FCM[2]	50–381	22–2940	0.18–0.29	75–16,000

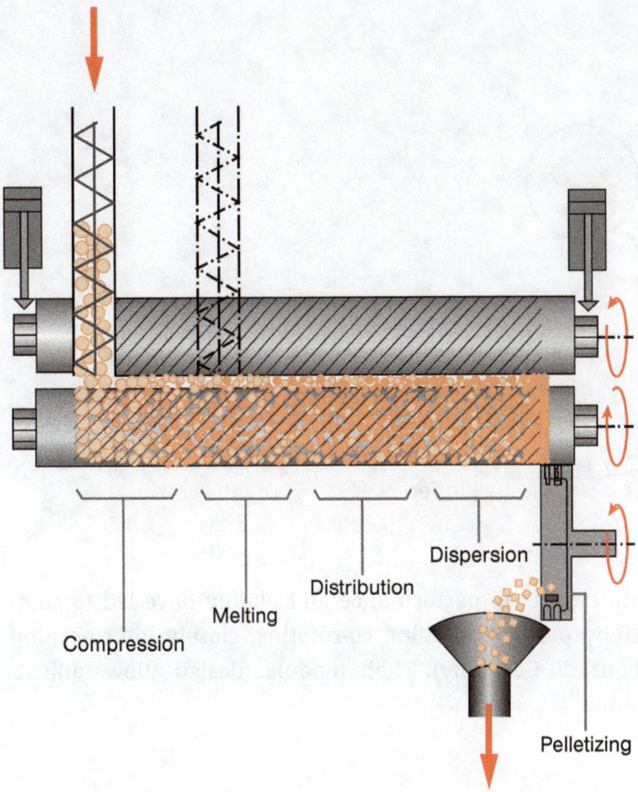

Figure 4.8 Processing steps during shear roll mixing

The roll separation width can be adjusted so it decreases from the beginning to the end of the rolls to 0.3 to 0.5 mm in order to increase the shearing effect. The introduction of additives, such as fillers, fibers, or pigments in any consistency, can be achieved through the nip region of the rolls. The finished material is either removed from the end of the rolls in form of a tape, or it is forced through small holes of a pelletizing barrel (Figure 4.9). The cylindrical nubs exiting on the inside of the barrel are cut and cooled by forced air. The throughput of current equipment ranges between 5 and 650 kg/h.

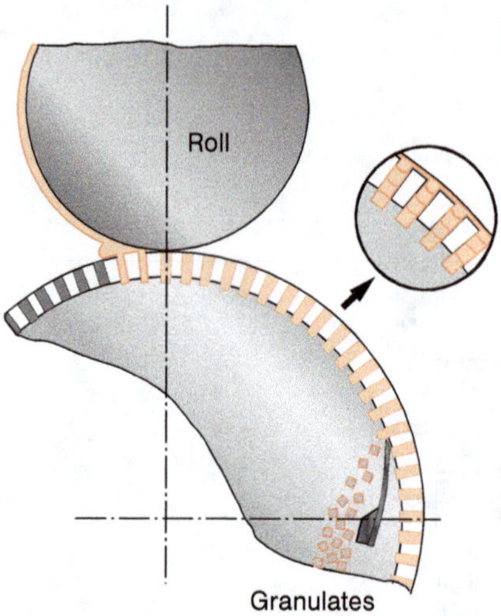

Figure 4.9
Flow into the nip region of the pelletizing cylinder

Multiple-Shaft Kneaders

Increased requirements with respect to performance and quality have led to compounders with, *e. g.*, 12 axially parallel, circular, co-rotating, closely intermeshing screws (Extricom GmbH, Lauffen, Germany). Their modular design allows adjustments for different applications.

Static Mixers

Static mixers are pressure-driven continuous mixing devices. One of the most commonly used static mixers is the twisted tape static mixer schematically shown in Figure 4.10. The polymer is sheared and then rotated by 90° by the dividing wall; the interfaces between the fluids increase. The interfaces are then re-oriented by 90° once the material enters a new section. The stretching/re-orientation sequence is repeated until the number of striations is so high that a seemingly homogeneous mixture is achieved.

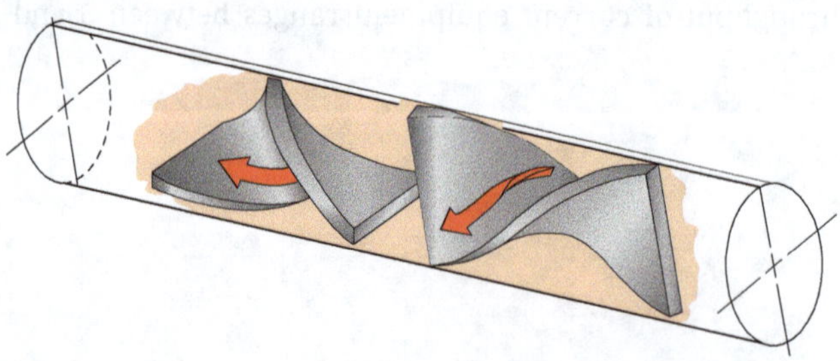

Figure 4.10 Schematic diagram of a Kenics static mixer

Figure 4.11 shows that in a static Kenics mixer the number of striations doubles with each mixing section. The number of striations can be computed as:

$$N = 2^n \quad (4.1)$$

Where

N = number of striations and n is the number of sections in the mixer

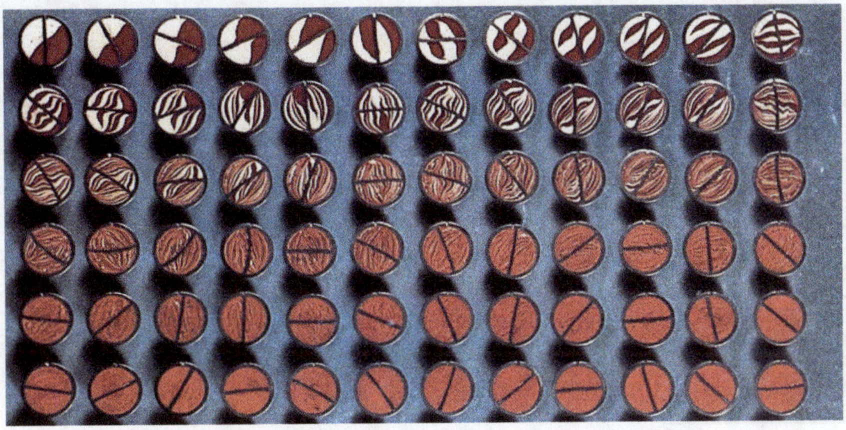

Figure 4.11 Experimental progression of the layering of colored resins in a Kenics static mixer [Source: Chemineer, Inc., North Andover, Massachusetts]

4.1.1.3 Other Mixers for Viscous Materials

Mixers both with and without rams have been developed for the compounding of rubber blends and are still used today for ABS (Figure 4.12).

Counter-rotating mixing rotors rotate in trough-shaped mixing chambers; here, intermeshing rotors introduce more energy into the blend than tangential rotor systems (Figure 4.13).

Twin trough kneaders are used to compound low-viscosity plastics both on a laboratory scale with volumes up to 10 l and on a production scale with volumes up to approx. 4000 l (Figure 4.14).

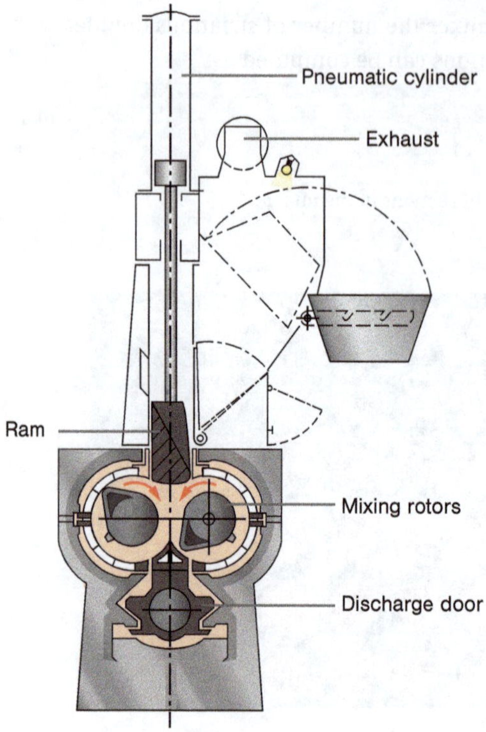

Figure 4.12 Ram kneader (internal batch mixer)

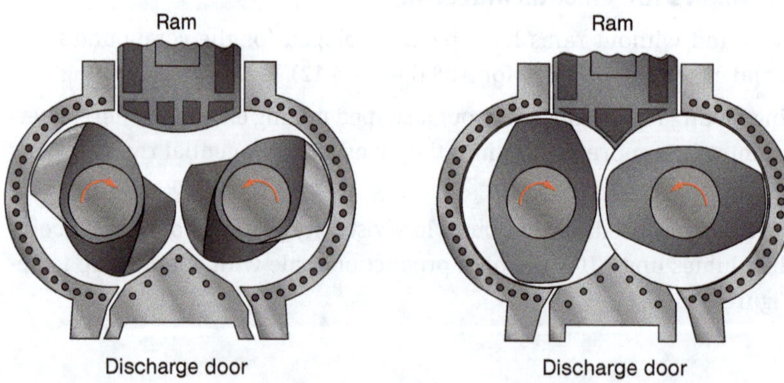

Figure 4.13 Geometries and rotor arrangements of internal batch mixers for discontinuous rubber compounding; tangential (left) and intermeshing (right) rotors

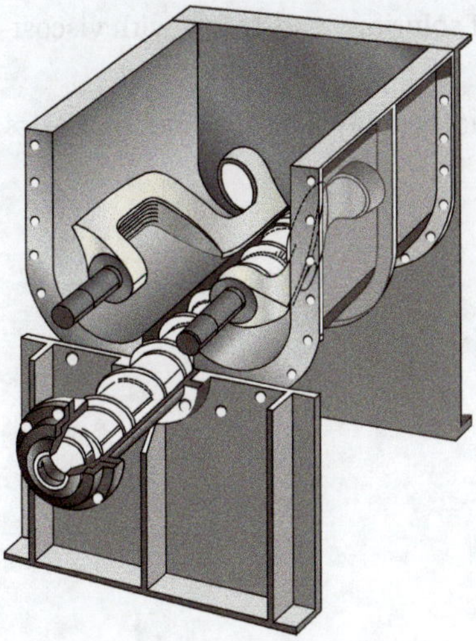

Figure 4.14
Twin trough kneader with sigma rotors and discharge screw

The continuous Conterna kneader (IKA-Maschinenbau, Staufen; Germany) consists of six (max. 12) mixing chambers arranged in series (Figure 4.15). This arrangement allows for the combination of multiple processing steps such as mixing, kneading, rolling, extruding, and adding of components.

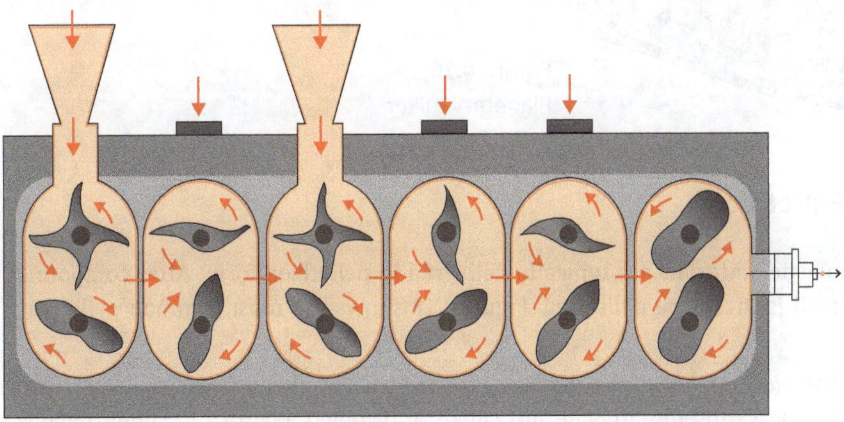

Figure 4.15 Continuous multi-chamber kneader; discharge can be combined with various discharge modules and/or cut-off devices and pelletizers, *e. g.*, gear pump, counter rotating twin screw extruder, single screw extruder

Planetary mixers are suitable for the preparation of solutions and pastes with viscosities up to approx. 150 Pa · s (Figure 4.16).

Roller mills, such as those used in rubber compounding, play a minor role in plastics processing operations.

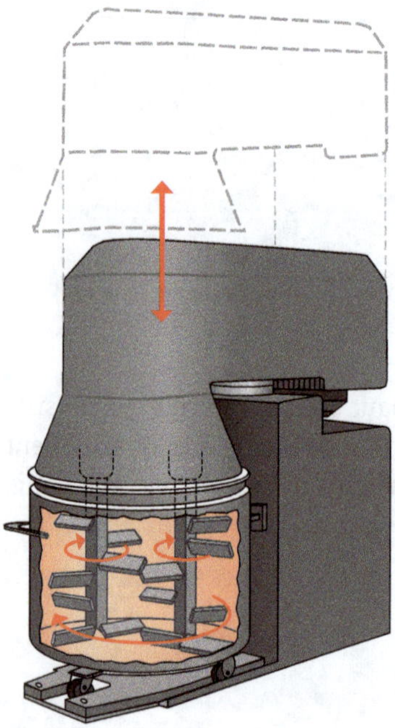

Figure 4.16
Planetary mixer

4.1.2 Pelletizers

Thermoplastic materials are typically delivered in pelletized form. After compounding the melt is therefore pelletized. Figure 4.18 shows the most common pelletizing methods.

During *hot plate or hot face pelletizing* the melt is fed through die plates to form strands that are subsequently cut into pellets and cooled. Figure 4.17 shows pelletizers that operate in air, air and water, and under water. Plastics with low tack, such as PVC, can be cooled in air, for all others the cooling medium of choice is water. After pelletizing, the material must be dried. Contamination by dust or debris during hot pelletizing is not an issue, because the cooling steps are performed in closed systems. Throughputs up to 75 t/h can be achieved.

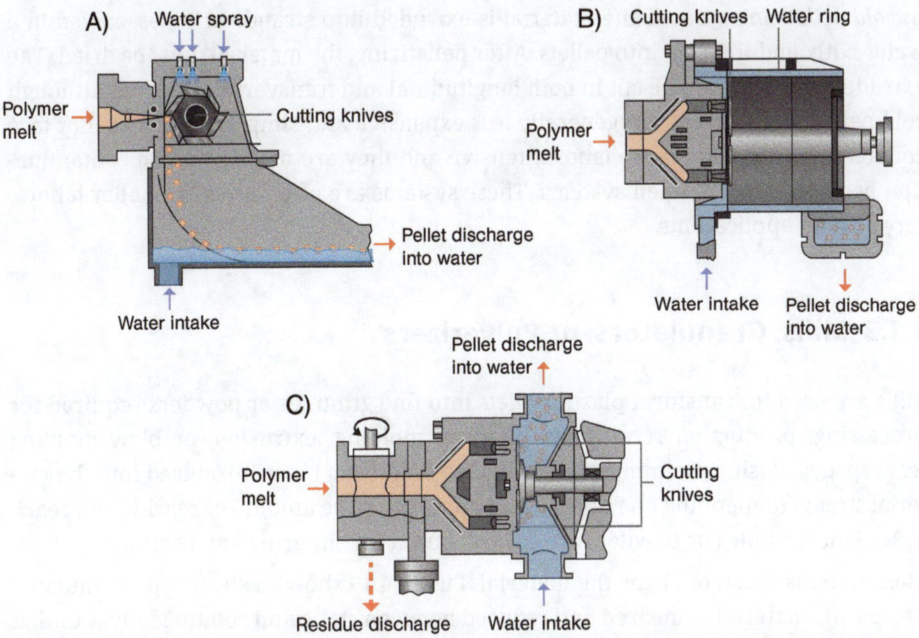

Figure 4.17 Rotary knife pelletizer (A), water ring pelletizer (B), and underwater pelletizer (C)

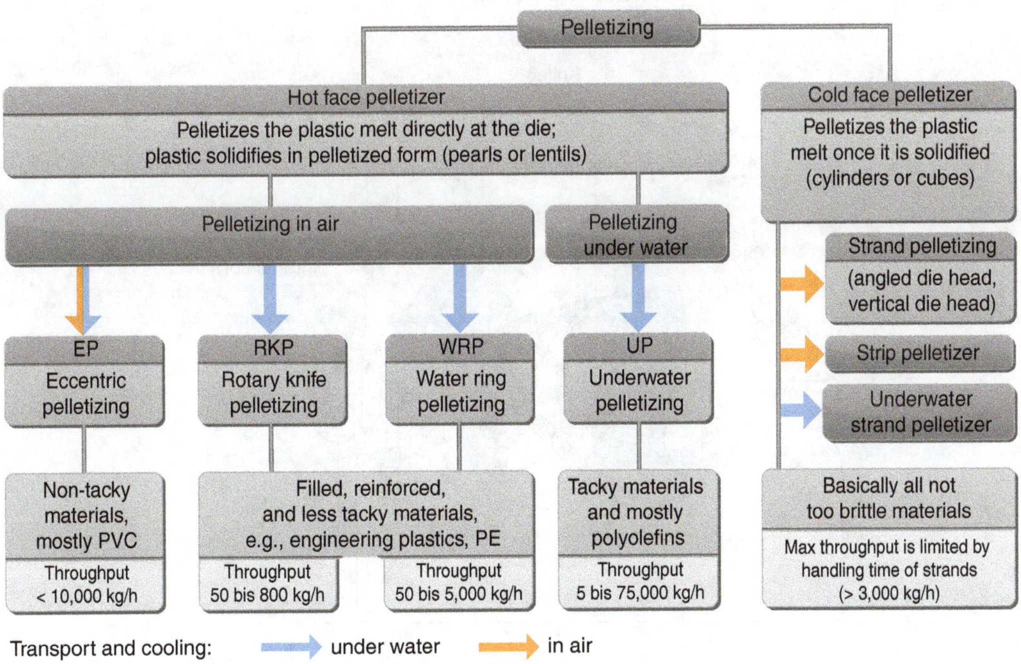

Figure 4.18 Pelletizing Systems

In *cold pelletizing systems* the material is extruded into strands or tapes, cooled in a water bath, and then cut into pellets. After pelletizing, the material must be dried. The extruded tapes have to be cut in both longitudinal and transverse direction. Although cold pelletizing systems are generally less expensive and simpler than their hot face counterparts, they are more labor-intensive and they are more prone to contamination because they are open systems. These systems are often used in smaller laboratory or pilot applications.

4.1.3 Mills, Granulators, or Pulverizers

Mills are used to transform plastic pellets into fine granules or powders required for processing: production scrap from injection molding, extrusion, or blow molding (*e.g.*, sprues, flash, or trimming) is ground so that it can be reintroduced into the material stream (depending on material and application the amount of regrind can reach 50%). Fine granules or powders are typically used for sintering and coating.

Most systems crush or shear the material. Figure 4.19 shows a shear-type granulator. The plastic material is sheared and reduced between static and rotating knives until it is fine enough to be discharged through the sieve. To avoid overheating caused by friction, pulverizing is often done in steps. When reducing the pellet size of ductile plastics, the required brittleness is achieved by pre-cooling using liquid nitrogen. Such a system is presented in Figure 4.20.

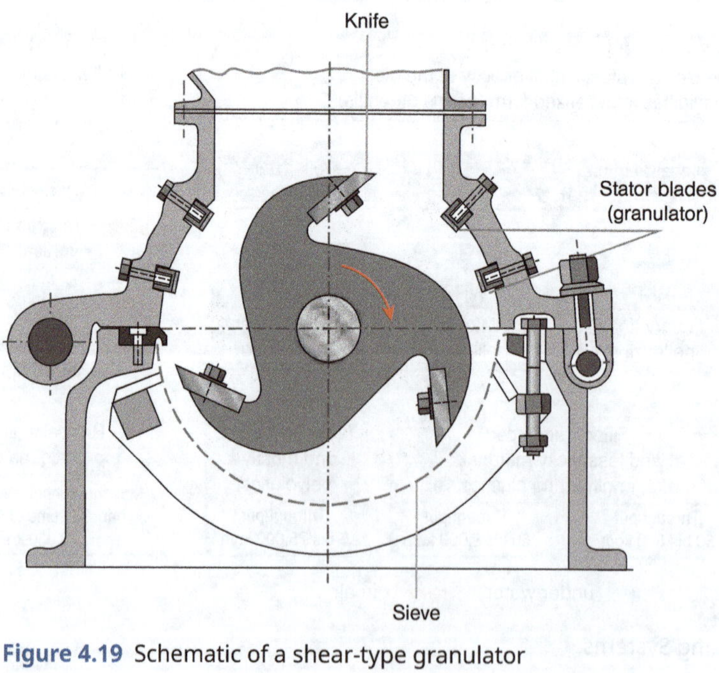

Figure 4.19 Schematic of a shear-type granulator

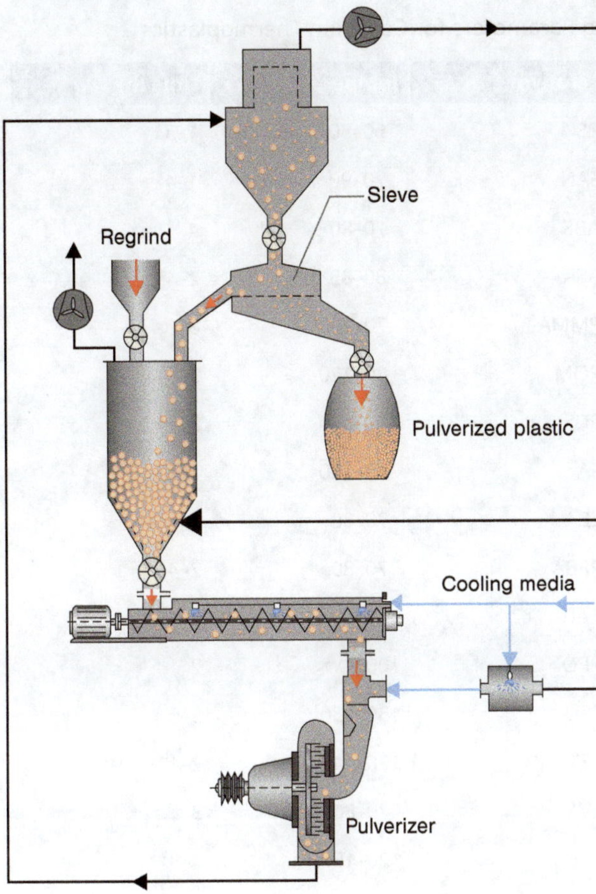

Figure 4.20 Schematic of a cooled pulverizing system

4.1.4 Dryers

To avoid hydrolytic degradation during melt processing, many plastics must be dried prior to processing. The most common dryer parameters are listed in Table 4.3. Hydrophilic engineering plastics in particular have to be pre-dried. Even hydrophobic plastics, such as polyolefins, may require pre-drying, *e. g.*, when their fillers or pigments are hydrophilic. In this case, the pellets' water content does not cause degradation; however, it acts like a blowing agent and increases the occurrence of shrink holes.

Table 4.3 Recommended Drying Parameters for Common Thermoplastics

Plastic material	Acronym	Temperature (°C)	Time (h)
Polystyrene and copolymers	PS	60–80	1–3
	SAN	70–90	1–4
	ABS	70–80	4
	ASA	80–85	2–4
Polymethylmethacrylate	PMMA	70–100	8
Polyoxymethylene	POM	80–120	3–6
Polycarbonate	PC	100–120	8
Polyamide	PA	80–100	16
Ethylene butyl acrylate	EBA	70–80	3
Polyether block amide	PEBA	70–80	2–4
Cellulose acetobutyrate	CAB	60–80	2–4
Polyphenylene oxide	PPO	100	2
Polysulfone or polyarylsulfone	PSU	130–150	3–6
Thermoplastic polyester	PET	120–140	2–12
Ethylene vinyl acetate	EVA	70–80	3
Polyarylate	PAR	120–150	4–8
Polyphenylene sulfide	PPS	140–250	3–6
Polyether ether ketone	PEEK	150	8
Thermoplastic elastomers	TPE	12	3–4

Dryer efficiency is determined by the actual temperature reached at each pellet and the relative humidity of the ambient air. If the temperature is too high, the pellets may agglomerate. Too little airflow will not be able to transport sufficient heat to the pellet and humid air away from the pellet. Note: drying time is cut in half with each 10 °C increase in temperature.

Removing moisture in these materials can easily be achieved by pre-heating the pellets with hot air in an oven fitted with removable baskets or trays. A more automated method uses a drying hopper that can be mobile, free-standing, or mounted directly on the throat of the processing machine. The capacity of the holding hopper is determined by the necessary drying time required at the temperature specified by the polymer supplier. For instance, an injection machine that processes 50 kg/h, with a determined drying time of 2 hours, requires a hopper holding at least 100 kg of material.

Hot air dryers: By heating the ambient air the relative humidity is reduced to a level that facilitates effective drying. These dryers are available as machine-side or top-mounted dryers and continuously provide the processing machine with pre-heated and uniformly dried material. Their drying efficiency is sufficient for many applications.

Dehumidified air dryers: Drying is achieved using pre-dried air at elevated temperatures, see Figure 4.21. These dryers generally operate with pre-dried rather than ambient air in order to achieve better drying results. This super-dry air is generated by injecting compressed air through a molecular sieve desiccant, such as crystalline aluminum silicate, which is contained in a pressure-sealed cylinder. To allow for continuous operations, these systems include two cylinders, one of which reduces the residual humidity in the dry air flow while the other cylinder regenerates. The advantage of high drying efficiency comes at the expense of higher energy consumption.

In automated material feeding systems the pre-drying process is integrated into the material conveying system.

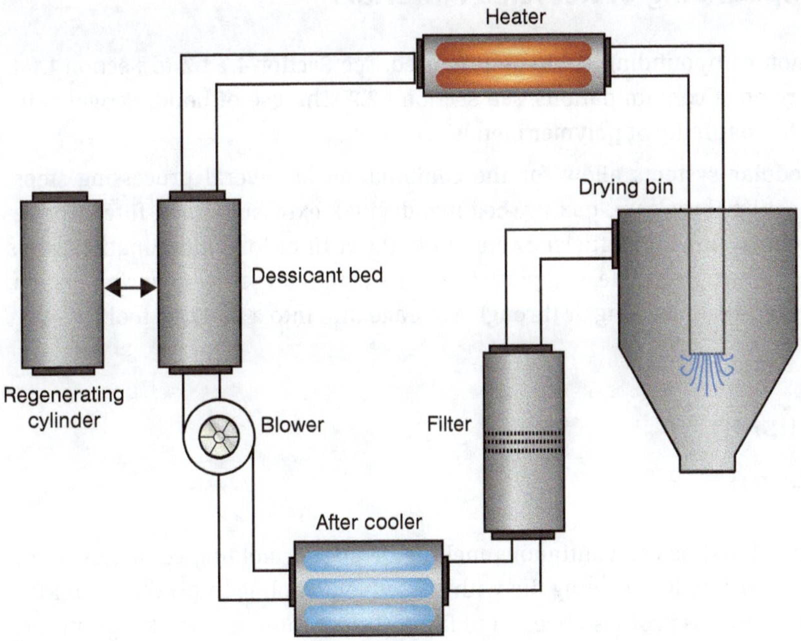

Figure 4.21 Schematic diagram of a dehumidifier drying system

4.1.5 Compounding of Thermosetting Materials

Thermosetting materials consist of a resin and powdered or fibrous fillers such as wood flour, glass fibers, paper, fabric, fiber strands, or chips. Powdered and short-fiber fillers are pre-mixed with the dry resin, if necessary with the addition of processing aids or pigments/dyes, and subsequently plasticized and homogenized in heated rolling mills or twin-screw extruders. At the same time, the resin is pre-condensed or pre-polymerized (B- or C-state) to reach the required viscosity for free flowing resins. The plastic sheet or extruded chips are subsequently reduced to uniform pellet size by milling and sieving.

Coarse-fiber materials and resin impregnated filler materials are predominantly manufactured by impregnation with liquid or dissolved resins and subsequent drying. Long-fiber reinforced materials are manufactured by impregnating continuous fiber strands and subsequently chopping them. Tapes for the manufacture of laminate materials are impregnated using impregnation machinery, see Section 4.8.5.

4.1.6 Compounding of Recycled Materials

Here, common compounding equipment is used, see Section 4.1.1.2 to Section 4.1.4. Melt filters remove contaminations, see Section 4.2.2. The use of bonding agents increases the compatibility of polymer blends.

Flexible, modular systems allow for the combination of several processing steps: blade compaction (break up, mixing, heating, drying), extrusion, melt filter (for reduced contamination), high-efficiency melt filter (laser filter for contamination levels up to 7%), devolatization zone, compounding zone with twin screw metering system, pelletizer with direct metering or through a storage unit into a molding tool.

4.2 Extrusion

Reinhard Schiffers

Extrusion is defined as the continuous melting of plastic molding compounds and their discharge through a forming die with subsequent cooling to produce semi-finished products such as profiles, sheets, and films. Output rates of a few kilograms per hour and up to more than 1000 kg/h can be achieved. The raw material can be plastics in granulate form, as a powder, or as a mixed recipe. In some cases, the extrusion systems also take over compounding tasks. These processes are often referred to as direct extrusion since a separate compounding step is no longer needed. There are various technologies for preparing the melt and building up pressure. The systems are equipped with sensors for the most important process parameters so that these can be monitored and controlled. The aim is to achieve consistent product quality. Increasingly, systems are being used for in-line recording of product quality and, where possible, the measured values are fed back for (indirect) control of the measured variable.

4.2.1 Extruder Designs

Common types and process variants of extruders are described below. Table 4.4 introduces the different system concepts with their relevant process engineering properties and shows their main areas of application.

4.2.1.1 Single-Screw Extruder

Single-screw extruders are the most frequently used machines in plastics processing due to the achievable cost efficiency; see Figure 4.22. Other designs are used whenever the process engineering requirements make it advantageous. The molding compound is fed via a hopper to a screw, which rotates in a cylinder (or barrel) with heated zones. There, the plastic is melted and possibly degassed mainly by dissipation of the torque introduced via the drive, rather than by heat conduction. The shear introduced homogenizes the melt. The conveying effect is caused by the frictional forces that the surfaces of the screw and barrel exert on the raw materials. The basic design of an extrusion screw with feed, transition or compression zone, and discharge or pumping zone is known as a three-zone screw and is shown in Figure 4.23. The specific geometry of the screw is adapted to the range of plastics to be processed in order to achieve optimum throughput with good melt quality and efficient operation.

Table 4.4 Process Engineering Properties, Design Principles, Sizes, Speeds, Throughputs, and Application Ranges of Extrusion Systems

	Co-rotating	Single screw	Co-kneader	Continuous mixer	Counter-rotating, intermeshing	Counter-rotating, non-intermeshing	Planetary roller	Multi-screw
Solids feed	+++	+	++	++	+++	++	++	+
Melting	+++	+	+++	+++	++	++	++	Melt feed
Distribution	+++	++	+++	++	+	++	++	+++
Dispersion	+++	++	+++	++	+++	+	+++	+
Self-cleaning	+++		+++		+++		+++	+++
Residence time distribution	Narrow	Wide	Narrow	Wide	Very narrow	Wide	Very narrow	Narrow
Degassing	++	++	++	+	+	++	++	+++
Pressure buildup	+	++			+++	+		++
Energy input (mechanic, drive)	+++	+++	++	+++	++	++	++	+
Cylinder heating/ cooling capacity	+	++	+	+	++	+	+++	++
Screw temperature control	Possible	Possible	Yes	Yes	Yes	No	Main screw	Main screw

		Co-rotating	Single screw	Co-kneader	Continuous mixer	Counter-rotating, intermeshing	Counter-rotating, non-intermeshing	Planetary roller	Multi-screw
Adaptation options without adaption of the process section		Speed Throttle Heating/cooling	Heating/cooling (Speed)	Speed Heating/cooling Discharge system Throttle	Speed Outlet flap Discharge system Heating/cooling	Heating/cooling (Speed) (Throttle)	Speed Heating/cooling	Speed Heating/cooling Discharge system	Speed Heating/cooling
Structure of the process section	Screw	Modular	One-Piece	Modular	One-Piece	One-Piece	Modular	One-Piece	One-Piece
	Cylinder	Modular	One-Piece	Modular	Modular	One-Piece	Modular	One-Piece	One-Piece
Screw size	[mm]	16 to 380	20 to 600	30 to 200 (600)	29 to 437	25 to 160	50 to 460	70 to 350	47 to 150
Screw speed	[min^{-1}]	100 to 1200	20 to 250	20 to 500	200 to 1800	5 to 80	100 to 500	10 to 60	20 to 200
Throughput range	[kg/h]	1 to 50,000	5 to 30,000	5 to 4000	5 to 44,000	1 to 2000	100 to 15,000	400 to 4000	50 to 5000
Main areas of application		Compounding Processing with powder or throughputs > 1 t/h	Processing of granulate Discharge extruder Degassing	Compounding Calender feeding (PVC, PP)	Compounding	Processing and compounding PVC dry blend	Degassing	Compounding and calendering of PVC	Degassing reactions with long residence times

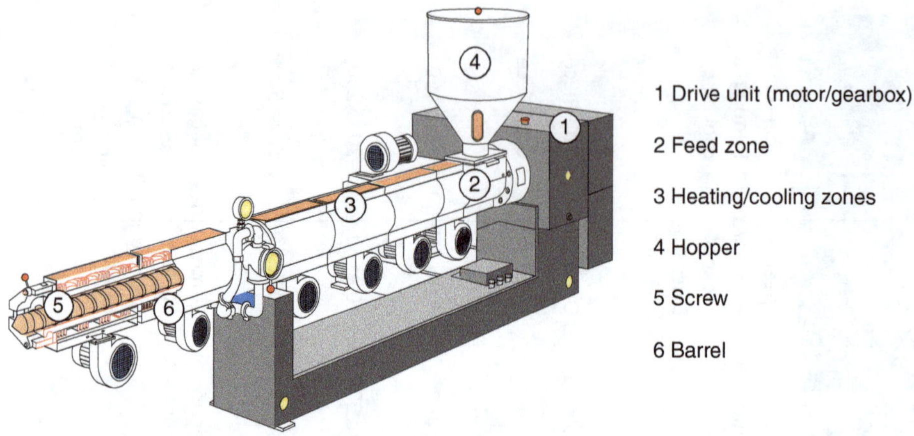

Figure 4.22 Single-screw plasticizing extruder

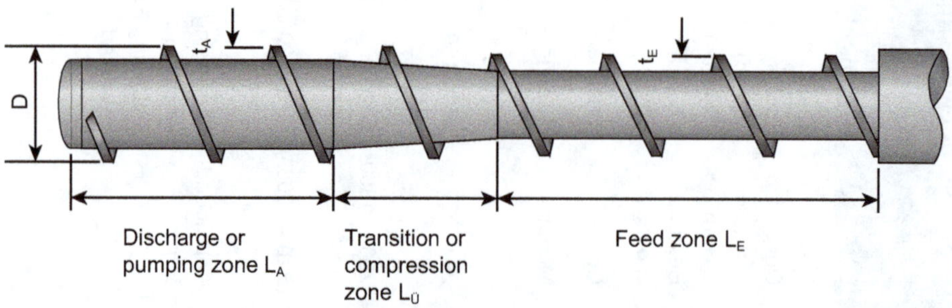

Figure 4.23 Zones of a standard three-zone screw

4.2.1.2 Cylinder Types: Smooth Bore and Grooved Liner

Single-screw extruders are divided into smooth-barrel (conventional) and grooved-barrel extruders. While conveying in smooth feed zones is dependent on temperature-related friction conditions, forced conveying occurs in a grooved feed zone in the barrel in the axial direction; see Figure 4.24. The grooves can have different geometric shapes (sickle-shaped, semicircular, rectangular). More relevant for the operating behavior is the ratio of the groove depth to the size of the granulate. They usually run in the axial direction of the cylinder, but can also be helical. Rectangular grooves running in the axial direction with decreasing depth up to the end of the feed zone are the most common.

The decisive effect of the grooves is that they make it more difficult for the solids to rotate with the screw. A grooved-barrel extruder can therefore convey more at the same speed than a conventional extruder. Ideally, there is an almost linear relationship between screw speed and throughput. This type of conveying results in a different operating behavior for the grooved-barrel extruder compared to the conventional extruder and, in particular, a different pressure profile along the screw.

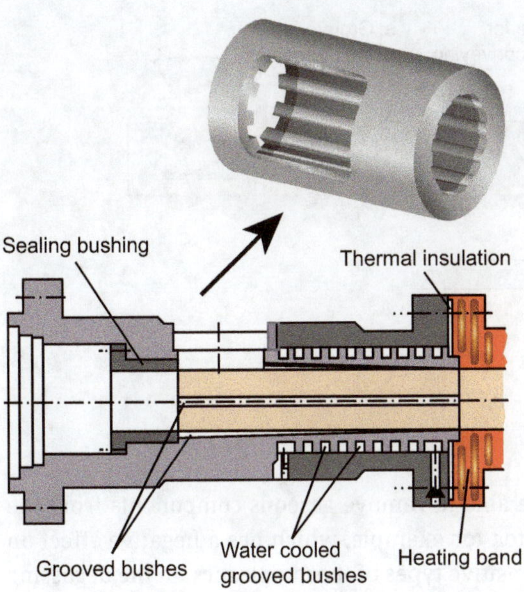

Figure 4.24
Solids-conveying principle of the grooved-barrel extruder

The tapered grooves at the end of the feed zone cause a strong compaction of the solids. The resulting frictional heat may already melt some of the solids at the end of the grooves. The stiffer the granulate, the greater the axial conveying of the solids. For example, the conveying of PE-HD is greater than that of PE-LD. In addition, plastics with a very high molecular weight can be processed.

The grooved area of the cylinder is temperature-controlled via a water or air unit to prevent the cylinder from overheating and building up a melt film. Furthermore, for reasons of energy efficiency, the grooved feed area is thermally decoupled from the rest of the process section.

4.2.1.3 Screw Concepts: Three-Zone, Barrier Screw

On the basis of the three-zone screw, various approaches to increase the throughput of extruders have been implemented in the course of development. This mainly concerns the melting capacity and thus the transition from solid to melt. One principle for increasing the throughput is solid/melt separation, which is used in barrier screws. These have an additional flight at the start of the melting zone, which has a greater pitch than the continuous screw flight and thus meets up with it again at the start of the compression zone; see the Maillefer screw (Figure 4.25). The already melted material is pressed into the screw flight, which increases in the conveying direction. This prevents unmelted pellets from being conveyed into the discharging zone. In addition, the melting capacity is increased and the homogenization of the melt is improved. Such screw concepts are available on the extrusion technology market under various names and are also increasingly being adapted for use in injection molding machines when enhanced melting capacity is needed.

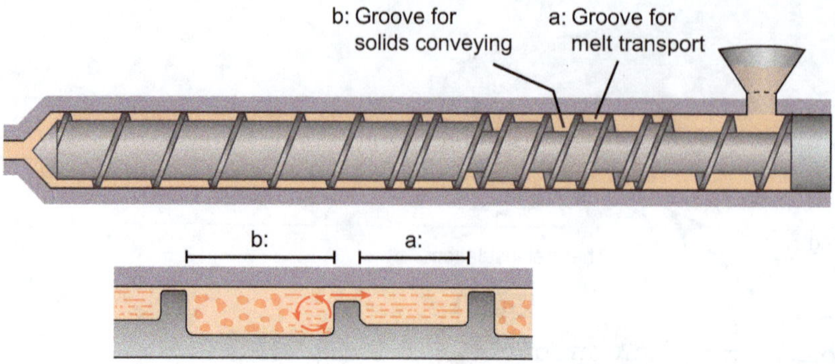

Figure 4.25 Barrier screw

4.2.1.4 Degassing Extruder

Degassing extruders (Figure 4.26) are able to remove gaseous components from the plasticized material. This may be water, for example, which has a negative effect on the product properties of moisture-sensitive types of thermoplastics. In the degassing extruder, after an initial plasticizing step, the screw flight depth is increased so that the pressure drops to atmospheric pressure and water vapor or other volatile components can escape through a degassing opening in the barrel. As a rule, the efficiency of the degassing is increased by applying a vacuum. The core task when designing such systems is to match the conveying capacity of the two screw sections to each other. If the second screw stage cannot convey the melt fast enough at all operating points, the degassing opening is "flooded" and the process comes to a standstill. The principle of degassing is not only used in single-screw extruders but also in other screw-based extrusion systems.

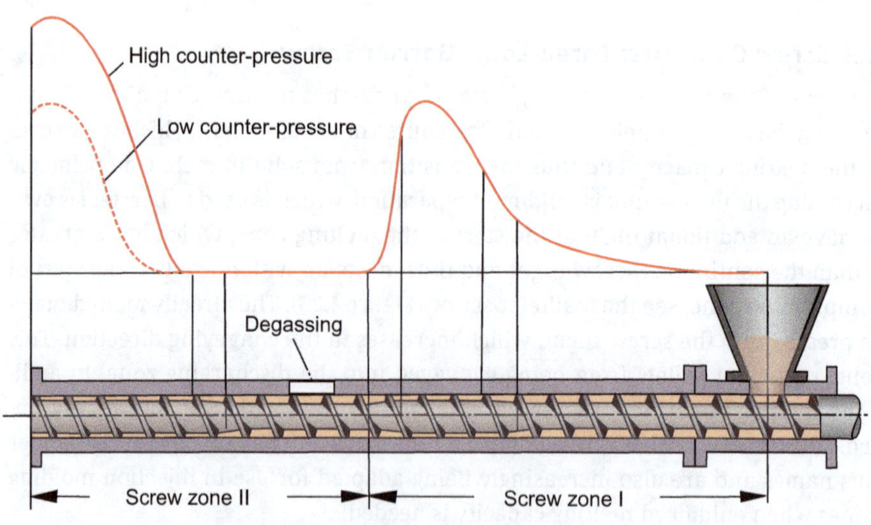

Figure 4.26 Schematic and pressure curve for a degassing extruder

4.2.1.5 Cascade or Tandem Extruder

Cascade or tandem extruders consist of a series connection of a plasticizing extruder, which prepares the plastic melt, and a melt extruder, which homogenizes the melt and builds up the pressure; see Figure 4.27. As both extruders have their own drive, the conveying capacities of the two screws can be easily coordinated for different operating points and raw materials, for example. In many cases, such systems also have the option of using a degassing opening (see Section 4.2.1.4). However, the mechanical engineering effort is significantly greater and these extruders can only be used effectively if the flexibility gained can be utilized.

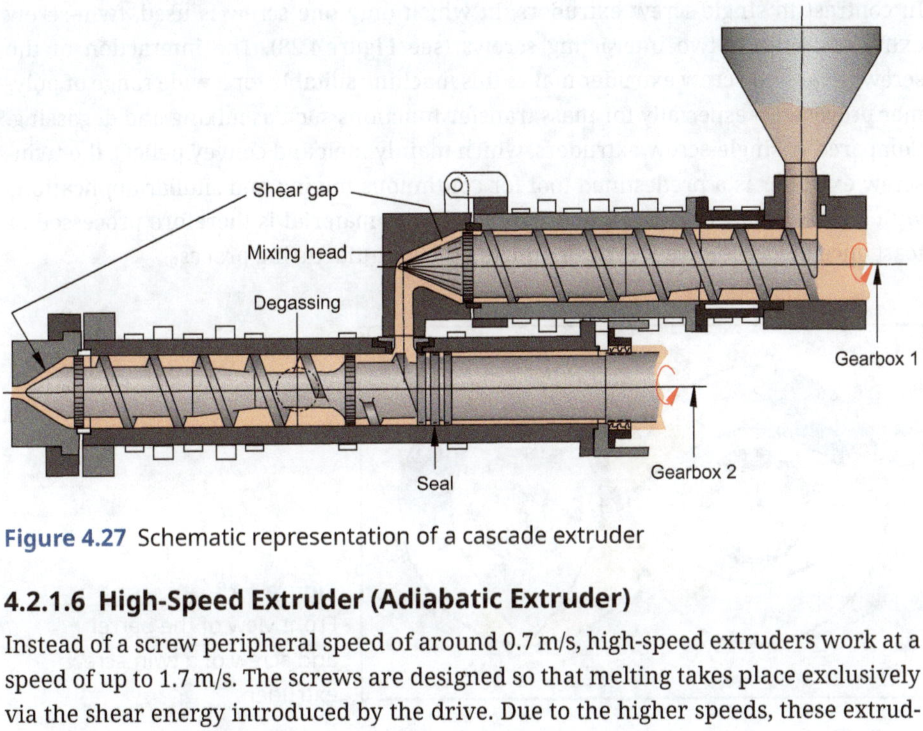

Figure 4.27 Schematic representation of a cascade extruder

4.2.1.6 High-Speed Extruder (Adiabatic Extruder)

Instead of a screw peripheral speed of around 0.7 m/s, high-speed extruders work at a speed of up to 1.7 m/s. The screws are designed so that melting takes place exclusively via the shear energy introduced by the drive. Due to the higher speeds, these extruders can often be operated directly (without a gearbox) on a drive. The axial bearing for absorbing the forces caused by the pressure buildup is integrated into the motor flange to save space, thus enabling a very compact design. Due to the process engineering properties and short residence times (a few seconds), high-speed extruders are used for plastics such as PE, PS, and PA, which are insensitive in this respect.

4.2.1.7 Planetary Roller Extruder

Planetary roller extruders exhibit very gentle plasticizing, good homogenization and dispersion and are therefore particularly suitable for compounding tasks. In the transition zone of the planetary roller extruder, the screw is equipped with special teeth, as is the barrel. In between, there are several planetary screws that mesh with these teeth. When the central screw rotates, the planetary screws rotate around their own axes and simultaneously revolve around the central screw. The granulate or powder conveyed from the feed zone is rolled out in a fine layer between the central and planetary screws. The resulting large surface area enables very effective degassing and temperature control in the system. Planetary roller extruders are also used for the post-condensation of polycondensates or in chemical or food process engineering.

4.2.1.8 Twin-Screw Extruder

Sebastian Fraas, Charlie Martin

In contrast to single-screw extruders, in which only one screw is used, twin-screw extruders employ two interacting screws (see Figure 4.28). The interaction of the screws in a twin-screw extruder makes this machine suitable for a wide range of polymer processes – especially for mass transfer functions such as mixing and degassing. Compared to single-screw extruders, which mainly melt and convey pellets, the twin-screw extruder is a predestined tool for continuous mixing and similar applications with a material transition. Almost every polymer material is therefore processed at least once in a twin-screw extruder during the manufacturing process.

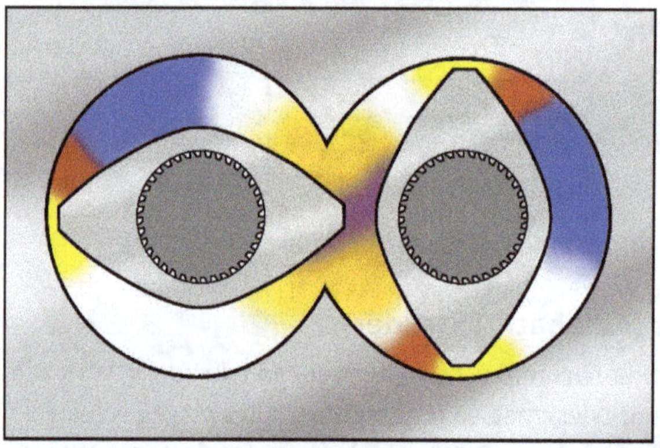

Figure 4.28
Front view of the barrel and screw of a twin-screw extruder

The smallest twin-screw extruders for research and development purposes can process batches of 50 g; in series production, the capacity ranges up to 50,000 kg/h and more. The length of a processing section is specified by the ratio of the screw length to the screw diameter (*L/D*). For example, if the screw diameter is 100 mm and the screw length is 4000 mm, the *L/D* ratio is 40/1. The motor and the gearbox are part of the drive train that applies shear energy to the materials to be processed inside the barrels via rotating screws.

Important process parameters during extrusion are the screw speed [min^{-1}], the throughput [kg/h], process temperatures [°C], and the vacuum level in degassing zones [mbar]. Continuous monitoring of mass pressure [bar], melt temperature [°C], torque [N·m], and in-line sensor data ensures consistent product quality. Modern twin-screw extruders use programmable logic controllers (PLC) with graphic HMI (human–machine interface) touchscreens that integrate data acquisition, process flow, and recipe management.

Raw materials in various forms such as pellets, granules, powders, fibers, gases, and liquids (that is, the "formulation") are fed into the twin-screw extruder. These materials can be premixed in the extruder (premix process) or fed separately via individual doses (split-feed process). The raw materials are metered as required using screw conveyors, vibrating feeders, belt weighers, and injection pumps. Table 4.5 compares the co-rotating and counter-rotating twin-screw extruders with regard to their most important properties.

Table 4.5 Comparison of the Core Properties of Co-rotating and Counter-rotating Twin-Screw Extruders

Twin-screw extruder design	Mixing effect	Degassing capacity	Pressure buildup	Residence time	Residence time distribution
Co-rotating, closely intermeshing twin screw	Very good	Good	Average	Short	Medium to wide
Counter-rotating, closely intermeshing twin screw	Good	Average	Very good	Short	Narrow, medium, or wide

A distinction is made between the following types of twin-screw extruders:

- Twin-screw extruder with high screw speed and high energy input in the design:
  - Uniformly (co-rotating) closely intermeshing
  - Counter-rotating, closely intermeshing
  - Counter-rotating, non-intermeshing or tangential
- Twin-screw extruder with low screw speed for mixing shear-sensitive formulations (for example, RPVC) and for generating high pressures:
  - Parallel, counter-rotating, closely intermeshing
  - Conical, counter-rotating, closely intermeshing

Each of these extruder types has technical advantages. The co-rotating, closely intermeshing twin-screw extruder is suitable for most compounding applications. The counter-rotating, closely intermeshing twin-screw extruder with low rotational speed dominates the market for the extrusion of RPVC profiles. These two types account for at least 90% of the world's twin-screw extruder lines.

Counter-rotating tangential and closely intermeshing designs with high rotational speed are used in special degassing processes and in reactive extrusion.

4.2.1.8.1 Co-rotating, Closely Intermeshing Twin-Screw Extruder

The co-rotating twin-screw extruder is the most widely used continuous-mixing device. Typical applications are masterbatch and compounding applications in which fillers, pigments, dyes, reinforcing agents, and additives are homogeneously mixed into a plastic matrix. High screw speeds (a speed of over 1000 min^{-1} is possible) facilitate the introduction of shear and energy into the process for efficient mixing. Due to the high surface renewal (rolling melt pools in the screw channels) and the large degassing surface, the co-rotating twin-screw extruder is also ideal for removing moisture and other volatile substances from the extrudate (degassing process). A controlled residence time and residence time distribution as well as intensive mixing enable optimum use in reactive extrusion processes.

The screws and barrels of a co-rotating, closely intermeshing twin-screw extruder have a modular design. By combining different barrel modules and screw elements, the extruder can be adapted to a variety of processes (such as the addition of fillers and fibers, degassing, reactive extrusion, etc.). The screw elements are mounted on shafts with gearing to transmit high torques.

Self-cleaning threaded elements convey the materials in the feed zones and the degassing/venting areas. These are called conveying elements. They also build up pressure in front of mixing zones and at the extruder discharge. Mixing elements are available in a wide variety of designs for optimum configuration. The most common are kneading elements; in this case, the degree of the dispersive and/or distributive mixing effect is determined by the width of the individual kneading disk and its offset angle. Depending on how many successive process steps are required for a particular application, the *L/D* ratio of the extruder is adjusted accordingly. An example of a screw configuration is shown in Figure 4.29.

Figure 4.29 Screw set of a co-rotating, closely intermeshing twin-screw extruder

The throughput and the screw speed are independent of each other. If the two parameters are varied, the filling level in the extruder can be changed. This allows the processing and degassing efficiency to be optimized. The pressure gradient in the twin-screw extruder is determined by the selection of the screw elements and the choice of process parameters. Conveying elements with a large pitch are strategically positioned so that there is no pressure in the extruder under the dosing and degassing zones. This ensures that material does not escape from the extruder unintentionally. To overcome pressure zones (mixing zones) and to generate pressure against an extrusion tool (die), conveying elements with a smaller pitch are generally selected. If material is to be added during the extrusion process after melting, a side feed (solids) or injection nozzle (liquids and gases) is often used.

The D_a/D_i ratio (outer/inner diameter of the screw) and the resulting screw flight depth are important design parameters of a twin-screw extruder. As the screw flight depth increases, the cross section of the screw shaft (and the corresponding maximum possible torque transmission) decreases and vice versa. For co-rotating, closely intermeshing twin-screw extruders, D_a/D_i ratios from 1.4 to 1.8 are common on the market. When designing the extruder, it is important to find the optimum compromise between free volume and torque, as these parameters often determine the achievable throughput rates.

4.2.1.8.2 Counter-rotating, Closely Intermeshing Twin-Screw Extruder with Low Screw Speed

The counter-rotating, closely intermeshing twin-screw extruder with low screw speed is characterized by a gentle, material-friendly melting process and a narrow residence time distribution at low screw speeds (less than 50 min^{-1}). At the same time, the system can build up high pressures at the end of the extruder. This combination is particularly useful when processing heat-sensitive PVC or similar formulations. The screws are one-piece and often have liquid cooling in the core. This is a preferred design feature for processing RPVC and similar temperature-sensitive processes. The barrels can be one-piece or modular.

If we look into the feed opening, we can see screws that rotate outwards. The direction of rotation makes it easier to feed material into the extruder. In the overlapping area of the screws, the screw flight of one screw penetrates the channel of the second screw. In this area, the movement directions of the two screws overlap – also referred to as the calender gap – creating a zone of shear, acceleration, and elongation of the material. This results in a highly effective extensional mixing effect.

A large part of the screw length essentially functions as a mixing device, as the materials constantly experience extensional mixing and shearing effects in the calender gap. The very high spreading forces in the calender area prohibit the use of high screw speeds and limit the achievable *L/D* ratios. Toothed mixing elements and blister rings can be used on the screws to improve distributive mixing and longitudinal mixing. Special, multi-start screw elements are placed under degassing openings to increase the surface area of the melt and thus the degassing performance. Elements with high displacement pumping efficiency are used upstream of the discharge nozzle to generate high pressures. An example of a screw configuration of a counter-rotating twin-screw extruder is shown in Figure 4.30.

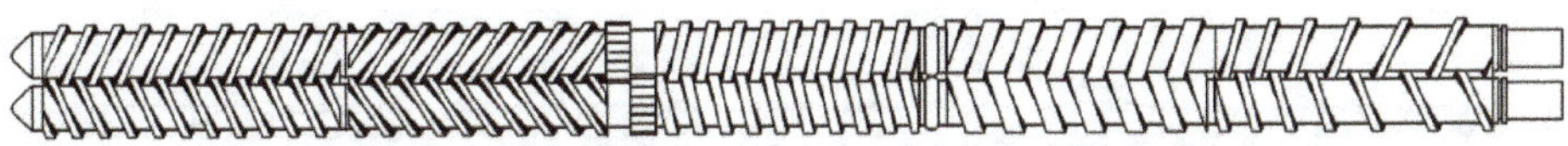

Figure 4.30 Pair of screws of a counter-rotating twin-screw extruder

An important difference with respect to co-rotating twin-screw extruders is that counter-rotating, closely intermeshing twin-screw extruders convey the extrudate in closed C-shaped, non-trailing chambers. This makes this device the only extruder design (apart from ram extruders) that can function as a positive displacement pump. Single-screw and co-rotating twin-screw extruders, on the other hand, convey the melt via drag flow.

There are two types of counter-rotating, closely intermeshing twin-screw extruders:

In *twin-screw extruders with parallel screws,* the diameter remains constant over the entire length. Parallel, counter-rotating twin-screw extruders are normally longer than extruders with a conical design. The screw geometry is based on geometric changes to the screw flights, the number of combs, and changes to the pitch in order to reproduce the desired process.

In contrast, with *conical screws* (Figure 4.31) the outer diameter decreases from the feed to the discharge. The large free volume in the feed zone of a conical twin-screw extruder is helpful when feeding powders with low bulk density. Another unique feature of the conical design is that the radial distance between screws and barrel can be changed to influence the mixing effect and the overall process.

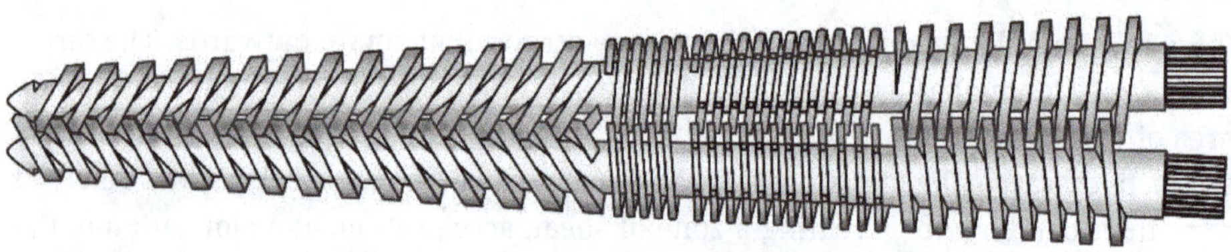

Figure 4.31 Counter-rotating, closely intermeshing, conical twin-screw extruder

4.2.1.8.3 Special/Niche Twin-Screw Extruders

These designs account for less than 5% of the total current twin-screw market and include:

Counter-rotating, non-sealed twin-screw extruders (Figure 4.32): As the name suggests, this design does not have sealed screws. The screws and barrels have a modular design. Most of the screw elements in one geometry are attached to the two shafts in mirror image. The screws can be designed for conveying and reverse conveying, have different pitches, and wide or narrow flight widths. Elements with different numbers of flights and other unique design features are also possible. A wide range of mixing elements is available.

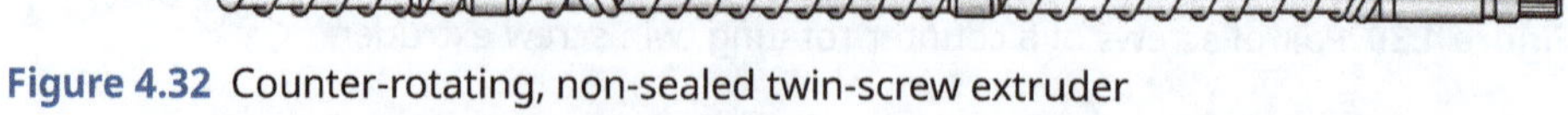

Figure 4.32 Counter-rotating, non-sealed twin-screw extruder

The screws rotate inwards, which can help to draw in materials in the feed opening and hold down the melt in degassing zones. One screw can be extended to simulate the pumping characteristics of a single-screw extruder at the material discharge. As there are no geometric restrictions due to a closely intermeshing screw profile, this extruder type can be designed for special processes that require a long residence time in process lengths of up to 100 *L/D* and more.

Counter-rotating, closely intermeshing, high-speed twin-screw extruders: In this design, the mixing mechanism is shifted from the calender gap to special mixing elements. This step enables speeds of up to 500 min^{-1} with minimal deflection of the screw. Up to six mixing chambers (Hexalobe Mixer; see Figure 4.33) are possible with a D_a/D_i ratio of 1.5. The use of these elements results in more stock partitions per screw revolution than with other designs. Counter-rotating, closely intermeshing, high-speed twin-screw extruders have proven to be particularly advantageous when mixing multiphase, shear-sensitive formulations.

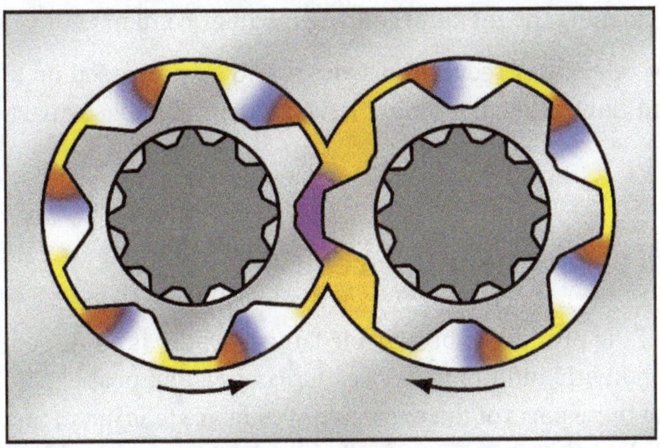

Figure 4.33
Example of a mixing element with six mixing chambers (Hexalobe Mixer)

Additional devices can be integrated into an extrusion system to map complex processes. These include filter systems, gear pumps, start-up diverters, and a wide range of devices that are individually tailored to the desired process.

The downstream equipment of a twin-screw extruder is mainly divided into pelletizing and direct extrusion processes. In compounding, pellets are generally produced as an intermediate product for processing on injection molding machines or single-screw extruders. Various pelletizing systems are available. These include strand cutting, in which the strands are cooled and cut, and hot surface cutting, in which molten strands are cut at the die plate in water or air. Alternatively, finished end products or semi-finished products such as films, sheets, filaments, fibers, profiles, pipes, and special parts can be extruded directly from the twin-screw extruder. Twin-screw extruders can also be combined with other extrusion systems in coextrusion lines.

4.2.2 Additional Equipment for Extruders

The following additional equipment is available for extrusion:

- *Dosing and tamping equipment* for uniform and controlled recipe feeding, even with varying bulk densities or raw-material properties
- *Heated vacuum hoppers* for pre-degassing and drying to increase output and product quality
- *Shearing and mixing parts* to increase the homogeneity of a melt – these are usually integrated into the screw concept of the extruder
- *Bushings or perforated plates* on the screw tip for specific adaptation of the melt pressure to the die resistance
- *Melt pumps* for a uniform and volumetrically adjustable mass flow to the forming die
- *Replaceable melt filters* for cleaning the melt or reducing the filler content
- Melt filters are particularly relevant if recycled plastics are to be processed on a system, as they may contain impurities that would have an unacceptable impact on product quality.

4.2.3 Extrusion Dies and Subsequent Devices

Depending on the final shape of the product to be extruded, the extruder is equipped with a specific die and combined with additional devices for cooling and post-extrusion product transport. Because these parts of the extrusion system are tailored to the specific final product, we will describe them in the following in the context of the final extruded product.

4.2.3.1 Pipes and Symmetrical Hollow Profiles

The melt stream is centrically expanded with the help of a mandrel (torpedo) and formed into a ring shape between die and die housing (Figure 4.34). The mandrel supports split the melt stream and create weld lines along the hollow profile that may weaken the extruded profile. To avoid this effect, lattice mandrel-, lattice basket-, or spiral dies are used. To calibrate the outer surface, the end of the profile is sealed with a drag plug attached to the mandrel by a rope or chain. The profile is provided with support air through the mandrel and pulled through a cooled calibration unit. If vacuum calibration is used, support air is not necessary (Figure 4.35).

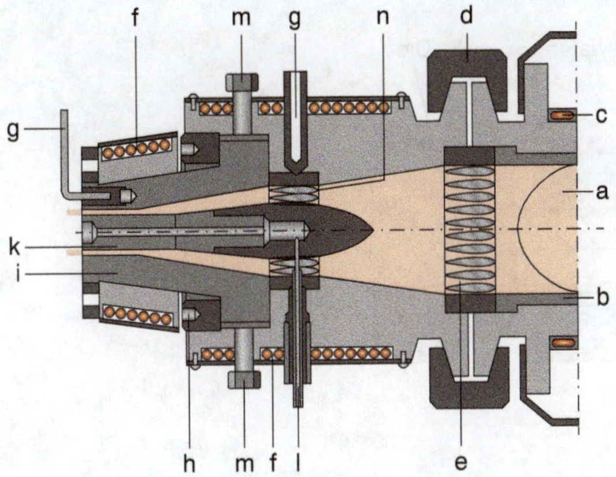

Figure 4.34 Hollow mandrel pipe extruder
a: Screw tip, b: barrel lining, c: barrel heating, d: die mounting with locking bolt, e: breaker plate with flow efficient drop-shaped holes, f: wound die heater, g: temperature measuring points with copper contacts, h: die screw connections, i: die, k: mandrel, parallel motion 30–40 times die gap, l: air supply, m: centering screws

4.2.3.2 Solid Profiles

For extruded products that do not exhibit a circular cross sections, die and extrudate cross sections are no longer similar because of non-uniform die swell. Therefore, it is necessary to empirically introduce corrections to the die. Once the die corrections have been introduced, to avoid jamming of the extrudate, solid profiles can only be calibrated from one side. The calibration gaps should be oversized by 15 to 20%. Furthermore, one may use a calibration surface that is compliant and able to yield in order to prevent jamming of the profile. With the cooling injector method, the outside of the extruded profile is calibrated immediately after exiting the die. To prevent void formation in the interior of the profile, melt is continuously injected into the profile core to compensate for shrinkage. The injection pressure is controlled by the screw speed. The Technoform precision profile drawing process monitors a representative dimension of the profile between die and calibration in order to control take-off. This approach not only prevents jamming of the profile, but also ensures tight manufacturing tolerances.

4.2.3.3 Hollow Chamber Profiles

These profiles are extruded through dies similar to pipes. Multi-wall sheets consist of thin top and bottom layers (and sometimes an intermediate layer) that are connected by webs to allow for shear transfer under bending loads. These sheets can reach widths of several meters, and therefore the extrudate exiting the extruder has to be spread to this width while ensuring that the average melt velocity over the entire profile cross section is uniform, preventing preferential flow in the middle; see also sheets and films. Vacuum is used for calibration (Figure 4.35).

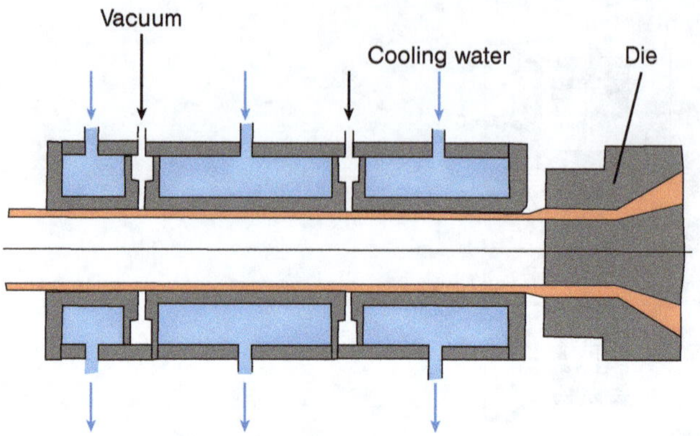

Figure 4.35 Vacuum calibration

4.2.3.4 Sheathing

Sleeve dies are used to sheath tubing with fabric reinforcement, for wire coating, or for coating of steel tubes (Figure 4.36).

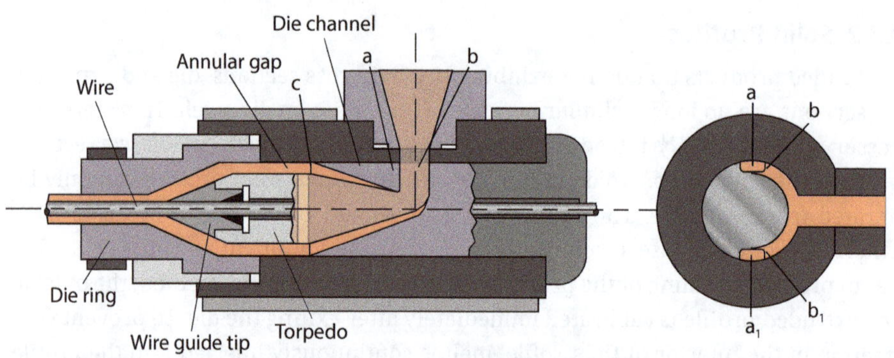

Figure 4.36 Sleeve die for pressure jacketing [Source: BASF AG]
(a, b, and c): flow around the mandrel (from two sides) (a, b and a_1, b_1 in the schematic on the right). Once (c) has been reached, the mandrel is surrounded by a molten layer

4.2.3.5 Sheets and Flat Films

Sheet and flat films with thicknesses ranging from approx. 0.3 to 30 mm and widths up to 4 m are produced with slit dies that are equipped with flexible lips (Figure 4.37). The centrically fed melt is distributed by a manifold so that the average velocity is the same at each point of the lip. This is achieved with so-called "coat-hanger dies" (Figure 4.38). Sheets with thicknesses below 0.5 mm are extruded horizontally and drawn off with a calender. When extruding films, the melt exits sloping downwards and is placed tangentially on a chill roll. Here, the films can be stretched to thicknesses of 8 to 15 µm in extrusion direction. Special stretching units allow for cold or hot stretching, uni- or biaxial stretching, and extremely high degrees of stretching. Using a two-stage stretching technology (Figure 4.39), the film is first longitudinally stretched between two rolls with different circumferential velocities. Transversal stretching is performed in a second step, in which the edges of the film are gripped by grips that are guided outward during the longitudinal transport of the film. During simultaneous biaxial stretching (Figure 4.40), the clips are guided outward and simultaneously in travel direction (with increasing velocity). Stretching and thermal conditioning increase stability, transparency, permeation resistance, low temperature resistance, and electrical properties.

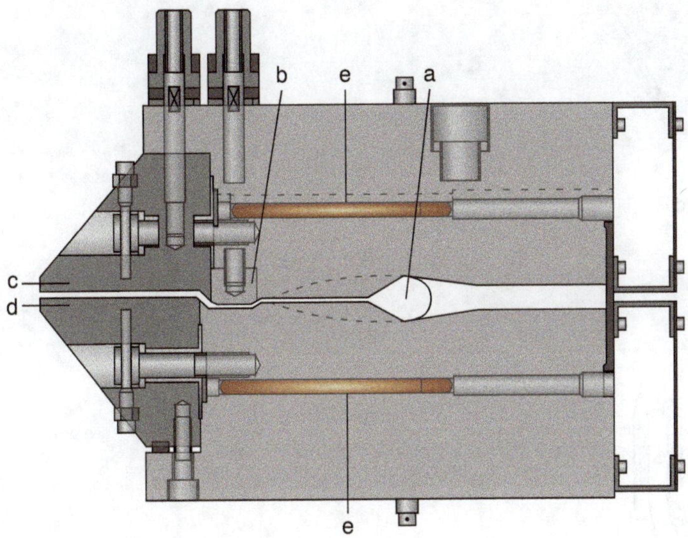

Figure 4.37 Slit die for sheet extrusion

a) Manifold; (b) choker bar; (c and d) lips; (e) heaters

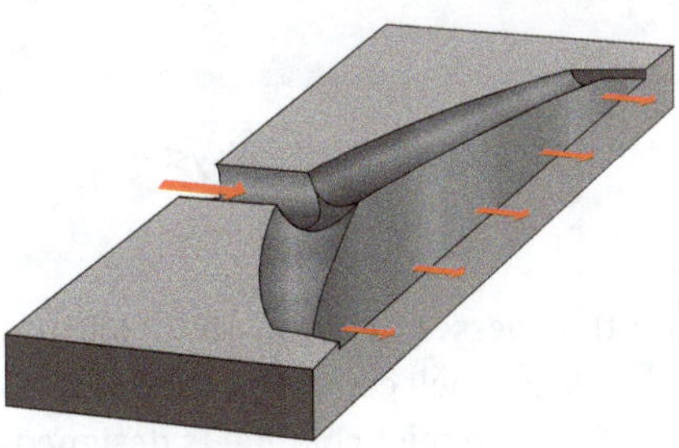

Figure 4.38 Coat hanger die

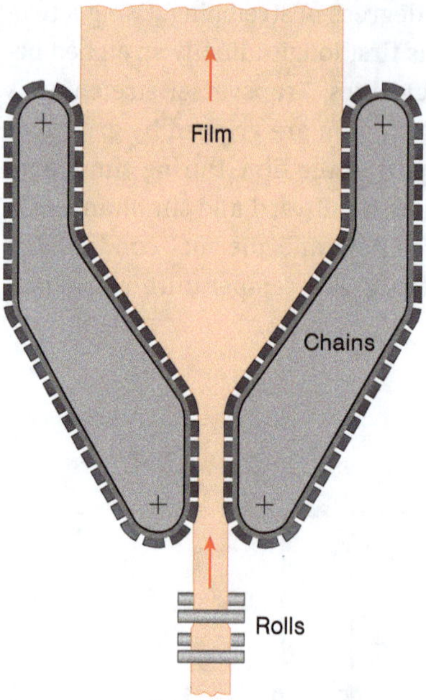

Figure 4.39
Two-stage biaxial stretching of flat films

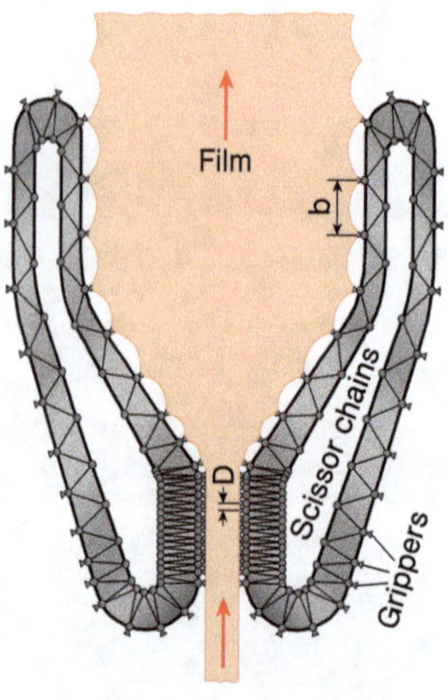

Figure 4.40
Simultaneous stretching with scissor chain systems

4.2.3.6 Blown (Tubular) Film

Blown films with circumferences up to 16 m and thicknesses between 10 to 300 µm are manufactured using annular dies. In mandrel dies, the melt completely surrounds the mandrel, which is fastened within a housing. Here, the inlet channel is designed so that a uniform flow in axial direction is achieved (Figure 4.41). However, the unavoidable weld line at the backside of the mandrel will be visible in the extruded product. This effect is reduced, but not eliminated, by using spider leg mandrels. Using downstream breaker plates and spiral mandrel distributors reduces the occurrence of weld lines (Figure 4.42). After the melt exits the annular die, the film is blown to the desired tube (or bubble) diameter with the help of support air, then cooled, and finally laid flat and wound up.

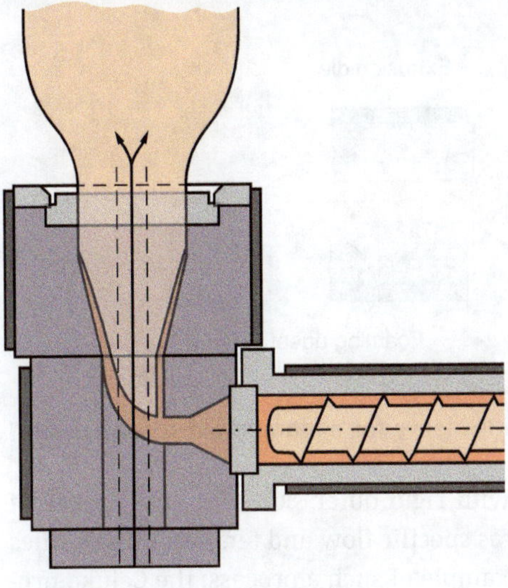

Figure 4.41
Mandrel die

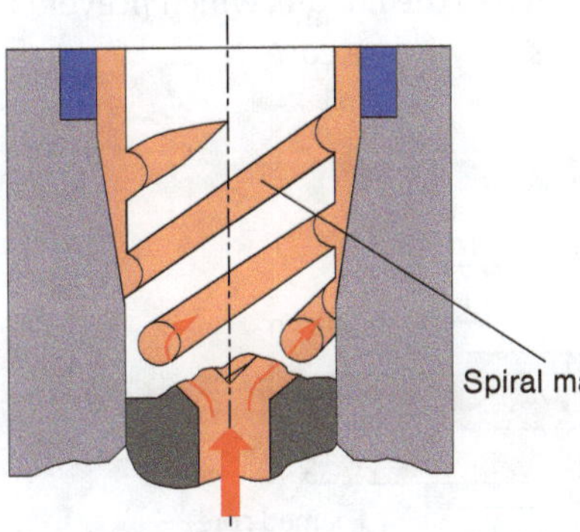

Figure 4.42
Spiral die

4.2.3.7 Foamed Semi-Finished Products and Profiles

The general principles of thermoplastic foam extrusion are detailed in Section 4.5.1. Physical foaming uses foaming agents in the high-pressure section of the extruder. The foaming agent, typically CO_2, is injected into the extruder under high pressure and molecularly dissolved in the melt – *e. g.*, in PS. When the melt exits the die and the pressure drops, the blowing agent falls out of solution and creates the foam structure. The pressure profile, but also the temperature profile and the level of foaming agent loading, determine the foam creation. These parameters essentially control the degree of foaming and the foam structure (open or closed cell).

Foamed semi-finished products and profiles are manufactured on conventional extruders. Figure 4.43 shows the foaming of the extrudate after exiting the die. The subsequent calibration or sizing determines the final dimensions and thus the degree of foaming.

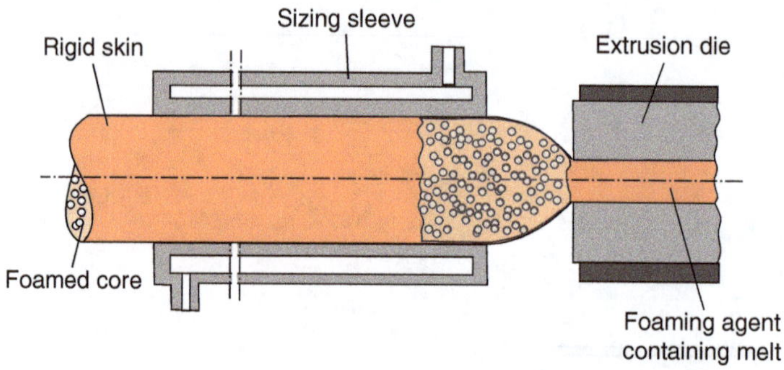

Figure 4.43 Schematic diagram of a calibration sleeve for a foamed extrusion profile

Generally, a sandwich-like foam structure with rigid outer skin and high degree of foaming in the core is preferred. This requires specific flow and temperature profiles in the foaming zone. Figure 4.44 shows an example of such a process: the Celuka process. Here, the cold calibration sleeve is directly flanged to the hot die, which prevents foam formation at the surface so that foaming occurs only in the core.

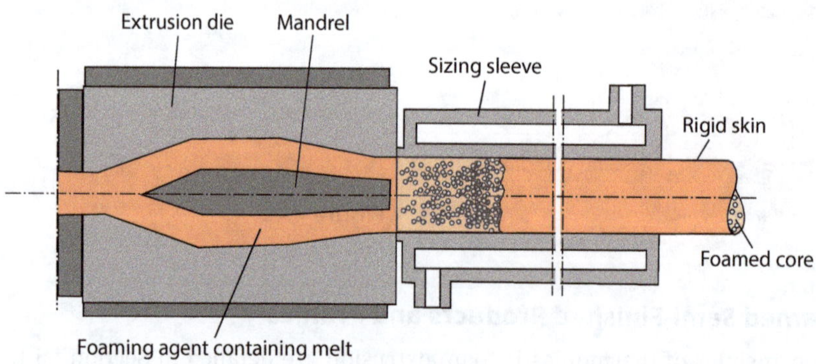

Figure 4.44 Core foaming with Celuka-type die

For the manufacture of foamed sheet, the sheet can be stabilized and pre-sized by guiding it through a pre-cooling unit after it exits the die. The final sizing occurs in a subsequent 3-roll calender.

When processing PVC-U (rigid PVC without plasticizer), force-fed twin-screw extruders are often used instead of single-screw extruders.

When thermoplastic vulcanized (TPV) profiles are manufactured on conventional single-screw extruders, a water releasing compound (WRC) is used to facilitate the manufacture of foams with densities as low as 200 kg/m^3.

4.2.3.8 Monofilaments, Tapes, Fibers

The extruders used to manufacture monofilaments, tapes, and fibers are typically equipped with melt pumps to allow for the creation of continuous melt flows. *Monofilaments* are extruded downwards through die plates with multiple drill holes and subsequently cooled in a water bath before being pre-stretched between rollers driven at various speeds. Additional stretch rollers will subsequently draw the filaments to stretch ratios of 1 : 5 up to 1 : 10 before the filaments are annealed in hot air.

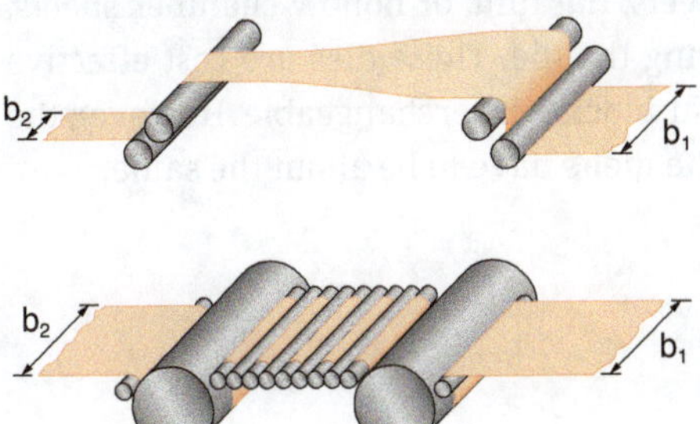

Figure 4.45
Principle of uniaxial stretching (top) and of transversely restricted stretching (bottom); b_1: initial width, b_2: final width

Tapes with widths of several mm and thickness between 20 and 200 µm are manufactured by either cutting stretched film or by cutting film and then stretching it. During transversely restricted stretching of films (Figure 4.45), lateral shrinkage is constrained which reduces the tendency of stretched tapes to splice out (applications: sacks, packing tapes). High splicing tendency is desirable for the manufacture of ropes and textile applications. Such fibrillated fibers are manufactured by one of three methods: the *Barfilex*-technology, where films are extruded with grooved thin sections in longitudinal direction. When stretched, the films will splice in these thin locations; the *roll-embossing technology*, where smooth films are contoured longitudinally with embossing rollers. If a meshed structure is embossed, stretching will result in a net after the thin sections rip. The third method uses knife- or needle rollers to cut the film after stretching. This fibrillation results in products that resemble natural fibers.

Fibers are extruded downwards through a spinneret, cooled in air, and then post-processed like monofilaments.

4.2.3.9 Co- and Multi-Layer Extruded Products

Co- and multi-layer extrusion are technologies to manufacture products that consist of several layers of similar or dissimilar plastic materials. These technologies are used to produce profiles, sheets, hollow chamber sheets, flat film, blown film, and extrusion blown film. Five to seven layers can be produced economically to obtain desired product characteristics by implementing gas or water vapor barrier layers, UV-absorbing layers, or by adding very thin adhesive layers. Layered product designs are also used to increase the economic viability of products by adding internal layers made of secondary or recycled material. Cost-effective, stiff water and sewer pipes with diameters between 32 and 630 mm are produced from PVC with stiff outer layers and foamed cores. Here, one or two twin-screw extruders are used for both outer layers. Coextrusion systems merge the melt streams from several extruders in one die. Figure 4.46 shows a blowing die for 2-layer films in which the melt streams exit the die separately and are then brought together by support air. Adding gas between the layers may increase adhesion. In multiple-layer dies, the melt streams are merged immediately after they exit the die (Figure 4.47). Like in the blowing die shown in Figure 4.46, the melt streams can be individually temperature controlled. In adaptor dies, which are typically used to produce sheets, flat film, or hollow chamber sheets, the melt streams are merged prior to entering the die. These dies are cost effective and compare to single-layer dies. The modular block is interchangeable. However, the processing temperatures and viscosities of the melts have to be about the same.

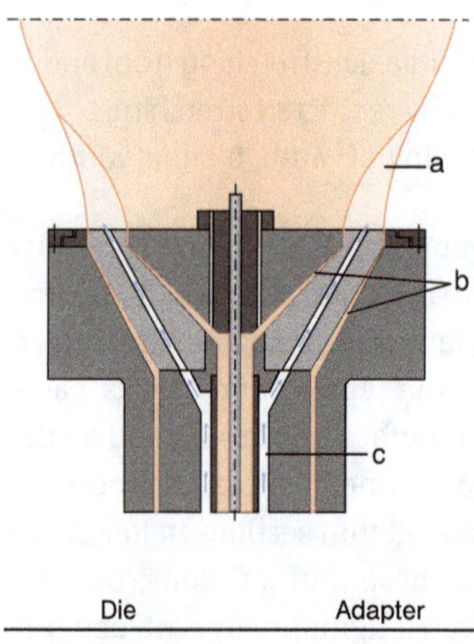

Figure 4.46
Blowing die for 2-layer film with melt separation and support air/gas
a: Air/gas support chamber,
b: Melt streams,
c: Air/gas supply

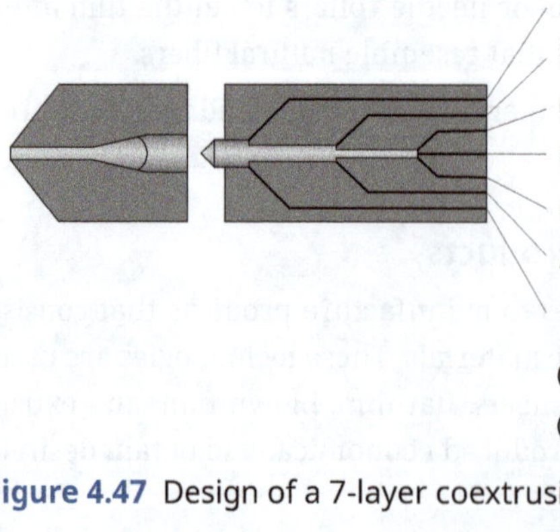

Layer (1)	Layer (2)	Layer (3)	Application examples (1)	Application examples (2)	Application examples (3)
A	A	A	PP	PS-HI	PP
B	B	B	HV	SCRAP	HV
C	C	C	EVOH	HV	PA
B	D	D	HV	EVOH	EVOH
D	C	C	SCRAP	HV	PA
A	B	B	PP	SCRAP	HV
-	A	E	-	PS-HI	PE-LD

(1), (3) Asymmetric layer structure
(2) Symmetric layer structure

Figure 4.47 Design of a 7-layer coextrusion adaptor

Figure 4.48 shows a coextrusion die where the breaker bar has been replaced by a flexible membrane to regulate the melt stream.

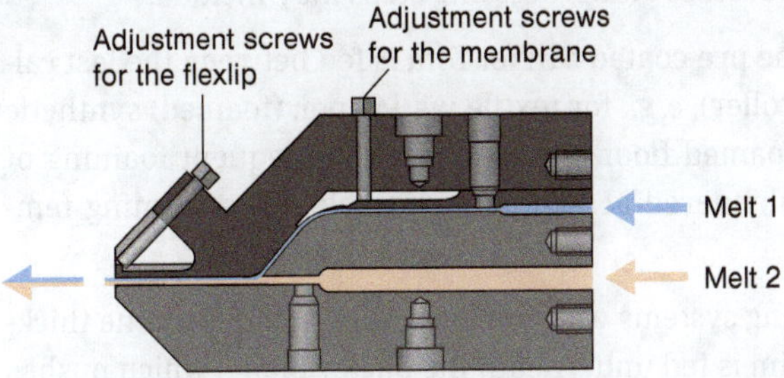

Figure 4.48 Coextrusion die without breaker bar

Greiner Co., Austria, offers a post-coextrusion process (PCE) in which a flexible lip seal can be added to the calibrated/sized profile with a coextruder in one processing step. Good adhesion is achieved by heating the joining surface. For a detachable connection, the seal is coextruded into a groove in the main profile.

4.2.3.10 Multi-Layer Films

4.2.3.10.1 Composite Films

Composite films are produced by depositing PE melt layers extruded from slot dies. Embossing calenders are used to laminate two similar film layers, forming a nonporous composite. In some systems pre-heated thicker films are laminated, *e.g.*, for multi-layer flooring. Here, a continuous circumferential pressure band made from steel pushes the layers against a driven cylinder (Figure 4.49).

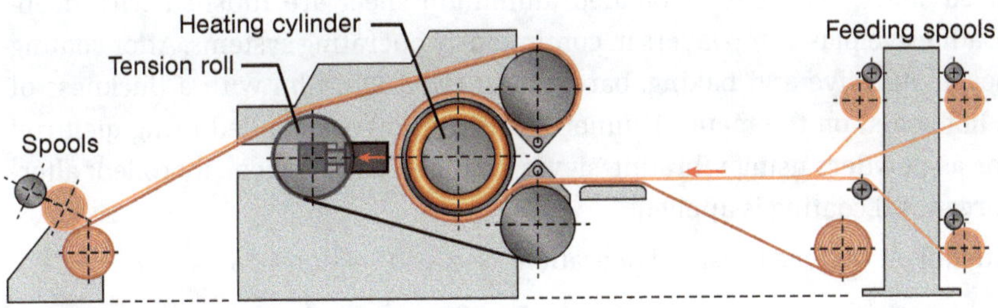

Figure 4.49 Laminating of multi-layer films

4.2.3.10.2 Laminates and Coatings

Flexible laminates made from thermoplastic and carrier films from paper or textile/non-woven fabrics (also based on mineral or glass fibers) are used in a variety of applications. They include packaging materials, tarps, construction and roofing sealing material, textile construction and flooring materials, textile wall paper, synthetic leather, and shoes. Often, successive layers of bonding agent, one or more (possibly foamable) layer, and a cover layer are continuously applied. Multi-color printing and embossing units are also often implemented. With additional units, such as drying, gelling, and cooling sections, these systems are highly sophisticated and specialized large-scale systems. Individual processes and processing steps may include:

- Laminating by calendering: the pre-coated carrier film is fed between the last calender rolls (or an additional roller), *e. g.*, for textile wall paper, (foamed) synthetic leather, or as base layer for foamed floor coverings with subsequent foaming of the core layer (which was calendered at temperatures below the foaming temperature).
- Coating on conventional coating systems with doctor blades that determine thickness. The supported carrier film is fed underneath the doctor blade, which pushes the viscous coating substrate.
- Reverse-roll coater (Figure 4.50) for higher coating speeds and thicknesses.

The two latter processes require liquid coating substrates (solutions, dispersions, pastes, PUR-precursors) and subsequent drying and gelling channels.

- Melt roll coating systems (Figure 4.51): The coating material (in pellet or powder form) is pre-plasticized in the extruder or by an additional roller and fed between the melt rollers.
- Hot melt processes without subsequent gelling or drying channels are economical and environmentally friendly. They include extrusion lamination using slot dies (Figure 4.52).
- Laminating systems (Figure 4.53): Films/sheets made of different materials are coated with solvent, dispersion, or melt adhesives and joined to form a laminate.

Phosphatized sheet steel and chromated aluminum sheet are finished with deep-drawable, adhesive plastic top layers in continuously operating systems. After coating with a special adhesive and baking, barely plasticized PVC film with a thickness of 0.2 mm is hot rolled on the metal. Thinner layers (50 µm) are applied using melt roll systems, or as powders using vibrating sieves and then homogeneously rolled; alternatively, organosol coating is applied.

For more detail on surface finishing operations, see also Section 4.15.

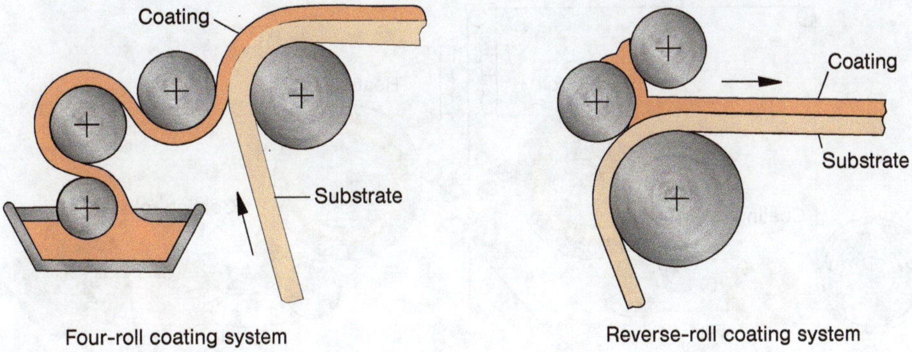

Figure 4.50 Schematic of a 4-roll coating system with forward-coating rolls (left) and a reverse-coating rolls system (right)

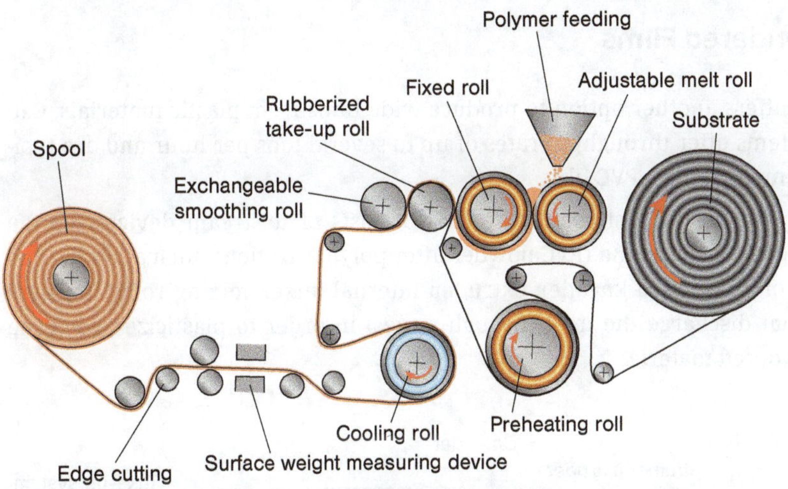

Figure 4.51 Schematic of melt roll coating system

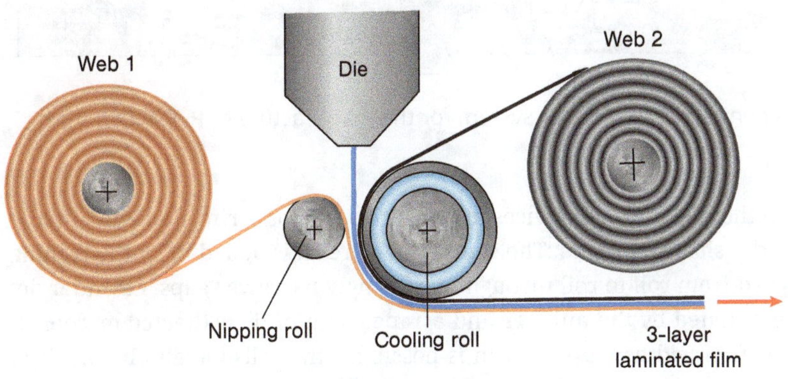

Figure 4.52 Hot melt 3-layer laminating system

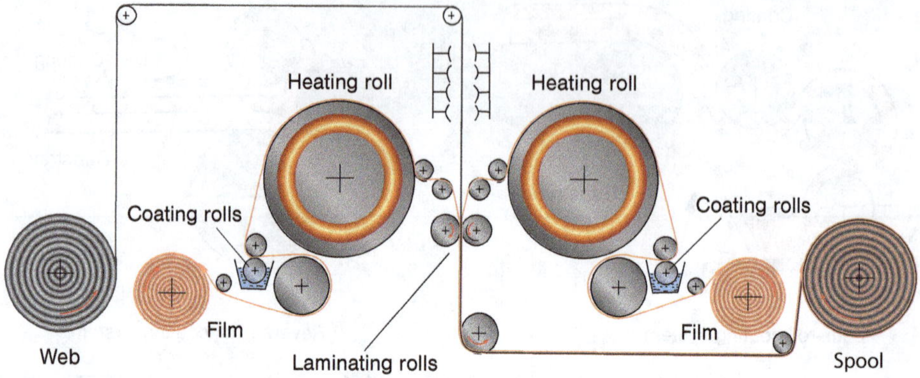

Figure 4.53 Three-layer laminating system

4.2.4 Calendered Films

Calendering offers another option to produce wide films from plastic materials. Calendering systems offer throughput rates of up to several tons per hour and are typically used to manufacture PVC film.

Such a PVC calendering system (Figure 4.54) consists of upstream devices for the compounding of the dry blend (PVC powder after polymerization), such as a mixer to process the formulation, a kneader, extruder, internal mixer, mixing rolls, strainers (extruders that discharge the melt through sieves) in order to plasticize or gel the typically powdered material.

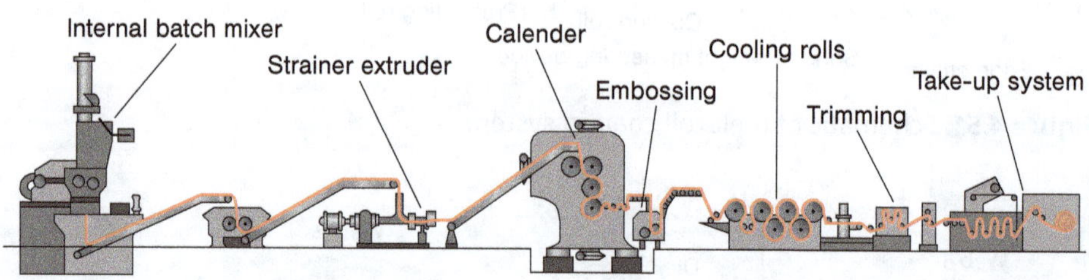

Figure 4.54 Schematic of a calender system for the manufacture of PVC-P film (flexible PVC)

Downstream of the calender are devices to cool, draw, emboss, trim, and wind up the film. The calender shapes the film. The compounded resin is pulled in by the feeding rolls and dragged from roll to roll through increasingly narrower gaps. Calender designs are differentiated by the number and arrangement of their (heated or cooled) rolls. For larger film widths – up to 2.8 m is possible – the rolls have to be profiled, bent, or their axes are misaligned by a slight angle in order to ensure uniform film thickness.

Table 4.6 shows the correlation between calender design and its respective area of application. Table 4.7 details some special calender designs and their respective applications.

Table 4.6 Calender: Standard Designs [Source: Krauss-Maffei, Munich]

	Design	General applications
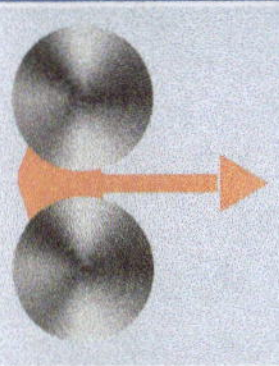	2-roll calender, -type	Floor coverings, vinyl tiles, texturing of plates and plates for shoe soles
	3-roll calender, I-type	Floor coverings, rubber sheets, frictioning
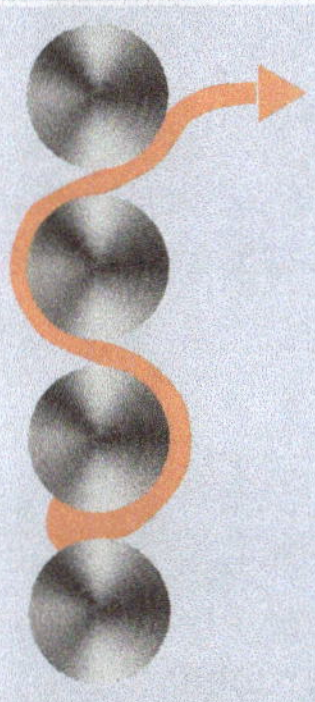	4-roll calender, I-type	PVC-U films, rubber sheets, single- and double-sided coating of fabrics, frictioning
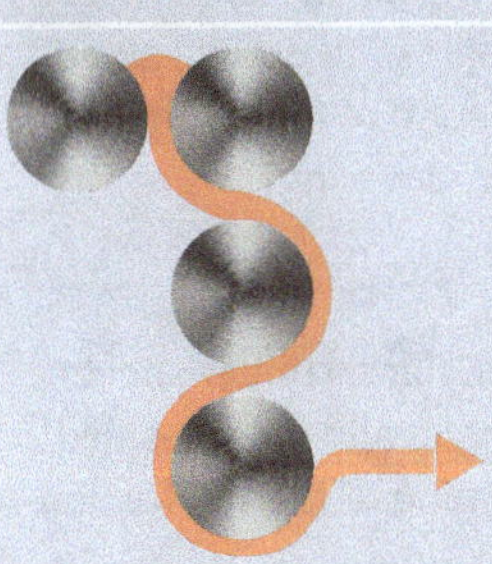	4-roll calender, F-type	PVC-P films, rubber sheets, single- and double-sided coating of fabrics, frictioning, doubling

Table 4.6 Calender: Standard Designs [Source: Krauss-Maffei, Munich] *(continued)*

	Design	General applications
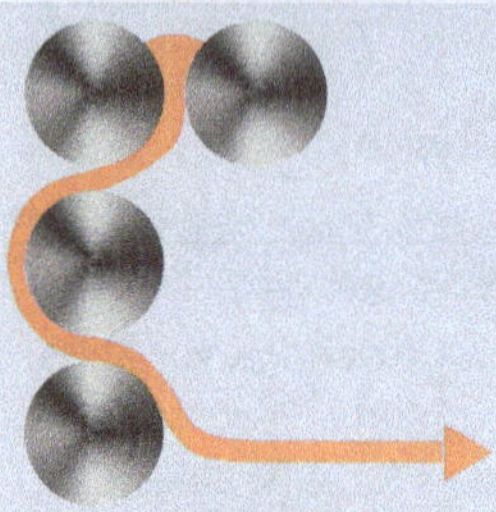	4-roll calender, F-type	PVC-P films, rubber sheets, single- and double-sided coating of fabrics, frictioning, doubling
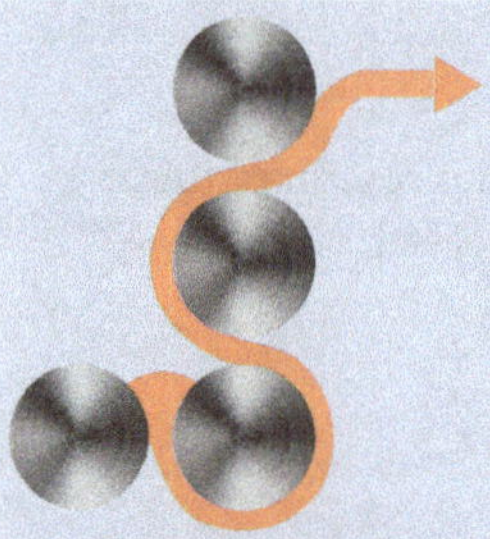	4-roll calender, L-type	PVC-U films
	5-roll calender, L-type	PVC-U films
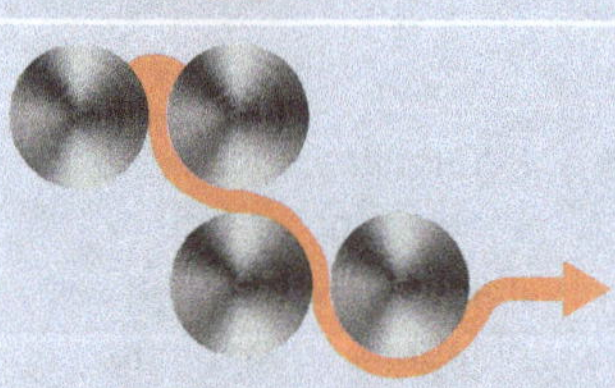	4-roll calender, Z-type	PVC-P films, rubber sheet, single- and double-sided coating of fabrics, frictioning, doubling
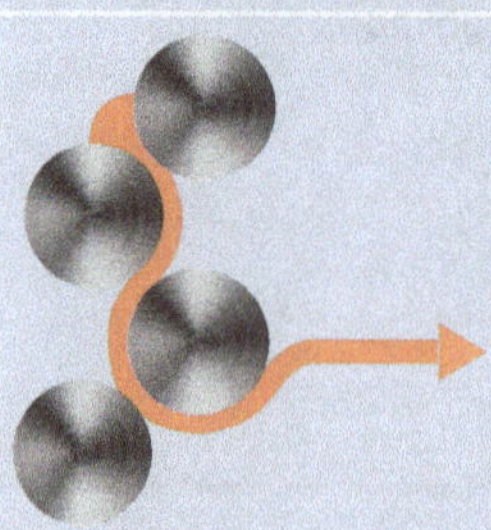	4-roll calender, S-type	PVC films, rubber sheets, single- and double-sided coating of fabrics, frictioning, doubling

Table 4.7 Calender: Special Designs [Source: Krauss-Maffei, Munich]

	Design	General applications
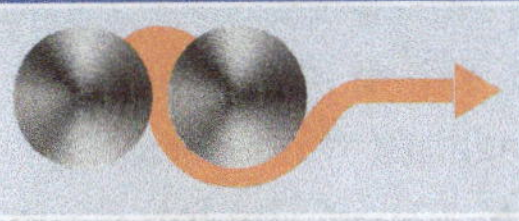	2-roll calender, horizontal type	Rubber sheets, floor coverings
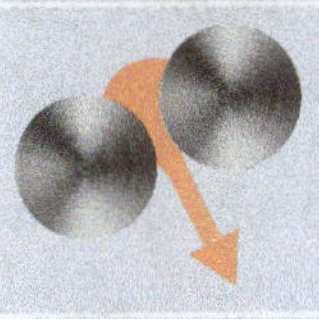	2-roll calender, sloped type	Rubber sheets, linoleum
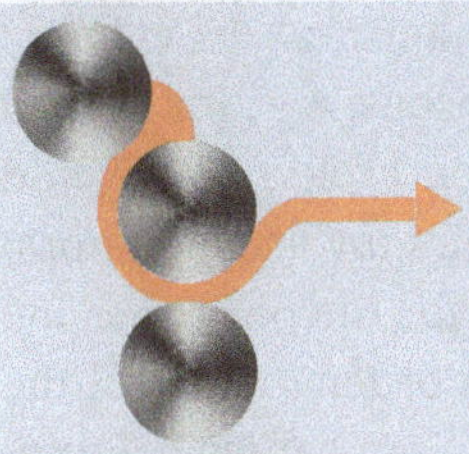	3-roll calender, sloped type	Rubber sheet, single-sided coating of fabrics, frictioning
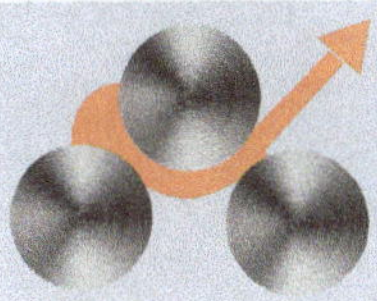	3-roll calender, A-type	Pre-films for PVC-U pressed sheet
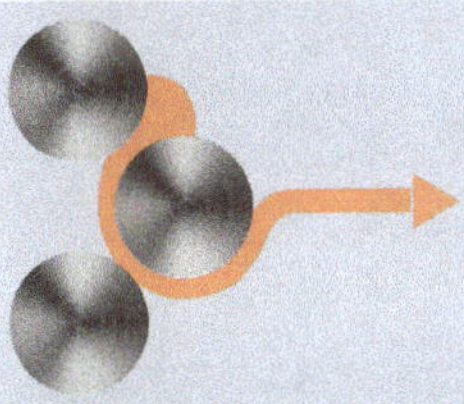	3-roll calender, triangular type	Rubber sheet, single-sided coating of fabrics, frictioning
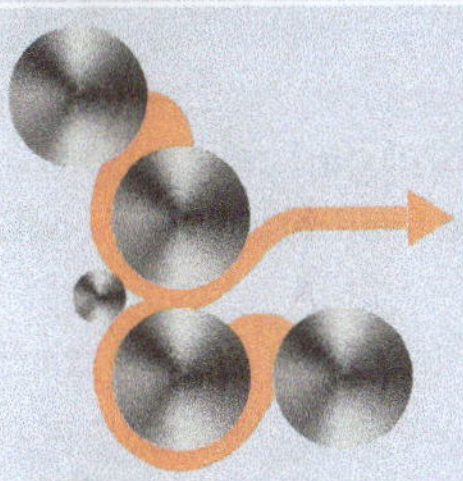	4-roll calender, Z-type with raised front roll	Rubber sheets, single- and double-sided coating of fabrics, frictioning, doubling
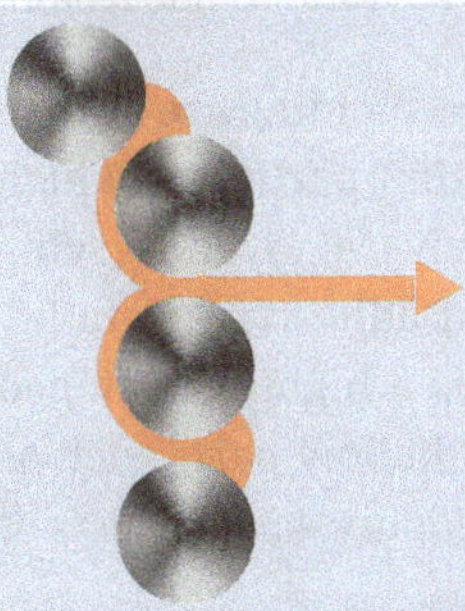	4-roll calender, I-type with sloped top roll	Rubber sheets, single- and double-sided coating of fabrics, frictioning, doubling

4.2.5 Extrusion Blow Molding of Hollow Parts with Variable Cross Sections

An extruder is used to manufacture tubular parisons, which are then placed between the two movable halves of a blow molding tool and pressed against the interior surface of the mold by forced air with pressures ranging from 5 to 10 bar. The forced air is introduced through the blowing mandrel which also sizes the opening of the hollow part (Figure 4.55). In order to provide biaxial orientation, the parisons are stretched by stretching mandrels or stretch-grips, or they are first pre-inflated and subsequently inflated at optimum stretching temperature in a larger mold to reach their final shape. Parisons with weights up to 1 kg (for hollow parts with approx. 30 l volume) can be continuously extruded. For larger, freely suspended parisons, *e. g.*, for fuel oil tanks with volumes up to 10,000 l, the relatively long extrusion cycle times would cause the parison to elongate and cool. In these cases, reservoirs (or accumulators) are continuously filled by one or more extruders and the melt for the parison is hydraulically extruded in a fraction of the plasticization time. In order to control the wall thickness of the finished part, the wall thickness of the parison can be continuously adjusted during extrusion.

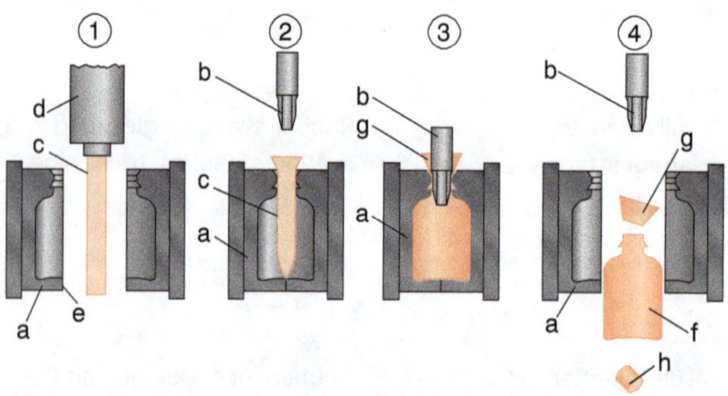

Figure 4.55 Extrusion blow molding processing steps (with mandrel gap calibration)
a: Blow mold (half), b: sizing blowing mandrel, c: parison (tubular), d: parison head, e: pinch-off edge, f: blown part, g: neck flash, h: bottom flash
1: Parison extrusion, 2: closing of blow mold; the lower end of the parison is closed and sealed by the pinch-off edge of the mold, 3: pushing in of sizing mandrel; sizing of bottle neck and blowing up of parison to form hollow part; cooling, 4: demolding of hollow part, trimming of neck and bottom flash

In many cases, the hollow parts manufactured via extrusion blow molding are used to store fuels or solvents, and therefore these containers must exhibit particularly high permeation resistance. This can be achieved by co-extruding several layers serving different purposes: load-bearing layers, diffusion barriers, and adhesive layers between bearing and barrier layers. Coextrusion blow molding is a highly developed technology that allows the large-scale manufacture of complex hollow parts such as automotive fuel tanks. The 3-layer accumulator head shown in Figure 4.56 consists of three concentric tubular ram accumulators that are fed by a separate extruder each. The layers are formed when the melt is discharged.

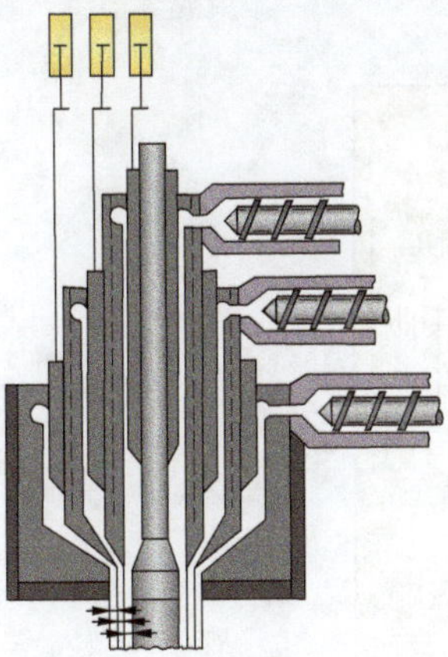

Figure 4.56
Coextrusion accumulator head with concentric tubular ram accumulators

To increase stiffness and/or thermal insulation of blow molded parts made from PE, a physical or chemical foaming agent is added to the middle of three layers, and this layer is foamed. Recycled material is often used for the core layer.

In order to reduce the length of the pinch-off welds in multi-dimensionally curved parts (*e. g.*, pipes for air ducts), the tubular parisons are picked up by special grabbers (robots) during or after extrusion and then placed according to the part geometry in the open blow mold that is tilted to take advantage of gravitational forces.

"Sequential coextrusion" combines different plastic materials in the longitudinal direction during parison production, allowing the manufacture of parts with more or less flexible sections.

In-mold labeling (IML) facilitates the decoration of blow molded parts. Printed films are positioned in the mold where the residual heat of the parison seals them in during the blowing step.

The Selar® RB-process uses blends of PE-HD and PA 6 with a bonding agent to blow mold automotive fuel tanks. During plasticization, the PA 6 pellets are fused and distorted by shearing forces in the extruder to form thin platelets. These platelets are aligned parallel to the surface of the fuel tank and thus increase the permeation resistance.

4.3 Injection Molding

Injection molding is the most important process used to manufacture products from thermoplastic materials. Finished parts with weights ranging from 1 mg to more than 10 kg can be manufactured within cycle times from mere seconds to several minutes and with a minimum of post-processing effort.

4.3.1 The Injection Molding Process

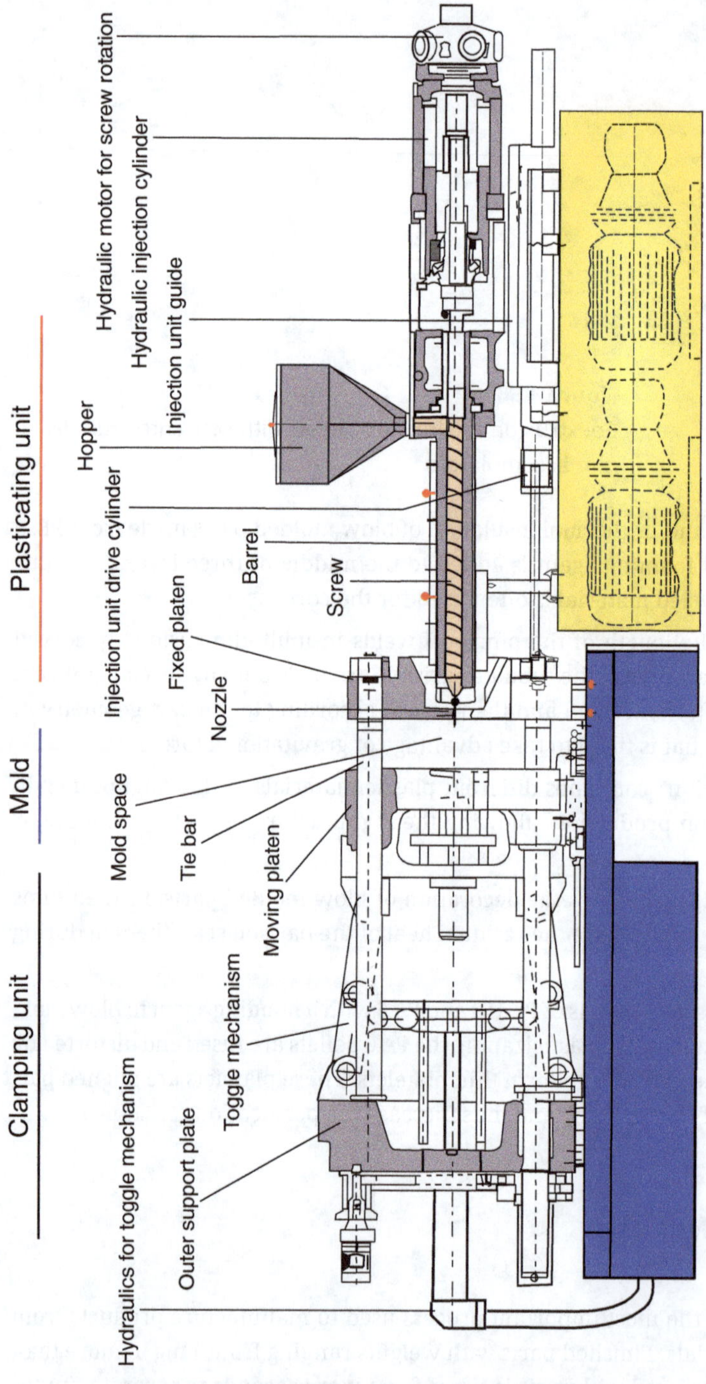

Figure 4.57 Schematic of an injection molding machine

Figure 4.57 shows the general design of a hydraulic injection molding machine. The essential parts of the system are the plasticizing unit, the clamping unit, and the mold. The mold contains the shaping cavity, which defines important technical parameters such as part volume, required melt pressure to fill the cavity, and required locking pressure for the clamping unit.

These parameters are used for an internationally valid classification system for injection molding machines:

MANUFACTURER *T/P*

where *T* is the clamping force in metric tons and *P* is defined as the product of the maximum shot size in cm^3 and the maximum injection pressure in bar.

The sequence of events during the injection molding of a plastic part, as shown in Figure 4.58, is called the injection molding cycle. The productivity of an injection molding process is determined by its cycle time. For parts with wall thicknesses above 1.5 mm the cycle time is determined by the cooling time, which increases with the square of the wall thickness.

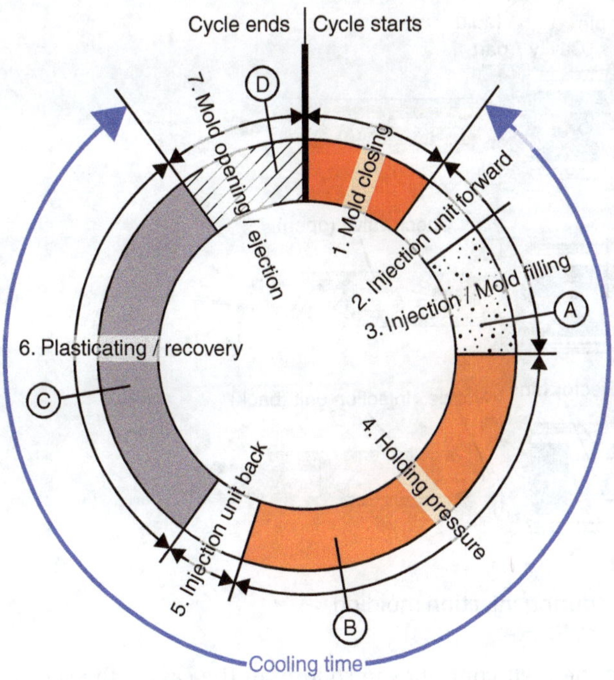

Figure 4.58 Injection molding cycle: the cooling time typically determines the cycle

The fundamental steps during injection molding operations are shown in Figure 4.59:

- **Injection:** The forward movement of the screw in the barrel of the plasticizing unit pushes the melt through the nozzle and sprue into the mold cavity. The built up pressure typically reaches several 100 bar. A non-return valve ensures that melt does not flow back into the screw during injection.
- **Holding pressure:** The melt solidifies in the cooler mold and contracts. To minimize voids and warpage, holding pressure (typically 600 to 1500 bar) is applied until the melt is sealed in the sprue.

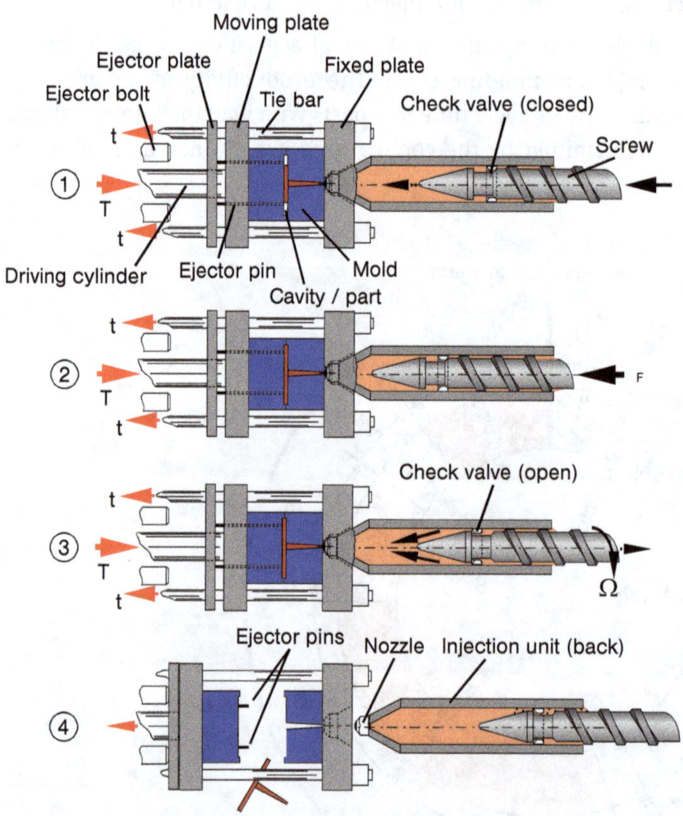

Figure 4.59 Basic steps during injection molding

- **Plasticizing:** While the melt continues to solidify in the cavity, the screw plasticizes new melt for the next cycle. The free-flowing material (in pellet or sometimes in powder form) is transferred from a hopper to the heated barrel with the help of a rotating screw. Here, the material is conveyed to the screw tip while being melted by friction and heat transfer. As the melt is conveyed to the screw tip, it opens the no-return valve. A reservoir of molten plastic is formed in front of the screw tip, which pushes the screw backward. Once there is enough molten plastic to fill the cavity and form the component, screw rotation stops and the plasticizing process is finalized. The parameters determining the plasticizing step are back pressure, screw speed, and maximum screw stroke.

- **Mold open and part ejection:** Ejection devices are used to ensure that the molded parts are demolded without damage. These devices are typically metal pins that eject the molded part from the movable mold half. The cavity has to be designed appropriately to make sure that the molded part is placed on the ejection side of the mold.

The mold is mounted in the clamping unit and electromechanically or hydraulically closed, counteracting the force created by the cavity pressure. In conventional designs these forces are taken up by four tie bars (for details see Figure 4.57).

To prevent return flow of melt into the screw channel during injection and holding, no-return valves are added to the tip of the screw (Figure 4.60). Because these valves cause a pressure loss during the metering phase, they should be designed to enhance flow and exhibit a free flow cross section of at least 80% of the free ring gap at the end of the screw.

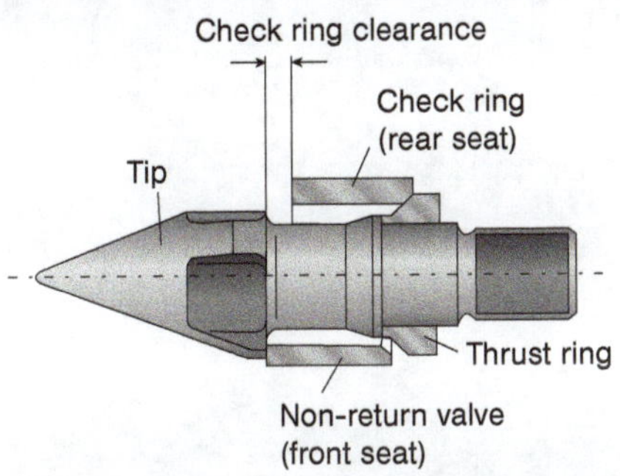

Figure 4.60
Non-return valve for thermoplastic processing

The nozzle is part of the plasticizing unit and represents a tight fit connection between barrel head and mold sprue bushing.

It is always preferable to use open nozzles because they allow for shorter melt residence times (Figure 4.61). Shut-off nozzles are used for cold runner sprues and plastic materials that tend to drip and string. Figure 4.62 shows two shut-off nozzles that open automatically when the nozzle touches the sprue bushing. Other nozzles use a needle valve that may be actuated by a spring (see Figure 4.63), or externally by mechanical or hydraulic controls.

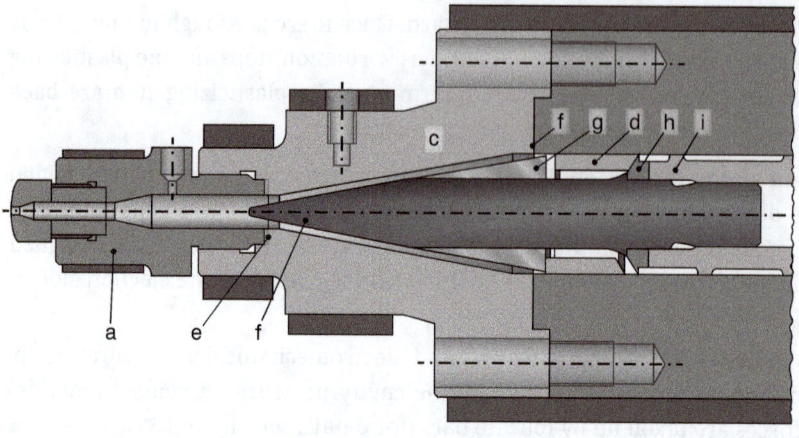

Figure 4.61 Design example of screw tip with non-return valve and nozzle

a: Nozzle, b: screw tip, c: barrel head, d: non-return valve, e: seal face nozzle, f: barrel head seal face ($p = 400\ N/mm^2$), g: four plated supports, h: pressure ring, i: screw

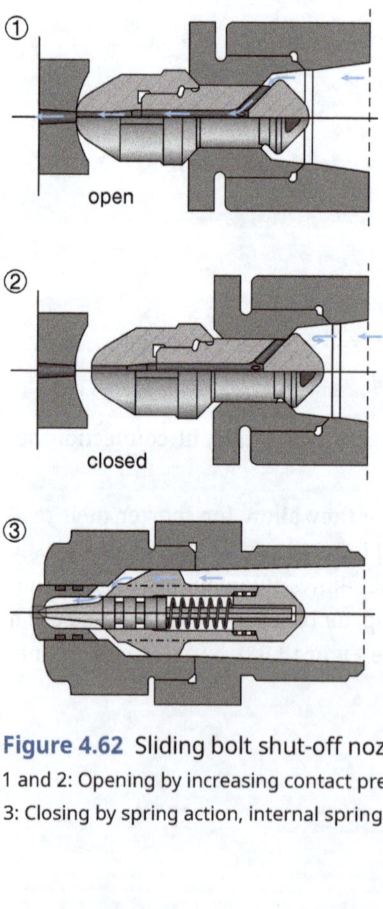

Figure 4.62 Sliding bolt shut-off nozzle

1 and 2: Opening by increasing contact pressure of nozzle

3: Closing by spring action, internal spring, opening actuated by injection pressure

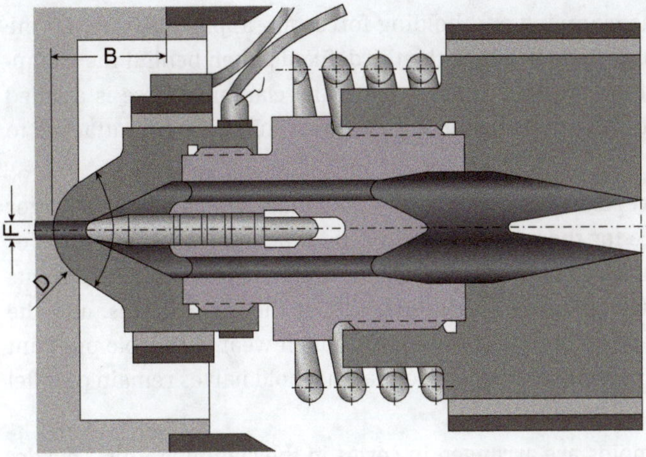

Figure 4.63 Needle valve with spring

B: nozzle retraction depth, D: nozzle radius, F: nozzle bore diameter

When a non-return valve is in place, the screw can be retracted from the mold as soon as the end of the hold phase is reached in order to interrupt the heat flow from the hot nozzle to the cold mold. When open nozzles are combined with cold runner systems, the material has to be plasticized first before the screw can be retracted from the cold mold. Prior to the separation of nozzle and mold the screw has to be slightly retracted (relief) in order to relieve melt pressure and prevent drooling. Hot runner systems are generally run with open nozzles.

Modern injection molding machines are equipped with multiple sensors. They monitor and control, *e.g.*, the temperature of the hopper, screw barrel, and mold, hydraulic pressure at the screw tip and in the mold, the screw stroke during injection, and cycle, injection, and hold times. High-quality parts require on-line quality control measures. If parameters exceed admissible tolerances, defective parts or inadequate quality may result. In this case, production must be interrupted to determine the cause for the deviating injection parameters.

The fundamental principle of an injection molding machine can be modified in many ways, both in terms of machine design and processing technology.

4.3.2 Design of Injection Molding Machines

4.3.2.1 Clamping Unit

The most common machine design is the *three-platen machine* (Figure 4.57): the movable platen moves between the fixed stationary and mounting platen. The clamping forces are applied either hydraulically or via a toggle lever. To avoid shifting of the mold halves in the clamping unit under the high applied forces, the clamping unit is mounted between four horizontal tie bars (or tie rods).

In *two-platen machines* the clamping and holding forces are applied directly by the mold mounting platen. There is no need for the third, fixed platen behind the clamping unit, which makes these machines shorter. Here, the clamping force is applied between the tie bars and the mounting platen, *e.g.*, via hydraulic cylinders attached to the four tie bars.

More recent machine concepts use tie bar-less clamping units, which allow for larger and easier accessible space for the mold. These injection molding machines transfer the clamping forces via a C-frame. In rigid machine designs the bending moments typically occurring here would cause a misalignment of the mold halves, and the resulting high bending loads on the mold would cause high wear. A flexible pin joint can mitigate this negative bending movement so that the mold halves remain parallel despite the bending of the C-frame.

In *tandem machines* two molds are arranged in series in the clamping unit. This design allows for cost savings of up to 25% for the injection molding process.

4.3.2.2 All-Electric Injection Molding Machines

In addition to hydraulically driven machines, there is also an established market segment for electrically driven machines. Here, the functions are either partially (hybrid machines) or completely controlled by electrical drives (all-electric injection molding machines). Currently, they come at a higher machine price (10 to 15%); however, they lower energy and maintenance costs, facilitate precise positioning and reproducible axis movements, and decrease noise levels.

4.3.2.3 Plasticizing Unit

The role of the plasticizing unit is to compound the melt and inject it into the mold at high pressure. Compared to the extruder, the plasticizing unit exerts higher loads on the melt because it operates discontinuously. Additional melt stressors are the fact that the screw traverses in the barrel (causing constantly changing flow conditions during plasticizing) and the typically longer residence times.

The size of the plasticizing unit must be matched to the shot volume because excessive residence times or plasticizing strokes will negatively affect the characteristics of the final parts. The metering stroke is the distance travelled by the screw; it determines how much melt (shot volume) is available for injection.

For the appropriate selection of a plasticizing unit it is important to understand the relationship between screw diameter D, metering stroke S, melt density ρ_S, and shot weight W_S (component including sprue):

$$S = \frac{W_S}{\rho_S \cdot \pi \cdot \left(\frac{D}{2}\right)^2}$$

If the melt density of a material is not known, it can be estimated from the density at room temperature by multiplying it with a factor of 0.85 for melts without fillers, or with 0.95 for highly filled (60%) melts.

The ratio of maximum screw stroke to screw diameter for three-zone screws with L/D-ratios of 18 : 1 to 22 : 1 ranges between 1 and 3 D to allow a shot size that is sufficiently large to produce the parts. However, larger screw strokes than necessary can result in longer melt residence times in the plasticizing unit, resulting in negative impact on part quality. Metering strokes of less than 1 D may result in inclusion of unmelted pellets, viscosity variations, or warpage. Here, a plasticizing unit with a smaller screw diameter should be selected. Metering strokes exceeding 3 D may cause a decrease of molecular weight, surface defects, or burnt molded parts. Here, a larger screw diameter should be selected.

In-Line Compounding Plasticizing Unit

In-line compounding plasticizing units are widely accepted. The major advantage of in-line compounding is the cost savings during raw material production, because there is no need for a second melting process – compounding and processing occur in one step.

Compounding in the plasticizing unit requires considerable effort in order to compensate for the rather poor plasticizing conditions compared to extrusion processes.

In general, two basic concepts are applied:

For the manufacture of blends in the injection molding machine or in order to mix easily miscible additives, screws with mixing zones or additional static mixers are applied (see also Section 4.1.1.3).

For the manufacture of more challenging compounds, conventional twin-screw compounders (see Section 4.1.1.2) that continuously compound the melt are combined with a melt buffer (accumulator) and an injection plunger. This arrangement combines the advantages of continuous compounding with the requirements for the typical discontinuous operation of injection molding plasticizing units.

Figure 4.64 shows the schematic design of the so-called injection molding compounder introduced by Krauss-Maffei. These machines directly compound long glass fiber reinforced molding materials from thermoplastic pellets and glass rovings (see Section 7.3.1.1), facilitating high reinforcement in the final injection molded part.

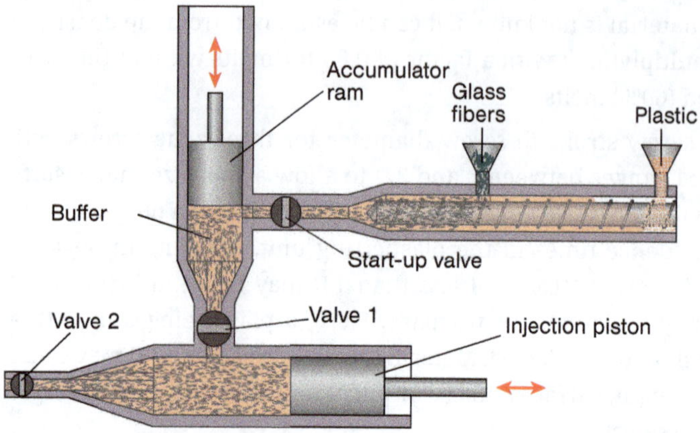

Figure 4.64 Plasticizing unit of injection molding compounder (Krauss-Maffei concept IMC)

Plunger Plasticizing Unit

In the early days of injection molding, machines used plungers rather than screws to meter and convey the pellets and melt, respectively. Today, these machines are used only for the manufacture of micro parts.

In a special design for a micro-precision injection molding machine (see Section 4.3.5.3) a screw plasticizing unit conveys the melt into the barrel of a piston injection unit, which then injects the melt into the mold.

Degassing Screws

This type of screw is advantageous for the processing of hygroscopic plastic materials, which are prone to damage/degradation by water at high processing temperatures. In these machines the flight depth is increased at approx. half screw length in order to lower the melt pressure to atmospheric pressure and allow water vapor and other volatiles to escape from the barrel through a venting port (Figure 4.65).

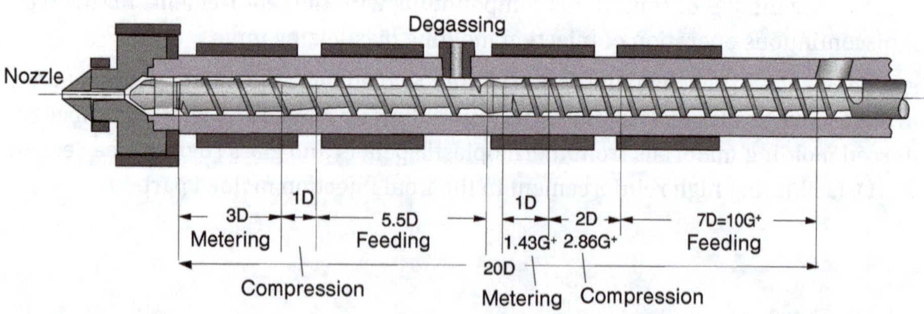

Figure 4.65 Degassing plasticizing unit for injection molding

The advantages of barrier screws, typically used in extrusion (see Section 4.2.1.3), are increasingly utilized in injection molding operations.

Extrusion-Injection

Extrusion-injection is used when manufacturing injection molded components with volumes that far exceed the maximum metering stroke of the screw. Here, the screw plasticizes during the injection phase and conveys the melts into the cavity; however, the process is determined by the low conveying pressure of the screw and its slow plasticizing rate compared to the much faster injection rate of conventional injection molding processes. Therefore, this technology requires large gates and allows for only small flow path/wall thickness ratios.

4.3.3 Injection Molding – General Information

During the *injection molding cycle* of a screw injection molding machine, injection typically takes less than 1 s, while the holding time until the solidification of sprue or gate can reach several seconds. The determining factor for the overall cycle time is the cooling time until the part can be demolded/ejected, which increases by the power of 2 with the wall thickness of the part. A part with a wall thickness of 2 mm will take between 4 and 10 s to cool, which may provide a reference value. While the part is cooling, the reciprocating screw is feeding new material.

The optimum processing parameters (melt temperature, mold temperature, residence time) vary for each specific plastic material. It is therefore advisable to adhere to the processing parameters recommended by the material manufacturers or to determine an optimum process window using set-up trials.

The latest engineering thermoplastic materials require injection units for temperatures up to 450 °C. Thin-walled parts with long flow paths require higher melt pressures than thicker walled parts.

Higher *mold temperatures* (within the given limits) result in higher surface gloss in polished molds and in lower residual stresses and frozen-in orientations. In some cases, they also improve the crystalline structure of the molded parts. However, they do increase cycle times. When selecting the gate position and controlling the injection process, care must be taken to make sure that the melt progresses from the gate while adhering to the mold walls, or otherwise jetting may occur (Figure 4.66) and result in insufficient connection of the following melt and in surface defects.

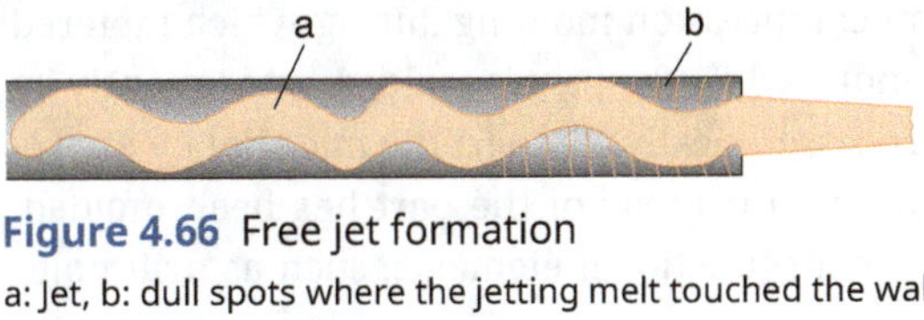

Figure 4.66 Free jet formation

a: Jet, b: dull spots where the jetting melt touched the wall and then – after slightly cooling – proceeded

Injection Molding of Filled Molding Compounds

Mineral filled and/or reinforced thermoplastic molding compounds generally require slightly higher injection molding temperatures and pressures than unfilled compounds. The injection time has to be short, because the melts quickly solidify. To ensure damage-free ejection, the molded parts should exhibit ejection draft angles of 1 to 2°, in particular for textured surfaces. It is difficult to manufacture components with wall thicknesses smaller than 2 mm from filled compounds. Hot runner molds help prevent premature solidification in the gate and minimize sprue waste.

Injection Molding of Thermosets

Screw injection molding machines for thermosetting materials do not differ significantly from machines for thermoplastics. To avoid excess shear heating of the crosslinking compounds, the screws exhibit shallow flight depths and a low flight depth ratio (Figure 4.67). They consist of only two zones, their compression is low, they do not have a non-return valve (except for low viscosity unsaturated polyester processing), and they operate with open nozzles. The screws are shorter, the L/D ratio ranges between 12 to 15 D; for wet compounds it is 10 D. Wear protection is of particular concern.

An essential difference is their temperature control. While thermoplastics are processed at high barrel and low mold temperatures, the barrel temperature for thermosets must not exceed 60 to 90 °C (depending on type) in order to prevent crosslinking in the barrel. Temperature in the barrel is typically controlled by water, while the molds are heated electrically to temperatures between 150 and 200 °C to achieve fast curing of the molding compounds. The crosslinked components can be demolded at elevated temperatures because they exhibit high heat deflection characteristics. In the molten state the compounds typically are of low viscosity, which allows for high flow path/wall thickness ratios. However, the high viscosity also requires a tight fit between machine and mold to avoid flash. The available injection pressure should be approx. 300 bar higher than for thermoplastics to facilitate the processing of easily crosslinking molding compounds.

Appropriate ventilation is important to ensure that water, mold release agents, and other volatiles can be removed. A proven approach is the ventilation of the mold, *i. e.*, after injection of the metered volume the mold is quickly ventilated before the curing process sets in.

Another process often used is injection compression molding during which metered melt is injected into a slightly opened mold, while the part is molded subsequently by closing the mold. Core compression processes locally compact the material by appropriate movement of the core after the main contour of the part has been molded; these processes are also used to implement functional elements, such as drill holes etc., without creating flow lines.

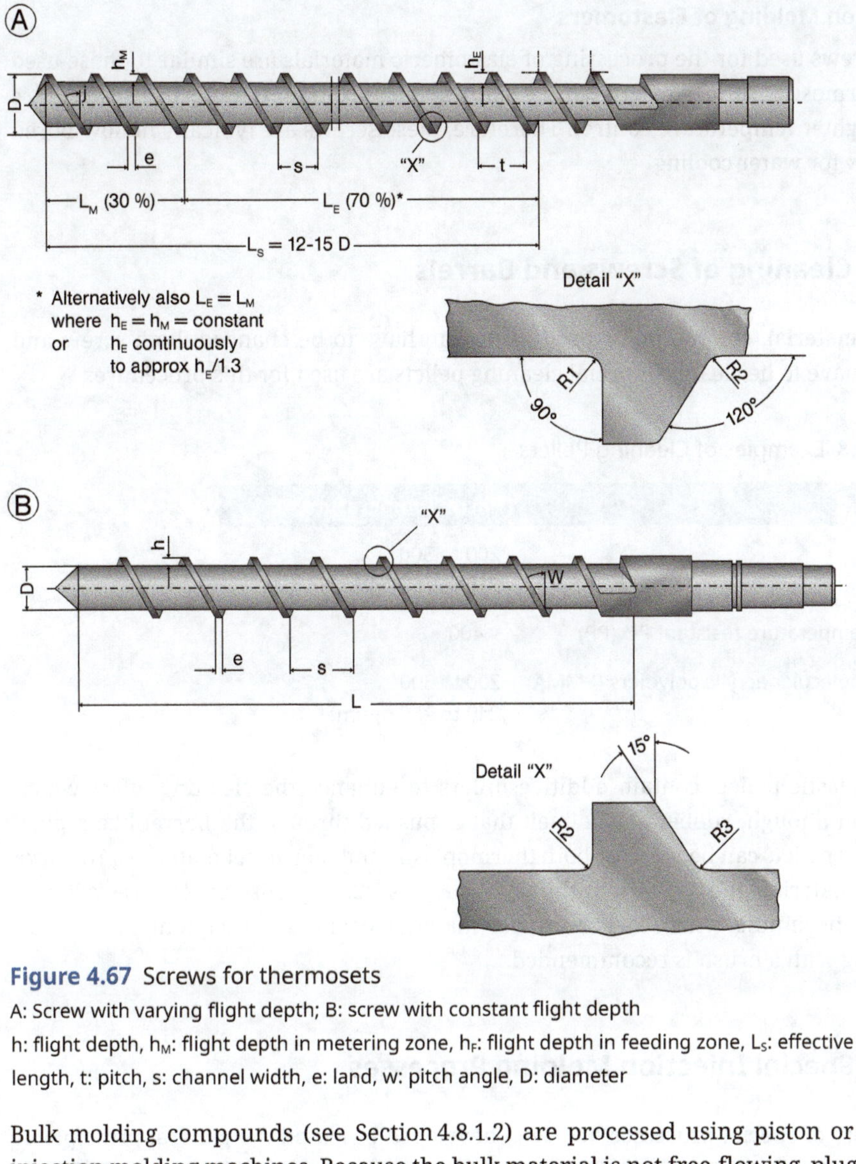

Figure 4.67 Screws for thermosets

A: Screw with varying flight depth; B: screw with constant flight depth

h: flight depth, h_M: flight depth in metering zone, h_F: flight depth in feeding zone, L_S: effective screw length, t: pitch, s: channel width, e: land, w: pitch angle, D: diameter

Bulk molding compounds (see Section 4.8.1.2) are processed using piston or screw injection molding machines. Because the bulk material is not free-flowing, plug screw loaders or piston feeders are employed. They maintain feeding pressures of 5 to 30 bar during the feeding phase of the injection molding machine. Screws with large flight depths and non-return valves and large sprue diameters minimize glass fiber breakage. The *BMC-injection process* uses metered bulk molding compounds that are fed into the plasticizing barrel of a specialized injection molding machine, which in turn feeds the melt into a second barrel acting as an injection plunger. This approach allows for bulk molding compounds to be fed into the mold without additional damage to the fibers.

Injection Molding of Elastomers

The screws used for the processing of elastomeric materials are similar to those used for thermosets. In order to avoid premature vulcanization, these processes require even tighter temperature control. Therefore, these screws are typically hollow-bored to allow for water cooling.

4.3.4 Cleaning of Screws and Barrels

When material, coloration, or production run have to be changed, both screw and barrel have to be cleaned. Special cleaning pellets are used for this procedure:

Table 4.8 Examples of Cleaning Pellets

Material	Barrel temperature, °C
PE	200 to 300 210 to 260 optimal
High temperature resistant PO (PP)	< 400
High molecular acrylic polymers (PMMA)	200 to 300 240 to 250 optimal

These plastic pellets contain additives/fillers to enhance the cleaning effect which relies on a tough, rubber-elastic melt that is pushed through the barrel like a plug. Cleaning pellets can be used for both thermoplastic and thermoset materials. To purge plastic materials that adhere to the steel, such as PC, step-wise procedures are employed, beginning with PE. For critical material changes subsequent mechanical cleaning with a brush is recommended.

4.3.5 Special Injection Molding Processes

A variety of special processes has been developed based on the general principle of the injection molding process. They are used to process many different materials in a whole range of arrangements and designs. These special injection molding processes come in a wide array of variations and are continuously developing, making categorization and systematical identification difficult. The authors consider the processes covered in this section as currently relevant. They are identified by technically appropriate labels or those used commercially when available (synonymous descriptions will be added in parentheses). It should be noted though that machine manufacturers often use different descriptions to establish unique selling point arguments.

In industrial practice, several special processes may be realized in a single production line. This follows the basic principle that the appropriate production technology is designed and built to meet the requirements of the component to be produced. The driving forces for the development of novel injection molding special processes or for the combination of existing processes to form a new variation are therefore the requirements of the component to be produced.

4.3.5.1 Injection Compression Molding

This process variation has proven advantageous for the manufacture of precision components, *e. g.*, optical applications such as data storage media. Here the volume of the tool cavity is reduced, *e. g.*, by a hydraulic sliding core or by moving of a movable tool half, to compensate for the material's thermal contraction upon cooling (Figure 4.68). Injection compression molding also minimizes orientations in the material during manufacturing and lowers injection pressures. The main effect of compression injection molding, *i. e.*, controlling mold pressure by moving mold surfaces, can often be achieved by using adjusted clamping force (breathing tools).

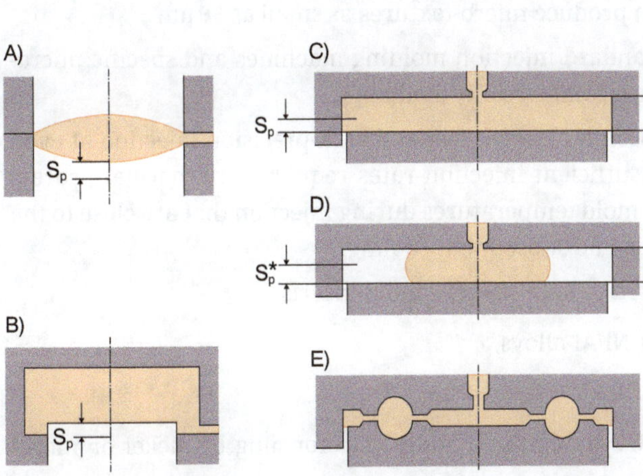

Figure 4.68 Variations of compression injection molding; left: using moving cores; right: using movable mold half

A) acting on cross section, B) acting on partial areas, C) complete mold filling, D) partial mold filling, E) with pinch-off edges

S: gap, S_p: compression gap compensating for volume contraction caused by cooling, S_p*: Open cavity gap > S_p

4.3.5.2 Powder Injection Molding

Powder injection molding (PIM) was developed for the manufacture of metal and ceramic components with sizes ranging from 0.2 to 0.5 cm. For metal injection molding (MIM), metal powders (iron, stainless steel, tool steel, tungsten carbide, titanium, Fe_2Ni, WCu, or MoCu) are mixed with a binder consisting of polymers and waxes and then processes into green parts on injection molding machines designed specifically for this application. Ceramics injection molding (CIM) follows the same approach with powders of alumina, zirconia, silicon nitride, silicon carbide and others. The binder is subsequently removed by a pre-sintering process (chemical-catalytically and/or thermally) without causing dimensional changes. The final sintering process takes place at temperatures between 1200 and 2000 °C.

To optimize the yield of the sintering process it is essential for the green parts to be manufactured with minimal residual stress and a homogenous filler distribution.

4.3.5.3 Micro-Injection Molding

Micro-injection molding processes are able to produce parts with: wall thicknesses of less than 20 mm, texture details as small as 0.2 mm, flow path/wall thickness ratios of 10 : 1 to 200 : 1 depending on material, and cycle times of approx. 5 s. Both reinforced and non-reinforced thermoplastic materials (PMMA, PC, PSU, PE, PE, POM, PA, PEEK, LCP) and powders (green parts) with binders based on POM and polyolefin/wax (carbonyl iron, steel, carbide metal WC-Co, Al_2O_3- and ZrO_2-ceramics) can be processed. Powder injection molding can produce micro-textures as small as 50 µm.

Both retro-fitting units for standard injection molding machines and specific micro-injection molding machines are commercially available.

Key to successful micro-injection molding is fast and high-precision injection of even smallest amounts of melt. Insufficient injection rates require variothermal process control with long cycle times, mold temperatures during injection that are close to the melt temperature of the plastic and subsequent cooling.

The following methods are used to manufacture mold inserts:

- Micro-chipping of brass or Ni/Al-alloys,
- Spark erosion,
- Electron beam, X-ray or UV lithography and electroforming of nickel or nickel alloys (NiCo, NiFe),
- Etching or reactive ion etching in silicon, and
- Laser texturing, *e. g.*, in carbide metals.

4.3.5.4 Injection Molding with Multiple Injection Units

4.3.5.4.1 Multi-Component Injection Molding, Multi-Color Injection Molding, Overmolding, Assembly Injection Molding

Here, we describe injection molding processes that use pre-fabricated preforms or parisons onto which additional plastic component(s) are molded in a subsequent process step. This results in parts that are composed of sections from different plastic materials or colors. Figure 4.69 shows a machine configuration typically used for such applications. Two or more injection units are employed successively, injecting into one mold. The molds are specifically designed for these processes. The size of the mold cavity can be changed by pulling cores or by changing specific parts of the mold. After the first injection unit has filled the initially small cavity, the cavity is increased accordingly to allow for material injection through the second unit. Both preform fabrication and injection molding can be performed simultaneously in appropriate molds (such as rotary table molds) so that a preform and a finished part are produced during each cycle.

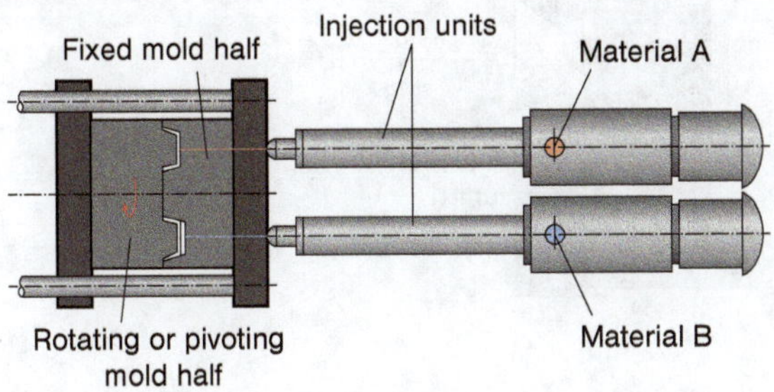

Figure 4.69 Machine with two injection units for two-component injection molding

Depending on the selected material and process parameters, the two components may form an adhesive, positive, or force fitting bond. The selected materials and parameters also determine the degree of the force fitting, making it possible to integrate components in the part that can move with respect to each other. A much quoted example for force fitting injection molding is the toy figure and its different manufacturing stages shown in Figure 4.70. Manufacturing and assembly of the individual injection molded parts occur simultaneously.

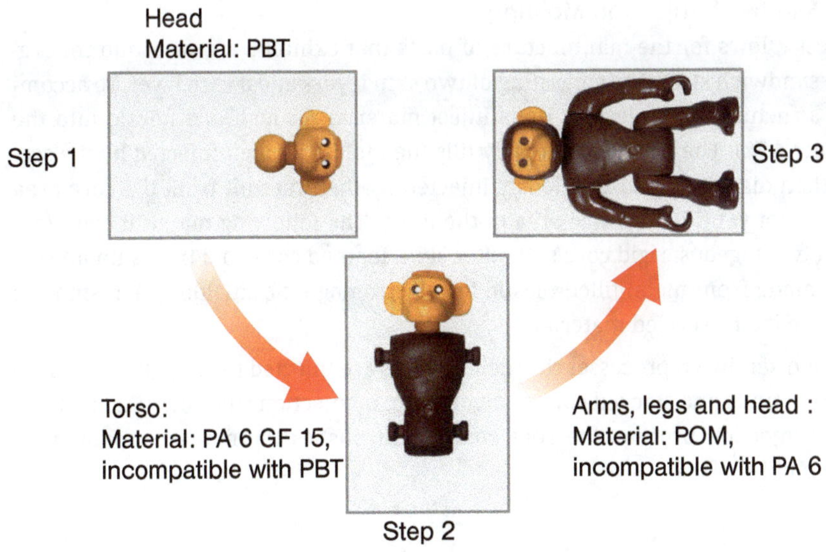

Figure 4.70 Assembly via multiple injection molding

Special molds for multi-component injection molding include those that pull a core (core-back technology), transfer the preform using a handling system, turn a part of the mold (typically the moving half), and use a rotating center platen (Figure 4.71).

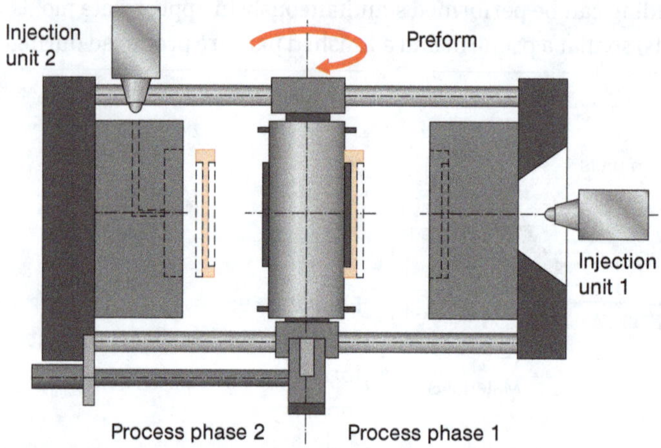

Figure 4.71 Multi-component injection molding with rotating center platen

Using multi-component injection molding, mechanical and electrical components can be integrated into a three-dimensional injection molded part. The so-called molded interconnect device process allows for metal coating of one of the components integrated during multi-component injection molding. Example: PA 12 + (PA 6 – GF 15 electroplatable).

4.3.5.4.2 Sandwich Injection Molding

This process allows for the manufacture of parts that exhibit a cross-section that resembles a sandwich structure consisting of two skin layers and a core layer. To accomplish this structure, two injection units inject material through one nozzle into the mold (Figure 4.72). The first unit partially fills the mold. The melt injected by the second unit then displaces melt previously injected by the first unit from the core area towards the not yet filled surface area of the mold. The following material combinations are advantageous: rigid core/soft outer skin; foamed core/void free, smooth surface; core made from metal-filled plastic for electromagnetic shielding/skin smooth; or core made from recycled material.

During mono-sandwich processes the melt streams are injected through two separate plasticization units and stacked consecutively into an injection chamber (Figure 4.73). During the injection process, the core component pushes the main component towards the mold surface.

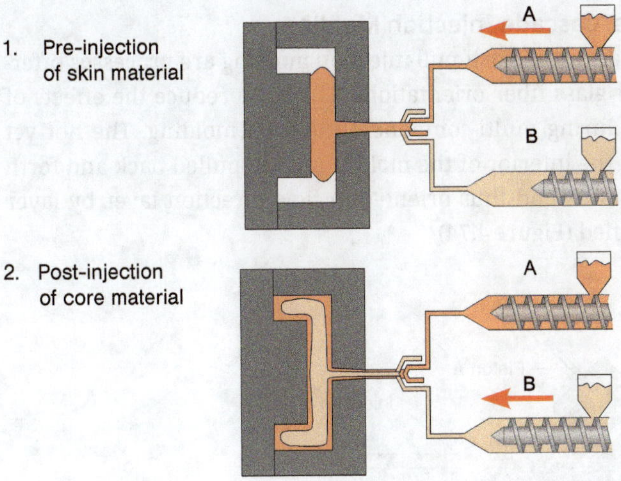

Figure 4.72 Sandwich injection molding

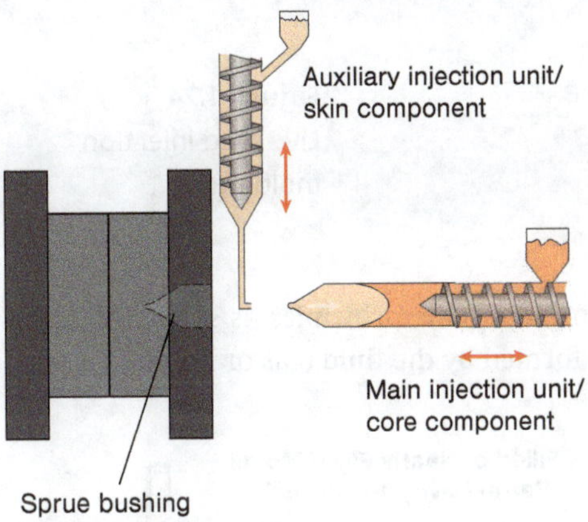

Figure 4.73 Schematic depiction of mono-sandwich process

4.3.5.4.3 Sequential Process

The sequential process is a variation of multi-component injection molding processes. It is characterized by the fact that the flow front of one melt stream is disrupted by the following melt, *e.g.*, one with a different coloration. This allows for the manufacture of decorative surface effects.

4.3.5.4.4 Interval Injection Molding

During interval injection molding processes (also called marbling process) according to Arburg, two differently colored plastic materials are introduced into a mixing nozzle prior to being injected into the cavity. The optical effects are created by the converging and mixing of the two different melts. This mixing nozzle can also be used for sandwich injection molding processes.

4.3.5.4.5 Oscillating Melts, Cascade Injection Molding

Multi live feed injection molding and push-pull injection molding are processes offering higher molecular and/or glass fiber orientations. They also reduce the effects of flow lines that are created during multi-component injection molding. The not yet solidified part of the melt in the interior of the mold is pushed/pulled back and forth by two pistons or injection units and thus oriented in flow direction layer by layer until the entire melt is solidified (Figure 4.74).

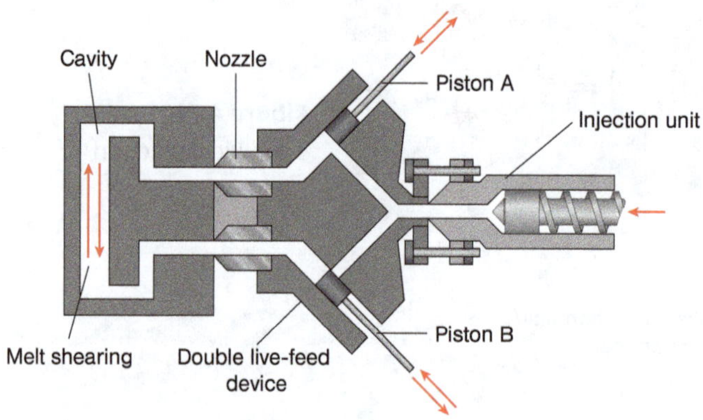

Figure 4.74 Live feed injection molding

4.3.5.5 Fluid Injection Technology

Fluid injection technology generally employs the same principles as sandwich injection molding. Here, the core component is formed by the fluid (gas or liquid).

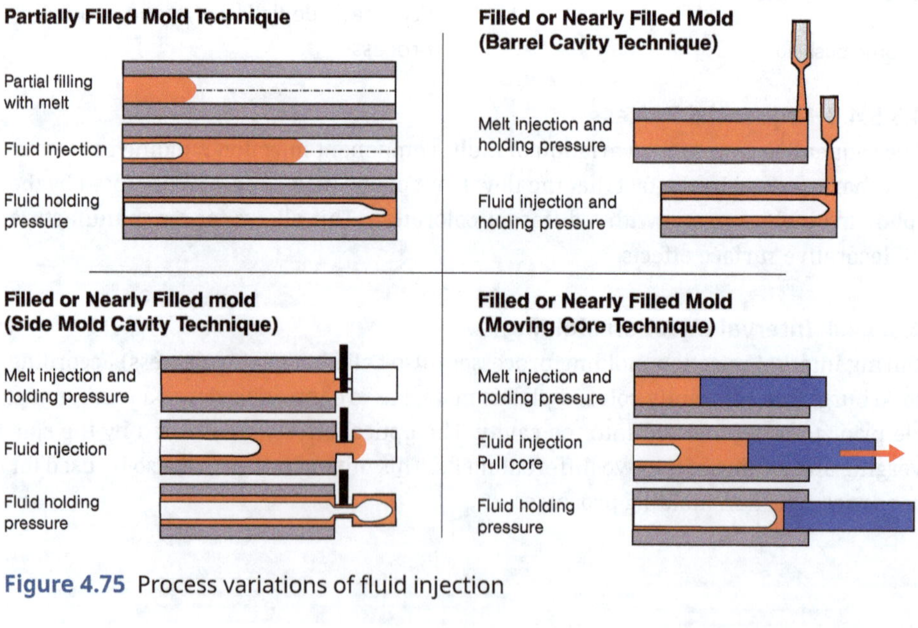

Figure 4.75 Process variations of fluid injection

The result is a hollow part. Figure 4.75 shows process variations for the molding of the part. The fluid is injected through a nozzle in the gate or the cavity at a place of mass accumulation with a pressure of up to 300 bar, which displaces the melt. Assuming laminar, wall-adhering flow, the fluid always flows in the center of the part. Because the fluid does not solidify, holding pressure can be applied until the plastic melt is completely solidified. The manufactured parts exhibit low residual stresses and good reproduction accuracy. The advantage of fluid injection molding processes becomes apparent for thick-walled components, where significant material savings and reductions in cooling time can be achieved. It is important to note that the fluid pressure must be released prior to demolding; when liquids are used, they have to be removed from the component interior.

When employing fluid injection molding technology it is also important to remember that the component's remaining wall thickness is entirely determined by fluid mechanical and thermal processes during molding. Therefore, the final wall thickness is not a design characteristic, but rather the result of effective parameters during the molding process (see also Figure 4.76).

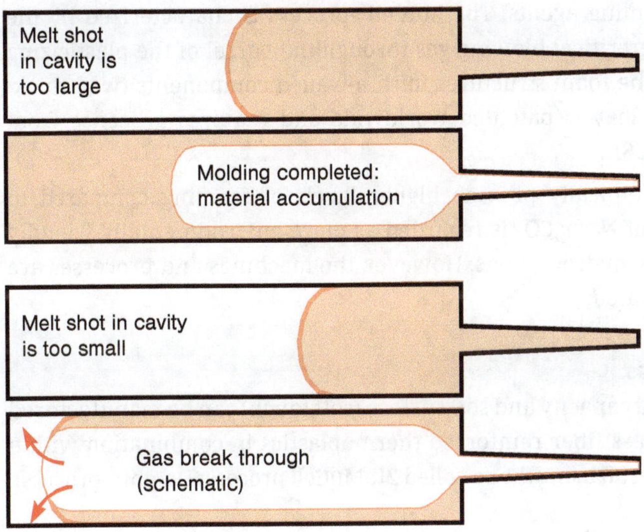

Figure 4.76 Gas bubble formation during gas injection molding (incorrect melt metering)

Today, gas injection technology has found widespread application using nitrogen as the fluid of choice. Water injection technology offers reduced cooling times compared to gas injection molding. However, the injection system is far more complicated as is the issue of water recycling after molding. Therefore, the majority of fluid injection molding applications employs gas injection.

Sink marks on visible surfaces can be avoided by creating gas pressure on the opposite surface of the component in the mold, *e.g.*, as in gas counter pressure processes.

4.3.5.6 Foam Injection Molding

4.3.5.6.1 Thermoplastic Foam Casting

The first foamed injection molded parts were manufactured from thermoplastic materials containing foaming agents. It was discovered early on that the process made it possible to produce thick-walled parts without sink marks/voids. Foaming agents of choice are low-boiling liquids or additives that decompose at high temperatures and thus cause foam formation. The results are integral structures characterized by compact exterior skins and cellular cores. When the melt is injected too slowly, it will foam during mold filling and create striations on the component's surface. By employing gas counter pressure foam casting technology, the formation of these striations is prevented by high gas pressure in the mold (N_2, ca. 80 bar) at the beginning of the filling phase. Abrupt pressure release at the end of the filling phase creates fine cellular foam structures because of the high nucleation density.

4.3.5.6.2 Injection Molding with Physical Foaming Agents, MuCell™, Optifoam™

Particularly fine foam structures can be achieved by employing foam injection molding with CO_2 or N_2 as foaming agents. The MuCell™-process is characterized by the precise dosage of the supercritical blowing gas through the barrel of the plasticizing unit and the achievable fine foam structures in thin-walled components (wall thickness < 3 mm). This technology is patented worldwide and requires a license from Trexel Inc., Woburn, MA, USA.

Physical foaming agents typically provide higher levels of foaming compared to chemical agents. The use of N_2 or CO_2 is regarded as more environmentally friendly than the use of low-boiling hydrocarbons. However, the machines and processes are significantly more complicated.

4.3.5.6.3 Dolphin Process

Components with high load capacity and soft surface (soft touch) can be manufactured by injection molding of glass fiber reinforced thermoplastics in combination with a foamed thermoplastic material using the so-called 2K-MuCell process (Dolphin-process).

4.3.5.7 Injection Molding of Liquid Silicone Rubber

Injection molding of liquid silicone rubber (LSR) has advanced significantly over the past 40 years. Because LSR is a thermosetting material, the injection molding process requires specialized handling. This includes thorough distributive mixing while keeping the material cool until it is injected into a heated mold cavity where vulcanization occurs. Figure 4.77 provides a schematic overview of the LSR injection molding process. LSR is typically supplied in barrels or hobbocks. Thanks to its low viscosity, it can be easily pumped through pipelines to the vulcanization equipment. The process involves two components, labeled A and B in the figure, which are delivered to a static mixer using a metering pump. One of these components contains a platinum-based catalyst. Before entering the static mixer, additives such as color pastes can also be incorporated. In the static mixer, the components are thoroughly combined, producing a highly homogeneous material. This consistency ensures uniform quality both within individual parts and across production batches. This is a key advantage over solid silicone rubber, or high-temperature vulcanized (HTV) silicone rubber, which is supplied premixed and partially vulcanized. Solid silicone rubbers are typically processed by transfer molding, which provides less material uniformity and higher part variability. Additionally, HTV materials require higher processing temperatures and longer curing times. After mixing, the LSR compound enters the cooled metering section of the injection molding machine. It is then pushed through cooled sprue and runner systems into a heated cavity where vulcanization takes place. The use of cooled feed lines minimizes material loss, enabling near-zero waste production. This eliminates the need for trimming operations and reduces material costs significantly. Silicone rubber belongs to the family of thermoset elastomers characterized by a backbone of alternating silicon and oxygen atoms, with methyl or vinyl side groups. Silicone rubbers make up roughly 30% of all silicone materials, representing the largest subgroup. These materials retain their mechanical properties over a wide temperature range and are extremely hydrophobic due to their methyl side groups. LSR is ideal for manufacturing high-precision components such as seals, sealing membranes, electrical and multi-pin connectors, and infant-care products like bottle nipples. Its smooth surface and biocompatibility also make it suitable for medical devices and kitchenware, including baking pans and spatulas (see also Section 6.23.6).

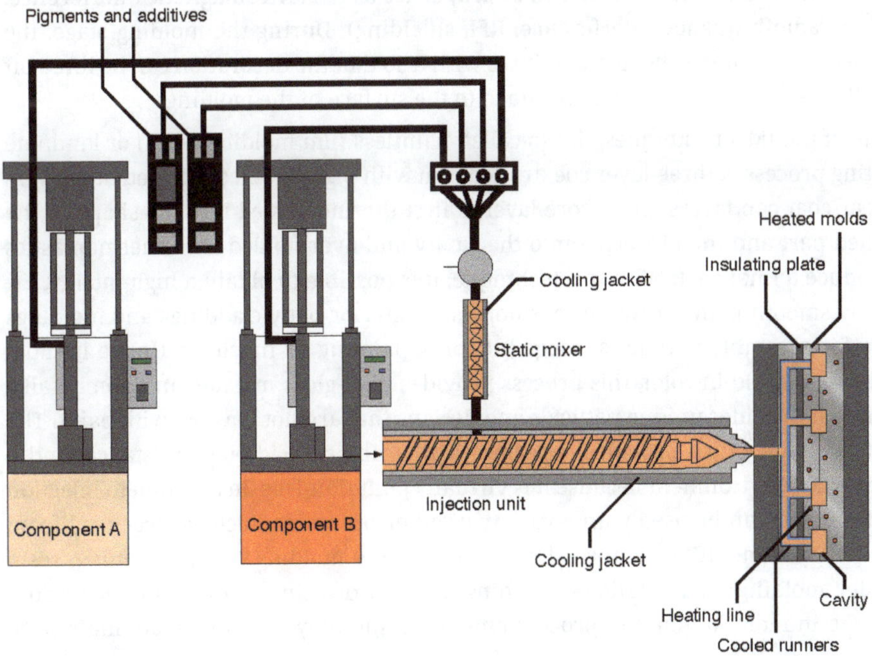

Figure 4.77 Schematic diagram of a liquid silicone rubber injection molding system

4.3.5.8 Injection Molding with Inserts

4.3.5.8.1 Overmolding of Inserts

To accommodate high local loads in a component, inserts are placed in the mold and then overmolded. A reversal of this approach is the injection of functional plastic elements into (metal) carriers inserted in the mold, the so-called outsert technology. Similarly, deep-drawn metal parts or glass fiber reinforced parts can be inserted in the mold and injected or overmolded with plastic (hybrid technology, *e. g.*, in automotive applications). This approach takes advantage of the high stiffness and Young's modulus of the insert and the high degree of design freedom of plastic materials for the manufacture of large-scale, stiff components that exhibit high energy absorption capacity. Instead of metal inserts continuous fiber reinforced semi-finished plastic parts are integrated into injection molded components to increase their lightweight design properties.

4.3.5.8.2 In-Mold Decoration and In-Mold Lamination

In-mold decoration (IMD) comprises insertion of a pre-decorated film or foil into the cavity, followed by injection of polymer melts on the inner side of the insert to produce a part with the final finish defined by the decorative film/foil. *In-mold lamination process* resembles IMD except that a multilayered textile laminate is used. These two processes provide a cost-effective way to enhance and/or modify the appearance of a product for marking, coding, product differentiation, and model change without costly retooling. The laminated component can have desirable attributes (*e. g.*, fabric or plastic skin finish with soft-touch) or properties (*e. g.*, electromagnetic interference, EMI, or radiofrequency interference, RFI, shielding). During the molding stage, the polymer melt contacts the film and fuses with it so that the decoration can be lifted off from the carrier film and strongly attach to the surface of the molding.

In one of the IMD techniques, the so-called paintless film molding (PFM) or laminate painting process, a three-layer coextruded film with pigment incorporated into layers of clear-coat cap layers and a core layer is first thermoformed into the shape of the finished part and then inserted into the cavity and overmolded with thermoplastics to produce a final part. With this technique, it is possible to obtain a high-quality, extremely smooth paint finish on thermoplastic exterior body claddings and moldings ready for assembly without subsequent spray painting or finishing. Unlike in-mold coating and mold-in-color, this process provides high-gloss metallic and nonmetallic finishes. It provides unique patterns and designs that are not feasible with paint. The paint laminate finish provides superior weatherability, acid etch resistance, and a safe worker environment because it is virtually pollution-free. In addition to injection molding, IMD can be used with a variety of other processes, such as structural foam injection molding, ICM, compression molding, blow molding, thermoforming, resin transfer molding, and rotational molding. Instead of using a thin film/foil as does IMD, the in-mold lamination process employs a multilayered textile laminate positioned in the parting plane to be overmolded by the polymer melt on the inner side. The decorative laminate can be placed in the mold as a cut sheet, pulled from the roll with needle grippers, or by means of a clamping frame method. By means of a thermoforming operation, the clamping frame method allows defined predeformation of the decorative laminate during the mold-closing operation. In-mold lamination is also known as *laminate insert molding* or *fabric molding* for manufacturing automotive instrument panels and interior panels, respectively. For in-mold lamination, the outer, visible layer of the decorative laminate can be made of polyester, PA, PP, PVC, or ABS film, cotton textile (woven, knitted, tufted, or looped fabrics), or leather. This outer layer typically comes with a variety of features to create a specific appearance or feel. In general, low surface texture is preferred because an ironing effect typically occurs with high pile or fiber loops. To provide the product with a soft-touch effect, there is typically an intermediate layer of polyurethane PU, PP, PVC, or PES foam between the top layer and the liner layer. Underneath the foam layer is the liner layer, which is used to stabilize the visible layer against shear and displacement, prevent the penetration of polymer melt into the intermediate layer, and provide thermal insulation against the polymer melt. To avoid damage or undesirable folding of the laminate during molding, low injection pressure and low temperature are desirable. This makes low-pressure injection molding, compression molding, and cascade injection molding with sequential valve-gate opening and closing suitable candidates for in-mold lamination. Although existing tools can often be adapted for IMD and in-mold lamination, it is generally recommended to design the tool specifically for these processes. In-mold decoration has been used to produce automotive rooftops, bumper fascia, and exterior mirror housings, as well as automotive lenses and body-side molding with mold-in colors to help automakers eliminate costs and environmental concerns associated with painting. In-mold lamination has been used widely for automotive interior panels and other applications such as automotive interior panels and shell chairs, both featuring textile surfaces.

4.3.5.8.3 Lost Core Technology

Injection molded parts with complex geometries or openings that cannot be manufactured using demoldable cores can be produced using so-called lost core inserts. These inserts are subsequently removed (melted). Core inserts made from low-melting metal alloys such as tin-bismuth are commonly used. This method is rather complex and involved, and therefore competes with other methods used to injection mold hollow parts such as die-slide injection molding with ultrasound welding. Here, the two halves of the hollow part are injection molded first. Then the mold is hydraulically shifted to position one component half over the other. In a second injection step plastic is molded around the seal area of the two parts, welding them into a single hollow part. However, lost core technology is used when edges are undesirable or when there are high requirements on the precision of the interior surface of the component.

4.3.6 Injection Blow Molding

Containers, in particular for packaging applications, are manufactured by first injection molding a cylindrical preform or parison with integrated threading for the cap. In the following processing step this parison is inserted into another mold and blow molded to its final contour (Figure 4.78). There are several variations of this process: the blow molding step may be performed immediately after injection molding in the same machine or the parison may be transferred to a separate blow molding station. Increasingly, the parisons are manufactured using multi-component injection molding processes, see Section 4.3.5.4.1; example: PET/recycled PET/PET.

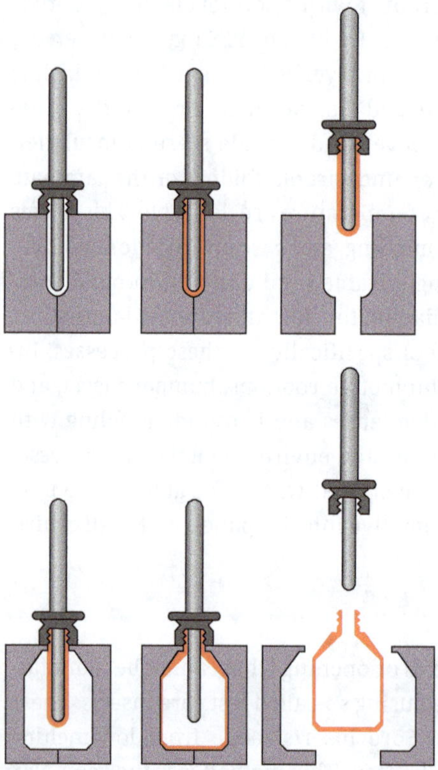

Figure 4.78 Principle of injection blow molding

4.3.6.1 Stretch Blow Molding

This is a variation of the injection molding process that employs bi-axial stretching to increase strength and decrease permeation (Figure 4.79). Here, the retractable mandrel stretches the component in axial direction while the blowing air creates the inflation of the parison in the circumferential direction of the container.

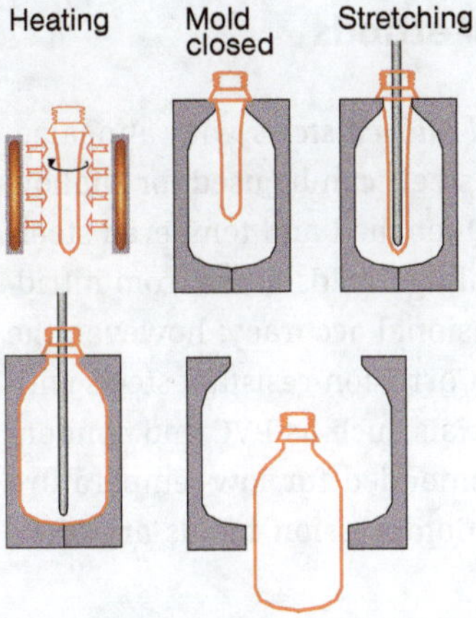

Figure 4.79
Stretch blow molding processing steps

4.4 Molds

Ruben Schlutter

4.4.1 General Introduction to Mold Design

Processing machines and their respective tools (or molds) are essential for the successful manufacturing of plastic compounds into final components (by injection molding, casting, or blow molding). The required high pressures and closing forces inside the cavity typically necessitate special mold steels. The mold materials have to provide not only mechanical strength but also sufficient stiffness. The requirements for the molded components rarely allow for mold deformations/deflections exceeding 0.01–0.02 mm despite the high pressures in the cavity (300–1000 bar or more), depending on the required tolerance for the molded components.

Today, plastic molded parts are often designed using injection molding simulations. The first step is to ensure that the molded part has been designed to be sufficiently plastic-compatible and can, in principle, be properly filled. Subsequent steps involve designing and optimizing the entire injection mold within the simulation. Initially, the number of molded parts to be produced, the respective gate position on the molded part as well as the type and geometry of the gate system are integrated into the simulation. Next, the cooling system can be designed to enable efficient and robust production. In the final step, the entire mold block, including inserts and cores, is considered. With each additional step, the production process is more accurately modeled, allowing for better insights into process robustness and optimization of target parameters such as minimal warpage or reduced clamping force.

4.4.2 Molding Tool Steels and Treatment Methods

The mold cavity is typically made from *case-hardened alloyed steels* with annealed, polished surfaces and tough cores. Through-hardened steels can be used for shallow textures or parts of the mold subjected to high wear. Quenched and tempered steels are used for large molds that may distort during annealing. Molds made from nitriding steel meet the highest demands in terms of dimensional accuracy; however, the thin, hard edges require appropriate careful handling. Corrosion-resistant steels with high chromium content are required for plastic materials such as PVC and condensation polymerization materials. They are also recommended for low-temperature cooled molds (condensation causing rust formation). Compression molds are often hard-chrome-plated.

The injection molds are mechanically designed to withstand the loads from both the clamping force (see Figure 4.80, right) and the internal pressure (see Figure 4.81, left). As a guideline, mold plates should not deflect by more than 0.02 mm (see Figure 4.81, right) to ensure a reliable production and minimize wear [1, 2].

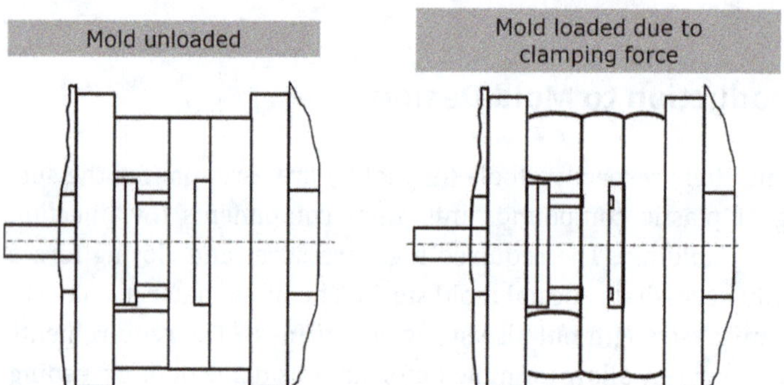

Figure 4.80 Load on the injection mold due to the clamping force

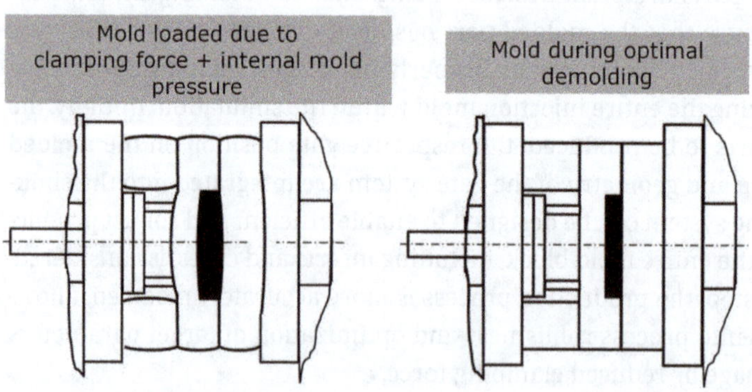

Figure 4.81 Load on the injection mold due to internal pressure

Physical vapor deposition provides molds and mold components made from hardened tool steels with hard, oxidation- and corrosion-resistant surfaces; see Table 4.9.

Table 4.9 Industrially Manufactured PVD Surface Coatings

Property	Unit	TiN	CrN	TiCN	AlTiN	WC/C
Hardness	HV	2500	2100	3000	2800	1000
Oxidation resistance up to:	°C	500	600	400	750	300
Coefficient of friction against steel 100 Cr 6	-	0.65	0.50	0.45	0.55	0.2
Density	g/cm^3	5.2	6.1	5.6	5.1	
Coefficient of linear expansion	10^{-6} K^{-1}	9.4	9.4	9.4	7.5	
Typical layer thickness	µm	2 to 4	3 to 8	2 to 4	2 to 4	1 to 4
Color	-	Gold	Silver	Red/brown to gray	Blue to black	Black/gray

The Lamcoat® process mechanically applies a tungsten-disulfide-based soft coating with a thickness of 0.5 µm to a maximum of 1.5 µm at room temperature. The coating reduces friction by approximately 70% and thus decreases the fill pressure and increases the service life of the mold.

Steel molds are typically machined by chip-removing processes; often, the actual mold contours are created by electro-discharge machining. Grains and other textures can be applied to steel molds by acid etching of photo-optically transferred patterns.

Molds cast from *zinc alloys* withstand temperatures up to approximately 100 °C and can be used for simple injection-molded components and for blow molds. Molds made by *light metal casting* exhibit porous surfaces and are therefore suitable only for blow molds. Alloys made of *copper* and cobalt and beryllium, which offer high thermal conductivity, are used for hot runner nozzles or other mold components that require intensive temperature control.

The use of hard-nickel mold inserts that are electrotyped from a master can be economically advantageous for molded components with finely textured, true-to-original surfaces (for example, technical toys).

A methodical approach to the *design of molds* includes the determination of parting line locations, number and arrangement of cavities, mold type, and applicable standards as well as planning for temperature control, ejection and ventilation systems, and the selection of the sprue type and gate locations for injection molds. Computer-based injection molding simulation is used to accomplish these tasks (see also Section 5.5.2).

4.4.3 Injection Molds and Sprue Types

Figure 4.82 shows the general design of an injection mold with only one parting line. The *sprue bushing* (17) of the mold is fitted to the nozzle of the injection unit (not depicted). The radius of the nozzle seat should be slightly larger than the nozzle radius and the sprue bushing should be wider than the nozzle to allow for the separation of the sprue runner and the molded component after demolding. In general, the *sprue* is attached to the molded component when it is demolded by the ejector system. The molded part is separated from the sprue at the gate (the narrowest part of the sprue runner) only after demolding. The oldest sprue design for single-cavity molds is the *sprue* gate, which ends in the cavity with its largest radius (Figure 4.83, left). This design is not economical for thin-walled parts because of its longer cooling times and the higher amounts of scrap it produces. It may also create sink marks on the opposite side because of shrinkage in the high sprue mass. Today, *pin point* gates with *melt chambers* (Figure 4.83, right) are commonly used. They represent a precursor to the current hot runner designs. By insulating the melt chamber from the filled cavity by air gaps, the plug retains a "plastic core" so that it does not have to be pulled between injection cycles but rather can be injected with the next shot.

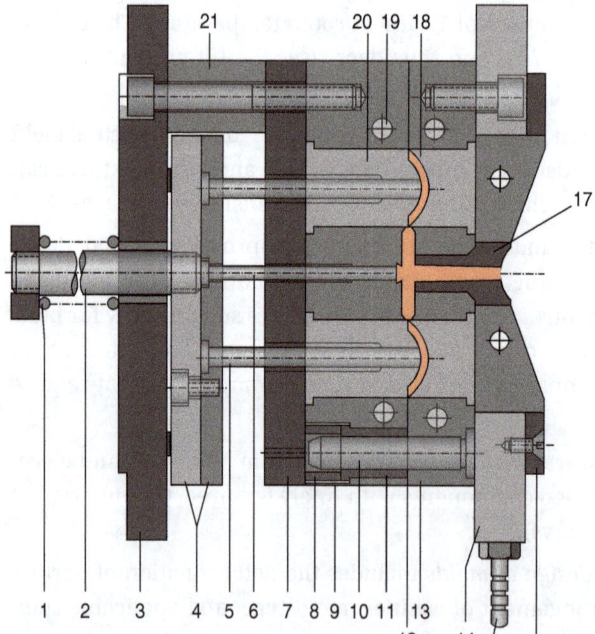

Figure 4.82 Identifying parts of an injection mold

1: Compression spring 12: Mold parting line

2: Ejector pin 13: Mold base

3: Movable platen 14: Stationary platen

4: Ejector plate 15: Tube connector for cooling

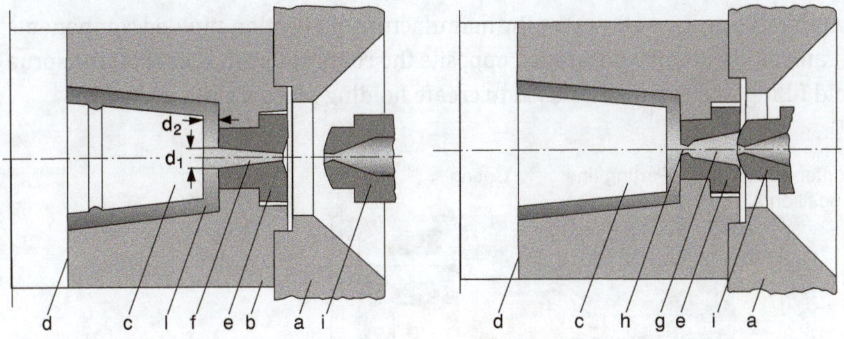

Figure 4.83 Center gate for cup-shaped components

Left: sprue gate, right: pin point gate with melt chamber

a: mold clamping plate, b: female mold, c: core, d: mold parting line, e: sprue bushing, f: sprue gate, d1 > d2, g: chamber, h: pin point gate, i: injection nozzle

This pin point design (for scrap-less injection molding) has been increasingly replaced by *hot runner systems* with multiple drops (pin points) (Figure 4.84). Here, the melt is delivered to the gates in heated channels. Both externally heated channels and drops with internal heating systems are employed.

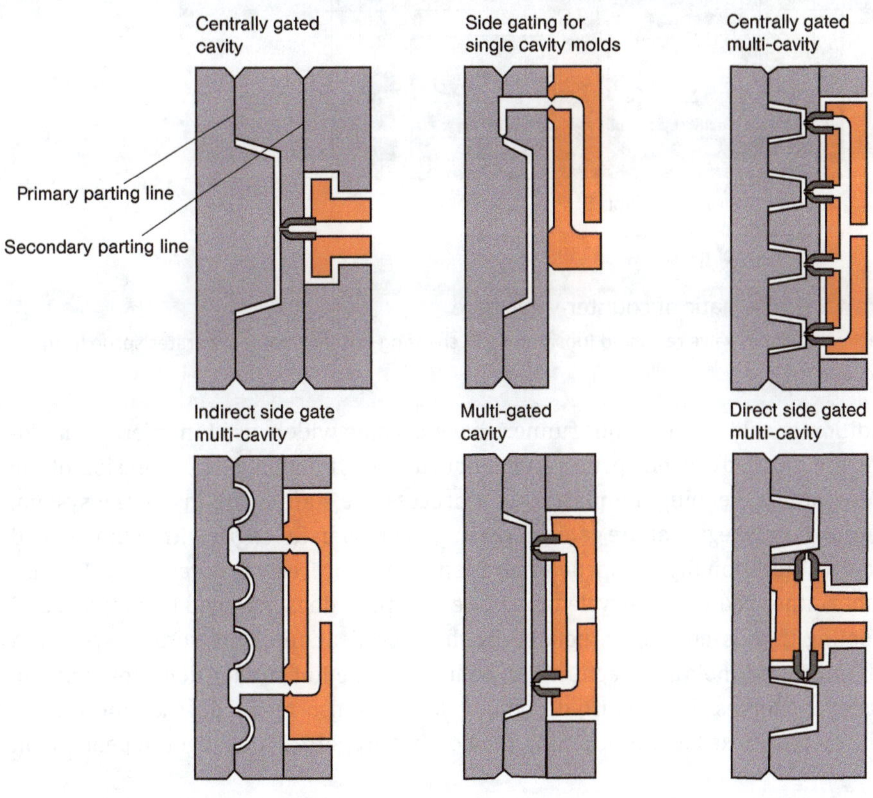

Figure 4.84 Main options for melt delivery in hot runner molds

The counter-valve process allows for the manufacture of injection-molded components without gate marks. A valve positioned opposite the runner system shears off the sprue after mold filling and can also be used to create holding pressure (Figure 4.85).

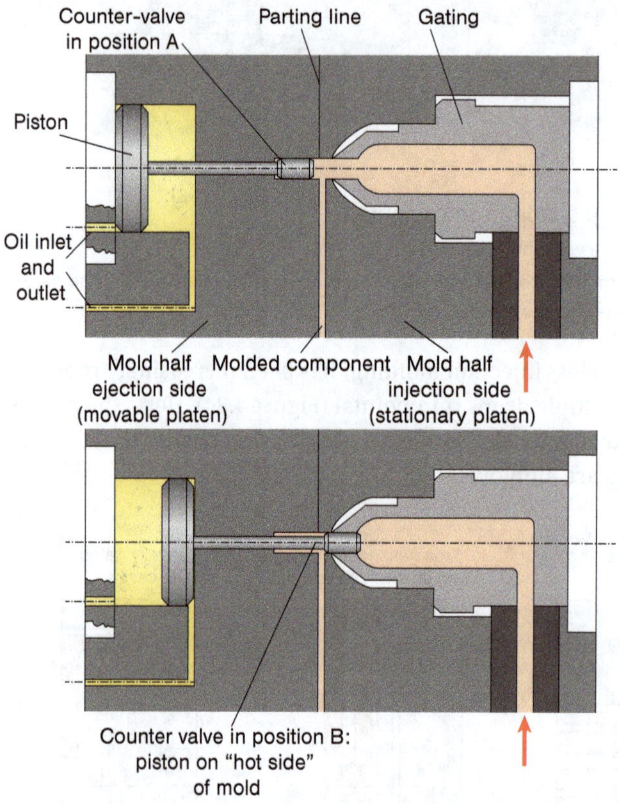

Figure 4.85 Schematic of counter-valve gate
A: Injection position, valve retracted (position A), B: shearing position, valve separates sprue from component

In addition to cold runners, hot runners have become widely used in injection molding. As the most advanced sprue system, hot runners function as an extension of the machine nozzle, keeping the plastics in a processable state while inside the system. Mold filling only begins at the gate, increasing the flow path length within the molded part cavity. Additionally, no sprue separation or post-processing is required. The use of a hot runner reduces the cycle time since the sprue does not need to be filled, and the cooling time is no longer dictated by the solidification of the runner system. A smaller injection molding machine can be used, as the hot runner does not generate any buoyant forces. The required opening stroke is also reduced, since there is no need to eject a sprue bar. Additionally, the holding pressure is no longer dependent on the sprue design, allowing for a longer holding time. The sprue can also be balanced by adjusting the temperature of individual zones [1, 3].

However, hot runners come with disadvantages, including a more complex mold design and higher costs. The system is also more prone to malfunctions due to potential leaks and electrical defects.

For detailed design guidelines on gating systems and gates for both cold and hot runners, refer to specialized literature such as [1, 3–5].

Sequential injection molding processes use needle valves to make sure that weld lines occur at noncritical component locations only. Here, needle valves are opened and closed at different times during the injection phase in order to shift the flow fronts.

Cascade injection molding processes eliminate weld/flow lines entirely. Here, the needle-valve gates are controlled such that only one melt front exists at any given time. Each hot runner gate opens only after the melt front has passed the gating position. The warpage of elongated components is reduced. This gating technology is used when long flow paths require several gates but at the same time weld/flow lines are not desired.

To make sure that the runners are uniformly filled in *multi-cavity molds* for small components and in molds with multiple gating locations (Figure 4.86), the runners are designed so that they exhibit the same length (from the nozzle of the injection molding machine to the gate at the cavity) and cross section. To adjust different lengths in flow paths, the runner cross sections are rheologically "balanced.

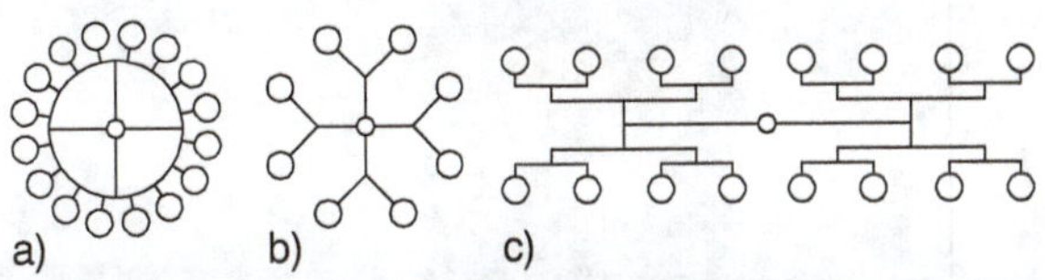

Figure 4.86 Different runner designs for multi-cavity molds
a: Circular runner, b: star-shaped runner, also multi-staged, c: balanced, multiple-stage gating; (a) not recommended because of different length in flow paths, (b) and (c) are preferred

When *angular components* or pipes are gated at only one edge, the melt has to flow around the core. A weld or flow line is created when the two now-cooled melt streams merge again. This line is a weak spot in the component and may also be visually undesirable. Centrally gated *large-scale* components may warp because the shrinkage in the flow direction is higher than in the one perpendicular to the flow. Film gates, serial pin point gates, or ribs are recommended for these types of components. Figure 4.87 shows a variety of sprue and gate designs.

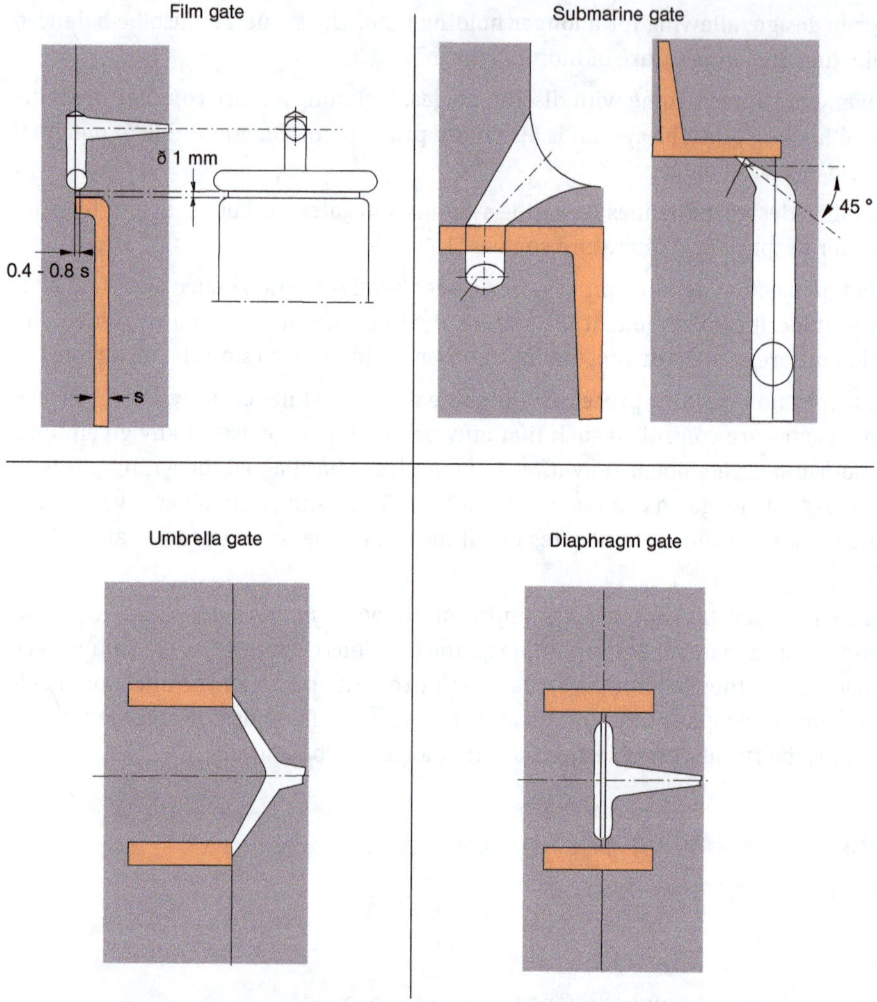

Figure 4.87 Different types of sprues and gates

For tunnel gates, the sprue is located in the opposite mold half; see Figure 4.87. On mold opening, the sprue is separated at the pin point sprue cross section and is pulled out by the runner. *Three-platen molds* (Figure 4.88) are used for multi-cavity molds that do not allow for tunnel gating and for large-scale, complex components that require multiple sprues. Here, the sprue is demolded together with the runner in the secondary parting line.

In order to increase the productivity of injection molding machines, *stacked molds* are also used for thin, large area components. Other than in regular molds, here multiple (typically two) cavities are arranged vertically "back to back" in the center of the mold and filled simultaneously. The high scrap caused by the branched runner system is a disadvantage that can be eliminated by hot runner systems.

For wall cutouts and exterior undercuts, *slides* and *stripper rings* are used. They glide outwards transversely with dovetail guides during demolding. They are actuated either by tension straps or hydraulically. Internal threads require unscrewing cores, which are manually unthreaded from the component after demolding. They are actuated as other movable mold components.

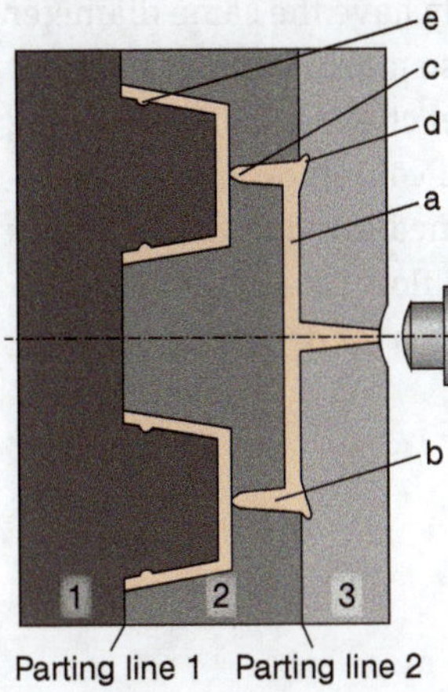

Figure 4.88
Three-platen mold, sprue with runner and mold separate on opening of plates 1, 2, and 3
a: Runner/sprue, b: runner/gate, c: gate, d: undercut for sprue, e: core undercut

4.4.4 Cooling Systems

The cooling system controls the temperature of the injection mold, significantly affecting the cycle time and, consequently, the efficiency of the injection molding process. It functions as a heat exchanger by removing thermal energy from the molten plastic until the molded part reaches a dimensionally stable state [1].
The temperature control unit defines the flow rate of the cooling medium, which is typically directed through drilled cooling channels and guided by plugs. Two primary configurations have been established: parallel and serial cooling channel layouts, as shown in Figure 4.89.

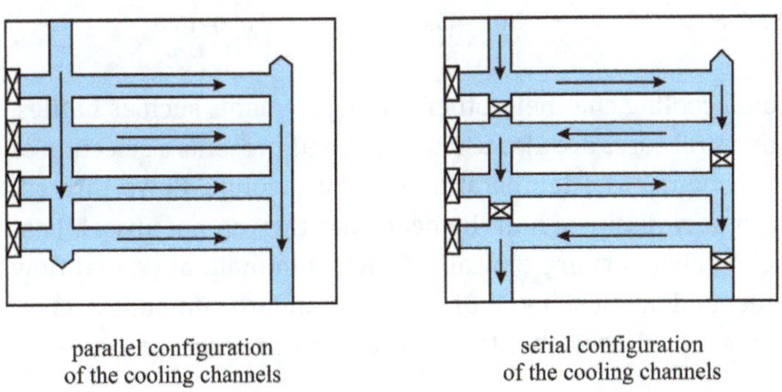

Figure 4.89
Configuration of cooling channels (own illustration based on [6])

In the parallel configuration (see Figure 4.89, left), the flow rates through the individual cooling channels may vary due to differences in flow resistance, pressure losses, or increasing blockages by deposits or corrosion. These variations in the flow velocities lead to inconsistent cavity wall temperatures, preventing uniform molding and cooling [6].

The serial configuration (Figure 4.89, right) is the recommended and most commonly used arrangement. Here, the cooling channel bores typically have the same diameter to maintain a turbulent cooling-medium flow. This setup also makes it easier to detect damage, such as blockages or leaks, as there are no parallel channels for the cooling medium to escape through. The disadvantages of the serial configuration are the increased pressure loss and greater heating of the cooling medium due to the longer flow path. The higher pressure loss reduces the achievable flow rate of the temperature control unit. Figure 4.90 and Table 4.10 provide guidelines for designing an effective cooling system [6, 7].

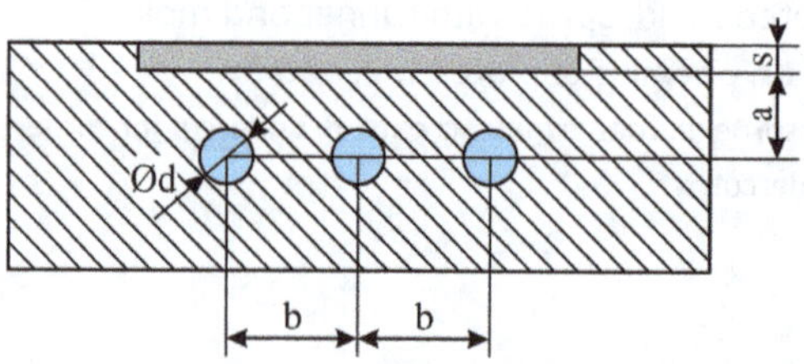

Figure 4.90 Cooling channel design (adapted from [7])

Table 4.10 Guidelines for the Dimensioning of the Cooling System

s [mm]	a [mm]	b [mm]	d [mm]
0–1	11.3–15.0	10.0–13.0	4.5–6.0
1–2	15.0–21.0	13.0–19.0	6.0–8.5
2–4	21.0–27.0	19.0–23.0	8.5–11.0
4–6	27.0–35.0	23.0–30.5	11.0–14.0
6–8	35.0–50.0	30.5–40.0	14.0–18.0

In addition to conventional cooling channels, other cooling elements such as baffles, baffle plates, bubblers, or spiral cores are also used. Figure 4.91 presents a selection of these different cooling elements. The temperature-control medium flows through them as well, but they can only dissipate half the heat since they do not absorb heat on the inside. These cooling elements are typically designed to maintain equal flow resistance on both the inlet and outlet sides [1, 6]. Various standard-component manufacturers already offer these cooling elements as part of their product range.

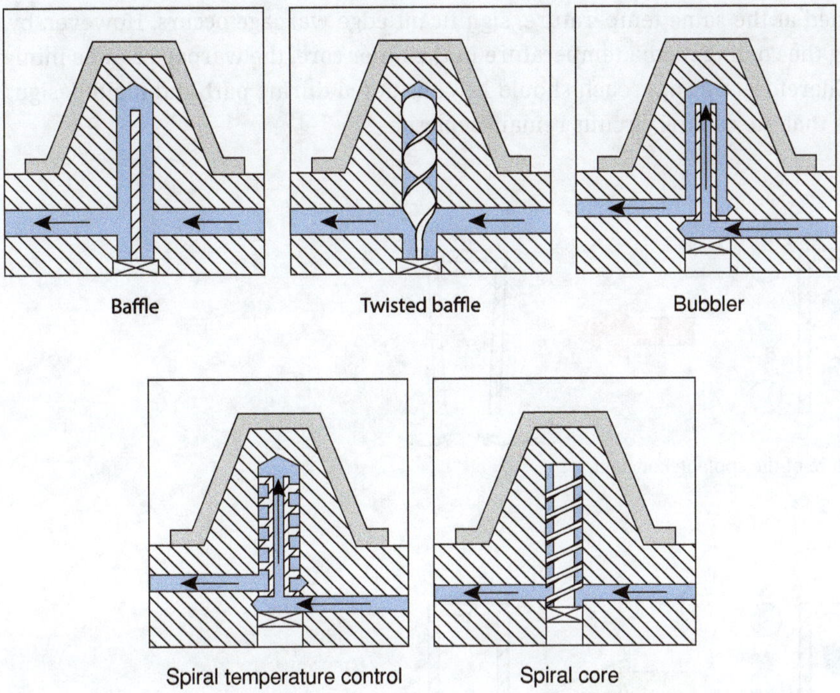

Figure 4.91 Cooling elements (adapted from [1, 6])

Cooling cores can also be used, especially for slim internal geometries (see Figure 4.92). To prevent excessive heating, these cores are made from materials with high thermal conductivity. Commonly used materials include copper–beryllium alloys and high-thermal-conductivity steels, with a thermal conductivity of approximately 100 W/(m K).

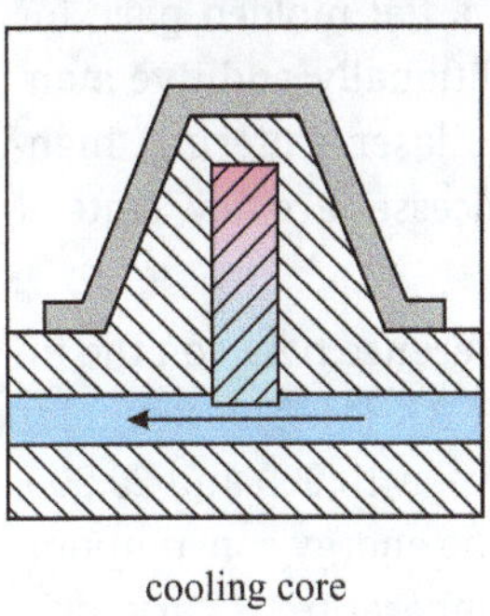

Figure 4.92 Cooling core (adapted from [6])

To minimize the warpage in unfilled plastics, multiple separate cooling circuits can be incorporated into the filling simulation. Figure 4.93 illustrates this effect using a molded housing part as an example (see also Figure 4.92). If all the cooling circuits are maintained at the same temperature, significant edge warpage occurs. However, by adjusting the cooling-media temperature in the inner core, the warpage can be minimized. Therefore, this approach should be considered during part and mold design, ensuring that the cooling circuits remain separate.

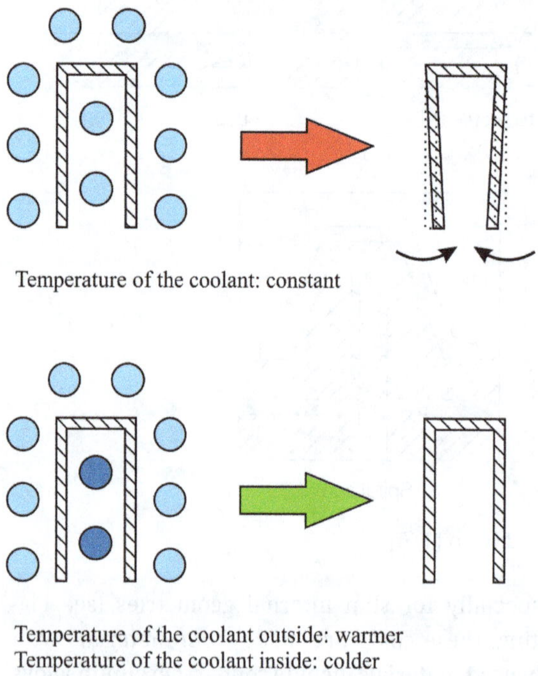

Figure 4.93 Minimization of warpage with separate cooling channels

Close-contour cooling systems are increasingly used for manufacturing complex molded part geometries. To achieve surface cooling beneath a flat molded part, for example, multiple inserts are manufactured and stacked. Additionally, additive manufacturing technologies such as laser casting, laser melting, laser sintering, high-temperature vacuum soldering, and various laser welding processes are now state of the art [1].

For effective and robust cooling, a maximum temperature difference between the inlet and the outlet of the cooling channels should not exceed 3 K [1]. The Reynolds number within the cooling system should be maintained between 10,000 and 20,000. Beyond 20,000, heat transfer improvement is minimal, while the energy consumption of the temperature control unit increases [8]. If the coolant temperature is very high, the outer surface of the mold should be insulated to prevent the cooling system from unintentionally acting as a heater in the peripheral areas of the mold [1].

4.4.5 Demolding Systems

Once the injection-molded part has cooled and solidified, it gains sufficient mechanical strength to be removed from the mold. Ideally, the part detaches from the mold cavity upon opening or separates from the cores due to gravity. In practice, however, molded parts often remain adhered to the mold cavity due to undercuts, adhesive forces, or internal stresses. A properly designed demolding system is required to release and eject the part effectively.

The demolding system is typically integrated with the ejector system of the injection molding machine and controlled via the interface of the machine. It is usually actuated electrically or hydraulically, creating a linear movement of the ejectors. In cases where this standard actuation is insufficient, additional pneumatic or hydraulic systems can be added, for example, to manage undercuts or threads. Most often, the ejectors are housed in the moving mold half. To ensure proper ejection, the mold and part design must encourage adhesion to the movable side. This is commonly achieved by incorporating undercuts or internal cores around which the melt flows, causing the part to shrink onto the core. At the same time, careful consideration must be given to draft angles and surface finishes on the internal part geometries to prevent excessive adhesion during demolding.

Demolding on the nozzle side is a less common approach and is only used when the part cannot adhere to the moving mold half. These solutions are typically costly and require custom mold designs, making them less desirable.

The demolding system plays a critical role in determining the structure and complexity of the injection mold. Consequently, molds are often given names based on their demolding system. The most common ejector systems include ejector pins, jaws, scrapers, sliders, and unscrewing elements for threaded parts.

At the final stage of part development, the geometry and mass of the part are established, allowing engineers to calculate the required demolding forces. To do this, the position of the molded part in the tool must be known. On the one hand, knowledge of the expected demolding force is important for the design of the demolding system with regard to the type, number, and positions of the ejector elements. On the other hand, it is essential to assess whether reliable demolding is feasible or if modifications to the mold layout or ejector system are necessary. In some cases, minor changes to the part design or the process parameters can significantly reduce the demolding forces.

There are two primary types of forces involved during demolding:

- *Opening forces*, which arise when the molded part is stuck in the mold due to insufficient shrinkage or excessive deformation.
- *Demolding forces*, which can be further categorized as:
  - *Breakaway forces*, which are present in all molded parts with cores. They result from the shrinkage of the part onto the core. Thin, slender ribs with insuf-

ficient draft angles may also experience breakaway forces, which can lead to breakage of the rib-forming tool elements.

- *Ejection forces*, which occur when the part remains in frictional contact with the core due to an inadequate draft angle.

While opening forces are generally less problematic when correctly accounted for, demolding forces are a key factor in ensuring a stable ejection during production. Figure 4.94 shows the factors influencing demolding forces.

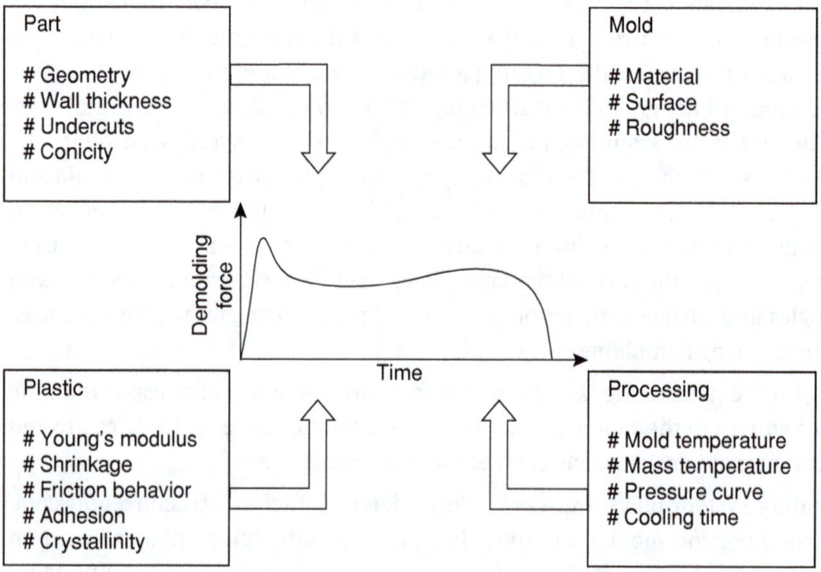

Figure 4.94 Influences on the demolding force (adapted from [9, 10])

In general, four key factors influence the level of demolding force, as illustrated in Figure 4.94. These factors primarily affect either the coefficient of static or dynamic friction, μ, or the cavity pressure. Due to the numerous influencing variables, complex physical interactions, and dependencies on external conditions such as pressure and temperature, accurately simulating demolding forces remains challenging and highly uncertain. As a result, mold designers usually rely on experience rather than precise calculations. In practice, the largest feasible draft angle is selected for cores, particularly when processing high-shrinkage molding compounds such as semi-crystalline thermoplastics. This draft angle accounts for the linear shrinkage of the internal part geometries. When undercuts or significant surface roughness are present, the draft angle must be further increased. As a practical guideline, an additional 0.5% to 2% per 2/100 mm of surface roughness may be required, depending on both the shrinkage of the plastic during cooling and the viscosity of the molten material.

4.4.6 Mold Standards

The only comprehensive summary of standards for injection molds can be found in the DIN standards of the German Institute of Standardization. For example, DIN ISO 12165:2019-11 provides normative references for all essential components of injection molding tools (both in text and figures).

The following is a list of characterizing labels:

- *WA* = Tool elements
- *P* = Undrilled platens
- *FP* = Drilled platens
- *FE* = Guide/centering element
- *BS* = Dies integrated in FP/WA
- *Af* = Aluminum mold design (mono-block)
- *F* = Springs

Examples of elements of the mold-filling system for thermoplastic materials that minimize sprues are:

- *HKB* = Hot runner manifold block
- *D1* = Mold, externally heated, with central gate
- *D5* = Mold, externally heated, with de-central gate
- *D8* = Mold, externally heated, with lateral gates

De-central and central single gates as well as 2- to 32-gate cold runner units are standardized. Open nozzles and needle-valve systems to balance the melt flow are available.

Mold elements can be heated using electrical heating cartridges *HP*, heating sleeves, and rigid and flexible coil heaters. Liquid media (oil, water) are used for both heating and cooling of mold platens, cores, and inserts. Here, temperature control channels *T*, turbo coils, and deflection lands are the preferred means. Miniature cooling channels together with high-pressure temperature control units, heat conducting pins, heat conducting tubes, alternative heat transfer media such as CO_2, N_2, compressed air, and liquid cooling media with phase changes are used to facilitate cooling of small areas with core diameters up to 1.5 mm. Interfaces for liquid cooling media are located at the outer mold surfaces, preferably equipped with recessed nipples *N* and couplings *K*.

Directly acting (part-contacting) demolding aids are labeled *Aw*. Ejector pins and sleeves are offered with different cross-sectional areas, lengths, and materials. Demolding aids not touching the product to be demolded, such as retraction devices, 2-step ejectors with single-stroke and multi-stroke, pneumatic safety clutches, and latch conveyors (round or angular) are indispensable, in particular to ensure the safe demolding of components with complicated functional sequences.

References for Section 4.4

[1] Menges, G., Michaeli, W., Morhren, P., *How to Make Injection Molds*, 3rd ed., Hanser, Munich · Vienna (2001)

[2] KIMW, *Guide to the Initial Setting of Injection Moulding Machines*, 4th ed., Horschler Verlagsgesellschaft mbH, Unna (2014)

[3] Johannaber, F., Michaeli, W., *Handbuch Spritzgießen (Injection Molding Handbook)*, 2nd ed., Hanser, Munich · Vienna (2004)

[4] Beaumont, J., *Runner and Gating Design Handbook*, 3rd ed., Hanser, Munich · Vienna (2019)

[5] Dangel, R., *Injection Molds for Beginners*, 3rd ed., Hanser, Munich · Vienna (2023)

[6] Shoemaker, J., *Moldflow Design Guide*, 1st ed., Hanser, Munich · Vienna (2006)

[7] Zöllner O., Optimised Mould Temperature Control. ATI Application Technology Information 1104 d, e Leverkusen: Bayer AG (1999)

[8] Pokorny, P., Raschke, F., Do Not Forget the Cooling. In: *Plastics Insights* 25 (2015) 6–7, pp. 43–46

[9] Korres, M., Optimierung eines Messwerkzeuges für das Entformungsverhalten im Spritzgießprozess (Optimization of a measuring tool for the demolding during injection molding process). Bachelor thesis, University of Applied Sciences South Westfalia (2016)

[10] Schlutter, R., *Injection Mold Coating Technology*, 1st ed., Hanser, Munich · Vienna (2024)

4.5 Foaming

Holger Ruckdäschel

Foams are frequently produced *in situ* during molding, either from the molten state, in the case of thermoplastics, or through chemical reactions, for thermosets and elastomers. These manufacturing approaches are considered sub-processes within their respective processing technologies; for example, foam extrusion within extrusion (see Section 4.2.3.7), foam injection molding within injection molding (see Section 4.3.5.6), and reactive foaming within polyurethane (PUR) processing (see Section 4.7). Additional technologies, particularly those related to bead foams, are covered in this section. To establish a solid foundation, the fundamental principles of polymer foaming are first discussed here.

4.5.1 Foaming Principles

Polymer foams are generated through the release of blowing agents, following either their prior homogeneous dissolution in the polymer matrix or their formation via chemical reaction. Upon release, the blowing agent forms gas nuclei that grow and stabilize into foam cells.

Foaming processes are generally categorized as either chemical or physical. In chemical foaming, the blowing agent is generated *in situ* through the thermal decomposition of additives (see Section 7.1.9). In contrast, physical foaming involves the use of externally introduced blowing agents – typically gases, supercritical fluids, or low-boiling liquids – that are dissolved into the polymer melt or solid and released upon a pressure drop or temperature increase.

Physical foaming is driven by a rapid change in thermodynamic conditions, through either a sharp temperature rise or a sudden pressure drop. This triggers phase separation and gas expansion, leading to foam formation.

In temperature-induced foaming, volatile blowing agents – such as hydrocarbons (*e.g.*, isobutane, pentane), carbon dioxide, alcohols (*e.g.*, ethanol), or nitrogen – are dissolved in the polymer. Heating causes these substances to vaporize, increasing the internal pressure and initiating the foam formation. Due to environmental concerns, many older blowing agents, such as chlorofluorocarbons (CFCs), have been phased out.

In pressure-induced foaming, the polymer melt is first impregnated with a blowing agent under elevated temperature and pressure. A rapid pressure drop – achieved, for example, by opening a valve or extruding through a nozzle – reduces the solubility of the blowing agent. This step initiates the phase separation of the polymer–blowing agent mixture and subsequent foaming.

Chemical blowing agents are typically solid compounds that decompose at elevated temperatures, releasing gas. Their decomposition temperature must match the processing window of the polymer, and a uniform dispersion within the matrix is critical. Common chemical blowing agents include azo compounds, *N*-nitroso compounds, and sulfonyl hydrazides. These can release between 100 mL and 300 mL of nitrogen per gram of blowing agent at temperatures ranging from 90 °C to 275 °C. The decomposition onset of azodicarbonamide (~230 °C) can be reduced to 150–200 °C with the use of initiators. For polyethylene (PE) foams, gas-releasing cross-linking agents are also employed.

In thermosetting and elastomeric foams, blowing gases may also arise from chemical reactions between resin components. These foams are typically manufactured using continuous belt foaming systems (*e.g.*, phenolic (PF), PUR, melamine-formaldehyde (MF)) or formed as blocks or molded parts (*e.g.*, PUR, PF, urea-formaldehyde (UF)); see also Section 4.7.

Depending on the foaming process, the resulting part exhibits a characteristic density profile across its cross section (see Figure 4.95). Foam injection-molded parts typically display profiles similar to the one in Figure 4.95(B) or (C), indicative of structural or integral foams (see Section 4.3.5.6). In contrast, extruded foams show a thin, compact outer skin, depending on processing, with profiles between the ones in Figure 4.95(A) and (B) (see Section 4.2.3.7). Bead foams, however, exhibit an almost-uniform density throughout the cross section, also illustrated in Figure 4.95(A).

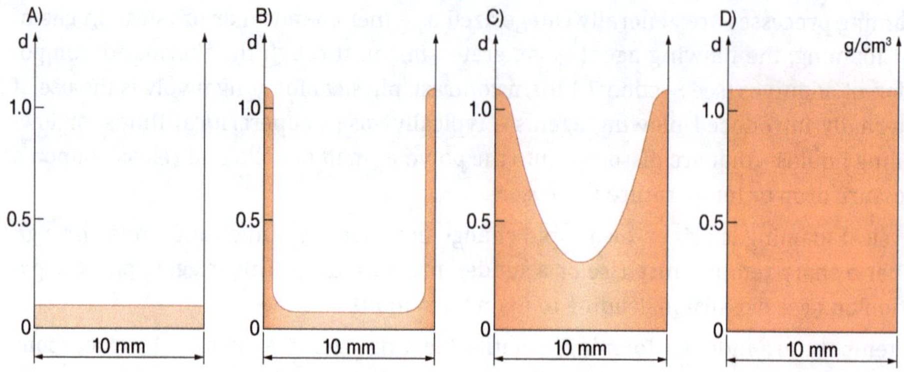

Figure 4.95 Example of density profile of structural foams over the cross section

A: Foam without or with minimal skin, density $d < 0.1$ g/cm^3; B: structural foam, bulk density: 0.2 g/cm^3; C: structural foam, bulk density: 0.6 g/cm^3; D: compact material, bulk density: 1.1 g/cm^3

4.5.2 Particle Foams

4.5.2.1 Classification

Particle foam parts are manufactured by molding foamed beads into specific shapes. These pre-foamed particles offer significant flexibility in designing complex geometries, while maintaining a low component weight and a uniform density.

As raw materials, both expandable and expanded particles can be used:

Expanded particles are already-foamed beads and can be produced from a wide range of polymers. The most common example of this type is expanded polypropylene (E-PP).

Expandable particles, on the other hand, contain a physical blowing agent. These are first pre-foamed by the user to the desired density using a pre-foamer, and then molded into near-net-shape components. However, the range of suitable polymers is limited, as both effective storage of the blowing agent and sufficient expansion during foaming must be ensured. As a result, expandable particle foams are often amorphous polymers with a glass transition temperature above 100 °C. Key advantages of expandable materials are their high bulk density during transport, allowing significantly more material mass to be shipped per unit volume, and their ability to adjust the density at the molder. The most widely used commercial example of this category is expanded polystyrene (E-PS).

4.5.2.2 Expandable Particle Foams

The production of parts made from expandable particle foams typically involves three main steps: (1) manufacturing of blowing-agent-loaded, expandable beads; (2) pre-foaming of these beads into foamed particles; and (3) molding of the foamed beads into final parts.

Two primary processes are employed for the production of expandable beads. The dominant method, in terms of production volume, is suspension polymerization of styrene into polystyrene (PS) with the addition of blowing agents, primarily pentane. This process yields bead sizes ranging from 0.2 mm to 2.0 mm. Alternatively, bead foam extrusion is used, wherein the polymer is first melted and then impregnated with a blowing agent under high pressure. The mixture is subsequently pelletized via pressure-assisted underwater granulation. The elevated water pressure and the simultaneous cooling during this step prevent premature foaming and minimize blowing agent loss.

Following bead production, the expandable beads undergo pre-foaming. This is typically achieved using steam, particularly for E-PS, at temperatures between 80 °C and 110 °C. Expansion ratios exceeding 80 can be attained. Depending on the process design, either batch (discontinuous) or continuous pre-expanders are used. Water vapor diffuses into the polymer matrix, aiding the foaming process. Recently, infrared (IR) heating has emerged as an alternative foaming method, enabling foaming at higher temperatures and without the use of water.

In the final step, the pre-foamed beads are molded into parts or blocks, which can reach dimensions of up to 1.25 m × 1.0 m × 8.0 m and bulk densities as low as 10 kg/m^3. Molding is typically carried out using steam, with temperatures for E-PS ranging from 110 °C to 120 °C and pressures from 0.5 bar to 1.5 bar. Under these conditions, the residual blowing agent inside the beads is activated, causing further expansion and facilitating the fusion of the beads. This results in dimensionally stable, low-density molded parts.

4.5.2.3 Expanded Particle Foams

Expanded particle foams differ from expandable foams in that they omit the pre-foaming step. Consequently, the production process consists of only two stages: (1) manufacturing of expanded beads and (2) molding of the foamed beads into final parts.

The direct production of foamed beads enables the use of a wide variety of polymer materials. Generally, two distinct manufacturing routes are employed. The first is a batch-wise autoclave process, which begins with solid microbeads. Commonly used blowing agents include alkanes, carbon dioxide, and nitrogen. During this process, the beads are saturated in a stirred, water-filled autoclave under elevated temperature and pressure. Subsequent rapid depressurization into a secondary vessel induces foaming of the beads.

The second route involves continuous extrusion. As in the production of expandable beads, the polymer melt is impregnated with a blowing agent under high pressure and then pelletized through underwater granulation. However, in this case, the lower water pressure and the controlled thermal conditions promote *in situ* expansion and stabilization of the beads during pelletization, resulting in foamed particles.

Expanded polypropylene (E-PP) is the most prominent material within this category of particle foams. E-PP is predominantly processed via the autoclave method. A characteristic feature of E-PP is the formation of a dual melting peak in its crystalline structure due to thermal treatment in the presence of physical blowing agents. This crystalline morphology facilitates subsequent molding into finished parts. In addition to E-PP, a growing number of expanded bead foams are now available, produced either through autoclave or extrusion techniques. These include expanded polylactic acid (E-PLA), polyethylene terephthalate (E-PET), thermoplastic polyurethane (E-TPU), and polyamide (E-PA).

The final molding step for expanded beads mirrors the one used for expandable bead foams. Depending on the specific polymer and application requirements, either steam-chest molding or, more recently, radio-frequency (RF) welding is employed to fuse the beads into shaped parts or blocks. The processing parameters vary significantly depending on the polymer type, target part density, and design specifications.

4.6 Casting

Casting is a pressure-less process for the manufacture of components or semi-finished products. For cast films, solutions, dispersions, or pastes are cast through a wide slit either onto a revolving belt of bare metal (*e. g.*, nickel) or onto a turning roll and subsequently directed past a heating unit to remove the solvent or to support the gelling of the paste.

When casting into larger molds, the material is filled with the help of gravity or centrifugal forces. Among thermoplastic materials, only semi-crystalline plastics can be cast, because they form low-viscosity melts. Here, the melt is extruded into an open mold and cooled without additional external pressure.

Solid, semi-finished goods, such as sheets, pipes, or rods are manufactured by polymerization inside the mold. Examples are PMMA semi-finished goods, *e. g.*, sheets that are cast between glass plates and polymerized to very high molecular weight, thus providing improved characteristics such as higher hardness and stress cracking resistance. However, the polymerization time may reach up to one week. Another example is cast PA 6 which can be processed pressure-less in light molds by anionic polymerization of caprolactam to form semi-finished goods. It can also be molded into solid components such as marine propellers, rollers, or gear rings.

Rotational casting is used to manufacture hollow thermoplastic components. The thermoplastic powder is filled into a pre-heated mold where it melts and forms an adhering layer at the mold wall. The hollow component can be demolded after cooling and opening of the mold. Rotational casting is suitable for large components and small production runs.

4.7 Polyurethane (PUR) Processing Technology

4.7.1 General Principles

Depending on their pre-polymerization, polyurethanes can be processed as thermoplastic or reactive materials, resulting in products with properties ranging from very rigid (in other words resembling thermosets) to rubber-elastic. In this section we will introduce a special technology that was developed for the processing of reactive precursors into polyurethane components (polyurethane (PUR) processing technology).

Technically, polyurethane "processors" are manufacturers of a plastic material, where the liquid precursors undergo a chemical reaction. Either all components react simultaneously ("one-shot process") or the reaction proceeds successively in two steps ("prepolymer process"). Accordingly, production lines for the manufacture of polyurethanes include the characteristic elements of a chemical reactor: vessels with mixing devices, pumps, pipelines, valves, among others. Construction and operation of PUR production lines, including transport and storage of the raw materials, are subject to regulatory control and permission. In addition, such production lines require safety equipment, *e.g.*, metal vats for storage tanks and other containers, overfill safeguard, exhaust and temperature control systems, as well as personal protection equipment (safety glasses, gloves, coveralls).

Figure 4.96 schematically shows the general design and operation of a PUR production line. The principal components A and B (polyisocyanate and polyol) are transferred from storage to working tanks, heated to the required temperature, and metered into the mixing head. The mixture is discharged from the mixing head onto a substrate or in a mold where it cures.

Solid raw materials (*e.g.*, for PUR elastomers) have to be melted, dehydrated, and vented in the working tank first. One of the additives C–F always is a crosslinking agent or a chain extender.

The PUR manufacturer can meter and add different additives separately, which allows for control of the formulation and thus of the properties of the manufactured product. Addition can occur with either a low- or high-pressured metering unit. Direct metering into the mixing head is also possible.

PUR production lines can be adapted to extreme processing conditions:

- Processing of raw materials that are liquid at room temperature with viscosities ranging from 5 to 20,000 mPa·s or of molten polyester polyols,
- Metering and mixing at ratios ranging from 1 : 100 to 1 : 1 for part weights from as little as few grams to 100 kg,
- Flow rate adjustment of the metering unit to match the reactivity of the system,

- Discharge technologies for continuously working lines and feeding systems for batch processing,
- Processing of reaction components filled with granular, flaky, or fibrous additives.

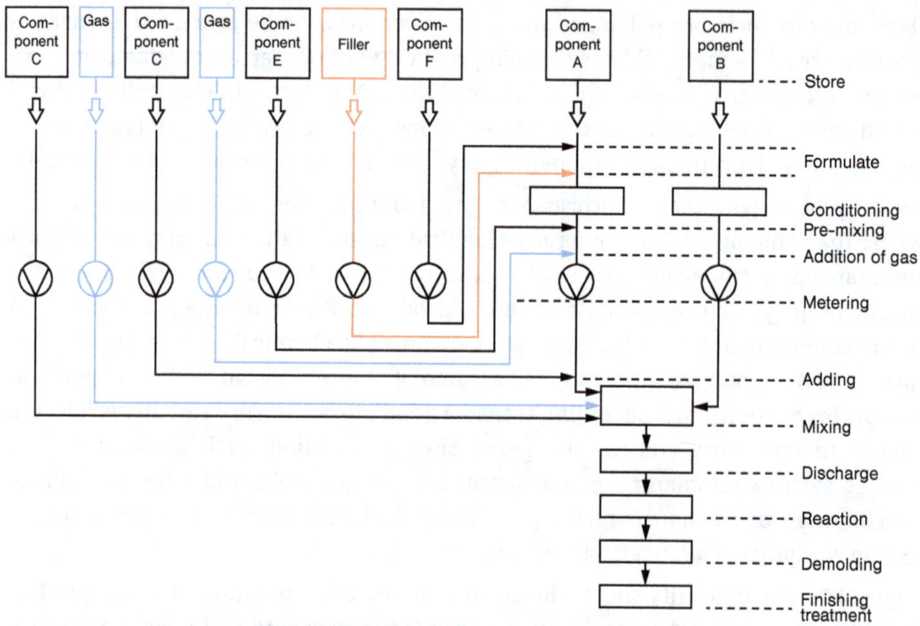

Figure 4.96 Scheme of a production line for PUR systems

The addition of 1.5 to 2 parts CO_2 to the reaction mixture is used to reduce bulk densities by approx. 10 to 20%, in particular for flexible foams (continuous and batch productions). In *inline-metering* systems, liquid CO_2 is added to one of the main components right before it enters the mixing head, where it goes into solution. In CO_2-*batch processes*, gas is added in liquid or gaseous form to one of the main components in a separate processing step. The mixture has to be maintained at equilibrium pressure of approx. 20 bar.

The bulk density of molded components can also be reduced by controlled evacuation of the specifically designed mold.

4.7.2 Manufacturing Steps

Table 4.11 summarizes PUR systems and their respective processing technologies. The first column lists the major PUR-based materials, while the following columns list the processes and process steps required for their manufacturing. The type of required machine systems is determined by viscosity and reactivity of the polyurethane precursors.

Table 4.11 PUR-Systems and Processing Technology

Processing steps → PUR-system ↓	One-shot process	Pre-polymer process	Pre-mixing	Intro-duction of gas	Adding	High-pressure-injection compounding	Low-pressure stirring mixing	Continuous manufacturing	Batch processes	Block manufacturing lines	Twin conveyor lines	Mold base	RIM	Demolding
Flexible foam														
Block	x				x		x	x		x				
Form	x					x			x			x		x
Fill	x					x			x			x		x
Rigid foam														
Block	x				x		x	x		x				
Hollow	x		x			x			x			x		x
Sheets	x		x	x		x		x			x			
Pipes	x		x			x			x					
Spray foam	x					x			x					
Integral foam rigid + flexible	**x**		**x**	**x**		**x**			**x**			**x**	**x**	**x**

Table 4.11 PUR-Systems and Processing Technology *(continued)*

Processing steps → PUR-system ↓	One-shot process	Pre-polymer process	Pre-mixing	Introduction of gas	Adding	High-pressure-injection compounding	Low-pressure stirring mixing	Continuous manufacturing	Batch processes	Block manufacturing lines	Twin conveyor lines	Mold base	RIM	Demolding
Elastomers														
Hot and warm cast		x					x		x			x		x
Cold cast	x						x	x	x			x		x
Spray	x						x	x						
Casting resins	x						x		x			x		x
PUR rubber		x							x			x		x
Thermo-plastic	x	x					x	x						

4.7.2.1 Low- and High-Pressure Machines

Slow rotating gear pumps are used to meter highly viscous polyester polyol melts. Because of their low operating pressures (max. 40 bar), these machines are called *low-pressure machines* (Figure 4.97). Mixing occurs in mechanical stirring devices (Figure 4.98).

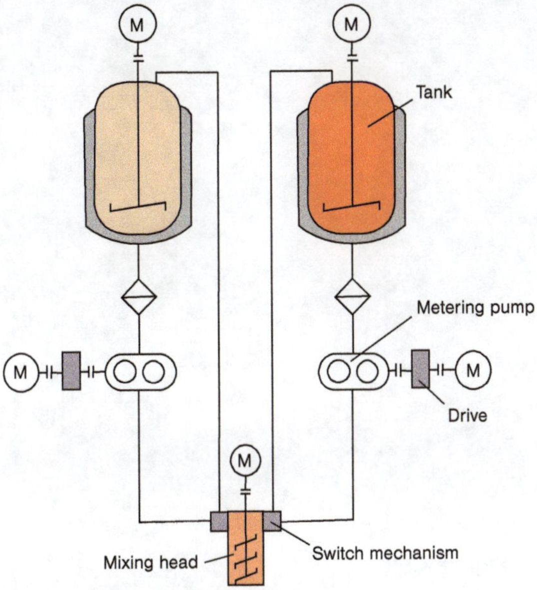

Figure 4.97
Flow diagram for a low-pressure machine

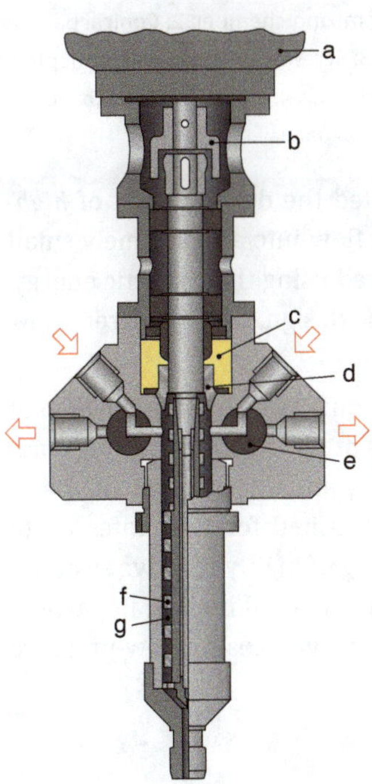

Figure 4.98
Agitator device for intermittent operation
a: Drive motor
b: Drive coupling
c: Lubricant
d: Mechanical seal
e: Switch
f: Paddle agitator
g: Mixing chamber

An alternative is the *low-pressure high-frequency mixer* (n = 150 to 250 s^{-1}), see Figure 4.99. In the cylindrical part of the screw mixing chambers are arranged as grooves or bores into which additional components can be metered laterally through valves (it is possible to add more than two components).

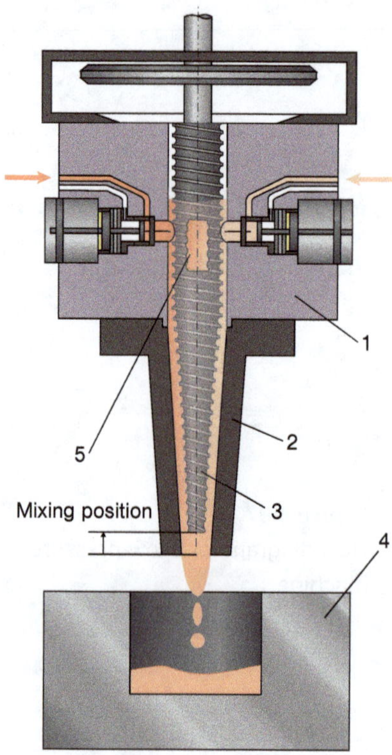

Figure 4.99
Working principle of friction mixer
1: Cylindrical part of mixing chamber, 2: Conical part of mixing chamber, 3: Screw with shallow flight depth, rotating at high frequency, 4: Open mold, 5: Mixing chambers in cylindrical part

The availability of low-viscous polyether polyols initiated the development of *high-pressure machines* (Figure 4.100). Here, the components flow into an extremely small mixing chamber at high rates and are continuously mixed using their kinetic energy. Figure 4.101 schematically shows one particular machine design, the concurrent flow mixing head.

The reaction components are metered into the injection mixing head with the help of piston pumps of variable design and working pressures of 100 to 300 bar. Critical advantages of high-pressure processes are exact metering and shot weight for parts with weights up to 50 kg; these processes are also well-suited for flow rates up to 400 l/min. These technologies also allow for the processing of highly reactive systems, minimal material losses, and last but not least less environmental impact, in particular when self-cleaning injection mixers are employed. The lower costs of low-pressure machines do not offset these advantages.

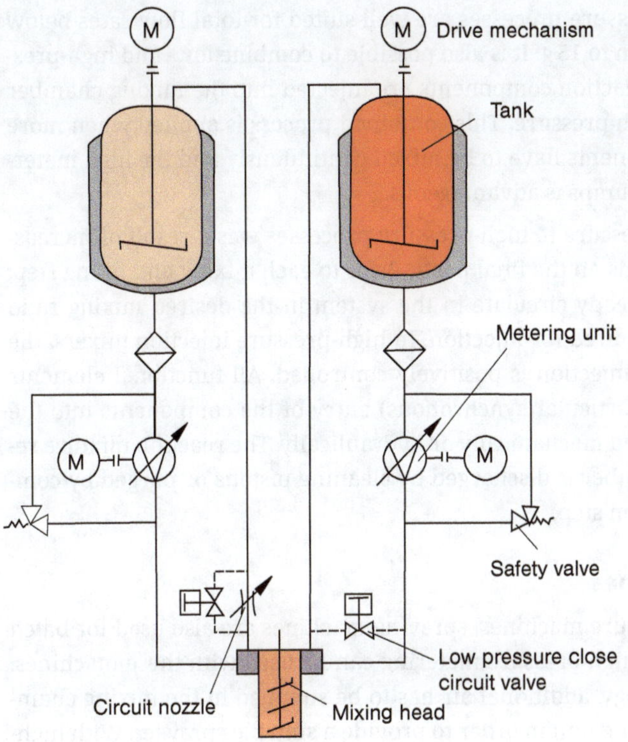

Figure 4.100 Flow diagram for a high-pressure machine (loop system)

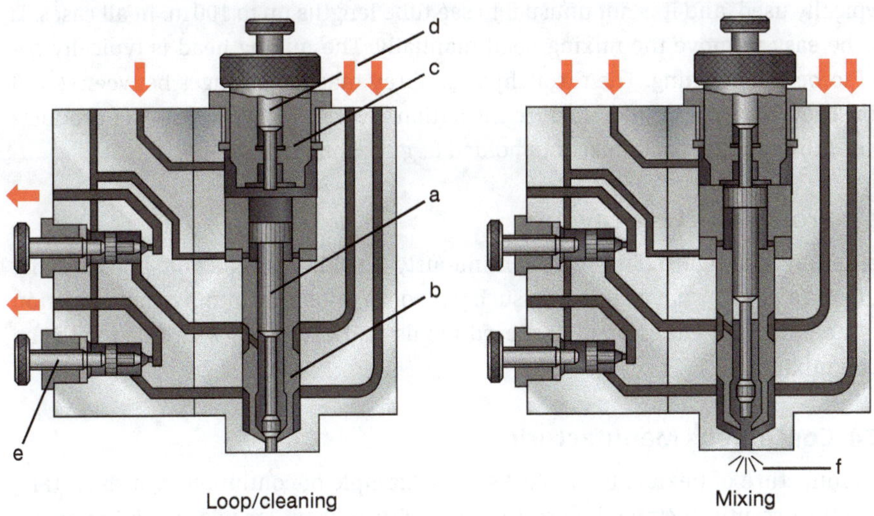

Figure 4.101 Schematic of concurrent flow mixing head (Design: Hennecke)

a: Internal valve pin, b: Hollow nozzle, c: Adjusting device for hollow nozzle, d: Adjusting device for internal valve pin, e: Restrictors, f: Mixing chamber

On the other hand, low-pressure processes are well-suited for total flow rates below 2 l/min or part weights down to 15 g. It is also possible to combine low- and high-pressure processes. Here, the reaction components are injected into the mixing chamber by piston pumps under high pressure. This combined process is applied when more than four individual components have to be mixed continuously and the high metering accuracy of the piston pumps is advantageous.

The transition from low-pressure to high-pressure processes was a result of increasingly higher quality demands on the final parts. Prior to each mixing and filling step, the component streams already circulate in the system in the desired mixing ratio and under the pressure required for injection. In high-pressure injection mixers, the switch from circulation to injection is positively controlled. All functional elements required for a proper (in particular synchronous) entry of the components into the mixing chamber are actuated mechanically or hydraulically. The reaction mixture remaining in the mixing chamber is discharged by cleaning pistons or purged by compressed air after the injection step.

4.7.2.2 Spraying Machines

Besides low- and high-pressure machines, spraying machines are also used for batch production. Both low- and high-pressure metering can be used with these machines. With low-pressure technology, additional air has to be supplied in the mixing chamber or at the exit of the mixing unit in order to provide a suitable spray jet. With high-pressure technology, the mixing pressure created by the high-pressure pumps can also be used to spray the components. Pneumatically driven one-cylinder piston pumps are typically used, and it is not unusual to see tube lengths up to 100 m. In all cases, is has to be easy to move the mixing head manually. The mixing head is typically designed to be self-cleaning. The throughput of spray machines ranges between 2 and 7 l/min. They are used to manufacture insulation layers as well as skin-like products that are subsequently back-injection molded (*e. g.*, dashboards).

4.7.2.3 Production Line Layouts

We generally distinguish between continuously working production lines for the manufacture of semi-finished goods, such as blocks or sheets, and batch production lines. The latter use either open or closed molds that are batch-wise filled with the reaction mixture.

4.7.2.4 Continuous Manufacturing

The manufacture of flexible foam blocks is an example of continuous processes (Figure 4.102). Currently, rectangular cross section with approx. 2 m width and 1 m height are state-of-the-art. Bulk densities of the foam blocks ranging from 15 to 60 kg/m^3 and desired conveyor rates up to 10 m/min result in reaction mixture throughput rates from 50 to 600 l/min. Depending on formulation and curing time, this requires conveyor lengths of 30 to 100 m. Production lines also include conveyor systems with subsequent cross cutting and trimming systems as well as finishing and packaging machines.

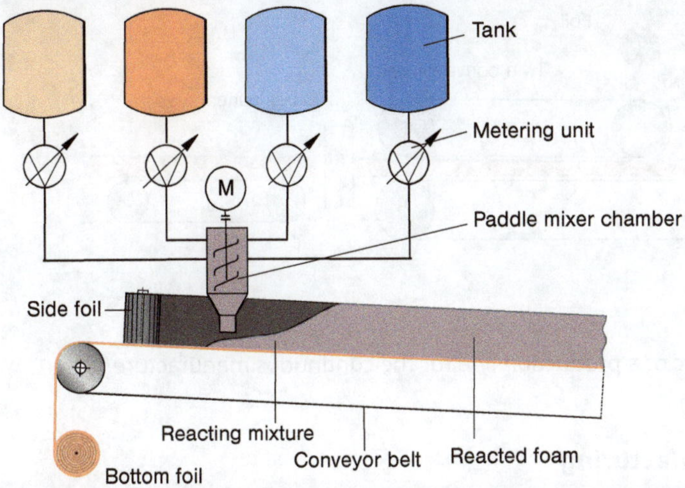

Figure 4.102 Schematic of a production line for the continuous manufacture of foam blocks

Another continuous manufacturing process is the production of rigid foam sheets on so-called *twin-conveyors* (Figure 4.103). In contrast to flexible foam production, during which large amounts of reaction mixture have to be discharged, continuous rigid sheet production requires the uniform distribution of extremely thin layers. The "wet unit" of the process is followed by reels (for rollable face layers), feeding stations (for rigid face layers), or cold roll forming lines for metal face layers. Rigid foam sheets with widths from 0.5 to 1.30 m and thicknesses of 5 to 200 mm are common. Production rates on modern twin conveyors reach up to 20 m/min. Bulk densities of 20 to 60 kg/m^3 result in throughputs of 3 to 40 l/min and twin conveyor lengths of 12 to 30 m.

A continuous free foaming process allows for the 1-step application of rigid foam layers (reinforced by glass beads with diameters of 10 to 60 µm) on pipes and other products. The compressive strength of these layers is approx. 15 MPa.

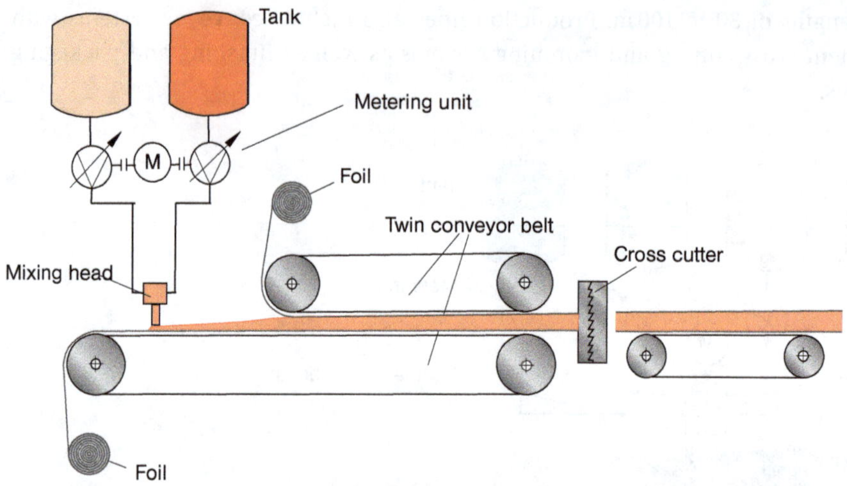

Figure 4.103 Schematic of a production line for the continuous manufacture of laminated foam sheets

4.7.2.5 Batch Manufacturing

In batch production lines the reaction mixture is filled intermittently into a mold cavity to create a molded part. Critical parameters for the design of these production lines include batch size, dimensions, and shape of the part. For the production of small and very small parts, the molds are placed on mounting plates or smaller clamping units which then move forward either continuously or intermittently. Both rotary tables and conveyor belts are used (Figure 4.104). The reaction mixture is filled into the open mold from no more than two mixing heads and deposited either in punctiform or linear patterns. The mixing head is carried along during the mixing phase when operating with continuous mold carriers. With intermittently moved mold carriers that operate according to the so-called *stop-and-go-principle*, the reaction mixture can be deposited either linearly or curvilinearly by robots or other manipulators to achieve improved wetting of the mold. A 24-position rotary table with two molds per palette has a production rate of 600 parts per hour. Both operating systems commonly use high-pressure machines.

Large and particularly complex parts are produced more economically with stationary production lines. Here, molds that are installed in a spatially close layout can either be filled with one mixing head using flexible pipelines or by moving the machine along the molds. Stationary machines are also common for the production of profile-like, elongated parts.

The filling rates for highly reactive systems are high and their curing times short so that removal of the mixing head and subsequent mold closure are no longer feasible. In these cases, reaction injection molding (RIM) technology is employed, using high-pressure metering systems with self-cleaning, add-on injection mixing heads. Each mold with a clamping unit is assigned a designated add-on mixing head (Figure 4.105). One metering unit may serve several mixing heads. Closed circular pipelines distribute the reaction component streams. Initially, the components circulate past the molds. When the system indicates that a mold is ready to be filled, the respective valves actuate the respective mixing head.

Similar to *multi-component injection molding* or *gas injection molding technology* for thermoplastic materials, PUR materials can also be manufactured into layered structures of different PUR systems or into hollow parts, see Section 4.3.5.4 and Section 4.3.5.5. *Overmolding or back injection molding* is also possible, in particular for textiles, see Section 4.3.5.8.

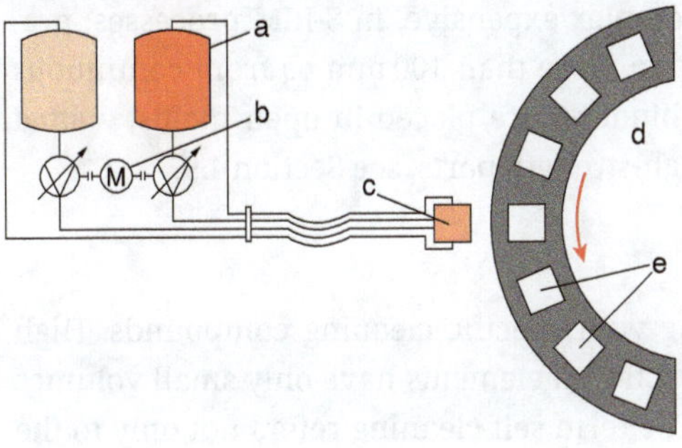

Figure 4.104 Schematic of a production line with movable molds and a mixing head that can be adjusted in charging/feeding direction

a: Working tank

b: Metering unit

c: Mixing head

d: Rotary table

e: Molds

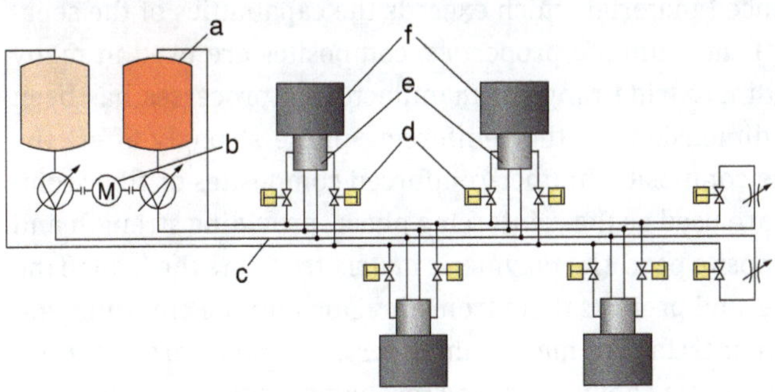

Figure 4.105 Schematic of a production line with stationary molds, add-on mixing heads, and multi-point dispensing unit (RIM-production line)

a: Working tank, b: metering unit, c: closed loop network, d: sequence valves, e: add-on mixing heads, f: mold clamping units

4.7.2.6 Manufacturing of Fiber-Reinforced Components

In *reinforced reaction injection molding (R-RIM) processes*, approx. 30% short glass fibers (0.1 to 0.3 mm) are added to the reaction mixture. When foaming systems are used, the processes are referred to as LD-R-RIM (low density R-RIM) processes. Here, the achievable strength is relatively low.

In *long fiber injection* or natural fiber injection processes, fiber bundles (2.5 to 10 cm length) are fed into the discharge pipe via a cutting unit in a special processing head, which can be added to any RIM metering unit. The programmed processing head distributes the PUR-fiber mixture according to the desired wall thickness. Some machines allow the simultaneous addition of other powder/granular additives, such as wood flour, or CO_2. This process is advantageous for multi-dimensional parts for which *structural RIM processes* are too complex/expensive. In S-RIM processes, preforms with filler fiber lengths from 10 up to more than 100 mm or from continuous fiber mats (typically with thermoplastic binders) are placed in open molds, wetted with PUR compounds, and molded into high-strength parts, see Section 4.9.

4.7.2.7 Cleaning

Mixing chambers are cleaned by purging with specific cleaning compounds. High pressure mixing chambers and their downstream elements have only small volumes and are therefore cleaned by forced air. The term self-cleaning refers not only to the cleaning of the mixing chamber (typically by rods), but also the cleaning of all downstream elements.

4.8 Manufacturing of Fiber-Reinforced Plastics

Composite materials are materials generated by the combination of individual components to achieve an enhanced material which exceeds the capabilities of the separate components. Because of their unique properties, composites are used in many industries and a correspondingly wide range of manufacturing processes has been developed. The shape and dimensions of the reinforcing phase strongly affect the mechanical properties of the composite. In fiber-reinforced composites or fiber-reinforced plastics (FRP), fibers are used as the reinforcing phase, providing strength and stiffness to the finished composite part. The polymeric matrix transfers the load to the fibers, maintains their shape, and protects them from abrasion and the environment. The load is transferred at the interface of matrix and fibers, and therefore the interfacial bond between the constituents determines the mechanical properties and performance of the composite. Besides the bonding properties of fiber and matrix, the type and quantity of matrix and reinforcing fibers determine the final properties. Glass fibers are the most common type of reinforcing fibers because of their low cost and high strength. However, a number of alternative fibers are available, including carbon, aramid, and natural fibers. More information about reinforcing fibers and their properties can be found in Section 7.3.1. Polyester resins are the most commonly used thermoset matrix resins, followed by epoxies, phenolic, and polyimide resins. Common metering, mixing, and dispersion devices are used for the manufacture of reactive resins, see Section 4.1. The most commonly used thermoplastic resins are polypropylene, polyamides, and other engineering plastics. The most commonly used processing machines are injection molding machines (Section 4.3) and presses (Section 4.9). Depending on the application and the process, highly-filled composites can have fiber contents up to 70% by weight. Because of their light weight and their high energy absorption during crashes, fiber-reinforced plastics are one of the fastest growing manufacturing materials.

With respect to processing and material classification purposes, FRPs are divided into discontinuous and continuous fiber composites, Figure 4.106. Discontinuous fiber composites are categorized into short and long fiber-reinforced composites, where the ratio of fiber length to fiber diameter, or *aspect ratio*, is an important property. Long fibers are those whose critical length is significantly longer than the critical length required to transfer the load from the matrix to the fiber. Long fibers are considered to exceed an aspect ratio of 100, whereas short fibers have an aspect ratio of less than 100. Continuous fiber-reinforced composites are used in structural applications where increased strength and stiffness are required. By stacking plies of continuous fibers, a composite laminate is formed. The fibers in each ply have a defined orientation and the arrangement of the plies is designed to enhance the strength of the composite in the primary load direction.

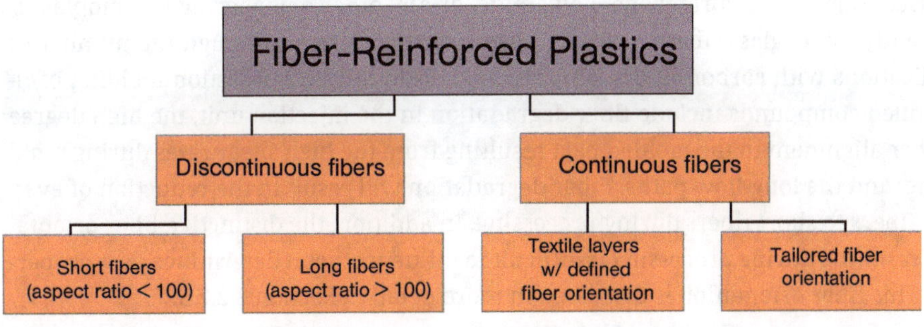

Figure 4.106 Classification of fiber-reinforced plastics

A variety of FRP manufacturing processes have been developed for different industries. The processes range from highly automated production lines in the automotive industry to small batch productions that require hand layup techniques for the aerospace industry. In general, the processes can be classified by the following criteria:

- Process principle,
- Matrix and/or fibers used,
- Fiber length,
- Freedom of design,
- Performance,
- Production volume/cycle time,
- Degree of automation.

The following sections summarize the main composite manufacturing processes with respect to fiber length (from short to continuous), matrix system, and increasing degree of automation (also reflecting higher production volumes).

4.8.1 Short-Fiber Reinforced Composites

4.8.1.1 Injection Molding

Discontinuous fiber-reinforced plastics have been used in injection and compression molding for a long time. *Injection molding* of FRPs is the most common process to date. The process is explained in detail in Section 4.3. Depending on the application, the matrices may be polypropylene, polyamide, or any other engineering thermoplastic. Currently, short glass fibers have the biggest market share, although the number of applications with carbon fibers is increasing. Issues limiting injection molding of fiber-filled compounds include fiber degradation in the injection unit, the high degree of fiber alignment in the molded part resulting from the high shear rates during mold filling, and the long flow paths. Fiber degradation will result in the reduction of even long fibers to short fibers during processing. In addition, the distinctive fiber orientation resulting during processing is difficult to influence, yet it determines part properties. This fiber orientation is discussed in more detail in Section 3.2.2.5.

4.8.1.2 Bulk Molding Compounds

Injection-compression molding is used to process thermoset matrices reinforced with discontinuous fibers. The material used here is called *bulk molding compound (BMC)*. BMC is composed of an unsaturated polyester resin reinforced with a 20–25% volume fraction of randomly oriented, 10 mm long fiber bundles. BMC is either injection molded or injection-compression molded. The injection-compression molding of BMC is very similar to the transfer molding process (see Section 4.8.4.1).

BMC materials are compounded in an internal batch mixer and subsequently extruded into the shape of a large strand, also known as a *BMC charge* (see Section 4.3.3). This charge is then fed into a cylinder above the BMC press. The cylinder injects the charge into the mold of the press. Closing the press causes the material to flow and fill the mold cavity, Figure 4.107.

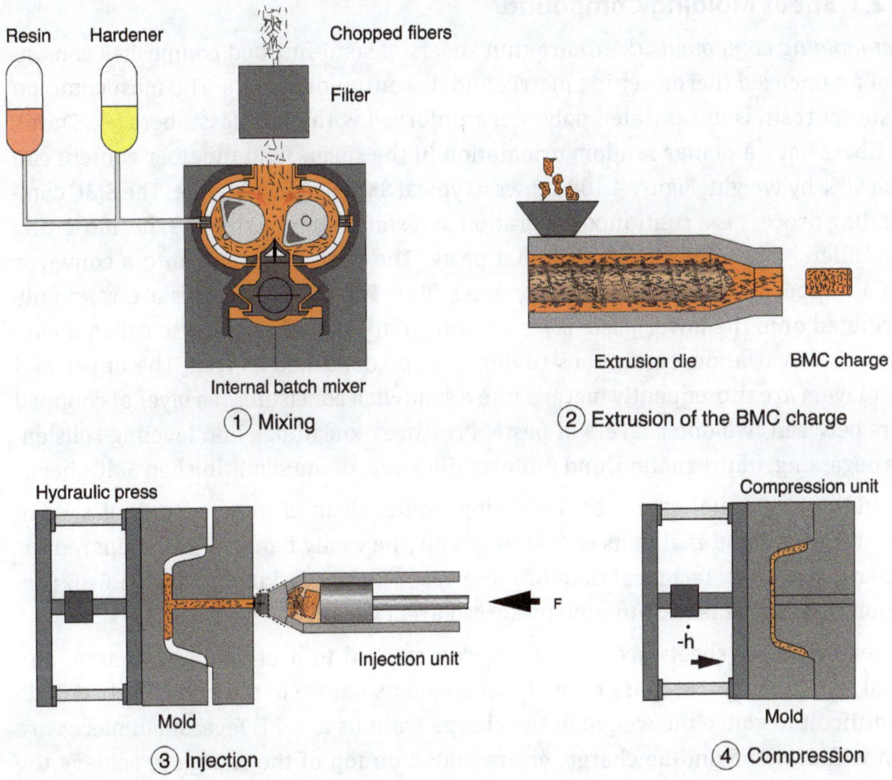

Figure 4.107 BMC injection-compression molding process

The main benefit of injection-compression molding over compression molding is its potential for automation, which makes it a suitable process for large batch production. In addition, the combination of injection and compression molding leads to a lower degree of fiber alignment and fiber attrition compared to injection molding. BMC is used in the production of smaller items, such as housings, valve covers, and other engineering applications that require strength.

4.8.2 Long-Fiber Reinforced Composites

Thermoset resins reinforced with long fibers exhibit higher strength than those reinforced with short fibers. In addition, here it is possible to create desired anisotropies by aligning the long fibers in the preferred direction.

4.8.2.1 Sheet Molding Compound

Sheet molding compounds (SMC) are thin sheets of semi-finished composites consisting of an uncured thermosetting matrix and discontinuous fibers. The most common thermoset resin is unsaturated polyester reinforced with long glass fibers (~ 25 mm). The fibers have a planar random orientation in the sheets, and the fiber content can reach 50% by weight. Figure 4.108 shows a typical SMC production line. The SMC compounding process is a continuous operation and starts with mixing the thermosetting resin, fillers, and other additives into a paste. The paste is spread onto a conveyor with an upper and lower polymeric carrier film. The chopped fibers are randomly distributed onto the lower paste layer forming a dry bed of fibers, also called a *mat*. In a process variation, continuous rovings can be deposited as well. The upper and lower layers are subsequently merged into a sandwich consisting of a layer of chopped fibers between two outer layers of paste. Pressure-, kneading-, and leveling rolls ensure degassing, impregnation, and uniform thickness of the semi-finished SMC sheets.

The addition of metal oxides as thickening agents ensures storage times of several days at 30–50 °C and that mats can subsequently be easily handled and prepared for further processing. Technical data of these resin mats: working width 1–1.6 m, production rate 2–10 (max. 20) m/min, preferred areal weights of 3–6 (max. 12) kg/m^2.

The semi-finished sheets are cut into pieces, stacked to a pre-defined charge, and placed into the mold. Because of the heterogeneous nature of the semi-finished SMC, it is difficult to gauge the weight of the charge from its area. Hence, small pieces are often withdrawn from the charge, or are added on top of the charge to achieve the desired weight of the final part. The charge usually covers 30–70% of the mold surface. The mold is closed rapidly first until both mold halves touch the SMC charge. Then, closing slows down, causing the material to flow and fill the mold cavity. Once the mold is completely filled, a constant force is maintained on the material as the curing of the thermosetting resin progresses. When the reaction is almost completed, the part is removed and allowed to cool. The cycle is repeated when a new charge is placed on the heated mold surface. Preheating the charge speeds up the curing process and is therefore desirable for some applications.

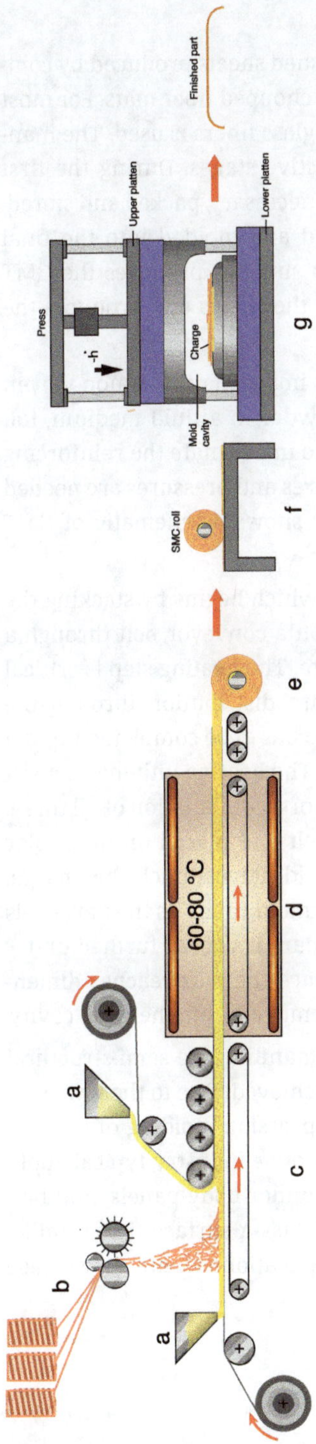

Figure 4.108

Continuous manufacture of resin mats

a: UP-resin/filler compound, b: Roving, c: Resin mat unit, d: Curing zone, e: Winding, f: Cutting table, g: Press

4.8.2.2 Glass Mat-Reinforced Thermoplastics

Glass mat-reinforced thermoplastics (GMT) are semi-finished sheets produced by combining a thermoplastic resin with continuous, woven, or chopped fiber mats. For most applications, a combination of polypropylene resin and glass fibers is used. The manufacturing process of GMTs is divided into two distinctive stages. During the first stage, the semi-finished GMT sheets are produced and, if necessary, packed and stored. During the second stage, the GMT sheets are re-heated and molded into the final shape using a compression molding process. Usually, a supplier produces the GMT sheets and sells them to a manufacturer that processes the sheets and produces the final parts.

The semi-finished GMT sheets can be produced by melt impregnation of non-woven glass mats, or by mixing sliced fibers with polymer powder in a fluid medium, followed by straining, drying, and consolidation. In order to impregnate the reinforcing fiber mats with the thermoplastic resins, high temperatures and pressures are needed because of the high viscosity of the resin. Figure 4.109 shows a schematic of GMT sheet production via melt impregnation.

The second stage is the actual molding of the material, which begins by stacking the GMT sheets up to a charge. The charge is then guided on a conveyor belt through a heating unit, which heats the material to the molten state. The heating step is critical in a GMT process because a homogeneous temperature distribution through the charge is required for molding. However, the heating step has to be completed rapidly to maintain short cycle times for a cost-efficient process. The heating unit may consist of radiative heating, contact heating, convective heating, or a combination of all three. After heating, the charge is lifted from the conveyor belt and placed on the cooler mold surface. The mold is closed rapidly until both mold halves touch the charge. Then, the mold slowly compresses the charge, filling the mold cavity as the part cools down. When the mold cavity is completely filled, the part is cooled further under pressure, allowing it to attain its structural integrity. Once the part reaches dimensional stability, the mold opens and the finished part is removed from the mold cavity.

GMT compression molding is a widely used process to manufacture semi-structural parts because of the mechanical properties that can be achieved. Due to the low cycle time and the high automation of the process, GMT compression molding offers economic advantages in large batch operations. In the automotive industry, typical applications are front ends, bumper beams, seat structures, under-body panels, and battery trays. However, parts made of GMT do not offer a class-A surface finish unlike parts manufactured of SMC. Thus, GMT is limited to applications where the surface finish is not important.

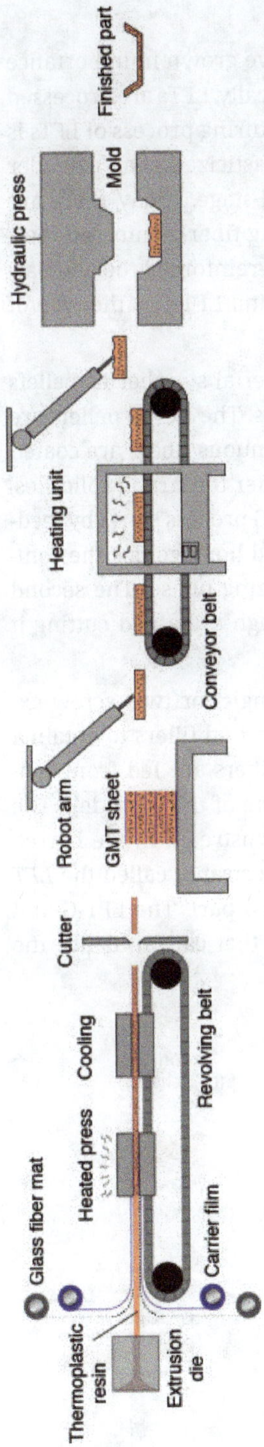

Figure 4.109
GMT sheet production and compression molding process

4.8.2.3 Long Fiber-Reinforced Thermoplastics

In recent years, *long fiber-reinforced thermoplastics (LFT)* have grown in importance in the field of discontinuous fiber-reinforced composites. Typically, LFTs are processed by compression molding and injection molding. The manufacturing process of LFTs is divided into two stages. In the first stage, the compound is plasticized in an extruder and, subsequently, molded into the final shape in the second stage. Today, there are various systems ranging from in-line direct processing of long fiber-reinforced thermoplastic (D-LFT) to the extrusion of prefabricated long fiber-reinforced thermoplastic granulates (LFT-G). The main distinction between D-LFT and LFT-G is the type of material initially used.

In the LFT-G process, the base material is supplied by a material supplier as pellets that consist of a thermoplastic matrix and incorporated fibers. The LFT-G pellets are usually produced in a pultrusion process, during which continuous fibers are coated with the polymer matrix by pulling the fiber through a die. After the strand solidifies, it is chopped into pellets that are 6–30 mm in length. The LFT-G process starts by feeding the semi-finished pellets into an extruder to plasticize and homogenize the compound. If necessary, additives are added during the plasticizing phase. The second stage of the process includes extruding the compound through a die and cutting it into a charge for the molding stage.

In the D-LFT process, the unreinforced resin is fed into a single- or twin-screw extruder, where it is melted and mixed with additives, pigments, and fillers to obtain a homogeneous material as shown in Figure 4.110. The glass fibers are fed from continuous rovings and added after the transition or melting zone of the extrude r. The fibers are pulled in by the rotation of the screw and cut downstream in the barrel. After the mixing phase, the compound is extruded, cut into a strand, called the *LFT charge,* and placed into a mold where it is shaped into a final part. The LFT-G and D-LFT charges have a distinct fiber orientation distribution that can influence the properties of the final part.

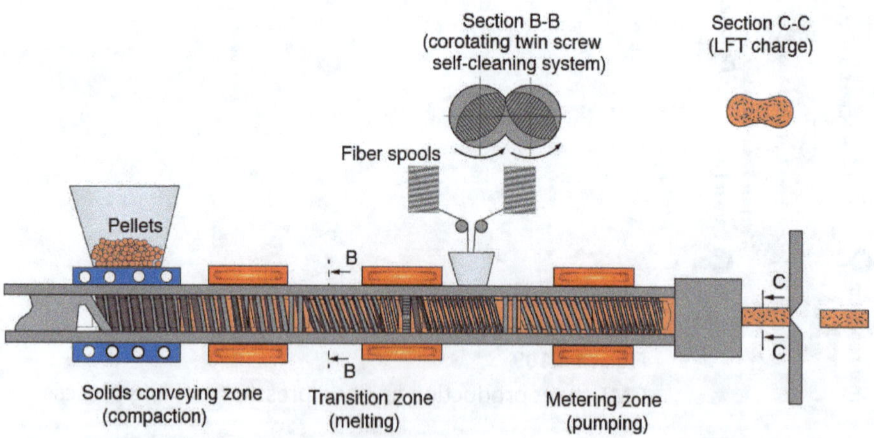

Figure 4.110 Direct processing of long fiber-reinforced thermoplastics (D-LFT)

4.8.3 Low Volume Liquid Composite Molding

4.8.3.1 Manual Techniques

Hand layup molding processes (hand lamination, layup, contact processes) are used to fabricate large components, such as boat hulls, with lengths up to 50 m, in low volumes. First, molds made from GFK, wood, sheet metal, gypsum, or other materials with non-porous surfaces are treated with a release agent; then the first, typically unreinforced, weather- and chemical-resistant resin layer (the so-called gel coat) is applied. Once this thin layer has reached gelation, fiber reinforcements in the form of mats, fabrics, or rovings are laid up and impregnated with resin so that the highest level of reinforcement is placed in the direction of highest loads in the final component. Another gel coat is applied as the final layer. Corrugated rolls are used to remove air from the layers and to press them together. Curing occurs at or slightly above room temperature over long periods of time.

The *fiber spray-up molding* process resembles hand layup processes. Cut rovings (20–50 mm in length) are mixed with pulverized resin and injected against a mold. Subsequently, they are manually compacted with rolls and de-aerated. Fiber orientation in the component can be influenced by the position and angle of the injection gun. Here, there is no need to cut reinforcing mats. Often, these two processes are combined. However, because of their high level of emissions and low reproducibility, both processes are increasingly replaced by RTM or vacuum-assisted processes.

4.8.3.2 Vacuum-Assisted Resin Infusion

The *vacuum-assisted resin infusion (VARI) technique* uses a continuous fiber reinforcement, such as woven or non-crimp fabric, that is impregnated with a low-viscosity liquid thermoset resin by forcing the resin into a fibrous structure trapped in a vacuum formed between the mold surface and flexible plastic film or vacuum bag. Similar to the hand layup molding technique, appropriate mold surface preparation is important to create a perfect seal and to render a part with a high quality surface. With the VARI molding technique, the fiber preform is cut and placed in layers over the mold surface. The fiber layers are then covered with three additional layers: 1) a treated release layer, sometimes referred to as *peel-ply*, 2) a flow medium that increases the permeability of the resin through the fibers, and 3) a vacuum bag (Figure 4.111).

Because the resin's resistance to flow is very high between the fibers, a flow medium is used. Its high permeability allows for long flow distances (typical are 500 mm), while the flow path through the fibers in thickness direction is very small. Taking advantage of this effect and using multiple inlets allows for large parts to be produced with this technique.

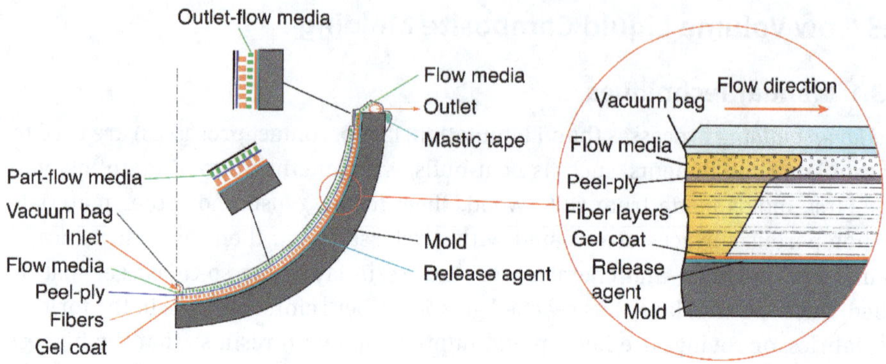

Figure 4.111 Schematic of the vacuum-assisted resin infusion process

4.8.4 High Volume Liquid Composite Molding

In contrast to the VARI technique for which a vacuum and only one mold half is used, resin transfer molding (RTM) and structural reaction injection molding (S-RIM) employ two mold halves to ensure controlled thickness and controlled mold surface on both sides. RTM and S-RIM are two similar liquid composite molding (LCM) processes that are well-suited for the manufacture of medium to large, complex, lightweight, and high-performance composite components primarily for the automotive and aerospace industries. In these processes, the fiber mat preform is placed in a closed mold to be impregnated by a low viscosity, reactive liquid resin in a transfer or injection process. The process characteristics of RTM and S-RIM differ in the resins used, mixing and injection set up, mold requirement, cycle time, fiber volume fraction, and suitable production volume. However, both LCM processes are recognized as the most feasible and structurally efficient approaches to mass-produce lightweight, high-strength, and low-cost structural composite components.

The preform is an assembly of dry (un-impregnated) reinforcement media, normally pre-shaped through a draping process and assembled with urethane-foamed cores into a three-dimensional skeleton of the actual part. A wide variety of reinforcements are available, such as woven and non-woven fiber products, die-cut continuous strand mats (CSM), random mats consisting of continuous or discontinuous fibers, knit braiding, two- and three-dimensionally braided products, or hybrid preforms made of layers of different types of media. The selection of the preform architecture depends on the desired structural performance, processability, long-term durability, and cost.

4.8.4.1 Resin Transfer Molding

The classic resin transfer molding (RTM) process uses a wide range of thermally activated resins, but because of their low cost, polyesters are the most commonly used resins in RTM. Other resins used include epoxies, vinyl esters, acrylic/polyester hybrids, acrylamide resins, and methyl methacrylate vinyl esters. These thermally activated resin systems rely on the heated mold walls to accelerate the chemical reaction. As a result, the filling times for RTM can be as long as 15 min, and the cycle time can reach one hour or longer, depending on the resin and application. Slow injection and reaction rates of some RTM systems can be used to achieve excellent impregnation of complex shaped parts. In addition, the low viscosity of the resin and the slow injection rate result in low injection-pressure and clamp-force requirements. As a result, RTM allows the use of so-called soft tools (*e.g.*, wood backed epoxy or aluminum molds) that require significantly lower investments compared to competitive manufacturing systems. Because of the relatively long cycle time, RTM is generally limited to low-volume production runs (*i.e.*, less than 10,000 parts).

4.8.4.2 Vacuum Assisted Resin Transfer Molding

The main challenges with RTM and S-RIM (see Section 4.8.4.1 and Section 4.8.4.4, respectively) technologies are the control of resin flow and fiber impregnation. Poor fiber impregnation can lead to voids between fiber and matrix, resulting in localized stress concentrations and possible crack initiation sites. One way to overcome these challenges is to use an assisting vacuum to force the resin through the network of fibers. This technique, schematically depicted in Figure 4.112 is referred to as vacuum-assisted resin transfer molding (VA-RTM). In contrast to the classic RTM process, only one mold half is necessary and complimented by a vacuum bag. However, it is also possible to evacuate the classical RTM mold.

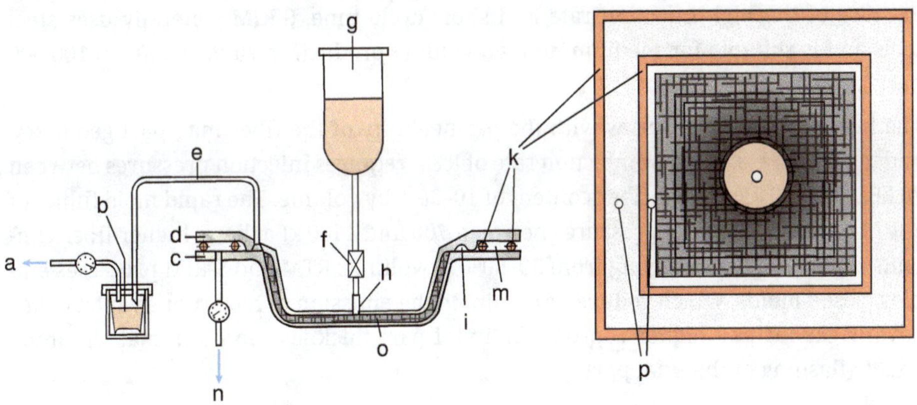

Figure 4.112 Vacuum-assisted resin transfer molding process

Left: cross section; Right: top view

a: Vacuum of 660–800 mbar, b: Excess resin container, c: Lower mold half, d: Upper mold half, e: Vacuum line to resin runner, f: Shut-off valve, g: Resin container, h: Injection point, i: Vacuum resin channel, k, m: Rubber seals, n: Vacuum of approx. 130 mbar, o: vacuum seal, p: GFP-component

4.8.4.3 Compression Resin Transfer Molding

In contrast to RTM and VA-RTM, compression resin transfer molding (C-RTM) or high-pressure RTM (HP-RTM) technologies use partially opened molds. The gap between the mold halves is sufficiently large to inject the resin into the cavity. Once a specific amount of resin is injected, the mold gap is closed, forcing the resin into the porous preform structure. Once the mold is closed, the preform is fully impregnated, and the resin is allowed to cure before demolding occurs. The main advantage of this process is the reduced time for resin injection and impregnation, facilitating the use of fast curing systems.

4.8.4.4 Structural Reaction Injection Molding

Structural Reaction Injection Molding (S-RIM) operates similar to RTM. However, the resins used in S-RIM are fast curing and mixing activated PUR-2K systems. Common resins used for S-RIM include urethane, acrylamate, and dicyclopentadiene. The chemical reaction is activated by impingement mixing of two highly reactive components in a special mixing head under high pressure, see also Section 4.7.2.6 and Figure 4.97. The mixture is subsequently injected into the mold at a lower pressure. The resin starts to cure as it impregnates the preform and forms the matrix of the composite. Because of the fast reaction rate and rapid increase in viscosity, the cavity has to be filled within a few seconds. In addition, the cycle time is as short as 1 min. Flow distances for typical S-RIM applications are therefore limited to 0.6–0.9 m from the inlet gate. In addition, the volume fraction and construction of the reinforcement have to be selected carefully to facilitate fast, complete filling before gelation occurs. Because of the high injection rate and short cycle time, S-RIM generally uses steel tools and is suitable for medium- to high-volume production runs (10,000 to 100,000 parts).

The injection pressure varies with the permeability of the fiber mat, part geometry, and injection rate. The low injection rate of RTM requires injection pressures between 70 kPa and 140 kPa, and a fiber content of 10–20% by volume. The rapid mold filling of S-RIM requires injection pressures between 700 and 1400 kPa; here, higher fiber contents can be reached, varying from 30–50% by volume. RTM and S-RIM processes employ closed molds, which reduce or eliminate the emission of hazardous vapor. Other advantages include higher reproducibility of part thickness and minimal trimming and de-flashing of the final part.

4.8.5 Filament Winding

Filament winding is used to manufacture high-strength, hollow structures, as well as other strong components. During filament winding, a fiber or roving is passed through a resin bath where it is coated or impregnated with a polymer, often a thermoset such as liquid epoxy resin that is mixed with an initiator, pigments, UV protectors, and other additives. Guides located before the resin bath control the tension experienced by the fibers. After impregnation, the fibers are usually pulled through a wiping device that removes any excess resin before gathering the fibers into a flat band. Figure 4.113 shows variations of impregnation bath configurations.

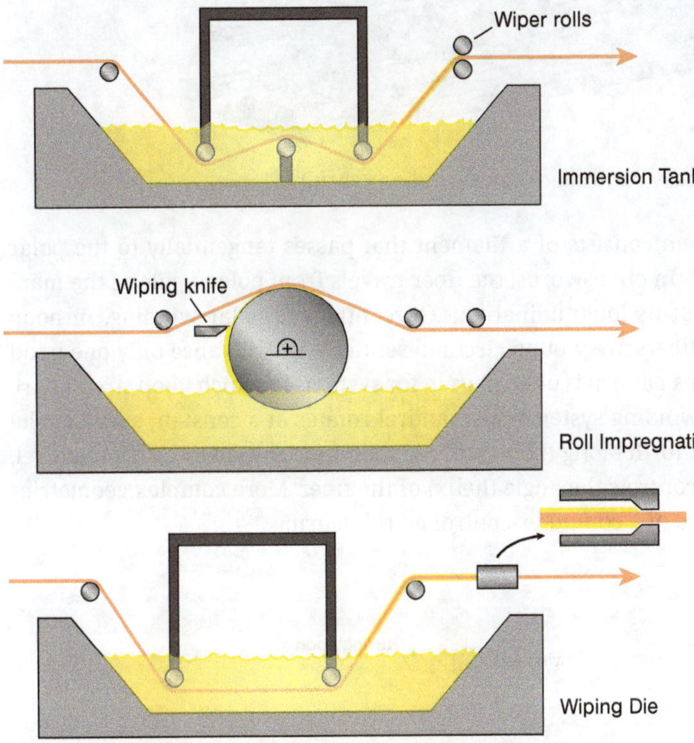

Figure 4.113 Impregnation bath configurations

The band is wound onto the body with an orientation that reflects the principal stresses and loads that the final product will experience. The final thickness is attained after multiple lateral displacements of the flat band are wound around the body. Once the part is fully wound, the flat band is cut and the part is taken to the next station where it is allowed to cure. The curing typically occurs in an oven because most of the resins used in filament winding are heat-activated thermosets. In order to prevent dripping of the resin and to achieve even heating throughout the final product, the part is supported by a rotating mandrel while being cured inside the oven.

Typical items manufactured using this technique are pressure vessels, pipes, or other tubular geometries. Figure 4.114 shows the schematic of a polar winding machine and Figure 4.115 a polar filament winding system for a pressure vessel.

Figure 4.114 Numerically controlled 5-axis winding machine

A polar winding system consists of a filament that passes tangentially to the polar openings of the vessel. In other words, the fiber travels from pole to pole as the mandrel arm rotates about its longitudinal axis. In contrast to polar winding, in hoop winding systems the fibers are wound circumferentially and advance only one band width per rotation. This pattern is often chosen for systems in which hoop stresses are dominant. In helical winding systems the mandrel rotates at a constant speed as the fiber travels back and forth along the axis of the mandrel. The speed of the mandrel, or of the fiber guide, controls the angle (helix) of the fiber. More complex geometries are wound with the aid of a computer-controlled robot arm.

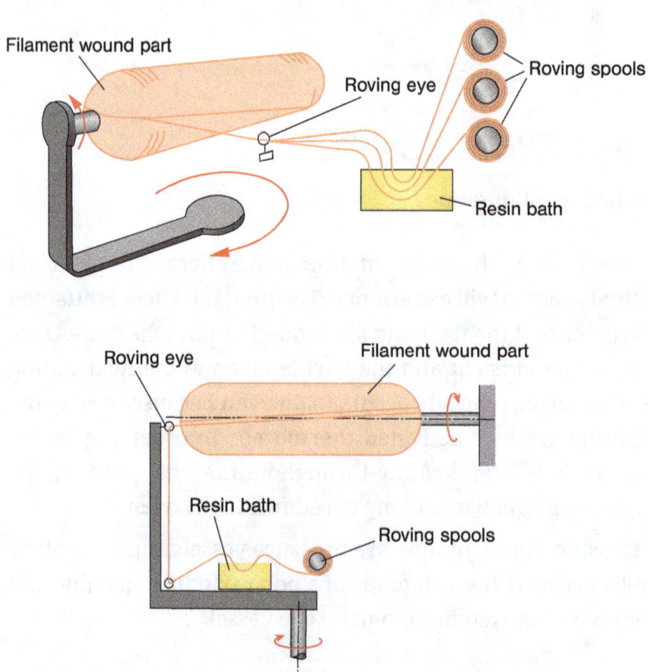

Figure 4.115 Polar filament winding system
Top: Tumbling system: Roving eye is fixed, mandrel moves around two axes
Bottom: Planet system: Roving eye describes circular path, mandrel can pivot around upper axis

4.8.6 Pultrusion

Pultrusion, schematically depicted in Figure 4.116, is a continuous, cost-effective process for the manufacture of high-strength profiles with controlled and uniform cross-sections.

Various types of reinforcing fibers are used in the pultrusion process, including fiber bundles, braided fibers, as well as woven and non-woven planar structures. During pultrusion, the fibers and fiber structures are passed through a thermosetting resin bath containing additives such as catalysts, initiators, and pigments. Next, the excess resin is wiped off the fibers and they are moved to a pre-forming vise to pre-shape the profile. The pre-forming vise often also removes excess resin. The pre-formed profile is then pulled through a heated die that gives the part its final shape and surface finish. Typical pultruded parts are hollow tubes, solid profiled I-beams, flat structures, handrails, and many more. Profiles are produced in thicknesses ranging from 1 to 30 mm and with pulling speeds of up to 5 m/min.

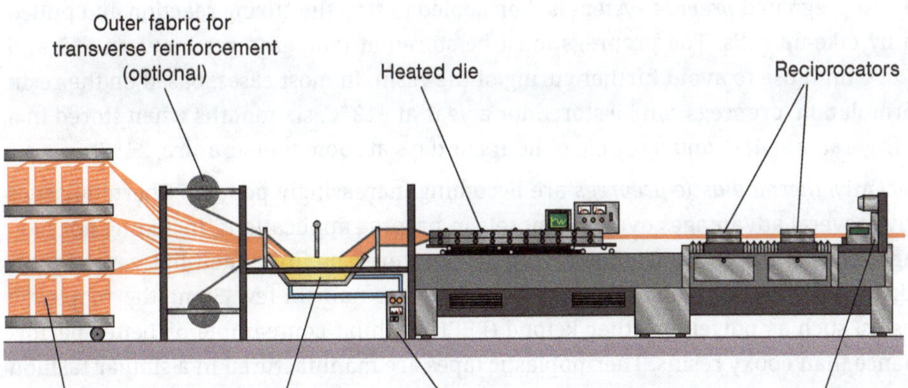

Figure 4.116 Pultrusion line

An alternative to the conventional pultrusion process uses an injection unit attached to the heating die, which replaces the resin bath. The principles of the process are maintained but volatile organic compound (VOC) emissions are eliminated and operators' direct contact with the resin is minimized. The biggest advantage, however, is the application of fast-curing resins, such as EP and PU, that could not be processed with the pultrusion process before.

4.8.7 Prepreg Layup

Fiber structure laminates that are pre-impregnated with a thermosetting or thermoplastic resin are widely used in the composites industry. These sheets or tapes commonly come with a unidirectional continuous fiber orientation and can be stacked on top of one another at predetermined orientations, resulting in composites with tailored properties. Compared to fabric reinforcements, the fiber strength is not diminished by fiber undulations.

4.8.7.1 Prepregging

The manufacture of these pre-impregnated unidirectional laminates begins by aligning the fibers that are fed from roving spools into their final planar arrangement using combs with adjustable spacing. The planar fiber arrangement is then sandwiched between two silicone-impregnated films, one of which has been coated with the polymer resin. The resin and fibers are consolidated between two heated rollers. The pressure generated by the two rollers causes the resin to fully impregnate the fibers. The impregnated *prepreg* system is then cooled to stop the curing reaction and pulled in by take-up rolls. The prepregs must be stored at temperatures between −15 and −20 °C until use to avoid further curing of the resin. In most cases, based on the resin formulation, prepregs can be stored for a year at −18 °C, six months when stored in a refrigerator at 4 °C, and a couple of hours or days at room temperature, 24 °C.

Recently, *thermoplastic prepregs* are becoming increasingly popular. Thermoplastics have several advantages over thermosets in prepreg applications, including not having to store the prepreg at sub-freezing temperatures, no finite shelf-life, ease of recycling, and higher fatigue and impact resistance, to name a few. Some thermoplastic resins, such as polyether ether ketone (PEEK), exhibit comparable or better performance than epoxy resins. Thermoplastic tapes are manufactured in a similar fashion as thermoset tapes, using hot melt (powder) impregnation and cooling before spooling the tape onto the take-up roll.

These prepregs are used for the manufacture of aircraft wings, among other parts, where pre-cut thin sheets of continuous carbon fiber-epoxy prepregs are layered at 0–90 degrees and ±45 degree orientations over a concave or convex mold surface that is relatively flat or slightly curved. Similar to the vacuum bagging techniques (see Section 4.8.3), peel-ply is laid on top of the laminate for easy removal, which is followed by a perforated air film and a layer of breather fabric. Once the layers are stacked, they are covered with a vacuum bag. The vacuum between the film and the mold surface forces the laminate layers against the hard metal mold surface. The whole mold-laminate structure is then placed in an *autoclave* where it is subjected to a pre-programmed thermal heating cycle. Once the epoxy resin is cured, the finished part is allowed to cool.

4.8.7.2 Automated Tape Layup

Automated tape layup (ATL) and automated fiber placement (AFP) are processes that use computer-guided systems to stack layers of fiber tape or tows onto a three-dimensional concave or convex mold surfaces to produce composite parts. Thermoset or thermoplastic prepregs for automated placement technologies are usually manufactured in wide formats and rolled up into parent rolls, which are later cut into narrow tapes or tows for ATL and AFP, respectively. Standard slit tow widths range from 1/8" to 1/2" and tapes from 3" to 12". By using unidirectional materials, the mechanical performance of the composite part can be optimized, as this ensures very high fiber content, a defined and homogeneous amount of resin throughout the part, and a very precise and optimized definition of the ply structures.

Both technologies allow a reduction in scrap and load-path optimized fiber orientation compared to other prepregs and semi-finished textiles (*e. g.*, woven or non-crimp fabric), as well as better precision and increased deposition rates compared to manual lamination. The use of ATL and AFP is increasing for aerospace applications, and they are the only techniques where the addition of layers is selectively managed to add strength and stiffness in a specific direction or position. It is therefore considered an additive manufacturing technology. Typical applications for these novel technologies include aircraft fuselages and wing skins. Selecting one process over another will always be a compromise between part surface geometry (AFP) and manufacturing throughput (ATL). The ATL process allows the deposition of both individual plies directly on the mold and the stacking of multiple plies in the desired orientation prior to placement on the tool. Because of the bigger width of the material as well as the deposition system, this process can be used to layup simple contours or flat parts at high deposition rates.

Compared to ATL, the draping process of the AFP technology differs in the simultaneous, but independent deposition of several tows, which can be independently cut perpendicular to the fiber direction. Contrary to the ATL process, these features allow higher material drapability and therefore the manufacture of highly complex, near net-shape three-dimensional geometries.

Because of the flexibility of the ATL method, designers often deviate from the standard 0–90° and ±45° layup orientations. In addition, the tape can be laid in a curved pattern, only restricted by potential wrinkle formation when the radius of curvature becomes too small. The in-plane draping (so called *steering*) of prepregs is constrained mainly by the tension/compression stiffness of the longitudinal fibers and the stiffness of the resin.

4.9 Compression Molding

There is a wide range of compression-molded parts and corresponding process variations. In many applications, injection molding has largely replaced compression molding, which is now primarily used for fiber-reinforced materials. One of the main reasons for this shift is the low flowability of compression molding compounds compared to that of injection molding materials. As a result, compression molding is often chosen for applications where a flow-induced orientation of the fibers or fillers would be detrimental to part performance.

In addition to the processes already described in Section 4.8 for the manufacturing of fiber-reinforced plastics such as SMC, BMC, and RTM, three additional compression molding processes are introduced below.

4.9.1 Laminating

On heated single- or multi-platen presses with more than 40 plates (Figure 4.117), flat resin-impregnated structures are compressed to form high-strength laminates. Resin carriers may consist of different raw materials, including natural and synthetic fibers, fibers made of glass, graphite, and wood. They are processed in the form of paper, fabrics, veneers, nonwovens, mats, films, sheets, etc.

The carrier materials, typically in web form, are impregnated or coated with binders using solutions, dispersions, casting resins, or thermoplastic films in dedicated impregnation systems. During the pressing process, smooth surface finishes are achieved by incorporating resin-rich surface layers and placing polished metal sheets between the laminate stack and the press platens.

Decorative laminate panels are produced by placing a patterned or colored sheet beneath a transparent top layer. After curing, this results in a high-quality decorative surface with enhanced visual and mechanical properties.

The required compression pressure varies depending on the type of resin and processing method. For cold-curing resins, pressures typically range from 5 bar to 10 bar, while hot-curing systems require 10 bar to 20 bar. In contrast, the processing of polyester premixes and prepregs can demand pressures of up to 300 bar. Presses with table surfaces of up to 15 m^2 and pressing forces as high as 200 MN are in use. While modern presses are often electrically heated, traditional systems use hot water or steam.

Pressing times can extend up to 100 min, depending on the material system and the part geometry. For boards that cannot be demolded while hot, integrated water cooling is necessary. An alternative approach is pressure cooling in a separate, cooled press unit. Decorative laminate panels can also be produced in a continuous process using double-belt presses, resulting in so-called continuous pressure laminates (CPL).

The key component of such a system is a heated drum, approximately 1300 mm in diameter, around which a pressure belt is guided. Typical process parameters for CPL production are a circumferential speed of up to 15 m/min, a temperature of up to 180 °C, a molding pressure of approximately 5 N/mm^2, a working width of up to 1.4 m, and a panel thickness from 0.3 mm to 0.7 mm.

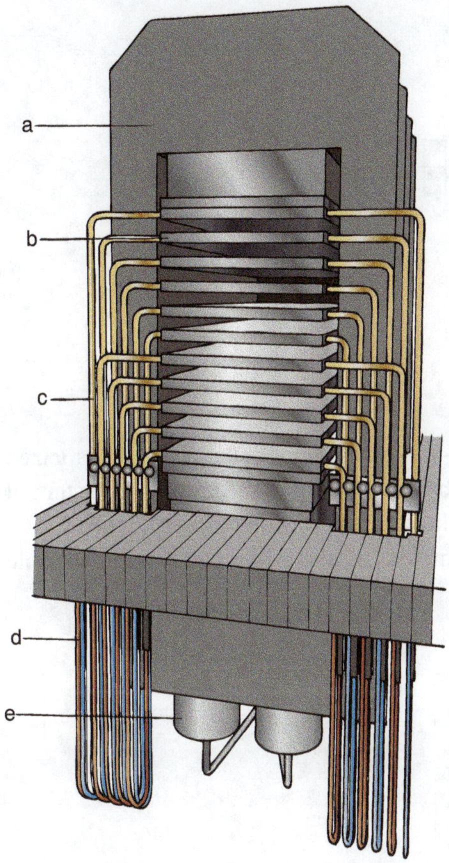

Figure 4.117
Multi-platen press; hydraulic vacuum frame press with 10 platens
a: Frame/rack, b: platens, c: pipes for heating and cooling, d: tubing for heating and cooling, e: operating cylinder

4.9.2 Extrusion Molding

Profiles made from thermosetting molding compounds can be manufactured using the extrusion molding process. This requires extruders designed to ensure that the material cures within the die, thereby generating the necessary counter-pressure to compact the molding compound effectively. In a batch (discontinuous) process, the molding compound is fed into the system and compressed by a reciprocating piston. Within the heated profile die, the material is plasticized and simultaneously cured.

The cured profile is then ejected in coordination with the back-and-forth motion of the piston, enabling controlled shaping and solidification of the final part (Figure 4.118).

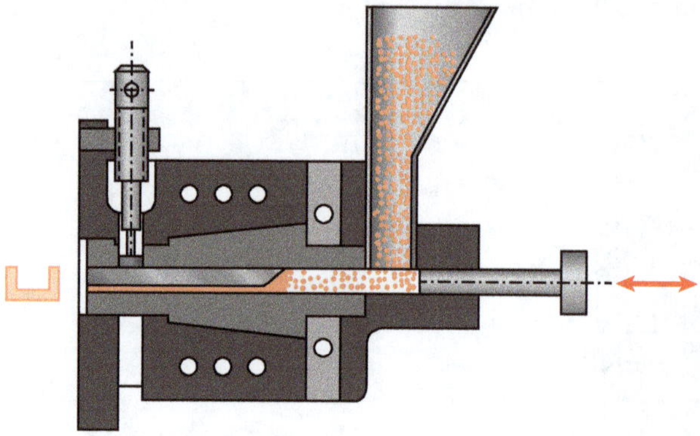

Figure 4.118
Extrusion molding

4.9.3 Transfer Molding

In transfer molding, the molding compound – metered and/or pelletized – is plasticized under heat and pressure in a transfer chamber and then pressed through a system of channels into a closed mold, which is maintained at curing temperature (Figure 4.119). This process allows for a precise filling of complex cavities and is particularly suitable for thermosetting materials.

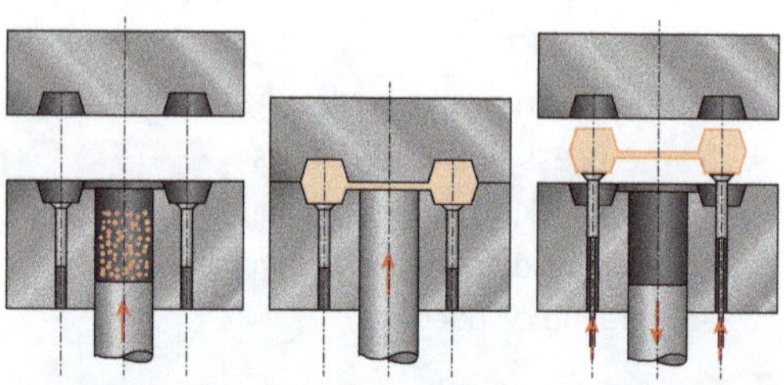

Figure 4.119
Transfer molding

Two-piston presses are commonly used: the larger piston is responsible for opening and closing the mold, while the smaller piston performs the injection of the molding compound into the mold cavity. Press designs vary in terms of the arrangement of the pistons and the method used to feed the injection cylinder. *Weigh feeding systems* use a hopper to introduce the molding compound from above into the injection cylinder. In contrast, *molding machines with pre-plasticization* first compound the molding material in a dedicated plasticizing unit before feeding it directly into the injection barrel, ensuring consistent material flow and improved cycle times (Figure 4.120).

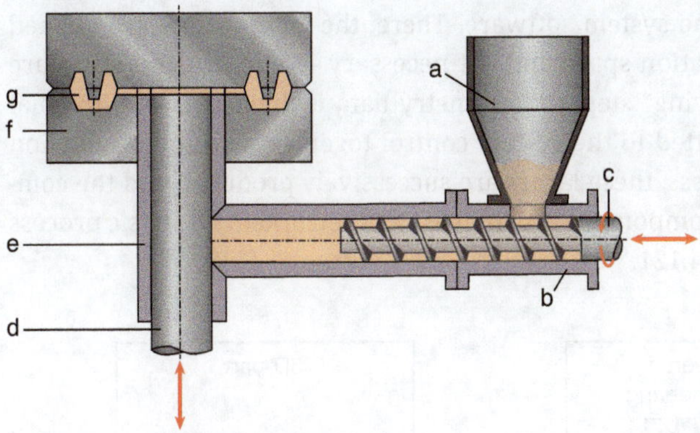

Figure 4.120
Transfer injection mold
a: Compound hopper,
b: plasticizing barrel,
c: plasticizing screw, d: transfer plunger, e: injection plunger,
f: mold, g: cavity

4.10 Additive Manufacturing Processes

Additive manufacturing is often referred to as the disruptive technology of the 21st century, primarily due to its ability to produce and functionalize geometrically complex components and even assemblies. Tool-free production can achieve a faster time-to-market, which can generate economic advantages compared to conventional manufacturing processes on the market.

Derivation of Terms

Additive manufacturing (AM), which is also referred to as 3D printing in the industrial or the popular environment, comprises those manufacturing processes that produce components by adding or joining volume elements (voxels) layer by layer. It can be understood as a 2.5D process, as the third dimension is not created as a continuous z coordinate, but by joining the individual layers together. According to DIN 8580, additive manufacturing is classified as a primary forming process. Due to the layer construction principle, complex structures can be realized with a very high degree of design freedom. This distinguishes it from conventional subtractive or formative processes. In subtractive manufacturing, the geometry is created by removing defined areas, such as in turning or milling. If materials are brought into a desired shape by mechanical forces or limiting mold cavities, it is a formative production process [1].

Procedure

A 3D CAD (computer-aided design) dataset serves as the starting point for the production of components using AM processes. In addition to direct digital design, it is also possible to digitize solid components whose data can be modified and adapted as required. This process is known as “reverse engineering”. The dataset is converted into a standardized STL (standard triangulation language or surface tessellation language) format and transferred to the system software. There, the components are aligned and positioned in the installation space and – if necessary – support geometries are created. In the so-called "slicing" step, the geometry data is divided into individual layer data, which is transmitted to the system control together with the production parameters. During the process, these layers are successively produced and the components are generated. The components are then usually reworked. The basic process sequence is shown in Figure 4.121.

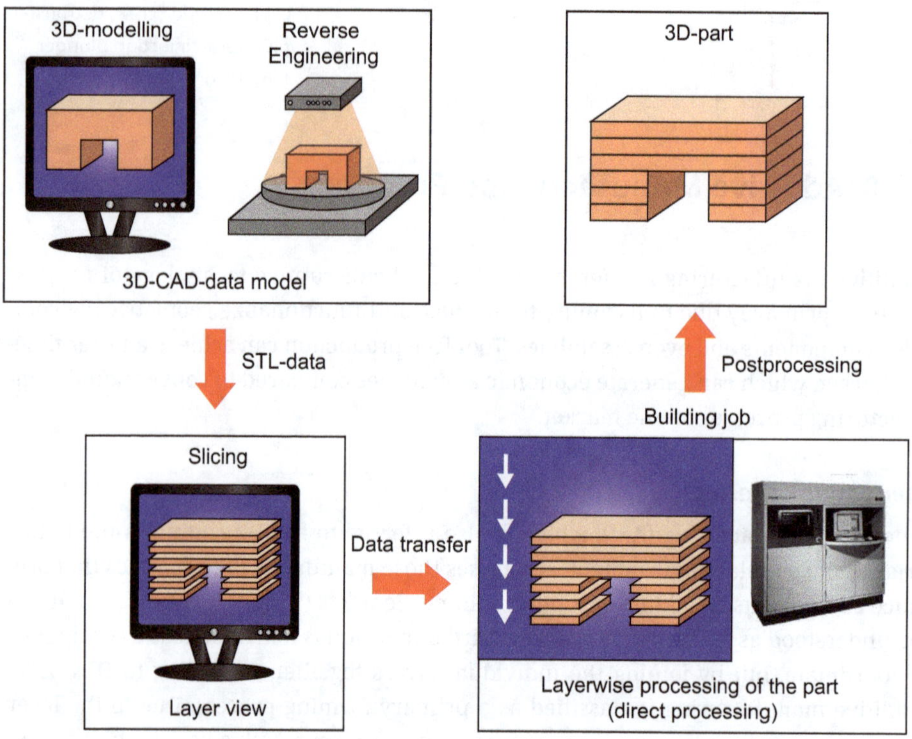

Figure 4.121 The concept of additive manufacturing: from virtual model to component

Intended Use

Depending on the intended use, AM processes can be divided into direct and indirect manufacturing processes. Prototypes (rapid prototyping) or final components (rapid manufacturing) can be produced using direct AM. Indirect AM is based on the use of AM in toolmaking (rapid tooling). Various areas of application have emerged for the use of AM processes. The most important ones are explained below.

Rapid Prototyping

Rapid prototyping refers to the fast and flexible production of prototypes or models. This allows the appearance (concept models), usability, shape and feel (geometry models), or even basic functional properties of the future product to be tested under thermal or mechanical stress (functional prototypes) at an early stage of product development. If the prototypes in pre-series already largely meet the requirements of the future series product (technical prototypes), final optimizations can still be made to the production of the series tool (for example, the injection molding tool) at this stage [1].

Rapid Tooling

Rapid tooling AM processes are used to produce tools or negative molds for the near-series molding of small or medium quantities. This allows the production of master models for vacuum casting, wax patterns for investment casting, or cores for sand casting. Additive manufacturing is particularly suitable for complex cast parts. Near-series materials or series materials are used in the molding process [1].

Rapid Manufacturing

If near-series components or end products are made using AM processes, this is known as rapid manufacturing. As no tools are required and the objects can be produced directly from CAD data, this enables fast, flexible, and efficient production. Rapid manufacturing is primarily used when only small batch sizes are required or when a component needs to have a geometry that is difficult or impossible to realize using conventional processes [1].

Fields of Application by Industry

There is a wide range of potential applications for AM processes in a variety of industries. Figure 4.122 lists the various fields of application. Thus, the main areas of application are in mechanical engineering, automotive engineering, aerospace, and (dental) medical applications [2].

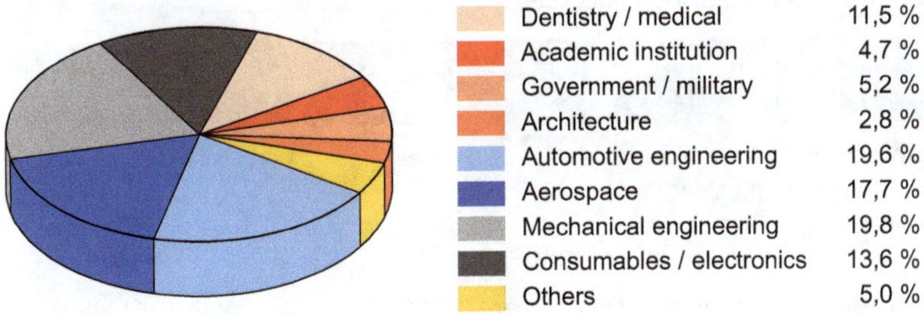

Figure 4.122 Fields of application of additive manufacturing processes according to Wohlers Report 2019 [2]

According to the illustration in Figure 4.123, AM processes are employed in particular for the production of components for end use and functional prototypes [2]. In this illustration, the material-specific differences in the processing of metals, plastics, and ceramics are not taken into account. For both plastics and metals, it can be assumed that the tendency is more towards end-use components.

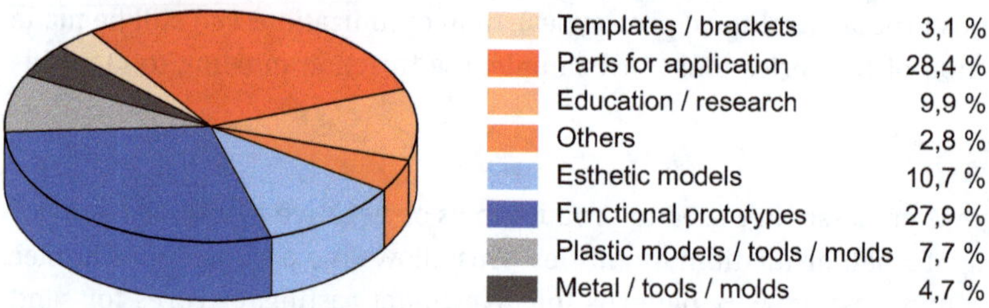

Figure 4.123 Areas of application for additively manufactured components according to Wohlers Report 2019 [2]

In the field of additive manufacturing, there are many different processes that can be classified by the aggregate state of the starting material (solid, liquid, gaseous) [1]. Processes that work with gaseous starting materials only play a subordinate role here.

In addition to the advantages mentioned, the layered structure also poses some challenges. For example, direction-dependent component properties (anisotropy) are typical for these manufacturing processes. There are also process-specific material and process requirements, which are explained below for the processes with the greatest industrial significance.

Table 4.12 provides a summary of AM processes categorized into the proposed ISO/ASTM categories and names for additive manufacturing. The table includes alternative names, often coined as brand names by the original inventors and companies that commercialized the various processes.

Table 4.12 Additive Manufacturing Processes

Default name[1]	Alternative name	Year of discovery	Patent or publication
Vat Photopolymerization (VPP)	Stereolithography (SLA or SL)	1984	US4575330
	Solid Ground Curing (SGC)	1986	US4961154
Digital Light Processing (DLP)[2]	Continuous Liquid Interface Production (CLIP); Digital Light Synthesis (DLS)	2015	WO2016140891

Default name[1]	Alternative name	Year of discovery	Patent or publication
Powder Bed Fusion (PBF)	Laser Sintering (LS); Selective Laser Sintering (SLS)	1986	US4938816
	Multi Jet Fusion		HP Brochure
	Selective Heat Sintering	2008	US9421715
Material Extrusion	Fused Deposition Modeling (FDM); Fused Filament Fabrication (FFF)	1989	US5121329
Sheet Material Lamination (SML)	Laminated Object Manufacturing (LOM)	1996	US5730817
Binder Jetting	Selective Binding; 3D Printing	1989	US5204055
Material Jetting (MJ)	Wax Jetting	1989	US5136515
	Polymer Jetting Freeforming	2013	EP2266782A1
Selective Thermoplastic Electrophotographic Process (STEP)[3]		2013 (commercial, 2017)	US8488994
Volumetric 3D Printing[3]	Computed Axial Lithography (CAL)	2017	WO2018208378
Xolography[3]		2019	DE102019115336 DE102019129868

1: ISO/ASTM 52900, Standard Terminology for Additive Manufacturing – General Principles – Terminology, *ISO/ASTM International* (2016)
2: VPP process variation with mask projection
3: Not standardized

4.10.1 Liquid Materials

An important group of AM processes is based on the use of liquid resin systems. The main components are acrylate monomers and oligomers, which are chemically cross-linked using a system of reactive photoinitiators to form solid structures.

4.10.1.1 Stereolithography

The process of stereolithography is the oldest of the AM processes, patented back in 1984 by Robert Hull and based, like the majority of them, on the successive formation of individual layers to generate a 3D model. Stereolithography is based on the principle of photoreaction, in which reactive photoinitiators are stimulated by UV radiation to react and cross-link the monomers and oligomers contained in the material to form a solid component. The materials used are mainly acrylates and methacrylates or acrylic-functionalized oligomers based on epoxy, polyester, polyurethane, and polyether. Due to the sensitizing and often toxic properties of these components in reactive form, it is generally important when using the stereolithography process or resin-based processes to provide personal protective equipment for the operator and to ensure adequate room ventilation. Figure 4.124 shows the process schematically.

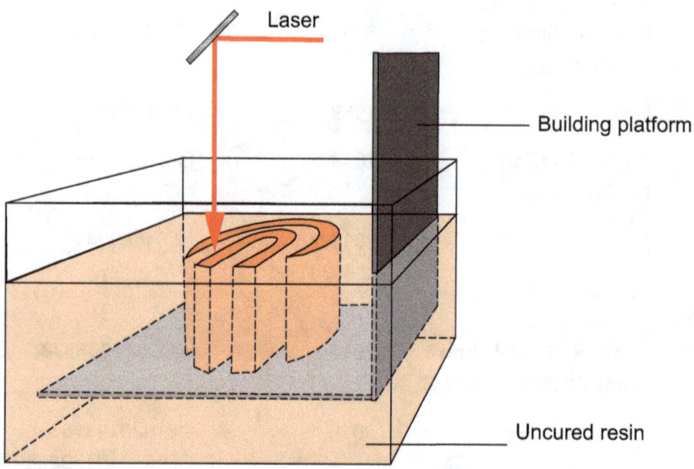

Figure 4.124 Stereolithography process principle (top-down)

In the process, one layer at a time is exposed to a UV laser whose beam is controlled via a movable deflecting mirror and thus stimulates the cross section of a component layer to react. The build platform is then moved downwards (top-down) or upwards (bottom-up) by one layer height, depending on whether the UV radiation is supplied from below or above. This process is repeated until the component is finished. The advantage of the bottom-up variant is that even large components can be produced with a very limited amount of material in the resin tank. The bottom-up method is mainly used in modern stereolithography systems. Incomplete cross-linking of the individual layers takes place during the exposure process. This step is essential as it maintains a sufficient number of reactive groups and ensures bonding between the individual layers. However, this incomplete cross-linking leads to a weakening of the mechanical component properties, which is why post-cross-linking is necessary after the printing process. In addition to post-cross-linking, the stereolithography process also involves further post-processing steps. As the resin cannot support overhangs, in contrast to other AM processes, the use of support structures (supports) is necessary. These are initially produced during the process and must then be mechanically removed in a post-processing step. In addition, excess, unreacted material must be removed by soaking in a suitable solvent, usually isopropanol.

Stereolithography is characterized by very high surface qualities and dimensional accuracy among the AM processes. For this reason, and due to the possibility of generating transparent components, it is primarily used in the field of rapid prototyping, as it allows detailed prototypes to be created directly from the CAD model in order to accelerate product development and secure it at an early stage. In addition, the generation of medical models to support diagnoses and the planning of operations is a strong focus of use, but the technology is also used mainly in the field of dental medicine.

4.10.1.2 Digital Light Processing (DLP)

In addition to classic stereolithography, there are several special processes that have become increasingly important in recent years. These differ primarily in the type of exposure, by which they can be differentiated from one another. While in classic stereolithography the UV radiation is applied to the material via a single laser spot, these approaches work with two-dimensional UV exposure, which can significantly accelerate the production process. In Digital Light Processing (DLP), the individual UV laser is projected onto a surface via a micromirror array. In LED technology, a LED or LCD display acts as the UV source. Another variant of DLP uses "Continuous Liquid Interface Production" (CLIP). This process is also based on a DLP projector and uses an oxygen-permeable membrane to create a so-called "dead zone" directly above it (see Figure 4.125), in which no local cross-linking can take place due to the natural oxygen inhibition of the photoreaction. This effect prevents the component from sticking to the bottom of the tank and promotes the flow of fresh resin. As a result, the component can be continuously pulled out of the resin pool, which accelerates the process and improves the surface structure.

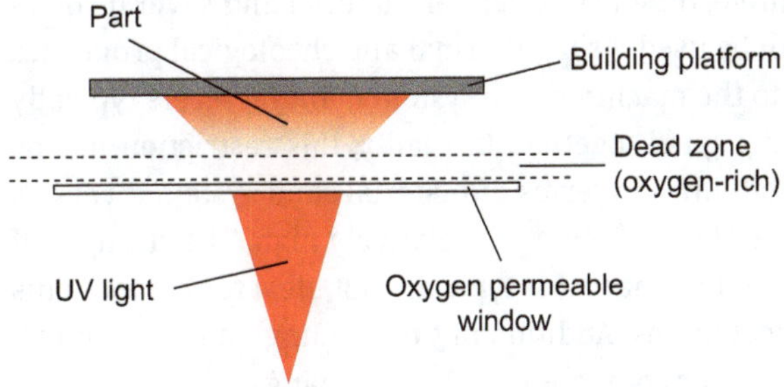

Figure 4.125 "Dead zone" in the CLIP process

4.10.1.3 UV-Assisted Direct Ink Writing (DIW)

Direct Ink Writing (DIW) is a relatively new 3D printing technology that enables the making of meter-sized and massive models (for example, for advertising purposes) as well as microfluidic chips in the micrometer range at an extraordinary production speed. The basic process is a hybrid additive manufacturing technology that combines extrusion-based AM and UV curing, as used in stereolithography. The liquid acrylic-based thermosets are extruded through a nozzle and then cured by an external UV light to form a solid structure. The highly viscous resin remains semi-solid or gel-like after the extrusion process. The external UV light then cures the structure completely in a short time. With certain machine types, the printheads can move at speeds of up to 1 m/s. This high speed makes it possible to produce large components in a significantly shorter time than conventional AM technologies can. The high or thixotropic viscosity of the materials used and the fast cross-linking allow the production of large overhangs without support as far as possible, which reduces the amount of material used, time in the printing process, and the amount of work required for post-processing. The costs for DIW are higher than for alternative extrusion-based AM processes (see Section 4.10.3). Both the material and the machine have a higher price, which is based on the need for a UV light and the composition of the resin. However, as the UV curing reaction is a chemical reaction and therefore forms a chemical bond between the layers, the resulting mechanical properties are better.

4.10.1.4 Material Jetting (MJ)

Material jetting (MJ) is a powerful manufacturing technology that enables the production of smooth and precise components. With a microscopically small layer resolution and an accuracy of up to 0.02 mm, thin walls and complex geometries can be produced with a wide variety of materials. Like a conventional 2D inkjet printer, the printhead sprays thousands of photopolymer droplets onto the build platform and solidifies them with UV light that follows the nozzle. Since MJ is based on a principle similar to that of inkjet printers, multiple nozzles with different colors and materials can be used in one process. This concept enables the production of MJ models with a colorful design and smooth surface as well as employing different material properties in one process step. In this way, for example, dissolvable carrier material and several colors or types of functional material can be used. As the thermal and rheological properties of the material must be matched to the machines, the system manufacturers typically bundle the materials they have designed together. In reality, this restriction means that MJ has relatively limited possibilities in terms of materials and system diversity, which in turn means that the costs of the process are relatively high. The strength of the parts produced is structurally so low that MJ is typically not ideal for components that will be subjected to mechanical forces. Additionally, depending on the manufacturing process, MJ requires supports for components with overhangs.

4.10.2 Powder-Based Processes

Powder-based AM processes are currently the most important class in industrial terms. The component can be produced by melting and solidifying or by bonding a fine, usually thermoplastic powder. What all processes in this class have in common is that no support structures are required to produce components with undercuts.

4.10.2.1 Laser Sintering (LS)

The first laser sintering (LS) systems were sold commercially in the early 1990s, making LS an established and mature technology. The overall LS process is divided into the preheating, construction, and cooling phases (Figure 4.126). The LS process is one of the so-called laser-based powder bed fusion (L-PBF) processes.

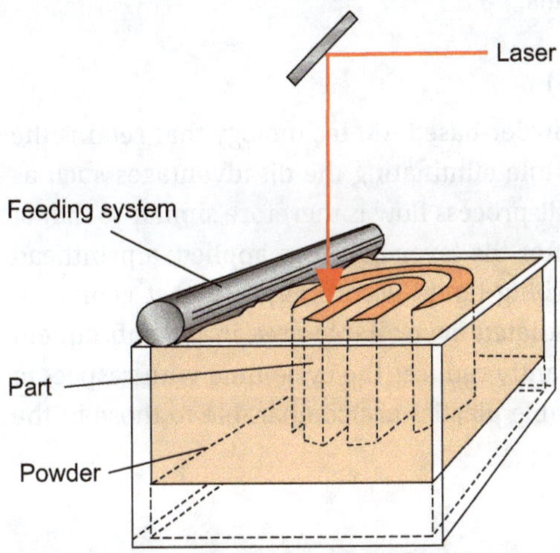

Figure 4.126
Laser sintering (LS) process principle

In the (pre)heating phase, the polymer powder provided by the powder storage chamber is applied via a system in the build chamber and preheated by radiant heaters to a temperature close to the melting range of the material. After the heating phase, further layers are applied, the so-called base layers. These layers are necessary to reduce the heat transfer between the exposed layers and the metal frame. In the construction phase, the desired geometry is selectively exposed with a focused laser beam via a mirror system (galvano-mirror). Typically, a CO_2 laser is used for the energy input, as the emitted wavelength in the range of 10.6 µm has a high absorption for most plastics. After exposure, the powder bed is lowered according to the set layer thickness, which is usually between 80 µm and 150 µm, and a new powder layer is applied. With repetitive geometry exposure, it is important that not only the particles of the current layer fuse, but that the individual layers also bond with each other in the *z* direction, that is, in the build direction. The process is repeated until all components have been completed. The unmelted powder serves as a support structure for the finished components. For thermal insulation, a defined number of empty powder layers are applied at the end of the construction phase. The heating systems are then switched off and the construction space cools down slowly and homogeneously. All three phases take place under continuous purging, using inert gas in order to prevent oxidation of the powder and thus ageing processes in the material. As soon as the powder cake has cooled down to room temperature, the components can be removed.

The LS process can be used to produce components with mechanical properties comparable to those from conventional processes. The yield strength and modulus of elasticity have similar characteristics to the ones obtained in injection molding, whereas the ductility is lower. Due to the good mechanical properties, components are also produced as functional parts in small series, in addition to being used in prototype construction. The most frequently used plastics include polyamides – especially PA12 – as well as PP, TPU, and filled systems.

4.10.2.2 High-Speed Sintering (HSS)

High-Speed Sintering (HSS) is a newer powder-based AM technology that retains the attractive properties of laser sintering while eliminating the disadvantages such as machine costs and process speed. The basic process flow is therefore similar to that of LS. The fundamental difference is that after the layer has been applied, a printhead wets the powder with an infrared-absorbing liquid according to the set geometry. This allows the build chamber to be irradiated over a large area in the subsequent step using infrared lamps, which significantly reduces the cycle time with respect to LS. The mechanical properties and available plastics are comparable to those of the LS process.

4.10.2.3 3D Printing (3DP)

3D printing (abbreviation: 3DP), also known as binder jetting, is another powder-based process. In contrast to laser sintering, the starting material is not melted in this process. The component is generated using a selectively introduced liquid binder, which bonds the powder particles together. This process principle means that theoretically all powdered materials can be processed as long as they can be bonded with a binder. Due to the low mechanical strength of the resulting composite material, this process is mainly used for large prototypes and design models. In the plastics sector, the process is also used for colored components, for example, made of PMMA. Other possible applications include the production of sand molds for cast parts or metallic components. The components produced in the process are "green" parts that have to be processed in further post-treatment steps. Due to the lack of thermal stress, the components can be freed from loose powder adhesions using compressed air.

4.10.2.4 Processes under Development

Selective thermoplastic electrophotographic process (STEP) represents a new approach in the field of powder-based AM processes. In contrast to the aforementioned processes, it does not work with a powder bed. The process principle is based on the functionality of a laser beam printer. A thermoplastic powder is electrostatically charged and selectively developed on a conveyor belt. To generate the desired components, this powder is melted with a flat energy input and transferred layer by layer to the building platform using pressure. This process enables very high build rates. At present, the range of materials is still very limited. Selective powder deposition should enable the production of multi-material components in the future.

There are two trends in the development of laser sintering of plastics that are being pursued by system manufacturers. On the one hand, the trend is towards filigree structures and thin-walled components. This requires new energy input concepts that can reduce the laser spot diameter. Additionally, material systems and system technology are being adapted in order to further reduce the minimum layer thickness, currently of approximately 0.08 mm.

On the other hand, in addition to detail resolution, system manufacturers are focusing on further increasing productivity. The use of new types of exposure modules should enable selective, flat exposure in the future. This will reduce temperature gradients in the melted area and the necessary exposure times. This technology should allow an increase in component reproducibility and productivity in the future.

Post-Processing for Powder-Based Processes

In all powder-based manufacturing processes, the resulting components must be freed from adhering powder residues. This unmelted powder serves as a support structure in the process and is essential for the geometry freedom of powder-based processes. The components are usually cleaned manually or automatically in a blasting process using glass beads and compressed air. One challenge is to remove powder residues even from small recesses and holes.

After cleaning, the components have a rough surface that depends on the particle system used. This can be smoothed by chemical post-treatment steps or vibratory finishing processes, for example. In addition, the components can be painted, dyed, infiltrated, or metallized in further processes. This creates unique surfaces that enable further functionalization of the additive components.

4.10.3 Filament/Granulate-Based Processes

The extrusion-based AM process is also classified as an additive manufacturing process. This process is known as fused layer modeling (FLM) and Melt Extrusion Manufacturing (MEM), among others [3].

In general, this process is described as melting a thermoplastic semi-finished product in a nozzle and then depositing it in the form of melt strands or droplets. After discharge, the material solidifies and forms the cross section traced with the nozzle. Therefore, this process differs significantly from the powder- and liquid-based processes presented above. These are based on the local melting and cooling of the starting material or its chemical cross-linking in a construction space completely filled with the starting semi-finished product. In contrast to other AM processes, this makes it easier to access the workpiece, which means that insertion operations can be integrated during component assembly, for example. In addition, a variety of temperature control strategies can be used in the installation space to influence the final component without affecting the semi-finished product. Finally, the principle of local material deposition enables resource-efficient production, even if, in contrast to laser sintering, support structures have to be generated for large overhangs. These can then be removed in a post-processing step using two different approaches. If water-soluble support material is used, it can simply be dissolved in a water bath to expose the final component. If the support material is not water-soluble, it must be removed using mechanical post-processing steps. Surface treatments in a post-processing step can improve the surface quality if needed. For example, a grinding process can be used to influence the visual and tactile properties of a component. Coating processes such as painting and metallization allow further functionalization of the surfaces. An overview of the general process principle is shown in Figure 4.127.

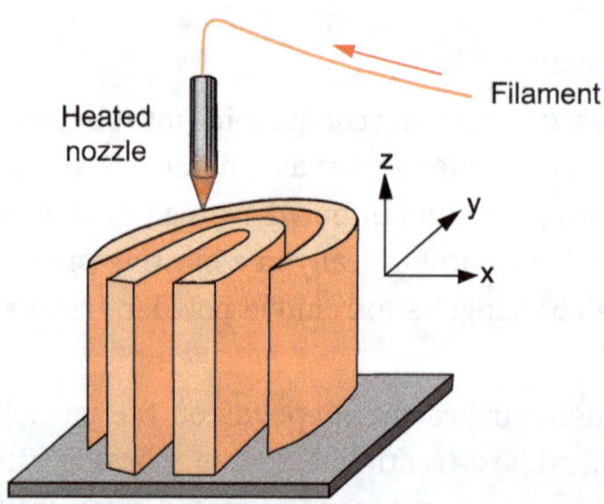

Figure 4.127
Process principle of the filament-based process of thermoplastics

The component properties that can be achieved and the processable materials that can be employed using the filament- or granulate-based process depend on various factors, where the shrinkage behavior and warpage are decisive for the processability of materials. For this reason, polymers with high shrinkage behavior (*e. g.*, unfilled PE and PP) are rarely used in extrusion-based additive manufacturing, despite their low prices and high market share in thermoplastic processing. The FLM materials can be classified as shown in Figure 4.128.

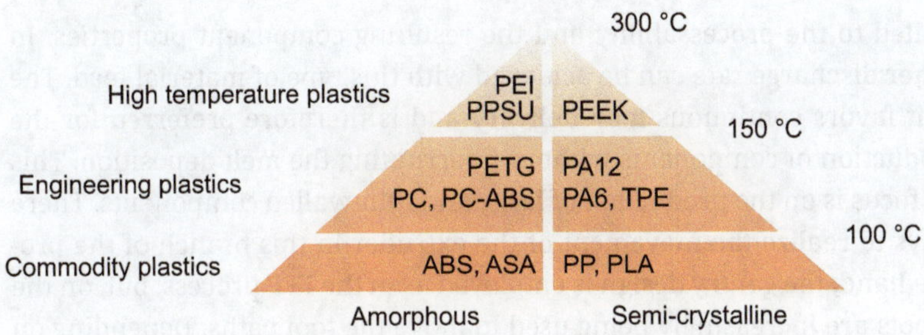

Figure 4.128 Classification of thermoplastic materials in the strand lay-up process

A variety of fillers can be used to make other materials accessible for extrusion-based additive manufacturing or to modify existing material systems. This influences not only the processing properties but also the mechanical, chemical, electrical, and optical properties of the FLM component. For example, the shrinkage behavior can be positively influenced with spherical fillers made of perlite or glass. Silicon carbide, on the other hand, can increase the wear resistance, while short carbon fibers increase the bending strength. Other fillers can include copper particles, blowing agents, talc, etc.

Current extrusion-based AM processes can be subdivided according to the semi-finished product supplied.

4.10.3.1 Fused Filament Fabrication (FFF)

The FFF process feeds the semi-finished product in the form of a filament. As this type of feed can be realized using simple and inexpensive machine technology and, in addition, material systems that are harmless to health can be processed, this is the most widely used AM process in the private sector. Feeding the semi-finished product in filament form results in a number of advantages and disadvantages. As the filament is moved towards the extruder by means of the gradual movement of gears, the melt discharge can be much better controlled than in granulate-based processes. Additionally, it is possible to pull back the material and thus to set down the die. However, the material variety is limited by this feeding mechanism, as the semi-finished product must achieve certain mechanical properties for successful material feeding. Furthermore, the discharge capacity is limited by the relatively small commercially available filament diameters, from 1.75 mm to 2.85 mm. Finally, the choice of a filament-based process results in higher material costs, as the semi-finished filament product has to be produced in an upstream process step.

4.10.3.2 Pellet-Based Strand Placement Process

The limitations with regard to the mechanical properties of the semi-finished product and the discharge volume are overcome by a pellet-based strand discharge process. As this process uses a screw to convey the semi-finished plastic product, the choice of material is limited to the processability and the resulting component properties. In addition, a higher discharge rate can be achieved with this type of material feed. The plasticizing unit favors continuous melt delivery and is therefore preferred for the continuous production of components without interrupting the melt deposition. This means that the focus is on the production of large and thin-walled components. There are various ways to realize the movement of the extruder in this branch of the process. On the one hand, the gantry design is employed as in the FFF process, but, on the other hand, robots are increasingly being used to move the tool paths. Depending on the size of the installation space, it is also temperature-controlled in order to counteract the anisotropic component properties.

4.10.3.3 Variants

Another granulate processing method is plastic freeforming. Here, the material is not applied as a continuous strand, but in the form of droplets via a valve gate nozzle, as in ballistic particle manufacturing.

The wide variety of materials and machines offers a broad spectrum of applications for FLM technology. The majority of FFF machines are used for hobby and prototyping applications. Here, functional prototypes, demonstration objects, and customized machine components can be manufactured in small quantities. In addition, FLM components made from high-performance plastics and fiber-reinforced plastics are increasingly being used directly in aerospace and medical technology. Prototype tooling or rapid tooling applications of the FLM process are also used in almost all branches of industry. These include molding tools, assembly aids, and lost cores for the metal casting process.

References for Section 4.10

[1] Gebhardt, A., *Additive Fertigungsverfahren: Additive Manufacturing und 3D-Drucken für Prototyping – Tooling – Produktion*, Hanser, Munich (2016)

[2] Wohlers Report 2019 – 3D Printing and Additive Manufacturing – State of the Industry (2019)

[3] Turner, B.N., Gold, S.A., A review of melt extrusion additive manufacturing processes. II. Materials, dimensional accuracy, and surface roughness, *Rapid Prototyping Journal*, 21 (2015) 3, pp. 250–261

4.11 Fiber Production

Dirk W. Schubert, Michael Redel

Organic fibers can be classified according to the way they are produced, the origin of their material, or their fiber diameter. In general, a distinction is made between staple fibers, continuous fibers, and fibers of limited length. A further subdivision can be made into synthetic fibers, usually made from petrochemical materials, and natural fibers, made from polymers harvested from plants or gathered from animals. A classification of chemical fibers according to their use as fillers is shown in Figure 7.2. The third classification relates to the fiber cross section and the associated properties. Extruded fibers, which are stretched via a so-called "haul-off", can have diameters in the range of a few millimeters (mm) and are used, for example, in toothbrushes. Conventional fiber spinning from the melt enables the production of fibers in the range of a few micrometers (5–100 µm); these are used as nonwovens in protective layers for masks or hygiene products. Finer fibers can be obtained using the meltblown process, which can produce fiber diameters in the range 1–20 µm. Nonwovens made from these fibers are typically used in filters due to their small pore sizes. Electrospinning offers the possibility of producing even finer fibers, which can achieve diameters of less than 1 µm (30–1000 nm). Potential areas of application for such fibers are biomedical materials for the support or, after colonization with cells, the reconstruction of tissue.

4.11.1 Textile Definitions

The most important characteristics of a polymer fiber are determined not only by its mechanical properties but also by the fiber diameter. This dimension is usually specified in micrometers. However, for historical reasons, the fineness *F*, which is specified in tex or denier (den), is also widely used.

The quantity "tex" has the unit g/km. Therefore, it indicates how many grams of material – in this case, polymer – are required to produce a fiber with a length of 1 km. The unit "den" has its origins in the French industry and refers to a scenario where 1 g of material yields the length of 9 km. The following, therefore, applies:

$$1\ \text{den} = \frac{1\ \text{g}}{9\ \text{km}} = \frac{1}{9}\ \text{tex} \tag{4.2}$$

Assuming a round fiber cross section and a polymer density ρ, the following equation provides a simple conversion of the fiber fineness F into the fiber diameter d and vice versa:

$$F = \frac{m\ (\text{mass})}{l\ (\text{length})} = \frac{\rho \cdot \pi \cdot d^2}{4} \tag{4.3}$$

$$d = \sqrt{\frac{4 \cdot F}{\pi \cdot \rho}} \tag{4.4}$$

Equation 4.3 shows a proportionality of the fineness *F* and the cross-sectional area of the fiber. Therefore, the following applies to the two important fiber polymers PP and PET:

PP:

$$d \cong 38\ \mu m \cdot \sqrt{F/_{g/km}} \tag{4.5}$$

PET:

$$d \cong 30\ \mu m \cdot \sqrt{F/_{g/km}} \tag{4.6}$$

The fineness F is currently used, among other things, in the evaluation of tensile tests and the differentiation of fiber types. One example of this is the tensile strength of fibers in tensile tests. This is often not specified in force per area (N/mm^2 = Pa) but in newtons per tex (N/tex), as the fineness is proportional to the cross-sectional area of the fiber. Furthermore, microfibers can be defined by their fineness and not only by the fiber diameter: They are fibers with a fineness of less than 1 dtex. This means that 1 g of polymer is spun into a fiber of at least 10 km.

4.11.2 Fiber Spinning from the Melt

Melt spinning is the most widely used industrial process for the production of polymer fibers. The main base materials used for production are polyamides (PA), polyesters, polyacrylic acids, cellulose-based polymers, and polyethylene (PE) as well as polypropylene (PP). A list of these materials and their characteristics relevant to the fiber spinning process is discussed later, in Section 4.11.5.

In polymeric fiber production, in addition to the base material, also the manufacturing process is decisive for the properties and areas of application. One method of differentiation in fiber production is the fiber elongation process. A distinction can be made between mechanical (godets) and aerodynamic (cold/warm air) stretching of the fibers.

Due to its high industrial relevance, aerodynamic stretching will be discussed first in the following, as the main field of application for fibers spun from the melt is spunbond nonwovens. The processes that use the aforementioned stretching principle can be divided into two widely used methods, which at first glance only differ in one small detail. These are the “classic” fiber spinning with aerodynamic stretching by cold air (15–80 °C) and the meltblown process, which uses warm air (150–300 °C) instead.

4.11.2.1 Classic Fiber Spinning

Classic fiber spinning is used to produce single fibers or nonwoven fabrics, so-called nonwovens. In this process, one or more fiber strands are extruded from polymer granulate through a die head and then stretched. A possible structure of such a system is shown schematically in Figure 4.129.

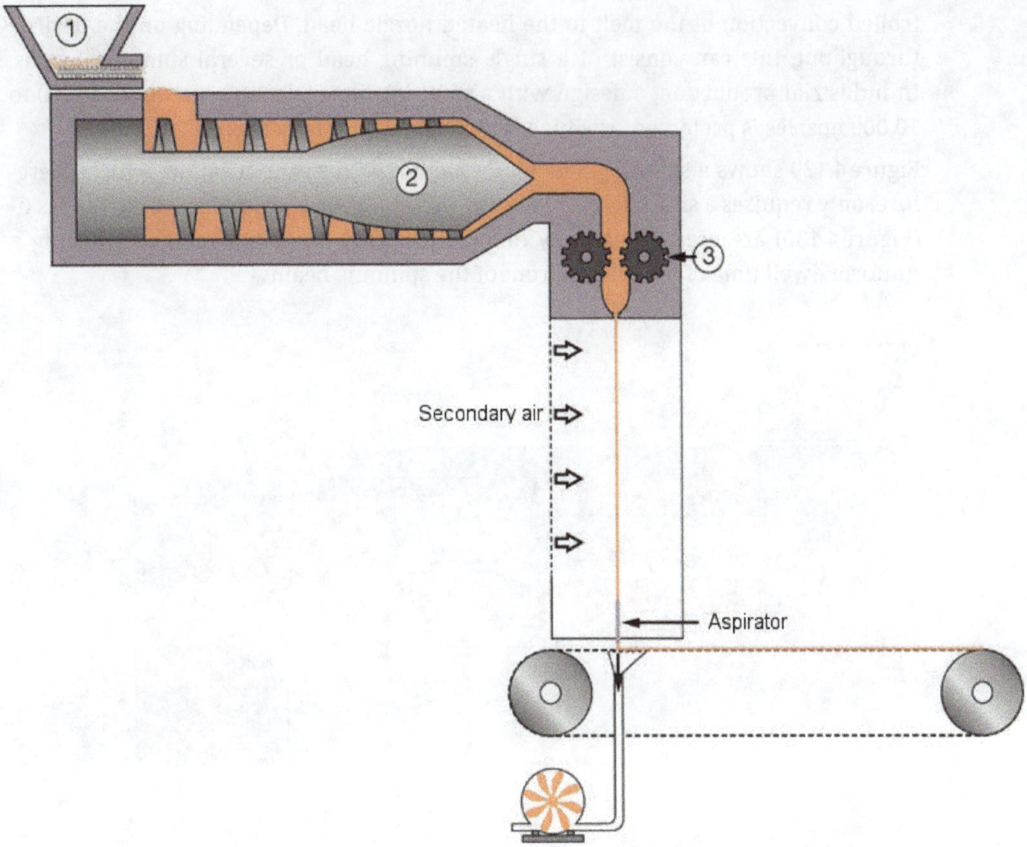

Figure 4.129 Schematic drawing of a conventional fiber spinning plant with deflection head, aspirator, and sieve belt

The fiber spinning system consists of a feeder ①, in which the material is pre-dosed and conveyed into the extruder ②. The extruder is the component responsible for melting the pellets, conveying them to the melt pump, and ensuring a continuous material flow before the pump ③. In practice, single-screw extruders are mainly used for this purpose, as the compounding capability of a twin-screw extruder is only very rarely necessary for fiber production.

However, due to the size of the industrially used fiber spinning plants, it is – as shown in Figure 4.129 – necessary to change the conveying direction of the melt stream towards the die using a deflection head. This has the advantage that the feeder and the extruder can be set up horizontally on a raised platform, but the fibers can be stretched vertically in the direction of gravity.

In classic fiber spinning systems, the melt filters and pumps are located after or just before the deflection head. These melt filters are used to extract foreign particles that could clog the nozzles, and the pumps allow for a continuous, volumetrically controlled convection of the melt to the heated nozzle head. Depending on the desired throughput, this can consist of a single spinning head or several spinning beams. In industrial production, a design with a spinning bar typically consisting of 1000 to 10,000 nozzles is preferred, as a higher throughput can be achieved.

Figure 4.129 shows a spinning head with one nozzle for simplification, which, therefore, only requires a single feed of polymer melt. So-called coat-hanger distributors ① (Figure 4.130) are used industrially to supply several nozzles while guaranteeing a uniform dwell time of the melt in front of the spinning beams.

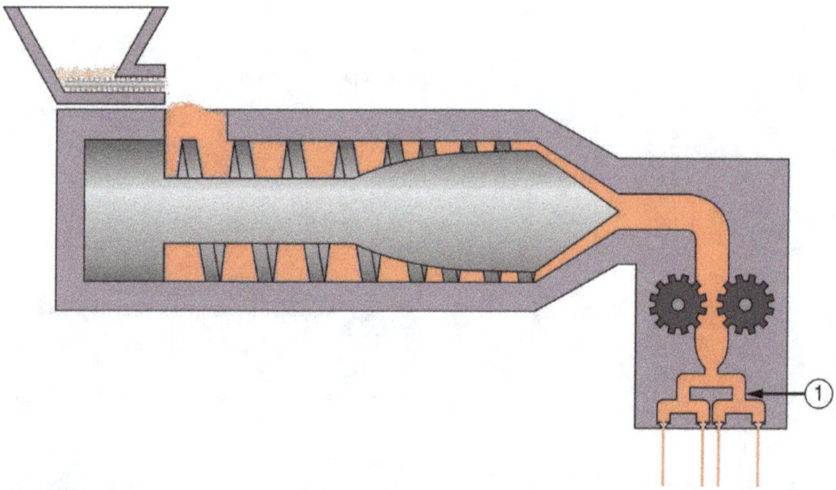

Figure 4.130 Schematic drawing of an extruder with melt pump and coat-hanger distributor

The continuous delivery of the polymer through the melt pump leads to a pressure buildup in front of the nozzle which, in the best case, is constant throughout production and ensures uniform flow through the individual nozzle holes. The emerging polymer strands are then aerodynamically stretched in a blown air shaft. The so-called "secondary air" applied to the shaft can be used to control the temperature and humidity of the air surrounding the strand during stretching. This serves to cool the resulting fiber in a controlled manner and thus achieve the optimum stretching parameters and desired fiber properties.

The aerodynamic stretching of the exiting strand can be achieved using aspirators (see Figure 4.129) or haul-off shafts (used, for example, in Reicofil® systems). Depending on the throughput, material, and take-off speeds, this usually enables the production of polymer fibers in a fiber diameter range of 10–100 µm.

For the production of a nonwoven described, at this stage, these fibers are deposited on an air-permeable sieve belt and transported towards further processing steps or windup. In order to avoid turbulent air flows, the air is usually removed via an extraction system. The fibers are deposited on the screen belt with the aim of producing fiber layers as homogeneous as possible with low cloudiness and the desired basis weight.

4.11.2.2 Meltblown Process

The meltblown process is similar to classic fiber spinning in many respects. However, it is employed in industrial applications to achieve finer fibers and, consequently, nonwovens. A possible design of such a meltblown system is shown in Figure 4.131.

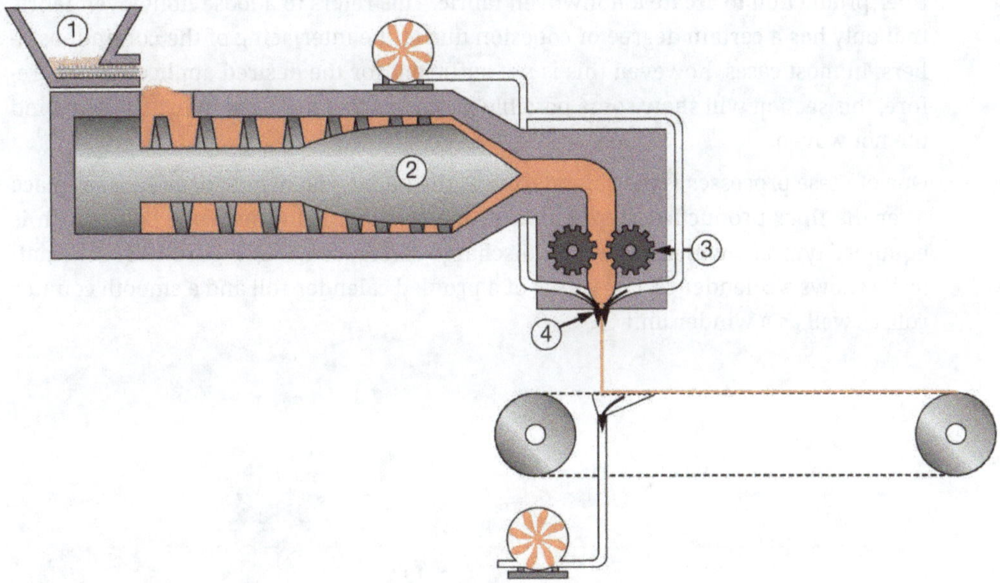

Figure 4.131 Schematic drawing of a meltblown system with sieve belt

The system shown also has a feeder ①, an extruder ② with a deflection head (only necessary for the vertical process), and a melt pump ③ as basic components for continuously conveying the polymer melt. In contrast to classic fiber spinning, however, no air stretching with cold air, and only rarely a closed blowing shaft, is used.

As shown in Figure 4.131, the polymer strand emerging from the nozzle is directly drawn off and stretched by blowing air onto it ④. In order to achieve the longest possible stretching time, the temperature of the draw-off air in the meltblown process is usually the same as or higher than that of the polymer melt. Furthermore, as described in Section 4.11.5 from the parameters of the polymers, less viscous polymers are preferred for this process. The combination of low viscosity with the longest possible stretching time should enable the production of polymer fibers in the range of only a few micrometers.

These very thin fibers (1–10 µm) are mainly produced for filtration processes. This is achieved either in a vertical process using a sieve belt (as shown in Figure 4.131) or in a horizontal process using a suction drum. As the name of the suction drum suggests, the hot air must be extracted from the system via a suction unit to prevent the belt or drum from heating up. This would lead to changes in the structure of the individual fiber and influence the adhesion of the fibers to each other.

4.11.2.3 From Single Fiber to Processed Nonwoven

The classic fiber spinning and meltblown processes can be used in the first step of fiber production to create a nonwoven fabric. This refers to a loose nonwoven fabric that only has a certain degree of cohesion due to the interlacing of the continuous fibers. In most cases, however, this is not sufficient for the desired application. Therefore, this section will show some possible processes that are used industrially to bind the nonwoven.

One of these processes is the calendering of the nonwoven, which usually takes place after the fiber production step and after the removal from the sieve belt, which is equipped typically with a process air discharge ① (Figure 4.132). Figure 4.132 schematically shows a calender ②, consisting of a profiled calender roll and a smooth counter roll, as well as a winder unit ③.

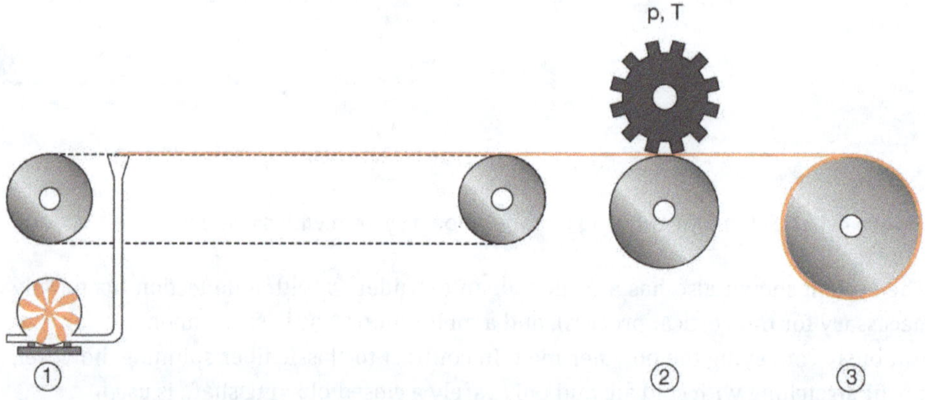

Figure 4.132 Schematic drawing of a sieve belt with calender and winder unit

The calendering process makes it possible to insert welding points between the continuous fibers. A suitable calender pattern, which consists of many small, usually round or oval elevations, is used for this purpose. The two rollers of the calender are heated to the desired temperature T before use and pressed together with a fixed pressure p during the process. This results in a local fusion of the fibers in a so-called calender point. The point acts as a kind of cross-link in the nonwoven and thus improves the cohesion and mechanical properties of the nonwoven.

Industrially relevant for the optimal calendering of the nonwoven are not only the already mentioned variables of roll temperature and pressure but also, above all, the calender pattern. The size, orientation, and flank angle α of the profiled roll have reportedly shown a strong influence (see Figure 4.135) on the welded surface and the properties of the nonwoven.

Two further possible post-processing steps, displayed in Figure 4.133, are the hydroentanglement ① and the Omega dryer (air-through dryer, ATD) ②. These differ greatly in their application and are only shown one after the other on the same sieve belt for illustrative purposes.

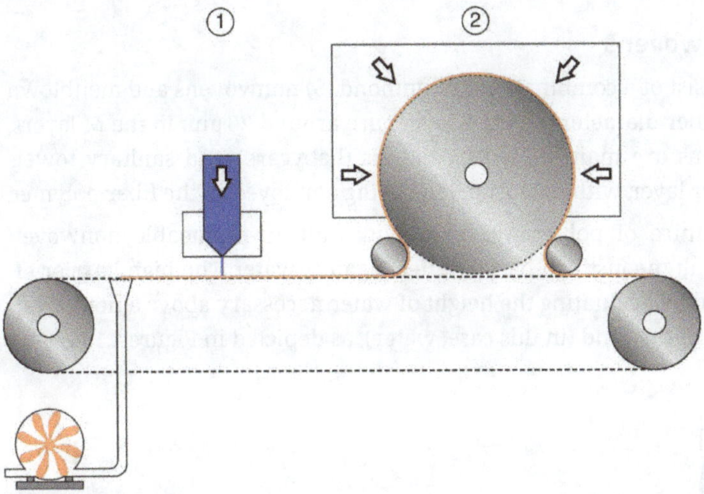

Figure 4.133 Schematic representation of a hydroentanglement and an Omega dryer

Hydroentanglement can be used to strengthen the nonwoven fabric conveyed on the sieve belt. For this purpose, water is pressed through a nozzle strip and shot onto the nonwoven at high speed. A sieve with an application-specific pattern is fitted underneath the nozzle strip to allow the water to run off. This setup causes the fibers to be "pulled along" by the water jet and then deflected by the sieve fitted below. The fibers are thus reoriented in a similar way to the needling process. This leads to entanglement and compaction of the fibers in the nonwoven and thus to an improvement in the mechanical properties. On the other hand, the second possible application of hydroentanglement is the splitting of two-component fibers, which is described in more detail in Section 4.11.2.5. In this application, the water jet is used to separate the interfaces of weakly bonded polymers.

The Omega dryer is also employed for the purpose of bonding nonwovens, creating a bond between the individual fibers by flowing hot air through the nonwoven. This approach is exemplified by the use of core-sheath fibers made of PET and co-PET in order to bond the material, which consists of a PET core and a sheath made of a PET-based copolymer. This combination has sufficient compatibility of the materials to ensure spinnability. Furthermore, the copolymer has lower crystallinity and lower glass and melting temperatures (T_g, T_m) in order to achieve melting under the conditions in the ATD. Contacting fibers can thus stick together and the nonwoven subsequently has a better bond between the continuous fibers.

The post-treatment steps described enable the optimization of individual nonwovens and their mechanical properties. Combinations of these processes can also be applied to alternately deposit layers of spunbond and meltblown nonwovens on a sieve belt and then join them using a calender. With the right production systems, the manufacture of so-called SMS nonwovens becomes feasible.

4.11.2.4 SMS Nonwovens

SMS nonwovens consist of a combination of spunbond (S) nonwovens and meltblown (M) layers. Typical fiber diameters in the S layers are around 20 µm; in the M layers, 3 µm. SMS nonwovens are mainly used in diapers (babycare) and sanitary towels (femcare) as a barrier layer, with polypropylene being employed as the fiber polymer.

The hydrophobic nature of polypropylene provides an air-permeable nonwoven structure with a simultaneous high barrier effect against water. The high barrier effect can be quantified by evaluating the height of water necessary above a nonwoven to cause penetration of the fluid (in this case, water), as depicted in Figure 4.134.

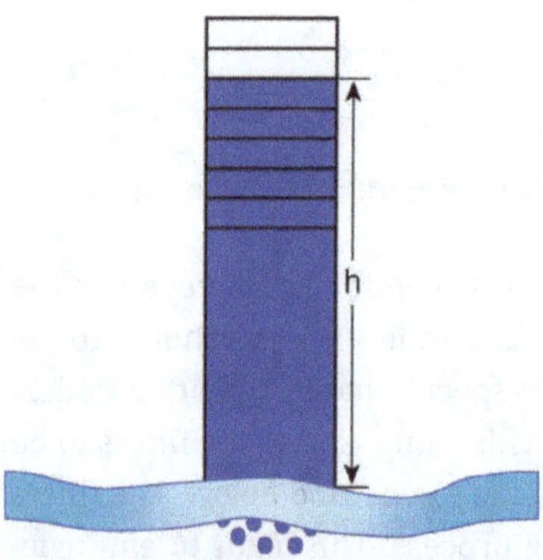

Figure 4.134 Schematic representation of the "hydrohead test" to determine water penetration in a fabric

The water of height h causes a hydrostatic pressure that must exceed the counteracting capillary pressure in order to strike through. The height of the water column at which the strike-through occurs is inversely proportional to the pore diameter. Since the pore diameter is, in turn, proportional to the fiber diameter, it follows:

$$\text{Hydrohead } h \sim \frac{1}{\text{Fiber diameter } d} \tag{4.7}$$

It can thus be concluded that the meltblown layer comprising the smallest fiber diameters is the most significant functional layer in determining the barrier properties of the SMS nonwoven. The S layers are essentially protective and carrier layers of the M layer and are bonded to it in a calendering process (described in Section 4.11.2.3), resulting in point sealing. The proportion of the welded area is typically between 3% and 30%. Figure 4.135 shows typical sealing patterns.

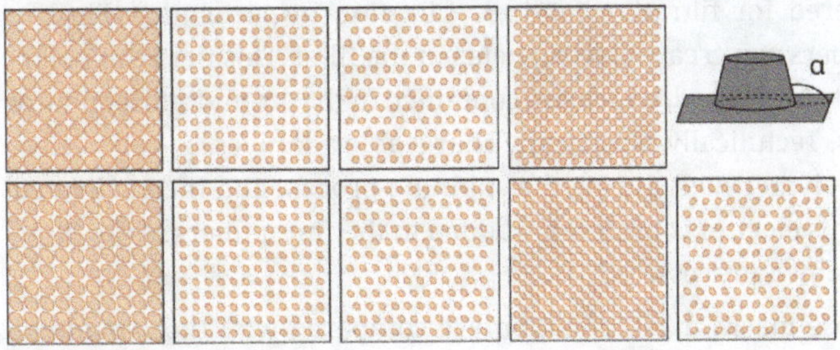

Figure 4.135 Schematic representation of various calender patterns and geometries with flank angle α

In special applications, the SMS structure is stacked multiple times, for example, as SMSMSMS or SMMMS. Statistically occurring errors in the individual M layers are therefore compensated by the multiple superimposed M layers.

In addition to the barrier function in baby- and femcare applications, the M layer allows filter functions in vacuum-cleaner bags and masks. The fine pores are used to extract particles that are blown/breathed through the nonwoven with the air flowing through. In this case, the M layer causes a higher flow resistance or breathing resistance, in the case of the mask. Technically, this is expressed as the pressure loss Δp. For a better understanding, an analogy with electrical engineering and Ohm's law can help (see Table 4.13 and Figure 4.136).

Table 4.13 Analogy: Filter and Electrical Resistance

Electrical engineering (resistance R)	Current I	Voltage U
Nonwoven fabric (filter)	Volume flow rate $\dot{V}$	Pressure loss Δp

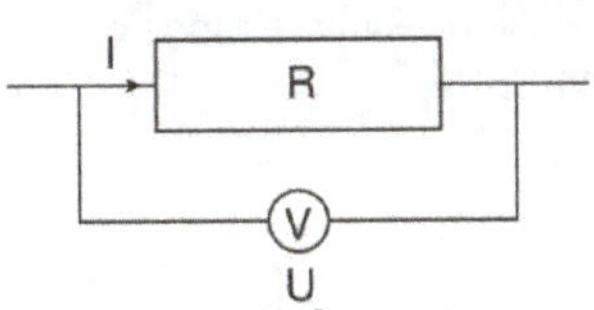

Figure 4.136
Measurement of electrical resistance and analogy to nonwoven fabric with regard to current I, voltage U, and resistance R

In principle, filter systems can be calculated like circuits with resistors if the relationship between the volume flow rate $\dot{V}$ (which corresponds to the current *I*) and the pressure loss *Δp* (which corresponds to the voltage *U*) is known.

$$\dot{V} * R = \Delta p \tag{4.8}$$

Given that finer pores would increase the resistance *R* or reduce the volume flow rate at a constant pressure difference, it is evident that additional effects beyond the "sieve effect" must be considered for filtration (and air filtration in particular). Electric charges in and on the fibers can create electric fields in the filter. This field can polarize the particles that are to be filtered, which are contained in the air, and subsequently attach them to the fibers. Technically, electrically chargeable additives, so-called electrets, are added to the polymer melt during the spinning process and then activated by high-voltage fields. This electret is preferably added to the finest fibers of the filter, that is, used in the meltblown process.

In Table 4.14, mouth–nose masks made of cotton are compared with those made of SMS fleece to also discuss the influences on relative filtration efficiencies, shown in Figure 4.137.

Table 4.14 Protective Effect of Mask Types

Mask type	Typical pore size	Structure	Water barrier	Protection
Cotton mask	20 µm	Sewn	No (hydrophilic)	Reduces the release of aerosols when coughing, speaking, and breathing
OP mask	M layer approx. 3 µm	SMS M layer without electret 20–40 g/m^2	Yes (hydrophobic)	Reduces the release of aerosols when coughing, speaking, and breathing No penetration of water droplets
FFP2/FFP3 mask	M layer approx. 3 µm	SMS M layer with electret 50–70 g/m^2	Yes (hydrophobic)	Reduces the release of aerosols when coughing, speaking, and breathing No penetration of water droplets Filters viruses in the range of 100 nm

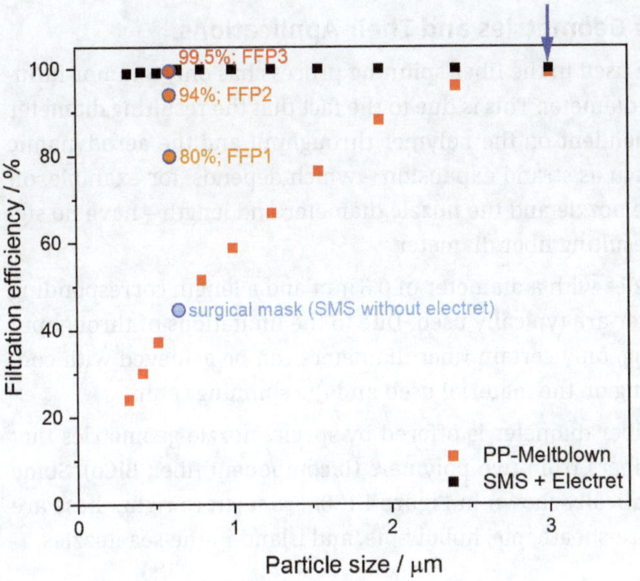

Figure 4.137 Filter efficiencies of nonwovens for relevant protective mask applications

It can be seen that without the additional electrostatic equipment provided by the electret, an almost 100% filter efficiency is only achievable for particles larger than 3 µm (blue arrow in Figure 4.137), which corresponds to the pore size and thus the "sieve effect". For nonwovens with electret, a filter efficiency of almost 100% is possible down to the nanometer range. Besides the possibility of introducing additives to the fibers to optimize the filter properties, the fiber diameter can be reduced and thus the "sieve effect" improved. However, a reduction in the fiber diameter always means an increase in breathing resistance or pressure loss.

The area A through which breathing takes place, the base weight BW, the fiber diameter d in the mask, and the pressure difference Δp caused by breathing are relevant variables for the design of mask or filter nonwovens. Therefore, the volume flow rate (V per time t) and the resistance R when exhaling can be calculated according to the following equations:

$$\frac{V}{t} \sim \frac{d^2 A}{BW} \Delta p \tag{4.9}$$

$$R \sim \frac{BW}{d^2 A} \tag{4.10}$$

In principle, the area A of the mask should be at least 150 cm^2 for a filter material suitable for FFP2/3.

4.11.2.5 Special Nozzle Geometries and Their Applications

The diameter of the nozzle used in the fiber spinning process has only a minor influence on the resulting fiber diameter. This is due to the fact that the resulting diameter of the fibers is mainly dependent on the polymer throughput and the aerodynamic stretching. Other effects such as strand expansion – which depends, for example, on the pressure in front of the nozzle and the nozzle diameter and length – have no significant influence on the resulting fiber diameter.

In industrial contexts, nozzles with a diameter of 0.5 mm and a length corresponding to 3 to 10 times the diameter are typically used. Due to the limitations of throughput and aerodynamic stretching, only certain fiber diameters can be achieved with conventional nozzles, depending on the material used and the spinning setup.

One way of reducing the fiber diameter is offered by special nozzle geometries that enable the production of fibers from two polymers (bicomponent fiber, BiCo). Some possible results are schematically shown in Figure 4.138. From left to right, these are fibers from side-by-side, core-sheath, pie, hollow-pie, and island-in-the-sea nozzles.

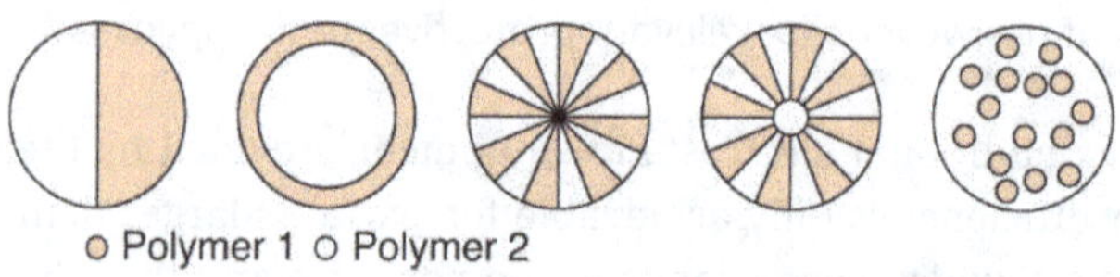

Figure 4.138 Schematic representation of various industrially relevant BiCo fiber structures

The side-by-side, pie, and hollow-pie nozzles are intended to make it possible to create an interface between two polymers, which is split in a post-processing step connected to the fiber production. This produces from one fiber two or more fiber segments that are smaller than the initial fiber diameter. It is important for processing that the polymers have an overlapping processing window. In the case of pie structures, it is essential to consider the surface tensions and viscosities of the two materials, ensuring that they are not too different. Otherwise, this can lead to the low-viscosity component enclosing the entire fiber and thus preventing splitting.

The core-sheath and island-in-the-sea nozzles are also intended to enable the simultaneous processing of two materials. The production goal is, in contrast to the previous ones, to combine two desired properties of the individual materials. The focus lays upon optimizing the fiber properties instead of reducing the fiber diameter.

The basic idea of splitting the pie and hollow-pie fibers as well as the island-in-the-sea fiber is to obtain very fine fibers, which would not be possible with a conventional spinning setup. In the case of the island-in-the-sea fiber, the "sea" material would have to be dissolved away or melted. In consequence, a large-scale industrial implementation of the island-in-the-sea process has not been successful yet.

4.11.3 Special Fiber Production Processes

The manufacturing and further processing methods of polymer fibers/nonwovens presented in the previous sections related exclusively to the established processes with aerodynamic stretching. The present section will provide some insights into less common processes that also have industrial relevance. The focus here is on electrospinning, a newer (compared to conventional fiber spinning) industrially relevant process.

In the electrospinning process, endless nanofibers (fibers with a diameter of less than 1 µm) are produced. A polymer melt or solution is supplied to an electrode using a dosing device. From there, the polymer, or the polymer solution, is accelerated and flies towards the collector electrode (Figure 4.139).

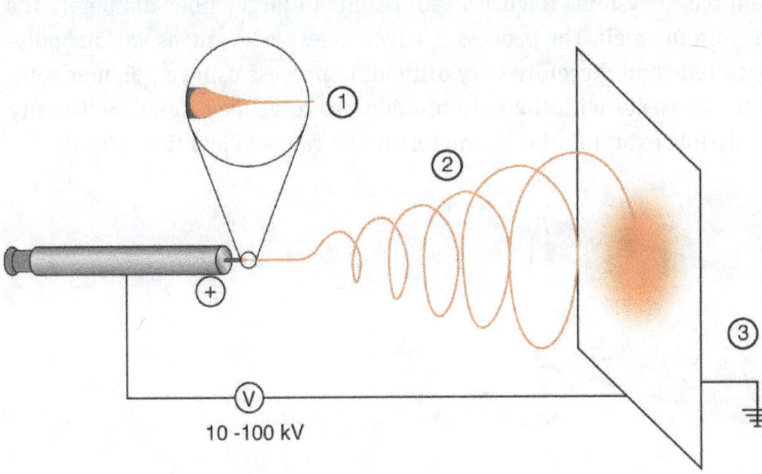

Figure 4.139 Schematic representation of the e-spinning (electrospinning) process

In the electrospinning process, a large number of collector electrode configurations are employed for this purpose, which, among other things, allow for deposition and/or windup of highly oriented nanofibers. At least as wide a variety can be used for the spinning electrodes. Nozzle-free electrodes can also be employed, exploiting the instabilities of liquid surfaces in the electric field. Potential setups use liquid films on wires, bubbles, spheres, cones, rollers, or even free surfaces.

As shown in Figure 4.139, the formation of a so-called Taylor cone ① can be detected at the dosing electrode, regardless of its geometry. From this cone, the polymer/solution is accelerated towards the electrode ② and undergoes various processes as the fiber solidifies. It is deposited on an earthed collector ③. As the beam diameter at the moment of the initial beam formation is usually one to three orders of magnitude larger than the final fiber diameter, the values described in Section 4.11.7 can be used for estimating the fiber diameter.

Due to the structure and complexity of the process, fiber yields are generally orders of magnitude lower than in the meltblown process. For this reason, the production of ultrafine fibers is only technically relevant for the fiber diameter range smaller than 1 µm, using the electrospinning process.

4.11.3.1 Melt Electrospinning

Like in the meltblown process, polymers with high melt flow rates (MFRs) are used and have to be combined with additives for melt electrospinning. These additives are necessary to adjust the viscosity, prevent electrostatic effects, or increase conductivity.

The viscosity of the polymer melt is one of the decisive factors in the production of electrospun fibers. The high viscosity and the absence of solvent evaporation in comparison to solvent-based systems (Figure 4.140) results in larger fiber diameters for fibers produced from the melt. The process is nevertheless relevant, as various polymers are poorly soluble and therefore very difficult to process using a polymer solution. Modifying the viscosity with the help of additives to achieve smaller viscosity values makes it possible to attain fiber diameters in the range of less than 100 nm.

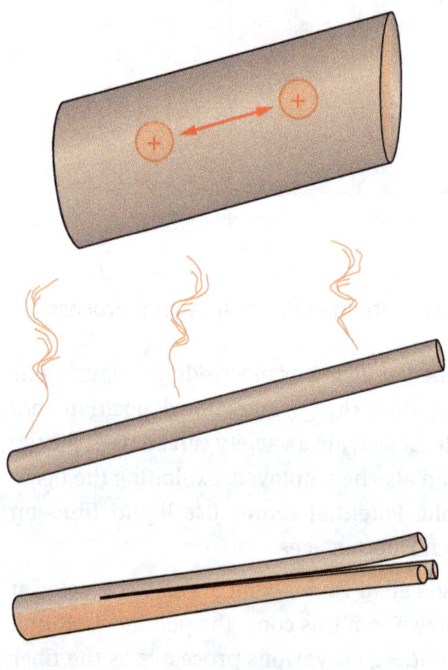

Figure 4.140 Schematic representation of effects during the stretching of fibers in the e-spinning process; from top to bottom: drawing by repulsion of like charges, evaporation of solvent, and occurrence of jet splitting

As insulators, nonconductive polymers (*e. g.*, PE/PP) tend to accumulate charge carriers on the surface. In the electrospinning process, this can lead to an inhomogeneous distribution of charges, resulting in an unstable process and, therefore, non-reproducible results. To avoid this, antistatic, surface-active additives are used.

In most applications, however, this homogeneous distribution is not sufficient to achieve the finest possible fibers. Increasing the conductivity is the easiest way to obtain a better processability and a further reduction in the final fiber diameter. In practice, this is usually achieved by adding ionic salts. Their admixture leads to additional charge carriers in the polymer, which can improve or enable the processing of nonconductive polymers (*e.g.*, PE/PP). Furthermore, the ions are used during fiber stretching and solidification (see Figure 4.139 ②) as additional charge carriers, which can neutralize repulsive forces (see Figure 4.140) on each other and thus generate increased stretching.

4.11.3.2 Solution Electrospinning

Even finer fibers are available in solution electrospinning. As the name implies, a polymer solution is used to produce the fibers. The liquids used for electrospinning are polymer solutions whose viscosities must be selected within a narrow range. They not only have an influence on the resulting fibers and their diameters but are, above all, important for the feasibility and reproducibility of nanofiber production.

Too-low viscosities lead to droplet formation ("electrospraying"), as the strand is unstable and will snap when pulled from the dosing electrode. Viscosities that are too high can lead to the nozzles sticking together and/or to undesirably large fiber diameters (larger than 20 µm). Ultimately, this means that the production of fibers with such diameters in other processes (see the meltblown process) is more cost-effective and, therefore, more industrially relevant.

4.11.4 Additives for Fiber Spinning

The industrial application of polymers as fiber or nonwoven materials requires properties in many areas that pure polymers cannot achieve. Some examples of this are mechanical or electrical properties but also the coloration of the product. In order to achieve the desired product properties, additives in the form of pigments (colors), electret, and organic or inorganic (*e.g.*, TiO_2) fillers are introduced in many applications (for more information, see also Chapter 7).

Pigments are intended to influence the coloring of the fiber and thus the product. Furthermore, these pigments can sometimes be used to enhance the UV stability of the polymer as a side effect. In the case of "pure" polymers without dyes, it may be necessary to add UV stabilizers, depending on the application.

Electret is used for the electrostatic finishing of polymer fibers. Fibers modified in this way are, for example, used in filtration applications. The fibers used in this case are themselves electrically charged after processing, with the help of charged electrodes. These charges enable an increase in filter efficiency, as small particles can be polarized while they flow through the electrical field of the fibers. The polarization results in an attraction to the fiber and an adhesion of small particles.

Organic and inorganic fillers can be added for various reasons. In theory, cost minimization could be achieved by adding inexpensive fillers. However, as the price per kilogram for most materials (PE, PP) used in mass production is less than 1 €/kg, this is rarely done. Of greater importance is the adjustment of the mechanical properties, which can be achieved by influencing the morphology and crystallinity of the fibers. A side effect of adjusting the crystallinity is a change in the transparency or opacity of the material. Surface-active additives can also be used for hydrophilization or hydrophobization by accumulating hydrophobic or hydrophilic end groups on the surface of the fibers. This is relevant, for example, in the use of nonwovens as a base material for diapers. The size and added quantity of inorganic substances is ultimately important for the processing of fillers. The particle size should not exceed $\frac{1}{10}$ of the fiber diameter and the filling level should remain below 0.5% by volume, as otherwise, instabilities may occur during the fabrication process.

4.11.5 Relevant Polymer Parameters for Fiber Spinning

In addition to the process parameters, the properties of the materials used are of great importance for the production of polymer fibers. Table 4.15, Table 4.16, and Table 4.17 show a classification of the MFRs (see Section 3.2.1.3) and their influence on the resulting fiber diameters for three industrially relevant polymers (PE, PP, and polybutene). Due to the different viscosities, which are caused by the architecture and chain length of the polymers, strong differences in the fiber diameter and the properties of the fiber/nonwoven result under the same process conditions.

The variation in the MFR of the polybutene allows a variation in the fiber diameter over an order of magnitude with otherwise identical process parameters. Compared to polybutene and PP, the available spinning grades for polyethylene are less common. Therefore, Table 4.15 only lists data for a few commercial materials.

The different PP spunbond and meltblown types mentioned (cf. Table 4.16 and Table 4.17) allow for a large variation in the fiber diameter for the same process parameters. The “fiber diameters” indicated by circles in the table clearly show the extent to which these would change with an otherwise identical process. When processing these materials, the molecular weight, or, as a result, the MFR, is therefore an important parameter in the production of polymer fibers.

Table 4.15 Typical Material Characteristics for Fiber Spinning Types Made of Polybutene and PE

Polybutene spunbond								PE spunbond			
MFR 190 °C / 2.16 kg (g/10 min)	**1.3**	**15**	**45**	**200**	**600**	**1200**	**1500**	**MFR 230 °C / 2.16 kg**	**11**	**18**	**26**
Fiber diameter	○	○	○	○	○	○	○	Fiber diameter	○	○	○

Table 4.16 Typical Material Characteristics for Available PP Fiber Spun Types

PP spunbond													
MFR 230 °C / 2.16 kg (g/10 min)	**2**	**12**	**15**	**18**	**19**	**25**	**27**	**30**	**32**	**34**	**36**	**37**	**38**
Fiber diameter	○	○	○	○	○	○	○	○	○	○	○	○	○

Table 4.17 Typical Material Characteristics for Available PP Fiber Spun Types

PP meltblown									
MFR 230 °C / 2.16 kg (g/10 min)	**125**	**450**	**500**	**800**	**1100**	**1200**	**1300**	**1500**	**1800**
Fiber diameter	○	○	○	○	○	○	○	○	○

Fibers Made from Regranulated Polymer Using the Example of rPET

The production of polymer fibers is often implemented in fields where direct reusability or multiple use is not possible (for example, babycare/femcare). For this reason, the recycling of polymers for fiber production, which is discussed in more detail using PET as an example, is also important in this field.

Recycled PET (rPET) is generally playing an increasingly important role in the world of spunbond fibers and nonwovens, with chips and pellets made from PET being used. In a study [1], virgin PET (vPET) and rPET were evaluated in detail with regard to spinnability and fiber properties. For this purpose, different aerodynamic take-off speeds were applied at processing temperatures of 270 °C and 280 °C (nozzle). It was found that fibers produced from rPET grades can achieve properties comparable to those of vPET. In particular, it was shown that a weight average M_W in the range of 40 kg/mol to 46 kg/mol at $M_W/M_n = 1.7–1.8$ provides the best fiber properties, where fiber diameters of less than 20 µm are possible.

4.11.6 Process Instabilities

Process instabilities in the production of polymer fibers can occur for various reasons. Some of the possible reasons and associated measures or calculations are explained below.

Critical shear stresses and shear rates at the nozzle can lead to instabilities in the strand and the resulting fiber breakage. One way to avoid these instabilities is to estimate the wall shear stress τ_W or shear rate $\dot{\gamma}_W$ (see Equation 4.11 and Equation 4.12). These make it possible to avoid critical shear stress or shear rate values during production. To calculate the shear stresses and rates, the nozzle length L, its radius R, and the volume flow rate $\dot{V}$ or the pressure before the nozzle Δp are required.

Estimation of shear stress and shear rate:

$$\tau_W = \frac{\Delta p}{L}\frac{R}{2} \tag{4.11}$$

$$\dot{\gamma}_W = \frac{4\dot{V}}{\pi R^3} \tag{4.12}$$

Snapping of the fiber during the stretching process is another possibility for a process instability. There can be several possible reasons for this. If the viscosity of the polymer melt is, for example, too low, the applied stretching force may exceed the melt strength, leading to fiber breakage. Possible countermeasures are an adjustment of the melt temperature or the draw-off air in order to achieve an adjustment of the viscosity or the draw-off force. When using different batches, fiber breakage can occur due to molecular differences. These result in different properties and can be detected by comparing GPC (gel-permeation chromatography) and light-scattering measurements of previously used "good" batches with those of the current batch. In particular, small amounts of branched molecules in the spinning polymer lead to strain hardening and thus to spin stabilization, but too many branched molecules lead to instabilities in the nozzle itself. The optimum branching ratio depends on the process conditions and the application.

In addition to problems in the production of single-component fibers using conventional processes, various problems can also occur in the production of BiCo fibers. Core-sheath fibers and split fibers, in particular, will be considered here.

Core-sheath fibers, as shown in Figure 4.138, have a polymer core which is to be enclosed by a second polymer in order to combine the properties of the polymers. Problems can arise here if the throughput of the sheath component is too low. This results in insufficient sheathing and, therefore, deformed (usually, out-of-round) and non-reproducible fibers. Instabilities can also occur during production if the core-sheath fiber is insufficiently formed due to large differences in viscosity between the two components. The sheathing of the core takes place in the spinneret during the production of these fibers. Excessive differences in viscosity can lead to instability when the second polymer coats the core. This can result in core fibers with a sheath that is too thin or sheathed fibers without a clear core.

Split fibers are a second type of BiCo fiber whose instabilities during production will be explained. In addition to the problems already mentioned due to differences in viscosity and fiber breakage caused by excessive pull-off forces, problems can also occur during production when the fibers are split. The splitting process is usually carried out using a water jet, which was already shown schematically in Figure 4.133. With an optimally adjusted fiber spinning process with desirable fiber geometry (see Figure 4.141, left), the water jet (or several hundred of these jets) can be used to split the segments at the interface between the incompatible polymers. Instabilities during splitting can occur due to different surface tensions of the two polymers. During processing, these result in one of the components flowing around the second component. The result is a fiber that (shown schematically in Figure 4.141, right) cannot be split as the water jet cannot break the polymer coating. The occurrence of this effect can be avoided by reducing the temperature of the flowing component (in this case, "Polymer 1"). If the described "reflowing" occurs after the use of different batches, the molecular architecture of the new batch can also have an influence on the viscosity. Information on this can be obtained, as discussed before, via GPC/light-scattering measurements and compared with that of "good" batches. Measurements of intrinsic viscosity or other parameters are not sufficiently meaningful.

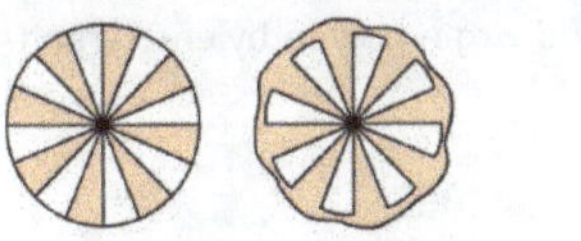

Figure 4.141
Desirable geometry (left) and incorrect creation (right) of a split fiber

4.11.7 Simple Correlations in the Fiber Spinning Process for the Practitioner

To summarize the previous sections, the fiber spinning process is an industrially very relevant method for the production of polymer fibers. Continuous and reproducible production is particularly important for the application. Since the fiber diameter d

plays a decisive role in the resulting nonwoven properties, an estimation of d in fiber spinning processes with aerodynamic stretching is shown below. Using the viscosity η of the polymer melt, the following relationship can be postulated:

$$d \sim \frac{\eta^{\frac{1}{3}} \cdot \dot{V}^{\frac{1}{2}} \cdot D^{\frac{1}{3}}}{\Delta p^{\frac{2}{3}}} \tag{4.13}$$

The exponents in Equation 4.13 are of a semi-empirical nature, can be adjusted in individual cases for better data description, and may deviate by up to 10%. The material parameters used here are the viscosity η (the so-called zero viscosity η_0 can be used here) and the volume throughput $\dot{V}$ of the polymer melt, the diameter D of the spinneret, and the air pressure difference Δp, which occurs or is used in every form of aerodynamic stretching (spunbond or meltblown).

A comparable estimation of the fiber diameter d in electrospinning processes from the polymer solution follows the laws shown here:

Without jet splitting

$$d \sim c^{1/2} \eta^{1/3} p_V^{1/6} U^{-2/3} \tag{4.14}$$

With jet splitting

$$d \sim \eta^{4/9} \cdot c^{1/2} \cdot \gamma^{1/3} \cdot p_V^{2/9} \cdot U^{-14/9} \tag{4.15}$$

These use the electrical voltage U between the electrodes, the viscosity η of the polymer solution (the so-called zero viscosity η_0 can also be used here), the concentration c of the polymer solution, the vapor pressure p_V of the solvent, and the surface tension γ of the polymer solution (the surface tension of the solvent can be used here). In the case of melt electrospinning from the polymer melt, the contribution of the concentration dependence is eliminated.

References for Section 4.11

[1] Qin, Y., Qu, M., Kaschta, J., Schubert, D. W., Comparing recycled and virgin poly (ethylene terephthalate) melt-spun fibres, *Polymer Testing*, 72 (2018), pp. 364–371

4.12 Rotational Molding

Rotational molding, also known as rotational casting, is a primary molding process for the production of hollow plastic components. Due to the possibility of generating virtually stress-free components with a constant wall thickness and great design freedom as well as comparatively inexpensive plant and tool technology, it competes with processes such as blow molding and injection molding and has established itself as a standard process, especially for the production of large-volume hollow components in small quantities. Typical product examples include tank containers, kayaks, boat hulls, machine housings, furniture, and toy parts.

The rotational molding process sequence can generally be divided into four phases, which are shown in Figure 4.142 schematically. In the first step (a), the open mold is filled with the polymeric starting material, which is usually in powder form. After the mold is closed, what follows is the heating phase (b), in which the mold and the plastic powder on the inner wall of the mold are heated. From a polymer-dependent temperature, the powder begins to adhere to the mold wall and gradually melt. A uniform melt layer is generated by multi-axis rotation of the mold, which causes the plastic powder to be evenly distributed within the cavity. Continuous rotation is followed by the cooling phase (c) with the accompanying solidification of the plastic. In a final step (d), the hollow plastic molded part is demolded.

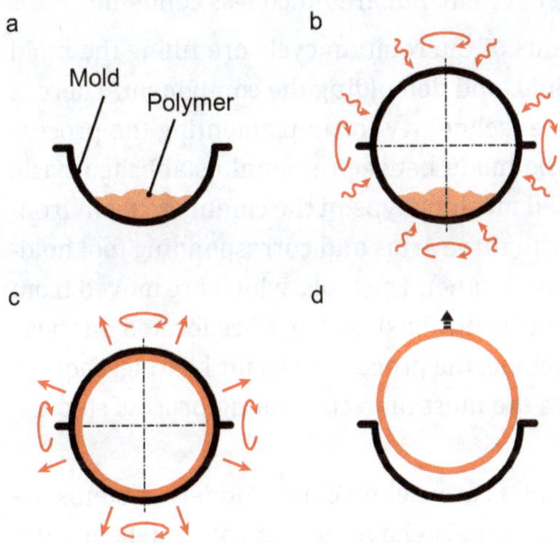

Figure 4.142

Rotation cycle (schematic)

a: Filling, b: heating phase, c: cooling phase, d: demolding

As the process generally takes place without pressure, rotary tools usually only have low wall thicknesses and are comparatively inexpensive to manufacture. In comparison with competing processes, uniform wall thicknesses can be achieved across the entire component cross section and the desired thickness can be controlled via the amount of material added and the tool rotation, without the need for subsequent tool adjustments. The process offers a great deal of design freedom and virtually waste-free production thanks to the complete utilization of the material. In special processes derived from the basic principle, rotational molding also allows the production of rotationally molded foam parts as well as of multilayer components in which the individual layers can fulfill different functions.

In industrial terms, the numerous potentials that rotational molding offers are offset by the long cycle times – as the biggest challenge – which reduce the cost-effectiveness of the process, especially with increasing quantities. They also result in a limited selection of raw materials, as only a few polymers are able to meet the resulting thermal requirements. Another limitation is the high cost of grinding the raw materials, which are generally only available as granules, in order to generate a free-flowing powder for the rotation process. The industrially established trend of using microgranulates in the rotation process represents an economic and technological potential, so that complex grinding processes can be eliminated and a more sustainable value chain can be created.

Both thermoplastic and thermoset materials can be processed in the rotational molding process. The most established material in the thermoplastic sector and in general is polyethylene, due to its high thermal resistance and availability as a free-flowing powder at low prices. In addition, the use of polypropylene is steadily increasing with higher product requirements. Other typical thermoplastic polymers for rotational molding are polystyrene, polyvinyl chloride, and nylon. Thermoset systems such as polyurethanes, unsaturated polyesters, or phenols are mostly used in the form of liquid resins in combination with a hardener system, but are much less common.

As already described, the basic components of the rotation cycle are filling the mold with material, heating and cooling the mold, and demolding the component. There is a wide range of industrially used machine technology for implementing the process principle, from which a distinction can be made between several established basic machine designs. The most commonly used machine type in the commercial environment is the so-called carousel machine, with three arms and corresponding tool holders, which are fixed at an angle of 120° to each other. The tools, which are moved from one production station to the next when the central axis rotates, are located on these rigid arms. For an optimum use of the machine, the process times for heating, cooling, and filling or demolding should match, as the most time-consuming process step determines the total cycle time.

Figure 4.143 shows a schematic structure of a carousel machine. Modern systems deviate from a rigid fixation of the arms and can also have four or five independently operating arms so that dwell times at the respective production stations can be adjusted as required, thus increasing the flexibility of the rotary systems. In the course of increasing modernization and automation, the use of robots in the area of tool handling is also growing significantly.

Rotational molds are generally made of steel, nickel, or aluminum alloys and typically have only thin walls, as the forces exerted by the rotational process are low in contrast to those in injection molding or blow molding, and rapid heat transfer to the mold and through it to the plastic must be ensured. Therefore, the tools largely correspond to a shell construction around the desired molded part geometry and are usually divided into two halves along a parting line. Rotary tools make it possible to reproduce complex molded part geometries and also offer a high level of functional integration, including the option of integrating inserts or metal inserts into the tool.

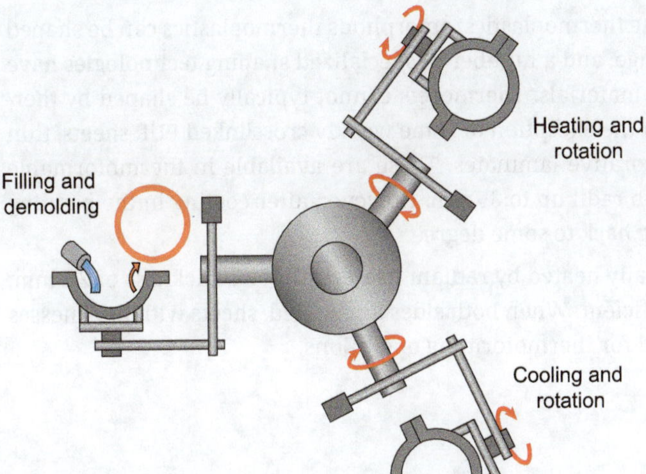

Figure 4.143 Carousel machine (schematic)

4.13 Secondary Shaping

Secondary shaping methods are described as chip-less processing methods, in particular of semi-finished products, but also of molded components. This technology takes advantage of a specific property of thermoplastic materials: their Young's modulus decreases by approx. one order of magnitude in the glass transition zone (for amorphous plastics) and in or slightly above the melting temperature (for semi-crystalline plastics). In this so-called rubber-elastic state, these materials can be shaped using little force. However, these shaping operations introduce orientations in the macromolecules that would lead to a shape recovery once the deforming force is removed. Therefore, the formed parts have to be either stored under constraint and at elevated temperatures in order to relax the molecular orientations, or cooled under constraint in order to freeze these orientations. However, this introduces changes in characteristics (anisotropies), see Section 2.4.2, and at temperatures close to the softening range, these parts would revert back to their original shape. Rubber-like plastics with glass transition temperatures T_g below service temperature can therefore typically not be thermoformed.

The resilience that thermoformed thermoplastics exhibit when re-heated is utilized in the production of *heat-shrink tubing* for the sheathing of rods or handles and the production of *shrink-wrap films* for tight-fitting packaging applications such as palettes.

Other than semi-crystalline thermoplastics, amorphous thermoplastics can be shaped in a wide temperature range, and a number of specialized shaping technologies have been developed for these materials. Thermosets cannot typically be shaped by thermoforming methods, with the exception of some weakly crosslinked PUR sheets, thin laminated paper, and decorative laminates. These are available in thermoformable types and can be bent with radii up to 35 mm. However, after cooling in the bending mold, they typically spring back to some degree.

Sheets and films are typically heated by radiant heaters. Up to a thickness of 2.5 mm, single-sided heating is sufficient. When both sides are heated, sheets with thicknesses up to 10 mm can be heated for thermoforming operations.

4.13.1 Bending

Semi-finished products, such as rods, pipes, and sheets are shaped by heating the bending area using hot air or gas, liquid baths, or radiant heat. Similar to metal shaping operations, pipes and tubes are internally supported by a heated rod, a silicone rubber rod, a fabric-sheathed metal spiral, or an inflated hose in order to maintain the circular cross section. Larger pipes are manufactured from sheet goods by bending over a roller and subsequent welding of the abutting edges. Pipe couplings are shaped on a counter pipe or by using calibrating mandrels, where casing pipes are heated and shrunk on a base with the use of expansion mandrels. Pipe ends are heated and fit into ring gauges and collar bushings, beaded, or in case of polyolefins, compressed.

Break bending of sheets is performed in fixtures that allow small bending angles. Sharp bends are achieved through a combination of bending and welding (Figure 4.144).

Integral hinges with good resistance to alternating loads can be manufactured in a similar way. A heated sword with flattened cutting blade is pressed into the plastic material thus stretching the material in the area of the blade.

Large-area components that require only minimal shaping operations, such as acrylic covers, are manufactured using *drape forming* processes. Here, the heated sheet is placed over a positive mold (typically covered with a friction-reducing material) and secured in a hold-down device.

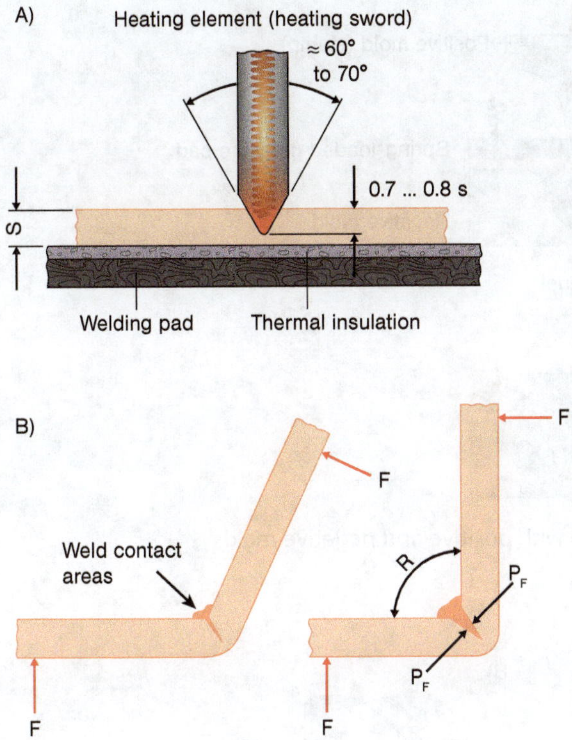

Figure 4.144 Break bending/welding

A: The hot heating element is pressed into the plastic sheet;

B: Break bending of the sheet;

F: Bending force;

P_F: Welding pressure

4.13.2 Mechanical Forming

Mechanical forming operations utilize positive and negative molds (Figure 4.145) where the upper part can also be replaced by a water-filled or foamed rubber cushion. Because the pre-cut plastic sheet is allowed to slip over the mold edge during shaping, it is possible to achieve almost uniform wall thicknesses. Plugs and draw rings are used for the mass production of simple, typically rotational symmetric packaging containers, see Figure 4.146. Mechanical forming uses temperatures below the glass transition for amorphous and below melting temperature for semi-crystalline thermoplastics *(cold thermoforming)*. The drawing forces required are only 40 to 60% of the forces required for the drawing of metals. Mold costs are 25 to 35% of those for injection molding. Automation and use of multi-cavity molds allow for production rates of > 10,000 per hour.

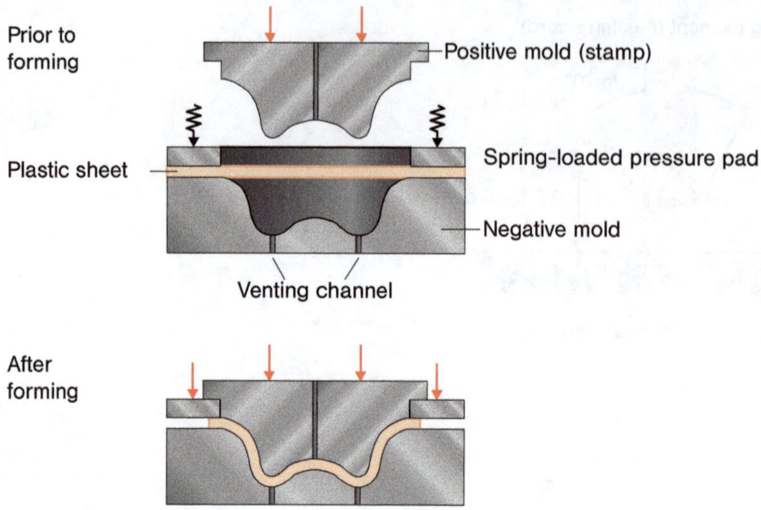

Figure 4.145 Mechanical forming with positive and negative molds

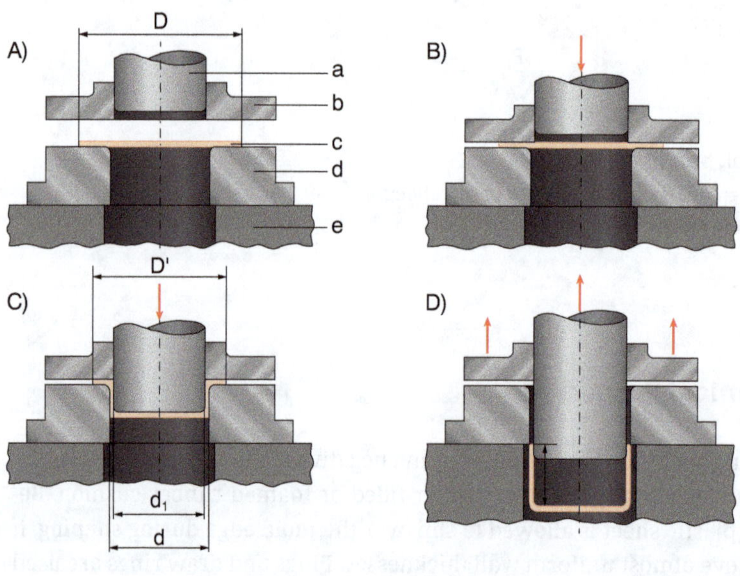

Figure 4.146 Forming (A to D) of cylindrical parts

a: Core plug, b: Pressure pad/clamp, c: ABS sheet, d: Draw ring, e: Press bed

4.13.3 Pressure Forming

During pressure operations, the edges of the pre-cut plastic sheet are clamped so that the shaping of the part will cause a decrease in wall thickness. The shaping forces are applied via vacuum. On a large scale, these processes are used to manufacture plastic film packaging and to produce large-area parts. Thermoforming machines offer mold surfaces ranging in size from approx. 250 mm × 350 mm to 3000 mm × 9000 mm and mold depths up to 2500 mm; they allow for the manufacture of up to 5000 pieces per hour. Mold cost reach only approx. 20% of comparable injection molds. Molds made from gypsum, Stonex clay, or wood are suitable for small production runs up to approx. 50 parts. For larger production runs or for parts requiring higher-quality surfaces, molds are typically cast from light metal, brass, or resins with a filler content of approx. 60% Al-powder, all of which exhibit good thermo-conducting properties and can be cast with additional cooling channels. Development targets of this processing technology are both minimizing of wall thickness variations and of processing scrap (usually the edges of the thermoformed components have to be trimmed).

Figure 4.147 shows the *vacuum process* with a negative mold (vacuum thermoforming). The areas of the sheet that adhere to the mold first are no longer drawn, and therefore the process significantly decreases the wall thicknesses in corners, edges, and in deeper parts of the component. This effect can be mitigated by pneumatically pre-stretching the heated sheet and subsequently placing it in the mold using a plug assist device. Figure 4.148 shows an example of the automation of this processing technology in the packaging industry. Both the filling station for the good to be packaged and sealing station (*e. g.*, via heat sealing) are integrated.

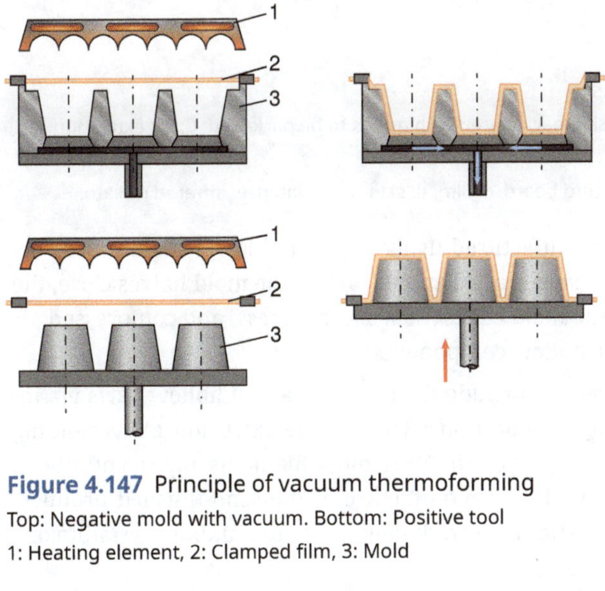

Figure 4.147 Principle of vacuum thermoforming

Top: Negative mold with vacuum. Bottom: Positive tool

1: Heating element, 2: Clamped film, 3: Mold

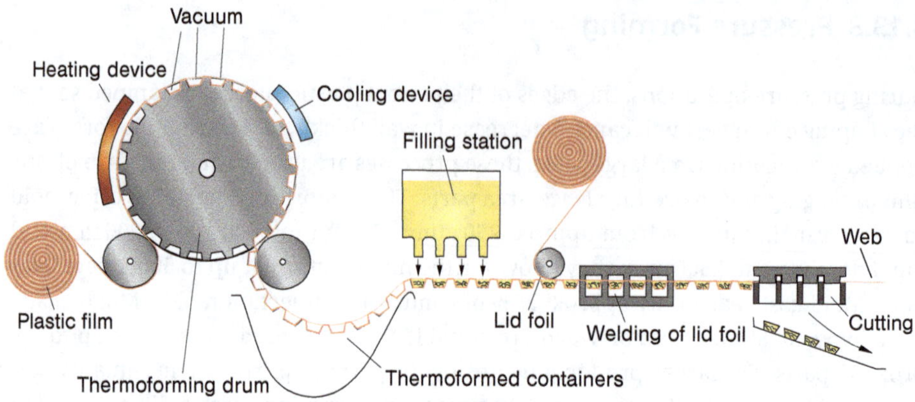

Figure 4.148 Simultaneous vacuum thermoforming, filling, and sealing of packaging

Processes in which the heated plastic sheet is pre-stretched above the mold and then pulled into mold by vacuum are called *positive processes*. A variation of this process is the so-called skin-pack process: here, the goods to be packaged are placed on a special, heat sealable card board with perforations for the vacuum and used as a positive mold (Figure 4.149). Blister or bubble processes use films (typically transparent plastics) with integrated shells that are put over the goods to be packaged and then welded to the sealable base.

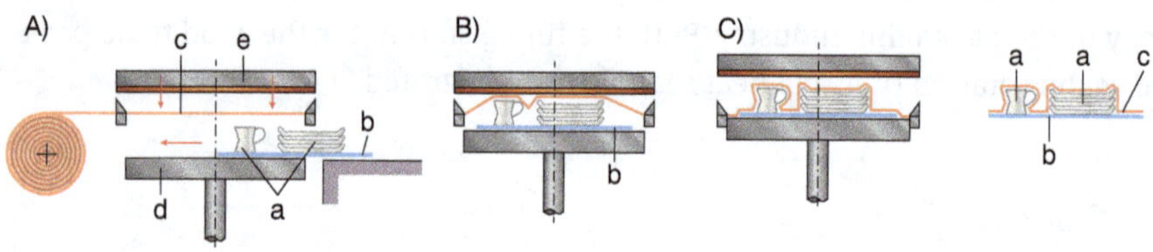

Figure 4.149 Skin-pack process

A: Pre-heating of film. B: Raising of skin-pack platen with goods to be packaged. C: Air extraction and cooling

a: Goods to be packaged, b: special card board, c: film, d: skin-pack platen, e: infrared radiator

Gas tanks or surf boards are manufactured using upper and lower mold halves between which two heated plastic sheets are clamped. When the mold halves close, the sheets are welded together at the mold edges *(twin sheet process)* and compressed air or vacuum is used to create the hollow components.

Other secondary forming operations include the manufacture of hollow parts via injection or stretch blow molding (see Section 4.3.6) and via extrusion blow molding (see Section 4.2.5), and the drawing of plastic films, monofilaments, tapes, and fibers. They can be integrated in-line with the manufacture of the semi-finished products and thus improve product properties by introducing molecular and/or crystallite orientation in the direction of drawing. It is also possible to proceed with thermoforming operations immediately after the manufacture of plastic sheets or films, thus eliminating a heating process.

4.14 Plastics Joining

Michael Gehde

4.14.1 Classification of Manufacturing Processes

The manufacturing processes available for the production of goods and commodities are divided into six main groups in accordance with DIN 8580 (see Figure 4.150). Each main group is characterized by the type of change in the material composition.

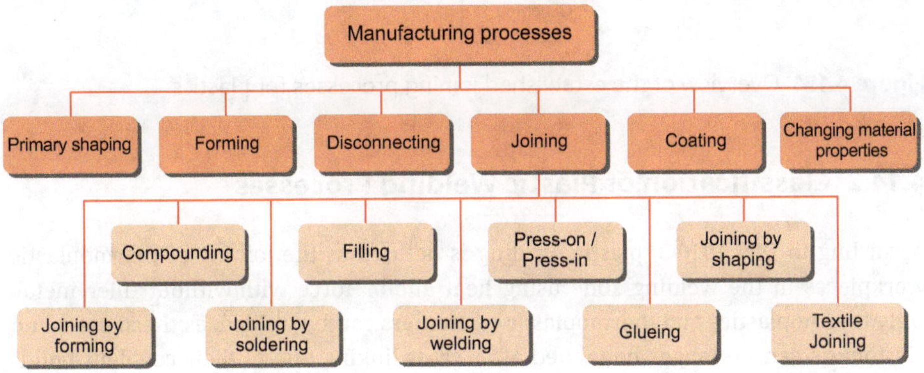

Figure 4.150 Classification of manufacturing processes according to DIN 8580

The purpose of the "joining" manufacturing process is the permanent joining of two or more workpieces of a specific geometric shape or the permanent joining of similar workpieces with shapeless material. As shown in Figure 4.150, the individual joining processes are divided into nine further subgroups in accordance with DIN 8580, described in detail in DIN 8593. However, not all of the joining processes listed in DIN 8593 are suitable for joining plastics, which is why Figure 4.151 shows a classification of the joining processes established for plastics. The classification is based on the physical principles of the joining method.

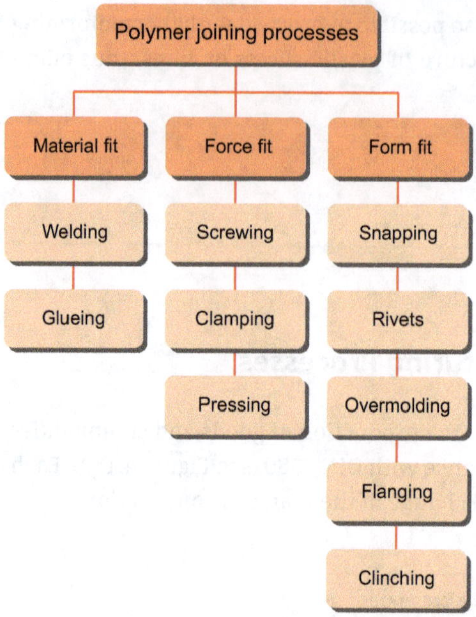

Figure 4.151 Overview of the established joining processes for plastics

4.14.2 Classification of Plastic Welding Processes

According to DIN 1910-3, plastic welding is defined as the joining of thermoplastic workpieces in the welding zone using heat and/or force with/without filler metal. Only thermoplastics and thermoplastic elastomers can be welded, as thermosets and elastomers can no longer be melted after cross-linking due to their covalent bonds between the macromolecule chains.

The processes for joining by welding are generally classified in DIN 8593-6. In contrast, the welding processes suitable for plastics are classified and presented in DIN 1910-3. Figure 4.152 shows the classification of plastic welding processes according to the type of heat input.

Figure 4.152 Classification of plastic welding processes according to DIN 1910-3

4.14.3 Welding

During welding, the workpieces are joined in the welding zone and welded using force with or without filler material. The required joining force is applied manually, mechanically, via an interference fit, or as a result of thermal expansion of the workpieces.

The welding process is usually divided into three process phases. In phase 1, the component surfaces are heated to a temperature above the melting temperature and joined in phase 2 under pressure and the resulting flow movement of the melt. The flow processes in the weld seam area are influenced in particular by the viscosity of the melt and the joining pressure and result, on the one hand, in expansion flows, which promote the diffusion process in the joining plane, and, on the other hand, in shear flows, which lead to structural notches in the material. At the end of the welding process, the melt cools down in phase 3, where it solidifies and the formation of the material structure begins as the joining pressure continues to act. Statements on weld seam quality and process parameters adjustment can be made, in particular, on the basis of the thickness of the residual melt (weld seam thickness) and optimized on this basis. Different types of plastic can only be welded together if there is a minimum chemical compatibility and their melting temperatures are not too far apart.

An evaluation of the welded joint must include both the tightness (pipes, containers, bags, packaging) and the strength. While tightness is usually examined in practical tests on welded molded parts, test specimens are suitable for determining the short- and long-term welding factors (quality factors). They reflect the ratio of the strength of the welded joint to that of the base material and are in the range of approximately 0.3 to 1. When specifying welding factors, the test method must always be included, as this can strongly influence the level of the factors. Common test methods are tensile, bending, folding bending, and impact bending or impact tensile tests as well as creep tests, for example, under internal pressure on pipe welded joints.

A welded joint can have a notch effect on the molded part. The local heat input in the weld seam leads to local changes in residual stresses, which have a significant influence on the service life of a welded joint. This may make it necessary to test the long-term properties of the molded part. The quality of a weld seam is also influenced by the material and processing quality (residual stress, crystallization, orientation, etc.) of the components to be joined.

4.14.4 Series Welding Processes

The term "series welding process" is used to describe welding processes that are employed in large-scale production in particular. The requirements for minimum cycle time, high process reliability, and maximum automation capability are the main focus here. The industrially established series welding processes for plastic applications are presented in more detail below.

4.14.4.1 Sequential Butt Welding Processes

Hot Plate Butt Welding

Hot plate butt welding is one of the dominant joining techniques in the series production of molded parts made from thermoplastics. It is suitable for complex three-dimensional joining surfaces and can be used for a wide range of plastics.

A typical welding arrangement for hot plate butt welding is shown in Figure 4.153. In the process, the joining surfaces of the molded parts or semi-finished products are heated by direct contact with the temperature-controlled heating element (approximately 180 °C to 300 °C, up to approximately 600 °C for high-temperature applications) and then welded without filler material under a pressure of approximately 0.1 N/mm^2 to 2 N/mm^2. As a result of the separate heating and joining process, hot plate butt welding is a multistage process. The process is divided, as shown in Figure 4.154, into four process phases: equalizing, heating/melting, transposing, and joining. During the equalizing phase, the joining surfaces make full contact with the heating element under pressure. This is followed by the formation of a fusion layer under almost pressure-less heating. At heating element temperatures of up to 260 °C, the heating element should be provided with a PTFE nonstick layer, if possible, in order to minimize adhesion effects between the plastic melt and the heating element. After heating, the heating element is removed (transposing phase) and the components are welded under pressure. The cooling already starts in the transposing phase. After the required joining distance has been covered and part of the melt has been ejected, the final cooling time takes place under pressure, and the residual melt solidifies. The welding process can be pressure-controlled or path-controlled. The characteristic process parameters include the heating element temperature, the heating time, the joining pressure, and the cooling time. The process times required for hot plate butt welding are in the medium to high range.

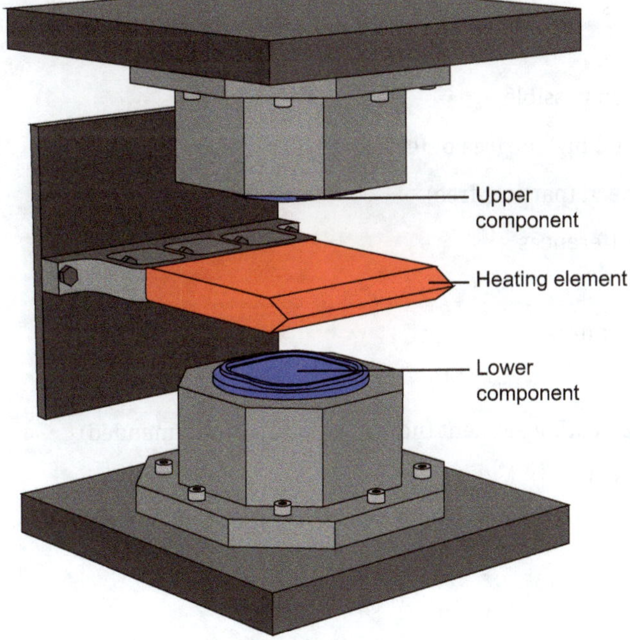

Figure 4.153
Schematic diagram (example) of a welding arrangement for hot plate butt welding [Source: KT-Chemnitz, Gehde]

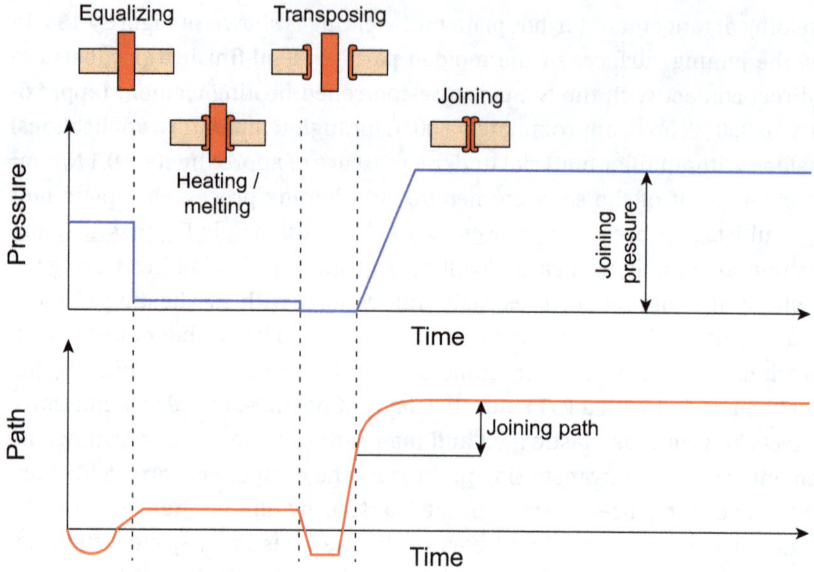

Figure 4.154 Schematic representation of the process sequence for (pressure-controlled) hot plate butt welding

The hot plate butt welding process is characterized by the advantages and disadvantages listed below. The designs shown in Figure 4.155 are suitable for this process. Detailed information on the process description and design can be found in the DVS Guideline 2215, among others.

Advantages:

- High degree of automation possible
- Simple process control and high degree of technological maturity level
- Homogeneous welding bead (particle-free)
- Tolerance for material differences

Disadvantages:

- Complex tools for 3D structures
- Long cycle times
- Material may stick to the heating element (nonstick coating recommended)
- High running costs (energy)

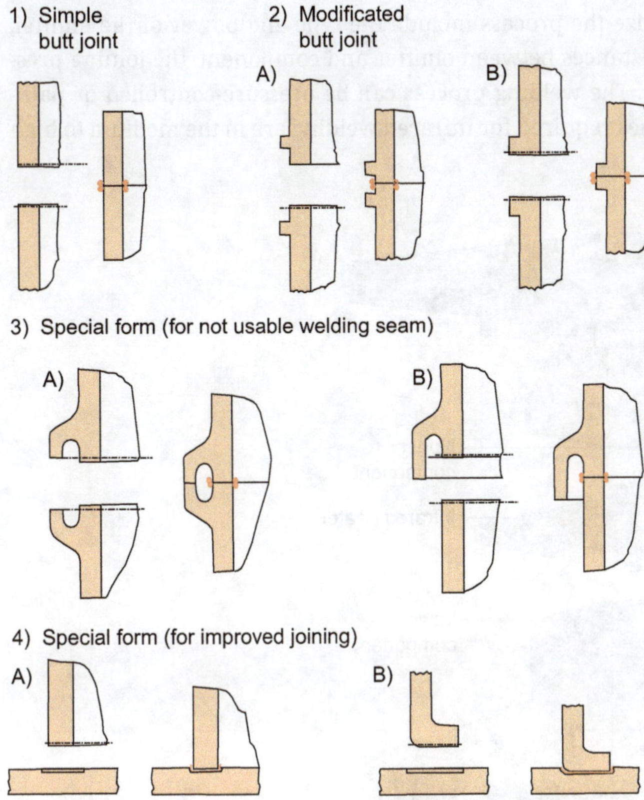

Figure 4.155 Joint seam design for hot plate butt welding

Infrared Welding

The process sequence for infrared welding is similar to that for hot plate butt welding. However, in infrared welding, the heat input takes place without contact by means of radiation and the process phase of adjustment is omitted. The radiation emitted by the infrared heater is absorbed by the plastic to be welded. As a rule, the short-wave (0.78 µm to 1.4 µm) and medium-wave (1.4 µm to 3 µm) infrared radiation of the light spectrum is used. To achieve a high degree of efficiency, the emission spectrum of the emitter should be matched to the absorption spectrum of the plastic. It should be noted that any additives, such as soot or color pigments, change the absorption capacity. A distinction must be made between volume and surface absorption. When using short-wave infrared emitters with a generally higher surface power, attention must be paid to possible thermal damage to the component surface.

The process control for infrared welding is similar to that for hot plate welding. A typical welding arrangement is shown in Figure 4.156. After noncontact heating of both components, the infrared emitter is removed from between the components. These are then welded under a pressure of approximately 0.1 N/mm² to 2 N/mm². The parameters that characterize the process include the type and power of the emitter, the irradiation time, the distances between emitter and component, the joining pressure, and the cooling time. The welding process can be pressure-controlled or path-controlled. The process times required for infrared welding are in the medium to high range.

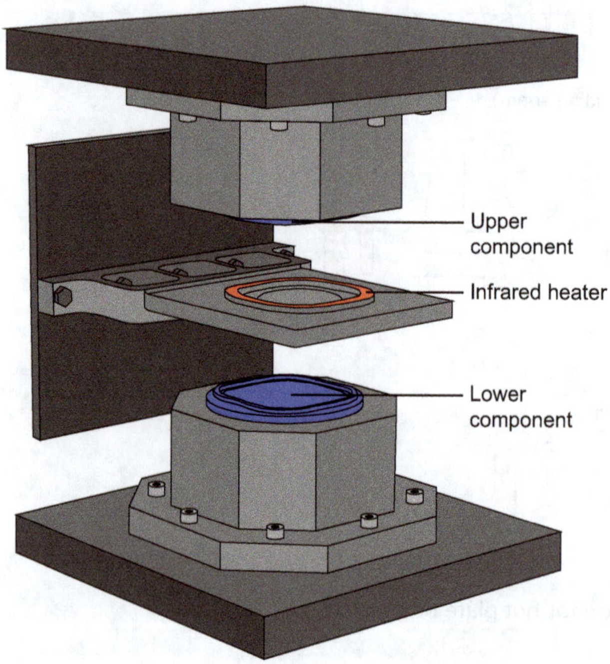

Figure 4.156 Schematic diagram (example) of a welding arrangement for infrared welding
[Source: KT-Chemnitz, Gehde]

Infrared welding is used across all industries and is very well suited for applications with high purity requirements, but also for joining high-temperature thermoplastics with and without fiber reinforcement (no risk of adhesion or abrasion). Depending on the emitter system used, complex 3D geometries can also be welded. Infrared technology can also be combined with other welding processes. A typical application is vibration welding in combination with infrared preheating to reduce particle formation (see Section 4.14.4.2 on vibration welding).

The infrared welding process is characterized by the advantages and disadvantages listed below. The designs shown in Figure 4.155 are suitable for this process as well. Detailed information on the process description and structural design can be found in the DVS Guideline 2228, among others.

Advantages:

- Contactless heating
- High design freedom of the joining surfaces and the infrared radiator
- Homogeneous welding bead (particle-free)
- Low-stress connection

Disadvantages:

- Complex emitter-material dependency
- Coordination of the emission and absorption behavior
- Energy input outside the joining zone possible
- Possibly complex spotlight construction
- Possibility of local material overload

Hot Gas Butt Welding

Hot gas butt welding is a fully automated welding process that is generally suitable for all thermoplastic molded parts. It offers the possibility of heating three-dimensional joining surfaces without contact and welding them reliably. Examples of applications include, in particular, components from the automotive industry where complete freedom from partlines is required. Intake and charge air pipes as well as fuel and hydraulic oil tanks but also headlights can be joined using this welding process.

The process control for hot gas butt welding is similar to that for infrared welding. In contrast, however, the heat input takes place by means of convection and conduction. A typical welding arrangement is shown in Figure 4.157. The joining surfaces are plasticized by a hot gas jet that exits through a hot gas nozzle system adapted to the joining part. The choice of process gas (*e. g.*, air, nitrogen, carbon dioxide, argon) can be made freely, but it is selected depending on the oxidation susceptibility of the plastic to be welded. The gas temperature, which is significantly higher than the melting temperature, can lead to thermal or thermal-oxidative damage to the plastic. As with hot plate and infrared welding, the heating and joining phases are carried out separately. Once the desired fusion layer has been created, the hot gas tool between the components is removed and these are welded under pressure. The characteristic process parameters are the gas temperature, the gas volume flow, the distance between the nozzles and the component surface, the heating time, the joining pressure, and the cooling time. The joining process can be pressure-controlled or path-controlled. The process times required for hot gas butt welding are in the medium to high range.

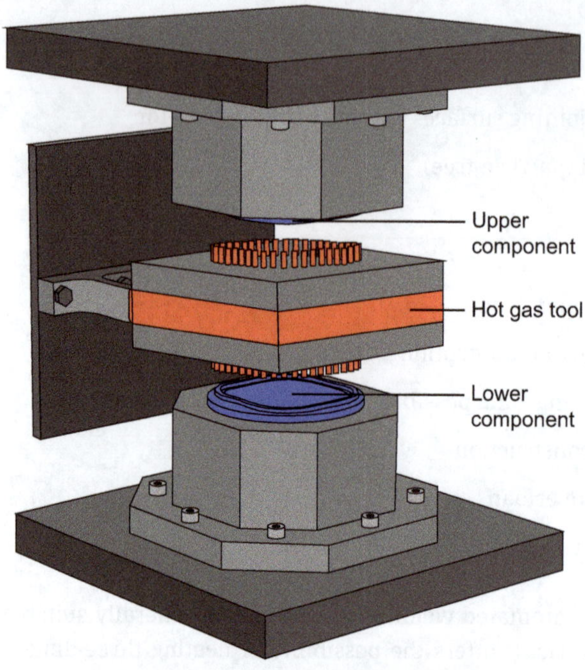

Figure 4.157
Schematic diagram (example) of a welding arrangement for hot gas butt welding [Source: KT-Chemnitz, Gehde]

The hot gas butt welding process is characterized by the advantages and disadvantages listed below. At the same time, the designs shown in Figure 4.155 are suitable for this process as well.

Advantages:

- Contactless heating
- Homogeneous welding bead (particle-free)
- Suitability for complex component geometries
- Suitability for high-temperature plastics

Disadvantages:

- Risk of thermal/thermal-oxidative damage to the plastic
- Energy input outside the joining zone possible
- Comparatively high operating costs of hot gas generation

4.14.4.2 Friction Welding Processes

In friction welding processes, the heat required to weld the parts to be joined is generated by a relative movement of the workpieces on the joining surfaces. Industrially established friction welding processes include vibration welding and rotary friction welding.

Vibration Welding

In vibration welding, different process variants can be used, in the form of linear vibration welding, biaxial vibration welding, and angle welding. The most common industrial application is linear vibration welding, as this process variant only requires a single degree of freedom. Vibration welding is suitable for welding injection-molded, extruded, and blow-molded components in series production. Different materials can also be welded together (*e.g.*, plastic with elastomers, textile fibers, or wood-based materials). The main areas of application are the automotive and household appliance industries.

In linear vibration welding, the plastic is heated by dissipating frictional energy as a result of interfacial friction. The parts to be joined are moved with respect to each other in a horizontal direction under the influence of a joining force (with an amplitude on the order of 0.25 mm to 2.5 mm). The upper machine table is set into a horizontal oscillation via electromagnetic spring systems (with frequencies of 100 Hz to 300 Hz, depending on the system concept). The onset of plasticization creates a melt layer, which is subjected to alternating shear stress due to the vibration movement. This causes an additional energy input. The joining pressure applied (approximately 0.5 N/mm² to 6 N/mm²) simultaneously forces melt into the bead.

Vibration welding is a single-stage welding process that is divided into four process phases (Figure 4.158). Particles escaping during welding can have a detrimental effect and must be taken into account when designing the process with regard to cleanliness requirements. In order to minimize unwanted particle formation during solid friction (phase 1), infrared preheating of the joining surfaces can be provided before the friction movement begins.

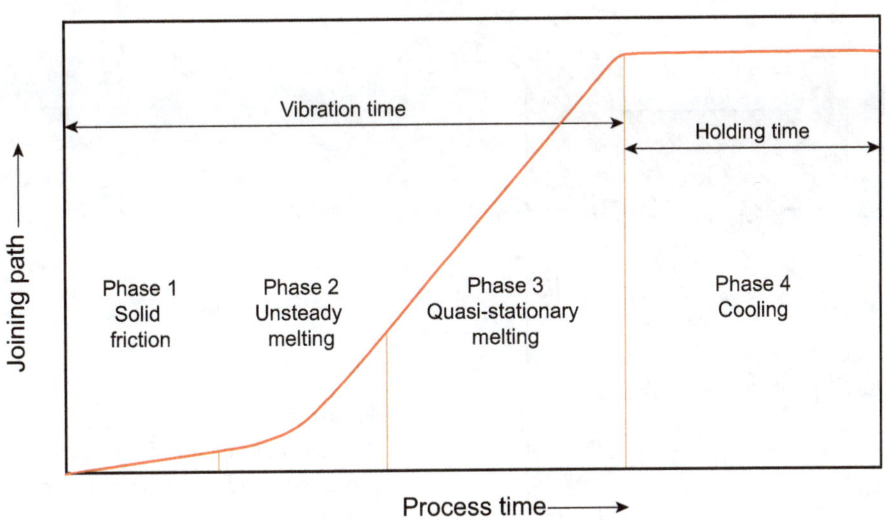

Figure 4.158 Process sequence and temporal phasing in linear vibration welding

With regard to process control, a distinction is made in linear vibration welding between the longitudinal and the transverse (longitudinal or transverse to the main direction of expansion of the parts to be joined) vibration directions. The characteristic process variables are the vibration frequency, amplitude, and duration, the joining pressure, the joining path, and the cooling time. The welding process can be time-controlled or path-controlled. The process times required for vibration welding can be estimated to be short. It is also possible to further reduce the process time by using the process control strategy of the so-called high-pressure approach.

The process of vibration welding is characterized by the advantages and disadvantages listed below. At the same time, a typical seam design as shown in Figure 4.159 is recommended for a functional bond. Detailed information on the process description and structural design can be found in the DVS Guideline 2217, among others.

Advantages:

- Short cycle times
- High weld seam quality
- Robust system technology
- Tolerance for contamination

Disadvantages:

- Limited suitability for components with integrated electronic components
- Particle abrasion (lint formation)
- Limited suitability for flexible components

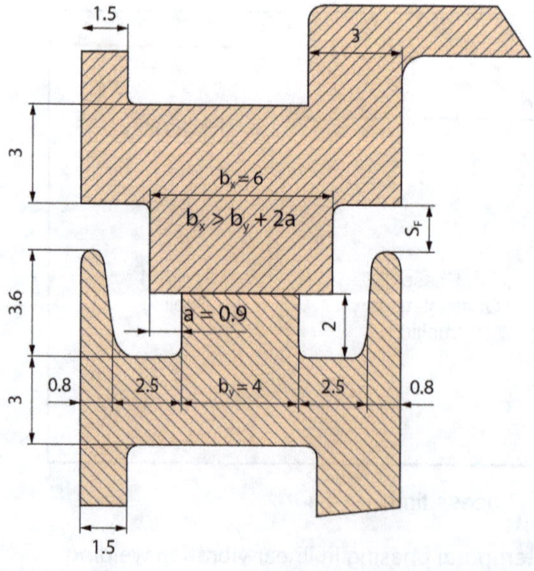

Figure 4.159
Typical seam design for linear vibration welding in accordance with DVS Guideline 2217-2 [Source: DVS RL 2217-2]

Rotary Friction Welding

Rotationally symmetrical parts such as rods, tubes, hollow cylinders, tube bundles, and round parts that are welded (in)to other bodies can be joined using rotary friction welding. On lathes or special rotary friction welding machines, one joining part is locked against rotation while the other joining part rubs against it at a circumferential speed of around 0.5 m/s to 8 m/s. The welding process is divided into four typical phases: start-up, friction, compression, and holding. Figure 4.160 and Figure 4.161 show a schematic diagram of the process principle and an example of a welding program, respectively. As a consequence of the interfacial friction, the component surfaces are plasticized during the friction phase (dissipation) and the joining pressure applied (approximately 0.1 N/mm² to 2.0 N/mm²) results in a joining distance being covered. When the set value is reached, the rotational movement must be braked as quickly as possible (approximately 0.5 s). If the joining pressure continues to act, then the joint cools down. If the parts have to be joined at a specific angle to each other, it is possible to stop the welding position at precisely this position. The characteristic process variables are the speed, the friction time, the joining pressure, the stop position, and the cooling time. The welding process can be time-controlled or travel-controlled. The process times required for rotary friction welding can be estimated to be short.

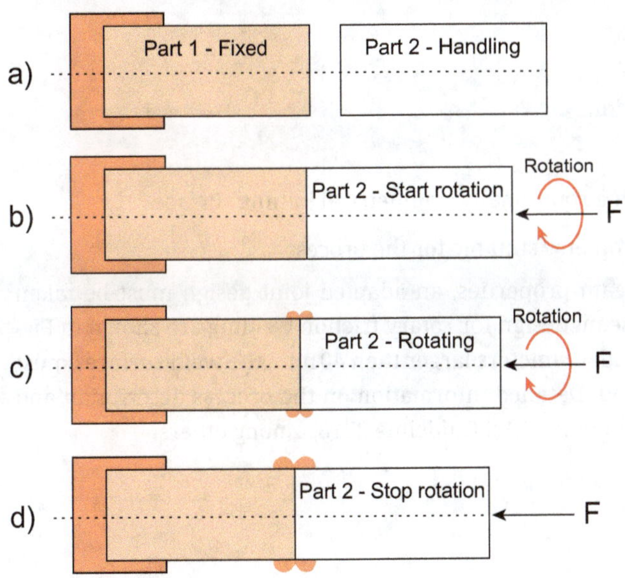

Figure 4.160 Process principle for rotary friction welding (schematic diagram)

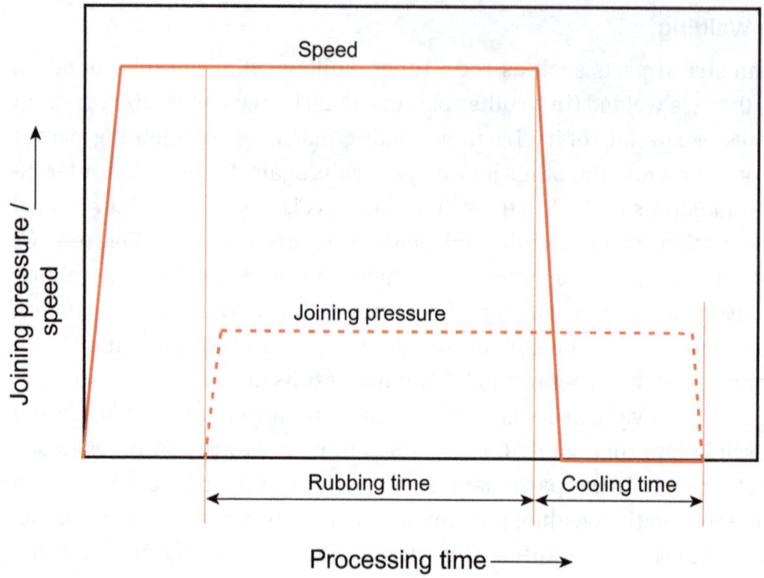

Figure 4.161 Example of welding program for rotary friction welding

Advantages:

- Short process times
- Robust system technology
- Tolerance for contamination

Disadvantages:

- Only for components with a rotationally symmetrical joining surface
- Special joining surface geometry suitable for the process

To achieve optimum weld seam properties, an adapted joint design must be taken into account. Examples of a seam design for rotary friction welding are shown in Figure 4.162. For butt welding with diameters larger than 40 mm, the weld surfaces must be crowned or slightly tapered. Detailed information on the process description and structural design can be found in the DVS Guideline 2218, among others.

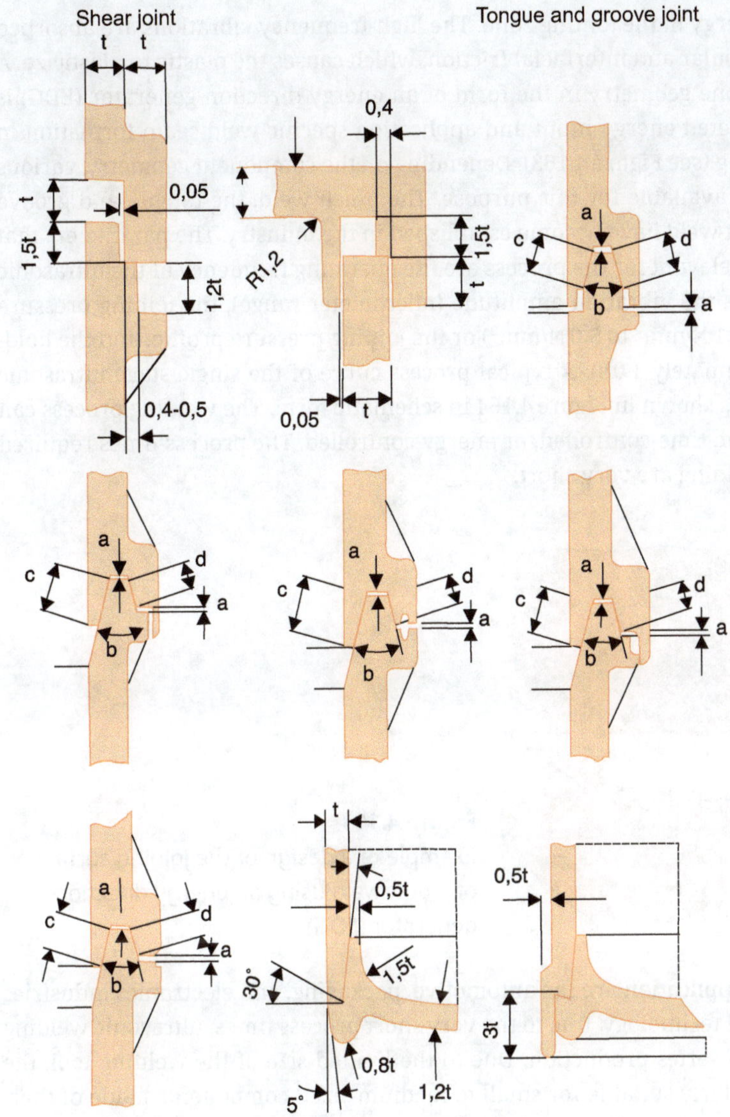

Figure 4.162 Seam design for rotary friction welding

4.14.4.3 Ultrasonic Welding

In ultrasonic welding, the high-frequency energy from a generator is converted into high-frequency mechanical vibrations by means of a sound transducer (converter) and transferred to the component by the welding tool (sonotrode). The so-called oscillating system, consisting of a converter, a transducer, and a sonotrode, works in resonance. The operating frequency for ultrasonic welding is typically between 20,000 Hz and 40,000 Hz. The heat required for welding is generated by dissipating the mechanical vibration energy in the joining zone. The high-frequency vibrations are absorbed and lead to molecular and interfacial friction, which causes the plastic to plasticize. A special joining zone geometry in the form of an energy direction generator (EDG) is required for targeted energy input and application-specific weld seam formation in ultrasonic welding (see Figure 4.163). Depending on the component geometry, various seam shapes are available for this purpose. The pinch weld, the tongue and groove joint, and the butt weld have become established in the industry. The parameters that are particularly relevant for the process are the operating frequency of the ultrasonic welding machine, the vibration amplitude (micrometer range), the joining pressure (approximately 0.1 N/mm^2 to 5.0 N/mm^2) or the joining pressure profile, and the holding time (approximately 1.0 s). A typical process curve of the single-stage ultrasonic welding process is shown in Figure 4.164 in schematic form. The welding process can be path-controlled, time-controlled, or energy-controlled. The process times required for ultrasonic welding are very short.

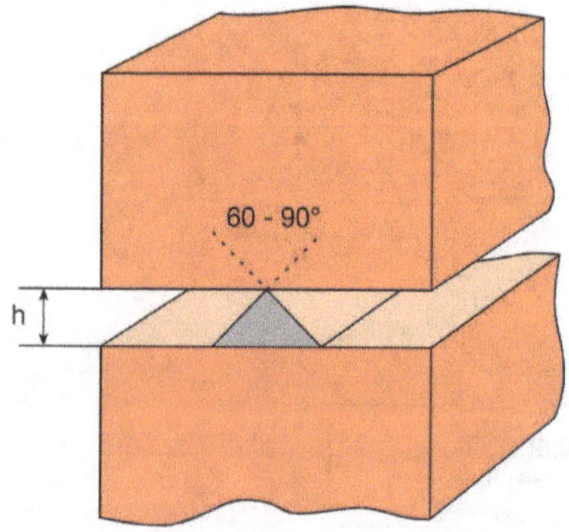

Figure 4.163
Example of a design of the joining surface of a butt weld using an energy direction generator (EDG)

Typical areas of application are the automotive, packaging, and electronics industries as well as medical technology. Due to the very short process times, ultrasonic welding is ideal for use in series production. Due to the limited size of the welding tool, the process is particularly suitable for small to medium-sized components made of thermoplastics and for welding films and nonwoven materials.

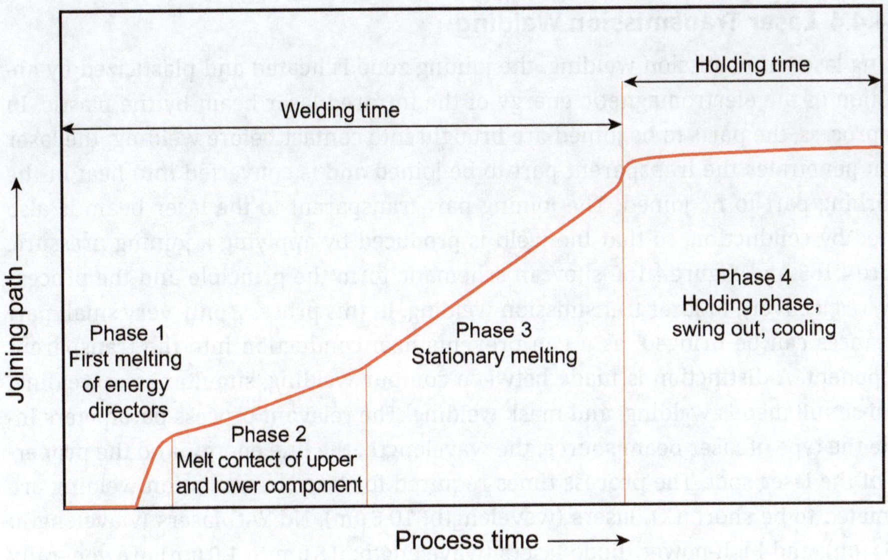

Figure 4.164 Process sequence and temporal phasing in ultrasonic welding with energy direction generator (EDG)

The ultrasonic welding process is characterized by the advantages and disadvantages listed below. Detailed information on the process description and design can be found in the DVS Guideline 2216, among others.

Advantages:

- Very short process times
- High degree of automation possible
- Energy-efficient and resource-saving
- Low investment costs

Disadvantages:

- Special joining surface design (EDG)
- Vibration input into the component
- High material and geometry dependency
- Particle abrasion

4.14.4.4 Laser Transmission Welding

During laser transmission welding, the joining zone is heated and plasticized by absorption of the electromagnetic energy of the infrared laser beam by the plastic. In this process, the parts to be joined are brought into contact before welding; the laser beam penetrates the transparent part to be joined and is converted into heat in the absorbing part to be joined. The joining part transparent to the laser beam is also heated by conduction, so that the weld is produced by applying a joining pressure. Figure 4.165 and Figure 4.166 show in schematic form the principle and the process flow, respectively, of laser transmission welding. In this process, only very small part tolerances can be bridged, as a gap prevents heat conduction into the transparent component. A distinction is made between contour welding, simultaneous welding, quasi-simultaneous welding, and mask welding. The relevant process parameters include the type of laser beam source, the wavelength, the line energy, and the properties of the laser spot. The process times required for laser transmission welding are estimated to be short. CO_2 lasers (wavelength: 10.6 µm), Nd:YAG lasers (wavelength: 1.064 µm), and high-power diode lasers (wavelength: 0.8 µm to 1.0 µm) are generally used for plastics.

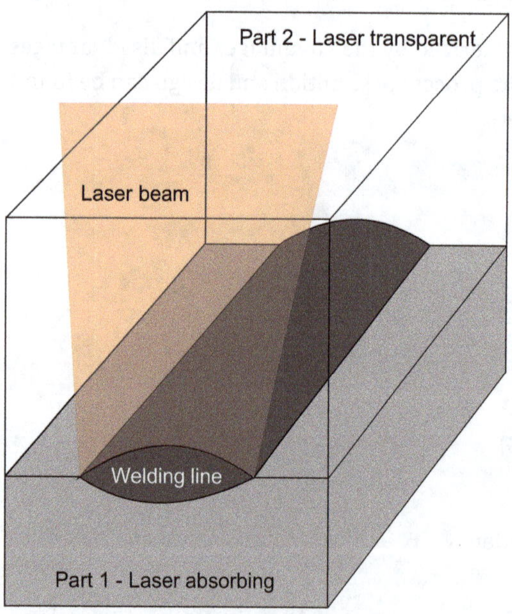

Figure 4.165
Principle sketch of laser transmission welding (schematic representation)

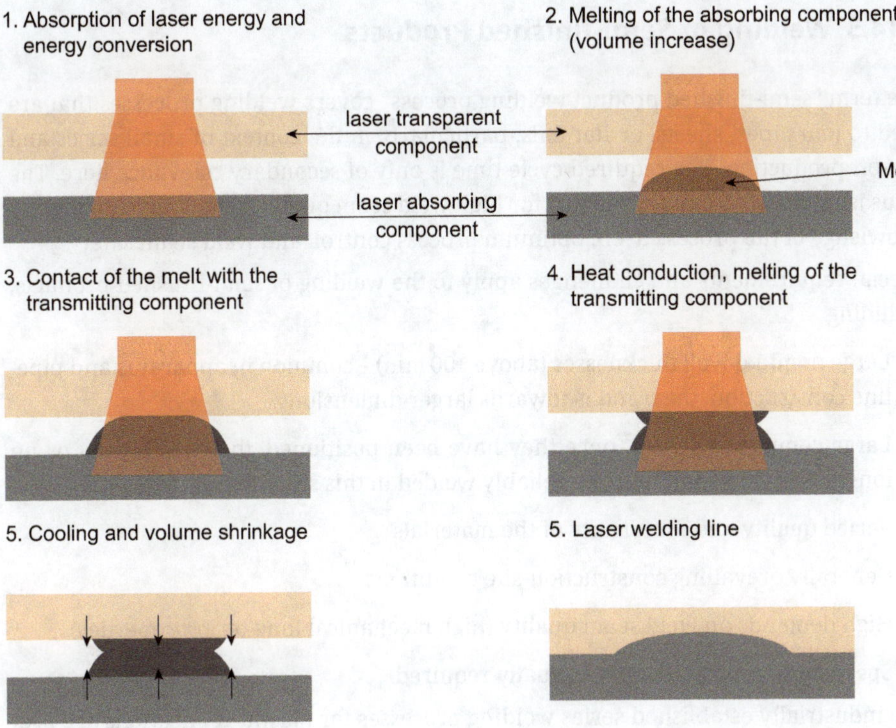

Figure 4.166 Schematic process sequence for laser transmission welding

The main areas of application for laser transmission welding are in the electronics industry, medical technology, and the automotive industry. In addition to similar plastics, different types of plastic can also be welded together as long as their melting temperature ranges overlap sufficiently and the two materials are chemically and physically compatible.

Advantages:

- Small, easily controllable heat-affected zone
- Optically high-quality weld seams (no squeeze flow during contour welding)
- Short cycle times
- Can be used for very small and precise components

Disadvantages:

- Possibility of bridging gaps to compensate for manufacturing tolerances
- Cost-intensive system technology
- Elaborate protective measures
- Radiopacity of a joining part is necessary

4.14.5 Welding of Semi-finished Products

The term "semi-finished product welding process" covers welding processes that are used to join pipes, sheets, or flat foils, particularly in the context of small series and one-off production. The required cycle time is only of secondary relevance here. The focus is rather on the requirements for high weld seam quality, which involve precise knowledge of the process itself, optimum process control, and weld seam safety.

Special requirements and challenges apply to the welding of semi-finished products, including

- Large nominal wall thicknesses (above 100 mm) – common in apparatus and pipeline construction, the trend is towards larger dimensions
- Large components – *i. e.*, once they have been positioned, they can hardly or no longer be moved and must be reliably welded in this situation
- Varied quality and uniformity of the materials
- Generally prevailing construction-site conditions
- High demands on weld seam quality (high mechanical long-term properties)
- Specially trained personnel – usually required

The industrially established series welding processes for plastic applications are presented in more detail below.

4.14.5.1 Direct Hot Plate Welding

Hot Plate Butt Welding

The hot plate butt welding of semi-finished products is identical to hot plate butt welding in series production in terms of process control and the advantages and disadvantages (see Section 4.14.4.1). The suitability of hot plate welding in the area of semi-finished products focuses in particular on polyolefin semi-finished products. Typical applications of hot plate welding are

- Welding and bending of panels
- Welding of pipes
- Pipe-fitting welds
- Welding components into plates or pipes (Y pipes)
- Mitre welds, for example, of PVC window frames

Detailed information on the process description and design can be found in the DVS Guidelines 2207 and 2208, among others.

Hot Plate Sleeve Welding

In hot plate sleeve welding, pipe and pipeline parts are welded overlapping, primarily in the area of domestic installations. The process principle is shown in Figure 4.167. The joining parts are heated radially. Filler metals are not used in this manual or automated process. Two pipes are always joined using a fitting, for example, an elbow or a T piece. While hot plate sleeve welding is very easy to install, attention must be paid to the need for a nonstick coating on the heating plates. The application of the process is also limited to small pipe diameters, of approximately 20 mm to 110 mm, in terms of tool technology. Further information on the process description and design can be found in the DVS Guideline 2207-5, among others.

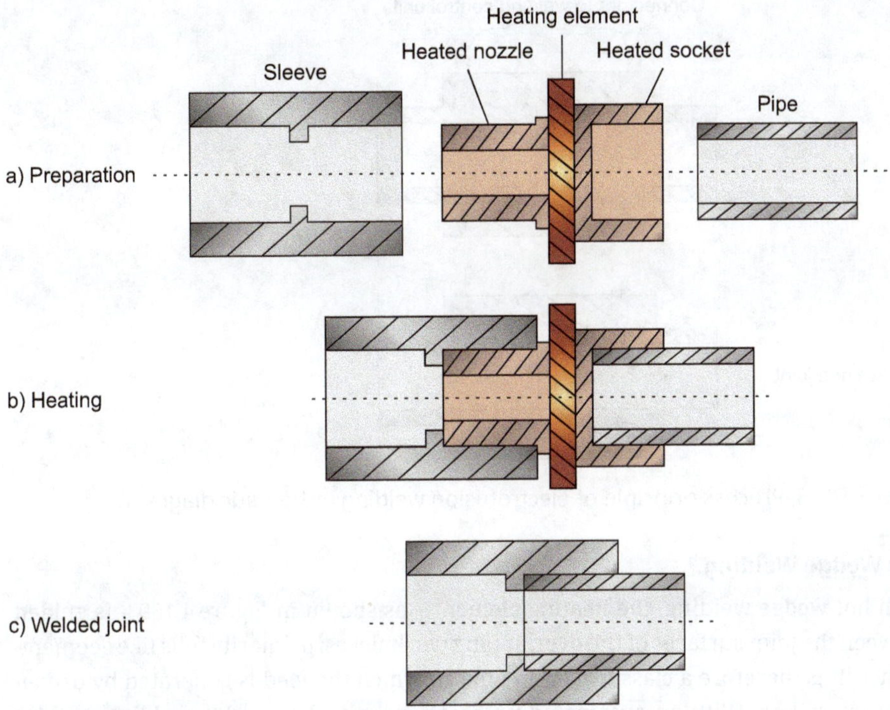

Figure 4.167 Principle of hot plate sleeve welding (schematic diagram)

Electrofusion Welding

A special form of hot plate butt welding is sleeve welding with electrofusion fittings, known as electrofusion welding. In this process, the joint surfaces are heated by electric current through resistance wires (electrofusion coils) embedded in the sleeve (see Figure 4.168). The thermal expansion of the plastic generates the joining pressure. The other welding parameters are based on the diameter and the nominal pressure of the pipe. Typical areas of use for electrofusion welding are applications in gas and water supply with plastic pipes with an outer diameter in the range of 32 mm to 1200 mm. Detailed information on the process description and design can be found in the DVS Guideline 2207-1 – Supplement 1.

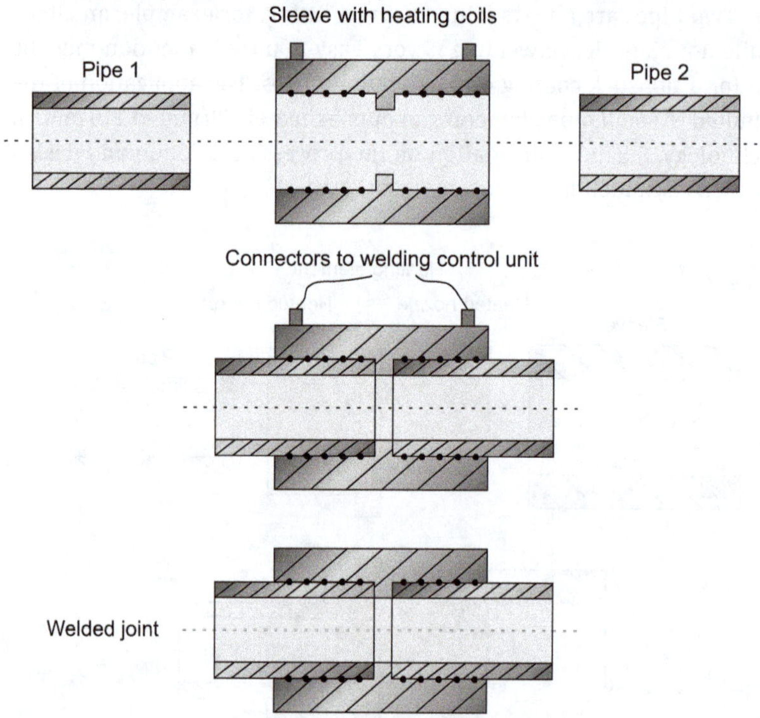

Figure 4.168 Process principle of electrofusion welding (schematic diagram)

Hot Wedge Welding

With hot wedge welding, the heating element – as shown in Figure 4.169 – is guided between the joint surfaces of the overlapping workpieces, primarily foils like geomembranes. It is therefore a classic overlap joint in which the feed is generated by driven pressure rollers. With an automated process sequence, hot wedge welding enables comparatively simple quality control. However, this requires plane-parallel joining surfaces. The advantages of this welding process lie in the high achievable strength and the equally high achievable welding speed. The disadvantage is that it is not possible to join geomembranes to solid structures in construction. Overlap joints of soft geomembranes for roof and landfill sealing are produced continuously using heating wedges, over which the sheets are guided and then welded using pressure rollers. Detailed information on the process description and structural design can be found in the DVS Guideline 2225, among others.

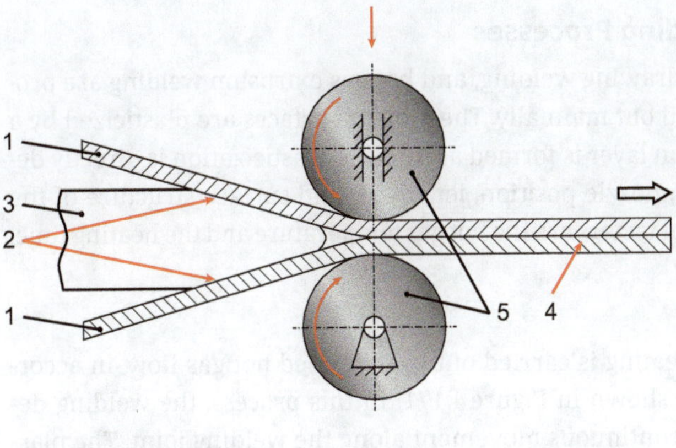

1 Joining components
2 Effective heating area
3 Heating wedge
4 Welding seam
5 Transport and pressure rollers

Figure 4.169 Principle sketch for hot wedge welding (schematic diagram) [Source: DIN 1910-3]

4.14.5.2 Indirect Hot Plate Welding

In indirect hot plate welding, the joining surfaces are heated from the back of the workpiece. Therefore, heating takes place through the workpieces. Indirect hot plate welding processes include hot plate heat contact welding, shown in Figure 4.170, hot plate heat impulse welding, and hot plate rolling tape welding. These processes have been specially developed for films in the packaging industry. The films to be joined are overlapped between two pressing systems, one or both of which are heated.

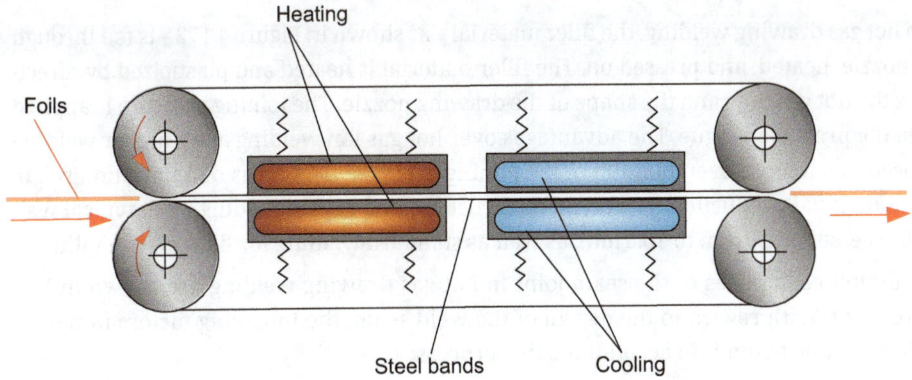

Figure 4.170 Heat contact welding for flat seams using circumferential welding bands with cooling under pressure

4.14.5.3 Convective Welding Processes

Hot gas fan welding, hot gas drawing welding, and hot gas extrusion welding are processes that are usually carried out manually. The joining surfaces are plasticized by a hot gas jet and a defined fusion layer is formed as a result. Plasticization is directly dependent on the nozzle shape, nozzle position, jet speed, and surface structure of the part to be joined. The dominant factors are the hot gas temperature and the heating time.

Hot Gas Fan Welding

In hot gas fan welding, the heating is carried out by a defined hot gas flow, in accordance with the arrangement shown in Figure 4.171. In this process, the welding device is guided manually in a continuous movement along the welding joint. The plasticized welding rod, which is also guided manually, is brought into the weld joint while the joining pressure is applied. The main applications are hard-to-reach areas and root layers, preferably made of PVC. Further information on the process description and design can be found in the DVS Guideline 2207-3, among others.

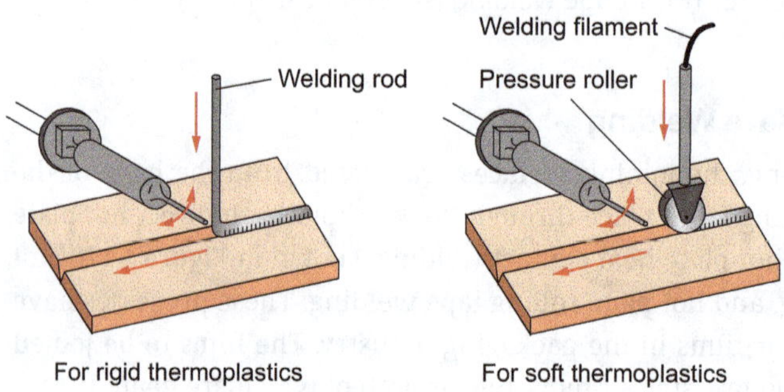

Figure 4.171 Welding arrangement for hot gas fan welding

Hot Gas Drawing Welding

In hot gas drawing welding, the filler material – as shown in Figure 4.172 – is fed through a nozzle, heated, and pressed on. The filler material is heated and plasticized by directing the hot gas flow and the shape of the drawing nozzle. The joining pressure is applied via the pressure tongue. The advantages over hot gas fan welding are a higher welding speed and a more even buildup of the joining pressure. The areas of application relate to individual production, prototype construction, and repair welding, with typical wall thicknesses of 2.0 mm to 10.0 mm, as well as standard welding for linings with inliners.

Common geometries of the seam joint in hot gas drawing welding are shown in Figure 4.173. With regard to the design of the weld seam, the following factors in particular must be taken into account for this process:

- Avoidance of cross impacts
- Avoidance of wall thickness cracks
- Free accessibility of the joining surfaces

Detailed information on the process description and design can be found in the DVS Guideline 2207-3, among others.

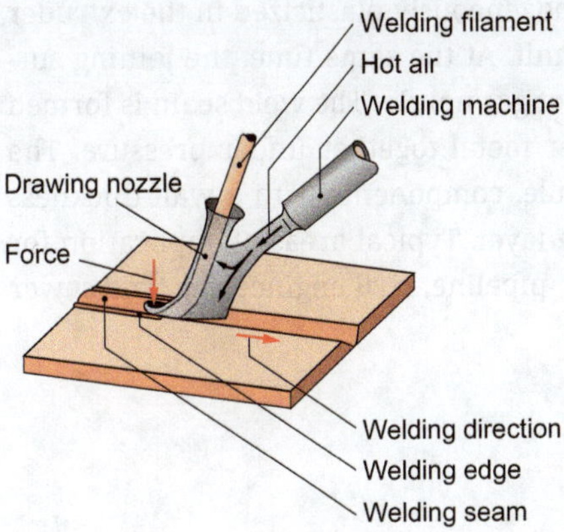

Figure 4.172 Welding arrangement for hot gas drawing welding [Source: DVS 2207-3]

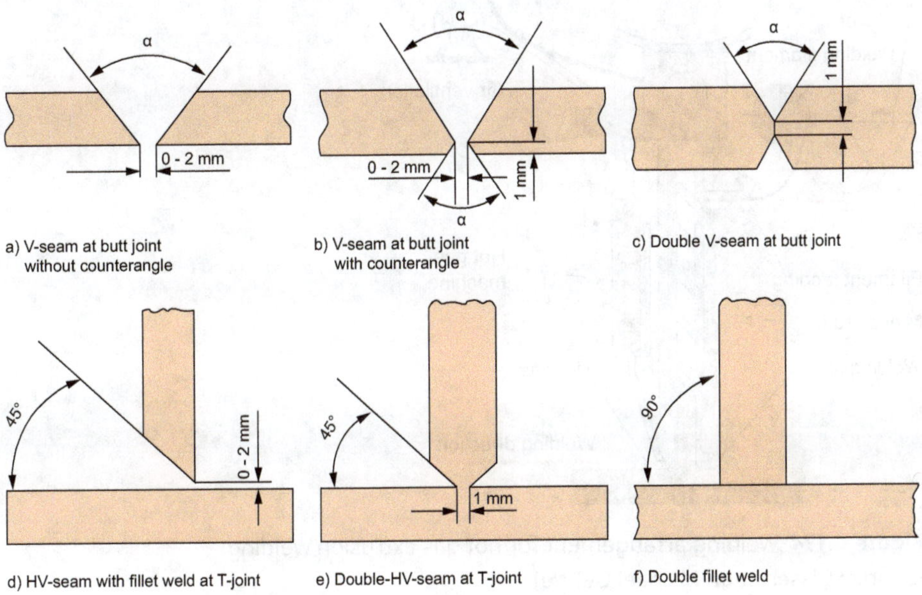

Polymer	Weld seam opening angle α
PE-HD[1)], PP[2)], PVC, PVC-C, PVDF, E/CTFE	60°-70°
MFA, FEP, PFA[3)]	0°-30° (depending on material thickness)

Figure 4.173 Common geometries of the seam joint in hot gas drawing welding [Source: DVS 2207-3]

Hot Gas Extrusion Welding

Hot gas extrusion welding is a manual welding process that uses a continuously fed filler metal in the form of a plastic granulate or a welding rod (Figure 4.174). In this process, the filler metal is completely and homogeneously plasticized in the extruder and emerges as a strand from a plasticizing unit. At the same time, the joining surfaces are also heated and plasticized using a hot gas nozzle. The weld seam is formed by bringing the joining surfaces and the filler metal together under pressure. The weld seam is cooled without pressure. As a rule, components with a wall thickness of up to 30 mm can be joined with one joining layer. Typical areas of application for hot gas extrusion welding are apparatus, tank, pipeline, civil engineering, and sewer construction.

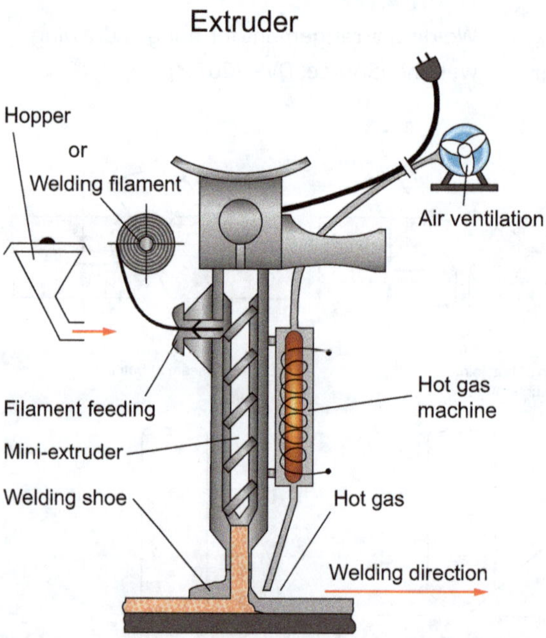

Figure 4.174 Welding arrangement for hot gas extrusion welding [Source: Dissertation Michael Gehde]

The most important influencing variables in hot gas extrusion welding are the successive process steps of seam preheating and the introduction of the filler material. This process is shown in Figure 4.175. Detailed information on the process description and design can be found in the DVS Guideline 2207-4, among others.

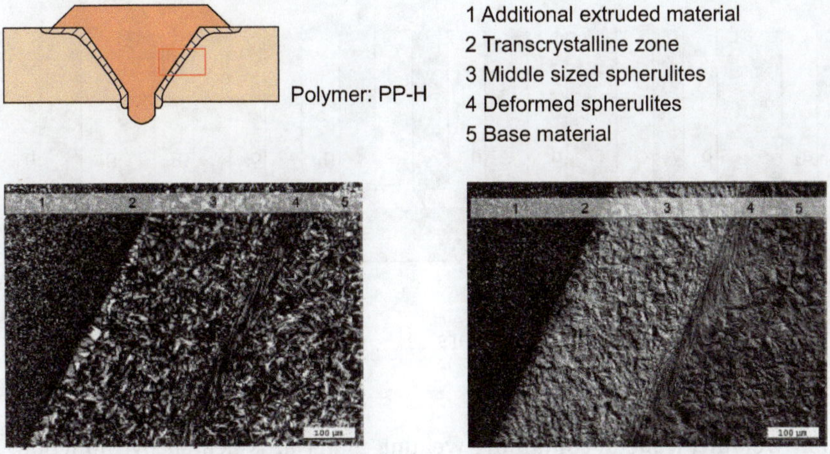

Figure 4.175 Structural seam design for hot gas extrusion welding [Source: KT-Chemnitz, Gehde]

4.14.6 Welding with an Alternating Electromagnetic Field

4.14.6.1 High-Frequency Welding

Plastics with a sufficiently high dielectric loss factor, tan δ > 0.1 (*e.g.*, PVC, PUR, PVDC, EVA, PET, ABS, CA), can be heated in a high-frequency (HF) alternating electric field and thus welded under pressure. Plastics whose loss factor increases with temperature can be made weldable by preheating or preheating electrodes. Interposing a weldable film or a corresponding composite film can also be a solution.

An HF welding machine consists of an HF generator (welding frequency: usually 27.12 MHz) and a pneumatic or hydraulic welding press, which applies the required welding pressure via the electrodes (the upper electrode is usually a shaped one made of brass; the lower electrode is an insulated plate-shaped counter electrode). Depending on the main field of application of HF welding (production of bags, book covers, upholstery, inflatable articles, automotive equipment parts, etc.), different shaped electrodes, including three-dimensionally shaped ones, are used; see Figure 4.176. High-frequency welding presses are usually installed in production lines. Further information on process control and design can be found in the DVS Guideline 2219, among others.

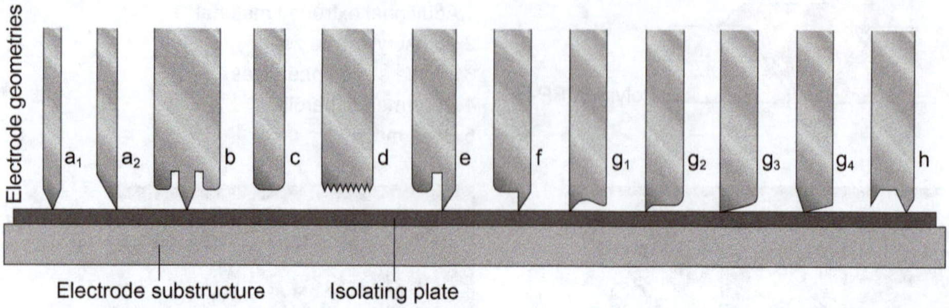

Figure 4.176 Different types of electrode bars

4.14.6.2 Induction Welding

In induction welding, a metal-containing welding material is heated by induction. Oxide particles are used for this: they are provided as electromagnetic additives in the base material of the parts to be joined or placed between the parts in a plastic intended as a welding filler. The principle is shown in Figure 4.177. Induction welding is suitable for welding most thermoplastics and, in particular, fiber composites with conductive fibers (*e.g.*, carbon-fiber fabric).

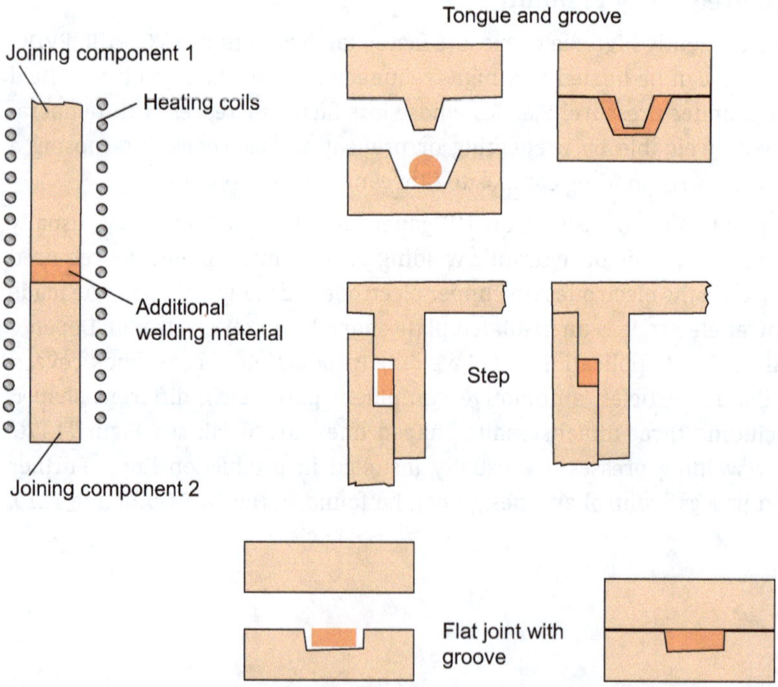

Figure 4.177 Principle and joint design for induction welding

4.14.6.3 Microwave Welding

In microwave welding, the joining zone is melted by dielectric heating as a result of an applied alternating electromagnetic field. Depending on the dielectric properties of the material, direct or indirect microwave welding can be used.

While in direct welding the parts to be joined are heated directly by microwaves, in indirect welding a filler metal is used: it plasticizes the joining zone through its own heating and the resulting heat conduction effects. The filler metal (*e. g.*, carbon black and polyaniline-modified compounds) has significantly higher dielectric losses than the polymer to be welded, causing it to heat up locally. The indirect method is used in particular for nonpolar polymers (*e. g.*, PP) with low dielectric loss. This method also offers the possibility of joining concealed welding geometries (inside the parts to be joined) by irradiating the entire component with microwaves, but only the joining zone is heated by the filler metal.

Direct microwave welding, on the other hand, requires a thermoplastic polymer with a polar character and sufficient dielectric losses. Special coupling systems are used to direct the microwaves locally to the joining zone, causing it to plasticize. However, the direct process principle is accompanied by a restriction of the possible welding geometry. In general, the advantages of microwave welding lie in the noncontact melt formation, the short heating times, and the low energy consumption.

4.14.7 Glueing

Glueing is a process in which components are bonded together by using a glue as an additional layer between the components. The glue is often referred to as adhesive.

Glueing is used to join polymer parts as well as assemblies with polymer parts and other materials. Polymers very often were themselves employed as an adhesive for joining many different materials of all classes.

The adhesive has the task of mechanically bonding the joining partners and transferring forces. In addition, depending on the requirements, it can also reduce stress peaks through deformation and perform a protective function for the joining partners, for example, against media. The bonding of the adhesive can be based on physical, chemical, and/or mechanical bonding principles.

Adhesive forces, when bonding plastics, can be attributed essentially to secondary valence, dipolar, and dispersion forces. The plastics to be bonded differ not only in terms of their surface energy (wettability) but, in particular, in terms of their chemical structure, which enables the formation of these forces. The notoriously difficult bonding of nonpolar polyolefins (PE, PP) demonstrates the strong influence of polarity properties. Nonpolar plastics can therefore only be bonded – with low strength – after surface treatment; see also Section 4.15.1.

In many cases, it is the solvency or diffusion behavior of thermoplastics that makes diffusion bonding possible in the first place. The solvents listed below are primarily used to dissolve or swell the individual thermoplastics:

- PVC: tetrahydrofuran, cyclohexanone
- PS: toluene, xylene
- PMMA: methylene chloride, methyl ethyl ketone
- POM: hexafluoroacetone sesquihydrate
- PC: methylene chloride, tetrahydrofuran
- CA: methyl ethyl ketone, methyl alcohol
- PPE: chloroform, toluene
- PA: formic acid
- PET: benzyl alcohol

In contrast to solvent-based adhesives, solvent-free reactive adhesives – provided they do not contain monomers that dissolve the parts to be joined – do not change the parts to be joined. Adhesives based on epoxy resins (EP), polyurethanes (PUR), methyl methacrylates (MMA), and unsaturated polyesters (UP) are mainly used. Cyanoacrylates are also widely used for small-area plastic and elastomer bonding. Diffusion bonding by swelling or loosening of the joining surfaces is therefore possible with thermoplastics, with the exception of PE, PP, POM, and polyfluorocarbons, for example. It leads to weld-like bonds but can promote stress cracking. Adhesive bonding through physical and chemical bonding processes between bonding surfaces and adhesive is possible with almost all plastics.

Examples of Bonded Joints

- In craftsmanship, acrylic glass, for example, is bonded with solvent-based adhesives for low stresses. Bonding agents that polymerize and are applied in a thicker layer, for example, as a V-seam, are optically perfect and weather-resistant. Similarly, acrylic glass is bonded with silicate glass; otherwise, with pressure-sensitive adhesives. The bonding of PVCs can be found in DIN 16970. For fiber-reinforced plastics to be processed, PVC is treated with a dissolving UP resin as a bonding agent.
- PIB membranes are bonded to concrete with bitumen-plastic hot-melt adhesives and to metals with special contact adhesives.
- Contact adhesives based on natural or synthetic rubber are suitable for bonding plastics, especially in the form of sheets or panels, to impermeable substrates (metals, concrete, stone, glass). They are usually applied to both surfaces, which are joined together by rubbing or tapping after the solvent has largely evaporated.

Good contact adhesives can absorb considerable shear forces while remaining flexible. Elastic cross-linking two-component adhesives can also do this at higher temperatures. Polychloroprene adhesives solidify through gradual partial crystallization.

- Solvent-free dispersion adhesives are suitable for bonding plastics, especially films, to porous materials (paper, cardboard, felt, textiles, leather, wood). Fresh, liquid adhesive can be removed with water; dried adhesive cannot. The bonds are largely resistant to moisture.
- Thermoset fittings are bonded to each other and to other materials with similar cold- or hot-curing synthetic resins. Phenolic resin laminates (roughened or with a paper-gray back that can be glued) also bond with carbamide resin glues. In addition to these, the abovementioned contact and dispersion adhesives are also used for decorative laminates. Vulcanized fiber and synthetic horn can be bonded to each other and to wood using standard wood glues.
- Solvent-free, pressure-less setting reaction resin adhesives are generally used for high-strength adhesive bonds between components made of fiber-reinforced high-performance materials or with components made of other materials. Cyanoacrylate single-component reactive resin adhesives are frequently used in precision engineering.
- Adhesive films made from thermoplastic hot-melt adhesives, which soften reversibly when heated, are primarily used for bonding flat structures such as textiles, paper, film sheets, leather, and wood products. In addition to "irons", heated molding presses and calender-type laminating systems are also used – in the latter, the adhesive film is melted continuously by radiators, flame treatment, or heated rollers.
- Adhesive films made from PUR-based reactive hot-melts consist of PUR prepolymers with free, terminal isocyanate groups that react with water to form polyureas. To achieve the best bond, the adhesive film must absorb the required amount of water from the air before joining with the preheated surfaces to be bonded (application: lamination of window profiles).

4.14.8 Screws, Rivets, Snaps

Other possible assembly processes include screwing, riveting, and snapping. The corresponding fastening elements must be designed to suit the plastic and integrated into the molded part; see Section 5.7. In that section, information on the dimensioning of screw and snap connections can also be found. Designed to be suitable for plastics, these assembly processes do not require any further post-treatment of the plastic components. For example, the insertion of threaded holes and the cutting of threads can be omitted, as either self-tapping plastic screws are screwed into the screw tubes integrated into the injection-molded component or, in the case of metric screw connections, the thread is inserted into the injection mold in the form of a metal insert.

4.15 Surface Treatments

Surface treatment and design include processes that can be integrated in the manufacturing process of components and semi-finished products, see the respective sections: multi-color/multi-component injection molding (Section 4.3.5.4); back injection molding/in-mold decoration/labeling (Section 4.3.5.8); co- or multi-layer extrusion (Section 4.2.3.9); laminates and coatings (Section 4.2.3.10.2); decorative laminated sheets (Section 4.9.1); and in-mold coating (Section 4.9). In this section, we will cover only those processes that are applied after the initial forming process.

4.15.1 Pre-Treatment of Surfaces

The durable surface treatment (coating, painting, adhesive bonding) of plastic components requires *pre-treatments* to remove surface defects or contamination introduced during the manufacturing process or by the environment. Difficult to wet surfaces of polyolefins, fluoro-plastics, or polyacetals have to be chemically activated prior to adhesive bonding, painting, printing, or metallizing.

The required chemical reactivity of the processes also poses the danger of potentially emitting toxic compounds. Additional costs are accrued with the disposal of the etching chemicals used for wet chemical treatments. With these facts in mind, it is not surprising that plasma-based processes have become increasingly important.

The following processes are used for cleaning and chemical activation:

4.15.1.1 Wet Chemical Treatments

- Wet chemical treatments take place in etching baths (*e. g.*, for PE, PP in aqueous chromosulfuric acid at 70 °C; for PTFE in naphthalene or sodium in tetrahydrofuran at 20 °C; for POM in 85% phosphoric acid). Here, the surface of the plastic component is typically oxidized.
- Another surface activation technique similar to wet chemical treatments is the use of diluted adhesive solutions, etch-, or halogenating agents as primers. For example, PVC-U is activated using 1-K-PUR-adhesive; PP is activated using cyanoacrylate.

4.15.1.2 Vapor Phase Treatment

Here, the surface to be treated is exposed to a gas consisting of N_2 and 0.1 to 10 vol.-% F_2 for a period of 10 s up to 10 min. During this exposure, the H in the plastic material is substituted by F; application for reaction adhesive bonding of PP, PP-GF, EPDM, PBT, PBT + ABS, PET, POM, PPS.

4.15.1.3 Flame Oxidation

Oxygen groups are chemically incorporated into the surface of plastic components by short-term exposure to an oxidizing gas flame and heating to 300 to 400 °C. This turns the surface polar and ensures wettability. The thermodynamic imbalance on the surface is released through molecular movements (the oxygen groups are partially moved from the surface into the material), thus making the surface activation non-permanent. However, if the activation is utilized immediately (*e.g.*, through painting/coating operations), the polar coupling mechanism is retained permanently

4.15.1.4 Radiation Treatment

UV-radiation is a suitable energy source for the initiation of oxidation reactions. UV-laser treatment allows for spatially exact surface activation.

4.15.1.5 Corona Treatment

Here, a corona discharge is created between two high-voltage electrodes and a dielectric. Voltages between 5 and 10 kV at a frequency of 6 to 100 kHz are common. The corona discharge creates a reactive plasma that is used to clean and activate the plastic surface of thin-walled products, such as films.

4.15.1.6 Plasma Treatment

The term "plasma" refers to the fourth state of aggregation of matter. It was first introduced in 1926 by Langmuir who used it to describe ionized gases. Today, the term is used to describe materials whose properties are mainly determined by the existence of electrically charged particles. Defining plasma as the fourth state of aggregation of matter can be justified by the fact that matter is generally in the state of plasma at temperatures of several thousand degrees Kelvin.

Exposure to electromagnetic fields can be used to excite a gas and turn it into plasma. Here, it is mainly the charged particles of the gas that absorb energy from the electromagnetic field. At high frequencies of the electromagnetic field, the energy absorption occurs through the lighter electrons. Because of their significantly smaller ratio of charge to mass, the ions absorb considerably less energy through the external field. However, the electrons are able to transfer energy to the heavier particles by inelastic impacts.

The level of energy transfer is determined by the number of impacts and/or the mean free path of the electrons, where the density of the gas plays a decisive role. By reducing the gas pressure it is therefore possible to achieve a lower impact probability. This in turn results in little interaction between the electrons, which are rich in thermal energy via the external field, and the heavy particles of the plasma. Now, in this state a hot electron gas is located in a cold gas formed by ions, molecule(-ion)s, and neutrons, where the heavier particles in the plasma carry the predominant part of the heat capacity. In this state, the plasma is not in thermodynamic equilibrium; this is also called a "cold plasma".

Such plasmas combine high reactivity with low temperatures and are therefore highly suitable for the treatment of plastic surfaces. We distinguish between low-pressure plasma and atmospheric plasma treatments.

Low-pressure plasma treatments take place in cylindrical processing chambers. After evacuation to 0.1 to 10 mbar, an electromagnetic field with a frequency in the GHz range is established in these discontinuously operating devices to create the plasma. Exposure of plastic surfaces to the plasma has a cleaning and activating effect.

The specific cleaning and activation properties can be tailored by adding certain process gases (O_2, N_2, fluorinated gases, inert gases). Exposure cycles typically take only several minutes, making the process rather economic in terms of unit cost despite the high investment cost. The treated plastic components will be heated to 60 to 100 °C, depending on the energy input in the plasma reactor.

In atmospheric plasma treatment processes the created plasma exits through the nozzle of a jet head at high flow velocities. The significantly higher pressure causes the plasma to dissipate quickly after exiting the nozzle. However, a plasma jet length of 2 to 5 cm is sufficient to activate the treated surface. Compared to low-pressure plasma, this process has the advantage of operating at ambient pressure, *i. e.*, without reactor chambers. It is possible to treat large areas by traversing the jet nozzle across the surface.

4.15.1.7 Mechanical Pre-Treatment

Mechanical pretreatment of parts with large porous surface areas or streaky appearance by sanding or blasting is labor-intensive. However, at times this effort is mandatory in order to roughen surfaces to promote adhesion or in applications with high surface quality requirements such as Class A surfaces in automotive applications. Residual release agents are removed by washing with solvents, in special facilities with solvent vapor, or by ultrasound in solvent baths. Dust attracted by electrostatic charges is blown off using deionized air.

4.15.2 Polishing

While molded components are occasionally polished in order to remove flash or gate marks, machined components are routinely polished. Special sanding/shining waxes are used; for manual operations, polishing mops, or better folded polishing rings, are recommended (sanding: 15 to 25 m/min, polishing: 25 to 30 m/min, shining: without polishing agent). Small parts are tumbled (each operation cycle 8 to 12 h, 20 to 25 U/min, 1/3 fill with two parts polishing agent (cubes or spheres) for one part goods to be polished. It is important to ensure sufficiently low temperatures. Antistatic trimming granulates and polishing agents reduce annoying dust.

4.15.3 Coating

The surfaces of high-quality plastic components often require painting to meet optical and haptic specifications as well demands for light and weather resistance. The hardness, expansion and temperature characteristics of the paint layer have to match the characteristics of the component's material. If the paint layer is too hard, the plastic component's impact strength may be significantly reduced. Another aspect to be carefully considered is making sure that the plastic component can withstand exposure to painting/lacquering conditions (temperature, solvents). Often, the cost of painting/lacquering a plastic component is as high as its manufacturing cost.

Using reaction resins to paint structural foam and glass fiber reinforced molded components typically requires surface pre-treatment by wet sanding and priming, unless the reaction resin can be applied in the mold. Black-pigmented lacquer is used as UV-filter (Transfer-Electric) to protect polyolefins against photochemical degradation. Abrasion-resistant conductive lacquers are used as anti-static coatings of gas tanks, and, when filled Ag, Ni or Cu, for radio-frequency shielding of electronic devices.

The scratch resistance of PMMA and PC is increased by applying special clear lacquers (acrylate-, siloxane-, polyurethane- and nano-clay lacquers). Applying aqueous dispersions based on 3,4-polyethylene dioxythiophene by spraying, compression molding, or casting results in *conductive polymer coatings*. Spraying, flow or dip coating are used to apply scratch-resistant coatings based on *organically modified ceramics* (ormocers) to transparent plastics, facilitating their use as automotive glazing. Special surface effects with thermoplastics, such as mother-of-pearl with fish silver, Iriodin® pigments, or two-tone grain of synthetic leather, require lacquers that are based on similar base materials and special solvents.

The manufacture of coated or impregnated sheets is detailed Section 4.2.3.10.2.

Dipping processes immerse the component to be coated in a plastic solution, paste, or dispersion. When the component is slowly extracted, a skin is formed on the component, and subsequently this skin is solidified/cured using standard processes (evaporating of solvent or water, gelling, cooling, vulcanization). The skin can also be peeled off the produced components, *e.g.*, gloves.

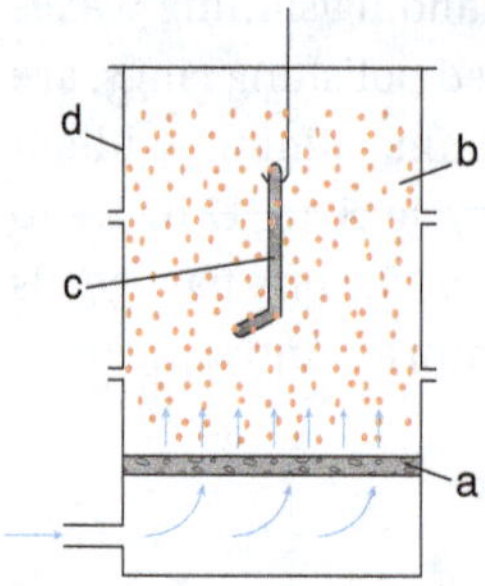

Figure 4.178
Fluidized bed coating; arrows indicate supply of compressed air
a: Porous plate
b: Powder bath
c: Metal part to be coated
d: Multi-part container

A variety of processes is used for *powder coatings*. For fluidized bed coatings, air or nitrogen is blown into the device through porous bottom plates (pore size < 25 µm) at an excess pressure of 0.1 bar to suspend plastic powder particles with sizes ranging from 50 to 300 µm, Figure 4.178: the bed is fluidized. When pre-heated metal components (200 to 400 °C) are introduced, a non-porous plastic coat is melt-deposited on the surface within 2 to 5 s, which, if required, can be oven-cured and finished in a continuous cycle. Depending on immersion time, layer thicknesses ranging from minimal 75 to 500 µm are achievable.

Powder slash processes use heated, rotating metal molds in which powders (PVC-P, TPU, PP, PA) are melted and deposited in layer thicknesses from 0.8 to 1 mm and subsequently removed and used, *e.g.*, as interior panels or lining in automotive applications.

Atmospheric plasma spray (APS) processes deposit sub-micron powder dispersions in water or solvents like lacquers via high-pressure or airless spray guns.

For *electrostatic powder spray processes*, plastic powders with particle sizes ranging from 40 to 100 µm are electrically charged via a (non-hazardous) high-voltage field of 50 to 90 kV when exiting a compressed air gun. The charge allows powder deposition on grounded metal components (even on their backsides) where it adheres for extended periods of time. Subsequently, the coated parts are processed in hot air ovens at approx. 200 °C, where the plastic melts, forming coating layers of 40 to 150 µm, max. 300 µm, before the components are cooled.

During *electrostatic fluidized bed coating* (Brennier process), charged powder is deposited on grounded components where it adheres.

Textiles are reinforced or lined using PA or PE powders that are ironed on the surface.

When *flame spraying* is used, an air gun directs the powder through the hot gases of a flame ring nozzle to melt it and then deposit it on the pre-heated surface to be coated.

4.15.4 Printing, Labeling, Decorating

Thermoset components are rarely printed. Planar melamine-formaldehyde or glass fiber reinforced plastic components are decorated by embedding printed or painted papers or patterned fabrics in a clear top coat; compression- and injection molded components are decorated by inserting decorative films in the mold, see Section 4.3.5.8.

Films are printed off the roll using conventional rotary printing presses; cut film is sheet-fed printed (also pull-printed for subsequent thermoforming) using machines that are modified to facilitate guiding and drying of plastic material to be printed. Special machines are available to facilitate multi-color printing of components. Printing ink manufacturers provide inks for plastic applications.

Of the common printing processes, *relief printing* with plastic or metal printing plates is used for small runs using sheet-fed printing and for printing of planar components. Aniline- or *flexo-printing*, used for roll-to-sheet printing, employs printing plates made from rubber that can print several colors. Indirect relief printing, also called *dry-offset printing* is used for sheet-fed printing and for the printing of molded components. Here, the printing ink (sometimes more than one color for one print run) is transferred from the relief printing plate to a rubber blanket. While *gravure printing* facilitates the typographic perfect reproduction of the master on films and spherical parts, it is economically feasible only for large print runs because of its high capital investment cost. Multi-color machines (even for spherical parts) are available for the universally suitable screen printing process, originally a manual process for small print runs.

The *thermo-diffusion printing process* for polyolefins, PA, and POM employs printing inks that are applied in a thick layer that diffuses into the surface of the printed components after exposure to elevated temperatures (100 to 150 °C). This allows the products to be thermoformed afterwards. The *therimage process* transfers the relief print on wax-coated paper to a hollow thermoplastic part where it is heat-fused. Pre-printed films that attach to the component are used for blow molded components employing a conformal printing process; for injection molded components, the *in-mold labeling (IML) process* is used, where the film is automatically inserted into the mold. Irregular and large parts that are not suitable for printing can be decorated using stencils or screens and *spray coating processes*.

These traditional processes are increasingly replaced by non-contact processes.

Thermoplastic components are durably labeled using electromagnetically controlled Nd-YAG-lasers (also neodymium-doped yttrium-aluminum-garnet laser) or CO_2-*laser beams* and a mask. Here, the plastic surfaces are labeled with high-contrast lettering by engraving, discoloration, foaming, or charring, *laser labeling*. Labeling rates range from 0.3 to 1 m/s. Suitable pigments can be added to materials with insufficient absorption properties, and their resulting color changes can create certain contrast effects as well as multi-color labels. Appropriate additive batches are commercially available.

Ink-jet technologies are also used with labeling devices.

In order to apply irregular or multi-dimensional decorations, a special immersion bath technology is used (marble or wood effect). The pattern is printed on a film that is adhered to a carrier, then placed on the surface of a water bath and sprayed with a solvent, which dissolves the carrier and leaves the thin film floating on the water. When an injection molded part is dipped into the water, the decorating film adheres to it. After drying the part, a clear coat is applied to increase abrasion resistance and optical depth effect.

4.15.5 Embossing, Hot-Embossing/Stamping

Local *blind- and color embossing* is applied using heated dies. For PVC-films, RF-blind embossing (see Section 4.14.6.1, High-Frequency Welding) and melt application of embossed films are often combined. Multi-colored labels are fusion-bonded by embossing.

Hot-embossing films, consisting of a PET carrier film with a layer design as described in Figure 4.179 and a thickness of 12 to 75 µm, are used for the application of scales, lettering, or decoration using positive or negative hot stamping (also by counterstamping). In a hot-embossing machine, the film is pressed to a substrate with a heated die under high pressure. The decorative layers are released from the carrier only where the die and the substrate make proper contact.

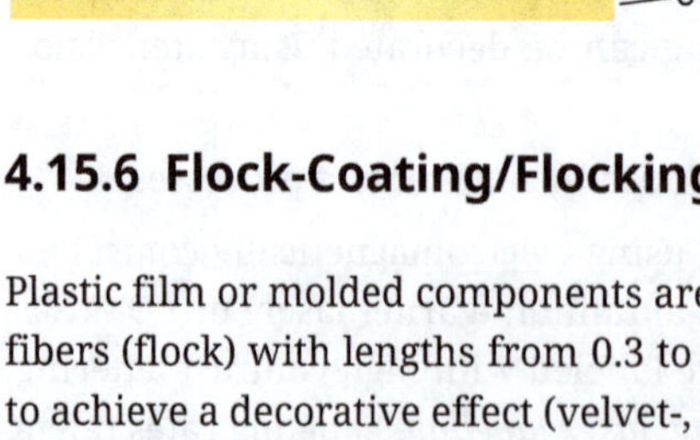

Figure 4.179 Composition of embossed film
a: Carrier film, b: interlayer, c: protective lacquer, d: metallization, color layer, e: adhesive layer; layers b through e form the embossed substrate

4.15.6 Flock-Coating/Flocking

Plastic film or molded components are flock-coated (or flocked) with short-cut textile fibers (flock) with lengths from 0.3 to 5 mm, or textile fiber dust or similar materials to achieve a decorative effect (velvet-, plush- or fur-like textures) or for technical purposes (*e. g.*, automotive window slide strips, friction elements in precision engineering, case linings, sound dampening, or condensation water collection). The flocks are sifted on a substrate covered with an adhesive layer and either blown on by compressed air or electrostatically deposited. Figure 4.180 shows a device in which the substrate is flocked from the top. While exiting the screen, the flock particles (which were treated with suitable aqueous solutions to impart sufficient electric conductivity) are electrically charged, oriented in the direction of the electric field, and pushed in the direction of the film at a rate of up to 200 cm/s so that they are anchored in the adhesive at right angles to the substrate. Non-anchored flock particles are re-charged and attracted by the high-voltage electrode or removed by a suction column.

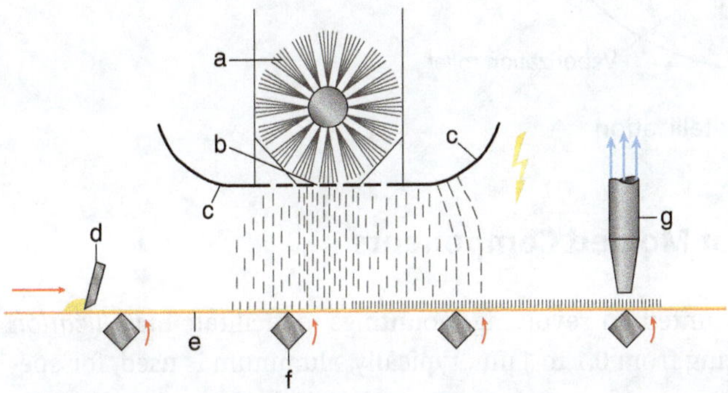

Figure 4.180 Electrostatic flocking from top

a: Flock hopper with brush metering, b: metering screen, simultaneously high-voltage electrode, c: pre- and post-flock zone, d: adhesive, e: adhesive covered substrate, f: beater bars, g: suction column for removing excess flock

4.15.7 Metallization of Film

A semi-continuous process is used to rewind rolled film in a vacuum installation while simultaneously applying super purity aluminum with layer thicknesses ranging from 0.03 to 0.04 µm via vapor deposition, see Figure 4.181. This process requires a high vacuum of 10^{-4} mbar. Carrier films are PET (minimum thickness 9 mm), biaxially stretched PP (minimum thickness15 µm), or PC with widths ranging from 0.3 to 2.4 m. Areas of application: capacitor films, water vapor and oxygen barriers, decoration.

Transfer- or transmission metallization processes are used to metallize films that cannot be directly metallized because of their volatile components' content. A coated carrier film is metallized on the coated side; the film to be metallized is laminated to the metallized side of the carrier film, and the carrier is subsequently removed.

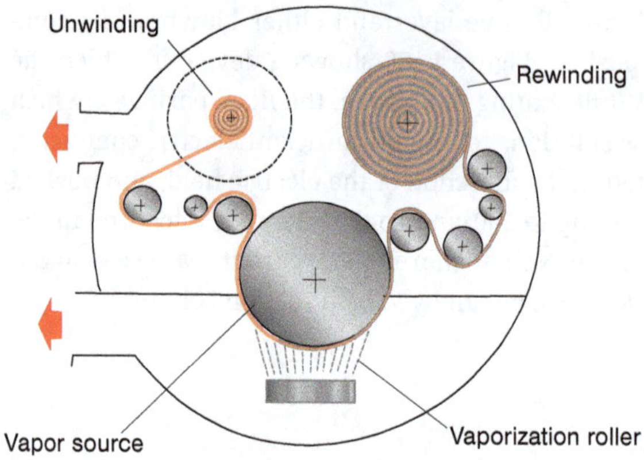

Figure 4.181 Principle of metallization

4.15.8 Metallization of Molded Components

Molded components are mounted on revolving mountings to facilitate *metallization* with layer thicknesses ranging from 0.1 to 1 µm; typically, aluminum is used, for special applications also copper or noble metals that are electrically heated and vaporized. For layer thicknesses exceeding 0.2 µm, the metal layers are opaque, thicker layers may be additionally vapor deposited. Plasticizer-containing plastics require a base coat, the reflective metal layers a protective layer.

Low temperature melting metals are applied to plastic surfaces using metallizing guns; gold, silver, and copper layers are precipitated by chemical reduction (*e. g.*, with formaldehyde) from metal salt solutions.

To ensure permanent adhesion of metal surface coatings applied by *vapor deposition* with thicknesses from 50 to 100 µm, plastic components are first immersed in etching baths in order to chemically roughen their surface. Suitable etching agents are available for most thermoplastics, PF laminates, EP resins, and integral PUR foams; some plastic materials can be galvanized without pre-treatment. The roughened surfaces are then activated in solutions of precious metal salts, which allows the use of electroless methods to deposit a firmly adhering copper layer from copper baths onto plastic surfaces. In turn, the copper layer is further electro-plated with copper and then nickel- or chrome-plated. These processes result in composite components with increased mechanical and thermal characteristics (Table 4.18). Surface metallization is also an important approach when shielding electronics against electromagnetic fields.

Table 4.18 Comparative Characteristics for Galvanized Molded Components

	ABS		PP	
	Non-galvanized	Galvanized	Non-galvanized	Galvanized[1)]
Limiting bending stress N/mm²	390	520	450	830
Young's modulus N/mm²	22,000	63,000	12,000	61,000
Ball indentation hardness N/mm²			540	860
Dimensionally stable up to °C	90	130	148	> 170

[1)] after pre-treatment

4.15.9 SiO_x-Coating of Films

Another development is the coating of films with SiO_x, where $1.5 < x < 1.7$ to ensure sufficient transparency while achieving adequately low oxygen permeability. So far, both chemical separation from the gas phase (based on organosilicon compounds) and vaporization (sublimation) of SiO_x by thermal energy have proven inefficient.

Currently used are electron beam coating processes, typically for PET, that work following the principles of metal vapor deposition. A computer-based system controls the vaporization of SiO_x via electron beam guns. In order to protect the brittle SiO_x layer, the process is almost exclusively used for composite films.

4.15.10 Rubbing

Plastic components are rubbed with anti-static agents in order to prevent the attraction of dust; rubbing with silicone oil increases gloss and scratch resistance; germicides can be used to stabilize some plastics, such as PVC, against microorganisms. However, most of these coatings wear off over the service life of the components.

4.15.11 Fluorination, Gas-Phase Fluorination

When PE, *e. g.*, in the interior of automotive gas tanks, or PP are exposed to fluorine-containing gases, the reactivity of the fluorine causes a more or less statistical substitution of hydrogen atoms in the PE chain through fluorine atoms. CF-, CF_2- and also (undesirable) CF_3-groups are the result. The latter indicate a break in the PE chain. Fluorination results in the hydrophilization of the surface and thus decreases adsorption and sorption of hydrophobic fuels.

4.15.12 Plasma Polymerization/Chemical Vapor Deposition (CVD)

Plasma polymerization has become a commonly used coating process (Figure 4.182). When monomers are exposed to a low-pressure plasma (completely or partially ionized gas, Section 4.15.1.6), the resulting impacts of electrons with monomer molecules create temperatures sufficiently high to break chemical bonds and initiate reactions that otherwise occur only at very high temperatures. The plasma temperature ranges from 60 to 150 °C so that the thermal load on the substrate to be coated is comparatively low. The appropriate choice of monomer and processing parameters facilitates the creation of a variety of surface effects: hydrophilic or hydrophobic layers, anti-fog layers, adhesive and non-stick layers, UV-protection layers, layers with tailored refractive indices, diffusion barriers, or scratch-resistant layers (0.1 to 8 µm thick SiO_2-compound layer).

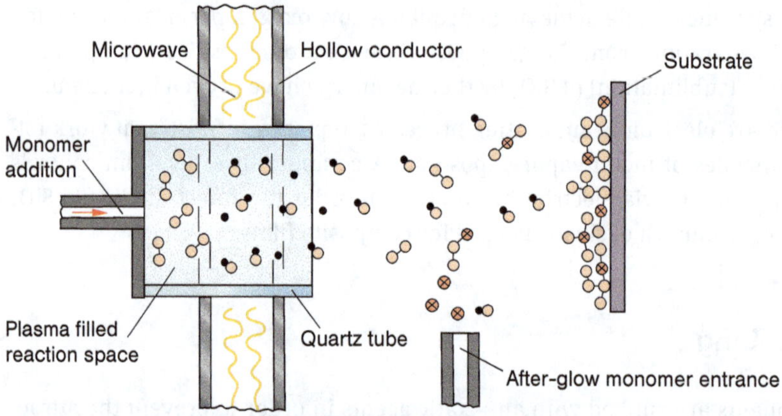

Figure 4.182 Principle of plasma polymerization

When applying scratch-resistant surface treatment to plastics in automotive glazing applications, organosilicon compounds (hexamethyldisiloxane) and oxygen are excited and fragmented by radio-frequency or microwaves. The fragments react with each other and precipitate on the plastic surface, *e. g.*, PC, forming a highly crosslinked layer; layer thickness: 0.1 to 8 µm.

4.15.13 Thermal Spraying

Metals, plastics, metal and plastic foams, ceramics, and other materials can be coated at high rates (10 min/m^2 for 0.1 mm layer thickness) with thermoplastics with coating thicknesses typically ranging from 0.1 to 0.5 mm. Inert gases are available to melt, convey, and protect the coating materials.

4.16 Other Treatment Processes

4.16.1 Machining

Forming and shaping operations are the appropriate processing methods for plastic materials. However, in the following cases machining is expedient or required:

- For small production runs because the cost for the molds would be too high,
- For applications with high requirements in terms of dimensional accuracy,
- For plastics that are difficult to form or shape, such as PTFE; here, blocks are shaved into veneer-like films or machined using mills or lathes,
- Finishing of thermosets and laminates,
- For the subsequent addition of bores/holes in order to keep molds simple,
- De-flashing and removing of sprues,
- For the manufacture of test specimens from molded or semi-finished components.

Compared to other materials, such as metal or wood, plastics exhibit a number of specific characteristics that need to be considered when machining them:

- Thermoplastics soften when heated and tend to stick; thermosets are prone to thermal or chemical degradation. Therefore, it is recommended to use small chip cross sections and high cutting rates with good cooling using compressed air or inert gas. Lubricants and liquid cooling agents may cause stress cracking.
- Their low thermal conductivity promotes heating during machining.
- Their coefficient of thermal expansion is high; therefore, dimensional checks should be performed only after cooling.
- The Young's modulus of plastics is more than one order of magnitude smaller which leads to elastic deformations under load that will recover with a certain time delay and may cause dimensional inaccuracy: holes are often smaller than the drill diameter.

- Semi-finished products and molded components may exhibit residual stresses, and machining potentially disrupts their equilibrium, leading to warpage and in extreme cases to crack formation. Annealing/tempering may be required prior to machining.
- Plastics with inorganic fillers cause higher tool wear; therefore, it is recommended to use carbide metal, oxide ceramics, or diamond tools. The filler-containing dust created during machining should be suction-extracted. Stainless steel tools are sufficient for unfilled thermoplastics. Superabrasives (diamond tools such as those used in wood milling machines) are successfully used for the machining of carbon- and fiber reinforced plastics (manufacture of test specimens from semi-finished products). Milling under water is advantageous.

Table 4.19 shows an assessment of plastic processing and machining technologies.

The required *de-flashing* can be achieved in different ways: for *thermosets*, tumbling with thermoplastic granulates at room temperature is suitable, while *elastomers* are shock-frozen to −130 °C using nitrogen so that only sprues and gates embrittle and can be removed.

Elastomers (NBR, Shore A = 80) can be easily machined in the entropy-elastic range (room temperature) using high cutting rates (60 m/min), feed rates of approx. 0.1 mm, and high-positive tools ($\alpha = 12°$, $\gamma = 25°$, cutting edge radius < 10 mm).

Table 4.19 Comparison of Processing/Machining Technologies

	Severing		Machining		Removing	
	Cutting	Stamping	Drilling	Milling	Water jet	Laser beam
Mechanical load	Low	Medium	Medium–high	Medium	Low	Very low
Thermal load	Very low	Very low	High	Medium–high	Very low	Very high
Harmful substances in air	Low	Low (fibers)	Medium	High (dust, fibers)	Low (easy aerosol formation)	Very high (smoke, gasses)
Noise	Low	Medium–high	High	Very high	High–very high	Low
Kerf width	–	–	–	Medium (tool)	Low	Low
Smallest possible radii	–	–	–	Tool diameter	0.5 to 1 mm	0.5 mm

	Severing		Machining		Removing	
	Cutting	Stamping	Drilling	Milling	Water jet	Laser beam
Symmetry of cut edge	No dimensional stability	Parallel/ concave to stamping direction	Perpendicular to tool axis	Perpendicular to tool axis, slight expansion	Parallel to jet direction, slight expansion	Parallel to beam direction
Tool wear	Medium	Medium-high	High	High	Low	Very low
Cutting speed	–	–	0.2 m/min	6 m/min	1 to 5 m/min (pure high pressure)	7 to 15 m/min (1500 W)
Surface roughness R_z	40 µm	–	–	Up to 10 µm	Up to 4 µm (abrasive) 50 µm (high pressure)	40 µm
Flexibility	High	Low	High	High	High	High
Personnel requirements	Low	Low	Medium	High	High	High

4.16.2 Cutting, Machining

Circular and straight knives as well as cutting rolls are used to prepare pre-cut parts from semi-finished products such as films, sheets, and profiles. Cut-outs/apertures are created by punching. Heating of rigid sheets will reduce the cutting forces and thus avoids rough edges. Thick sheets are cut using circular or band saws. Diamond cutting machines, such as those commonly used for glass or stone, are employed to wet-cut/separate fiber reinforced resins.

Thermoplastics can also be cut/separated by local melting *(cut and seal)*, see also Section 4.14.2. Electrically heated blades or wires are used as cutting tools to ensure smooth cutting surfaces. *Incandescent wire cutting* is of particular importance for the separation and shaping of PS- or PUR-foams. Spatial cuts with desired surface profile can be created when flexible foams are compressed (laminar or between rolls) during cutting operations, *e. g.*, for the packaging of delicate goods (Figure 4.183).

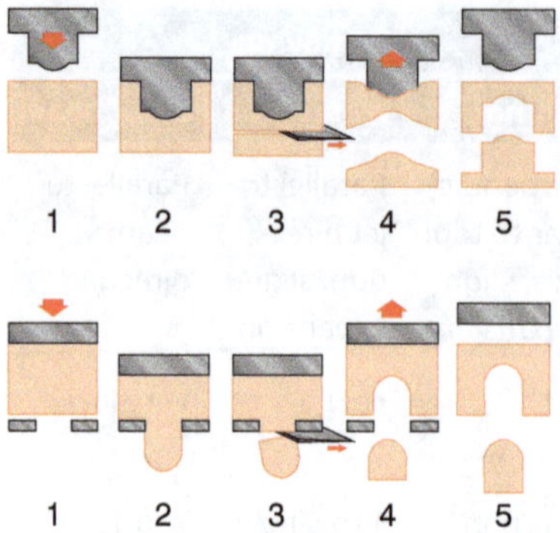

Figure 4.183
Compression cuts:
Top: indentation process, Bottom: push out process

Minimal kerf widths (0.1 mm) and smooth cutting edges in plastics can be achieved using *laser beams*. The high energy of laser radiation causes the degradation and/or vaporization of the plastic. This process is particularly suitable when using numerically controlled machines (robots) for the manufacture of irregular molds or to drill precision holes. *High-pressure water jet cutting* (2500 to 4500 bar) operates without generating potentially hazardous degradation products. The water jet exits at several times the speed of sound and immediately flushes out the removed material. Although the process is suitable for fiber reinforced materials, the achievable cut quality does not match that of diamond cutting tools.

Lasermicrojets use the light of a solid state laser (λ = 1064 nm) coupled with a laminar water jet with a diameter of 0.1 mm and a pressure of to 200 bar. Advantages: cooling and cleaning effects, large working length of the laser beam facilitated by guidance through the water jet, distance between processing head and component to be cut: 1 to 100 mm, minimum kerf 0.05 mm.

4.16.3 Radiation Crosslinking

When exposed to high-energy radiation, plastic materials will crosslink or degrade, depending on their macromolecular structure. The addition of "crosslinking agents" facilitates the crosslinking of plastics that would otherwise be degraded by high-energy radiation. This approach allows for plastic materials to be processed like thermoplastics and subsequently be crosslinked.

A well-known application is the radiation crosslinking of PE cable sheathings which is achieved using electron beam accelerators immediately post-extrusion, Figure 4.184. This process requires specific safety measures because ozone, nitrous fumes, and X-rays are created.

The crosslinking ensures that these cables will not melt during a potential fire, thus offering increased fire safety. Other PE applications include shrink tubing (crosslinking prior to drawing locks in the molecular structure, and when re-heated the tubes/

hoses shrink back to their original dimensions) and pipes (crosslinking increases their heat and crack resistance).

Increasingly, other engineering plastics are also radiation crosslinked, for example to increase their mechanical characteristics at elevated temperatures.

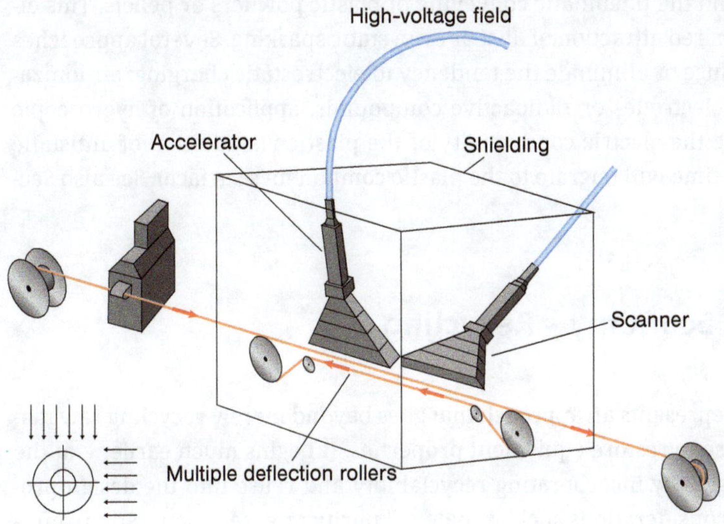

Figure 4.184 Radiation crosslinking with two accelerators (90 °C offset) for cable sheathing

4.16.4 Heat Treatment

By exposing plastics to elevated temperatures over a certain period of time (tempering) the following effects can be achieved:

- Post-curing of thermosets,
- Increasing the crystalline portion (and thus the strength and stiffness) of semi-crystalline plastics,
- Reducing or eliminating residual stresses by storing at temperatures in the melt-freezing range (for amorphous plastics) or below the melting range (for semi-crystalline plastics). Slow cooling after storing is required because otherwise the local temperature gradients would again create residual stresses. The molded component has to be retained in a fixture during storage at elevated temperatures to avoid warpage.

4.16.5 Elimination of Electrostatic Charges

The surfaces of plastic materials with high specific electric resistivity ($\rho > 10^{12}\ \Omega \cdot cm$) will charge electrostatically when they are rubbed or when they disconnect from another surface. Examples include the de-molding of injection molded parts, the unwinding of films, and the pneumatic conveying of plastic powders or pellets. This effect can cause undesired attraction of dust or even static sparking. Several approaches are available to reduce or eliminate the tendency to electrostatic charging: air ionization via discharge electrodes or radioactive compounds, application of hygroscopic layers or increasing the electric conductivity of the plastic via addition of antistatic agents (which over time will migrate to the plastic component's surface). See also Section 7.1.3.

4.17 Circular Economy – Recycling

Circular economy represents an approach that goes beyond merely recycling raw materials after their use to restore equivalent properties. It begins much earlier – at the product design stage – by incorporating recyclability and reuse into the design process. This includes considerations such as material purity or ease of separation, durability, repairability, and the potential for material recovery through recycling.

Although the economic feasibility of recycling fluctuates significantly due to its dependency on oil prices, the early 2020s have seen increased regulatory actions and societal pressure, particularly in Europe and to some extent globally, which are driving a transformation in the plastics processing industry. This shift is especially notable in the plastic packaging sector, which accounts for approximately 50–60% of the global plastic production.

Key legislative measures by the European Union include the following:

- EU Directive 2019/904, which targets the reduction of single-use plastics and typical marine litter such as fishing gear
- EU Directive 94/62/EC, on packaging and packaging waste
- The Circular Economy Action Plan, also known as the EU Green Deal, which outlines ambitious targets for waste and pollutant reduction across sectors ranging from automotive to battery production and plastics processing

As a result, many European countries have enacted corresponding national legislation. Examples include the German Packaging Act (2019) and the plastic taxes introduced in the UK and Italy (2020, 2021).

These regulatory frameworks have, in turn, stimulated collaborations and innovations among polymer producers, converters, and industry associations. These initiatives focus on the development and evaluation of novel processes, product certification, and the logistics of material flows.

Even established plastics recycling methods are undergoing transformation. Plastics recycling refers to the reprocessing of both pre-consumer waste (also called post-industrial waste) from production (*e.g.*, sprues, scrap) and post-consumer waste, which is collected from industrial and municipal sources. At present, most recycled material still comes from packaging waste.

The core concept of recycling is to break down used products into their material components and reuse these materials to manufacture new products. The recycling process and resulting materials can vary greatly depending on the source and composition of the waste. Generally, a plastic can only be recycled a limited number of times before it becomes too degraded or contaminated for further use.

Pre-consumer waste is often shredded on-site and reintroduced into the production process by blending with virgin material in proportions ranging from 30 vol % to 70 vol %. However, post-consumer plastic waste is currently recovered to a much lesser extent for this type of reuse. This is largely due to lack of material purity, such as in multilayer films, and contamination in mixed collection streams.

Recycling of mixed and contaminated plastic waste is significantly more complex. When plastics are dirty or mixed with other types, the resulting recyclate quality is lower and often only suitable for downcycling, that is, use in lower-value applications.

Three general categories of recycling are distinguished today: mechanical (material) recycling, including primary and secondary processes, and chemical (feedstock) recycling. The choice of method depends on the molecular structure of the polymer and the level of contamination in the waste.

4.17.1 Mechanical Recycling

Among the available options, mechanical recycling is the most desirable approach due to its low cost and high reliability. In this method, the molecular structure of the polymer is largely preserved. After grinding the plastic waste, the key processing step is remelting, which limits this method to thermoplastic polymers.

Because remelting leads to polymer degradation (*i. e.*, chain scission and molecular breakdown), recyclates are often blended with virgin materials to minimize performance loss in the final product. Additives such as UV stabilizers must also be reintroduced to ensure processing and product stability.

A distinction is made between:

- Primary recycling, which deals with clean, homogeneous pre-consumer waste of known composition, requiring minimal additional processing

- Secondary recycling, which handles post-consumer waste that is often contaminated and requires sorting, cleaning, and reprocessing steps

Table 4.20 presents an overview of plastic waste types and recycling shares in the United States in 2015 [1].

Table 4.20 Plastic Waste and Recycling Share in 2015 in the US

Type of plastic	Share of plastic waste [%]	Recycling share [%]	Recycled products
PET	14.8	18.4	PET bottles
PE-HD	17.5	10.3	PE-HD bottles
PP	22.4	0.9	Screw caps
PE-LD/PE-LLD	23.0	6.2	Slides
PS	6.9	1.3	Styrofoam
Other plastics	12.8	22.5	Electronic housing, etc.
Total	100	9.1	

Sorting methods vary by plastic type. One basic method is manual sorting, in which trained personnel separate materials on the basis of visual cues such as recycling codes, shape, color, or product category. This explains the relatively high recycling shares for bottles and caps. However, manual sorting is labor-intensive and error-prone, and visual distinctions are often insufficient for an accurate identification.

Automated sorting technologies are increasingly used to improve efficiency and accuracy. Common systems include:

- Sink–float separation, which utilizes density differences to separate plastic types in water tanks; for example, PET bottle flakes can be separated from PP caps,
- Near-infrared spectroscopy (NIR), widely used in high-speed sorting facilities: this technique detects characteristic absorption bands in the infrared spectrum based on the chemical composition of the polymer; however, NIR is limited in detecting dark-colored plastics, adhesives, residues, and additive-laden materials.

After sorting and cleaning, the recyclates are reprocessed. At this stage, it is crucial to understand the degradation behavior of polymers. Degradation occurs due to heat, mechanical stress, oxygen, and UV exposure, causing changes in the structure of the polymer – typically, chain scission, which alters material properties.

Depending on the polymer structure and processing conditions, degradation can occur to varying degrees. General degradation effects include:

- Changes in molecular weight and molecular weight distribution, affecting viscosity,
- Formation of cross-links and branched chains,
- Creation of oxidized or unsaturated compounds.

A solid understanding of polymer degradation kinetics is essential for designing recycling processes and manufacturing high-quality products from recyclates. During recycling, the melt viscosity and flow behavior of polymers change, potentially leading to challenges in processing. With respect to virgin materials, recyclates may also exhibit reduced mechanical properties, though thermal stability is usually unaffected.

Degradation mechanisms can be classified as:

- Mechanical degradation,
- Thermal degradation,
- Thermo-oxidative degradation,
- Photochemical degradation.

These are influenced by the polymerization method, processing conditions, and polymer properties. In practice, the blending of virgin material with recyclate can significantly reduce or even eliminate many degradation-related issues.

4.17.2 Chemical Recycling

Chemical recycling, also known as feedstock recycling, is primarily used for cross-linked polymers or thermoplastics that cannot be mechanically recycled due to contamination, degradation, or mixed compositions. Unlike mechanical recycling, which retains the polymer structure, chemical recycling breaks down polymers into low-molecular-weight compounds or their original monomers, which can be reused to produce virgin-like plastics or serve as petrochemical feedstocks.

This category includes several distinct processes:

- Solvent-based purification: It dissolves polymers in specific solvents to remove additives and contaminants without altering the polymer backbone. It enables the production of high-quality recyclates but has limited scalability due to energy-intensive solvent recovery and toxic solvent concerns.
- Depolymerization: It chemically breaks down condensation polymers like PET, PA, PU, or PC into their monomers, which can be purified and repolymerized. Techniques include glycolysis, hydrolysis, methanolysis, and enzymatic depolymerization. These are under active research to improve catalyst efficiency and energy use. For addition polymers like PE and PP, depolymerization remains a challenge due to their chemical stability.

- Pyrolysis: It thermally degrades plastics in the absence of oxygen to produce pyrolysis oil and gases. While capable of handling mixed waste, it requires further refining and is energy- and emission-intensive, raising sustainability concerns.
- Gasification: It operates at even higher temperatures with limited oxygen to produce syngas, which can be used for energy or chemical synthesis. It handles highly contaminated waste but is currently used more for energy recovery than for material circularity.

Despite its potential, chemical recycling faces major challenges, including high energy demand, infrastructure requirements, and regulatory inconsistencies. The environmental benefits are still under debate, with life cycle assessments (LCAs) complicated by data gaps and varying methodologies.

Nonetheless, chemical recycling is a rapidly developing field, driven by the need for higher-quality recyclates and more circular plastic systems. While it cannot replace mechanical recycling, it provides valuable solutions for waste streams that cannot be processed mechanically. Future progress depends on technical improvements, policy support, and a shift towards better product design for recyclability.

For further reading, see [1].

References for Section 4.17

[1] Rudolph, N., Kiesel, R., Aumnate, C., *Understanding Plastics Recycling*, 3rd edition, Hanser, Munich (2025)

5 Material- and Production-Oriented Design and Simulation

The term "design" covers all activities with which the plans for a plastic product are created. In addition to the description of the geometry of the individual components (design drawings), these include, in particular, the definition of the requirements for a product, the development of solution principles taking into account possible manufacturing processes, the manufacturing and operating costs, the selection of suitable materials, and the assessment of risks in production and application as well as the ecological compatibility.

In the case of plastics, the design also includes the consideration of the manufacturing processes, as special system components such as injection molding tools or handling equipment usually are required to manufacture a product. The interactions between the selected material and the design and processing adapted to the material must also be considered (see Figure 2.16).

Design is an element of product development. In addition to design activities, this also comprises the functions of technology and innovation management in terms of the development and provision of production technologies, material technologies, and design tools; see Figure 5.1. The question of the appropriate mastery of computer-aided design tools is also of great importance.

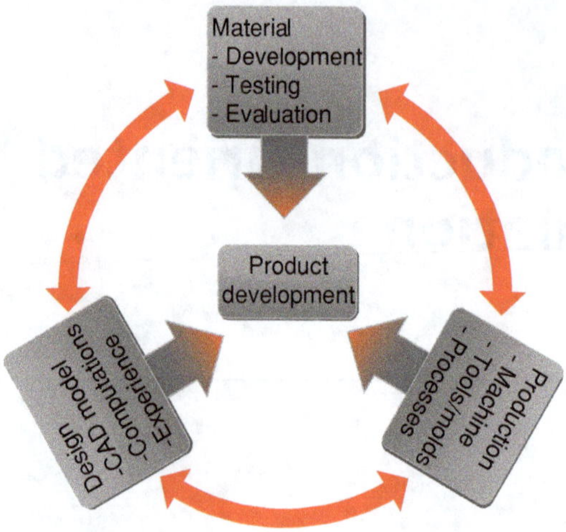

Figure 5.1
Elements of product development

5.1 Systematic Design of Plastic Products

Systematic design aims to master the product development process reliably, even with complex requirements. It is intended to ensure that all essential aspects are considered in the development processes.

An essential element of systematic design is the definition of the various activities in the design process and their interlinking, as schematically depicted in Figure 5.2. In addition, it must be noted for the field of plastics that the manufacturing tools required for production must also be developed alongside the actual product. This means that the process described in Figure 5.2 must be run through at least twice, namely for the molded part and the production tool.

Other plastic-specific aspects of systematic design are:

- Consideration of special material properties of plastics (*e. g.*, temperature dependence, long-term behavior, and aging),
- Production-oriented design,
- Consideration of the dependence of component properties on the manufacturing process (*e. g.*, anisotropy, shrinkage, residual stresses).

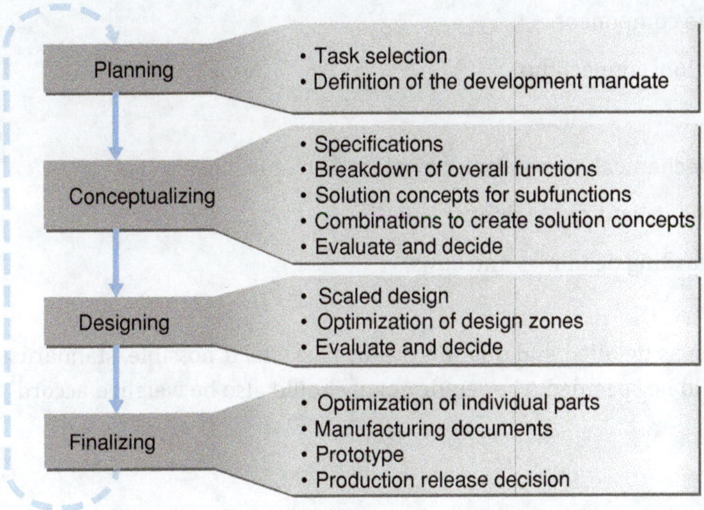

Figure 5.2 Procedure for developing technical products according to the German Association of Engineers (VDI)

5.1.1 Requirement Lists and Specifications

Requirement lists are used to define the properties that the product should fulfill. A distinction must be made between necessary and desirable requirements, in order to enable a balance to be struck between cost and benefit in the subsequent design phase. The requirement lists allow the derivation of requirements for the material and the manufacturing process and are the basis for the material selection.

The requirements specification summarizes the characteristics of a product that are required and desired by the customer. The functional specification is therefore part of a list of requirements. If products are to be developed on behalf of a supplier, the client formulates the specifications. In this specification sheet, the requirements for the product are formulated from the user's perspective. Therefore, the requirements specification and the functional specification also describe the requirements for a product, but from the different perspectives of the user and the manufacturer of the product.

The comprehensive and systematic creation of requirement lists or specifications for a product to be developed is the prerequisite for target-oriented product development. Requirements that are formulated too late can call into question the development process that has been carried out up to that point or mean a great deal of effort when subsequently considering the requirements that then need to be reformulated.

A list of questions for the systematic creation of a list of requirements for plastic products consists of the following topics:

- Information on the component,
- Designation, function, connections,
- Legal framework,
- Information on mechanical, thermal, electrical, and chemical loads,
- Material requirements and recyclability,
- Requirements regarding deadlines and costs,
- Other requirements.

All answers should be as detailed and quantitative as possible. If possible, standards and regulations should be specified. The requirement should also be weighed according to

- Fixed demand,
- Minimum requirement,
- Wish.

Documentation of the source of each piece of information must also be included.

The questions included here are only a first step towards finalizing the list of requirements. During the design process, questions may be changed (clarified or added). The recorded requirements must also be constantly scrutinized and may be changed during the design and testing phase. Each of these steps must be documented.

5.1.2 Design Catalogs

An important element of systematic design is the use of design catalogs to select and compare different solution principles, as such catalogs help to consider the different aspects of a solution principle. Table 5.1 shows a design catalog for the function of joining plastics. Standard solutions, so-called functional elements, can be developed for the individual solution principles, which can be used repeatedly for a wide variety of designs. A large number of such functional elements have been developed especially for designing with plastics; see Section 5.7.

5.2 Material Selection

In general, the material costs for plastic products represent the largest cost block; see Figure 5.3. The choice of material is therefore of great importance, not only from a technical but also from an economic point of view. An essential step in the material selection process is the implementation of the requirements through quantified material characteristics. A search profile is created from such defined requirements for the material characteristics. This can then be used to select the optimum material from a material database. It is important to ensure that the material characteristics in the database are comparable (for example, determined according to the same test standards, test specimens manufactured under comparable conditions). These requirements are met by the CAMPUS database, which most raw material manufacturers have joined.

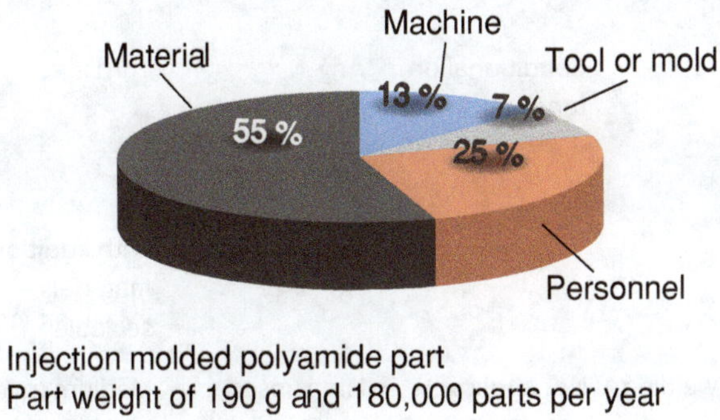

Figure 5.3 Breakdown of production costs

A major problem when selecting materials is that the requirements for the product are often not specified precisely enough in the early phase of product development. These uncertainties frequently are compensated for by excessive requirements, which ultimately leads to an overpriced product. It is therefore advisable to identify several alternative materials in a material preselection and only make the final specification in a later phase of product development – for example, after the first prototype tests have been completed.

Table 5.1 Design Catalog: Joining Plastics

	Type of conclusion		Material locking			Interlocking connection	
		No.	1	2	3	4	5
Main part	Designa-tion	1	Adhesive bond	Welded joint	Fabric-locking connecting elements (film hinge)	Overmolding	Rivet connection
	Sketch	2	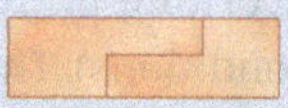	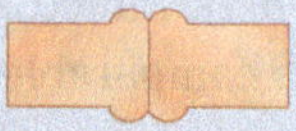	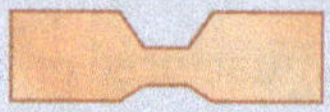	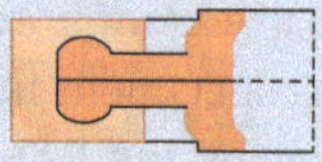	
Access part	Detachable	1	Not detach-able	Not detach-able	Not detachable	Not detach-able	Not detachable and detachable
	Type of load capacity	2	Thrust	Any	Conditional on tension	Any	Thrust
	Media per-meability	3	Impermeable			Impermeable	With additional materials – sealable
	Assembly effort	4	High (complex pre-treat-ment)	High (welding machine required)	Not applicable	High (insert parts individu-ally)	Medium (riveting tool required)
	Additional connecting materials	5	Adhesive or solvent	Welding material	None	Overmolding material	Rivet
	Material pairing	6	Both materi-als can be bonded	Same thermo-plastics	Made of PP, PE, PA, possibly also made of ABS or SB	Any	
	Miscella-neous	7					

Interlocking connection			Frictionally engaged		
6	7	8	9	10	11
Sliding connection	Snap connection	Form-fit connecting element (feather key, pins)	Press connection	Screw connection	Frictionally engaged connecting element (ring clamping element)
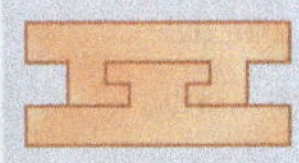	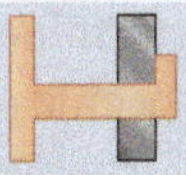	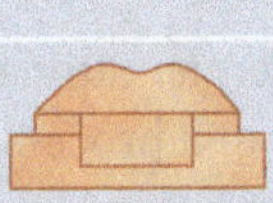	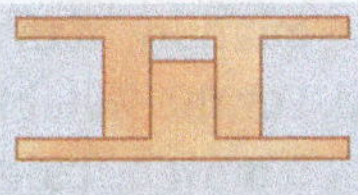	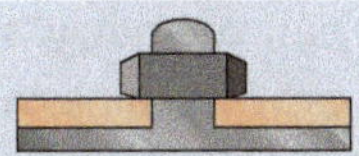	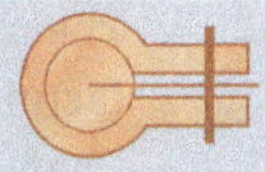
Detachable	Not detachable and detachable	Detachable	Conditionally detachable	Detachable or conditionally detachable	Detachable
Different, depending on the type of connection	Conditional on tension	Different, depending on the type of connection	Can be sealed with additional materials		
	Can be sealed with additional material		Low	Medium (for threaded inserts, larger)	Medium
Low	Low	Medium	None	Screw, nut (threaded insert)	Ring clamping element or similar
None	None	Feather key, pin, ...		Any	
	At least one viscoplastic material	Any			

5.3 Production-Oriented Design

The concept of production-oriented design covers all aspects of design that take appropriate account of the interaction between the manufacturing process and the product properties (see Section 2.4).

5.3.1 Selection of the Manufacturing Process

The material, geometry, and quantity of the product ultimately determine the manufacturing process to be used. For the suitability of a process for a specific product, see Chapter 4. Particular attention must be paid to the strong interaction between process and product shape. The geometry often has to be modified so that it meets the requirements of a specific manufacturing process.

The manufacturing process frequently also has to be modified, for example, to meet the geometric requirements of the component. The development of many special injection molding processes thus can be derived from the requirement for a specific product shape. The example of a component with complex spatial structures illustrates the difficulty of determining a suitable manufacturing process. Such a component can be blow-molded as a geometrically simple hollow body or consist of several individual shell-shaped parts that are joined together in a further process to form a component (multi-shell technology) or manufactured using melt core technology. An analysis of today's plastic intake systems for internal combustion engines, for example, shows that all the processes mentioned here are ultimately used. In addition to the manufacturing costs, which usually differ depending on the quantity and the geometric complexity of the component, the decision in favor of a particular process also depends on which processing technology is available or better mastered by the respective manufacturer.

5.3.2 Design

The properties of plastic components are influenced to a considerable extent by the processing; see Section 2.4. In addition, the selected manufacturing process imposes restrictions on the design of the component geometry. Simple examples illustrate this relationship well for injection molding:

- Large wall thicknesses result in long cycle times. Doubling the wall thickness quadruples the cooling time. Typical wall thicknesses of injection-molded parts are 2 mm to 4 mm.

- With small wall thicknesses and long flow paths, the cavity can no longer be filled.
- The design should be such that the molded part can be demolded when the mold is opened. Suitable draft angles must be provided for this purpose (Figure 5.4). In the case of structured surfaces, the demolding slope must be adapted to the roughness depth of the structure (see Section 3.6.5). The separation of the mold must be designed in such a way that sliders are avoided as much as possible (Figure 5.5).
- Smaller undercuts, such as those resulting from the circumferential bead, shown in Figure 5.6, can be forcibly demolded if the elongation imposed on the component remains sufficiently small. Such small undercuts are often designed into the molded part so that the part remains on the ejector side of the cavity when the mold is opened.

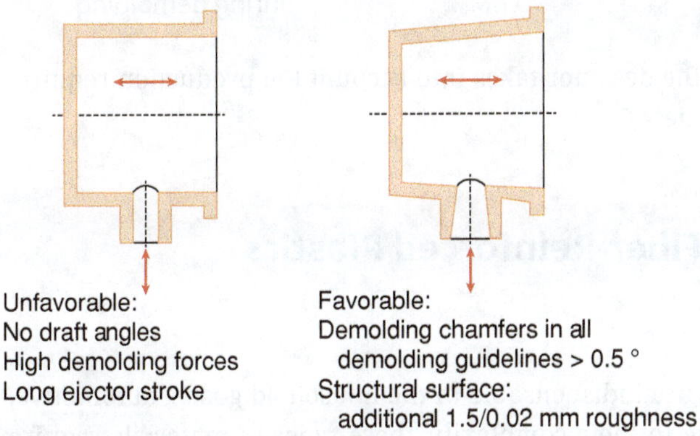

Figure 5.4 Notes on draft angles

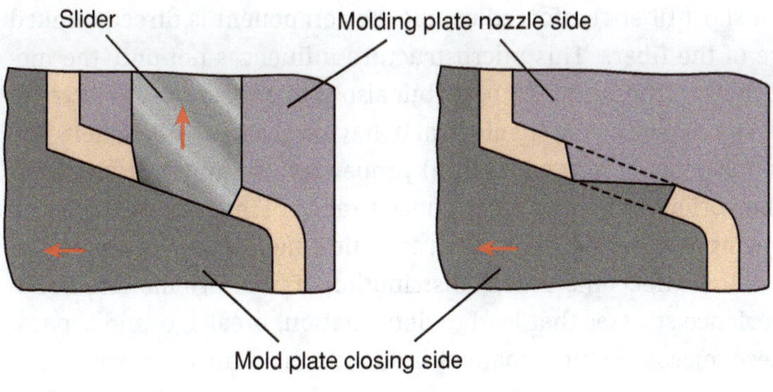

Figure 5.5 Avoidance of undercuts

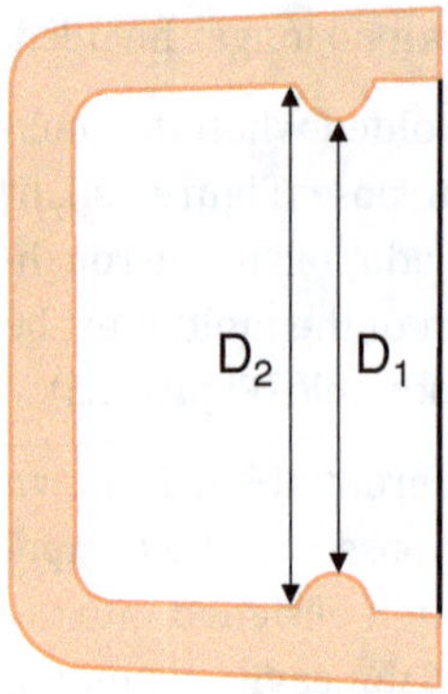

Polymer	Maximum allowable strain [%]
PS	< 0.5
PS-HI	< 1.0
ABS	< 1.5
PC	< 1
PA	< 2
POM	< 2
PE-LD	< 5
PE-HD	< 3
PVC hard	< 1
PVC soft	< 10
PP	< 2

$$\varepsilon = \left[\frac{D_2}{D_1} - 1\right] \cdot 100\,\%$$

Figure 5.6 Permissible strains during demolding

By designing for production, the designer takes into account the production requirements for the component geometry.

5.4 Designing with Fiber-Reinforced Plastics

Sebastian Goris

Fiber-reinforced plastic parts are indispensable in the household goods, automotive, and aerospace industries. Due to their complexity, these types of materials require special attention and are therefore discussed in detail in this section. A summary of common fiber types can be found in Section 7.3.

The performance of a short-fiber-reinforced composite component is directly linked to the microstructure of the fibers. This microstructure influences not only the mechanical behavior, such as stiffness and strength, but also properties such as electrical and thermal conductivity as well as the rheological behavior of a fiber-filled melt. The three most important changes in microstructural properties during processing concern fiber length, fiber orientation, and fiber concentration. Fibers embedded in a deforming matrix during processing, such as in injection molding or compression molding, tend to twist; this affects the overall distribution of fiber orientation. In addition, they also experience stresses that lead to deformation, breakage, and separation of the fibers. These microstructural changes in the fibers, in turn, determine the mechanical behavior of the manufactured component. In this section, fiber abrasion, fiber alignment or orientation, and fiber–matrix separation are presented individually to help the reader gain some understanding of the fundamental aspects of microstructural evolution during processing and to provide a basis to fully understand the underlying physics.

5.4.1 Fiber Length Distribution

The superior mechanical capability of fiber-reinforced composites is achieved by transferring the applied load from the matrix to the reinforcing fibers. Consequently, the interfacial adhesion and the length of the fibers embedded in the matrix determine the maximum load that can be transferred. Keeping a sufficiently great fiber length for load transfer while maintaining economical processes is of great interest to the industry.

"Fiber breakage" refers to the fact that fibers break during processing and their length inevitably decreases from their initial value. While the initial length of the fibers is uniform and can be up to 30 mm for long-fiber thermoplastics (LFTs), fiber abrasion leads to a fiber length distribution in the finished part. This distribution ranges from powdery fibers (length < 0.1 mm) to a small proportion of fibers that have retained their original length. In general, the degree of fiber attrition is influenced by the choice of processing method, processing parameters, material selection, and tool design. Since the mechanical properties depend on the fiber length, it is desirable to preserve the length of the fibers as much as possible by making the processing conditions gentler. Along the entire process line, the fibers are exposed to severe stresses, including hydrodynamic effects within the highly viscous matrix, fiber–fiber interactions, and fiber–mold interactions, which can reduce the average fiber length to as little as 0.5–1.0 mm in the molded part, even with LFT materials.

The fiber length in a finished part is highly dependent on abrasion during processing. Plasticization is the critical phase, followed by abrasion in the mold cavity, which can depend on gate design, mold design, and flow management. Fiber breakage during plasticizing can be reduced by optimizing process parameters such as screw speed, pressure, and temperature. A screw design with variable pitch and geometry can also help to reduce fiber attrition. In most cases, it is a balancing act between production throughput and reducing fiber abrasion.

The length of each fiber can vary from 0 to L_{max} and the total sample consists of N total fibers. The entire population can be represented by a histogram with n bins, where each bin contains N_i fibers of length l_i. The distribution obtained is usually summarized by calculating representative averages, such as the number-averaged length L_N and the weight-averaged length L_W, to obtain a concise understanding of the fiber population as a whole. The weight-averaged (or length-averaged) fiber length L_W is an average value of the fiber length distribution that highlights the proportion of long fibers in the distribution. This is of particular interest for LFT components as they typically have a length distribution with a distinct peak followed by a long tail.

Figure 5.7 shows a discrete fiber length distribution function $\mathfrak{n}(l)$ and a length-weighted distribution $\mathfrak{w}(l)$ in the form of histograms, with a maximum fiber length L_{max} of 15 mm. By integrating $\mathfrak{n}$ over the length, the cumulative length distribution $\mathfrak{N}$ is obtained: it describes the proportion of fibers that are less than or equal to a certain length l_j; $\mathfrak{W}$ presents the cumulative length-weighted distribution. The cumulative distributions $\mathfrak{N}$ and $\mathfrak{W}$ are represented by continuous curves. The graphs show the corresponding mean values as well as the median values from the number distribution and the length-weighted distribution ($L_{N,50}$ and $L_{W,50}$).

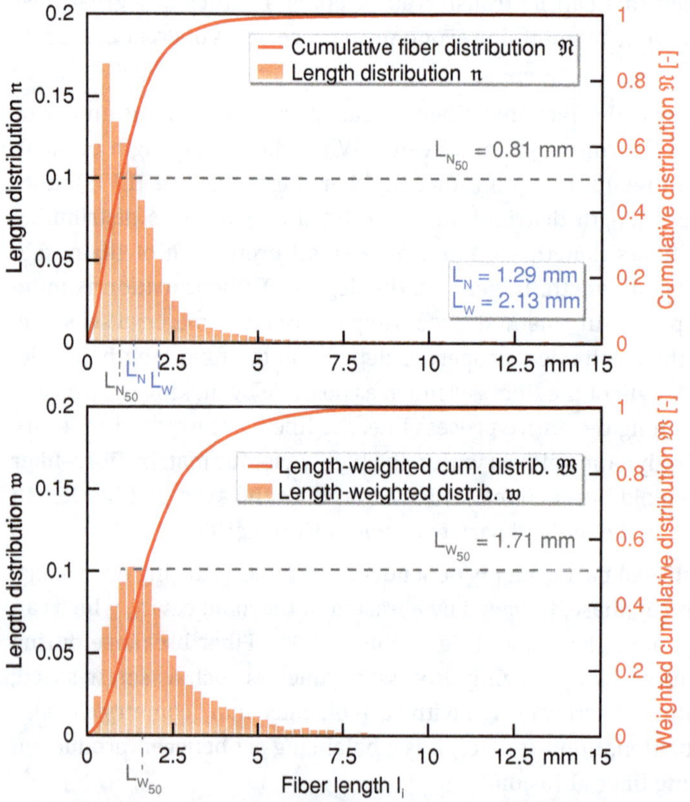

Figure 5.7 Example of a fiber length distribution function $\mathfrak{n}$ and cumulative fiber length distribution $\mathfrak{N}$ (top) and a length-weighted distribution $\mathfrak{w}$ and cumulative length-weighted distribution $\mathfrak{W}$ (bottom). Both distributions represent the same dataset

5.4.2 Fiber Orientation Distribution

Discontinuous fibers embedded in a matrix rotate and shift due to the deformation of the suspension during processing. Internal shear and tensile stresses cause a distribution of fiber orientation that leads to anisotropic properties of the molded part. The development of fiber orientation in a highly filled material is determined by the processing conditions, the design of the cavity, and the material properties.

In general, injection-molded parts have a distinct orientation pattern across the thickness, known as the skin–core–skin layers structure, as shown in Figure 5.8. The core layer in the middle of the part consists of fibers that are predominantly oriented in the transverse direction, while the fibers in the shell layers are oriented in the flow direction. The surface (skin) layer is very thin and usually has a random fiber orientation. This specific orientation pattern is caused by the fountain flow effect during thermoplastic processing (see Figure 5.9), which realigns the polymer molecules or fibers at the advancing flow front.

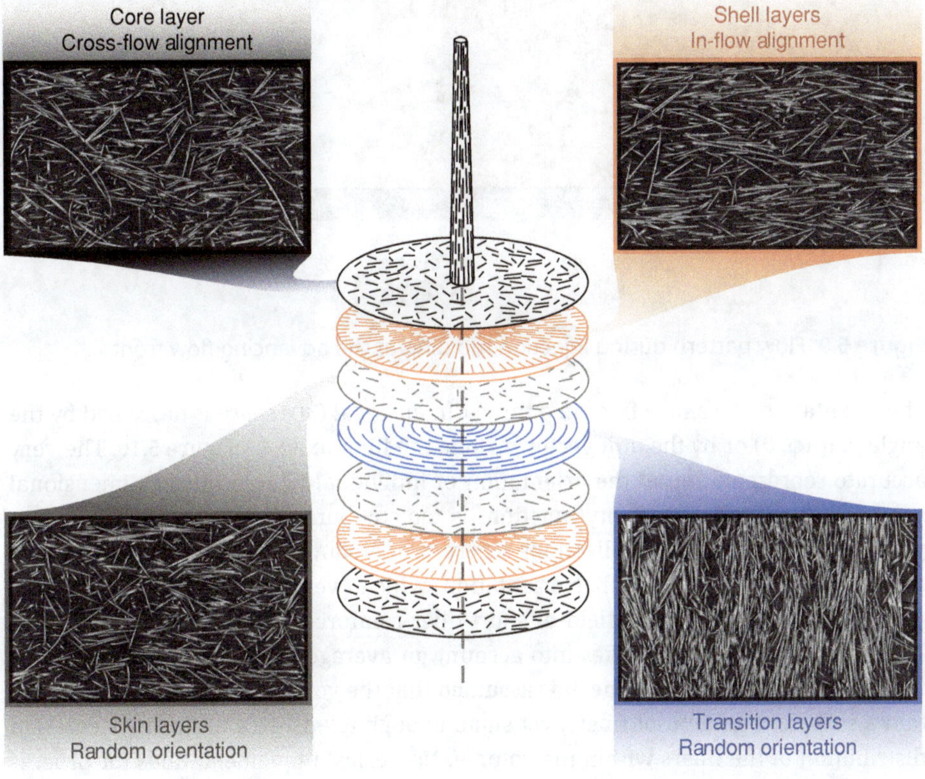

Figure 5.8 Illustration of the typical fiber orientation distribution across the thickness of an injection-molded part, including micro-computed tomography (μCT) images of a PP sample filled with 40 wt % glass fibers

The material at the flow front is forced to move from the center of the cavity outwards onto the mold surfaces. The material freezes on contact with the cold mold and forms a solidified edge layer. As additional material enters the cavity, the advancing flow front is continuously stretched and rolls onto the cold mold wall, where it solidifies immediately. The high shear region at the solidified layer and the strain in the advancing melt core cause strong fiber orientation in the shell layers; seven layers can be identified, and each layer has its own characteristic fiber orientation distribution, as shown schematically in Figure 5.8. The thick core layer consists of fibers that are predominantly oriented in the cross-flow direction. Two transition layers with a random fiber orientation surround the core layer. The fibers in the thicker shell layers are predominantly oriented in the direction of flow. The two thin edge layers on the mold surfaces have a random fiber orientation.

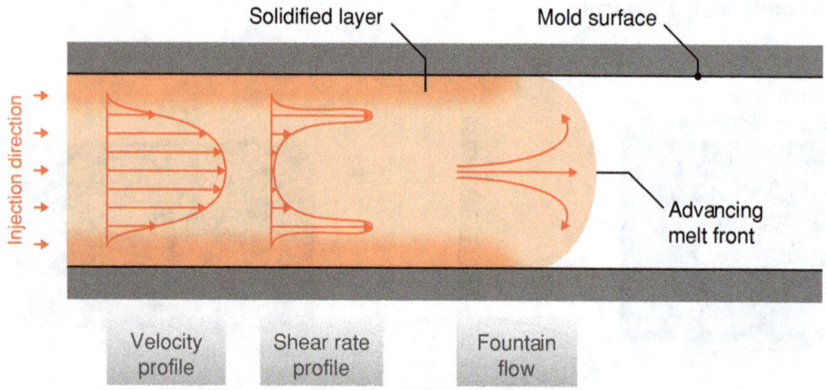

Figure 5.9 Flow pattern during injection molding at the advancing flow front

The orientation of a single fiber in a three-dimensional (3D) space is described by the angle pair (ϕ, ϑ) or by the unit vector $\boldsymbol{p}$ along the fiber axis; see Figure 5.10. The very accurate representation of the orientation of individual fibers in three-dimensional space using probability density functions is too computationally intensive to be implemented in simulation applications. Therefore, Advani and Tucker [1] proposed a tensorial representation of fiber orientation, which eventually became the conventional representation in the field of short-fiber-reinforced composites. The Advani–Tucker orientation tensor takes into account an average orientation property of all fibers within a discrete volume. It is assumed that the volume is large enough to contain a sufficient amount of fibers, yet small enough to ensure a uniform orientation distribution of the fibers within the volume. Numerical implementations for process simulation and structural analysis usually employ the tensorial representation.

In very simplified terms, the orientation can be described by the three tensor components A_{11}, A_{22}, and A_{33}, which describe the degree of orientation in relation to the defined coordinate system. The reference coordinates are usually defined in such a way that the *1* direction represents the flow direction; the *2* direction, the cross-flow direction; and the *3* direction, the thickness direction. A completely random orientation (Figure 5.11 (A)) is described by $A_{11} = A_{22} = A_{33} = 1/3$. A value of $A_{22} = 1$ means that all fibers are oriented in the *2* direction (Figure 5.11 (B)). A value of $A_{22} = 0$, on the other hand, means that no fibers are oriented in the *2* direction. A random planar orientation, as seen in sheet molding compound (SMC) blanks, is represented by $A_{11} = A_{22} = 1/2$ and $A_{33} = 0$ (Figure 5.11 (C)).

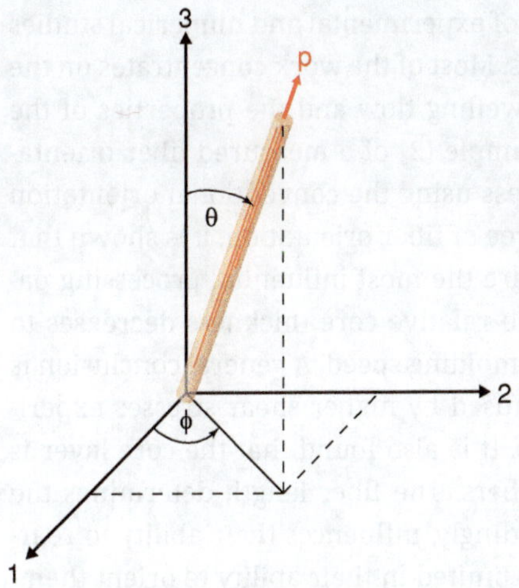

Figure 5.10 Schematic representation of the orientation of a single fiber in 3D space

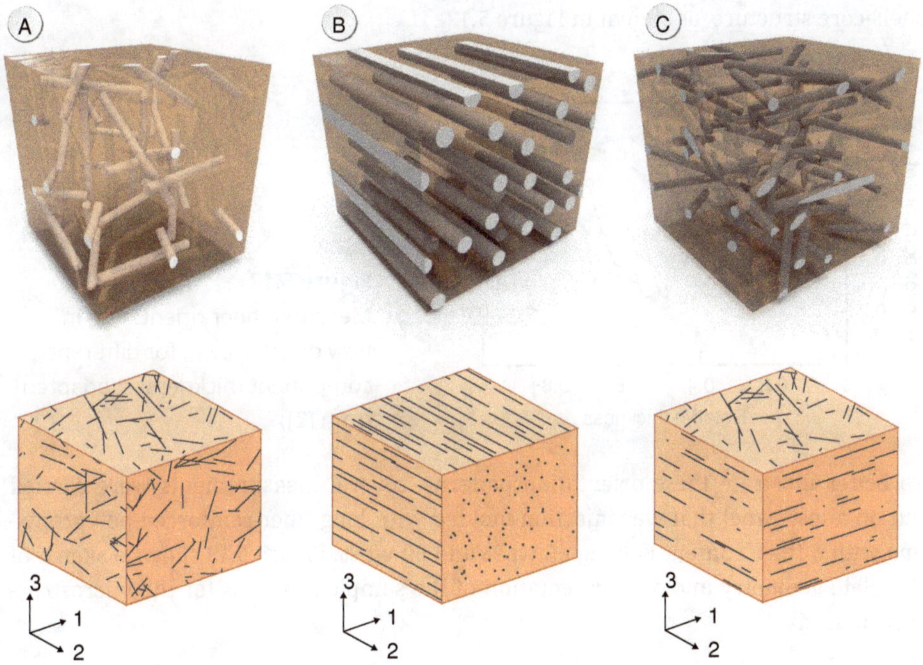

Figure 5.11 Orientation of a fiber population within a volume: randomly oriented fibers (A), fibers oriented in the *2* direction (B), and random fiber orientation in the plane (C). The images in the top row illustrate the 3D fiber configuration and the bottom row shows the 2D projection onto the corresponding plane

The process-related fiber orientation is a focus of experimental and numerical studies in the field of short-fiber-reinforced composites. Most of the work concentrates on the fiber orientation pattern resulting from the swelling flow and the properties of the shell–core structure. Figure 5.12 shows an example [2] of a measured fiber orientation distribution plotted along the part thickness using the conventional orientation tensor representation. With respect to the degree of fiber orientation, it is shown that the injection speed and the holding pressure are the most influential processing parameters. At low injection molding speeds, the relative core thickness decreases to half of the core thickness at standard injection molding speed. A general conclusion is that the decrease in core layer thickness is caused by higher shear stresses experienced during the mold filling of thinner parts. It is also found that the core layer is wider with longer fibers than with shorter fibers. The fiber length determines the mobility of the fibers in the matrix and, accordingly, influences their ability to rearrange at the flow front. Long fibers are clearly limited in their ability to orient themselves. Further scientific work indicates a strong correlation between the shell–core structure and the thickness of the part. The thicker the part, the more pronounced the shell–core structure, as shown in Figure 5.12.

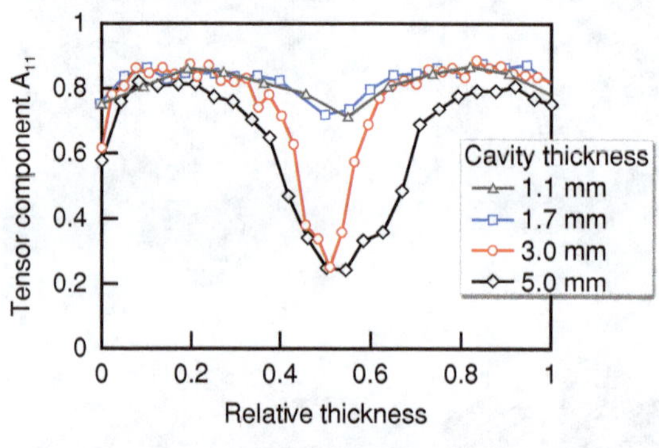

Figure 5.12
Measured fiber orientation in flow direction, A_{11}, for different component thicknesses (adapted from [2])

To better illustrate these orientation patterns, several measurements were carried out on a test panel that was injection-molded with long-fiber-reinforced polypropylene with a fiber content between 5 wt % and 60 wt %. Figure 5.13 shows a sketch of the plate geometry and a representation of the sample positions for the microstructure analysis.

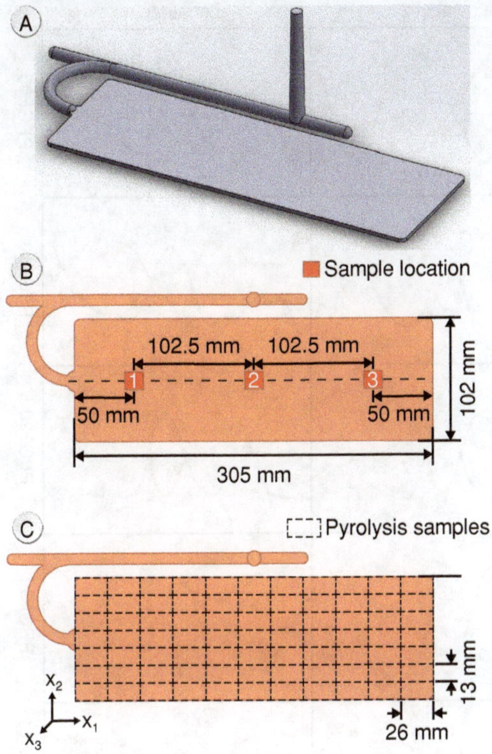

Figure 5.13
Sketch of a plate geometry and illustration of the sample positions for a microstructure analysis

Figure 5.14 summarizes the measured fiber orientation at measuring points 1, 2, and 3 for all fiber fractions (PPGF05 to PPGF60). The diagrams show the fiber orientation along the flow direction, A_{11}, and in the cross-flow direction, A_{22}. The fiber orientation in the thickness direction, A_{33}, is not shown for reasons of clarity ($A_{11} + A_{22} + A_{33} = 1$). Three samples were analyzed for each location and the low standard deviation of the measurement results indicates a high reproducibility of the molded samples.

The tests with higher concentrations (PPGF20 to PPGF60) show the expected core–shell–skin pattern. The thick core layer consists of fibers that are predominantly oriented along the cross-flow direction (A_{22}), while the fibers in the shell layer are oriented along the flow direction (A_{11}) due to the swelling flow. The orientation in the thickness direction (A_{33}) is low, with average values of less than 0.06 and evenly distributed across the thickness in all tests. The measurements of the samples with 5 wt % and 10 wt % nominal fiber concentration (PPGF05 and PPGF10) show a changing fiber orientation across the thickness of the sample, which is particularly pronounced at position 2 (see Figure 5.14).

In all tests, the samples near the gate (position 1) show a wider core layer than the one shown by the samples at positions 2 and 3, while the degree of orientation in the shell or skin layer is lower. This can be explained by the radial flow field near the gate, which is created by the edge cut.

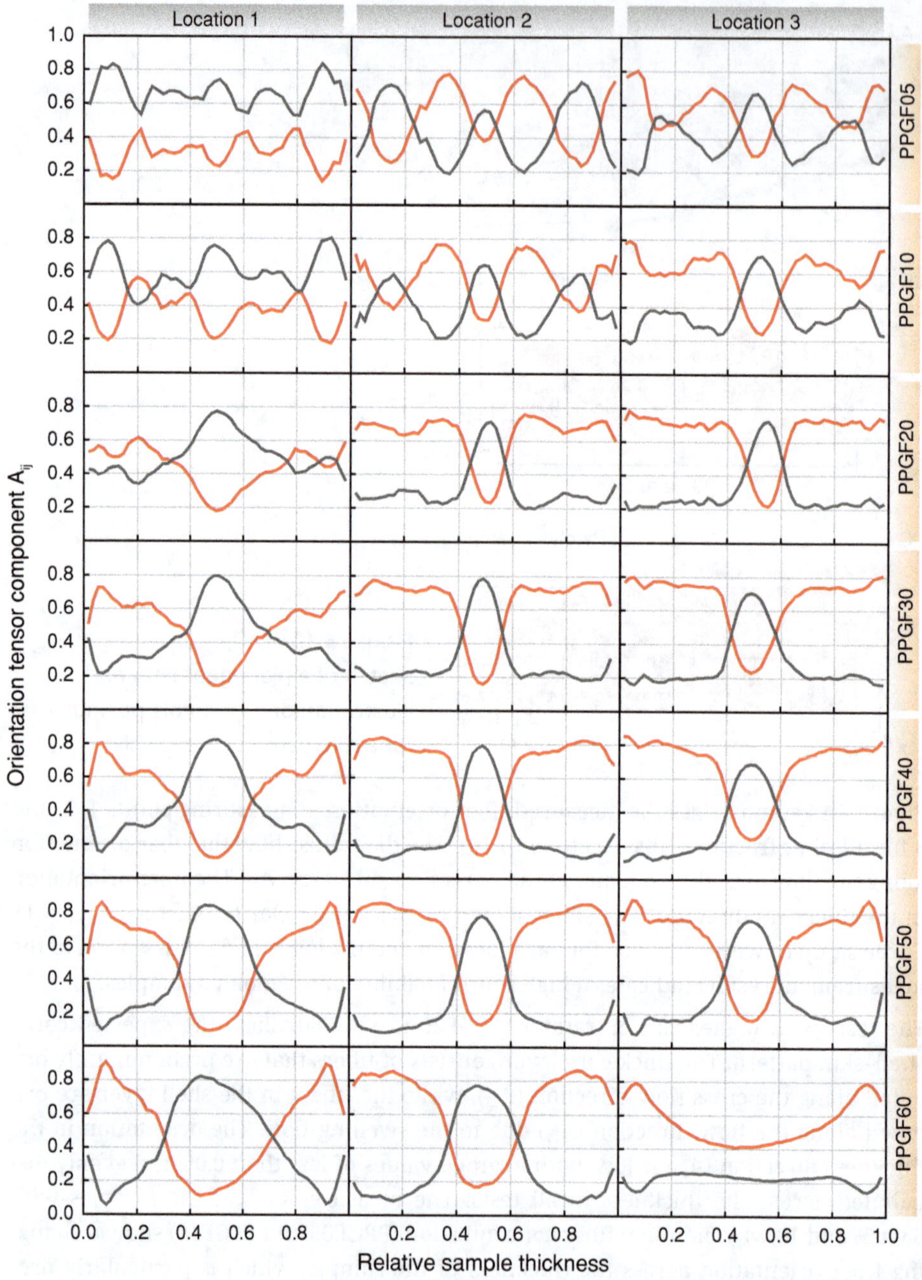

Figure 5.14 Measurements of fiber orientation at measuring points 1, 2, and 3 for all tests: diagrams with A_{11} (red) and A_{22} (gray) as a function of thickness (A_{33} is not shown for reasons of clarity)

The measurements for PPGF60 at location 3 indicate a large core layer and an overall random fiber orientation, including an overall larger standard deviation in the replicate measurements. This suggests that the high fiber content of 60 wt % may have caused a nonuniform flow front at the end of the flow process due to the high viscosity of the suspension.

The measurements show that the nominal fiber concentration influences the degree of orientation in the core and shell layers as well as the width of the core layer. The distribution of fiber density is discussed in the next section.

5.4.3 Fiber Density Distribution

The reinforcing effect of fibers is directly related to the fiber concentration or fiber density in a material. The complex flow properties that occur in injection molding during mold filling can lead to spatial differences in the fiber concentration in the finished part. The phenomenon of fiber–matrix separation describes the process-related variation in fiber concentration. Early work has shown that fiber migration occurs on two levels during processing: variations in *local fiber concentration along the thickness of a molded part* and the *global fiber concentration gradients* along the flow path.

These studies dealt with the process-related concentration gradients in molded parts made of filled thermoplastics. When investigating fiber–matrix separation, it becomes clear that the separation effects are more pronounced with spheres than with fibers, which indicates an influence of the shape and dimensions of the fillers. In addition, the mold geometry and the filler concentration are important parameters that determine the degree of filler migration during injection molding. Other processing parameters such as mold temperature, injection speed, matrix material, and screw speed do not appear to have a significant influence on filler–matrix separation in their tests. For example, the work of Ogadhoh et al. [3], in Figure 5.15, shows the filler concentration along the flow length for glass bead diameters from 11 µm to 203 µm at a nominal concentration of 40 wt %. It is clear that the larger particles are more susceptible to separation behavior.

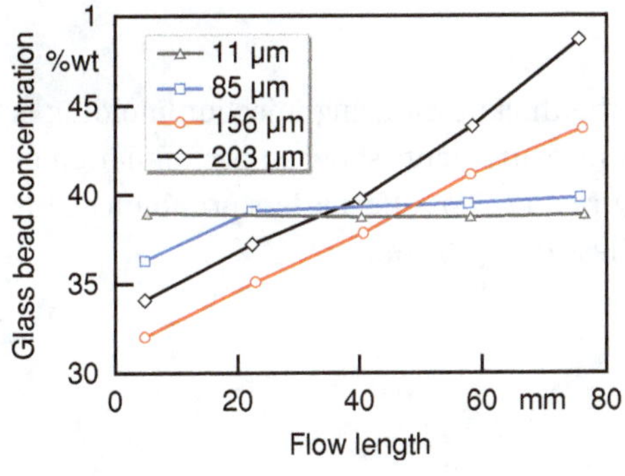

Figure 5.15
Glass bead demixing along the flow path for different glass-bead sizes (nominal filler concentration: 40 wt %) (figure adapted from [3])

From many studies and observations, it can be concluded that the fiber concentration increases towards the end of the flow path, indicating that a greater amount of filler is present at the last filled point. Furthermore, the processing conditions appear to play a much smaller role than the filler properties in assessing the degree of fiber–matrix separation. The concentration and the shape of the filler are mentioned as influencing factors, as the migration effect is generally more pronounced for longer fibers with higher concentrations than for shorter fibers with lower concentrations. The global variation of the fiber concentration in the manufactured panels was determined by ashing; see Figure 5.13. The results for all tests are displayed in Figure 5.16. The measurements show areas of low and high fiber concentration for each test. PPGF05 has a maximum with up to 20% higher fiber concentration in the middle of the plate. PPGF10 to PPGF60 show a tendency towards a low fiber concentration near the gate and an increased fiber concentration at the end of the flow path.

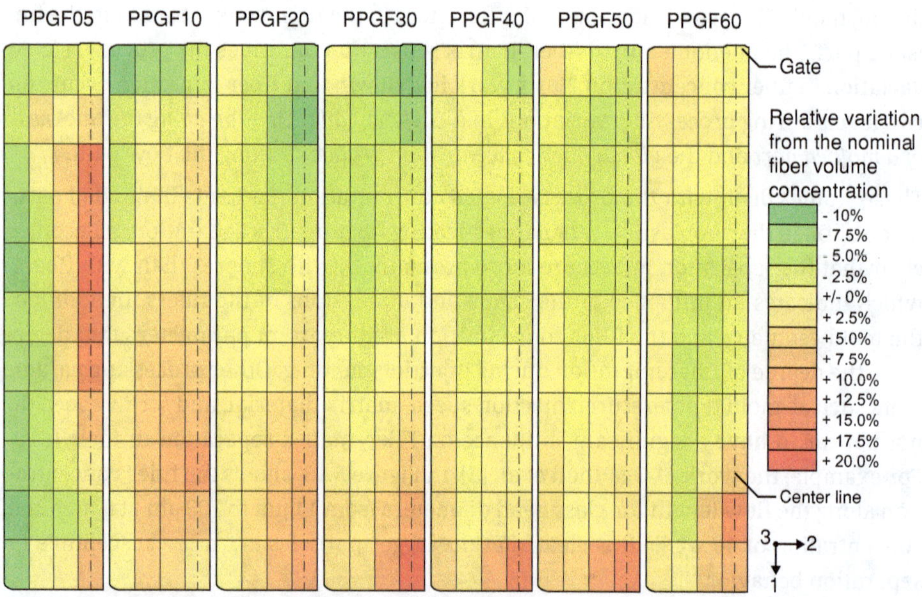

Figure 5.16 Results of the analysis of the global concentration gradient in the entire molded plate for all experiments

The gradient of filler concentration across the thickness during injection molding is less frequently investigated. However, most measurements show an accumulation of fibers in the core layer, as can be seen in Figure 5.17. This effect is less pronounced for short fibers with an average fiber length of less than 0.6 mm.

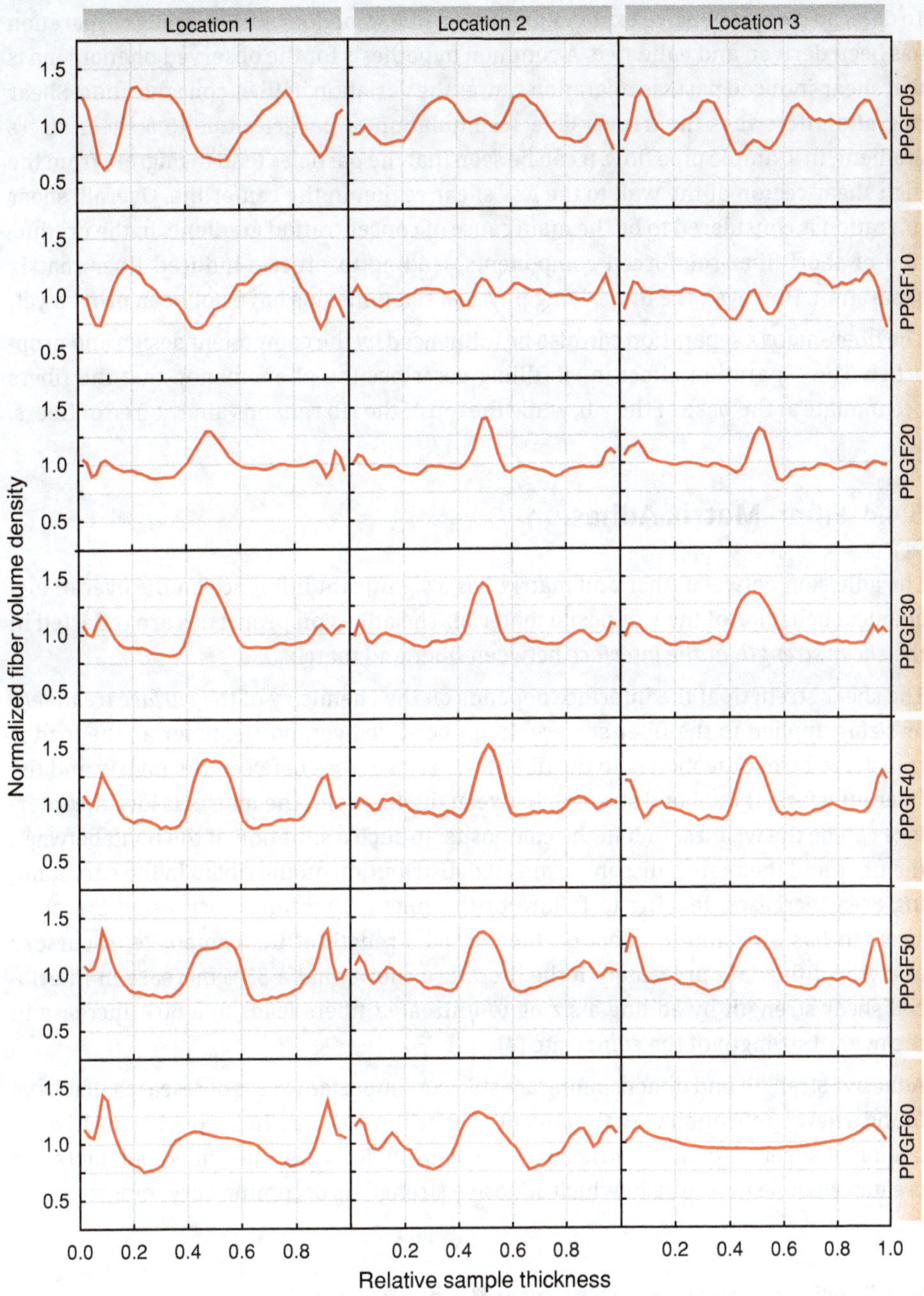

Figure 5.17 Measured fiber concentrations across the thickness in positions 1, 2, and 3 of Figure 5.13 for fiber weight fractions between 5% and 60%

To date, no comprehensive explanation for the phenomenon of fiber–matrix separation has been derived and validated. A common hypothesis for the observed phenomena is that shear-induced particle migration causes the variation in fiber concentration. Shear migration describes the irreversible accumulation of particles due to a shear stress gradient. In a simple pipe flow, it can be seen that the particles tend to migrate from the high shear region at the wall to the low shear region in the centerline. Overall, shear migration is considered to be the main cause of concentration gradients in the production of short-fiber-reinforced components, leading to process-induced fiber–matrix separation. However, the underlying physical relationships have not been proven yet.

The fiber–matrix separation can also be influenced by the component design and properties. The separation effect in rib filling describes the phenomenon that the fibers accumulate at the base of the rib, while the top of the rib remains almost free of fibers.

5.4.4 Fiber–Matrix Adhesion

The adhesion between fiber and matrix has a significant influence on the overall mechanical behavior of the composite material. The adhesion properties are reflected in the *shear strength at the interface* between fiber and matrix.

The shear strength at the interface depends on the chemistry of the *surface treatment* or *sizing* applied to the fiber surface. In a fiber-filled composite material, the load is transferred from the matrix to the fiber by shear stresses between the matrix and the fiber interface. If the bond strength between the fiber and the matrix is low, the interface can be the weakest link in the composite. In such a situation, if the bond between the fiber and the matrix dissolves, the load distribution on the fibers in the surrounding area increases, leading to failure of the fibers. Therefore, increasing the bond strength has a profound effect on the overall strength of the composite. A cursory analysis of the data presented in the literature shows that a 35% increase in interfacial shear strength by adding a sizing to untreated fibers leads to a 60% increase in the overall strength of the composite [4].

Adhesive strength and dimensioning are still very important areas of research that have found a new application with the advent of hybrid structures. In particular, the hybridization of sheet metal geometries that are overmolded with fiber-filled polymers has created a whole new area in which adhesive strength is becoming very important.

5.4.5 Microstructure–Property Relationship

The condition of the fibers – namely their orientation, length, and dispersion – in the finished component depends on the process parameters used. As a reinforcing phase in the composite material, the condition of the fibers determines the mechanical performance of the finished product; see Figure 5.18. The anisotropy introduced by the orientation of the fibers has a significant influence on the mechanical properties. The tensile strength is higher in the direction of the fiber orientation and lower in the transverse direction. Therefore, anisotropy is an important aspect in the design of fiber-reinforced components, as process-related fiber orientation is unavoidable during injection molding. Similarly, the concentration and the length distribution of the fibers also play an important role in the relationship between microstructure and properties. Longer fibers tend to increase the strength and stiffness of the finished component. But there are also interactions: longer fibers tend to interact with each other, resulting in resistance to alignment along the direction of flow. This, in turn, means less anisotropy and less warpage. In a similar way, a higher fiber concentration improves the mechanical properties; however, if the processing conditions are not optimal, fiber bundles may occur in the component. Agglomerates are the condition in which the fiber bundles are not completely dispersed during melting; they often lead to a deterioration in mechanical properties. The dispersion of fibers is still an area of research that is not yet fully understood.

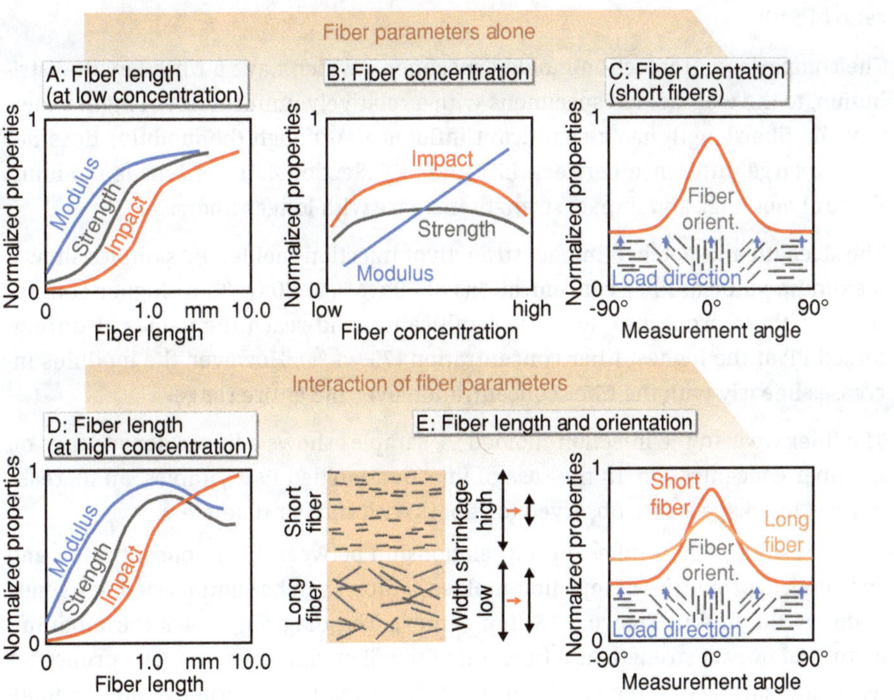

Figure 5.18 Summary of the process–microstructure–property relationship for discontinuous fiber-reinforced thermoplastics: effects of individual fiber parameters (top) and generally observed effects of fiber parameters interactions (bottom) [5–11]

Since the 1980s, extensive work has been carried out [5–11] to determine the structure–property relationship of discontinuous fiber-reinforced thermoplastics. Extensive mechanical testing has been performed, including tensile, flexural, and impact tests, and the local fiber orientation and fiber length have been characterized. The main conclusions from this work are:

- A linear increase in tensile strength with fiber concentration is observed for cast polypropylene (PP) samples. It is also shown that the tensile strength depends on the average fiber length up to 3 mm. However, the increase in strength flattens out for all material combinations with a fiber length of more than 3 mm, which together with a fiber diameter of 11 µm corresponds to an aspect ratio of 270.
- The tensile modulus of fiber-reinforced PP increases linearly with the fiber concentration up to 40 wt %, although the rate of increase is significantly lower at higher concentrations. Measurements indicate that the modulus of the composite from 0.5 mm fiber length (aspect ratio of 45) is independent of its length.
- The impact strength increases with increasing fiber concentration. The strength also improves with the fiber length but converges at lengths above 6 mm (aspect ratio of 540).
- The comparison of injection-molded specimens, which have a fiber length distribution, to the wet-molded specimens with a relatively uniform fiber length shows that the fiber length has a significant influence. Although the modulus does not show a large difference between LFT and SFT (Section 4.8), the tensile modulus, flexural modulus, and impact strength increase with longer fibers.
- The strength and notched impact strength of injection-molded PP samples show a maximum value at fiber concentrations of 40 wt % to 50 wt %. At higher concentrations, these properties decrease significantly and reach the values of unreinforced PP at the highest fiber concentration (73 wt %). However, the modulus increases linearly with the fiber concentration over the entire range.
- The fiber wear in the injection-molded PP samples shows a linear dependence on the fiber concentration. In the case of injection-molded PA6 samples, an increase in fiber breakage is also observed for fibers with smaller diameters.

The main results of the studies on the relationship between fiber concentration and the mechanical properties of injection-molded samples can be summarized in a single diagram, as the one in Figure 5.18 (top center). The diagram shows the modulus, strength, and impact strength as a function of the fiber concentration. The properties are each standardized to the values of the unfilled matrix material. While the modulus increases linearly with the fiber concentration, the strength and the impact strength show a maximum at 40 wt % to 50 wt %. A further increase in fiber concentration leads to a decrease in strength and notched impact strength. At a high filling value of over 60 wt %, these properties even drop to the same level as with unfilled PP.

The effect of the residual length of the fibers on the mechanical properties of a molded part is shown in Figure 5.19. The diagram illustrates the effects of the fiber cross section ratio on the tensile modulus, tensile strength, and notched impact strength. It can be seen that these properties reach saturation and that increasing the fiber length would not further improve the mechanical properties. The image also qualitatively illustrates the expected range of residual fiber aspect ratios for LFT injection molding. This expected range indicates a strong sensitivity to the fiber length of the tensile strength and the impact strength of injection-molded LFT parts. While the diagram in Figure 5.19 provides a qualitative overview of the relationship between fiber aspect ratio and mechanical properties, the quantitative interpretation of this diagram should be made with caution, as universal validity has not been proven yet. For example, the underlying experimental data does not account for variation in fiber orientation or local fiber concentration due to the change in fiber length.

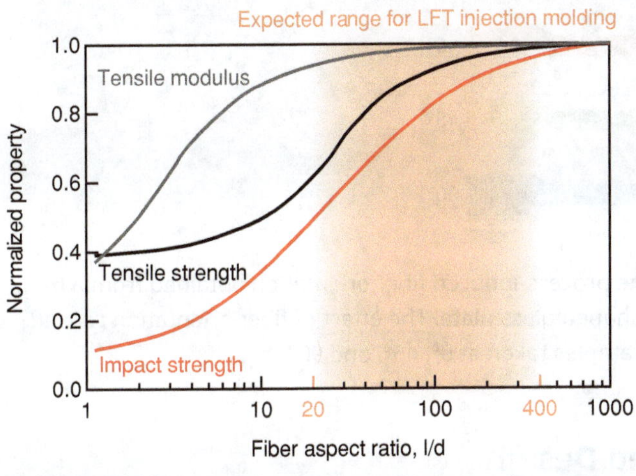

Figure 5.19 Standardized mechanical properties as a function of the fiber aspect ratio for tensile modulus, tensile strength, and notched impact strength [11]. The highlighted area illustrates the qualitative range of fiber aspect ratios expected in LFT injection molding

In the following, mold filling simulation tools widely used in the industry to predict the evolution of the microstructure during processing are presented. To illustrate this, the predicted anisotropy and its effects on the mechanical performance of injection-molded parts are summarized in Figure 5.20. The results of the process simulation show the effect of fiber orientation on the stress–strain behavior for a simple film sprue sheet geometry. In this plate design, the majority of the fibers are aligned in the flow direction. Consequently, the 0° specimen has a significantly higher stiffness than the 45° and the 90° specimens.

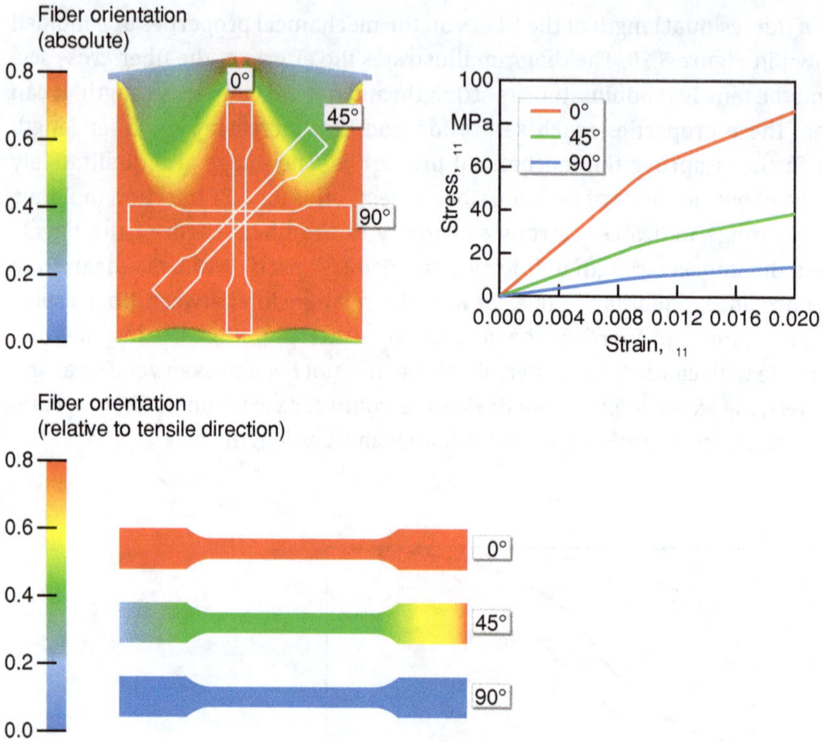

Figure 5.20 Illustration of the process-induced fiber orientation obtained from a basic process simulation of a fan-shaped glass plate. The effect of fiber orientation caused by mold filling is shown for samples taken at 0°, 45°, and 90°

5.5 Computer-Aided Design

The simulation of polymer processes and the behavior of plastic parts has made considerable progress since the mid 1980s. No other area of polymer processing is more affected by this than shaping processes such as injection molding and compression molding. Simulation is also widely used today in the design of extrusion tools and mixing heads and in processes such as thermoforming and blow molding. For basic considerations, universal computer programs are used for modeling and simulation, usually based on the finite element (FE) or the finite difference (FD) method.

The computer simulation of polymer processes offers the tremendous advantage of allowing designers and engineers to consider virtually any geometric and process option without the cost associated with prototyping molds or tooling, or the waste of material from time-consuming trial-and-error processes. The ability to try out new designs or concepts on the computer gives the engineer the opportunity to identify and fix problems before going into production. In addition, the process engineer can determine the influence of processing parameters on the quality and properties of the final part. For example, computer-aided engineering (CAE) gives the designer the flexibility to determine the effect of different gating scenarios, sprue designs, or cooling line locations when designing an injection mold.

However, process simulation is not a panacea. As with any modeling technique, there are limitations caused by assumptions in the constitutive material models or by geometric simplifications of the model cavity. For example, there is a tendency in the industry to continuously reduce the part thickness of injection-molded parts. Thickness reductions increase the pressure requirements during mold filling, with typical pressures of up to 2000 bar being reached. Such pressures have a profound effect on the viscosity and thermal properties of the melt – effects that are largely not taken into account in commercially available software.

5.5.1 Structural Simulation Using the Finite Element Method

The finite element method (FEM) provides the designer with a powerful tool for reliably predicting the mechanical properties of a component, even with very complex geometries and stress cases. In principle, the application of FEM for plastics is not different from that for other materials. However, the special features of the material behavior and the often-complex component geometry lead to typical processing approaches. The following generally applies to FEM:

- Complex geometries are represented by finite elements.
- The load can be applied without geometric restrictions (point, line, volume, surface, etc.).
- The loading can be changed in time.
- Contact problems, such as touching down on another body, can be depicted.
- Nonlinearities can be taken into account (see Figure 5.21).

The choice of elements for mapping the geometry is also decisive for the quality of the calculation. In Figure 5.22 various elements are shown. It is clear that the calculation results cannot conduce to making any statements about behaviors that are not represented in the model. For example, membrane elements are not suitable for calculating a bending moment in a shell-shaped geometry.

When creating the model, it is also essential to achieve a suitable mesh geometry and fineness. Highly distorted or too coarse elements lead to incorrect calculation results. Figure 5.23 uses the example of a snap connection to show how local refinement of the mesh can achieve a compromise between the necessary accuracy and the required calculation time.

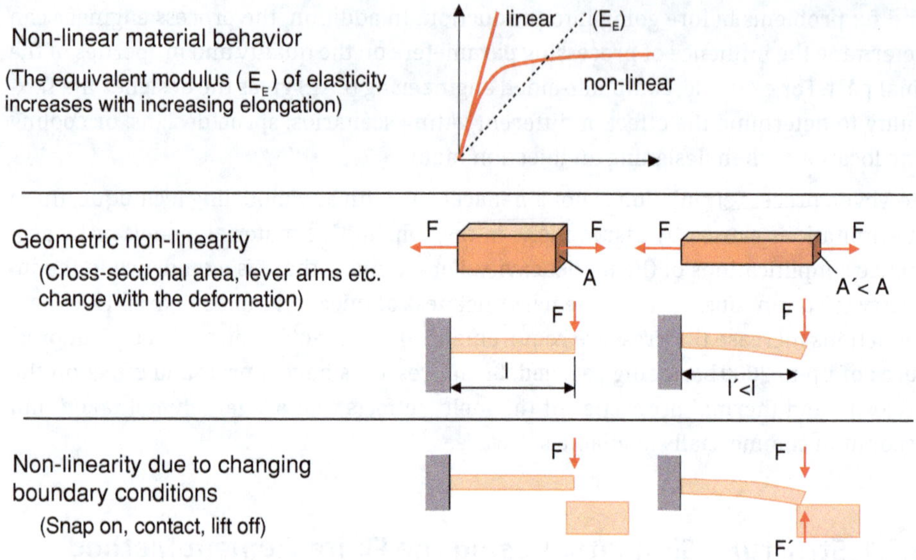

Figure 5.21 Nonlinearities in the application of the finite element method

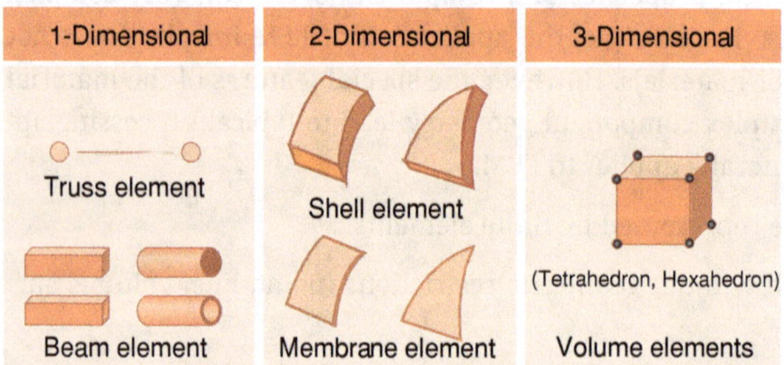

Figure 5.22 Various finite elements

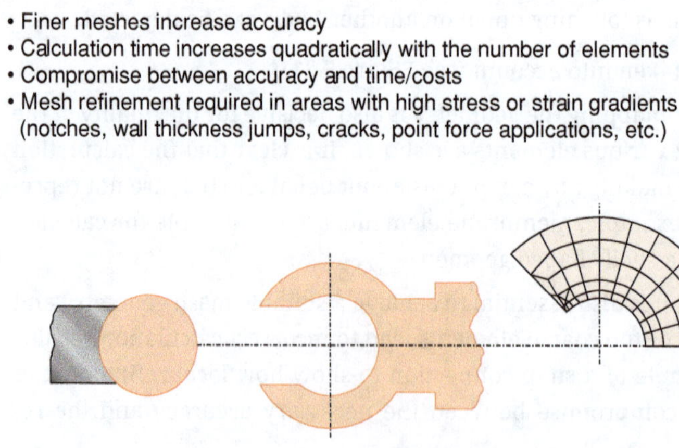

Figure 5.23 Fineness of the structural elements

The application of a finite element analysis results in the following typical workflow:

- Geometry creation,
- Preprocessing (model preparation: boundary conditions, loads, choice of material model),
- FE calculation (solution of the system of equations),
- Postprocessing (presentation of results),
- Interpretation using dimensioning parameters.

The designer pursues various objectives with the finite element analysis:

- Shape optimization,
- Proof of manufacturability and influence of production on component properties,
- Service life FEM.

FEM is therefore used to first develop the mechanical concept of a component. At this stage, the geometry of the component is changed several times in order to arrive at an optimum design. Here it is sufficient to calculate with simplified geometries (or partial geometries, in the case of symmetry) and material laws to reduce the effort involved.

Once the optimum component geometry has been found, the feasibility must be checked using process simulation. This can also lead to a change in the component geometry, which results in another change in the geometry of the FE mesh.

The dimensioning calculations are completed by a service life FEM. The aim is to predict the behavior of the component for specified parameters relating to geometry, material, manufacturing conditions, and load case. Here, it is usually necessary to consider the influences of production on the local material properties in the component by coupling process simulation and component simulation with FEM (integrative simulation) and thus arrive at a precise description of the component behavior. The influences of the stress and the manufacturing conditions on the material behavior are considered by attributing locally adapted material properties to each element.

Only once these calculations have been completed is the CAD model of the component – usually, a 3D solid model – finalized.

5.5.2 Process Simulation

The first step of CAE in process design and optimization is to create a solid model such as the one shown in Figure 5.24 and break it down into a finite element mesh that can be used by the simulation software package. The image shows the CAE model of the mold cavity and the positions of the mold cooling lines.

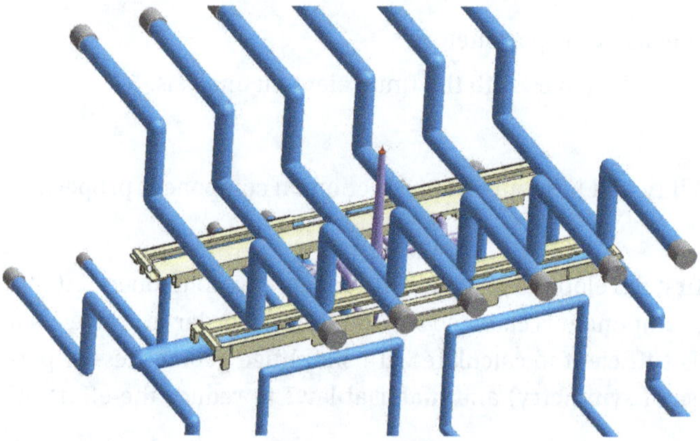

Figure 5.24 Mold cavity components (cream), sprue (purple), and positions of the mold cooling lines (blue) for the production of two contact image sensors of a scanner

As a first approximation, a three-dimensional geometric model is converted into a midplane model, which is essentially a two-dimensional geometry oriented in three-dimensional space. A finite element model is then generated on the midplane surface. Basically, the most common injection molding models use this approach to represent the geometry of the part. While most injection-molded parts are thin and planar and can be represented well with such a model, some injection-molded parts have a smaller aspect ratio or have three-dimensional features that make these models invalid.

In the case of the contact image sensor of the scanner, a three-dimensional model was used to simulate the mold filling process. Figure 5.25 shows the meshing strategy employed to represent the part. Prisms and hexahedral elements were used at the interface between the sprue, gate, and the components. The corners and the surface of the part are usually represented with prismatic elements, and tetrahedral elements are used for the inner elements.

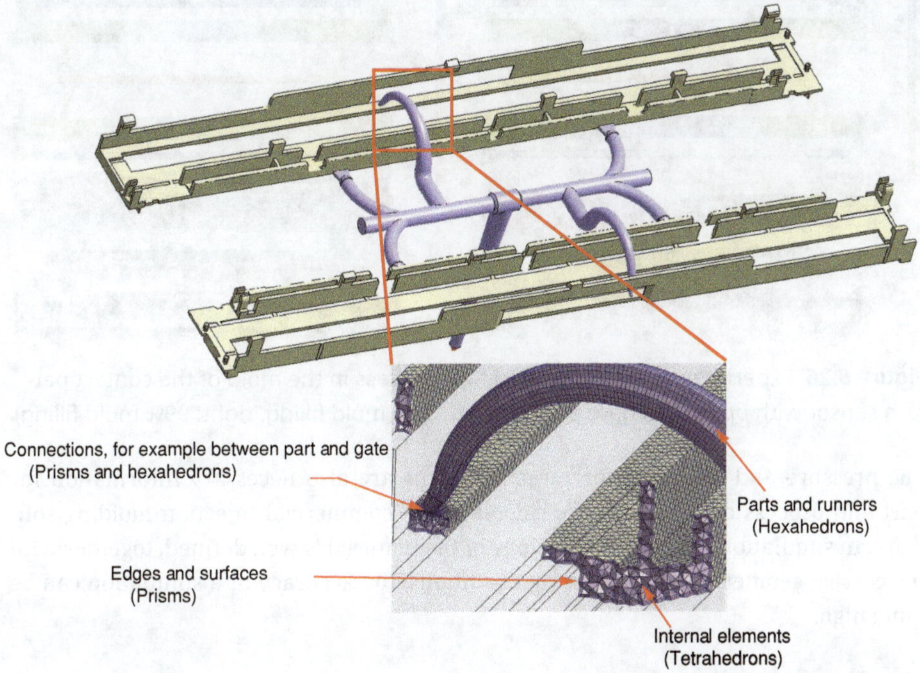

Figure 5.25 Mold cavity and finite element meshing strategy for the contact image sensor of the scanner

5.5.2.1 Mold Filling Simulation

Using a finite element mesh – as shown in Figure 5.25 – a simulation package, in conjunction with the control volume approach, solves a coupled energy and momentum balance and provides a mold filling pattern that not only takes into account the non-Newtonian effects present in the flow of polymer melts but also the effect of cooling on the melt flow within the mold cavity.

The mold filling analysis and the resulting filling pattern can be used to predict the formation of weld lines and gas inclusions. These can cause weak spots and surface finish issues that can lead to cracking and failure of the final part as well as aesthetic issues in the finished product. To minimize flow imbalances during mold filling and reduce warpage, the position of the gates was changed (optimized) from the original design in Figure 5.25. The best results were achieved by moving the gates 40 mm outwards, as shown in Figure 5.26. The figure displays the predicted 90% (left) and 99% (right) filled mold cavities for the optimized gate position.

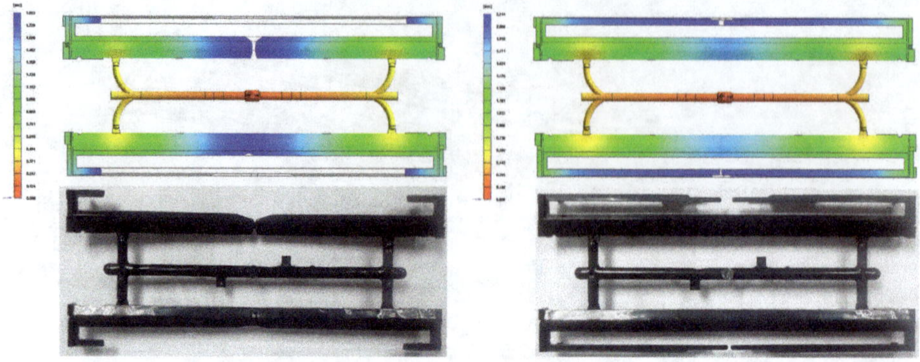

Figure 5.26 Experimental and simulated flow process in the mold of the contact pattern sensor with optimized gate position (left: 90% mold filling, right: 99% mold filling)

The pressure and clamping force requirements are also necessary information for part and process design. Both are calculated by commercial injection molding software. In simulations where the rheology of the material is well defined, together with the correct geometry and processing conditions, the accuracy of a simulation can be quite high.

5.5.2.2 Prediction of Shrinkage and Warpage

Shrinkage and warpage are directly related to residual stresses. These result from locally varying strain fields that occur during the hardening or solidification phase of a manufacturing process. Such strain gradients are caused by uneven thermomechanical properties and temperature fluctuations within the mold cavity. Shrinkage due to curing can also play a dominant role in residual stress development in thermoset polymers and becomes important in fiber-reinforced thermosets; shrinkage is also a problem when sink marks occur in thick part areas or ribs. When processing thermoplastic materials, shrinkage and warpage in the final product depend on the molecular orientation and residual stresses that form during processing. The molecular or fiber orientation and the residual stresses inside the part depend, in turn, on the flow and heat transfer during the mold filling, holding pressure, and cooling phases of the injection molding process. In order to predict the residual stresses in the finished part, modern software packages characterize the thermomechanical behavior of the polymer from the melt to room temperature using the *pvT* behavior of the material in conjunction with the temperature-dependent stress–strain behavior. Figure 5.27 shows the distorted part from Figure 5.25 with the original sprue position (left) and with the geometry shown in Figure 5.26 (right). Minimizing warpage is one of the biggest problems for the designer. This is sometimes achieved by changing the formulation of the resin. A further reduction in warpage can also be achieved by changing the number and position of the gates. Although trial-and-error solutions, which are still the most practical with today's technology, are often performed, computer optimization often leads to a reduction in costs.

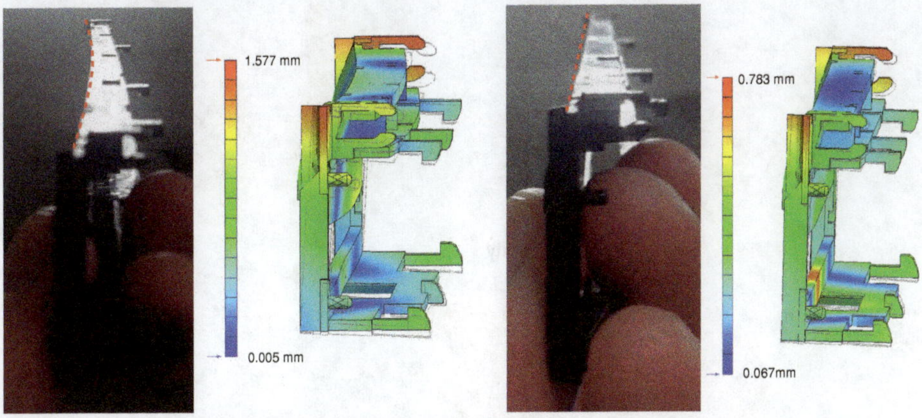

Figure 5.27 Experimental and predicted distorted geometry after demolding and cooling of the original model with the shape shown in Figure 5.25 (left) and the optimized sprue system shown in Figure 5.26 (right)

5.5.2.3 Orientation Predictions

Molecular and filler orientation have a profound influence on the properties of the finished part. The molecular orientation influences not only the mechanical properties of the polymer but also its optical quality. For example, birefringence is controlled by molecular orientation, which must be kept low for products that require certain optical properties, such as lenses. Figure 5.28 shows a comparison between predicted and experimental birefringence patterns. In the case presented in Figure 5.28, the birefringence pattern is directly related to the molecular orientation. Fiber orientation will be discussed in the next section, which deals with the fiber microstructure of discontinuous fiber-reinforced parts.

Like molecular orientation, fiber orientation is also determined by the flow field and the solidification process during mold filling. Much work has been done to develop models that predict fiber orientation in a discontinuous fiber-filled system with short and long fibers [1]. Various models for predicting fiber orientation distributions across a part have been successfully implemented in commercial software. Figure 5.29 shows a disk-shaped geometry with a central gate, a common geometry used in industry and research. Figure 5.30 compares the predicted and experimental fiber orientation tensors across the thickness of the part at three different locations along the radius of the disk. The predicted fiber orientation is the so-called iARD-RPR model (improved ARD and Retarding Principal Rate), which is implemented in the commercial software Moldex3D.

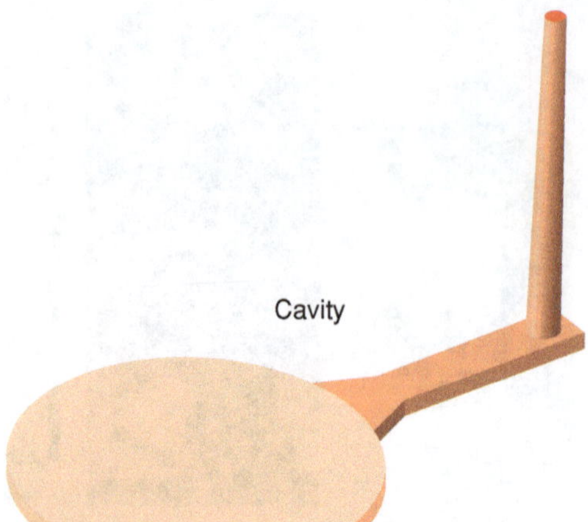

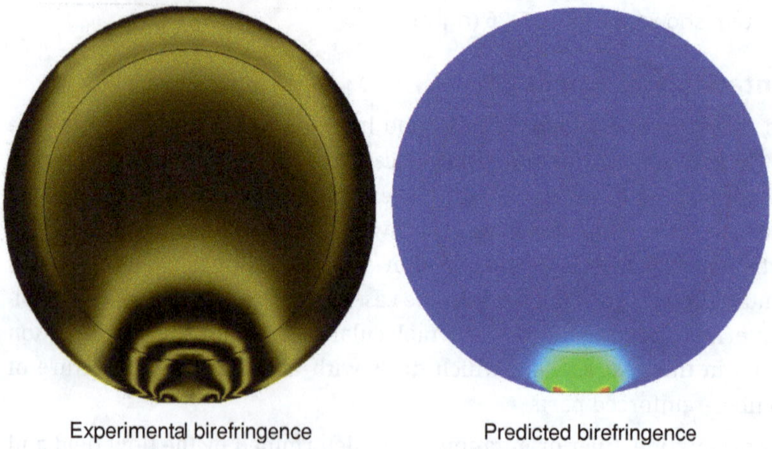

Figure 5.28 Experimental and simulated birefringence patterns in an injection-molded polycarbonate lens

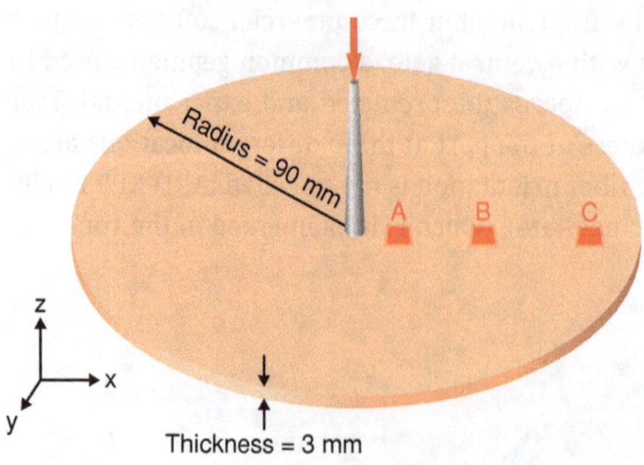

Figure 5.29 Geometry of the disk with central sprue, with the three areas A, B, and C for measuring the fiber orientation

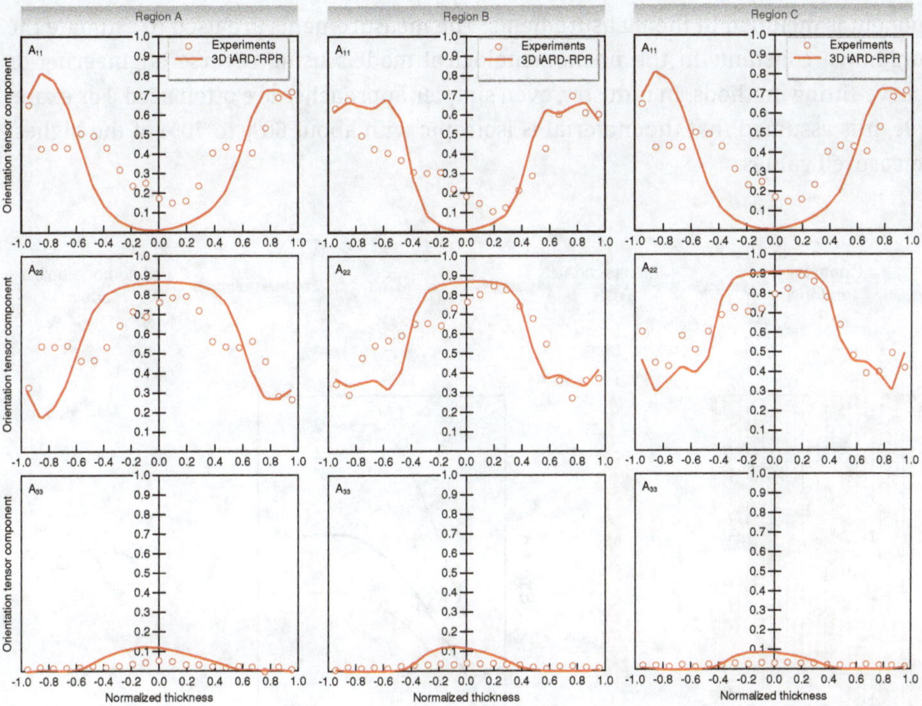

Figure 5.30 3D orientation predictions using the iARD-RPR model [12] compared with experimental data [13] for the three diagonal components of the orientation tensor, A_{11}, A_{22}, and A_{33}, through the normalized thickness in regions A, B, and C of the disk with central sprue

5.5.3 Modeling of Discontinuous Fiber-Reinforced Materials for FEM Implementation

The material models to be implemented in a FEM model can be roughly divided into two categories: (1) macro-level or phenomenological models and (2) micro-level or multiscale models. As the names suggest, the main difference between the two methods is the way in which the effects of each component are considered. The choice of method in a study is usually based on the availability of data, the need for accuracy, and the resources available. The details of each approach are discussed below.

In the past, macroscopic or phenomenological modeling was one of the most commonly used methods in the industry, as it was applied to fiber-reinforced plastics as well as unfilled ones. Essentially, the material is treated as homogeneous and orthotropic. Material properties such as stress and strain are measured using test specimens in different directions, such as yielding, transverse flow, and some additional directions in between (see Figure 5.31). It is assumed that the effect of the micro-component is included in the measurements. The measurements are used to estimate the unknown constants in the phenomenological models using reverse engineering or curve-fitting methods. In industry, even simpler approaches are often used. For example, it is assumed that the material is isotropic with about 60% to 70% of the highest measured values.

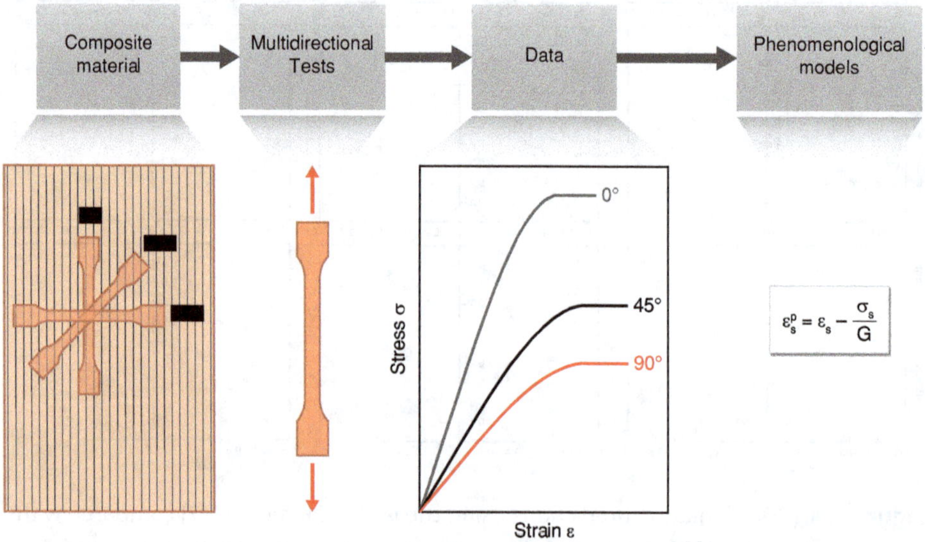

Figure 5.31 Macroscale approach for modeling

Modeling at the macro level represents a considerable simplification of the material properties of filled systems. The individual materials are only considered on a global level: their geometry and properties on a micro level are ignored. The fiber-reinforced plastic is considered as a single material with homogeneous properties, and the variation of these properties from place to place is also ignored. The accuracy of the finite element model is quite limited when using such coarse material properties. The main reason for its popularity lies in the simplicity of the model, the easy checking methods, and the fact that almost no specialized knowledge is required. Engineers familiar with finite element analysis of conventional isotropic materials can easily adopt and use this type of material model. The measured data can be employed to estimate the parameters for the constitutive material laws used in the finite element model.

Depending on the part under investigation and the manufacturing processes used, such an approach can work well in many cases for the calculation of stiffness and force deformation estimates. Furthermore, for large, complex assemblies, the macroscale is the only possible option due to limited computing power. In industry, it is quite common to use such material models in which the constitutive material behavior and the required parameters are based on years of measurements and know-how.

In a microscale model, one would ideally like to consider each component of the composite, namely the fibers, the matrix, the interface, and their geometrical arrangements. A finite element model of the microstructure that includes all fibers with their exact geometry and the matrix with voids can be considered. However, even in a cubic millimeter there are thousands of fibers. Modeling each individual fiber can lead to a very large number of elements in the model, which is a challenge for the calculations. In addition, constants such as interfacial properties are not precisely known. Such models, which we refer to as microscopic models in the true sense, are considered impractical for any useful engineering application. Over time, many concepts have evolved to overcome these challenges and to simplify the development of computational models. These are referred to as multiscale modeling. Multiscale modeling essentially considers the fiber details at the micro level (micrometers) and translates them into material models at the macroscopic level (millimeters). There are many different methods for converting the microscopic details into macro-level material data that can be used in a finite element model.

The basic principle of multiscale modeling is shown in Figure 5.32. There, the concept of the representative volume element (RVE) is used. The RVE is the smallest volume that can accurately capture the structural behavior of the effective properties of the composite material. In nature, most structures are heterogeneous, and when dealing with such heterogeneous materials, it is common to average the effects of the microstructure over a selected volume and refer to this as the effective properties. The size of such an RVE is debatable. If the microstructure is periodic, such as in woven or layered continuous fibers, the choice of the size of the RVE is straightforward. For discontinuous fibers, however, this can be a challenge. On one hand, we want the volume to be as small as possible to capture the effects of material variations throughout the part; on the other hand, it should be large enough to average the effective composite properties as a continuum material. In general, it is expected to be an order of magnitude larger than the microstructure details to be averaged [4].

The fiber-reinforced composite component can be divided into a series of RVEs. The constitutive material model for each RVE can be calculated individually on the basis of the fiber and matrix properties and the geometric details of the fibers, such as fiber length, orientation, and concentration. There are many approaches to calculating the material properties in an RVE. The simplest calculation is when they are assumed to be homogeneous and the averaged homogeneous material properties for the RVE are used in the finite element calculations. It is important to emphasize that the material properties for an RVE, if assumed to be homogeneous, can be – and probably are – anisotropic for discontinuous fiber-reinforced composites. Compared to the macroscopic approach described above, where the entire component was assumed to be homogeneous, the RVE approach subdivides the component into many smaller volumes (RVEs) on a macroscale. While each of these volumes is assumed to be homogeneous, the properties are expected to vary from one volume (RVE) to another. The gross properties at the level of the whole component may differ significantly from the properties of the RVE (macroscale). In addition, an RVE approach better represents the actual conditions in the material. Therefore, this can be a significant improvement in representing the overall material properties in the FE model.

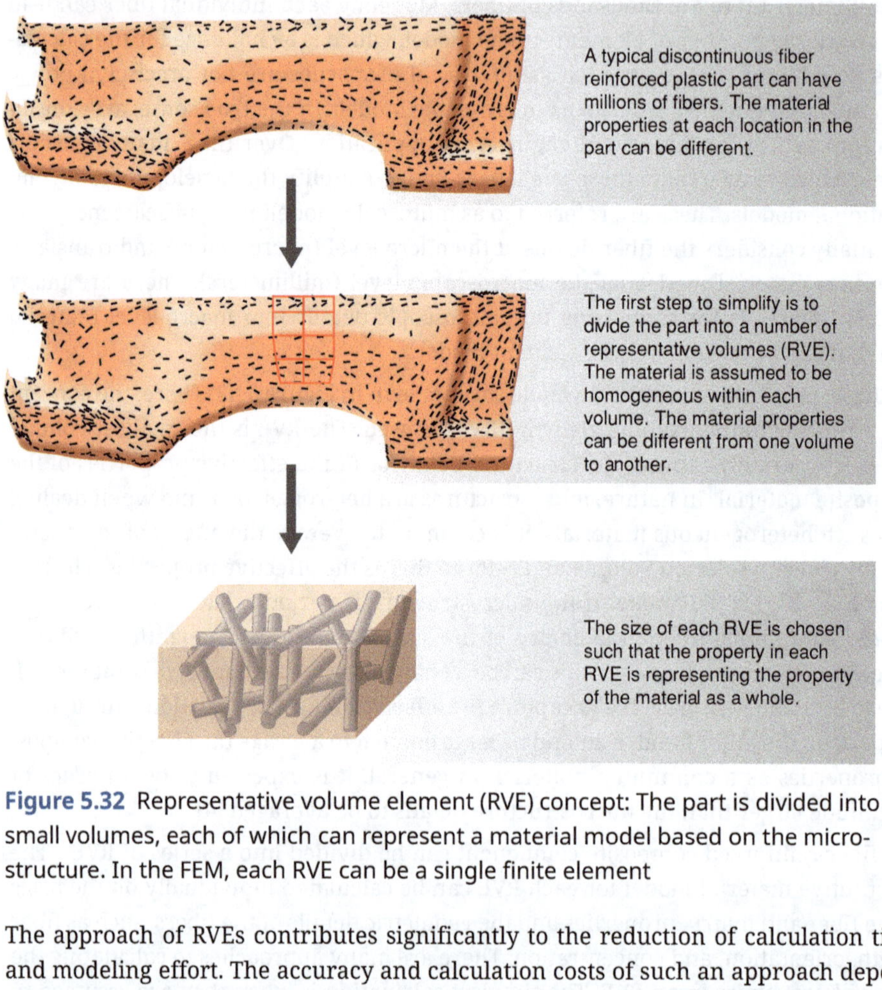

Figure 5.32 Representative volume element (RVE) concept: The part is divided into small volumes, each of which can represent a material model based on the microstructure. In the FEM, each RVE can be a single finite element

The approach of RVEs contributes significantly to the reduction of calculation time and modeling effort. The accuracy and calculation costs of such an approach depend on how the representative volume is chosen and which method is used to represent the stresses and strains within this volume. Once the finite element model is employed for the calculation, the stresses and strains estimated via the RVE need to be interpreted. This is important because the microstructure, which is assumed to be homogeneous in the RVE, is heterogeneous and consists of fibers and matrix. Depending on the fiber geometry and the material properties of the fibers and the matrix, the stresses and strains in the RVE are not homogeneous but vary from place to place. The average stress based on the homogeneous material will not be accurate. Therefore, it is important to translate the average stress and strain based on the microstructure into a meaningful stress and strain distribution within the RVE.

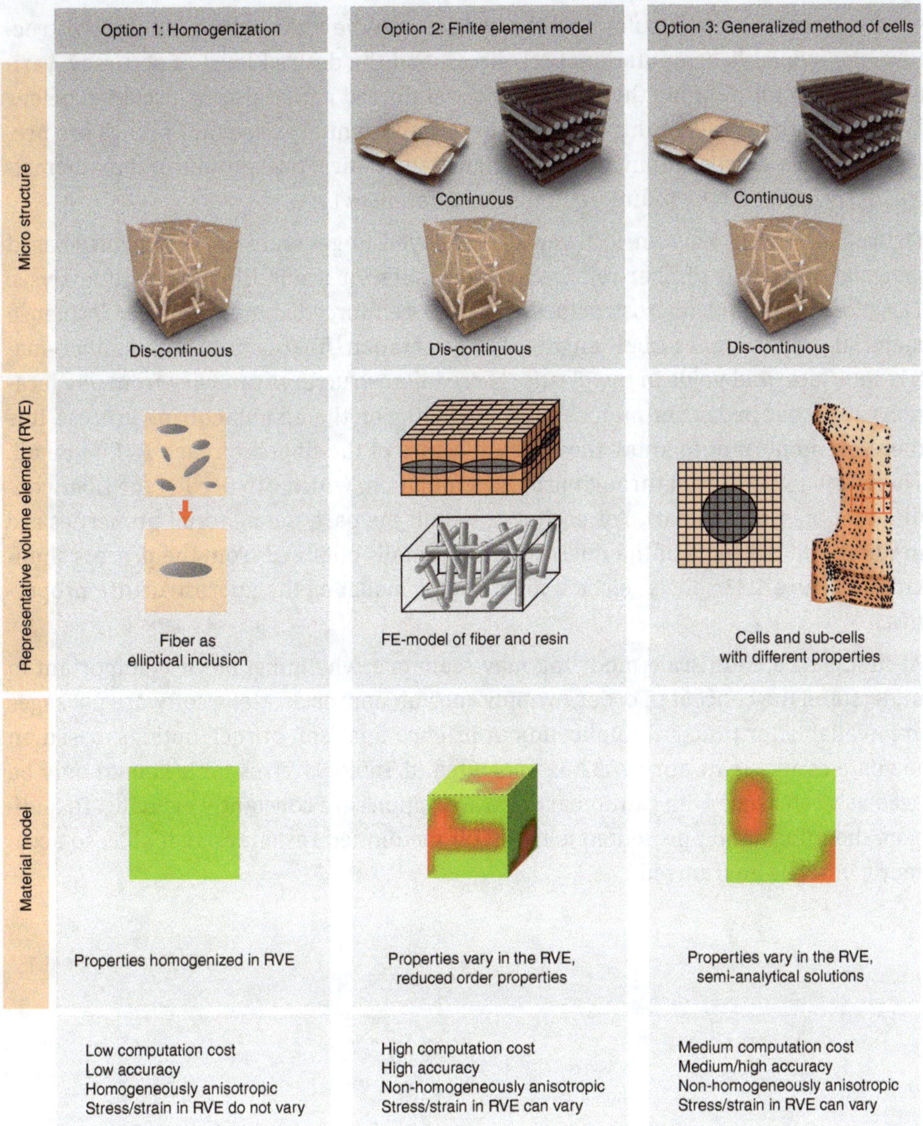

Figure 5.33 Methods for calculating the material properties based on the micro-structure for a representative volume element (RVE) for fiber-reinforced polymer composites

Since most materials found in nature are heterogeneous, that is, have two or more different materials at the microstructural level, the concept of using RVEs is widely recognized in physics. As shown in Figure 5.33, many different methods have been developed over the years to calculate the RVEs for fiber-reinforced materials. Each approach has its advantages and disadvantages. The selection of a representative solid element for a particular problem is based on the knowledge of the microstructure, the available computational resources, and the desired level of accuracy. Various methods for defining the RVE for fiber-reinforced materials are discussed below. To maintain the focus of this book, the most important aspects and findings are presented from a practical point of view. Detailed theoretical background and mathematical derivations can be found in numerous references [4].

Figure 5.34 shows how an RVE can be employed to generate microstructure-based material properties of fiber-reinforced composites for use in finite element analysis. The most important microstructures for fiber-reinforced composites that influence material properties are fiber length, fiber orientation, fiber concentration, fiber–matrix interface, and voids in the plastic. These microstructures can vary from one location to another in discontinuous fibers depending on the manufacturing process. It is therefore important to know the microstructure of the fibers for each RVE in detail. This is usually achieved through process simulation, which can predict the fiber condition in the finished part. For each RVE within the part, the material properties are predicted on the basis of the microstructure details obtained from the process simulation. (Figure 5.33 shows selected methods for modeling the microstructure properties.)

At first glance, multiscale modeling may seem overwhelming, but it is important to understand the concept in order to apply such an approach. Many software packages are available for practical application. A brief comparison of such options, based on the data available in April 2022, is presented in Table 5.2. This table should only be used as a reference with caution as software options are constantly evolving. In addition, the information presented is based on the limited resources available, so accuracy cannot be guaranteed.

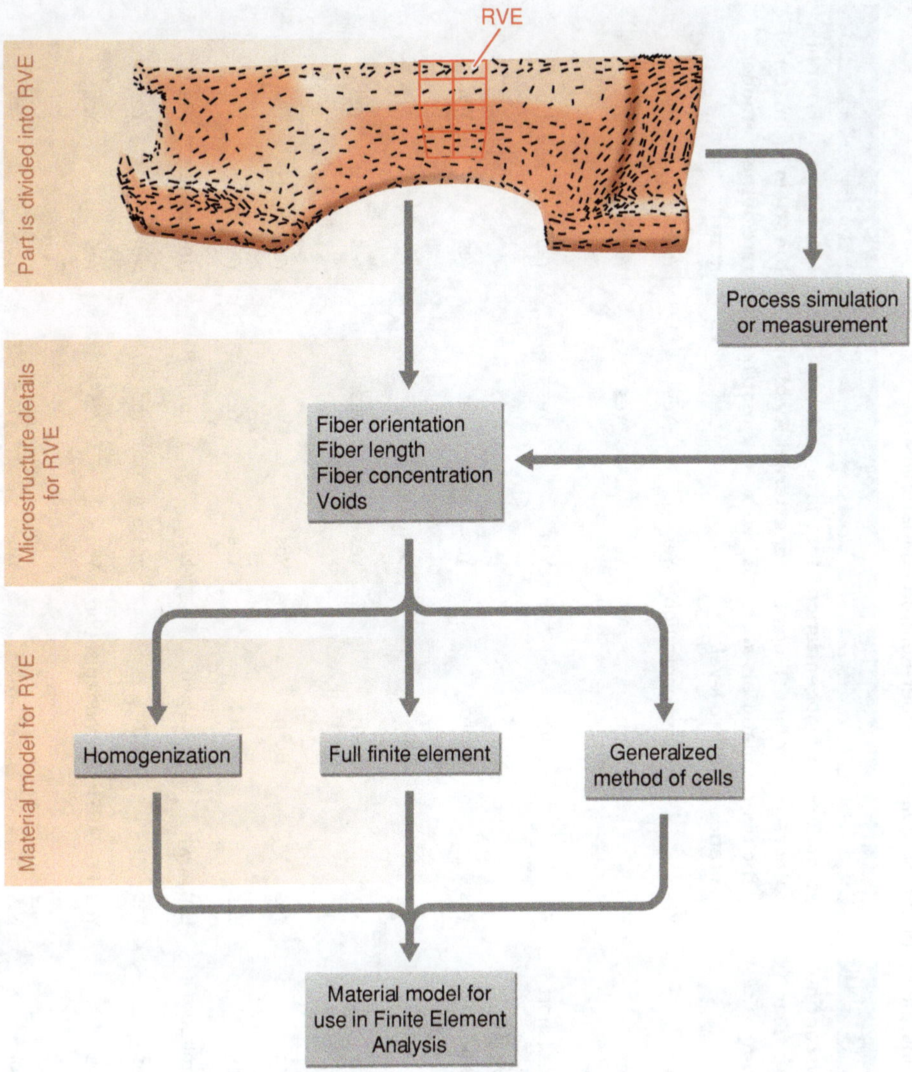

Figure 5.34 Approach for integrating material properties from RVE into finite element analysis

Table 5.2 Overview of Available Multiscale Modeling Software for Fiber-Reinforced Materials

Software	Application/features (April 2022 version)						
	Continuous fiber composite materials	Discontinuous fiber composites	Method for incorporating the fiber orientation	Homogenization or modelling of materials at micro level	Interface for external solvers	Nonlinearity of the matrix level	Modeling the strain rate of the matrix	Error in material model
Digimat™	Yes	Yes	Yes	Homogenization	Yes	Yes	Yes	Yes
AlphaSTAR™ Products	Yes	Yes	Yes	Both	Yes	Yes	Yes	Yes
Converse™	No	Yes (short fiber)	Yes	Both	Yes	Yes	Yes	No
NASA Generalized method of cells	Yes	Yes	Yes	Micro	Yes	Yes	Yes	Yes
ESAComp™ (Altair)	Yes	No	No	Both	Yes	Yes	Yes	Yes
Autodesk™	Yes	Yes	Yes	Both	Yes	Yes	Yes	Yes
Moldex3D™	Yes	Yes	Yes	Both	Yes	Yes	Yes	No
LMS™	Yes	Limited	Unknown	Unknown	No	Limited	Unknown	Yes
MultiMech™	Yes	Yes	Yes	Both	Yes	Yes	No	Yes
SwiftComp™ Micromechanics	Yes	Yes	Yes (PreVABS)	Homogenization	Yes	Yes	Yes	Yes
HyperSizer™	Yes	No	No	Unknown	Yes	No	No	Yes

5.6 Material-Compatible Design

The specific properties of plastics must be given due consideration during design in order to achieve technically sophisticated and economical products. In comparison to metallic materials, the following must be considered:

- The much lower modulus of elasticity (only about 0.1% to 1%),
- The lower strength (about 1% to 10%),
- The significantly higher elasticity (about 10 to 100 times),
- The significantly higher thermal expansion (around 10 to 100 times),
- The dependence of many properties on temperature and time.

Some generally applicable rules for plastic-compatible design can be derived from this:

- Bending-resistant components with a low modulus of elasticity can be achieved by ribbing.
- Mechanically highly stressed parts must be designed without notches and corners!
- The high malleability of plastics allows characteristic designs (film hinges, snap connections).
- The low strength favors the use of self-forming metallic screws as a cost-effective joining technique.

As the properties of plastics can be greatly influenced by adding fillers or changing their molecular structure, the modifiability of the plastic with regard to its area of application is a key characteristic of this group of materials.

5.6.1 Part Dimensioning Calculation

As with classic design materials, plastics are also dimensioned to clarify the issue: Can the component withstand the forces occurring during production and use without cracking or unacceptable deformation? The procedure is as follows:

- Definition of the boundary conditions:
  - Geometry of the component,
  - Acting component stress (loads, temperature, time, medium),
  - Dimensioning parameters (permissible material stresses, modulus of elasticity, etc.).

- Determination of component behavior under applied load (alternatively by):
  - Conventional dimensioning calculation,
  - Empirically supported dimensioning calculation,
  - Application of the finite element method for mechanical analysis,
  - Component tests on prototypes or models.
- Admissibility check by comparison with limit values:
  - Stiffness verification to check the permissible component deformation,
  - Strength verification to check the admissibility of the material stress.

5.6.2 Dimensioning Parameters

Dimensioning calculations require dimensioning parameters that allow a quantitative description of the influence of the material on the component behavior. The following can be regarded as the essential dimensioning parameters:

- Modulus of elasticity,
- Cross-contraction number,
- Permissible stress and permissible elongation.

In order to adequately consider the viscoelastic properties of the plastic, the dependencies of the dimensioning parameters on the operating conditions must be considered. Accordingly, the corresponding characteristic values must be redetermined for each load case. The usual procedure is to determine a “quasi-elastic equivalent characteristic value”. This considers the change in material stiffness, for example, due to creep, or an increase in temperature, by defining a “substitute material” to which the now-lower modulus of elasticity is assigned as an elastic property. The determination of the elastic equivalent modulus of elasticity as a function of the respective time, temperature, and stress level is carried out in the isochronous stress–strain diagram; see Figure 5.35. The equivalent modulus of elasticity is calculated as the gradient of the connecting line from the origin of the stress–strain diagram to the respective stress state under consideration. Figure 5.36 shows how a creep process is mapped using the model of the ideal elastic equivalent material.

Another characteristic value required for the stiffness analysis of the component is Poisson’s ratio ν. As no measured values are usually available for this, an approximate approach is used (see Section 3.3.5).

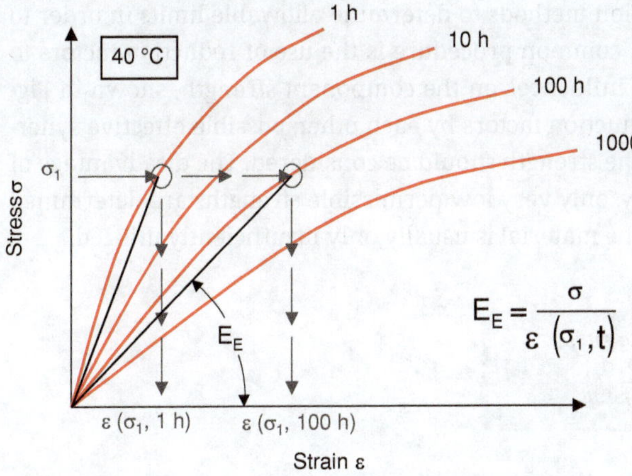

Figure 5.35 Determination of the elastic equivalent modulus of elasticity from the isochronous stress–strain diagram

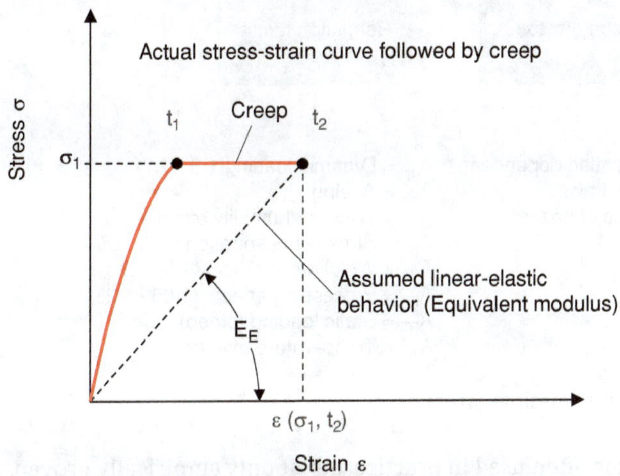

Figure 5.36 Description of creep process using the model of the elastic substitute material

The calculation of static mechanical stress states can be approximated using the formulas for elastic materials with the aid of the equivalent characteristic values determined here. The restrictions are the following:

- In these dimensioning calculations, the influence of the molecular or filler orientation on the mechanical properties is not considered.
- The behavior under dynamic, cyclic, or non-isothermal stresses can only be roughly estimated using the procedure described above. In such cases, it is necessary to use more complex models and material laws to consider the influence of the stress history when solving the mechanical problem. Only then is it possible to determine the energy converted into heat (dissipated) during cyclic loading.

While isochronous stress–strain diagrams are increasingly documented in the material database, there is practically no information on stress limit values. Therefore, the designer has to rely on estimation methods to determine allowable limits in order to avoid costly material testing. A common procedure is the use of reduction factors to take into account the different influences on the component strength, shown in Figure 5.37. By multiplying the reduction factors by each other, possible effective synergies of different influences on the strength should be considered. The disadvantage of this approach is that, ultimately, only very low permissible strengths are determined and therefore the potential of the material is usually only insufficiently utilized.

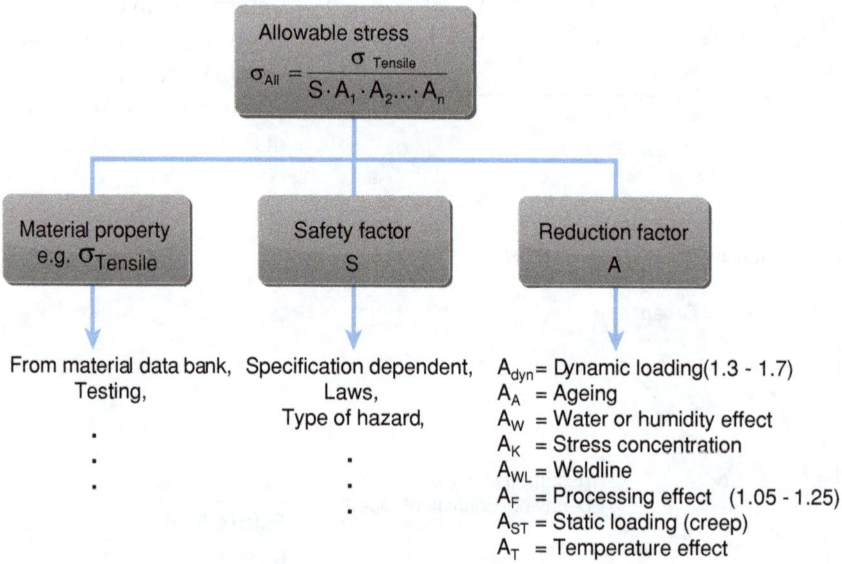

Figure 5.37 Determining the permissible stress

A limit for dimensioning that is often used in practice, albeit only empirically proven, is the critical elongation; see Table 5.3. It was developed by G. Menges, based on the theory that the first irreversible damage to plastics can only be observed when plastics are stressed above the critical elongation.

Table 5.3 Critical Elongation of Thermoplastics

Plastic	Critical elongation
Amorphous, unfilled thermoplastics	0.8%
Exception polystyrene	0.2%
Semi-crystalline thermoplastics (above T_G; otherwise, see amorphous thermoplastics)	2.0%
Impact-modified blends	0.8%

When using the critical strain as a dimensioning limit, it must be noted that the local strain of the matrix is always greater than the average macroscopic strain in plastics filled with hard fillers. Therefore, the critical strain must be reduced depending on the filler content. The following approximation formula can be used for plastics filled with short glass fibers, where φ is the weight fraction of glass fiber content:

$$\epsilon_{\text{crit,GF}} \approx \epsilon_{\text{crit}}(1 - 1.5\varphi)$$

Dimensioning with critical elongation is a deliberately simple and clear rule. It guides the designer to think in terms of strain. It allows an initial design of a component even if there are major uncertainties in the requirement profile. A prediction of failure states cannot be derived from the critical strain.

Both stress limits discussed above refer to the uniaxial stress state. If multiaxial stress states are present in the component, they must be transformed into the normal stress system (shear-free state) and compared with the permissible uniaxial stress using a failure hypothesis. When proceeding according to the critical strain, the corresponding permissible uniaxial stress is obtained by multiplication by the corresponding equivalent modulus of elasticity. Common failure criteria for estimating multiaxial stress states are the Mises criterion and the cone criterion.

5.6.3 Dimensioning Calculations

A design can be roughly calculated if it can be broken down into geometrically simple structures (members, beams, plates, shells, etc.): the load cases can be calculated individually and then superimposed. Rough calculations are helpful for estimating the magnitudes of stresses and deformations. The procedure is only permissible if the deformations remain sufficiently small, as, otherwise, the change in geometry and the force application points are not adequately considered. It is assumed that the equations of the general theory of elasticity are applicable. The solutions derived from this are used for the respective geometries (*e. g.*, bending of plates, buckling of bars, boiler formula for thin-walled tanks, etc.). With the development of the finite element method, conventional dimensioning is becoming less and less important. Nevertheless, it is a very useful tool for estimating the feasibility of a structure and the magnitude of the stresses in the initial design stages.

5.6.4 Empirically Supported Dimensioning Calculation

For functional elements in particular (see Section 5.7), the empirically supported dimensioning calculation is often used. The basic idea here is that certain functional principles are tested using sample components and the mechanical operating principle is then laid down in approximate equations. For example, design rules are used to design the tube of a bolted connection in such a way that failure due to the tube bursting open is avoided. Depending on the screw-in depth, the breaking force of the screw connection can then be determined from the strength of the screw, the strength of the tube cross section, or the shear strength of the sheared area of the thread screwed into the tube. It is therefore characteristic of the empirically supported dimensioning calculation that certain design rules are initially observed and that empirical knowledge is processed into these rules. Such empirical dimensioning (*e. g.*, for plastic pipes under earth pressure) can often be found for plastic components that are subject to building authority approval by the German Institute for Building Technology.

5.6.5 Component Tests on Prototypes and Models

For components with high requirements for safety against failure, component tests must be carried out to validate the dimensioning, which is always based on hypotheses. For the final release of a series component, these component tests should be carried out on samples that have been manufactured in the same way as the series components. However, this is not possible in the early phases of component development, as the molds for series production are typically not available at this stage.

However, in order to check the dimensions, it is advisable to use rapid prototyping (see Section 4.10). This can be done according to the method described in Section 5.5.1, which can be used directly to generate the prototype. The advantage of this, in addition to significant time and cost savings, is that any errors in geometry generation in the CAD system are immediately recognized on the model.

For large components, scaled-down prototypes must be produced, and they are tested under a correspondingly adapted load. The use of different materials for the series component and prototype can be considered by means of appropriate correction factors or a suitably modified test temperature.

The particular benefit of the component test lies in the additional information in conjunction with the FE calculations. The coupling of test and calculation provides engineers with an additional level of experience that allows them to develop improved designs and calculation methods.

5.7 Functional Elements

In terms of using tried and tested designs, it makes sense to avoid new designs as far as possible and instead come up with new solutions by modifying existing design methods. If this approach is broken down to the individual elements of a design, numerous examples of such recurring functional elements can be found. Some of these functional elements are described below.

5.7.1 Living Hinges

Living or film hinges make use of the high elasticity of plastics as well as the possibility of producing two molded parts simultaneously in the injection mold; the second part is connected to the first part by a film that is as thin as possible. This film then forms the film hinge of the now-connected parts (*e. g.*, component with locking element). A critical aspect of film hinges is a design that leads to unacceptably high local expansion; see Figure 5.38. When designing the rheological properties of injection molds, it is important to note that premature solidification of the flow fronts in the film hinge can occur due to the large jumps in wall thickness in the component. Therefore, care must be taken to ensure that the sprues in the first part are arranged in such a way that the film hinge is located at the end of the flow path of the first part as far as possible. Finally, the second part should be at least partially filled by the film hinge.

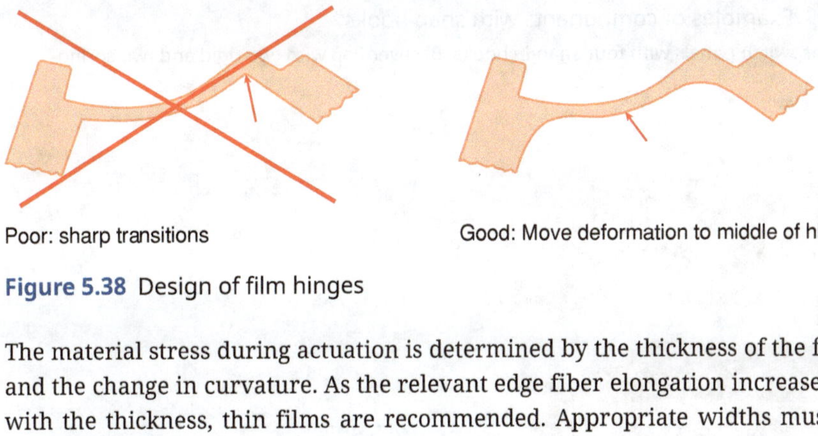

Poor: sharp transitions

Good: Move deformation to middle of hinge

Figure 5.38 Design of film hinges

The material stress during actuation is determined by the thickness of the film hinge and the change in curvature. As the relevant edge fiber elongation increases linearly with the thickness, thin films are recommended. Appropriate widths must be provided to absorb the tensile loads.

5.7.2 Snap-Fit Connections

In many cases, snap-fit connections are technically and economically favorable types of connection. Depending on the hook geometry, they can be designed to be difficult or easy to release or even conditionally releasable (latching in such a way that release is not possible by force but only by additional mechanical actuation). Because many plastics allow quite large elongation values for a short time without breaking or permanent deformation, and the shaping is usually achieved during injection molding with little additional effort, plastics are particularly suitable for this joining method. Figure 5.39 (A) and (B) show simple examples with snap-fit hooks; Figure 5.40 shows a chassis cover with two snapping tabs with high force transmission, which can be released by thumb pressure in the direction of the arrow; and Figure 5.41 shows an interrupted ring snap connection.

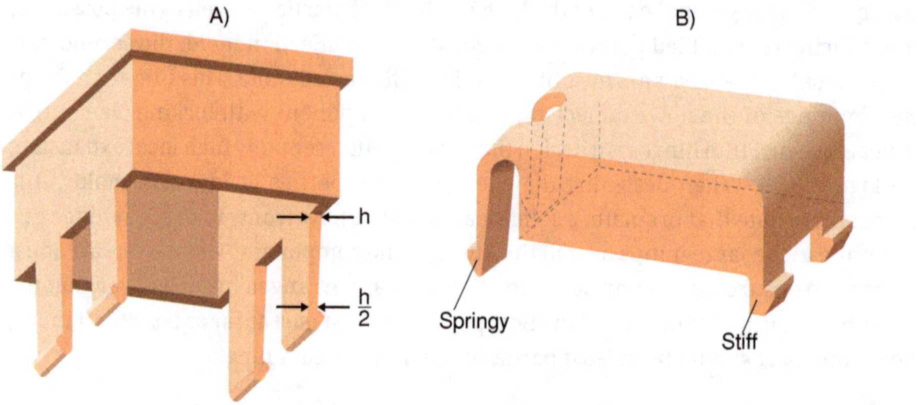

Figure 5.39 Examples of components with snap hooks

A: module for switch panels with four snap-in hooks, B: cover flap with two rigid and two spring-loaded snap hooks

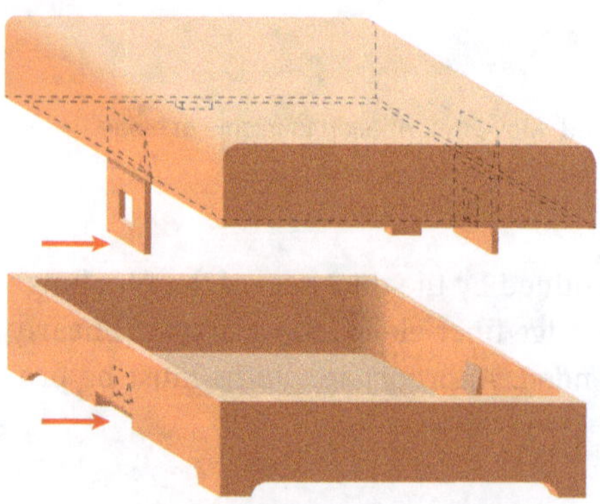

Figure 5.40
Detachable snap connection of a chassis cover

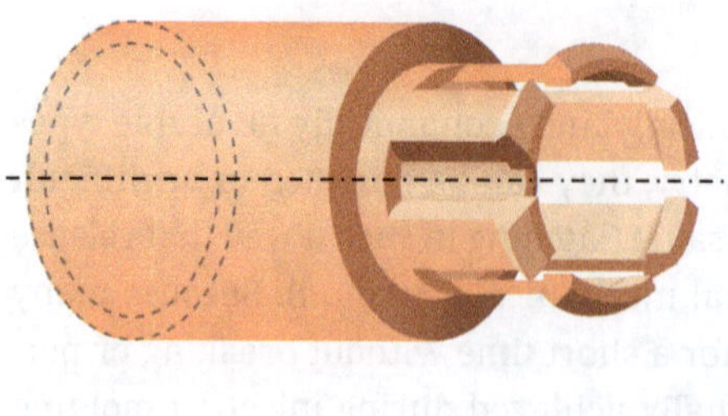

Figure 5.41
Interrupted ring snap connection

The most important basic forms for such connections are spring hooks, ring snap connections, and torsion snap connections. The calculation equations for three simply designed snap hooks are included in Figure 5.42.

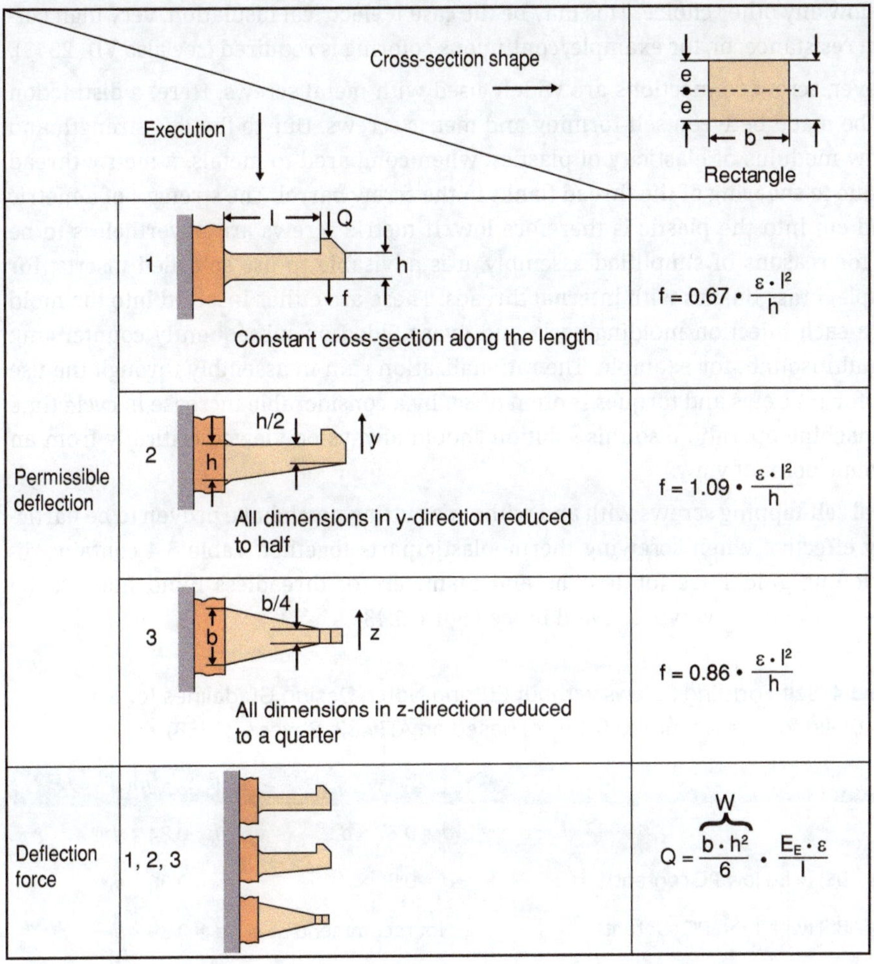

Figure 5.42 Calculation equations for snap hooks

f: (permissible) spring travel (= undercut), *ε*: (permissible) elongation in the edge fiber of the clamping point (in the formulas, *ε* is an absolute value: percentage/100%), *l*: arm length, *h*: thickness at the clamping cross section, *b*: width at the clamping cross section, *e*: edge fiber distance from the neutral fiber (center of gravity), *W*: axial moment of resistance ($W = I/e$; *I*: area moment of inertia), E_E: equivalent modulus of elasticity, *Q*: (permissible) deflection force

5.7.3 Screw Connections

Screws or nuts made of plastics as fasteners are only useful due to the generally low strength and/or rigidity of plastics if special requirements for the screw connections do not allow any other choice. This may be the case if electrical insulation, very high corrosion resistance, or, for example, continuous coloring is required (see also VDI 2543).

However, screw connections are widely used with metal screws. Here, a distinction must be made between self-forming and metric screws. Due to the low strength and the low modulus of elasticity of plastics, when compared to metals, a metric thread fails due to shearing of the thread flanks in the screw barrel. The strength of a metric thread cut into the plastic is therefore low. If metric screws are nevertheless to be used for reasons of simplified assembly, it is advisable to use so-called inserts; for example, brass bushes with internal threads. These are either inserted into the mold before each injection molding cycle and overmolded or subsequently countersunk using ultrasound, for example. The rationalization gain in assembly through the use of uniform screws and torques is often offset by a considerable increase in cycle time and machine operation, so this solution should always be viewed critically from an economic point of view.

Special self-tapping screws with and without a cutting notch have proven to be particularly effective when screwing thermoplastic parts together. Table 5.4 contains dimensioning guidelines for lengths and diameters of threadless blind holes to be formed for two screw types, A and B; see Figure 5.43.

Table 5.4 Self-Forming Screws without Cutting Notch Design Guidelines for Screw Studs (Valid for d = 2.9 mm to 5.1 mm) Based on ATI 482, Bayer G (1988)

Materials (examples of thermoplastics)	Screw type A[1)]	Screw type B[1)]
ABS	$d_i = 0.87 \times d$	$d_i = 0.84 \times d$
(PC + ABS) (with low PC content)	$d_i = 0.90 \times d$	$d_i = 0.86 \times d$
(PC + ABS) (with high PC content)	Not recommended	$d_i = 0.89 \times d$
PC		$d_i = 0.90 \times d$
PC, glass-fiber-reinforced		$d_i = 0.92 \times d$
PET	$d_i = 0.89 \times d$	$d_i = 0.85 \times d$
PET, glass-fiber-reinforced	$d_i = 0.91 \times d$	$d_i = 0.87 \times d$
PA 66, glass-fiber-reinforced, freshly sprayed	$d_i = 0.94 \times d$	$d_i = 0.90 \times d$
PA 66, glass-fiber-reinforced, conditioned	$d_i = 0.92 \times d$	$d_i = 0.87 \times d$
PA 6, freshly sprayed	$d_i = 0.90 \times d$	$d_i = 0.86 \times d$

Materials (examples of thermoplastics)	Screw type A[1)]	Screw type B[1)]
PA 6, conditioned	$d_i = 0.85 \times d$	$d_i = 0.82 \times d$
PA 6, glass-fiber-reinforced, freshly sprayed	$d_i = 0.92 \times d$	$d_i = 0.88 \times d$
PA 6, glass-fiber-reinforced, conditioned	$d_i = 0.90 \times d$	$d_i = 0.86 \times d$

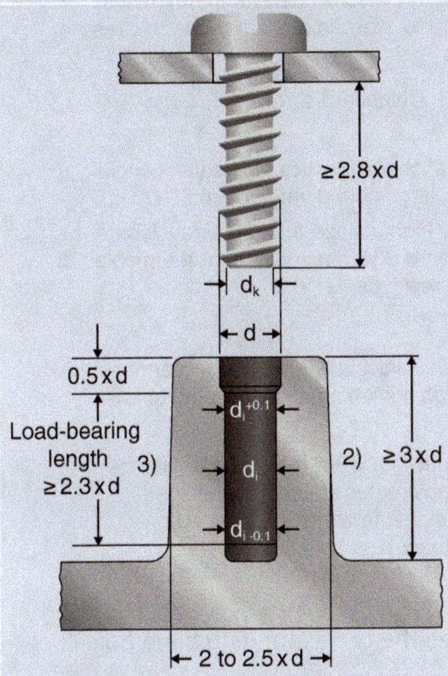

1) For screw types, see type A and type B in Figure 5.43
2) Load-bearing length ≥ 3 × *d* recommended to comply with VDE regulations
3) The greater the screw-in depth, the better the VDE regulations are met

It can be seen from the table that the diameter d_i can be smaller for tough, less crack-sensitive plastics (ABS, PA 6 conditioned) and that the screw type B is generally more favorable, especially for PC.

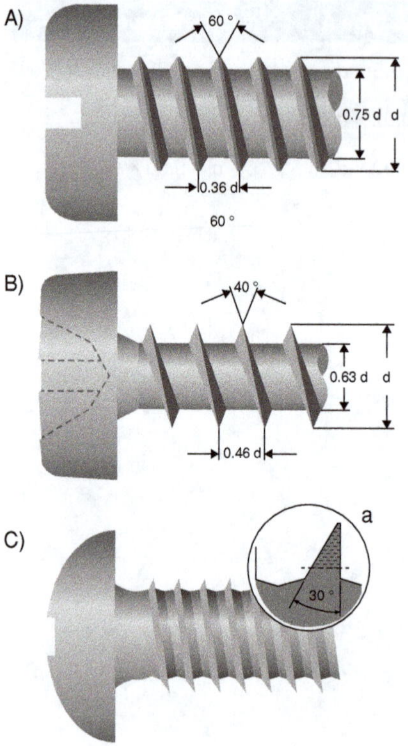

Figure 5.43 Examples of self-forming screws (see the corresponding blind hole dimensions in Table 5.4)

For a screw specially developed for thermoplastics, similar to type B in Figure 5.43, with a flank angle of 30°, the manufacturer (Ejot-Verbindungstechnik) recommends screw stud diameters of 1.8 *d* to 2.5 *d*, screw-in depths of 1.7 *d* to 2.2 *d*, and the following hole diameters:

- Approx. 0.70 *d*: PE-LD, POM, PP, PP with 20% talc,
- Approx. 0.75 *d*: PA 46, PA 6, PA 66, PBT, PE-HD, PET, POM-GF 30,
- Approx. 0.80 *d*: ABS, ABS+PC, ASA, PS, PVC-U, PA 46-, PA 66-, PBT-, PET-GF 30,
- Approx. 0.85 *d*: PC, PC-GF 30, PMMA, PPO.

When screwing thermoset plastic parts, self-tapping screws are favorable. They do not deform the plastic (as with tough thermoplastics) but cause it to be machined; see type C in Figure 5.43. An asymmetrical thread profile with a load flank angle of 30° is suitable for this purpose. The blind hole core diameter should be approximately 0.85 *d* and the chip space depth at least 0.8 *d*, preferably more, so that the bottom is not blasted off at the end of the blind hole when screwing in.

5.7.4 Rivets

The riveting of plastics to each other or to other materials can be done with soft rivets made of copper, brass, and aluminum and also with flanged hollow rivets. The impulse (mass × speed) of the riveting impact must be adapted to the susceptibility of the plastic to breakage or cracking. If one of the joining parts is a thermoplastic material, the end of the studs of the molded part in question can be formed into a rivet head by hot or cold forming (ultrasound); see Section 4.14.

5.7.5 Welded Joints

Thermoplastics can be joined by welding. For the various processes, see Section 4.14. If welding is to be used as an assembly process in series production, processes that allow correspondingly short cycle times (*e.g.*, friction welding, ultrasonic welding, hot plate welding) must be selected. The joining areas must be designed in accordance with the requirements of the welding process.

It should be noted that the strength of a welded joint generally does not reach the strength of the base material. This is mainly due to the notch effect caused by the resulting weld bead, the residual stresses caused by local heating during joining, and the generally unfavorable orientation of the molecules and reinforcing fibers. Depending on the suitability of the material for welding and the process selected, 40% to 80% of the short-term strength is achieved. The weakening is even greater with long-term exposure to load, dynamic stress, or media attack. Correspondingly high reduction factors must be considered when dimensioning welded joints.

The rules and regulations of the German Welding Society (*Deutscher Verband für Schweißtechnik*) set out the characteristic design values for some plastics and welding processes.

5.7.6 Press Fits

The press fit or connection is a simple and cost-effective way of creating a nonpositive connection between a shaft and a hub. It is mainly used in precision engineering for shaft–hub connections of gears, actuators, drivers, and couplings as well as for fastening fan and pump impellers.

Due to the lower modulus of elasticity of plastics, the undersize must be larger than in press connections between two metal components. The greater thermal expansion of the plastics (in the case of hydrophilic plastics, the swelling expansion as well) and the time-dependent mechanical behavior must also be considered. These influences lead to a partial relaxation of the stresses in the plastic, which reduces the pressing forces in the joint.

References

[1] Tucker, C. L., *Fundamentals of Fiber Orientation: Description, Measurement and Prediction*, Hanser, Munich (2022)

[2] Vincent, M., Giroud, T., Clarke, A., Eberhardt, C., Description and modeling of fiber orientation in injection molding of fiber reinforced thermoplastics, *Polym. Blends Compos. Hybrid Polym. Mater,* MACRO 2004, 46 (2005) 17, pp. 6719–6725

[3] Ogadhoh, S. O., Papathanasiou, T. D., Particle rearrangement during processing of glass-reinforced polystyrene by injection molding, *Compos. Part Appl. Sci. Manuf.*, 27 (1996), pp. 57–65

[4] Gandhi, U. N., Goris, S., Osswald, T. A., Song, Y.-Y., *Discontinuous Fiber-Reinforced Composites: Fundamentals and Applications*, Hanser, Munich (2020)

[5] Thomason, J. L., Vlug, M. A., Influence of fiber length and concentration on the properties of glass fiber-reinforced polypropylene: 1. Tensile and flexural modulus, *Compos. Part Appl. Sci. Manuf.*, 27 (1996) 6, pp. 477–484

[6] Thomason, J. L., The influence of fiber length, diameter and concentration on the impact performance of long glass-fibre reinforced polyamide 6,6, *Compos. Part Appl. Sci. Manuf.*, 40 (2009) 2, pp. 114–124

[7] Thomason, J. L., Vlug, M. A., Schipper, G., Krikor, H. G. L. T., Influence of fiber length and concentration on the properties of glass fiber-reinforced polypropylene: Part 3. Strength and strain at failure, *Compos. Part Appl. Sci. Manuf.*, 27 (1996) 11, pp. 1075–1084

[8] Thomason, J. L., Vlug, M. A., Influence of fiber length and concentration on the properties of glass fiber-reinforced polypropylene: 4. impact properties, *Compos. Part Appl. Sci. Manuf.*, 28 (1997) 3, pp. 277–288

[9] Thomason, J. L., The influence of fibre length and concentration on the properties of glass fiber reinforced polypropylene: 5. Injection moulded long and short fibre PP, *Compos. Part Appl. Sci. Manuf.*, 33 (2002) 12, pp. 1641–1652

[10] Thomason, J. L., The influence of fibre length and concentration on the properties of glass fibre reinforced polypropylene. 6. The properties of injection molded long fibre PP at high fibre content, *Compos. Part Appl. Sci. Manuf.*, 36 (2005) 7, pp. 995–1003

[11] Schemme, M., Long-fiber reinforced thermoplastics, *Kunststoffe*, 93 (2003), pp. 106–109

[12] Tseng, H.-C., Chang, R.-Y., Hsu, C.-H., Phenomenological Improvements to Predictive Models of Fiber Orientation in Concentrated Suspensions, *J. Rheol.*, 57 (2013), pp. 1597–1631

[13] Nguyen, N., Jin, X., Wang, J., Phelps, J. H., Tucker III, C. L., Kunc, V., Bapanapalli, S. K., Smith, M. T., Implementation of New Process Models for Tailored Polymer Composite Structures into Processing Software Packages, US Department of Energy, PNNL Report under Contract DE-AC05-76RL01830, PNNL-19185 (2010)

6 Plastic Materials

6.1 General Remarks

Comparisons between different plastic materials will be based on characteristics provided in Table 3.1 in Section 3.1 (see also CAMPUS database or ISO 10350). For example, in general, no values for impact strength will be provided, because these values typically do not contribute significant information to either part design or a part's load bearing capacity, see also Section 3.3.3. As a general rule of thumb it can be assumed that parts made from resin types with a lower modulus of elasticity and a higher elongation at yield (elongation at break) will be more ductile (tougher) under excessive loads.

Unless particularly noted, specific property values are given for generic unmodified material classes. Because in general there will be no data provided for specific trade names, the property values given should be considered an indicating range. Specific material types may vary widely from these ranges, particularly regarding classifications such as "resistant" or "non-resistant" to environmental conditions.

6.2 Polyolefins (PO), Polyolefin Derivatives and Copolymers

Polyolefins are polymers built from hydrocarbons with a double bond and the general chemical structure C_nH_{2n} (ethene, propene, butene-1, isobutene). They include polyethylene, polypropylene, polybutene-1, polyisobutene, poly-4-methylpentene as well as their respective copolymers. Today, the available homo- and copolymeric resins based on ethylene and propylene provide an extraordinarily broad range of properties.

The polyolefin architecture is determined in particular by the catalysts used, with *metallocene catalysts* gaining increasing importance. Polymeric resins made with metallocene catalysts provide a narrow molecular weight distribution and their polymerization allows for a particular sequence and order of their monomer building blocks. In addition, it is possible to incorporate building blocks into the polyolefin structure that could not be incorporated previously by copolymerization.

6.2.1 Polyethylene Standard Homo- and Copolymers (PE-LD, PE-HD, PE-HD-HMW, PE-HD-UHMW, and PE-LLD)

Polyethylenes (PE) are semi-crystalline thermoplastic materials. Their structure, molecular weight, crystallinity, and thus their properties depend to a high degree on the polymerization method used and on possible chain branching options (see Section 2.3). The relative molecular weight and its distribution can be influenced by high thermal or mechanical stresses, which is typically avoided. The degree of crystallinity is determined by the structure of the polymer and the processing conditions. For industrial applications, ISO 1133 categorizes polyethylenes by their different densities, which depend on the respective degree of crystallinity of each specific PE grade. ISO 1872 is used internationally and specifies PE resins mainly by "*density*" and "*melt flow rate (MFR)*"; in addition, it uses a system of abbreviations indicating application, processing method, additives, fillers, and reinforcements. However, the characterization by these designations is inadequate in describing the properties and application ranges for the numerous PE grades available and is therefore rarely used in practice.

6.2.1.1 Polymerization, Chemical Constitution

In general, polyethylenes are produced either by high-pressure processes in the presence of radicals (radical polymerization) or by medium- and low-pressure processes with the help of catalyst systems (anionic polymerization). In addition, polyethylenes are classified as suspension-, solution-, gas phase-, or mass polymerization grades depending on their state. High-pressure processes are used for highly branched homopolymers (PE-LD), while medium- and low-pressure processes are used to synthesize linear homo- and copolymers (PE-HD, PE-MD, PE-LLD).

High-pressure process: PE-LD (LD = low density) is synthesized by a *high pressure process* from ethylene ($CH_2{=}CH_2$) under a pressure of 1000 to 3000 bar at 150 to 275 °C with 0.05 to 0.1% oxygen or peroxides as catalysts; synthesis occurs either discontinuously in stirrer vessels or continuously in pipe reactors. The result is a highly branched PE with side chains of different lengths. Its crystallinity ranges from 40 to 50%, its density from 0.915–0.935 g/cm^3, and it has a molecular weight average up to 600,000 g/mol. With the help of high-performance catalyst systems, PE-LD synthesis equipment can also produce linear low density PE (PE-LLD).

Medium- and low-pressure process: PE-HD (HD = high density) is synthesized by either a *medium-pressure* (Phillips method) or a *low-pressure process* (Ziegler method), both of which are suspension processes. The Phillips method uses pressures from 30 to 40 bar, temperatures from 85 to 180 °C, and chromium oxide or aluminum oxide as catalysts. The molecular weights obtained are approx. 50,000 g/mol. The *Ziegler method* uses pressures from 1 to 50 bar, temperatures from 20 to 150 °C, and titanium halides, titanium esters, or aluminum alkyls as catalysts, obtaining molecular weights ranging from 200,000–400,000 g/mol. PE-HD is almost unbranched and therefore has a higher degree of crystallinity (60–80%) and higher densities (0.94–0.97 g/cm^3) than PE-LD.

PE-HD-HMW (HMW = high molecular weight) with a density of 0.942–0.954 g/cm^3 and *PE-HD-UHMW* (UHMW = ultra-high molecular weight) with a density of 0.93–0.94 g/cm^3 offer high molecular weights together with their high densities. These resins are produced with special catalysts in a low-pressure process. The average molecular weight ranges from 200–500 kg/mol for PE-HD, from 500–1000 kg/mol for PE-HD-HMW, and from 3000–6000 kg/mol for PE-HD-UHMW.

PE-LLD (LLD = linear low density) is polymerized with high-efficiency catalysts (metal complexes) in four different processes: a *low-pressure process* in the gas phase, in solution, in suspension, or in a *modified high-pressure process*. Copolymerization of ethylene with 1-olefins, such as butene-1 or hexene-1, creates short side chains. Compared to linear PE-HD, PE-LLD contains a higher ratio of comonomers. The higher molecular weight and the low number of side chains lead to improved properties of these resins.

PE-VLD (VLD = very low density) with a density of 0.905–0.915 g/cm^3 and *PE-ULD* (ULD = ultra low density) with a density of 0.890–0.905 g/cm^3 exhibit such a high degree of branching caused by their increased comonomer content that their densities decrease below 0.915 g/cm^3.

PE-(M) (polyethylene produced with metallocene catalysts) exhibits a narrow molecular weight distribution and can be produced in a broad range of densities. Although PE-MLLD (M) is of the same composition as "regular" PE-LLD, it shows a different sequence statistic. Transition metal compounds activated by methyl aluminoxanes are used as catalysts, typically with Ti- or Zr-centers linked with cyclopentadienyl residues.

The densities of PE-(M) grades range from:

- PE-MLLD: 0.915–0.930 g/cm^3
- PE-MMD: 0.930–0.940 g/cm^3 (medium density)
- PE-MHD: 0.940–0.995 g/cm^3

In addition, it is possible to produce PE-MVLD with densities from 0.863–0.885 g/cm^3 (polyolefin-elastomers) and 0.866–0.915 g/cm^3 (polyolefin-thermoplastics).

6.2.1.2 Processing

Processing of PE is non-critical. The broad variation of PE grades covers a wide range of processing conditions. Special grades with unique processing characteristics are used for particular applications and processes.

For injection molding, the melt temperatures for PE-LD range from 160–260 °C, for PE-HD from 260–300 °C, and the mold temperatures range from 50–70 °C and 30–70 °C, respectively. Easy flowing PE grades are used for mass production. The density and the shrinkage of parts made of these semi-crystalline resins are determined by their temperature history until demolding. Parts that were cooled quickly exhibit low crystallinity and minimal mold shrinkage, but also a high degree of post-mold shrinkage caused by post-crystallization at elevated temperatures. The results can be warpage and stress cracking caused by frozen-in stresses. This problem can be avoided when using PE grades with lower melt flow rates. Gating brittleness, resulting from strong molecular orientation, can be alleviated by an increase in melt temperature or grades with the highest applicable melt flow rate.

Only special grades of PE-HD-UHMW can be injection molded (parts up to 1 kg). Because of their low flow rates, they require machines with high injection pressures (≈ 1100 bar), no backflow valves, grooved feed zones if possible, and short flow distances. Processing temperatures range from 240–300 °C, mold temperatures range from 70–80 °C. PE-LLD is more difficult to process than PE-LD. As a rule, screw driving horsepower should be increased, output decreased; shrinkage properties are less favorable.

PE grades with higher melt viscosities and sufficient melt strength must be selected for blow molding so that the parisons do not tear under their own weight. Melt and mold temperatures range from 140 °C for PE-LD to 160–190 °C for PE-HD, respectively. PE-LLD is not well suited for blow molding because of its narrow molecular weight distribution and the resulting higher melt viscosity. However, it is well suited for rotational casting.

PE-LD is extruded at melt temperatures of 140–210 °C (films and pipes), 230 °C (cable sheathing), and 350 °C (coating). PE-HD requires a temperature increase by 20 to 40 K. These grades are also used to produce boards and monofilaments. High-pressure plasticization (2000–3000 bar) in twin screw extruders and ram extruders is used for PE-HD-UDMW profiles. Co-rotating twin screw extruders, running at approximately 10 RPM, are also suitable for profile extrusion of PE-HD-UHMW profiles (melt temperature 180–200 °C).

Extrusion of PE-LLD on equipment designed for PE-LD will result in a 20–30% reduction in output. This disadvantage is caused by the necessary reduction in screw length from 30 D to 25–20 D and by a reduction in RPM by 50%. Measures to compensate for the loss in throughput include: using screws of larger diameters, increasing the screw pitch, or increasing the die gap. Optimum melt temperatures range from 210–235 °C, for film extrusion from 250–280 °C.

Extremely tough reinforcement fibers can be obtained by progressive, controlled orientation (30-fold) of fibers under conditions that lead to almost single-crystalline orientation of the crystallites. These fibers exhibit a strength of 1–5 GPa, a modulus of elasticity from 50–150 GPa, and an elongation at break of approximately 5%.

Separation of PE from solution under shear results in cellulose-like fibers, so-called fibrides.

The Neopolen process uses pre-foamed particles, obtained by quenching from a melt containing foaming agent, to form pellets or particles by steam sintering PE-E (E = expanded). Simple parts are formed by compression molding at a pressure of 2–5 bar. Electrically conductive resins can easily be heated by a current flow. PE-HD and PE-LD can be compression molded at 105–140 °C.

Powder techniques (rotational molding, fluidized bed sintering) use PE powders with particle diameters of 30–800 µm, molded densities of 0.92–0.95 g/cm^3, and low volume flow rates. PE grades with higher flow rates are used for carpet backing and fusible reinforcement fabrics. Precipitated PE powders with uniform particle sizes of ≈ 50 µm are suitable for electrostatic coating of metals and fabrics. Still finer powders (8–30 µm) are dispersed in the beater in papermaking or in printing inks.

6.2.1.3 Post-Processing Treatment

PE is easily weldable with standard welding techniques: heated element-, friction-, hot gas-, ultrasonic-, and extrusion welding. Induction welding of conductive grades is possible; however, high-frequency welding cannot be used because of the non-polar character of polyolefins.

Adhesive joining and decorating of PE is difficult because of its non-polar character and low solubility. Section 4.14 provides more information regarding surface pre-treatments to increase bonding. Grafting PE-LLD with maleic anhydride (see Table 6.4, No. 31) provides excellent bonding strength and high heat distortion stability.

Parts made from plasma-treated PE-LLD-(M) powder can be coated with water-based lacquer or foamed with PU foam without pre-treatment when processed by rotational molding.

During machining of PE parts, it is important to make sure that the material does not get too hot, because it tends to become tacky as a result of frictional heating. Parts made of soft grades or with thin walls are easy to stamp.

6.2.1.4 Properties

Numerous grades of PE with widely differing properties can be produced by homo- and copolymerization and by the creation of low, medium, and high densities, low, medium, high, and ultra-high molecular weights, or narrow and broad molecular weight distributions, respectively. Low-molecular weight PE is used as an aid in plastics processing. Compared to other polymeric materials, all high-molecular weight polyolefins share low densities, relatively low strength and stiffness, high toughness and elongation at break, good friction and wear behavior, and very good electrical and dielectric properties (PE is non-polar). The tensile stress at yield increases approx. linearly with density. Water absorption and water vapor permeability are low. Permeation of oxygen, carbonic acid, and many aromatic and odiferous agents is substantial, but it decreases with increasing density. The maximum allowable temperatures for short-time exposure range from 80–120 °C (PE-LD-UHMW can be used from −268 °C to 150 °C, short-term) depending on grade; continuous service temperatures range from 60–95 °C (PE-LD-UHMW: 100 °C).

PE is resistant to water, saline solution, acids, alkalis, alcohols, and gasoline. Below 60 °C, PE is insoluble in all organic solvents, but it increasingly swells in aliphatic and aromatic hydrocarbons with decreasing density. Individual grades of high-density PE have been approved for applications such as containers for heating oil, gasoline, and vehicle fuel tanks. If the internal surfaces have been fluorinated (by fluorine-nitrogen blends) or sulfonated (by SO_3), these containers are impermeable to all kinds of fuels and hydrocarbons, see also Section 4.13.1.6 "Plasma Treatment". PE is not resistant to strong oxidizing agents, such as fuming sulfuric acid, concentrated nitric acid, nitration acid, chromium-sulfuric acid, and halogens as well as to some cleaning agents. Surfactants (washing and wetting agents) can induce environmental stress cracking in PE parts. PE-LLD and all PE grades with densities of ≈ 0.90 g/cm^3 provide the highest stress cracking resistance.

Addition of carbon black protects PE against photo-oxidation. PE can be crosslinked by high-energy radiation; it degrades in the presence of oxygen. PE burns like wax. Flame-retardant types, *e. g.*, for applications in construction, are available.

PE is odorless, tasteless, and physiologically inert. Most PE grades meet the current FDA guidelines for use in contact with food. The dependence of certain properties on structural parameters (density, molecular weight, and molecular weight distribution) is summarized in Table 6.1.

Metallocene PE grades (PE-M) distinguish themselves from conventional PE grades by a combination of properties:

- *Polyolefin-thermoplastics:* A new polymer class with high comonomer content offers improved penetration and perforation resistance, gloss, sealing capability, oxygen- and hydrogen permeability compared to PE-LLD (packaging of fresh foods).
- *PE-MD/HD-(M):* The modification of type and length of comonomer allows for tailoring of properties, such as toughness, stiffness, strength, and optical and organoleptic properties to meet particular requirements.
- *Polyolefin-elastomers* offer the following advantages compared to conventional polymers (in parentheses) (also as comonomers in TPO or PP): thermoplastic processing (crosslinked elastomers), shelf life (TPS), transparency, flexibility, and crack resistance (EVA, EMA), processability, cost effectiveness, and environmental

compatibility (PVC-P), and after crosslinking: long-term and temperature resistance (EPDM, IIR, SBR, NR).

Table 6.1 Structural Parameters and Properties of PE

Structural parameter	Density g/cm^3		Molecular configuration		MW-average		MW distribution M_w/M_n	
Limiting values	0.915	0.97	Highly branched	Linear, no or only short side chains	Low 20,000–60,000	High 200,000–400,000	Narrow	Broad
Degree of crystallization	−/+	++	−	++	−	+	+	−
MFI	▭		▭		++	−	▭	
Processability	+	−	+	−	+	−	-	+
Tensile and flexural strength	→		→		→		←	
Elongation at break	←		↔		→		←	
Stiffness and hardness	→		→		▭		←	
Impact strength	←		→		→		←	
Stress cracking resistance	↔		↔		→		←	
T_m and HDT	→		→		→		←	
Low temperature range	→		→		→		←	
Chemical and solvent-resistance	→		→		→		▭	
Vapor and gas diffusion resistance	→		→		▭		▭	
Transparency	←		←		▭		▭	

+, −: High and low values, respectively
→: Arrow indicates direction of positive effect
▭: No significant influence

PE properties can be customized for specific fields of application by the use of additives: glass fibers are used to improve rigidity and strength, anti-oxidants and UV-stabilizers for outdoor use, flame-retardants, foaming agents, anti-static agents, carbon black and other means to enhance electric conductivity (EMI, electromagnetic interference), slip agents, small amounts of fluoro elastomers to enhance melt strength and flowability, and pigments for coloring. A comparison of properties is provided in Table 6.2.

Table 6.2 Property Comparison of Polyethylenes

Property	Unit	Polyethylene					
		PE-LD	PE-MD	PE-HD	PE-UHMW	PE-LLD	PE-(M)**
Density	g/cm^3	0.915–0.92	0.925–0.935	0.94–0.96	0.93–0.94	ca. 0.935	0.904
Tensile modulus of elasticity	MPa	200–400	400–800	600–1400	700–800	300–700	75
Yield stress	MPa	8–10	11–18	18–30	ca. 22	20–30	7
Elongation at yield	%	ca. 20	10–15	8–12	ca. 15	ca. 15	–
Nominal elongation at break	%	>50	>50	>50	>50	>50	>50
Stress at 50% elongation	MPa	–	–	–	–	–	–
Ultimate stress	MPa	–	–	–	–	–	–
Elongation at break	%	–	–	–	–	–	–
Melting temperature	°C	105–118	120–125	126–135	130–135	126	100
Heat deflection temperature HDT/A 1.8 MPa	°C	–	30–37	38–50	42–49	ca. 40	–
Thermal coefficient of linear expansion, parallel (23–55 °C)	10^{-5}/K	23–25	18–23	14–18	15–20	18–20	–
Thermal coefficient of linear expansion, perpendicular (23–55 °C)	10^{-5}/K	–	–	–	–	–	–
Flammability UL 94 for 1.6 mm thickness	Class	HB*	HB*	HB*	HB*	HB*	–
Dielectric constant at 100 Hz	–	2.3	2.3	ca. 2.4	2–2.4	2.3	2.3

Property	Unit	Polyethylene					
		PE-LD	PE-MD	PE-HD	PE-UHMW	PE-LLD	PE-(M)**
Dielectric loss factor at 100 Hz	$\cdot 10^{-4}$	2–2.4	2	1–2	ca. 2	2	–
Specific volume resistivity	Ohm · m	$>10^{15}$	$>10^{15}$	$>10^{15}$	$>10^{15}$	$>10^{15}$	$2 \cdot 10^{4}$
Specific surface resistivity	Ohm	$>10^{13}$	$>10^{13}$	$>10^{13}$	$>10^{13}$	$>10^{13}$	–
Dielectric strength	kV/mm	30–40	30–40	30–40	30–40	30–40	–
Comparative tracking index CTI/A		600	600	600	600	600	–
Water absorption at 23 °C	%	< 0.05	< 0.05	< 0.05	< 0.05	< 0.05	–
Moisture absorption at 23 °C/50% rel. humidity, saturation	%	< 0.05	< 0.05	< 0.05	< 0.05	< 0.05	–

* Also available from V-2 to V-0 ** Produced with metallocene catalysts

6.2.1.5 Applications

Table 6.3 provides an overview of injection molding application of PE.

Table 6.3 Processing Properties and Applications for PE Injection Molding Resins

Density	0.92 g/cm³	0.93 g/cm³	0.94 g/cm³ [3] [2]	0.95 g/cm³	0.96 g/cm³
MFR 190/2.16: > 25 to 15[1]	Easy flow; mass products, not exposed to particular stress	Easy flow; large surface parts, little warpage, good gloss	Easy flow; impact resistant parts, no particular requirements on stiffness	Easy flow; little warpage, difficult to injection mold household goods	Easy flow; hard and stiff; bowls, sieves, dishes, transport cases[3] hard hats
15 to 5	Parts with higher strength, little surface gloss	Low stress parts with good surface gloss	Good impact resistance, little susceptible to stress corrosion, high performance technical parts	Easily processed, good impact strength, screw caps, closures, technical parts	Impact resistant, dimensionally stable, mechanically highly stressed parts, trash cans[3], seats

Table 6.3 Processing Properties and Applications for PE Injection Molding Resins *(continued)*

Density	0.92 g/cm³	0.93 g/cm³	0.94 g/cm³ 3 2)	0.95 g/cm³	0.96 g/cm³
ca. 1.5	Very good mechanical properties and resistance to stress corrosion		Good creep resistance, little susceptibility to stress corrosion, particularly stressed closures	Resistant to stress corrosion, good surface quality, high performance technical parts	
<1				Highly molecular, usually highly stabilized; pressure valves, pipe fittings	

[1] Extremely free-flowing PE-LD and -HD grades are available with MFR > 100
[2] Often PE-LD/HD blends; similar applications as for PE-LLD of low density
[3] Bottle cases and trash cans are subject to quality standards including limits for co-processing of recycled material

PE-LD: Main application areas are films for packaging, films for heavy-duty bags and sacks, shrink film, carrying bags, agricultural films, water vapor barriers in composite films, in which PE-LD copolymers are used as coupling agents with, *e.g.*, EVA, EAA, or EEA; pipes; sheets for thermoforming; sheathing for wires (also foamed and cross-linked PE); coating of steel pipes; flexible containers and bottles; canisters up the 60 l capacity; and tanks up to 200 l capacity. Blends with PE-LLD produce stretched films with higher extensibility.

PE-LLD: Films (extendable to 5 µm) with good optical properties, improved strength at low temperatures, resistance to tearing and penetration, and lower tendency for stress cracking than films made from PE-LD. Blown film made from blends with PE-LD (also as composite films) are increasingly substituting PE-LLD in rotational molding of canisters, canoes, and surfboards.

PE-HD: Housewares, storage and transportation containers, trash cans and trash containers up to 1100 l capacity, bottle cases, gasoline canisters, automotive fuel tanks. Special grades are used for high-pressure pipes, fittings for drinking water supply and waste water disposal (extruded diameters up to 1600 mm, even larger diameters for rolled pipes), sheets (also glass fiber reinforced) for devices for the chemical industry and in the automotive industry (lower cover for engine compartment). Fibrides are used as hydrophobic and hydrocarbon-binding additives in patching compounds for the absorption of oils, and as reinforcement for paper, fibers, and highly strengthened reinforcement fibers.

PE-HD-HMW: Surfboards up to 5 m in length; monofilaments for nets, cables, cords, ropes, and woven fabric; packaging film with a reduced thickness of 10 to 7 µm (from 20 µm) for carrying bags and with a thickness from 80–120 µm for inner linings of paper bags; sealing liners for landfills.

PE-HD-UHMW: Pressed blocks and filter-pressed sheets. Because of its excellent wear behavior, it is used to line hoppers and chutes for abrasive materials, machine elements such as screw conveyors, pump and slide elements, pulleys, gears, bushings, and rolls; sintered spacers with a porous core for lead batteries, surgical implants, prostheses, anti-friction coatings for skis.

PE-(M): Packaging films, heat sealing coatings, surface protection films, highly flexible gaskets and covers.

6.2.2 Polyethylene Derivatives (PE-X)

Crosslinked PE, PE-X

The linear PE macro-molecules can be 3-dimensionally *crosslinked (PE-X),* which will considerably improve creep resistance, low temperature impact resistance, and stress cracking resistance while hardness and stiffness are slightly reduced. Because PE-X behaves like an elastomer and will not melt, it can withstand higher thermal loads: short-term without additional mechanical loads up to 250 °C, long-term up to 120 °C. With increasing crosslinking density, the shear modulus increases even at elevated temperatures.

We distinguish between four types of crosslinking, all of which are used in extrusion operations: peroxide-(A)-, silane-(B)-, electron beam-(C)-, and azo-(D)-crosslinking. In injection molding, PE-HD is processed with peroxide crosslinking agents in the barrel within a precisely defined temperature range of 130–160 °C and then crosslinked in the mold at 200–230 °C. Injection molded and blow molded parts can also be crosslinked by irradiation.

Peroxide crosslinked PE-XA: The *Engel-process* uses a conveying machine with twin high-pressure plunger to continuously sinter pipes from a grit-like *PE-HD/peroxide mixture,* subsequently heats it in a heating barrel, molds it into its final shape, and crosslinks it at temperatures ranging from 200–250 °C, above the crystallization temperature. This creates a uniform network of macromolecules with low stresses and high flexibility. The material is softer and tougher than electron beam crosslinked PE-X. In a modified process, the *PAM (Pont-a-Mousson) process,* the extruded pipe is crosslinked in the sizing zone in a hot salt bath. The two-step *Daopex process* uses standard extruded PE-LD pipes and crosslinks them from the outside by exposing them to an epoxide emulsion under pressure and at temperatures above the crystallite melting temperature of PE. The Engel process achieves crosslinking degrees up to 99.5%.

Silane crosslinked PE-XB: With the Sioplas-, Hydro-Cure-, Monosil-, and Spherisil processes, silane-grafted PE compounds are processed with the addition of a silane crosslinking catalyst. The crosslinking process is started by a hot water-pressure treatment with the formation of Si-O-Si bridges.

Electron beam crosslinked PE-XC: PE parts made without crosslinking aids are crosslinked in special irradiation equipment at temperatures below the crystallization point under separation of hydrogen. Radiation sources are electron beam accelerators or isotopes (β- or γ-rays). The penetration depth can be adjusted (β-rays up to 10 mm, γ-rays up to 100 mm) so that the cores of thick walls or specified areas of a part will remain uncrosslinked.

The crosslinking of PE-HD compounded with an inhibited version of butyl perbenzoate is initiated immediately after extrusion or molding by *UHF radiation*.

Azo crosslinked PE-XD: Azo compounds are added to PE, which will crosslink in a subsequent hot salt bath with nitrogen under formation of crosslinks *(Lubonyl process).*

Application fields for PE-X are medium- and high-voltage cable coatings, pipes for hot water and radiant heating installation, parts for electrical engineering, chemical engineering, and automotive applications.

6.2.3 Chlorinated and Chloro-Sulfonated PE (PE-C, CSM)

Chlorination of polyolefins is achieved in solution, dispersion, or by direct exposure to chlorine gas. *PE-C* with a chlorine content of 25–30% is pliant to rubber-like soft and low temperature resistant. It is miscible with many plastics.

Applications: PE-C is blended with polyolefins to reduce flammability; with PVC to increase impact resistance; with PS for recycled commingled compounds. PE-C+PVC compounds with 70–90% PE-C content are used as unplasticized and bitumen resistant roofing and waterproofing material, films for liners, and profiles.

Several approaches can be used to crosslink PE-C and create a *PE-C elastomer*. Typically, PE-C is vulcanized by peroxides, because they provide higher valence forces than sulfur- or radiation crosslinking. The favorable price/property ratio has made PE-C elastomers competitive in cable *applications* and in the rubber industry, competing with CSM (chloro-sulfonated PE, Equation 6.1), CR (polychloroprene), EPDM (ethylene propylene diene rubber), and NBR (acrylonitrile butadiene copolymer). Its outstanding properties include: resistance to aging, weather and ozone resistance, flame resistance, oil resistance (even at elevated temperatures), wear resistance, low ductile-brittle transition temperature, and good processability.

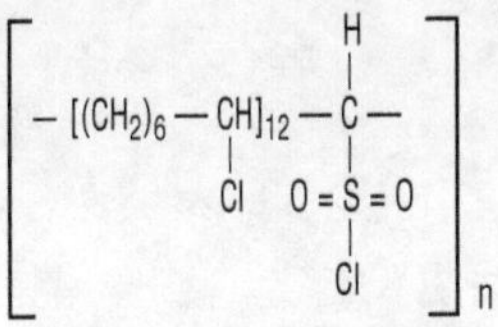

Chloro-sulfonated PE, CSM

(6.1)

6.2.4 Ethylene Copolymers (PE-ULD, EVAC, EVAL, EEA, EB, EBA, EMA, EAA, E/P, EIM, COC, ECB, ETFE)

Copolymerization of ethylene with propene, butene-1, vinyl acetate, acrylic acid ester, carbon monoxide, among others (for chemical structure of monomers, see Table 6.4) interrupts the linear structure of the methylene chain and thus reduces the degree of crystallinity. This results in a reduced melting temperature. The intramolecular forces and the glass transition temperature depend on the type and polarity of the comonomer. Table 6.5 compares selected characteristics.

Polar copolymers are exclusively produced with high-pressure processes. Copolymerization of PE with *non-polar* monomers, such as butene, hexene, and other higher α-olefins to produce linear PE of lower densities is carried out in low-pressure processes in gas phase-fluidized beds or in solution.

Table 6.4 Basic Structures of Olefin- and Vinyl Polymers and Their Copolymers (Terpolymers with PS, see Table 6.9)

	Chemical notation		Monomer structure	Homo-polymers	Copolymers with:		
No.		**R_1**	**R_2 (where -[ ]- is shown, the complete structure is given)**		**Ethylene**	**Styrene**	**Vinyl chloride**
1	Ethylene	–H	–H	PE		*	VCE
2	Propylene	–H	$–CH_3$	PP	E/P	*	
3	Butene-1	–H	$–CH_2–CH_3$	PB			
4	Octene	–H	$–(CH_2)_5–CH_3$				
5	Styrene	–H	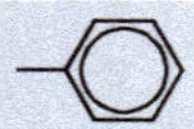	PS			
6	α-Methylstyrene	$–CH_3$	see 5	PMS		SMS	
7	para-Methylstyrene	–H	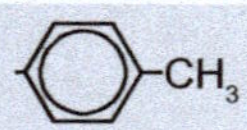				
8	Vinyl chloride	–H	–Cl	PVC			
9	Chlorinated ethylene	–H	–Cl	PE-C			VCPE-C
10	Vinylidene chloride		$-[C(Cl)(Cl)-C(H)(H)]-$	PVDC			VCVDC
11	Acrylic acid (acrylate)	–H	-C(=O)–O–H	PAA	EAA		

	Chemical notation		Monomer structure	Homo-polymers	Copolymers with:		
12	Methacrylate (acrylic acid methyl ester), (MA)	–H	$-C(=O)-O-CH_3$		EMA		
13	Ethyl acrylate (acrylic acid ethyl ester), (EA)	–H	$-C(=O)-O-CH_2-CH_3$		EEA		
14	Butyl acrylate (acrylic acid butyl ester), (BA)	–H	$-C(=O)-O-(CH_2)_3-CH_3$	PBA	EBA		
15	Octyl acrylate (acrylic acid octyl ester)	–H	$-C(=O)-O-(CH_2)_7-CH_3$				VCOA
16	Methacrylic acid	$-CH_3$	See 11	PMA			VCMAK
17	Methyl methacrylate (methacrylic acid methyl ester)	$-CH_3$	See 12	PMMA		SMMA	VCMMA
18	Ethyl methacrylate (meth-acrylic acid ethyl ester)	$-CH_3$	See 13				
19	Butyl methacrylate (methacrylic acid butyl ester)	$-CH_3$	See 14				
20	Vinyl acetate (VAC)	–H	$-O-C(=O)-CH_3$	PVAC	EVAC		VCVAC

Table 6.4 Basic Structures of Olefin- and Vinyl Polymers and Their Copolymers (Terpolymers with PS, see Table 6.9) *(continued)*

	Chemical notation		Monomer structure	Homo-polymers	Copolymers with:		
21	Vinyl alcohol	-H	-OH	PVAL	EVAL		
22	Vinyl ether	-H	-O–R ∣ R = $-CH_3$, $-(CH_2)_n-CH_3$				
23	Vinyl methyl ether	-H	$-O-CH_3$	PVME			
24	Vinyl pyrrolidone	-H	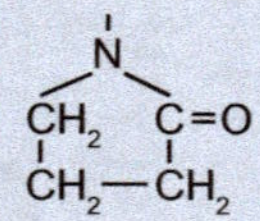	PVP			
25	Vinyl carbazole	-H	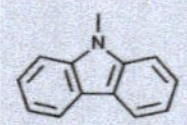	PVK			
26	Acrylonitrile	-H	-C≡N	PAN		SAN	
27	4-Methylpentene-1	-H	$-CH_2-CH(CH_3)_2$	PMP			
28	Vinyl butyral		$-[CH_2-CH-CH_2-CH]_m-[CH_2-CH]_n-$ (O–CH–O with $(CH_2)_2CH_3$; OH)	PVB			
29	Vinyl formal		$-[CH_2-CH-CH_2-CH]_m-[CH_2-CH]_n-$ ($O-CH_2-O$; OH)	PVFM			
30	Itaconic acid ester	CH_2COOR	$-C(=O)-O-CH_2-CH_3$				

	Chemical notation		Monomer structure	Homo-polymers	Copolymers with:		
31	Maleic anhydride		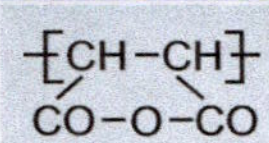			SMAH	VCMAH
32	Maleic imide		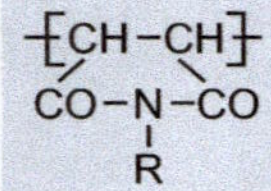 R = $-CH_3$, $-(CH_2)_n-CH_3$				VCMAI
33	Acrylic ester elastomer	–H	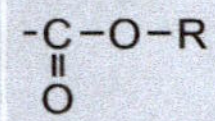 R = $-CH_3$, $-(CH_2)_n-CH_3$	ACM			
34	EPDM-rubber		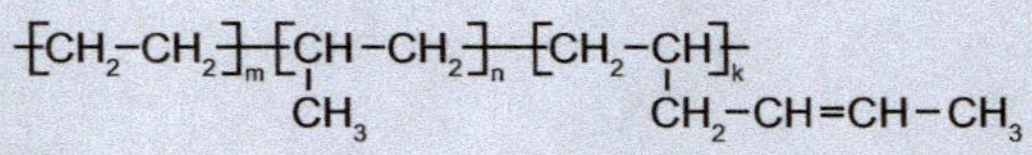	EPDM		SEPDM	
35	Butadiene rubber		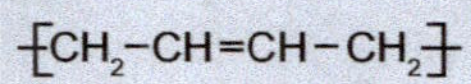	BR		SB	
36	Isoprene rubber		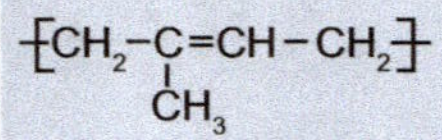	IR			
37	Thermoplastic PUR elastomer		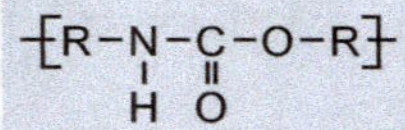	TPU			
38	Ethene butene		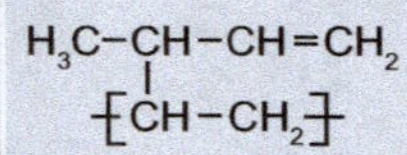				
39	Ethene propene		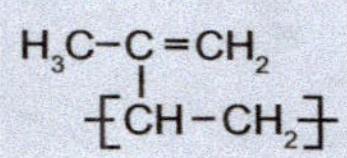				

Table 6.5 Property Comparison for Ethylene Copolymers and Other Polyolefins

Property	Unit	EVAC	EIM ionomer	COC 52% norbornene	PDCPD	EA	PB	PMP
Density	g/cm^3	0.93–0.94	0.94–0.95	1.02	1.03	0.925–0.935	0.90–0.915 (0.94)	0.83–0.84
Tensile modulus of elasticity	MPa	30–100	150–200	2600–3200	1800–2400	40–130	210–260 (420)	1200–2000
Yield stress	MPa	–	7–8	–		4–7	15–25	10–15
Elongation at yield	%	–	> 20	2–5	4	> 20	ca.10	>10
Nominal elongation at break	%	> 50	> 50	–		> 50	> 50	>10
Stress at 50% elongation	MPa	4–9	–	–		–	–	–
Ultimate strength	MPa	–	–	46–66	46	–	–	–
Elongation at break	%	–	–	2–10	25	–	–	–
Melting temperature	°C	90–110	95–110	80–180		92–103	125–130	230–240
Heat deflection temperature HDT/A 1.8 MPa	°C	–	–	75–170	90–115	–	55–60	40
Thermal coefficient of linear expansion, parallel (23–55 °C)	$10^{-5}/K$	ca. 25	10–15	ca. 6	8.2	ca. 20	13	12

Property	Unit	EVAC	EIM ionomer	COC 52% norbornene	PDCPD	EA	PB	PMP
Thermal coefficient of linear expansion, perpendicular (23–55 °C)	10^{-5}/K	–	–	–	–	–	–	–
Flammability UL94 for 1.6 mm thickness	Class	HB	HB	HB		HB	HB	HB
Dielectric constant at 100 Hz	–	2.5–3	ca. 2.4	ca. 2.4		2.5–3	2.5	2.1
Dielectric loss factor at 100 Hz	$\cdot 10^{-4}$	20–40	ca. 30	–		30–130	2–5	ca. 2
Specific volume resistivity	Ohm · m	> 10^{14}	> 10^{15}	> 10^{14}		> 10^{14}	> 10^{14}	> 10^{14}
Specific surface resistivity	Ohm	>10^{13}	> 10^{13}	–		> 10^{13}	> 10^{13}	> 10^{13}
Dielectric strength	kV/mm	30–35	40	–		30–40	20–40	
Comparative tracking index CTI/A		600	600	600		600	600	600
Water absorption at 23 °C, saturation	%	< 0.4	ca. 0.5	< 0.01		< 0.4	< 0.1	< 0.01
Moisture absorption 23 °C/50% rel. humidity, saturation	%	< 0.2	ca. 0.3	–		< 0.2	< 0.05	< 0.05

6.2.4.1 Ultra-Light Polyethylene (PE-ULD, PE-VLD)

PE-ULD and PE-VLD are co- and terpolymers of ethylene with up to 10% octene-, 4-methylpentene-1, and sometimes propylene, with densities ranging from 0.91–0.89 g/cm^3. They have a low degree of crystallinity, are transparent, and, with an elongation at break of > 900%, they are flexible in a wide temperature range.

Applications: penetration resistant stretch film for heavy loads, films for heavy-cargo bags, (multi-layer) barrier films and medical packaging, and parts manufactured by high-speed injection molding. These easily processable materials are also used to improve elastic properties and stress cracking of other PE grades. Butyl acrylate modified PE is similar to PE-LLD.

6.2.4.2 Ethylene Vinyl Acetate Copolymers (EVAC)

Gas permeability of PE increases with increasing vinyl acetate content and the materials become softer because the crystallinity decreases. Table 6.6 shows the properties and applications of copolymers with varying VAC content. Resins with more than 10% VA are more transparent, tougher, and easier to heat-seal than PE-LD; they are approved for use in food packaging. EVAC copolymers are grafted with vinyl chloride (VC), which makes them easier to process as plasticizers and impact modifiers.

EVAC is *processed* similar to PE-LD; however, the melt temperature depends on the VA content: 175–220 °C and mold temperatures of 20–40 °C. Extrusion temperatures range from 140–180 °C (for flat film up to 225 °C). To avoid separation of acetic acid, the melt temperature should not exceed 230 °C and residence time should be kept short. Flushing with PE-LD is recommended after production stops.

Printing and *coating* of EVAC is easier than of PE because of its polar structure. Pretreatment, usually with corona discharge, improves bonding strength.

Applications: Flexible pipe, profiles, wire- and cable coating, bags, gaskets, dust- and anesthesia masks, packaging and green house films, foldable and deformable toys, material for compounds.

EVAC (and ethylene/acrylic acid acrylate, EAA) are processed with paraffins, waxes, and synthetic resins and sometimes crosslinked with peroxides to manufacture polishing compounds, hot melt adhesives, and coatings. Modified and partially saponified EVAC is used as interlayer for compound glass.

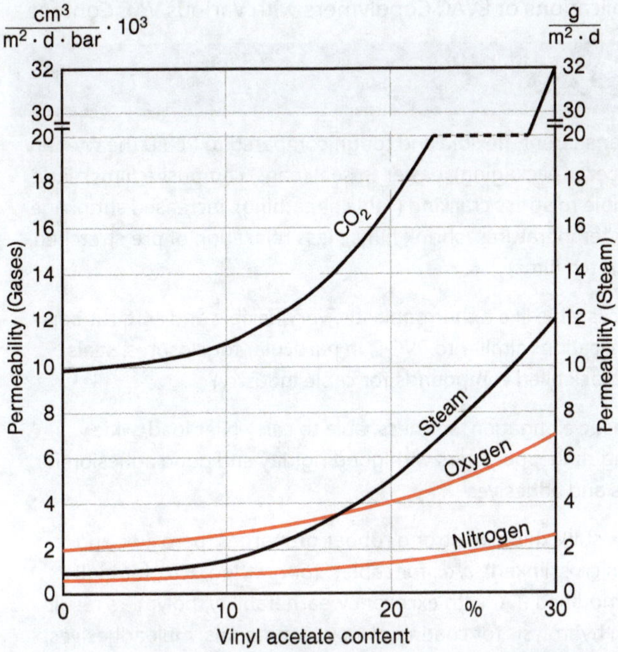

Figure 6.1 Water vapor and gas permeability of EVAC

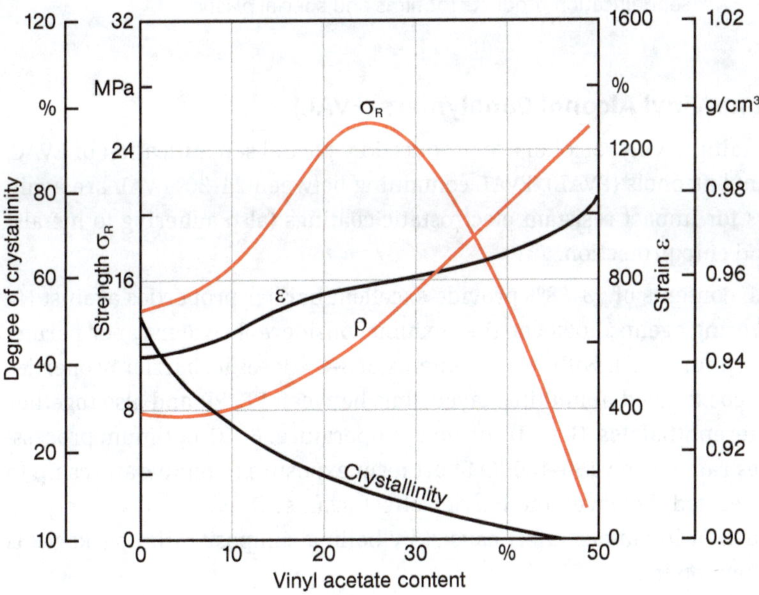

Figure 6.2 Properties of EVAC copolymers versus VAC content

Table 6.6 Properties and Applications of EVAC Copolymers with Various VAC Content

VAC-content of copolymers (wt.-%)	Properties and applications
1 to 10	More transparent, flexible and tough compared to PE-LD (heavy-duty films, freezer packaging), easier to seal (bags, composite films), little susceptible to stress cracking (cable sheathing), increased shrinkage at lower temperatures (shrink films), less relaxation of pre-stretched films (stretch films)
15 to 30	Still processable like a thermoplastic, very flexible and soft, rubber-like (applications similar to PVC-P, in particular for closures, seals, carbon black filled compounds for cable industry)
30 to 40	High elastic elongation, softness, able to carry filler load, wide softening range, polymers with good rigidity and good adhesion for coatings and adhesives
40 to 50	Products still exhibiting strong rubber properties (peroxidic and radiation crosslinked), *e. g.*, for cables; for grafting, *e. g.*, for high impact modified PVC with excellent weatherability; polymers resulting from hydrolysis for coating of woven materials, melt adhesives, thermoplastic processing for parts and films with high rigidity and toughness
70 to 95	Used as lattices for emulsion paints, paper coating, adhesives, and saponification products for films and special plastics

6.2.4.3 Ethylene Vinyl Alcohol Copolymers (EVAL)

Copolymers containing vinyl alcohol are produced by partial saponification of EVAC, just like poly vinyl alcohols (PVAL). EVAL containing between 24–30% VAL are available as powders for impact resistant electrostatic coatings (also adhering to metals) for corrosion and chip protection.

Grades with VAL contents of 53–68% provide excellent barrier properties against N_2, O_2, CO_2, and flavoring agents; however, they exhibit considerable water vapor permeability and -absorption. EVAL with water contents of 3–8% loses its barrier properties and is therefore coextruded in multiple-layer films between PE, PP, and also together with PA or polyterephthalates. Glass transition temperature: 66 °C, optimum processing temperatures range from 160–180 °C. Short-term exposure to temperatures up to 200 °C can be tolerated; however, the barrier effect against O_2 will be reversibly reduced. The reduction in barrier effect caused by boiling composite film packages is also reversed after cooling.

PVALs are semi-crystalline, water soluble polymers. Saponified grades dissolve at acceptable rates only in hot water. Partially saponified grades dissolve easier in cold than in hot water. Both PVAL types are used for the manufacturing of films: monofilms for water-soluble bags and as coextruded film because of their extremely low O_2 permeability.

6.2.4.4 Ethylene Acrylic Copolymers (EEA, EBA, EAA, EAMA, EMA)

Ethyl acrylate- (EA), butyl acrylate- (BA), and methacrylate- (MA) copolymers with PE are used at low temperatures for high-elasticity, stress crack resistant, highly filled packaging films and hot sealing layers. EMA with up to 20% and EEA with up to 8% comonomer content are approved for food contact applications. Carbon black-filled, semi-conductive films and tubes are used for microchip packaging, explosive packaging, and medical and other applications where static electricity is a hazard.

EAA, EAMA, and EMA, ethylene-meth(acrylic acid) copolymers, and terpolymers modified by acrylamide are used as intermediate bonding layers in multi-layer film, *e.g.*, between PE and PA, and coextruded for metal coating. They are basic components of ionomers and useful as compatibilizers in blends, for example of PE+PET.

Applications: EAA see also EVAC (Section 6.4.2.4).

6.2.4.5 PE α-Olefin Copolymers (PEα-PO-(M))

Copolymers of PE and α-olefins produced with metallocene catalysts (PE-LLD (M)) are polymers with a linear structure and narrow molecular weight distribution. They exhibit low densities (0.903–0.917 g/cm^3), favorable optical properties, and high toughness.

Applications: Packaging film for medical devices and food.

6.2.4.6 Cycloolefin Copolymers (COC, COP)

COCs are copolymers of linear (ethene) and cyclic olefins (technically usually norbornene).

They can be produced with metallocene catalysts in two categories: random (amorphous) and alternating (semi-crystalline) polymers, the latter exhibiting better solvent and chemical resistance. The characteristics for COCs can be tailored in a broad range during polymerization. For example, the glass transition temperature of COC can be adjusted from 0 °C for 12 mol-% to 230 °C for 80 mol-% norbornene content. COC polymers are highly transparent, blood compatible, sterilizable, metallizable, easy processable, and hydrolysis resistant. Injection molded parts show very little optical anisotropy, the index of refraction is 1.53, and the density is 1.02 g/cm^3.

Applications: CDs, other optical data storage media, optical, pre-filled transparent syringes, biaxially stretched condenser films, H_2O barrier films, blister films, and cosmetics.

Table 6.5 provides a comparison of characteristics.

Polymerization and subsequent hydrogenation of *dicyclopentadiene* produces an amorphous COC copolymer (COP) with high transparency, low birefringence, good chemical resistance, low moisture absorption, and temperature resistance from 120–130 °C (new developments: T_g = 100–165 °C).

Applications: Medical devices, such as syringes, and blister film, LCD, optical projection plates, condenser films.

6.2.4.7 Ionomers (EIM)

In contrast to conventional plastics, here ionic bonds (electrostatic bonds) exist between the molecular chains in addition to the secondary valence bonds. Ion contents up to 10% in otherwise non-polar materials define ionomers; materials with higher ion contents are called polyelectrolytes. Ionomers are typically synthesized by copolymerization of a functionalized monomer (*e. g.*, acrylic acid, methacrylic acid, or p-styrene sulfonic acid) with an olefinic monomer and subsequent salt formation. An important group are thermoplastic copolymers of ethylene with monomers containing carboxylic groups, such as acrylic acid, some of which are free carboxylic groups, while the rest is linked with metal cations of the 1st and 2nd group in the periodic table, facilitating a certain level of physical crosslinking, see Equation 6.2. At elevated temperatures, these crosslinks are dissolved so that ionomers can be processed by all standard thermoplastic processes at melt temperatures between 150–260 (330) °C. The melt has a high degree of elasticity, thus facilitating the production of sheet and film. Because of their polar groups, ionomers bond well to various carrier materials, and non-porous films of only 12 µm thickness can be extruded. These films are easily warm drawn with a favorable draw ratio.

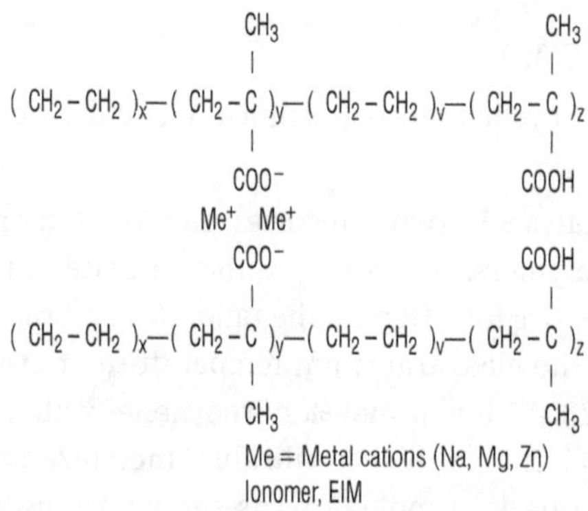

(6.2)

General properties: Ionomers are non-crystalline. Over the range of their service temperature (−40 to +40 °C) and above these materials are tough and crystal-clear. They are resistant to alkalis, weak acids, fats, and oils. Organic solvents will merely swell them. They are not resistant to oxidizing acids, alcohols, ketones, aromatic and chlorinated hydrocarbons, and their susceptibility to stress cracking is slight. Their permeability to water vapor, O_2, and N_2 is comparable to that of PE; they are less permeable to CO_2. Ionomers burn with a bright flame. There are grades available approved for food contact.

Applications: Crystal-clear pipes for drinking water, wine, and fruit juice; transparent films for fat-containing foods; parts for laboratory and medical applications; skin- and blister packaging; bottles for vegetable oil, liquid fats, and shampoos; shoe soles, ski shoe shells, tool grips, transparent coating, corona- and stress cracking resistant insulation.

Ionomers act as bonding agents between polymers of varying polarity in alloys and composites. Low-molecular polymers are useful additives for the homogenization of dispersions and to reinforce adhesive bonds.

6.2.4.8 Ethylene Copolymer Bitumen Blends (ECB, ECB/TPO)

ECBs are blends of ethylene copolymers (ethylene/butyl acrylate copolymer) and special bitumens. In ECBs, the polymer forms the continuous phase. Colored roofing and sealing sheets are available from ECB/TPO (thermoplastic polyolefins) blends.

Applications: Sealing films (also with glass- or polyester reinforcement) for flat roofs, tunnels, and in civil engineering, as additives to bitumen- and asphalt materials to improve structural stability.

Copolymers with fluoropolymers (ETFE) are presented in Section 6.5.2.

6.2.5 Polypropylene Homopolymers (PP, PP-H)

PP is polymerized from propylene. Like PE, it is a semi-crystalline thermoplastic material; however, it exhibits higher strength, stiffness, and crystalline melting temperature at lower densities (0.905–0.915 g/cm^3). A wide variety of PP grades is available (see Table 5.7). The product palette and area of applications for these materials is continuously expanding. Worldwide, 50 million tons of PP are consumed annually. New PP grades with tailored property profiles are produced with special catalyst systems. High-molecular homo- and copolymers, as well as block copolymers and elastomer-modified grades are of interest. Transparent grades, grades with low stiffness, grades with high stiffness and good heat distortion stability, and scratch resistant grades are available. In certain application areas, PP grades will replace higher priced materials, such as ABS, PS, PET, PC, and TPE.

6.2.5.1 Chemical Constitution, Polymerization

The CH_3 groups of PP (Equation 6.3) can be sterically arranged in different configurations and thus create different properties during polymerization, see Figure 2.6 in Section 2.3.1.2.

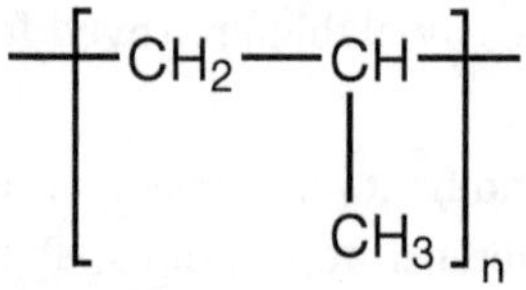

(6.3)

The majority of CH_3 groups in *isotactic PP* (PP-I) are arranged either on the same side of the C-chain or helically turned outward. In *syndiotactic PP* (PP-S), the CH_3 groups alternate on the opposing sides of the main chain, while in *atactic PP* (PP-A or PP-R for random) the CH_3 groups are arranged randomly on both sides of the chain. Atactic PP has the consistency of non-vulcanized rubber. Quantitatively, isotactic PP is the most important grade; it is technically characterized by its index of isotacticity, which indicates the percentage of polymer insoluble in boiling xylene.

The basic *synthesis* is low-pressure precipitation polymerization of propene gas at the surface of *Natta* (1955) organometallic catalysts. These are stereo-specifically effective Ziegler catalysts dispersed in hydrocarbons. A certain ratio of atactic PP is generated as a by-product, creating a softer and less temperature-resistant material. This by-product is heptane-soluble and can be separated. The newer gas-phase polymerization processes produce high yields of pure products (97% isotactic PP) using minimum amounts of selectively adjustable high-efficiency catalysts. The Spheripol process with its “Catalloy”-catalyst systems, which can be tailored to produce different homo- and copolymer grades, produces particles with 0.5–4 mm diameters that are processable without additional granulation. These gas-phase processes also facilitate the production of PP blends of normally incompatible amorphous thermoplastic materials, creating a continuous PP matrix with an evenly distributed amorphous phase (reactive blending).

The properties of PP can be specifically adjusted by polymerization with *metallocene catalysts* (PP(M)). With this technology, tacticity, molecular weight, molecular weight distribution, and comonomer content can be tailored to meet customer requirements. In addition, PPs with narrow molecular weight distribution and lowered melt viscosities (PP-CR, controlled rheology) are produced, *e. g.*, by adding organic peroxides during mixing or processing.

6.2.5.2 Processing

PP grades for *injection molding* cover a broad range of requirements, ranging from high-temperature resistant, rigid to elastic, to low-temperature impact resistant grades. The melt temperature ranges from 250–270 °C, the mold temperature from 40–100 °C. Moisture can condensate on the pellets’ surface in humid climates and should be removed by drying or with the help of vented extruders before processing. Blown films, flat films, sheet, pipe, blow molded parts, and monofilaments are *extruded* at melt temperatures ranging from 220–270 °C. Because of the high cooling demands during film production, chill-roll flat sheet die processes are preferred over tubular film blowing, which requires intensive water-cooling of the film tube. To achieve brilliant transparency in PP-H films, they must be shock-cooled below the crystalline melting temperature. Other than films for biaxial orientation, most films contain slip- and anti-blocking agents. Highly viscous PP-H grades are used for extrusion blow molding at temperatures from 190–220 °C.

Melt spin processes are used to produce staple fibers from low-viscosity, highly isotactic PP-H; *drawing* of blown and flat films produces woven film tapes and fibrillated fibers; precipitation of PP under shear from solution produces fibrides.

6.2.5.3 PP Foams (PP-E)

PP foams are produced by the following processes:

- Extrusion to produce flexible foams with small closed cells and extremely low densities (10 kg/m^3),
- High-pressure processes for rigid foams with densities from 50–120 kg/m^3,
- Injection molding with gas or chemical foaming agents to produce structural foams with densities from 400–700 kg/m^3,
- Extrusion of rigid profiles in the same density range.

6.2.5.4 Post-Processing Treatments

The narrow thermo-elastic temperature range of PP makes *thermoforming* of flat parts difficult. However, PP can be melted by solid phase pressure processes at temperatures just below the crystalline melting temperature (150–160 °C) and molded, pressed, or rolled at room temperature.

PP must be pre-treated prior to any *surface finishing* (coating, printing). PP can be vacuum-metallized; specific grades can be galvanized after surface activation by precious metal salts. In general, a nickel layer is applied first, and subsequently plated with nickel or chromium.

Similar to PE, PP can be *joined* by welding and adhesives. Contact adhesives on the basis of natural rubber or chlorobutadiene rubber, as well as adhesives based on silicone, epoxy resins, or polyurethanes are generally used. Surfaces are pre-treated with chromium sulfuric acid. Diffusion bonding is not possible.

Machining of PP is easier than of PE because of its greater hardness; however, stamping is usually not possible.

6.2.5.5 Properties

The variety of available PP grades is wider than that of most other plastic materials. Molecular structure, average molecular weight (200,000–600,000 g/mol), molecular weight distribution, crystallinity, and spherulite structure can be varied over a broad range, thus determining the properties.

Stiffness and hardness range between those of PE and engineering plastics, such as ABS, PA, and others. The dynamic load capacity is relatively high. With a glass transition temperature of 0 °C, all PP-H grades embrittle at low temperatures. The crystalline melting point ranges from 160–165 °C, higher than that of PE. Therefore, maximum service temperatures are also higher: short-term 140 °C, long-term 100 °C. The electrical properties compare to those of PE and are not affected by exposure to water. The dielectric constant and the dielectric loss factor are largely independent of temperature and frequency (see Table 6.7).

PP exhibits only minimal water absorption and permeability. Grades approved for food contact can be hot-filled and hot sterilized. Gases, in particular CO_2, low-boiling hydrocarbons, and chloro-hydrocarbons diffuse through PP. PP swells in contact with chloro-hydrocarbons. Because of its non-polar structure, PP is chemically highly resistant: up to 120 °C resistant to aqueous solutions of salts, strong acids and alkalis, as well as brines. High-crystalline PP grades provide particularly good resistance to polar organic solvents, alcohols, esters, ketones, fats, and oils. Only special grades are resistant to fuels at elevated temperatures. Strong oxidants, such as chlorosulfonic acids, oleum (disulfuric acid), concentrated nitric acid, or halogens attack PP even at room temperature.

For outdoor exposure PP needs weatherproofing; yet, it still cannot compete with PE with regard to weather resistance. PP continues to burn with a light luminescent flame after removal of the ignition source; flame-resistant grades are available.

Parts made from PP are translucent. Orientation below the crystalline melt temperature increases crystallinity, as does the addition of nucleating agents to create a fine grain crystalline structure.

Compared to conventional PP, the properties of *syndiotactic PP* can be modified: these include the possibility to lower the melting point to ≈ 150 °C, increase gloss and transparency while maintaining stiffness (thus competing with PS because of the higher impact resistance), decrease of melt viscosity, and improve barrier properties.

Table 6.7 Comparison of Properties for Special PP Grades

Property	Unit	Polypropylene								
		PP-H Homo-polymer	PP-R Random polymer	PP-B Block-copolymer	PP + EPDM	PP-T 20 Talcum	PP-T40 Talcum	PP-GF30 Glass fiber	PP-GFC30 Glass fiber, chem. coupled	PP-B25 Barium
Density	g/cm^3	0.90–0.915	0.895–0.90	0.895–0.905	0.89–0.92	1.04–1.06	1.21–1.24	1.12–1.14	1.12–1.14	1.13
Tensile modulus of elasticity	MPa	1300–1800	600–1200	800–1300	500–1200	2200–2800	3500–4500	5200–6000	5500–6000	1850
Yield stress	MPa	25–40	18–30	20–30	10–25	32–38	30–35	–	–	26
Elongation at yield	%	8–18	10–18	10–20	10–35	5–7	3	–	–	–
Nominal elongation at break	%	> 50	> 50	> 50	> 50	> 20	4–10	–	–	–
Stress at 50% elongation	MPa	–	–	–	–	–	–	–	–	–
Ultimate strength	MPa	–	–	–	–	28–30	30	40–45	70–80	–
Elongation at break	%	–	–	–	–	15–20	3–15	3–5	3–5	–
Melting temperature	°C	162–168	135–155	160–168	160–168	162–168	162–168	162–168	162–168	–

Table 6.7 Comparison of Properties for Special PP Grades *(continued)*

Property	Unit	Polypropylene								
		PP-H Homo-polymer	PP-R Random polymer	PP-B Block-copolymer	PP + EPDM	PP-T 20 Talcum	PP-T40 Talcum	PP-GF30 Glass fiber	PP-GFC30 Glass fiber, chem. coupled	PP-B25 Barium
Heat deflection temperature HDT/A 1.8 MPa	°C	55–65	45–55	45–55	40–55	60–80	70–90	90–115	120–140	53
Thermal coefficient of linear expansion, parallel (23–55 °C)	10^{-5}/K	12–15	12–15	12–15	15–18	10–11	8–9	6	6	0.7
Thermal coefficient of linear expansion, perpendicular (23–55 °C)	10^{-5}/K	–	–	–	–	10–11	8–9	7	7	
Flammability UL94 for 1.6 mm thickness	Class	HB	HB	HB	HB	HB	HB	HB	HB	HB
Dielectric constant at 100 Hz	–	2.3	2.3	2.3	2.3	2.4–2.8	2.4–3	2.4–3	2.4–3	ca. 2.6

Property	Unit	Polypropylene								
		PP-H Homo-polymer	PP-R Random polymer	PP-B Block-copolymer	PP + EPDM	PP-T 20 Talcum	PP-T40 Talcum	PP-GF30 Glass fiber	PP-GFC30 Glass fiber, chem. coupled	PP-B25 Barium
Dielectric loss factor at 100 Hz	$\cdot 10^{-4}$	2.5	2.5	2.5	2.5	7–10	12–15	10–15	10–15	ca. 20
Specific volume resistivity	Ohm · m	$>10^{14}$	$>10^{14}$	$>10^{14}$	$>10^{14}$	$>10^{14}$	$>10^{14}$	$>10^{13}$	$>10^{14}$	$<10^{15}$
Specific surface resistivity	Ohm	$>10^{13}$	$>10^{13}$	$>10^{13}$	$>10^{13}$	$>10^{13}$	$>10^{13}$	$>10^{13}$	$>10^{13}$	$<10^{13}$
Dielectric strength	kV/mm	35–40	35–40	35–40	35–40	45	45	45	45	ca. 28
Comparative tracking index CTI/A		600	600	600	600	600	600	600	600	600
Water absorption at 23 °C, saturation	%	< 0.2	< 0.2	< 0.2	< 0.2	< 0.2	< 0.2	< 0.2	< 0.2	–
Moisture absorption 23 °C/50% rel. humidity, saturation	%	< 0.1	< 0.1	< 0.1	< 0.1	< 0.1	< 0.1	< 0.1	< 0.1	< 0.01

6.2.5.6 Applications

Their special properties provide an exceptionally wide range of applications of PP, a fact that the resin suppliers pay tribute to in their formulations.

Amorphous atactic PP and other atactic α-olefinic polymers are flexible to rigid plastics at temperatures as low as −30 °C. They are used as (melt) coatings for paper packaging and carpet tile backing, automobile insulation, corrosion protection, paint for road markings, hot-melt adhesives, sealing compounds, bitumen blends, and aging-resistant roofing and weather strips.

Injection molding of isotactic PP: Sound-absorbing automobile interiors that meet requirements of head impact tests, ventilation systems, dashboards, center consoles, front- and rear light housings; mass-produced articles such as cups and food containers, tool boxes, suitcases, transport and stacking boxes, housings and functional parts, such as living hinges, for household appliances, such as coffee makers, toasters, heat blowers, dish washers, dryers, and washing machines; filled and reinforced grades for automotive body parts, automotive expansion tanks for coolants (for 2.3 bar at 125 °C), lawn mowers, electric tools, parts for submerged pumps, electric installations, lawn furniture.

Blow molding: highly viscous grades (PP-Q, polypropylene with high impact resistance) for bottles for cosmetic and medical powders; somewhat less viscous grades for mass production of containers up to 5 l – also biaxially stretched; hot water containers, automotive air ducts, cases with living hinges for tools, sewing machines, electronic parts (antistatic), motorcycles, and surfboards.

Extrusion: high-pressure warm water and waste water pipes, profiles, sheet, cable sheathing, unstretched and biaxially stretched (oriented) film (PP/BO, transparent) for packaging, insulation, and as composite film, foamed sheet, packaging tape, woven film tape, non-woven fabric for geo textiles and filters, cords for agricultural applications (degradable), sacks, glass mat reinforced sheet (GMT) for hot-press molding.

Preferred applications for *PP(M)*: transparent packaging (injection molded, flat sheet), thin-wall applications, carpet yarns, staple fiber; for *PP-CR*: packaging (thin-walled cups), coatings, and fibers for filter- and hygiene applications.

6.2.6 Polypropylene Copolymers and Derivatives, Blends (PP-C, PP-B, EPDM, PP+EPDM)

In addition to using additives, the characteristics of PP can also be modified in a wide range by copolymerization and blending with other polymers.

6.2.6.1 Chlorinated PP (PP-C)

PP-C is of less importance than PE-C. It is used for chemical- and corrosion-resistant coatings.

6.2.6.2 PP Copolymers (PP-B)

Ethylene, butene-1, and higher α-olefins are used as copolymers (see Table 6.4, No. 1, 3, and 27). Sometimes, PP-B grades are also called block-copolymers, although they are heterophasic blends of homo- and copolymers. The insertion of PE interrupts the molecular chain; however, the crystallinity of PP is maintained up to a PE content of 20%. PE lowers the glass transition temperature by 5 °C. The melt temperature of a random PE copolymer is substantially lowered, even at low PE contents. If PE is inserted as a block, the melt temperature is not substantially lowered; 10% PE content significantly improves the impact resistance at low temperatures (−30 to −40 °C); 20% PE results in flexible products, such as PE-LD with a melting point above 160 °C. It is used for hot-sealing layers, flexible pipes, and transparent injection molded and blow molded parts.

6.2.6.3 Ethylene-Propylene (Diene) Copolymers (EPDM)

Copolymerization of PE, PP, and ethylene/norbornene (for terpolymers) in hexane with Ziegler catalysts yields EPDM. Norbornene is obtained by synthesis of ethylene and cyclopentadiene; it is a raw material for synthetic rubbers and is pressed in bale form, while PP-EPDM compounds are provided as pellets.

Random copolymers (amorphous EPDM): Macromolecular mobility of linear PE freezes only at temperatures below −100 °C. However, this property does not result in the expected flexibility because PE-HD is semi-crystalline. Hence, in order to achieve an amorphous product, some H-atoms in the PE molecule must be replaced by random distributed polar groups. Copolymers with α-olefins, such as PP or butene-1, with up to 70% PE are amorphous, 3-dimensionally cross-linked, exhibit extremely low densities of 0.86–0.87 g/cm^3, and a glass transition temperature sufficiently below room temperature. These products can be processed like rubbers. Prerequisite for being vulcanizable with sulfur and for predetermining the degree of crosslinking is the insertion of dienes. E/P copolymers (EPM) can only be chemically crosslinked with the help of peroxides; however, this makes optimization of their crosslinking degree and -density impossible. EPDM with a PE content of more than 50% cannot be processed like a thermoplastic; it is used in automotive and construction applications and in the cable industry.

Sequential copolymers (semi-crystalline EPDM) are physically cross-linked. The cross-linking bonds are formed between crystalline or glass-like solid parts of the polymer chain. In addition, they can be vulcanized. With a PE content of at least 70%, the PE sequences are long enough to form such domains. These materials can be processed like thermoplastic; however, the physical bonds have the disadvantage that, with rising temperatures and depending on block structure, they will start to break, thus losing the elastomeric character of the material. Both vulcanized and unvulcanized materials exhibit good weather resistance and weldability and are therefore used as sealing membranes for roofs and floors. Because they are easily filled with heavy fillers, they are used as sound insulating layers, *e.g.*, in automotive applications.

6.2.6.4 PP+EPDM Elastomer Blends

The ideal structure of a polyolefin elastomer consists of blocks of amorphous ethylene and propylene sequences in a random distribution together with fixed PP blocks. These structures do not necessarily have to be connected to each other in a chain, but can be achieved by blending PP with EPDM. These products exhibit high stiffness and softening temperatures, are easily modified by copolymerization of PP, and are compatible with EPDM.

UV stabilization is achieved by the addition of either carbon black or sterically hindered amines, if light-colored products are required. There is also the possibility of coating with flexible PUR to achieve UV resistance. Stiffness can be enhanced by chalk or glass fibers.

The properties of PP + EPDM elastomers depend on their blend ratio. 90% PP results in the properties of conventional PP with slightly lower stiffness and softening temperature, but also with increased impact resistance at −40 °C. Blends with 40% PP exhibit the typical properties of thermoplastic rubbers. Other determining factors are crystallinity, molecular weight, and molecular weight distribution of the PP. In addition, it is important whether a homo- or a copolymer, random or sequential PP is used. It is also possible to create blends with PE.

All processing methods used for PP can be used with PP-elastomer blends. Highly viscous grades are extruded, blown, or pressed at ≈ 250 °C. Less viscous grades are injection molded at melt temperatures from 220–260 °C and mold temperatures of ≈ 60 °C.

Main application area is the automotive industry: bumpers, spoilers, coverings for wheel houses and trunks, mud flaps, dashboards, consoles and other interior parts, steering wheel covers; flexible tubes and pipes in construction; shoe-, sport-, and toy industry.

6.2.6.5 Polypropylene Blends

Blends of PP-I and, *e.g.*, methyl methacrylate or styrene are produced by reactive blending. They have the following advantages: low density (0.91–0.96 g/cm^3), weather resistance, scratch resistance, little processing shrinkage (little warpage) and moisture absorption.

Compounds with 3–6% hydrocarbon resins, for example, hydrogenated dicyclopentadiene (DCPD), increase the glass transition temperature of PP films by up to 25 K and thus the modulus of elasticity by up to 50%, while reducing water vapor permeability by up to 30%.

6.2.7 Polypropylene, Special Grades

Additives, such as nucleating agents, cause a finer spherulite structure and thus higher transparency and flexibility, but they also reduce stiffness and lower the heat distortion point. The addition of powdered peroxides during compounding or processing creates radicals that split hydrogen from the molecular chain, thus leading to a narrower molecular weight distribution (PP-CR). This reduces the melt viscosity and facilitates processing. Because copper ions catalyze thermo-oxidation, PP should be appropriately stabilized for applications such as insulation of copper wires. Applications in washing machines require stabilization against alkaline solutions and heat aging. Antimony trioxide in combination with halogen compounds and phosphoric acid esters are flame retardant (HB to V-0 according to UL 94). Grades suitable for plating contain pigments that cause finely cracked surfaces, which in turn increase the adhesion of the first metal layer during plating. Carbon black is used for weather stabilization; for colored applications, amines are used.

Modified grades with higher melt stability are used for extrusion and blow molding.

Talc is the most common filler used with PP. It improves stiffness, dimensional stability, heat resistance, and creep behavior, and it also serves as a nucleating agent. Disadvantages are the decrease in low temperature impact resistance, the reduction in weldability and oxidation resistance at elevated temperatures, and the formation of matte surfaces. Calcium carbonate has the same effect as talc but has additional advantages: easier to disperse, better flowing melt, higher UV- and oxidation stability, higher surface quality, less tool wear, and reduced cycle times for injection molding; 40% of mica increases stiffness to the same extent as 30% glass fibers at lower cost. Wood flour increases acoustic insulation. Calcium silicate increases impact resistance and electrical and thermal properties. Zinc oxide protects against microorganisms and increases UV-resistance.

Scratch resistant surfaces similar to those of ABS can be achieved with special coated fillers in combination with additives and mineral fillers.

Glass fiber reinforced PP is available with ground and cut glass fibers. Short glass fibers increase the stiffness and lower impact resistance, while longer glass fibers increase structural integrity, strength, and creep resistance. Chemically linked glass fibers enhance this effect. Particularly long glass fibers cause anisotropic shrinkage as a result of their fiber orientation (thus causing warpage), matte surfaces, and increased wear. The addition of *glass spheres*, also in combination with glass fibers, creates reinforced PP that exhibits higher stiffness and compressive strength as well as a lower tendency for warpage (see Table 6.8).

Table 6.8 Comparison of Characteristic Properties for GMT and Long Glass Fiber Reinforced PP (LFT) (base material: polypropylene; all characteristics were determined with test specimens)

Property	Unit	GMT		D-LFT	
		PP-GM 30	PP-GM 40	PP-GF 30	PP-GF 40
		Glass mat reinforced, needle punched		Reinforced w/long glass fibers, direct process	
Density	g/cm^3	1.13	1.22	1.13	1.22
Glass content	wt.-%	30	40	30	38
Tensile strength	MPa	70	90	70	80
Tensile modulus of elasticity	MPa	4500	5500	5000	6000
Elongation at break	%	1.8	1.6	2.0	1.8
Coefficient of expansion	10^{-6} K^{-1}	ca. 30	25–30	30–40	20–30

Glass mat reinforced PP sheet (GMT) for compression molding operations (see Sections 4.8.2.2 and 4.8.2.3) and long fiber reinforced compounds are used to produce components for high-load applications, particularly in the automotive industry.

Properties: Processing temperature 215 °C, heat distortion point ≈ 150 °C, short-term service temperature 140 °C, long-term 100 °C (see Table 6.8).

6.2.8 Polybutene (PB, PIB)

There are two forms of isomeric butene of technical interest: butene-1 and isobutene. The linear monomer butene-1 is used for thermoplastic polybutene-1 (PB, see Equation 6.4), while the branched isobutene is used for rubber-like polyisobutene (PIB, see Equation 6.5).

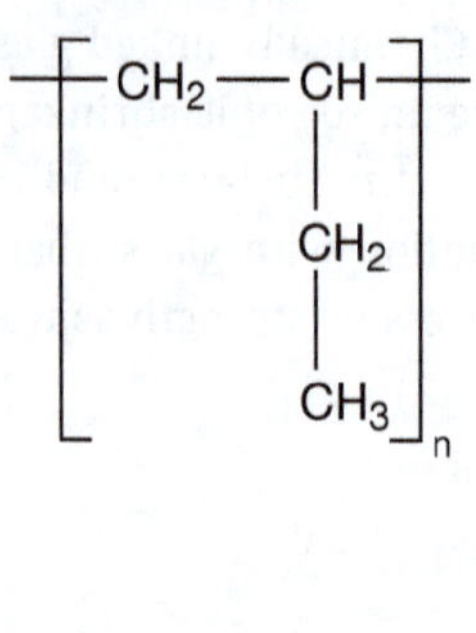

(6.4)

6.2.8.1 Polybutene-1 (PB)

Chemical Constitution

Polybutene-1 is created by stereospecific polymerization of butene-1 with specific Ziegler-Natta catalysts. PB is a mostly isotactic, semi-crystalline polymer with a high molecular weight of 700,000–3,000,000 g/mol and a low density of 0.910–0.930 g/cm³. During cooling, it first crystallizes to a degree of ≈ 50% in a metastable, tetragonal modification (density ≈ 0.89 g/cm³), forming a soft, rubber-like material. With corresponding shrinkage, this material transforms at room temperature within approx. one week into a stable, double hexagonal modification. Under higher pressures the transformation completes faster; at temperatures higher or lower than room temperature, the process is slower. During the transformation, density, elongation at break, and hardness increase. Atactic polybutene-1 is used as melt adhesive; syndiotactic polybutene-1 has no technical significance.

Processing

Hopper drying of pigmented grades is necessary for the major processing technologies (injection molding and extrusion). Melt temperatures for injection molding range from 240–280 °C, for extrusion from 190–290 °C. Mold temperatures should range from 40–80 °C. Note that PB recrystallizes after thermoplastic processing. This fact can be utilized when producing pipe bends: extruded pipes are wound over a drum and left there until the material transforms into a stable modification.

Properties

The mechanical properties of PB in its stable modification at room temperature range between those of PE and PP. Even at elevated temperatures it exhibits high creep rupture strength, low creep, and high stress cracking resistance because of its high molecular weight and the strong bonds between the crystalline blocks. These favorable properties are sustained even at carbon black loads of 20%. Particularly notable are low creep in combination with chemical resistance and excellent abrasion resistance, for example, for the transportation of slurry or diluted solid particles. PB pipes meet international standards for stress cracking resistance.

PB is resistant to non-oxidizing acids, oils, fats, alcohols, ketones, aliphatic hydrocarbons, and cleaning agents. It is non-resistant to aromatic or chlorinated hydrocarbons. Like all other polyolefins, PB burns easily and must be stabilized for outdoor use. Food contact is permitted, PB is physiologically harmless. Contact with and inhalation of butene-1 vapors should be avoided because they have anesthetizing effects.

Property comparison: see Table 6.5.

Applications

Hot water pipes, pipes for floor heating, fittings, extrusion blow molded hollow parts, containers for the chemical industry, telephone cable, two-layer blow molded or flat sheet film with PA/PE, PE, or PP as carrier and a peelable sealing layer made of PE/PB blends for hot- or cold-filled food or meat packaging. Adding 1–5% PB increases the extrusion speed of PE, PP, and PS. Blends of 25–30% PB with aliphatic adhesive components (tackifiers) und micro-crystalline waxes are used in special adhesive formulations that sustain long exposure times and high temperature resistance; 1–2% PB as processing aid for PE-LLD reduces melt fracture during film blow molding.

6.2.8.2 Polyisobutene (PIB)

Chemical Composition

PIB (raw density 0.91–0.93 g/cm^3) is a rubber-like thermoplastic material produced by cationic polymerization of isobutene (see Equation 6.5). Homo-polyisobutene is not vulcanizable because of its lack of functional properties (see also butyl rubber, IIR, Section 6.23.3).

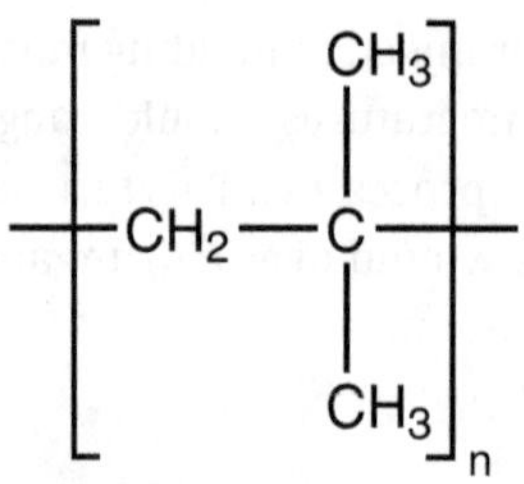

(6.5)

Processing

Solid rubber-like products are processed like rubber with kneaders, rolling mills, calenders, presses, extruders, and injection molding machines. Melt temperatures range from 150–200 °C. If the temperatures are too low, PIB will mechanically degrade. Dispersions for coatings and foams are available.

Properties, Applications

Depending on molecular weight, PIBs are viscous oils, tacky to soft/flexible, or rubber-like materials. Some applications rely on the high low-temperature flexibility (glass transition temperature −73 °C), others on the low gas permeability ($D = 0.081 \times 10^{-6}$ cm^2/s).

Applications: PIBs of all molecular weights are used as adhesives and sealers, because their properties do not change significantly between the glass transition temperature (−73 °C) and almost 100 °C. They are used for copolymerization with other polyolefins (to improve processability), styrenes, and other monomers. In rubber blends, PIB improves weather- and age-resistance and adhesive strength of tire treads, while reducing gas permeability.

Oily fluids, MW 300–3000 g/mol: electro insulation oils, viscosity improving additives to mineral oils, adhesion promoting additives for oriented films.

Highly viscous materials, MW 40,000–120,000 g/mol: blends for laminating waxes, sealing resins, chewing gum.

Elastic materials, MW 300,000–3,000,000 g/mol: polymer additives, electro insulation films, lining and sealing webs against acids and water pressure, for flat and conventional roofs (with light-stabilizing and mineral fillers).

High-molecular PIBs for membranes for acid protection (MW 4,000,000 g/mol) contain exclusively special grades of carbon black as inert fillers as well as graphite and are plasticizer-free. Chemically resistant to HCl 70%, HNO_3 50%, HF 40%, and H_2SO_4 70% at room temperature. Not resistant to mineral oils, and organic solvents. Used as liners and membranes under panels and acid-resistant stones in special cements. Physical characteristics: tensile strength: 4 N/mm^2. Elongation at break: 500%. Thermal service limits: > 85 °C. Excellent dielectric properties: spec. volume resistivity > 100 Ohm; surface resistivity < 1000 Ohm. Very low coefficient of water vapor diffusion at 70 °C: $0.05 \cdot 10^{-9}$ cm^2 s^{-1}. Typical applications: industrial flooring with high chemical resistance, waste water treatment facilities, collecting trays for substances hazardous to water, aluminum sulfate absorbers and dryers, pickling baths for stainless steel, electrically conductive floors, scrubbers for flue gas desulfurization facilities.

PIB is resistant to acids, alkalis, and salts, exhibits limited resistance to nitric and nitrided acids; it is not resistant to chlorine, bromine, chloride sulfuric acid; it is soluble in aromatic, aliphatic, and chlorinated hydrocarbons; it swells in butyl acetate, oils, and fats; it is insoluble in esters, ketones, and low alcohols. PIB must be stabilized against sunlight and UV radiation. It burns like rubber.

6.2.9 Higher Poly-(α-Olefins) (PMP, PDCPD)

6.2.9.1 Poly-4-Methylpentene-1 (PMP)

Chemical Composition, Properties

After polymerization, 4-methylpentene-1 is a highly branched, glass clear (90% light transmission), hard, semi-crystalline thermoplastic material with a low density in the raw state (0.83 g/cm^3), see Table 6.4, No. 27.

It has a degree of crystallinity of 65%; it is highly transparent because both amorphous and crystalline domains have the same index of refraction. Micro-cracks forming between the amorphous and the crystalline domains because of their different coefficients of expansion make PMP homopolymers slightly cloudy in appearance. The melting point is at 245 °C and can be lowered by copolymerization, which will also lower the tendency for micro-crack formation. Short-term service temperatures are at 180 °C, long-term at 120 °C. Electrical properties resemble those of PE-LD.

PMP is resistant to mineral acids, alkaline solutions, alcohols, cleaning agents, oils, fats, and boiling water; it is not resistant to ketones, aromatic and chlorinated hydrocarbons; it is also susceptible to stress cracking. Weather resistance, even of stabilized grades, is low. PMP will yellow and then lose its good mechanical properties. It burns with a luminescent flame, is approved for food contact, and physiologically inert.

For *property comparison*, see Table 6.5.

Processing

PMP is injection molded at melt temperatures ranging from 280–310 °C and mold temperatures of ≈ 70 °C. Because of the narrow melt temperature range, extrusion is difficult. Blow molding of hollow parts is possible at temperatures from 275–290 °C; however, the low melt strength limits the size of the hollow parts. Thermoforming is also possible. Like all other polyolefins, PMP is weldable and easier to join by adhesive bonds than PE, as long as the surfaces are roughened. Treatment with chromium sulfuric acid enhances bond strength; however, it is recommended to use plasma- or jet blast corona treatment instead to avoid waste disposal issues.

Applications

Sight glass or observation windows, interior light fixtures, sterilizable parts and films for medical applications and packaging, packing for instant meals, coloring spools, Raschig rings, transparent pipes and fittings, cable insulation flexible at low temperatures (−10 °C).

6.2.9.2 Polydicyclopentadiene (PDCPD)

Chemical Composition, Properties

Dicyclopentadiene is a by-product of the crack process used to produce gasoline. At 99% purity in a mixture with norbornene it is liquid above −2 to 0 °C. In the presence of catalysts, such as alkyl aluminum, it is polymerized to a crosslinked PDCPD by ring opening and splitting of the double bonds. Mixed with additives, DCPD can be processed on conventional RIM equipment in form of a 2-component system with reaction times of 15 s to 2 min.

Properties, Applications

Properties and applications compare to those of rigid PUR-RIM compounds. PDCPD exhibits higher strength, stiffness, and dimensional stability at elevated temperatures. Service temperatures range from −40 to 110 °C. Typical applications are body parts for agricultural- and construction machinery and short runs for medical devices.

6.3 Styrene Polymers

6.3.1 Polystyrene, Homopolymers (PS, PMS)

6.3.1.1 Polystyrene, PS; Poly-p-Methylstyrene (PPMS); Poly-α-Methylstyrene (PMS)

Chemical Composition

From a chemical standpoint, polystyrenes are actually polyvinylbenzenes, in which the phenyl groups are normally randomly distributed along the chain (Equation 6.6). Therefore, crystalline domains cannot form, so that conventional PS or PS-R (random) are amorphous, transparent thermoplastic materials.

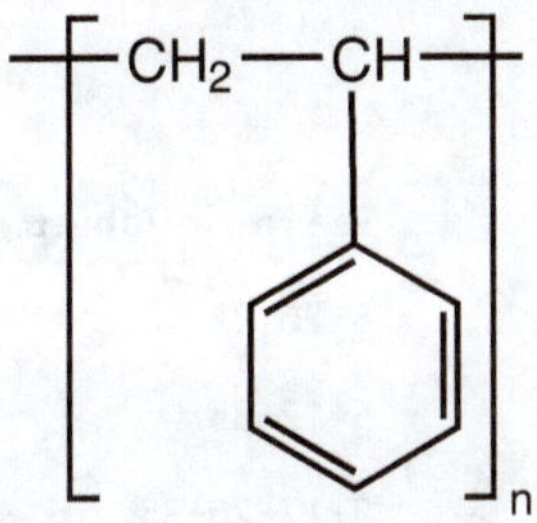

(6.6)

Suspension-, bead-, or mass-polymerization produce crystal-clear products. PS belongs to the most important materials for consumer goods that are mass produced and relatively inexpensive. The use of stereospecific Natta catalysts results in isotactic polymers, PS-I, in which all phenyl groups are in the same dimensional arrangement. These products are up to 50% crystalline and opaque when cooled slowly from the molten state or after exposure to temperatures of 150 °C.

PS polymerized with metallocene catalysts is syndiotactic (PS-S, PS-S(M)). It is crystalline and has a melt temperature of 270 °C; therefore, it is considered a high-temperature resistant plastic. Applications: injection molding and extrusion, electronics, heat-resistant household products.

Besides styrene, substituted styrenes such as p-methylstyrene and α-methylstyrene are polymerized (PS, PPMA, PMS, see Table 6.4, No. 5, 6, and 7). p-Methylstyrene is radical polymerized and exhibits higher heat distortion points and hardness than PS. α-Methylstyrene is difficult to radical polymerize into highly molecular products. Lower molecular poly-α-methylstyrenes are used as modifiers and processing aids, *e. g.*, for thermoplastic elastomers, PVC, and ABS.

6.3.2 Polystyrene, Copolymers, Blends

6.3.2.1 Chemical Composition

A wide variety of copolymers (also in combination) and blends were developed in order to modify specific properties, such as heat distortion resistance, stiffness, impact resistance, chemical resistance, stress cracking resistance, and to meet specific requirements. Table 6.4 lists the basic structures of these copolymers.

Table 6.9 lists selected co- and terpolymers as well as common blends. An important representative of this material class is SAN, which is used to produce ABS and ASA by addition of rubbers. These are further processed into blends, such as ABS + PC and ABS + PA. MABS and MBS are also members of the styrene-copolymer family, see also Section 6.6.2.3.

Table 6.9 Polystyrene Ter- and Block Copolymers, Blends

Terpolymers	Acronym	See No. in Table 6.4
Acrylonitrile butadiene styrene	ABS	5+26+35
Methacrylate butadiene styrene	MBS	5+12+35
Methyl methacrylate acrylonitrile butadiene	ASA	5+17+26+35
Acryl ester styrene acrylonitrile	ASA	5+26+33
Styrene butadiene methyl methacrylate	SBMMA	5+35+17
Acrylonitrile chlorinated PE styrene	ACS	5+26+9
Acrylonitrile EPDM rubber styrene	AES (AEPDMS)	5+26+34
Styrene maleic acid anhydride butadiene	SMAB	5+31+35
Styrene isoprene maleic acid anhydride	SIMA	5+26+31
Block copolymers, TPE-S		
Styrene acrylonitrile	SAN	5+26
Styrene butadiene	SB	5+35
Styrene butadiene styrene	SBS	5+35+5
Styrene isoprene	SIR	5+36
Styrene isoprene styrene	SIS	5+36+5
Styrene ethene propene	SEP	5+1+2
Styrene ethene butene styrene	SEBS	5+38+5
Styrene ethene propene styrene	SEPS	5+39+5

Blends of	with
PS	PPE
ABS	PC
	PC-blend
	PA
	TPU
	PVC
	SMA
	PSU
ASA	ABS
	AES
	PC

6.3.2.2 Processing

Most styrene polymers are *injection molded.* Melt temperatures range from 180–280 °C, mold temperatures from 5 °C (short-lived mass-produced articles) to 80 °C. Low mold temperatures and high injection rates cause frozen-in stresses and orientation and will reduce the long-term quality of the produced parts. Low shrinkage facilitates precision molding. Impact modifiers increase melt viscosity, thus lowering the melt flow rate. Grades with added blowing agents can be injection molded into structural parts.

In general, all styrene polymers can be extruded into profiles, films, and sheets (PS-S). SB and ABS in particular are used to produce sheets for thermoforming of large-area components. Injection and extrusion blow molding are common processing methods. Vacuum metallization of standard PS with aluminum results in mirror-like surfaces. ABS in particular is well suited for plating after special surface pre-treatment. Painting, printing, and laser-labeling are possible. Adhesion can be achieved with glues that fit the polymer type, or with 2-component adhesives. Common welding processes used for thermoplastic materials can be utilized.

6.3.2.3 Properties

PS homopolymers are crystal clear and glossy. Easily flowing grades are stiff and hard, but also brittle and susceptible to fracture which limits their suitability for use under dynamic load. Wear- and abrasion resistance and water absorption are minimal, dimensional stability is high. Short-term maximum service temperature ranges from 75–90 °C, long-term from 60–80 °C. PS exhibits very good electrical and dielectric properties. It is insensitive to moisture and resistant to brine, lye, and non-oxidizing acids.

Esters, ketones, aromatic and chlorinated hydrocarbons act as solvents and the solutions are used as glues. Gasoline, essential oils, and flavors (spices, lemon peel oil) cause stress cracking. Stress-cracking tendencies are high, in particular when residual stresses are frozen-in because of rapid cooling of the melt or because of disadvantageously positioned gates. While PS retains its gloss and transparency in-doors, it is not suitable for outdoor use. PS easily burns after ignition.

Syndiotactic PS-S (PS-S(M)) behaves like standard random PS of equal density; however, it is stiffer and exhibits a much higher heat distortion point – in particular glass fiber reinforced grades – and better melt flowability.

PS, as well as its copolymers and blends, is stabilized against degradation at high processing temperatures and yellowing caused by UV-exposure. The available grades range from easy flowing, glossy, antistatic, antiblocking, internally and externally lubricated, colored or pigmented (organic or inorganic), and electroplatable.

Fillers and reinforcements are rarely used with PS (particularly with pure PS), because they have little effect on hardness, brittleness, and stress crack resistance. Chalk, talc, and glass fibers/spheres are more commonly used with ABS and its blends.

The general effects of copolymer blocks are detailed in Table 6.10.

Table 6.10 Effect of Comonomers on the Properties of Polystyrene

Monomer	Line No. in Table 6.4	Influence on Polystyrene
α-Methylstyrene, MS	6	Higher heat distortion point
Chlorinated ethylene, PE-C	9	Impact modifier
Methyl methacrylate, MMA	17	Creates transparency in PS and ABS
Vinyl carbazole	25	Higher heat distortion point, but toxic
Acrylonitrile, AN	26	Higher stiffness, toughness, resistance, water absorption, decreased electrical properties, yellowish
Acrylic acid ester elastomer	33	Impact modifier, decreased heat deflection temperature, stiffness, rigidity, increased weatherability
EPDM rubber, EPDM	34	Same as acrylic acid ester elastomer
Butadiene rubber, BR	35	Same as EPDM, but decreased heat deflection temperature
Isoprene rubber, IR	36	Same as BR
Thermoplastic PUR elastomer	37	Increased impact and wear resistance

To enhance mechanical properties and heat distortion temperature, styrene groups can be replaced by methylstyrene. At 50% content, the glass transition temperature increases from 95 to 115 °C. *AN* and *MA* have the same effect. Examples: *SAN, ASA, AES, or SMAB*. AN increases toughness, resistance to oils and fats, but also water absorption and yellowing. SAN-modifications with polyvinyl carbazole (PVK is toxic!) result in extremely heat distortion resistant but toxic products. Modifications with carbonic acid dimethyl ester additionally result in higher chemical and UV resistance.

In addition to increasing heat distortion resistance, *PVK* is also used as a homopolymeric molding material: density 1.19 g/cm^3, modulus of elasticity 3500 MPa, short-time service temperature 170 °C, long-term 150 °C, brittle. Applications: insulation for high-frequency and television applications under high mechanical or thermal loads.

High impact resistant PS (PS-HI) is produced by graft copolymerization with, *e.g.*, 5–15% butadiene styrene rubber *(BR, SBR)*. Copolymers with 15–40% *BR* are used as processing aids in rubber manufacturing. There are numerous modifications of multi-phase engineering plastics (graft polymers and/or thermoplastic-blends) with *ACM, EPDM, PE-C*, or thermoplastic *PUR*-elastomers, exhibiting well balanced thermal, mechanical, and toughness properties (up to a leathery toughness).
Stiffness and strength of these products are reduced, in particular impact resistance at low temperatures, while stress crack resistance is improved. There are opaque and even transparent grades. Butadiene-containing grades are less UV-resistant than SAN copolymers and butadiene-free impact resistant grades.

ABS polymers have gained particular importance because of the broad range of variation in their properties. The two most significant manufacturing methods are:

- Graft polymerization of styrene and acrylonitrile on butadiene latex. The resulting graft polymer is blended with SAN, coagulated, and dried
- The graft polymer and SAN are produced separately, dried, blended, and then pelletized. By appropriate selection of the rubber component and substitution of styrene by α-methylstyrene it is possible to specifically control glass transition temperature and thus heat distortion point and impact resistance.

In addition to the standard opaque ABS, transparent plastics based on ABS with suitable rubber components are available.

With increasing PC contents, *ABS + PC blends* exhibit increasing heat distortion resistance; their stiffness and hardness compare to those of PC.

The unfavorable properties of PA, such as shrinkage and little impact resistance at low temperatures, are alleviated in *ABS+PA* (PA 6) blends. These blends offer good toughness, processability, chemical resistance, and a lower density compared to ABS + PC. Because of their excellent heat distortion resistance (180–200 °C), ABS + PA blends can be used for cathodic dip primer with subsequent drying at temperatures slightly above 200 °C.

Compared to ABS, *ASA* is distinguished by its weatherability and heat distortion resistance, caused by its acrylic ester rubber content. ASA in outdoor applications does not require a lacquer or paint coat. Like ASA, AES is used in similar applications as ABS; however, it meets higher requirements regarding light fastness and weatherability.

ASA + PC blends exhibit higher heat distortion resistance compared to ABS; in blends with PPE + PS this resistance can be increased up to 160 °C.

Polystyrene copolymers are used extensively as blend components for other thermoplastic materials. In the case of ABS, because of the polar nature of the CN bond and the relatively low melt viscosity of the SAN group. Special modifying resins are available for PVC. Coupling agents based on SB blends are used as bonding layers for composite sheet and coextruded films made of styrene copolymers with other polyolefins, PC, PMMA, or PA, and as coupling agents in mixtures of recycled plastic blends.

Table 6.11 and Table 6.12 compare the properties of some selected grades.

Applications

Standard PS is typically used for short-lived, non-technical applications, such as disposable packaging for food, pharmaceutical products, and cosmetics.

PS-S is supposed to substitute higher temperature resistant plastics such as PC, PT, LCP: headlight reflectors, electronic, medical, condenser film.

PS copolymers and blends, in particular ABS and similar products, are used in higher value products: insulation films and sheet for thermoforming, pipes, profiles; parts and housings for radios, TVs, notebooks, telephones, computers, printers, tools, housings for optical devices, household and gardening tools; plotters, luminescent screens, toys; inner linings for refrigerators and freezers; automotive: interior panels, dashboards, interior modules (*e. g.,* from SMA-GF10), ventilation systems, roof liners, battery housings. Weather- and UV-resistant grades: spoilers, radiators, fenders, and hubcaps for cars, exterior parts for tractors, trucks, motorcycles, RVs, boat hulls, and lawn furniture.

Table 6.11 Property Comparison of Styrene Homopolymers and α-Copolymers

Properties	Unit	Styrene Homopolymers and Copolymers								
		PS (PS-R)	PS-s-(M)	SMS	SB impact resistant	SB-T impact resistant, transparent	SB-HI highly impact resistant	SAN	SAN-GF35	SMAHB
Density	g/cm^3	1.04–1.05	1.05	1.05–1.06	1.03–1.05	1.0–1.03	1.03–1.04	1.07–1.08	1.35–1.36	1.05–1.13
Tensile modulus of elasticity	MPa	3100–3300	4500	3300–3500	2000–2800	1100–2000	1400–2100	3500–3900	10000–12000	2100–2500
Yield stress	MPa	–	–	–	25–45	20–40	15–30	–	–	37
Elongation at yield	%	1.5	1.4	–	1.1–2.5	26	1.5–3	–	–	–
Nominal elongation at break	%	–	–	–	10–45	20– > 50	40– > 50	–	–	–
Stress at 50% strain	MPa	–	–	–	–	–	–	–	–	–
Ultimate strength	MPa	30–55	–	50–60	–	–	–	65–85	110–120	–
Elongation at break	%	1.5–3	–	24	–	–	–	2.5–5	2–3	11–26
Melting temperature	°C	–	270	–	–	–	–	–	–	

Table 6.11 Property Comparison of Styrene Homopolymers and α-Copolymers *(continued)*

Properties	Unit	Styrene Homopolymers and Copolymers								
		PS (PS-R)	PS-s-(M)	SMS	SB impact resistant	SB-T impact resistant, transparent	SB-HI highly impact resistant	SAN	SAN-GF35	SMAHB
Heat deflection temperature HDT/A 1.8 MPa	°C	65–85	95[1)]	80–95	72–87	60–75	60–80	95–100	100–105	103–115
Thermal coefficient of linear expansion, parallel (23–55 °C)	10^{-5}/K	6–8	–	68	8–10	7–14	8–11	7–8	2.5–3	6–9
Thermal coefficient of linear expansion, perpendicular (23–55 °C)	10^{-5}/K	–	–	–	–	–	–		2	6–9
Flammability UL94 for 1.6 mm thickness	Class	HB*	–	HB*	HB*	HB	HB*	HB*	HB*	HB
Dielectric constant at 100 Hz		2.4–2.5	3	2.4–2.5	2.4–2.6	2.5–2.6	2.4–2.6	2.8–3	3.5	

Properties	Unit	Styrene Homopolymers and Copolymers								
		PS (PS-R)	PS-s-(M)	SMS	SB impact resistant	SB-T impact resistant, transparent	SB-HI highly impact resistant	SAN	SAN-GF35	SMAHB
Dielectric loss factor at 100 Hz	$\times 10^{-4}$	12	–	1–2	1–3	2–4	1–3	40–50	70–80	
Specific volume resistivity	Ohm × m	> 10^{14}	–	> 10^{14}	> 10^{14}	> 10^{14}	> 10^{14}	10^{14}	10^{14}	10^{13}
Specific surface resistivity	Ohm	> 10^{14}	–	> 10^{14}	> 10^{13}	> 10^{14}	> 10^{13}	10^{14}	10^{14}	> 10^{15}
Dielectric strength	kV/mm	55–65	–	55–65	45–65	40–50	45–65	30	40	25
Comparative tracking index CTI/A		350–500	–	350–500	350–550	600	350–550	400–550	600	
Water absorption at 23 °C, saturation	%	< 0.1	–	< 0.1	< 0.2	< 0.1	< 0.2	0.2–0.4	0.2–0.3	
Moisture absorption 23 °C/50% rel. humidity, saturation	%	< 0.1	–	< 0.1	< 0.1	< 0.1	< 0.1	0.1–0.2	0.1	0.2

[1)] PS-s-GF: 250

* also available in other flammability classifications

Table 6.12 Property Comparison of Acrylonitrile Styrene Copolymers and Blends

Properties	Unit	Acrylonitrile styrene copolymers and blends						
		ABS	ABS-HI	ABS-GF 20	(ABS + PC)	ASA	(ASA + PC)	ABS+PA$^{+)}$
Density	g/cm^3	1.03–1.07	1.03–1.07	1.18–1.19	1.08–1.17	1.07	1.15	1.07–1.09
Tensile modulus of elasticity	MPa	2200–3000	1900–2500	6000	2000–2600	2300–2900	2300–2600	1200–1300
Yield stress	MPa	45–65	30–45	–	40–60	40–55	53–63	30–32
Elongation at yield	%	2.5–3	2.5–3.5	–	3–3.5	3.1–4.3	4.6–5	
Nominal elongation at break	%	15–20	20–30	–	> 50	10–30	> 50	> 50
Stress at 50% strain	MPa	–	–	–	–	–	–	
Ultimate strength	MPa	–	–	65–80	–	–	–	
Elongation at break	%	–	–	2	–	–	–	
Melting temperature	°C	–	–	–	–	–	–	
Heat deflection temperature HDT/A 1.8 MPa	°C	95–105	90–100	100–110	90–110	95–105	105–115	75–80
Thermal coefficient of linear expansion, parallel (23–55 °C)	10^{-5}/K	8.5–10	8–11	3–5	7–8.5	9.5	7–9	9
Thermal coefficient of linear expansion, perpendicular (23–55 °C)	10^{-5}/K	–	–		5–6	–		

Properties	Unit	Acrylonitrile styrene copolymers and blends						
		ABS	ABS-HI	ABS-GF 20	(ABS + PC)	ASA	(ASA + PC)	ABS+PA+)
Flammability UL94 for 1.6 mm thickness	Class	HB*	HB*	HB*	HB*	HB*	HB*	HB
Dielectric constant at 100 Hz	–	2.8–3.1	2.8–3.1	2.9–3.6	3	3.4–4	3–3.5	
Dielectric loss factor at 100 Hz	$\cdot 10^{-4}$	90–160	90–160	50–90	30–60	90–100	20–160	
Specific volume resistivity	Ohm · m	10^{12}–10^{13}	10^{12}–10^{13}	10^{12}–10^{13}	$>10^{14}$	10^{12}–10^{14}	10^{11}–10^{13}	$2\text{-}10^{12}$
Specific surface resistivity	Ohm	$>10^{13}$	$>10^{13}$	$>10^{13}$	10^{14}	10^{13}	10^{13}–10^{14}	$3\text{-}10^{14}$
Dielectric strength	kV/mm	30–40	30–40	35–45	24			30
Comparative tracking index CTI/A		550–600	550–600	600	250–600	600	200–225	
Water absorption at 23 °C, saturation	%	0.8–1.6	0.8–1.6	0.6	0.6–0.7	1.65	1	
Moisture absorption 23 °C/50% rel. humidity, saturation	%	0.3–0.5	0.3–0.5	0.3	0.2	0.35	0.3	1.3–1.4

* Also available as V-2 to V-0, +) Conditioned

6.3.3 Polystyrene Foams (PS-E, XPS)

The market offers PS-, SB-, and ABS grades with chemical blowing agents, blowing agent-containing concentrates, and ready-for-use blends for the production of structural foams (some with flame-retardant treatment). The blowing agents, together with ≈ 0.2% butyl stearate or paraffin oil as coupling agent, are blended by the processor. Structural parts are typically injection molded (thermoplastic foam injection) from impact resistant modified grades with densities ranging from 0.7–0.9 g/cm^3; they usually have a wall thickness of ≈ 5 mm and can weigh up to 30 kg. The skin of these parts is too rough for some applications, so that post-treatments such as sanding and coating are necessary. Profiles for construction and furniture are extruded. PS grades with incorporated physical blowing agents, such as CO_2, are used for the extrusion of lighter foamed products, such as packing films with densities of 0.1 g/cm^3 or insulating sheets with densities of 0.025 g/cm^3; physical blowing agents can also be metered directly into the extruder. Extruded rigid polystyrene foam (XPS) is a closed cell foam with a dense skin. It exhibits good compressive strength, minimal water absorption, and low flammability. Compressive strengths from 0.20–0.70 MPa can be achieved, with raw densities ranging from 20–50 kg/m^3.

PS-E (PS particle foam, PS-expanded) are bead polymers with diameters ranging from 0.2–3 mm, which are polymerized with a low-boiling hydrocarbon, preferably pentane, as blowing agent. These grades are foamed in a 3-step process into rigid parts, *e.g.*, for packaging, or into blocks measuring up to 1.25 m × 1.0 m × 8.0 m. Styrene copolymer foams with maleic acid are temperature resistant up to 120 °C.

Applications of PS-E Foams

General insulation, shock- and sound insulation, packaging, transport palettes, safety gear (bicycle- and other protective helmets, safety seats), surfboard cores, and model airplanes. PSE parts are used in lost-core processes, *e.g.*, in the manufacturing of intake manifolds. Concrete filled with PS-E is lightweight with a density from 700–900 kg/m^3. Shredded PS-E parts or production waste (chips with mesh sizes from 4–25 mm) are used as insulating material for thermal insulation or for soil improvement (styrene mulch).

Incorporation of infrared absorbers or -reflectors into the foam structure reduces the radiation part in heat transmission and thus the thermal conductivity by 12–18%, depending on raw density. The pentane-containing, expandable, black PS pellets are processed into silvery-gray foam blocks or parts.

Incorporation of expandable hollow beads into TPS-SEBS pellets facilitates injection molding and extrusion of fine-pored foams with $\rho = 0.5$ g/cm^3 for the production of, *e.g.*, handles, wheels, and profiles.

6.4 Vinyl Polymers

6.4.1 Rigid Polyvinyl Chloride Homopolymers (PVC-U)

6.4.1.1 Chemical Composition

The major differentiating criteria for vinyl polymers are their different polymerization processes (emulsion polymerization, PVC-(E), suspension polymerization, PVC-(S), and mass polymerization, PVC-(M)) and their different properties (rigid grades without plasticizers, PVC-U), plasticized PVC (PVC-P), and PVC pastes.

Table 6.4, No. 8, 9, and 10, provides an overview of the basic structures of vinyl chloride (VC) and its copolymers. Today, VC is typically produced by combining chlorine and ethylene. The high content of available chlorine resulting from NaCl electrolysis (homopolymeric PVC contains 56.7% chlorine) determines both the economic importance and the special properties of VC polymers. PVC-(M) is extremely pure and exhibits a narrow distribution of grain sizes (≈ 150 µm). Processability and thermal stability are better than those of PVC-(S). Receptivity for plasticizers and liquid additives is very high.

6.4.1.2 Delivery Forms, Processing

While other thermoplastic materials are typically available as ready-to-use compounds, VC polymers are often mixed during processing from powdered raw materials and the additives required for processing and application. Unless granulation of the raw materials by extrusion is technically required, free-flowing powder mixtures (grain sizes should not be too small), agglomerates, or dry-blends are used. These processing techniques are very economical and prevent damage of the thermally sensitive PVC raw materials.

PVC-(E) primary particles are created with diameters of 0.1–2 mm. The final size and form of the powder grains is determined by the processing conditions (drying). This allows tailoring specific resin types for their respective applications: *micro-sized* resins for plastisol processing; *fine-grained* resins that are easily processed by calendering; coarse-grained, free-flowing resins with high bulk density for extrusion (PVC-U); and *coarse-grained,* porous resins for soft PVC applications (PVC-P, see Section 6.4.2). PVC-(E) contains up to 2.5% emulsifiers and sometimes other inorganic additives. Depending on the type and loading level of these additives, transparency, water absorption, and electrical insulation properties are typically inferior to those of PVC-(S) and PVC-(M). PVC-(E) possesses better processability, leading to finished products with smooth, non-porous surfaces, higher toughness, and low electrostatic chargeability.

Both *PVC-(S),* with a particle size distribution between 0.06 and 0.25 mm, and *PVC-(M),* with a particularly uniform particle size distribution, are very pure products, due to their manufacturing process. With little stabilization, both are suitable for crystal-clear, high-quality products with high mechanical and electrical load-carrying capacity and favorable corrosion and weathering resistance. Resins with porous grains (bulk density ranging from 0.4 to 0.5 g/ml) are suitable for PVC-(P); those with compact grains (bulk density ranging from 0.5 to 0.65 g/ml) are suitable for PVC-(U). Micro-PVC-(S) polymers can be encapsulated (grain size 10–1 µm).

Chlorinated PVC-(C), with a chlorine content increased by up to 60% by post-chlorination, is more difficult to process than PVC, but thermally stable to above 100 °C and shows even higher resistance, *e. g.*, to chlorine.

VC polymers cannot be processed without stabilizers that minimize discoloration and further damage both by oxidation and HCl separation during processing at high temperatures and by elevated temperatures and UV exposure during end use. Also used are flow enhancers, lubricants, and UV-absorbers. Finely ground calcium carbonate at a filler load of 5–15% facilitates extrusion and increases notched bar impact toughness.

PVC melts are highly viscous, shear- and temperature sensitive, but adhere less to the contact surfaces of the conveying equipment than other polymeric materials. Extrusion of pipes, profiles, and sheet and *calendering* and *compression molding* of sheet are the major industrial processing methods. *Injection molding* and *blow molding* are also used.

Calendering of rigid PVC sheet on an industrial scale is performed in several steps: metering and mixing of the compound ingredients, plasticization, typically with continuous screw kneaders, high-temperature calendering with temperatures increasing from 160–210 °C, until sheet thicknesses ranging from 0.02–1 mm (predominantly 0.1–0.2 mm for PVC-(S) and PVC-(M)) are obtained. The low-temperature process is used for PVC-(E) films with a high K-value (K-value, see Section 6.4.1.3). Here, temperatures decrease from 175 to 145 °C, the film is calendered, and subsequently stretched in one or both directions over a roller at 240 °C. Calendered film with a thickness of ≈ 0.5 mm is used to laminate stress-free panels and blocks in multi-platen presses.

Powdered or pelletized resin is used for *extrusion* of PVC-(U) at temperatures from 170–200 °C. Powdered mixtures require twin screw extruders or single screw extruders with a length of 20 D.

Only PVC-(S) or PVC-(M) resins with low K-values and appropriate processing aids are suitable for *injection molding* (melt temperatures from 180–210 °C, mold temperatures from 30–50 °C) or *blow molding* of bottles and other hollow articles. Processing has to be more careful than with other thermoplastics because over-heating must be avoided. It is particularly important to design the screws and molds for optimum flow conditions (no dead zones), to avoid degradation and subsequent separation of hydrochloric acid. In addition, corrosion resistant surface treatment of screws and molds is recommended.

Welding of PVC-(U) can be achieved by all standard processes used for thermoplastic materials: hot gas-, hot plate-, ultrasonic-, and high-frequency-welding. Solvent adhesives based on tetrahydrofuran (THF) or two-component adhesives based on EP, PUR, or PMMA are used for *bonding*. Adhesive bonding is achieved with PUR, nitrile-, or chlorobutadiene rubbers.

Thermoforming of panels, pipes, etc., is used on a large scale in the construction of chemical processing equipment.

6.4.1.3 Identification

VC polymers are identified according to their molecular weight by K-values or viscosity numbers, see Section 3.9.9. Both numbers are based on the relative viscosity in solution and are proportional to each other. There is a correlation between the following characteristic values: K-value, specific viscosity, viscosity number J, inherent viscosity (ASTM D 1234-T), melt flow index (ISO 1133 and ASTM D1238), and number and weight average molecular weight (see also resin standards ISO 1163 (PVC-U) and ISO 2898 (PVC-P)). PVC resins used for thermoplastic processing exhibit K-values between 50 and 80, according to ISO 1628. Higher K-values indicate better mechanical and electrical properties of the molded parts; however, higher K-values also indicate more difficult processing conditions for PVC-U. Table 6.13 shows typical K-values for standard processing methods. ISO 1060/1 provides classification and identification guidelines for VC polymers.

Table 6.13 Areas of Application for Various PVC Grades

PVC grades	PVC-U			PVC-P		
	E	S	M	E	S	M
Processing	**K-value**			**K-value**		
Calender (laminating calender, sheet presses)	(60–65)	57–65	57–65	70–80	65–70	70
Thermal films	78	–	–	–	–	–
Flooring	–	–	–	65–80	65–80	
Extrusion PVC(-U)						
Pipes	–	67–68	67–68			
Window profiles	–	68–70	–			
Furniture and construction profiles	60–70	60–68	60–68			

Table 6.13 Areas of Application for Various PVC Grades *(continued)*

PVC grades	PVC-U			PVC-P		
	E	S	M	E	S	M
Sheets and flat film	60–65	60	60			
Blown film	60	57–60	57–60			
Extrusion PVC(-P)						
General				65–70	65–70	65–70
Cable coating				–	70–90	
Preferred				–	70	70
Blow molding of hollow parts	–	57–60	57–60		65–80	60–65
Injection molding	–	50–60	56–60	–	65–70	55–60
Paste technology	–	–	–	65–80	(70–80)	

6.4.1.4 Properties and Applications

PVC-U is a thermoplastic material with high rigidity and a high modulus of elasticity, but with low abrasion resistance, impact resistance at low temperatures, and long-term alternating fatigue strength. Service temperatures are also relatively low: short-term up to 75 °C, long-term up to 65 °C. Appropriate selection of stabilizers and additives ensures excellent electrical properties, particularly in the low voltage, low frequency range. Higher frequencies cause heating because of the high electric loss factor. N_2-, O_2-, CO_2-, and air permeability are lower than for polyolefins; water vapor permeability is higher.

Up to temperatures of 60 °C, PVC-U is resistant to most diluted and concentrated acids, except for oleum-containing sulfuric acid, lactic acid and concentrated nitric acid, diluted and concentrated base and saline solutions. Gaseous chlorine forms a protective layer of chlorinated material; liquid halogens corrode PVC. PVC is resistant to alcohols, gasoline, mineral oils, other oils and fats; esters, ketones, chlorinated hydrocarbons, and aromatic hydrocarbons swell or dissolve PVC to various degrees. PVC is largely stress crack resistant. Appropriately stabilized PVC-U is suitable for outdoor applications (window profiles, etc.).

PVC-U is physiologically indifferent. Most PVC-U products are flame retardant, even without special flame inhibiting additives (up to V-0 according to UL 94). The flame will extinguish after the ignition source has been removed. Depending on the type of polymerization, PVC is translucent to transparent and can be easily colored.

In general, *fillers* are used only as extenders in PVC, and glass fibers are rarely used as reinforcement, see Table 6.14. With PVC-U in pipe and profile production, 1–2% fine-grain *chalk* (5–10 µm) is used as processing aid. A filler load of 5–15% can double notched impact toughness and more than 40% is used for pipes and profiles (pressure pipe, waste pipe, drainage pipe, sun shade profiles), depending on the mechanical loads to be sustained. Depending on application, PVC-P may contain up to 350 parts chalk. Chalk will absorb separated acids, thus improving thermal stability. Chalk also counteracts any plating effect (deposit of exuded particles on calenders and extruders).

Table 6.14 Properties of Filled and Reinforced PVC

PVC-U additive	Content by weight	Tensile strength	Strain at maximum load	Young's Modulus	Vicat temperature	Density
	%	MPa	%	MPa	°C	g/cm³
None	–	60	6–10	2700	85	1.36
Natural $CaCO_3$	30	46	8	3200	94	1.53
	100	–	–	–	116	1.78
Precipitated $CaCO_3$	15	30–47	6	3100	87	1.45
Chalk	20	34	6	3500	–	1.48
Silica flour	20	38		3100	–	–
Glass fibers	40	25	3	8000	85	–
Wollastonite	20	25	5.4	–	–	1.47

Because *kaolin* increases the volume resistance, it is used for PVC-P cable resins. *Silicates* reduce plate-out effects, increase thixotropy, and result in products with a matte surface. *Aluminum hydroxide* increases the flame retardance, *e. g.*, for carpet backing. *Barium ferrite* is used as a filler for magnetic sealing profiles for refrigerators, among others.

High impact resistant *PVC-HI grades* with 5–12% impact modifier content are 2-phase resins. *Polyacryl acid ester elastomers (ACM), PE-C,* and *EVAC* as a dispersed soft phase are typical impact modifiers. ACM is used either as graft copolymer with VC with 6–50% AN content or as copolymer with methyl methacrylate (MA) with 60–90% AN content. High-grade ACMs are mixed with PVC-S, -M, or -E to a common impact modifier concentration of 5–7% or above. In contrast to ACM, PE-C and EVAC are shear sensitive impact modifiers to a certain degree; that is, the achievable impact resistance depends on the processing conditions. *Polyacrylates* modified with polystyrene are used to manufacture transparent parts with improved impact resistance.

All PVC-HI are long-term weather resistant and therefore suitable for outdoor applications. *MABS* and MBS, particularly for hollow parts, are used as impact modifiers for crystal clear packaging materials, see Section 6.3.2 and Section 6.6.2.3. Because of their butadiene contents they are less suitable for outdoor applications.

Applications: see Table 6.13.

Comparison of properties: see Table 6.15.

Table 6.15 Comparison of Properties of Vinyl Chloride Polymers and Blends

Property	Unit	Vinyl chloride polymers and blends							
		PVC-U	PVC-C	VCA acrylate mod.	PVC + VCA	PVC + PE-C	PVC + ASA	PVC-P with DOP	
								75/22	60/40
Density	g/cm^3	1.38–1.4	1.55	1.34–1.37	1.42–1.44	1.36–1.43	1.28–1.33	1.24–1.28	1.15–1.20
Tensile modulus of elasticity	MPa	2700–3000	3400–3600	2200–2600	2500–2700	2600	2600–2800	–	–
Yield stress	MPa	50–60	70–80	45–55	45	40–50	45–55	–	–
Elongation at yield	%	4-6	3-5	4-5	4-5	3	3-3.5	–	–
Nominal elongation at break	%	10–50	10–15	35–>50	>50	10–>50	ca. 8	>50	>50
Stress at 50% strain	MPa	–	–	–	–	–	–	–	–
Ultimate strength	MPa	–	–	–	–	–	–	–	–
Elongation at break	%	–	–	–	–	–	–	–	–
Melting temperature	°C	–	–	–	–	–	–	–	–
Heat deflection temperature HDT/A 1.8 MPa	°C	65–75	ca. 100	72	74	69	65–85	–	–
Thermal coefficient of linear expansion, parallel (23–55 °C)	10^{-5}/K	7-8	6	8-9	7-7.5	8	7.5–10	18–22	23–25

Property	Unit	Vinyl chloride polymers and blends							
		PVC-U	PVC-C	VCA acrylate mod.	PVC + VCA	PVC + PE-C	PVC + ASA	PVC-P with DOP	
								75/22	60/40
Thermal coefficient of linear expansion, perpendicular (23–55 °C)	10^{-5}/K							–	–
Flammability UL94 for 1.6 mm thickness	Class	V-0	V-0	V-0	V-0	V-0	V-0	–	–
Dielectric constant at 100 Hz	–	3.5	3.5	3.5	3.5	3.1	3.7–4.3	4–5	6–7
Dielectric loss factor at 100 Hz	$\cdot 10^{-4}$	110–140	140	120–140	120	140	100–120	0.05–0.07	0.08–0.1
Specific volume resistivity	Ohm · m	$>10^{13}$	$>10^{13}$	$>10^{13}$	10^{13}	10^{12}–$>10^{13}$	10^{12}–$>10^{14}$	10^{12}	10^{11}
Specific surface resistivity	Ohm	10^{14}	10^{14}	$>10^{13}$	$>10^{13}$	$>10^{13}$	10^{12}–10^{14}	10^{11}	10^{10}
Dielectric strength	kV/mm	20–40	15	30				30–35	ca. 25
Comparative tracking index CTI/A		600	600	600	350–600	600	350–600	–	–
Water absorption at 23 °C, saturation	%	0.1	0.1	< 0.25	0.5	0.1	0.4–0.8	–	–
Moisture absorption 23 °C/50% rel. humidity, saturation	%	0.01	0.01	< 0.01	< 0.1	0.03	0.1–0.3	–	–

6.4.2 Plasticized Polyvinyl Chloride (PVC-P)

6.4.2.1 Chemical Composition

Hardness and brittleness of PVC can be influenced to a high degree by addition of plasticizers. They increase the distance or free volume between the PVC chain molecules, thus reducing the bonding forces. When a low glass transition temperature polymer, such as VCEVAC, is blended into the PVC molecule, the process is referred to as "internal plasticization", see also Section 6.4.3.

"External plasticization" is achieved by a low-molecular plasticizer penetrating the PVC and causing "swelling". This comes with the danger of exudation or seeping out, particularly at high temperatures and over longer time periods. Therefore, a distinction is made between primary plasticizers that are well gelled and do not exude, and secondary plasticizers that show little dipole effect and do not gel well. The latter are used in combination with primary plasticizers, which reduces migration tendencies and improves low-temperature toughness and extraction resistance. To reduce cost, primary plasticizers are often partially replaced by extenders, which are liquid fillers with low volatility and medium polarity. They do not act as gelatinizing agents; however, they can improve the rheological behavior of plastisols (fatty acid esters, alkylated aromatics, naphthenes) or improve the flame retardance (liquid chloroparaffins).

Polymeric plasticizers are particularly resistant to solvents and migration.

6.4.2.2 Delivery Forms, Processing

PVC-P resins, already equipped with stabilizers, lubricants, etc., are available as pellets for injection molding and blow molding of hollow parts and as free flowing powders for extrusion. Powder mixtures are used for dip coating and electrostatic coating.

PVC-P melts flow under comparatively low pressures; however, they should be *injection molded* at the maximum permissible melt temperature (170–200 °C). If lower processing temperatures are used, the parts will not display optimum mechanical and electrical properties; post-molding shrinkage will be irregular and they will exhibit uneven, matte surfaces. Rapid cooling, *i. e.*, a cold mold (mold temperatures from 15–50 °C) will result in similar effects. As with PVC-U, molds and machines should be corrosion resistant. Resins with a hardness of Shore A 60–80 are *extruded* at melt temperatures of 120–165 °C; harder resins require melt temperatures from 150–190 °C. *Blow molding* of hollow parts is common practice.

All *adhesion and welding methods* used for thermoplastic materials can be applied; high-frequency welding is used in particular for purses and bags.

6.4.2.3 Plasticizers

Table 5.16 provides an overview of commonly used types of PVC primary plasticizers and their characteristic properties.

Phthalate plasticizers (Group 1 in Table 6.16), primarily the all-purpose plasticizer DOP, account for 65–70% of all plasticizers consumed. For specific applications, phthalates of short-chained alcohols are used.

Esters of aliphatic dicarboxylic acids (Group 2) are predominantly used in blends with phthalates to improve low temperature impact resistance of PVC-P parts.

Phosphoric acid esters (Group 3) are favored for the manufacturing of technical parts with high flame-resistance.

Alkyl sulfonic acid esters of carbolic acid (Group 4), similar to DOP in their plasticizing behavior, exhibit minimal volatility. These esters provide good high-frequency weldability and weatherability, despite yellowing.

Citric acid esters (Group 5) are special plasticizers for products subject to FDA regulation for food contact.

Trimellitates (Group 6) are used for products subjected to elevated service temperatures over extended periods.

Epoxidized products (Group 7) are added to PVC-P primarily because of their co-stabilizing properties. Their exclusive use in significant quantities can lead to exudation.

Polyester plasticizers (Group 8) are available as oligomers or polymer plasticizers and offer a broad range of products, because they allow for the choice of esterification components and molecular weights ranging from 600–2000 g/mol and beyond. In addition to their low volatility, they are distinguished by their good extraction resistance against fats, oils, and fuels. Migration problems in contact with other substances can be resolved with these plasticizers, as long as the products are compatible.

Table 6.16 PVC Primary Plasticizers

Group No.	Name	Acronym	Characterization
1.	Phthalate plasticizers		
	Dioctyl phthalate	DOP	Special plasticizer for plastisols
	Di-iso-heptyl phthalate	DHP	Standard plasticizer for PVC, high gelling capacity, little volatile, balanced heat-, low temperature-, water resistance and electrical properties
	Di-2-ethylhexyl phthalate	DEHP	
	Di-iso-octyl phthalate	DIOP	
	(Phthalic acid mixture)		

Table 6.16 PVC Primary Plasticizers *(continued)*

Group No.	Name	Acronym	Characterization
	Di-iso-nonyl phthalate	DINP	Plasticizing effect, volatility, low temperature resistance decrease from DINP to DITDP (compared to DOP); heat resistance, electrical properties increase. High processing temperatures (addition of bisphenol A), special plasticizers for cable compounds
	Di-iso-decyl phthalate	DIDP	
	Di-iso-tridecyl phthalate	DITDP	
	C_7–C_9 phthalates	–	Mixed alcohol-esters, compared to DOP: low viscosity (for pastes), better low-temperature and water resistance, less volatile (important for artificial leather, floor coverings)
	C_9–C_{11} linear alcohols	–	
	C_6–C_{10} n-alkyl phthalates	–	
	C_8–C_{10} alkyl phthalates	–	
	Dicyclohexyl phthalate	DCHP	Limited applications, resistant to fuel extraction; good gelling, for foam pasted, floor coverings
	Benzylbutyl phthalate	BBP	
2.	Adipic, azelaic, and sebacic acid		
	Di-2-ethylhexyl adipate	DOA, DEHA	DOA: outstanding low-temperature-resistant plasticizer, light fast; more volatile and water sensitive than DOP
	Di-iso-nonyl adipate	DINA	DINA-DIDA: less low-temperature resistant and volatile than DOA
	Di-iso-decyl adipate	DIDA	
	Di-2-ethylhexyl azelate	DOZ	Less water sensitive than adipate, similar to DOS
	Di-2-ethylhexyl sebacate	DOS	Best low-temperature resistance, low volatility

Group No.	Name	Acronym	Characterization
3.	Phosphate plasticizers		
	Tricresyl phosphate	TCF	Flame resistant, for heavy-duty mechanical and electrical parts; not suitable for food contact;
	Tri-2-ethylhexyl-phosphate	TOF	flame retarded, light fast, less heat resistant than TCF; low-viscous (for pastes)
	Aryl alkyl mixed phosphates	–	similar to TOF, fuel resistant
4.	Alkyl sulfonic acid phenyl ester	ASE	Similar to DOP, less volatile than phthalates, tendency to discoloration, weather resistant,
5.	Acetyl-tributyl citrate	–	Similar to DOP, suitable for food contact
6.	Tri-2-ethyl hexyl trimellitate	TOTM	Little volatile, thermally highly resistant
	Tri-iso-octyl trimellitate	TIOTM	High price (cable compounds)
7.	Epoxidized fatty acid esters		Butyl-, octyl-epoxy stearate; low-temperature resistant, little volatile, synergistically stabilizing with Ca-Zn stabilizers
	Epoxidized linseed oil	ELO	ELO and ESO primarily for improving heat stability, resistant to extraction
	Epoxidized soy oil	ESO	
8.	Polyester plasticizers	–	Polyester from (propane-, butane-, pentane-, and hexane-) diols with dicarboxylic acids of group 1 and 2. Not volatile, little temperature dependent, largely resistant to extraction and migration
	Oligomer plasticizers	–	Viscosity < 1000 mPa · s, also mixed with monomer plasticizers for pastes
	Polymer plasticizer	–	Viscosity up to 300,000 mPa · s, suitable for extrusion and calendering

6.4.2.4 Properties and Applications

PVC-P resins contain 20–50% plasticizers. Low plasticizer content that does not allow sufficient homogeneous gelation will cause embrittlement; rigidity will increase and elongation at break may decrease. Beyond this limit, PVC-P behaves more rubber-like with increasing plasticizer content. It is therefore common to characterize PVC-P grades by their Shore A hardness. Shore A hardness values of 96 to 60 are common for extrusion, 85 to 65 for blow molding, and up to 50 for injection molding. Highly filled resins with Shore A values from 85 to 70 for floor coverings and cable sheathing exhibit low elongation at break and are less flexible at low temperatures than unfilled resins.

Low-temperature brittleness depends on the type and amount of plasticizer, see Table 6.15 and Table 6.17. At 40–60 °C, PVC-P exhibits distinct permanent deformation after higher mechanical loads for prolonged periods of time, which distinguishes it from rubber. The upper temperature limit for long-term use is at 80 °C, because above this limit plasticizer loss increases markedly while rigidity decreases. For more temperature resistant resins, for example, for cable sheathing, trimellitate- and polyester plasticizers should be employed, despite their unfavorable low-temperature behavior.

Table 6.17 Shore-A and -D Hardness of PVC-P

Shore-hardness range		General characteristics	Low-temperature brittleness range* (°C)
A	D		
98-91	ca. 60–40	Semi-rigid	0 to −20
90–81	39–31	Bend leather-like	−10 to −30
80–71	Not applicable	Taut rubbery	−10 to −45
70–61	Not applicable	Medium rubbery	−30 to −50
60	Not applicable	Very soft injection molded parts	−40 to −50

* measured by drop-hammer test

In general, the specific volume resistance will decrease with increasing plasticizer content. Addition of > 13% carbon black will lower the volume- and surface resistivity by ≈ 6 to 8 orders of magnitude.

On contact, monomer plasticizers can migrate into other resins, such as thermoplastics, rubber, and lacquers on contact. This will cause embrittlement of the PVC-P and potentially of the absorbing material, and may initiate stress cracks. Although most plasticizers are physiologically indifferent, the use of plasticizer-containing plastics is undesirable in food contact applications when there is a danger of plasticizer migration.

Blends of PVC-P with PVCEVAC-plasticizers, TPU, or NBR/AN copolymers exhibit excellent flexibility at low temperatures together with a general stability under continuous use in a wide temperature range. TPU- and NBR grades are also resistant to fats, oils, and fuels.

Comparison of properties: see Table 6.15.

Applications

Injection molding: seals, protective caps, suction cups, bicycle and motorcycle handles, boots, sandals, shoe soles, shock absorbers.

Extrusion: pipes, tubes, hand rail profiles, sealing profiles, cable- and wire sheathing, insulating tapes and tubes (limited to low frequencies because of the high dielectric constant); also by calendering: sheet, film for shower curtains, coatings for weather gear, conveyor belts; floor coverings.

Blow molding: armrests in automotive applications that are PUR foam resistant, dolls, balls, tubes, bottles.

6.4.3 Polyvinyl Chloride: Copolymers and Blends

Composition

Table 6.4 shows a list of vinyl chloride copolymers, Table 6.18 shows the influence of comonomers on the properties of polyvinyl chloride.

Table 6.18 Influence of Comonomers on the Properties of Polyvinyl Chloride

Monomer	Line No. in Table 6.4	Influence on PVC
Ethylene	1	Tougher at low temperatures, easier to process
α-Methylstyrene	6	Higher heat distortion point
Chlorinated polyethylene	9	Cl statistically distributed, increased impact resistance, more difficult to process
Vinylidene chloride	10	Lower heat distortion point, easier to process
Methacrylate	16	Easier to process, easier thermoforming
Vinyl acetate Ethylene vinyl acetate	20	Lower heat distortion point, easier to process

Table 6.18 Influence of Comonomers on the Properties of Polyvinyl Chloride *(continued)*

Monomer	Line No. in Table 6.4	Influence on PVC
Vinyl ether	22	Higher rigidity, more difficult to process
Acrylonitrile	26	Higher rigidity and impact resistance
Maleic (acid) imide	32	Higher heat distortion point, more difficult to process
Acrylic acid ester elastomer	33	Increased impact resistance

VCVAC copolymers are used as molding resins with VAC contents up to 50%. Their properties are similar to those of PVC; however, they are more sensitive to processing conditions and should be processed at lower melt temperatures; they exhibit lower heat resistance. Applications include thermoforming films and compression molded records (vinyls). Copolymers of vinyl ethers and PP can also be used for this purpose.

VCVDC copolymers with high VDC content are more temperature resistant then PVC; however, they are more difficult to process because of their tendency to split off HCl at elevated temperatures. PVDC homopolymers decomposes below melting temperature and is therefore not available commercially.

Modifier resins, mostly based on high-molecular methacrylate-(multi-)polymers are blendable in single-phase with PVC and result in crystal clear products. They improve the flow behavior of the melt and the thermoformability in the thermo-elastic range, facilitating the manufacture of impact and weather resistant products under gentle processing conditions. Modifier resins based on α-methylstyrene/acrylonitrile and styrene/maleic acid anhydride improve temperature resistance by 10–15 K.

Other blends are manufactured with ABS, NBR, PE-C, PMMA.

Comparison of properties: see Table 6.15.

Applications

PVC-U, copolymers, and blends: semi-finished products such as films, sheet, corrugated sheet, profiles, pressure pipes for water and chemical industry, sewage and drainage pipes, fittings, chemical apparatus, shutter elements, eaves, window and door profiles, siding, reflecting road markers, housings, packaging films, bottles and other blow molded parts for packaging, foam, barrier layers against O_2, H_2, and flavors in food contact (VCVDC), plasticizer-free soft films (VCEVAC), fibers, filaments, nonwoven fabric, nets.

6.4.4 Polyvinyl Chloride: Pastes, Plastisols, Organosol, Foams

6.4.4.1 Pastes, Plastisols, Organosol

PVC paste technology is a special method to manufacture PVC-P parts; it disperses fine-grain PVC-U particles (K-values between 56 and 80) in plasticizers. No reaction takes place between the PVC and the plasticizers at temperatures up to ≈ 35 °C. Gelation of the pastes will occur on heating to 150–220 °C; the material will then exhibit typical properties of PVC-P. VCVAC copolymers are used to lower the gelation temperature.

The composition of the pastes defines *plastisols*, which consist primarily of PVC and plasticizers and sometimes small amounts of extenders, stabilizers, pigments, and fillers. *Organosols* are plastisols with a higher content of volatile diluents, such as gasoline or glycols, to reduce viscosity, thus making them suitable for lacquers. *Plastigels* are made kneadable and gel-like by addition of high amounts of colloidal silica or metal soaps.

Regarding the selection of plasticizers, see Section 6.4.2. Monomer glycol methacrylate is added as a polymerizable plasticizer to lower the viscosity of the pastes. With additionally added catalysts, they polymerize during gelation to form materials with increased hardness.

Pastes can be manufactured in PVC/plasticizer ratios ranging from 50 : 50 to 80 : 20; they result in rubber-like to PVC-U-like materials. At higher plasticizer contents, the materials will turn jelly-like. Fast turning mixers with subsequent filtration and venting, particularly required for dipping- and casting pastes, are used for manufacturing. Temperature-sensitive resins are produced in slow moving mixers with subsequent passage over a water-cooled calender, see also Section 4.1.

PVC pastes can be further processed in a variety of methods: *spread coating, dipping methods, cast and spray processes* as well as *rotary screen printing*. Planar carriers (textile, paper, glass- and mineral non-woven materials, sheeting) are typically converted into artificial leather, textiles for rain gear, tarps, floor and wall coverings, and sidings by several subsequent spread coatings with either compact or foamed material. The *dipping method* is used to manufacture protective gloves and coatings on metallic carrier materials. Casting methods, in particular rotational casting, are used for hollow parts, such as balls, dolls, bellows, and so forth. *Sprayable plastisols* are typically used with airless methods for metal coatings, *e. g.*, for underbody coatings in automotive applications. With rotary screen printing, PVC pastes are applied to paper or textiles (among others) to manufacture tarps, artificial leather, and floor and wall coverings (wall paper).

6.4.4.2 Foams

Ready-to-use PVC-U resins containing blowing agents are available for extrusion of structural foamed pipes and other profiles and for thermoplastic foam casting. They are used for door frames, window sills, and furniture parts with overall densities from 0.7–0.9 g/cm^3. PVC-P compounds (*e. g.*, 100/70 PVC/DOP) with 0.5 parts foaming agent and blended with 20–30 parts nitrile butadiene rubber (NBR) or EVA are used for sealing profiles, shoe soles, bumpers, and shock absorbers.

At pressures from 200–600 bar, closed-cell blocks of PVC-U or -P are produced using the Airex technique. Cross-linking with diisocyanate increases the heat deflection temperature.

Pastes for foaming on carrier materials are formulated to allow for gelling and foaming in one step, *e. g.*, at 180–200 °C in the heating channel of a spread coating machine. With this method, single- and multi-layer foamed artificial leathers and carrier-free foamed film are produced. Chemically embossed cushioned vinyl floor and wall coverings are produced by printing an inhibitor (TMSA, benzotriazole) on the pre-gelled foam layer, which will locally prevent foaming during the subsequent foaming step at 190–220 °C.

6.4.5 Vinyl Polymers, Other Homo- and Copolymers (PVDC, PVAC, PVAL, PVME, PVFM, PVB, PVK, PVP)

The general chemical structure of the homopolymers is shown in Table 6.4.

6.4.5.1 Polyvinylidene Chloride (PVDC)

PVDC decomposes below melting temperature and therefore it is only used as comonomer with PVC.

6.4.5.2 Polyvinyl Acetate (PVAC)

PVACs (density = 1.17 g/cm^3, MM = 35,000–2,000,000 g/mol) are crystal clear, soft to rigid resins. Because of their insufficient temperature resistance, they are not suitable as molding materials; however, they are easily soluble in most solvents (except for aliphatic hydrocarbons and water-free alcohols). They form light-fast, gasoline-, oil-, and water-resistant films that slightly swell in water (water absorption up to 3%) and exhibit high pigment-binding properties and limited resistance to nitrocellulose and plasticizers. PVAC is susceptible to saponification.

Applications: Paints and coating compounds, also in the form of dispersions, adhesives, and finishes.

6.4.5.3 Polyvinyl Alcohol (PVAL)

PVAL is produced by saponification of polyvinyl acetate; it is a white-yellowish powder that does not dissolve in organic solvents. Partially saponified grades with ≈ 13% PVC content are easily water soluble, better than fully saponified PVAL.

Applications: clear films; with hydrophilic alcohols such as glycerin as plasticizer, PVAL can be processed like a thermoplastic material into age-resisting, leather-like products (fuel-, oil-, and solvent-resistant tubes, membranes, gaskets, and release sheets, *e.g.*, for processing of UP resins); UV radiation renders PVAL with added bichromate impermeable by water; it is used in graphic art.

Applications for solutions: protective colloids for dispersion and stabilization, thickeners for the plastics, cosmetics, and pharmaceutical industries, textile dressings and glazes, paper glues, binders for color pencil leads, paints and printing inks, adhesives, coatings.

6.4.5.4 Polyvinyl Methyl Ether (PVME)

PVMEs are soluble in cold water and almost all organic solvents, except gasoline-like hydrocarbons, ethyl ether, isobutyl ether, and decyl ether. Depending on the degree of polymerization, they are oils, or soft or rigid, non-saponifiable resins.

Applications: electric insulations, self-adhesive resins, blended resins for chewing gum and dental resins, octadecyl ether for high-gloss floor polish (V-wax). Applications for solutions, see PVAL.

6.4.5.5 Polyvinyl Butyral, Polyvinyl Formal (PVB, PVFM)

Polyvinyl acetals are created by conversion of polyvinyl alcohols with aldehydes. PVB and PVFM are solid resins soluble in organic solvents.

Applications: lacquers, (including hot-sealable lacquers, enamels, gold lacquers, printing inks, primers), impregnating resins, adhesives, shrink crown caps, peelable packaging, oil- and gasoline-resistant tubes and gaskets. High-molecular PVB is used for interlayers in safety glass; together with PVFM it is used for gasoline-resistant coatings for gasoline tanks; together with phenolic resins it is used for wire coating and heat-setting metal adhesive films.

6.4.5.6 Polyvinyl Carbazole (PVK)

PVK is a crystal clear thermoplastic material like PS; however, because of the voluminous side groups (see Table 6.4), it exhibits significantly higher thermal resistance: short-term 170 °C, long-term 150 °C. Copolymers with styrene are known. The electrical properties are very good. PVK is resistant to alkaline and saline solutions, acids (except concentrated chromate, nitric and sulfuric acid), alcohols, esters, ethers, ketones, carbon tetrachloride, aliphatic hydrocarbons, mineral oils with low aromatic contents, castor oil, water, and water vapor up to 180 °C. PVK is not resistant to dimethylformamide, fuels, aromatic hydrocarbons, chlorinated hydrocarbons, such as chloroform, methylene chloride, and tetrahydrofuran. Resistance against stress cracking and weathering are excellent. PVK is self-extinguishing; however; it must not be used in food contact.

Applications: PVK is used in the chemical industry because of its good thermal and chemical resistance; it is used in high-frequency applications with additional mechanical loads.

6.4.5.7 Polyvinyl Pyrrolidone and Copolymers (PVP)

PVP is available in water-soluble and in highly water-swelling modifications. PVP and its copolymers are used as harmless, non-skin irritating binders in the cosmetic and medical industry and for numerous other industrial applications.

6.5 Fluoro Polymers

6.5.1 Fluoro Homopolymers (PTFE, PVDF, PVF, PCTFE)

Chemical Composition

In fluoropolymers, the hydrogen atoms (H) in the main carbon chain have been completely or partially substituted by fluorine atoms (F), see Table 6.19. The F-atoms have a higher volume than the C-atoms and form a close, protective layer around the carbon chain. In addition, the F–C bond is very stable; therefore, these polymers exhibit excellent chemical resistance, even at elevated temperatures. They are weather-resistant without further stabilization, physiologically indifferent, non-flammable, and embrittle only at low temperatures. PTFE exhibits the highest heat deflection temperature of all engineering plastics: the use-temperature's upper limit short-term is 300 °C, long-term 250 °C. Stiffness (fracture strength below 50 MPa) and rigidity (modulus of elasticity 350–1800 MPa) are rather low, elongation at break is above 100%. Depending on molecular structure and processing, the degree of crystallinity can be up to 94%. Depending on fluorine content, the gross density can be up to 2.2 g/cm^3; therefore, PTFE is one of the plastic materials with the highest densities. Mold shrinkage is high, depending on degree of crystallization.

The following sections will describe the general properties of fluoropolymers and their areas of application. Table 6.19 provides information on service temperatures, and processing methods. Table 6.20 provides an overview of characteristic values.

Table 6.19 Fluorine-Containing Polymers

No.	Acronym	Chemical structure	Service temperature °C		Processing: Compression molding = C Injection molding = I Extrusion = E
			Short-term	Long-term	
1	PTFE	$[CF_2-CF_2]$	300	−270 to +260	C, sintering, ram extrusion
2	PVDF	$[CH_2-CF_2]$		−60 to +150	I, E, coating
3	PVF	$[CH_2-CHF]$		−70 to +120	(Semi-finished parts)
4	PCTFE	$[CFCl-CF_2]$	180	−40 to +150	I, rod extrusion
5	ECTFE	$[CFCl-CF_2][CH_2-CH_2]$	160	−75 to ca. +140	I, E, sintering
6	ETFE	$[CF_2-CF_2][CH_2-CH_2]$	200	+155 to −190	I, E
7	FEP	$[CF_2-CF_2][CF_2-CF(CF_3)]$	250	−200 to +205	I, E, C, sintering
8	PFA	$[CF_2-CF_2][CF_2-CF(O-R)]$ | $R=C_nF_{2n+1}$		PFA: −200 to +260 TFB: −100 to +130	I, E, coating
9	AF	$[CF_2-CF_2][CF-CF]$ (CF–CF ring-bonded via O, O to $C(CF_3)_2$)	300	−50 to +260	I, E, coating
10	TFEHFPVDF (THV)	$[CF_2-CF_2][CF_2-CF(CF_3)][CH_2-CF_2]$		−50 to +130	I, E, coating

Table 6.20 Property Comparison of Fluoropolymers

Property	Unit	Fluoropolymers							
		PTFE	PCTFE	PVDF	FEP	PFA	ETFE	ETFE GF 25	ECTFE
Density	g/cm^3	2.13–2.23	2.07–2.12	1.76–1.78	2.12–2.18	2.12–2.17	1.67–1.75	1.86	1.86–1.70
Tensile modulus of elasticity	MPa	400–750	1300–1500	2000–2900	400–700	600–700	800–110	8200–8400	1400–1700
Yield stress	MPa	–	–	50–60	ca. 10	ca. 50	25–35	–	ca. 50
Elongation at yield	%	–	–	7–10			15–20	–	
Nominal elongation at break	%	> 50	> 50	20–> 50	> 50	>50	> 50	–	> 50
Ultimate strength	MPa	20–40	30–40	–	15–25	20–35	40–50	–	40–50
Failure stress	MPa	–	–	–	–	–	–	80–85	–
Failure strain	%	–	–	–	–	–	–	8–9	–
Melting temperature	°C	325–335	210–215	170–180	255–285	305	265–270	270	240
Heat deflection temperature HDT/A 1.8 MPa	°C	50–60	65–75	95–110	–	45–50	70	210	ca. 75
Thermal coefficient of linear expansion, parallel (23–55 °C)	10^{-5}/K	15–20	6–7	10–13	8–12	10–12	7–10	2–3	7–8
Thermal coefficient of linear expansion, perpendicular (23–55 °C)	10^{-5}/K	–	–	–	–	–	–	> 3	–
Flammability UL94 for 1.6 mm thickness	Class	V-0	V-0	V-0	V-0	V-0	V-0	V-0	V-0
Dielectric constant at 100 Hz	–	2.1	2.5–2.7	8–9	2.1	2.1	2.6	2.8–3.4	2.3–2.6
Dielectric loss factor at 100 Hz	$\cdot 10^{-4}$	0.5–0.7	90–140	300–400	0.5–0.7	0.5–0.7	5–6	30–50	10–15

Property	Unit	Fluoropolymers							
		PTFE	PCTFE	PVDF	FEP	PFA	ETFE	ETFE GF 25	ECTFE
Specific volume resistivity	Ohm · m	$>10^{16}$	$>10^{16}$	$>10^{13}$	$>10^{16}$	$>10^{16}$	$>10^{14}$	$>10^{14}$	$>10^{13}$
Specific surface resistivity	Ohm	$>10^{16}$	$>10^{16}$	$>10^{13}$	$>10^{16}$	$>10^{16}$	$>10^{14}$	$>10^{15}$	10^{12}
Dielectric strength	kV/mm	40	40	40	40	40	40	40	40
Comparative tracking index CTI/A		600	600	600	600	600	600	600	600
Water absorption at 23 °C, saturation	%	< 0.05	< 0.05	< 0.05	< 0.05	< 0.05	< 0.05	< 0.05	< 0.05
Moisture absorption 23 °C/50% rel. humidity, saturation	%	< 0.05	< 0.05	< 0.05	< 0.05	< 0.05	< 0.05	< 0.05	< 0.05

6.5.1.1 Polytetrafluoroethylenes (PTFE)

Processing

PTFE undergoes a phase transition at 19 °C, associated with a 1.2% increase in volume, which has to be taken into consideration when dimensioning and machining parts (advisable at 23 °C). The volume increases by 30% when heated from 20 °C to the range of crystal melting temperature of approx. 327 °C, at which PTFE transforms into a clear gel-like material. This causes a corresponding, process-dependent anisotropic shrinkage when the melt is cooled.

Powdered PTFE suspension polymers are used to compression mold preforms at 20–30 °C, which are then processed above 327 °C by the following techniques:

- *Mold-free sintering:* At 20–30 °C and pressures of 20–100 MPa, simple preforms are compression molded in automatic presses; preforms with undercuts or hollow sections are compression molded under isostatic pressure with flexible tools. Subsequently, a pre-determined temperature profile is used to heat the free-standing molded parts in an oven, sintering them at 370–380 °C, and slowly cooling them. Mold-free sintered parts (density ≈ 2.1 g/cm^3) are porous.
- *Pressure sintering or sintering with post-molding pressure:* Sintering is achieved by applying pressure in the mold on parts made of conductively modified PTFE (easily heated by conduction), or by post-forming pressure on the heated parts that were formed in the mold without pressure and subsequent cooling under pres-

sure, or by blow forging of pre-sintered preforms. These processes result in non-deformable, non-porous parts of highest density and rigidity. Parts hot-formed slightly below the melting point have a tendency to revert, which is exploited for lip seals that adhere after warming.

- *Ram extrusion (powder extrusion)* of rods and thick-walled pipes: The entering molten resin is discontinuously pressed into tablets by a reciprocating ram in the first section of the long cylindrical barrel. The following section of the barrel is heated to 380 °C, where the tablets are sintered under counter pressure by heat expansion and barrel friction to form continuously discharging (if necessary additionally decelerated at the end of the sintering pipe) sintered rods or pipes.
- Films are peeled from sintered cylindrical block; they can be treated by rolling.

Long, thin-walled products, mainly tubes up to 250 mm diameter with wall thicknesses from 0.1–4 mm and cable sheathing, are produced by paste-extrusion of emulsion polymers – made kneadable by the addition of 18–15% naphtha solvent – in a ram extruder. In a downstream continuous oven, the lubricant volatilizes before the parts are sintered at 380 °C. Thread sealing tapes, which should remain porous, are not sintered after extrusion, but simply rolled and dried. PTFE dispersions are used to cast thin films or to impregnate glass fiber products or parts made from graphite or porous metals, which are subsequently sintered and compression molded at 380 °C. Antifriction coatings made of PTFE dispersions on metal or ceramic surfaces (if necessary with bonding agents) and subsequently annealed, are porous and therefore not suitable as anti-corrosion coatings.

Slight etching with solutions of alkali metals makes PTFE adhesive; there are special adhesives for temperatures up to 130 °C. Machining requires keen edge tools.

Forms of Delivery, Applications

PTFE is the most important fluoropolymer. Because of its high melt viscosity (melting point > 327 °C) it cannot be processed by standard thermoplastic processing methods. It is available as suspension polymerized powder for compression molding, sintering, and ram extrusion. Emulsion polymer powder is intended for paste extrusion, in dispersion for coatings and impregnations, or as an additive to other plastics to reduce dynamic friction.

PTFE is a plastic exhibiting low rigidity and stiffness. Its advantages are the broad range of use-temperature (−270 to 300 °C, PTFE embrittles only below −270 °C), universal chemical resistance, insolubility in all known solvents below 300 °C, weather resistance without stabilization, flammability rating, excellent electrical and dielectric properties, and the best slip and anti-stick properties of all plastics. Its wear properties are less outstanding; however, wear properties as well as stiffness and rigidity can be improved by modification (5–40 vol.%) with graphite, bronze, steel, MoS_2, or glass fibers.

Applications

Statically and dynamically loaded gaskets, bellows, pistons, and other machine elements, crucibles, coatings (for glass fabrics) and sheathings, carriers for printed circuit boards, films and other semi-finished products for subsequent machining.

6.5.1.2 Polyvinylidene Fluoride (PVDF)

Processing

Melt temperatures (injection molding and extrusion) range from 230–270 °C; mold temperatures range from 60–90 °C. Temperatures for coating with PVDF by spraying, dipping, or casting range from 190–215 °C. PVDF can be thermoformed at 180 °C. PVDF melts must not contact boron- or MoS_2-containing materials (screws, cylinders, certain glass fibers) to avoid spontaneous melt decomposition. In addition, manufacturers' recommendation for colorants, fillers, and reinforcing agents should always be followed. All standard welding processes and adhesion with glues or 2-component adhesives can be used.

More flexible modifications have a melting point at ≈ 165 °C.

Properties

PVDF contains 57% fluorine. Crystallinity depends on the thermal preconditioning of the parts: fast cooling of thin films results in transparent products, while annealing at 135 °C results in highly crystalline and rigid products. Service temperatures range from −60 to +150 °C. The polar structure excludes the use of high-frequency techniques. Weather exposure (radiation wavelengths from 200–400 nm) causes gradual decomposition. PVDF is highly resistant to chlorine and bromine, is UV resistant, and surpasses all other fluoropolymers in high-energy radiation resistance. It meets the highest requirements for purity and is therefore used for high-purity water conduits, packaging of chemicals (gas- and flavor-impermeable bottles), for chemical apparatus, and semi-conductor manufacturing.

Applications

Gaskets, membranes, pump and valve parts, pipes, shrink tubes, fittings, slide rails, linings, laminates for outdoor use, packaging film.

6.5.1.3 Polyvinyl Fluoride (PVF)

Properties

PVF is only available as film (crystal clear) and sheet. Lightly oriented or biaxially stretched films are on the market. Compared to PVDF, PVF exhibits higher rigidity and stiffness, lower density, ductility, and long-term service temperature. Because of its good UV-, IR-, weather-, and corrosion resistance, PVF is used for coatings of other materials in outdoor use. Thermoplastic processing is carried out at 260–300 °C. Semi-finished products are weldable and can be adhesively bonded with EP resins.

Applications

Roofing membranes, roof coverings, sun collectors, laminates for outdoor use, road signs, packaging films, shrink tubes.

6.5.1.4 Polychlorotrifluoroethylene (PCTFE)

Processing

PCTFE is generally processed by compression molding, injection molding, and extrusion. Melt temperatures range from 270–300 °C, mold temperatures range from 80–130 °C. Dispersions can be used to coat metal parts. Because copper- and iron-containing metals catalyze the decomposition of PCTFE, preparatory electro-less nickel plating is necessary when sheathing wires. High-frequency- and ultrasonic welding are possible; adhesive bonding requires pretreatment.

Properties

The symmetry of the macromolecules is disturbed by the integration of chlorine atoms because they are bigger than fluorine atoms. Therefore, the crystallinity is lower and the distance between chains is larger. Nonetheless, the higher polarity of the chlorine atoms results in higher stiffness and rigidity. However, high-frequency applications are limited. The chemical resistance is lower than that of PTFE. PCTFE has the lowest water vapor permeability of all transparent films; it is weather resistant; however, high-energy radiation causes the separation of chlorine; it does not burn, is physiologically inert, is used in medical applications, and can be used in food contact applications.

Applications

Fittings, tubes, membranes, crucibles, printed circuit boards, coil cores, insulating film, packaging.

6.5.2 Fluoro-Copolymers and Elastomers (ECTFE, ETFE, FEP, TFEP, PFA, AF, PTFEAF, TFEHFPVDF (THV, TFB), [FKM, FPM, FFKM])

Numerous copolymers with modified properties are available, in particular with improved thermoplastic processability, see Table 6.19. The nomenclature for these copolymers is not standardized; however, the names often include the abbreviation for the comonomer. Table 6.20 provides a comparison of properties.

6.5.2.1 Ethylene Chlorotrifluoroethylene Copolymers (ECTFE)

This block co-polymer belongs to the fluorine-containing polymers with the highest modulus of elasticity and highest stiffness, while also exhibiting high impact strength between −4 and +150 °C. It is available in pellet- and powder form for injection molding, extrusion, rotational molding, fluidized bed sintering, and foamed cable insulation. The melting temperature ranges from 260–300 °C. Welding and adhesive bonding are possible. The barrier properties against oxygen, carbon dioxide, chlorine gas, and hydrochloric acid are 10 to 100 times better than those of PTFE, which facilitates many applications in the chemical industry.

Applications

Cable sheathing in the electronics-, aero-, and oil-industry, radio chemistry, linings and coatings for containers in the chemical industry, packaging for pharmaceuticals, flexible printed circuits, films for lamination, fibers for filter cloths and fire-resistant upholstery, parts for clean-room applications.

6.5.2.2 Ethylene Tetrafluoroethylene Copolymers (ETFE)

Approximately 25% ethylene content in PTFE greatly improves thermoplastic processability: melt temperatures from 300–340 °C for injection molding (the most important processing method). However, the maximum service temperature is lowered by ≈ 100 °C. Stiffness and strength are improved, resistance is comparable. Stabilization against thermal and photochemical decomposition is necessary. Glass fiber reinforcement results in significant increase in stiffness and strength.

Applications

Gear wheels, parts for pumps, packaging, laboratory equipment, linings, cable insulation, blown film for transparent very durable roofing.

6.5.2.3 Polyfluoroethylene Propylene (FEP); Tetrafluoroethylene Hexafluoropropylene Copolymers (TFEP)

Processing

PTFE is copolymerized with 50–90% hexafluoropropylene to achieve thermoplastic processability. Injection molding and extrusion are carried out at high temperatures: melt temperatures from 315–360 °C and mold temperatures from 200–230 °C. Extrusion blow molding is possible. The barrels of the processing machines must be made of iron-free alloys, such as Hastelloy, Xaloy, Reiloy, or Monel. FEP is susceptible to melt fracture. Crystallinity can be increased up to 40–67% by annealing at 210 °C. FEP powders are used for fluidized bed sintering.

Properties

Compared to PTFE, FEP has a lower melt viscosity, higher impact resistance, lower stiffness and strength, lower long-term service temperature, but comparable chemical and weather resistance, flammability, and electrical properties.

Graphite and milled glass fibers are typically used as reinforcing agents to increase stiffness and wear resistance. The relatively high melt viscosity limits the filler content.

Applications

Cable insulation and coatings, lining of containers, flexible printed circuit boards, injection molded parts for the electric, electronic, and chemical industries, packaging films, impregnations, heat-sealing adhesives.

6.5.2.4 Perfluoropropyl Vinyl Ether Copolymer, Perfluoroalkoxy (PFA)

Copolymers of perfluoropropyl vinyl ether (PFA) with tetrafluoroethylene (TFE) reach the long-term service temperatures of PTFE while being thermoplastically processable because of their lower melt viscosity. Other properties are also comparable to those of PTFE. Melt temperatures reach 330–425 °C for processing, mold temperatures 90–200 °C. All melt conveying parts of the machine must be corrosion resistant; molds should be hard-chromium plated or made from nickel alloy. Because of the fluorine content of the copolymer, PFA exhibits the highest chemical resistance of all fluoropolymers processable in the molten state. Depending on the level of alkyl vinyl ether residual, the service temperatures can exceed +260 °C. The properties of PFA can be significantly influenced by varying the level of alkyl vinyl ether residual.

Applications

Linings, technical products, cable insulation, heat-sealing adhesives, chimney linings for energy efficient heating systems.

6.5.2.5 PTFE Copolymers with AF (PTFEAF)

PTFEAF is a copolymer of PTFE and 2,2-bis(trifluoromethyl)-4,5-difluoro-1,3-dioxolan (AF). While all fluorine-containing polymers described so far are semi-crystalline and therefore opaque to translucent and insoluble, PTFEAF is amorphous, transparent, and soluble, rendering it suitable for coating of substrates (anti-stick coatings). It is stiff and does not creep; the melt exhibits good flow properties and the lowest dielectric constant in the GHz range of all plastics (1.89–1.93). The service temperature of this fluoropolymer is below 360 °C; at temperatures above 360–500 °C, the polymer will degrade.

Applications

Anti-corrosion and anti-stick coatings, fiber optics, non-reflective coatings for optical parts, high-permeability membranes for gas separation, parts for electro- and electronic industries.

6.5.2.6 Tetrafluoroethylene Hexafluoropropylene Vinylidene Fluoride Terpolymers (TFEHFPVDF (THV, TFB))

TFB is a thermoplastic fluoroterpolymer that consists of *t*etrafluoroethylene (TFE) *h*exafluoropropylene (HFP), and *v*inylidene fluoride (VDF). This terpolymer has a melting temperature of 160–185 °C, low strength, a very low Young's modulus, and a very high elongation at break (even at low temperatures). Transparency in the visible spectrum is at 97%. The chemical resistance compares to that of PVDF, its flame resistance is better, and it holds up to weathering for 10 to 15 years. It can be processed like a thermoplastic material and is weldable by heat or high-frequency methods.

Applications

Fabric coatings (also with dispersions), fiber optics, solar cells; flexible tubes and pipes for the medical and chemical industries, cable insulation.

6.5.2.7 Other Fluoro Copolymers

Tetrafluoroethylene perfluoromethylene vinyl ether copolymer (MFA).

Vinylidene chloride hexafluoropropylene (VDFHFP).

Tetrafluoroethylene perfluoro vinyl sulfonic acid (TFE/PVS) (ionomer, applications: membranes).

Fluorinated rubbers; FKM, EPM, FFKM. See Section 6.23.4.

6.6 Polyacrylic- and Methacrylic Polymers

All polymers based on acrylic acid and methacrylic acid belong to the group of acrylate polymers, see Table 6.4, No. 11–16. They exhibit particularly good transparency and weatherability.

6.6.1 Polyacrylates, Homo- and Copolymers (PAA, PAN, PMA, PBA)

6.6.1.1 Polyacrylonitrile (PAN)

PAN is produced by polymerization of acrylonitrile, general structure see Table 6.4, No. 26. Major area of application is the production of fibers and use as copolymer with styrene (ABS, SAN) and butadiene (NBR; nitrile rubber). PAN has good barrier properties against most gases. Because pure PAN cannot be processed like a thermoplastic material, acrylonitrile copolymers are used as barrier layers in packaging films.

6.6.1.2 Polyacrylates, Special Products

Homopolymer acrylic acid esters (PAA), see Table 6.4, No. 11, are soft resins, exhibiting good resistance against light, oxidation, and heat. They are used for co- and terpolymerization with PS, PVC, VA, MA, AN, and acrylic esters because of their plasticizing effects and are available as solid resins, in solution, but mainly in dispersion. *Oxazolidine-modified acrylic resins* can be cross-linked with isocyanates. *Elasto-plastic copolymers* are used as base compounds for seals; those with > 20% acrylic acid are water soluble. *Polyhydroxyethylene methacrylate* saturated with ≈ 40% water is used for contacts and coatings (*e. g.*, for glasses) and encapsulation with controlled water absorption in medical and other technical applications. *Acrylic resins* soluble in the digestive tract are used for encapsulation of drugs. Mono-polymerization or polymerization of *crosslinked* two-component resins is used to produce heat- or radiation-curable paint resins. Rigid modified *methacrylate copolymers* are used as finishes for artificial leather and other fuel-resistant varnishes. *MMA-VC copolymers* are plasticizers for PVC. Unsaturated aliphatic polyurethane acrylate resins can be cross-linked with H_2O_2 to form tough, flexible glass fiber laminating resins.

6.6.2 Polymethacrylates, Homo- and Copolymers (PMMA, AMMA, MABS, MBS)

6.6.2.1 Polymethylmethacrylate (PMMA)

Chemical Composition

PMMA is the best-known acrylate grade. Block-, emulsion-, or suspension polymerization of methacrylic acid yields a thermoplastic material with high molecular weight; for basic structure see Table 6.4, No. 17. Block polymerization between glass plates or in molds produces semi-finished products such as sheet, solid profiles, or pipes of high strength, high modulus of elasticity, and good surface finish. Peroxides are used as polymerization initiators for hot-setting processes; the same peroxides with the addition of amino-activators or other redox-polymerization-starter systems are used for cold-setting processes. Addition of multi-functional components produces semi-crosslinked polymeric materials. Their further processing is only feasible by thermoforming, adhesion, or machining because of the extremely high molecular weight.

In order to produce resins for injection molding, blocks are polymerized in flat bags and subsequently milled. Uniform pellets are produced by extrusion and subsequent pelletizing.

Processing

PMMA resins can be injection molded and extruded; melt temperatures range from 200–230 °C, mold temperatures range from 50–70 °C. Pre-drying of the resin or use of vented screws is required. When printing or painting on PMMA surfaces it is important to ensure that the parts do not retain any residual stresses and that the appropriate kind of color pigment is selected in order to avoid stress cracking. PMMA surfaces can be metallized, both with and without pre-treatment. Hot air-, high-frequency-, and ultrasonic welding are possible, as is adhesive bonding with single- or two-component methyl methacrylate adhesives or with glues based on methylene chloride or chlorinated carbon hydroxides. However, glues containing solvents should not be used because of their environmental impact; in particular, given the fact that now adhesives based on epoxy resins, PUR, or cyanoacrylate are available that are adapted for polymeric materials.

Properties

PMMA is brittle with high strength, high modulus of elasticity, high surface hardness (scratch resistance), and, if polymerized between polished glass plates, with high surface gloss. It can be polished, is weather resistant without further stabilization, and resistant to weak acids and alkalis, non-polar solvents, fats, oils, and water. Stress crack formation may be problematic. PMMA is preferably used in the lighting industry because of its excellent light transmission and the fact that it can be colored to a high degree of saturation, see Figure 6.3 and Figure 6.4. For a *comparison of properties* see Table 6.21.

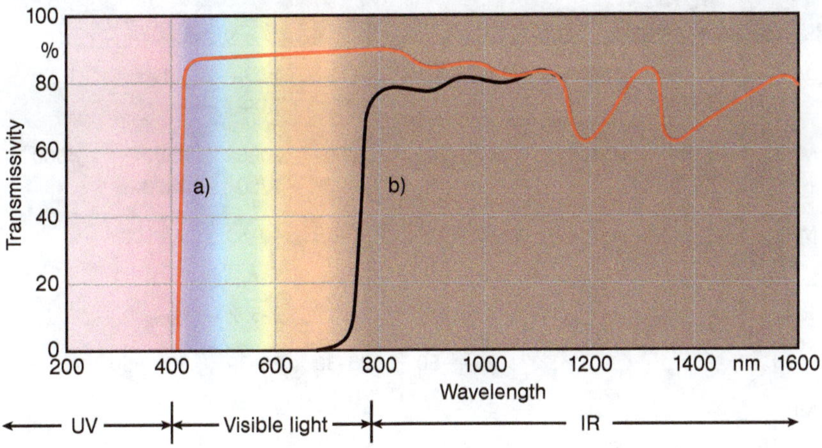

Figure 6.3 Spectral transmittance for crystal clear PMMA (a) and IR-transmitting black colored PMMA (b) (specimen thickness 3 mm)

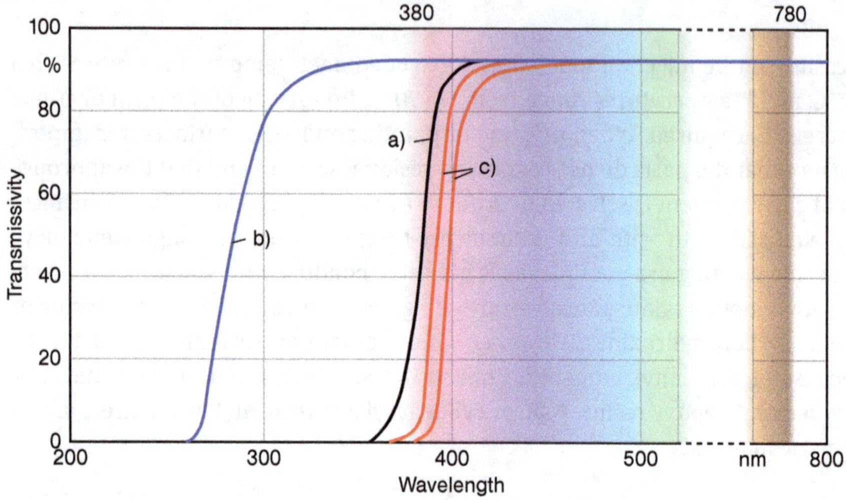

Figure 6.4 Spectral degree of transmittance of crystal clear (a), crystal clear, UV-transmitting (b), and two crystal clear, UV-absorbing (c) grades of PMMA (specimen thickness 1 mm)

Table 6.21 Property Comparison of Methyl Methacrylate Polymers

Properties	Unit	Methyl methacrylate polymers						
		PMMA resins	PMMA semi-finished products, high MW	PMMA-HI elastomer mod.	MBS	MABS	AMMA semi-finished products	PMMI
Density	g/cm³	1.17–1.19	1.18–1.19	1.12–1.17	1.05–1.16	1.07–1.09	1.17	1.22
Young's modulus	MPa	3100–3300	3300	600–2400	2000–2800	2000–2200	4500–4800	4000
Tensile modulus of elasticity	MPa	–	–	20–60	30–55	40–50	90–100	–
Yield stress	%	–	–	4.5–5	3–6	3–5.5	10	–
Elongation at yield	%	–	–	20– > 50	25–30	20–30	40– > 50	–
Stress at 50% strain	MPa	–	–	–	–	–	–	–
Failure stress	MPa	60–75	70–80	–	–	–	–	90
Failure strain	%	2–6	4.5–5.5	–	–	–	–	3

Properties	Unit	Methyl methacrylate polymers						
		PMMA resins	PMMA semi-finished products, high MW	PMMA-HI elastomer mod.	MBS	MABS	AMMA semi-finished products	PMMI
Melting temperature	°C	–	–	–	–	–	–	–
Heat deflection temperature HDT/A 1.8 MPa	°C	75–105	90–105	65–95	85	90	75	1.60
Thermal coefficient of linear expansion, parallel (23–55 °C)	10^{-5}/K	7–8	7	8–11	9	9	6.5–7	4.5
Thermal coefficient of linear expansion, perpendicular (23–55 °C)	10^{-5}/K	–	–	–	–	–	–	–
Flammability UL94 for 1.6 mm thickness	Class	HB	HB	HB	HB	HB	HB	HB
Dielectric constant at 100 Hz	–	3.5–3.8	3.5–3.8	3.6–4.0	3.0–3.2	2.9–3.2	4.5	–
Dielectric loss factor at 100 Hz	$\cdot 10^{-4}$	500–600	600	400–600	270–290	160–200	600	–
Specific volume resistivity	Ohm · m	$>10^{13}$	$>10^{13}$	$>10^{13}$	$>10^{13}$	$>10^{13}$	$>10^{13}$	–
Specific surface resistivity	Ohm	$>10^{13}$	$>10^{13}$	$>10^{13}$	$>10^{13}$	$>10^{13}$	$>10^{13}$	–
Dielectric strength	kV/mm	30	ca. 30	30	25–30	35–40	ca. 30	–
Comparative tracking index CTI/A		600	600	600	600	600	600	–
Water absorption at 23 °C, saturation	%	1.7–2.0	1.7–2.0	1.9–2.0	0.4–0.6	0.7–1.0	2–2.25	5
Moisture absorption 23 °C/50% rel. humidity, saturation	%	0.6	0.6	0.5–0.6	0.2–0.3	0.35	0.7–0.8	2.5

Applications

Molding resins: Piping/tubing elements, sanitation, glasses, lenses, housings, covers, viewing glasses, fiber optics, solar technology, cockpit and stadium window shields, light domes, filter for specific wavelengths, jewelry, optical data storage, "electronic paper" (60 µm thin film with finely distributed liquid crystalline drops: polymer dispersed LC, PDLC).

Casting resins: Gap fillers, safety glass, wood impregnation, integral casting of displays, glass fiber reinforced light displays, artificial stone tiles with, *e.g.*, marble inlays, two-component systems for highly resistant pavement coatings, concrete repair, road markings, special resins in rigid or rubber-elastic modifications for surgical, orthopedic, and dental prosthetics.

Modified, thermosetting, free-flowing, and highly reinforceable reaction resins based on MMA are used as bonding agents for BMC- and SMC molding resins, semi-finished products and pultrusion, resulting in highly moisture-, chemical-, and corrosion-resistant parts, some of which with increased heat distortion temperatures. Dispersions with fine-grain mineral fillers are used to cast and heat cure parts such as kitchen sinks, water basins, sanitary wares, and marble-like sheets that can be thermoformed or finished like wood.

6.6.2.2 Methyl Methacrylate Copolymers (AMMA)

Copolymers and blends with methyl methacrylate are used to increase heat distortion temperature, impact resistance, and stress crack resistance against alcohols.

Basic structure, see Table 6.4, No. 17.

Processing

High melt viscosity and a wide softening range render AMMA particularly suitable for blow molding and thermoforming. The pellets must be pre-dried for injection molding at melt temperatures not to exceed 230 °C.

Properties

Solvent resistance and heat distortion point, and at AN contents of more than 50%, also impact resistance, are increased. Because AN tends to form ring structures when heated, AMMA has a slight yellowish tinge. Light stabilization is required. Cast sheet can be produced by *in situ* copolymerization. Biaxial stretching up to 70% increases impact, burst, and tear resistance; AMMA is therefore suitable for heavy duty glazing applications, *e.g.*, in stadiums.

The low gas permeability of polyacrylonitrile determines the performance of (graft-) copolymers with methacrylate or styrene. These "barrier" plastics contain up to 70% nitrile and sometimes additional butadiene-containing, plasticizing components. Permeability decreases with increasing biaxial stretching; fracture strength and impact resistance increase.

AMMA is resistant to moderately concentrated acids and alkalis and most solvents. It swells in methanol and ketones; it is soluble in dimethylformamide and acetonitrile.

Comparison of properties, see Table 6.21.

Applications

Packaging for liquid and solid foods such as spices, vitamin supplements, instant soups, and meat products; car care products, and cosmetics. Beverages must not come in contact with nitrile-containing plastics; therefore, for these applications multi-layer packaging is used.

6.6.2.3 Methyl Methacrylate Acrylonitrile Butadiene Styrene Copolymers (MABS); Methacrylate Butadiene Styrene Copolymers (MBS)

Chemical structures, see Table 6.4, No. 5, 17, 26, 35.

MABS and MBS are impact resistant, clear, and translucent even at low temperatures. Because of their butadiene content they are not weather resistant; however, they are resistant to oils, fats, and fuels. They are suitable for sterilization by gamma radiation.

Processing

Injection molding, extrusion, blow molding, heat sealing, adhesive bonding, ultrasonic-, hot plate-, and friction welding.

Comparison of properties, see Table 6.21.

Applications

Bottles for cosmetic, spray, and cleaning products; hollow technical parts, disposable medical devices, controls, and packaging.

6.6.3 Polymethacrylates, Modifications and Blends (PMMI, PMMA-HI, MMA-EML Copolymers, PMMA+ABS)

6.6.3.1 Polymethacrylmethylimide (PMMI)

Technically, PMMI is a copolymer of methyl methacrylate (MMA) and glutaramide; however, it is produced by reacting PMMA with methylamine (MA) at high temperatures and pressures, see Equation 6.7.

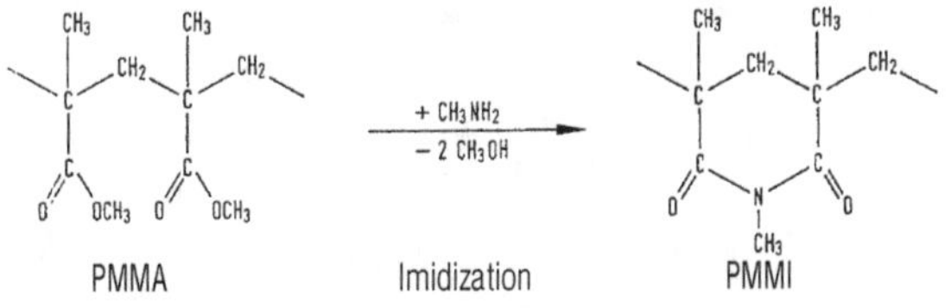

(6.7)

Colorless and with high transparency, crystal clear and free of cloudiness its properties closely resemble those of PMMA. The ring termination provides increased chain stiffness and dimensional thermal stability (heat distortion temperature). Depending on the degree of imidization, it is possible to obtain all intermediate property values compared to PMMA. PMMI exhibits low oxygen permeability and is less susceptible to stress cracking caused by ethanol, ethanol/water, and isooctane toluene mixtures than PMMA. These properties depend on the imide content.

Processing

PMMI can be injection molded (after pre-drying) with melt temperatures ranging from 200–310 °C and mold temperatures ranging from 120–150 °C.

Comparison of properties, see Table 6.21.

Applications

Headlamp diffusers for automotives, street light covers, fiber optics, blend components for fiber reinforced engineering materials.

6.6.3.2 Impact Resistant PMMA (PMMA-HI)

Impact-resistant resins are 2-phase systems manufactured by suspension or emulsion polymerization. The toughening disperse phase in the PMMA matrix is formed by, for example, styrene-modified acrylic elastomers (< 30%, with matched refractive index). This composition results in impact resistant molding compounds that can be blended as required with PMMA and that are as weather resistant and crystal clear as pure PMMA. Their stress cracking sensitivity is lower and their resistance to hot water is higher than PMMA's.

Processing

The compounds are either pre-dried or processed with vented screws by injection molding or extrusion at melt temperatures ranging from 210–230 °C and mold temperatures ranging from 60–80 °C.

Comparison of properties, see Table 6.21.

Applications

Anti-weathering coatings for construction profiles, for example PVC window profiles, household appliances, drawing and writing utensils; sanitary installations, light covers.

6.6.3.3 Methyl Methacrylate Exo-Methylene Lactone Copolymers (MMA-EML-Copol., MMAEML)

Compared to PMMA, copolymers of methyl methacrylate with exo-methylene lactones (EML), for example with methylene methyl butyrolactone, exhibit improved properties: at EML contents of 15–40% glass transition temperatures range from 140–180 °C, tensile strength from 46–82 MPa, modulus of elasticity from 3600–4000 MPa, elongation at break from 4.7–1.2%, and water absorption from 2.5–4.3%. The refractive index of 1.54 is higher than that for PMMA (1.49). Melt temperature ranges from 230–280 °C.

Applications

High hardness, optical brilliance, weather- and UV resistance, and paintability render these resins suitable for automotive headlights (diffusers and reflectors); fiber optics.

6.6.3.4 PMMA+ABS

This type of blend is used for automotive parts (body parts, housings, reflectors), sanitary installations (surface coatings for tubs and shower bases), and in electric applications. The material can be metallized, has good welding properties, and superior weather resistance and stiffness compared to ABS.

6.7 Polyoxymethylenes (Polyacetal, Polyformaldehyde) (POM)

Ursula Ziegler

6.7.1 Polyoxymethylene Homo- and Copolymers (POM-H, POM-C)

Chemical Composition

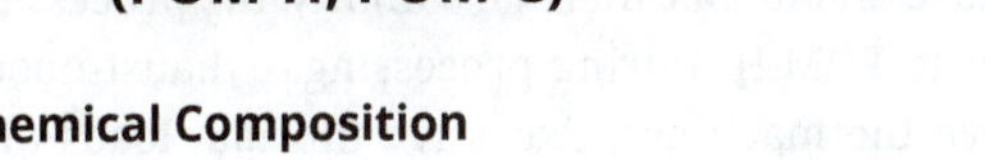

Acetal homopolymer, POM-H — Acetal copolymer, POM-C, (n is statistically distributed, m is 2 or 4) (6.8)

POM (polyoxymethylene, polyformaldehyde, polyacetal) are semi-crystalline materials that can be processed like thermoplastics. We distinguish between homopolymers that are formed by anionic polymerization of formaldehyde, and copolymers that are formed by ring-opening cationic polymerization of trioxane with various comonomers, see Equation 6.8. Because the polyacetal bonds are thermally and chemically relatively unstable, the goal of implementing comonomers is to stabilize the material for processing and end-use. Once exposure to thermal oxidatives or acids initiates the degradation of an acetal bond in the homopolymer, the entire polymer chain will degrade entirely. In the copolymers, the more stable ether bonds (typically with C2- or C4-comonomer units) in the chain act as interrupters that impede further degradation so that only a small part of the chain is subject to degradation, making sure that the characteristics of the material can be maintained. Because formaldehyde is formed during degradation of the acetal chain, its creation is limited with the implementation of comonomers, making processing and use of these products safer. Copolymers differ from homopolymers in their terminal groups: the latter contain unstable semi-acetal terminal groups that are capped by hydrolysis-prone ester groups. In contrast, due to their ether end-groups, copolymers are alkali-resistant and exhibit better hot water resistance. On the other hand, homopolymers exhibit a slightly higher melting point and mechanical properties than copolymers, because the comonomers contained in the copolymers cause a lower degree of crystallinity. After the initial commercial introduction of homopolymers in 1959, POM copolymers' higher stability led to a significantly higher market share compared to POM homopolymers.

The variation options during copolymerization of trioxane allow for the incorporation of several different comonomers. The incorporation of multi-functional comonomers allows for the creation of branched or even crosslinked terpolymers with increased melt strength that facilitate specific processing technologies, such as extrusion blow molding.

Processing

In general, POM materials can be processed with all common technologies used for thermoplastics, such as injection molding, extrusion, injection and extrusion blow molding, and compression molding. Pre-treatment is typically not required. Drying is recommended for materials that have absorbed moisture. Depending on material grade and the respective dwell time, melt temperatures should not exceed 220–240 °C. Because of its lower melting temperature and higher thermal stability, the processing window of POM-C is wider compared to POM-H. During processing, exhaust hoods should be installed immediately above the machines. Excessive thermal loads and dwell times cause the degradation of POM under formation of pungent smelling, gaseous formaldehyde that irritates mucous membranes. In addition, if the nozzle is closed or frozen, the pressure of the forming gaseous formaldehyde may damage the machine.

POM is not miscible with most other thermoplastics. Contamination with other materials, even in small quantities, will lead to non-homogeneous components. Particular caution is required with decomposition-causing thermoplastics, specifically PVC, because even at low concentrations they may initiate a severe degradation reaction. Therefore, POM materials even with traces of PVC contamination must not be processed.

The most common processing technology for POM is injection molding. The market offers a number of grades with different viscosities that allow the injection molding of components ranging from extremely delicate components to thick-walled, void-free parts. Particularly easy flowing injection molding grades are available for parts with particularly thin walls or long flow paths. POM's good flowability and fast crystallization make it suitable for both hot- and cold-runner molds. Mold wall temperatures should range between approx. 60 and 130 °C. Increasing mold wall temperatures lead to increasing processing shrinkage and decreasing post-shrinkage. Therefore, the mold wall temperature should be as high as possible for precision injection molding operations in order to ensure dimensionally stable molded components. For most commercially available POM grades, total shrinkage ranges between 1 and 3%, depending on modification with additives and wall thickness of the molded part. Smaller dimensional corrections can be made by changing injection or hold pressure. Components made from unmodified POM typically exhibit only little warpage; however, in fiber reinforced POM components the fiber orientation in flow direction causes higher warpage.

POM is also suitable for multi-component injection molding, *e. g.*, for the manufacture of rigid/soft composites. POM and elastomer manufacturers offer special elastomer grades that adhere to POM together with processing advice to ensure good bonding of the components. Components with adhesion modified TPS-SEBS or with thermoplastic polyurethane elastomers (TPU) or thermoplastic polyester elastomers (TPC, see Section 6.20) deserve particular emphasis.

POM is used for the extrusion of semi-finished products (rods, flat and hollow bars, and sheet) that can subsequently be machined into components. In addition to standard injection molding grades, the market offers higher viscous extrusion grades. For extrusion operations, the narrow melting range and fast freezing of POM needs to be taken into consideration. Low output rates are recommended to ensure that heat created by crystallization is sufficiently dissipated despite the low thermal conductivity of POM. Non-uniform cooling rates across profile cross sections create residual stresses that must be compensated by subsequent tempering. It is also possible to extrude pipes that are calibrated by vacuum combined with internal positive pressure.

To facilitate extrusion blow molding operations, POM grades with particularly high melt strength are required. Suitable here are branched terpolymer grades and modified grades in which a blend component provides improved melt strength. Hollow components up to 5 l volume can be manufactured without a melt accumulator. Larger hollow parts require machines with accumulator heads where plasticized material is accumulated and then extruded relatively quickly.

POM is also suitable for injection blow molding operations. This technology is used for the waste-free production of packaging containers that are under internal pressure such as aerosol containers.

Hot plate welding at temperatures ranging from 220–240 °C and friction welding are particularly suitable methods to join components made from POM. However, RF welding is not suitable because of the low dielectric loss factor of POM.

Because of its high solvent resistance, it is difficult to bond POM with common adhesives. To facilitate such bonds, surface activation and the selection of special adhesives is required.

Printing, painting, and metallizing of POM is also relatively complex and typically requires special pre-treatment of the components to ensure sufficient adhesion.

Properties

As a result of its high degree of crystallinity, non-reinforced POM is among the stiffest and toughest thermoplastic materials. The melting temperature of the homopolymer is approx. 178 °C, the melting temperatures of most copolymers range from 160 to 172 °C, depending on type and concentration of the comonomers used. The rate of crystallization is very high. Because of its high crystallinity, POM also exhibits a very high density of approx. 1.4 g/cm^3.

Even unreinforced, POM exhibits very good heat resistance; its short-term service temperature is approx. 150 °C, its long-term service temperature is approx. 110 °C. The material is impact resistant, and it embrittles only below −40 °C because of its low glass transition temperature of approx. −60 °C.

Table 6.22 provides a comparison of properties of POM homopolymers and POM copolymers and materials modified with them.

Figure 6.5 shows the shear modulus G and the mechanical loss factor δ of a typical POM copolymer. Figure 6.6 shows stress/strain diagrams of a high molecular POM homopolymer. At room temperature, POM exhibits a distinct elastic limit at approx. 7 to 15% strain. Below this limit, POM exhibits very good resilience characteristics, even under cyclic loads, which makes it particularly suitable for snap-fits and spring elements.

Table 6.22 Property Comparison for Polyacetals

Properties	Unit	Polyacetals				
		POM - H	POM - C	POM - H HI	POM - C HI	POM - C GF 25
Density	g/cm^3	1.40–1.42	1.39–1.41	1.34–1.39	1.32–1.39	1.6
Water absorption at 23 °C, saturation	%	0.9	0.7–0.8	0.8–0.9	0.6–1.0	0.9
Water absorption 23 °C and 50% rel. humidity	%	0.2	0.2	0.2–0.35	0.2–0.3	0.17

Properties	Unit	Polyacetals				
		POM - H	POM - C	POM - H HI	POM - C HI	POM - C GF 25
Tensile modulus of elasticity	MPa	2900–3400	2500–3200	1400–2500	900–2300	9000
Yield stress (failure stress for GF)	MPa	70–75	60–70	43–58	30–55	(135)
Elongation at yield	%	10–25	7–12	16–30	6–30	/
Nominal elongation at break	%	20–45	15–35	30– > 50	30– > 50%	3
Notch impact strength (Charpy) at 23 °C	kJ/mm^2	6–14	6–11	15–100	9–100	8
Notch impact strength (Charpy) at −30 °C	kJ/mm^2	6–13	5–9	12–20	6–15	8
Melting temperature	°C	178	162–172	178	162–172	166
Heat deflection temperature HDT/A 1.8 MPa	°C	93–108	90–107	60–80	55–87	160
Thermal coefficient of linear expansion, parallel (23–55 °C)	$10^{-5}/K$	10–11	11–12	13	12–14	4
Thermal coefficient of linear expansion, perpendicular (23–55 °C)	$10^{-5}/K$	10–11	–	12–14	–	8
Flammability UL94 for 1.6 mm thickness	Class	HB	HB	HB	HB	HB
Dielectric strength	kV/mm	32–33	35–40	23	28–40	40
Dielectric constant at 100 Hz		3–4	4	3–5	3–6	4
Dielectric constant at 1 MHz		3–4	4	3–4	3–5	4

Table 6.22 Property Comparison for Polyacetals *(continued)*

Properties	Unit	Polyacetals				
		POM - H	POM - C	POM - H HI	POM - C HI	POM - C GF 25
Dielectric loss factor at 100 Hz	10^{-4}	100–200	10–30	65	30–300	30
Dielectric loss factor at 1 MHz	10^{-4}	50–70	50	160	60–250	60
Specific volume resistivity	Ohm · m	$>10^{13}$	$>10^{12}$	$>10^{12}$	$>10^{11}$	10^{12}
Specific surface resistivity	Ohm	$>10^{15}$	$>10^{13}$	$>10^{14}$	$>10^{13}$	10^{14}
Comparative tracking index CTI/A		600	600	600	600	600

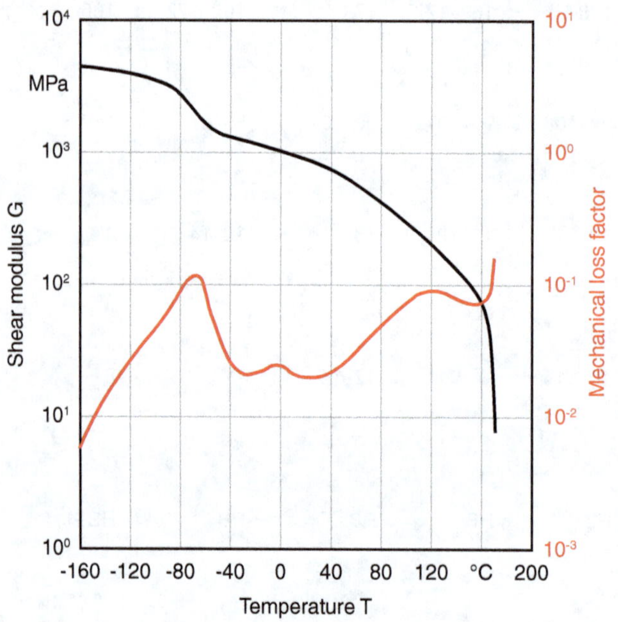

Figure 6.5 Shear modulus G and mechanical loss factor δ of a typical POM copolymer as a function of temperature; torsion vibration test according to DIN EN ISO 6721

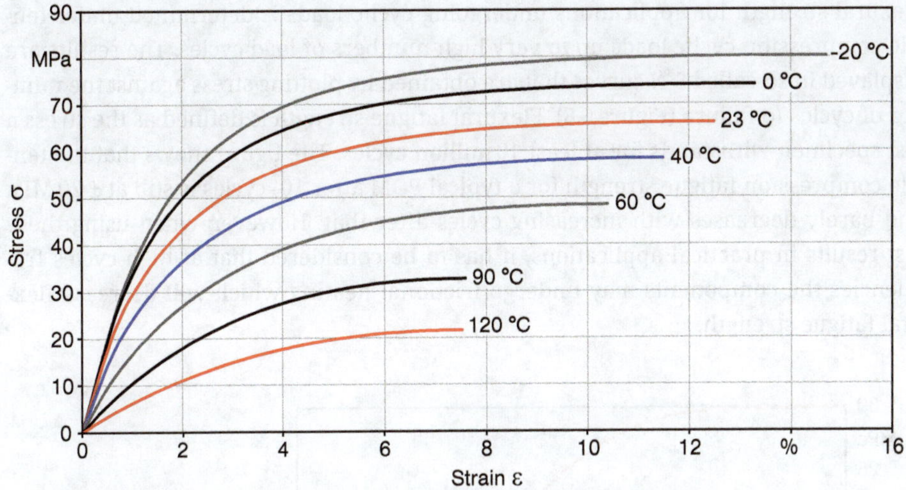

Figure 6.6 Stress-strain diagrams of high molecular weight POM homopolymer; measured at different temperatures according to ISO 11403-1

POM exhibits little tendency to creep under long-term loads and very good flexural strength under continuous cyclic loads. Figure 6.7 shows creep curves of a typical POM copolymer.

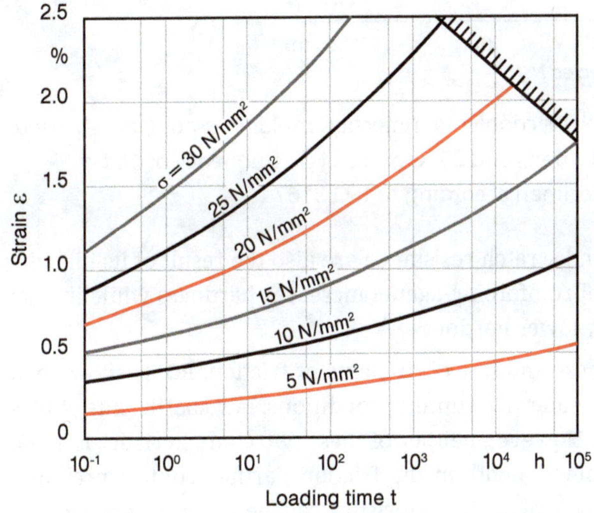

Figure 6.7 Creep curves of a typical POM copolymer, measured at 23 °C and different loads (according to tensile creep test specified in ISO 899)

Flexural strength for applications undergoing cyclic loads is determined under tensile-compression cyclic loads up to very high numbers of load cycles. The results are displayed in so-called S-N curves that are obtained by plotting stress against the number of cycles to failure (Figure 6.8). Flexural fatigue strength is defined as the stress a test specimen withstands for at least 10 million cycles. The figure shows that the tensile-compression fatigue strength for a typical POM after 10^7 cycles is still at ± 20 MPa and barely decreases with increasing cycles after that. However, when using these test results in practical applications, it has to be considered that at high cycles frequencies the components may undergo frictional heating which will decrease flexural fatigue strength.

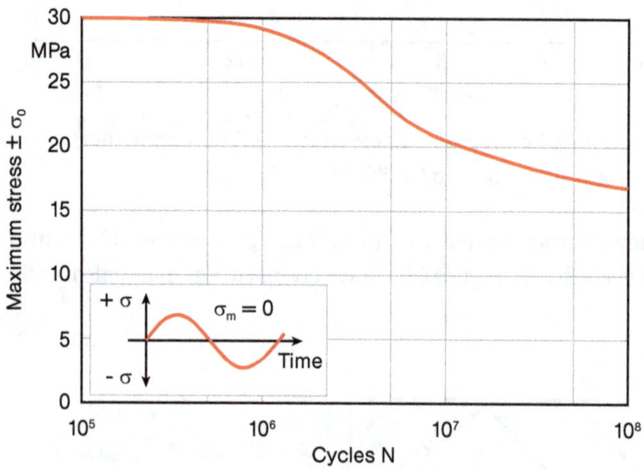

Figure 6.8 S-N curve of typical POM copolymer (injection molding grade); determined under tensile-compression cyclic loads and 23 °C and a cycle frequency of 10 Hz according to DIN 50100; test specimen according to ISO 3167

POM's high surface hardness and scratch resistance are also the result of its high degree of crystallinity. Addition of reinforcing agents increases hardness while impact modified grades exhibit slightly lower hardness.

Because of their surface hardness and low coefficients of friction, POMs show good slip and wear resistance. Even under dry running conditions, low coefficients of friction and wear can be achieved. However, neither of these two characteristics is a material characteristic because they depend on the friction partner, contact pressure, and sliding velocity. POM's tribological characteristics can be further improved by special additives that modify the surface. This approach will also reduce squeaking noise created by stick/slip effects, see Sections 3.8 and 7.6.

POM exhibits little water and moisture absorption so that, compared to other thermoplastics, the unique combination of stiffness, strength, and toughness is only little dependent on ambient conditions such as humidity. Water absorption in standard climate is approx. 0.2%; at saturation after immersion in water at 23 °C it is only 0.8% and the physical properties of the molded components undergo only minimal changes. Therefore, even when used under changing climate conditions, POM components exhibit largely uniform dimensions and application properties.

Both POM-H and POM-C are attacked by strong acids (pH 4) and oxidants. However, POM-C is alkali-resistant while POM-H is not. Even at elevated temperatures, POM copolymers exhibit good to excellent long-term chemical resistance against water, detergent solutions, aqueous solutions of salts, and most common organic solvents (alcohols, esters, ketones, aliphatic and aromatic hydrocarbons, fuels (also methanol- and ethanol-containing fuels, *e. g.*, M15, CM15, CM15A, CM15AP, E85, FAM-B, bio diesel), as well as fats and oils, and brake and cooling fluids. Some solvents and fuel components, in particular alcohols, such as methanol and ethanol, cause minor (reversible) swelling. Fuel permeability is also very low. Specially stabilized grades are resistant in hot diesel fuels up to 100 °C and in aggressive gasoline.

Permeation of gases and vapors, also of organic compounds, for POM is also low.

UV resistance of POM can be improved by UV stabilizers and/or addition of carbon black. Stabilization for colored grades with good hot light fastness and products with good weather resistance for outdoor applications are available.

At high doses, POM is not resistant to high-energy or ionizing radiation such as gamma-radiation.

Films made from unmodified POM are translucent. Molded components are more or less translucent to opaque/white, depending on their wall thickness. POM can be easily colored using pigments. Effect pigments (*e. g.*, metal pigments or fluorescent pigments) for special decoration applications are available. The addition of carbon black or additives also facilitates laser labeling. The degree of gloss of molded components depends mainly on the surface quality of the mold. Specially modified grades for molded components with matte surfaces are also available.

POM burns with a weak blue flame and drips. According to UL 94, POM is classified as HB. Classification as V-0 is not possible. For a wall thickness of 1 mm, burn rates typically range below 100 mm/min, the maximum rate allowed by FMVSS 302.

Their good insulation and dielectric properties are little temperature- and frequency-dependent. POM is not prone to static charging. Grades with tailored anti-static additives provide lower specific volume resistivity. Conductive grades with reduced specific volume and surface resistivity are also available.

POM is physiologically inert; for food applications and use in medical technology, appropriately highly pure grades are offered.

Applications

POM is used in particular for technical parts with specific requirements in terms of combinations of stiffness, strength, and good toughness even at low temperatures, low fatigue and creep properties, and high flexural fatigue strength. POM is particularly suitable for precision injection molding because of its high dimensional stability; applications are in precision engineering, *e.g.*, gear wheels and toothed rings, switches, levers, bearings, screws, and coils. Here, POM has largely substituted traditionally used materials, such as metals. In addition to a palette of basic grades with a variety of flow characteristics, there are specifically impact modified POM grades and reinforced grades.

In automotive engineering, *e.g.*, for functional parts in seat belts, airbags, power window motors and housings, door lock housings and electrical mechanism, door handles, loudspeaker housings, parts of car seats and head rests, switches, levers, push buttons and knobs in dashboards, in steering and control for stabilizer arms, ball sockets, connecting rods, sensor components, adjustment elements, *e.g.*, for air conditioning or sun roofs.

For functional parts and controls in machines, motors, pumps, for gas meters, parts of textile machines, office equipment, copiers, electrical tools, and DIY devices. For small household appliances, white goods, *e.g.*, dampers in washing machines, filters in dish washers, and vacuum components. For electric razors, motors of electric toothbrushes and waterpicks, consumer electronics and keyboards, *e.g.*, for personal computers. "Outsert" injection molding is used to print circuit boards with up to 100 functional POM elements. Keyboards and other visible controls can be laser printed when POM grades with specific modifications are used. Here, change-over from dark/black to light or from light to dark is possible.

Because of its good flexibility and low creep, POM is often used for snap-fit applications, such as fasteners, cable guides, clasps, and spring plates, in a wide variety of industries. Window, door, and furniture hardware and drawer guides are made from POM.

POM is also used for moving parts with low coefficients of friction, low abrasion values, and high long-term strength, such as rollers, sliding components (sliding and fastening parts for window treatments), chain links for conveyor belts, and zippers. Tailored grades are offered for many of these applications, combining frictions and wear properties for different loads (rate and pressure) and friction partner.

Because of its very good resistance to hydrocarbons, alcohols, and other solvents, POM is used in fuel-related applications, such as fuel pumps, flanges, filter housings, surge tanks, level gauges, tank cap, rollover valves, built-in components in diesel tanks (special hot diesel stabilized grades are available).

POM copolymers are used in sanitary applications, such as hot water carrying parts of shower heads and faucets, and for pipe fittings, such as for sprinkler systems.

Special UV-stabilized POM grades are used for outdoor applications. In automotive applications they are used for functional window wiper components, adjustment mechanisms for mirrors and headlights (*e.g.*, electro-mechanical, with Bowden cable), for gardening tools, and sport equipment (ski bindings, inline skates).

POM materials are also used for functional toys, functional packaging in the food and cosmetics industry (valves and metering systems). Because of their high purity, special products may also be used for medical applications, such as technical devices and packaging for pharmaceutical products (*e.g.*, spring and sliding element in insulin pens and inhalers).

6.7.2 Polyoxymethylene, Modifications and Blends (POM+TPU)

Basic Grades and Special Stabilizations

Most manufacturers offer a palette of basic grades with a variety of flow characteristics as well as grades with nucleating agents for injection molding and high molecular grades for extrusion and blow molding. Beyond these option are grades with specific stabilization for different areas of application.

Grades with added UV stabilizers (in particular HALS compounds) and UV absorbers deserve recognition. These additives are particularly important for outdoor applications. The effect of the chemical UV stabilizers is often amplified by the addition of carbon black or titanium dioxide as UV absorbers. For automotive interior applications, UV stabilized grades in interior colors are often used; they have to meet specific requirements for color fastness.

Many large manufacturers offer special grades with reduced formaldehyde emission for interior applications. These grades exhibit higher chain and terminal group stability during polymer manufacturing and compounding. In addition, special processing stabilizers, such as highly effective antioxidants, acid- and formaldehyde-scavengers are added to these compounds to reduce formaldehyde emission by the finished part.

Other special stabilizations include hot water resistance (for sanitary applications, heating water systems, coffee makers, and against acidic decalcifiers in household applications, such as bathroom faucets, or shower heads). Also on the market are grades with improved resistance against chlorine in water and alkaline bleaches. Grades that are stabilized against specific media include those for fuel applications with improved hot diesel resistance. These highly stabilized grades typically contain increased contents of antioxidants and acid scavengers.

Because of its low formaldehyde content, POM has very good microbial and fungal resistance. For sanitary or household applications, anti-microbial modified grades will further inhibit the growth of microbes and fungi. Modifiers here include zinc oxide or special silver salts and, depending on application, biocides.

Filled and Reinforced Grades with Increased Strength

Incorporation of glass fibers (in a range from 10 to 40 wt.-%), steel fibers, or other fiber types, glass spheres, or mineral fillers (such as talcum, chalk, wollastonite) is used to increase stiffness and strength. When using glass fibers, the resulting injection molded parts exhibit strong orientation and thus orientation-dependent properties. The shrinkage of these glass fiber reinforced grades is also orientation dependent and may lead to warpage of the molded parts. This phenomenon is not encountered when using glass spheres and other isotropic fillers and they are therefore used to manufacture less warpage-prone grades.

Incompatible fillers, such as glass fibers and glass spheres, cause a decrease in toughness because of the micro-notches they introduce in the polymer matrix.

Long-term creep is significantly reduced by the addition of glass fibers.

Impact Modified Grades for Applications with Increased Impact Loads

These grades are compounds with elastomers and exhibit higher impact resistance than basic grades. Typical elastomer loads reach up to 40 wt.-%. With increasing (flexible) elastomer content, strength, stiffness and hardness of the grades decrease, while their toughness increases. A number of testing procedures is available to determine impact resistance – for notched and unnotched specimens and also biaxial puncture tests for sheets. The failure behavior is determined by impact velocity and temperature. When designing components for impact load, structural strength, *i. e.*, the combination of toughness and strength, has to be considered.

As modifying elastomers, typically thermoplastic polyurethane elastomers (TPU) with low glass transition temperatures of the flexible component are selected for modification because of their compatibility with POM. Chemically coupled systems are also commercially available; they are produced by reactive compounding of functionalized POM and TPU together with reactive coupling agents. This chemical coupling improves impact resistance and weld line strength and decreases delamination phenomena of the POM/TPU blends. It is also possible to use particulate crosslinked core-shell impact modifiers, *e. g.*, based on polybutadienes, for POM impact modified grades with particularly good low-temperature toughness and weld line strength.

Tribologically Modified Grades for Sliding/Friction Applications

These grades are modified by special additives that improve dry-running behavior and/or wear resistance. These additives include special waxes, but also PTFE, silicone oils, polyethylene and modified polyolefins, molybdenum disulfide, and special, coated chalks. Loads are dependent on additive and application and typically range from several tenths of a percent for waxes up to 20–30 wt.-% for polymers.

Molded components made from POM offer good anti-stick properties; therefore, POM is often used for gear wheels, bearings, and sliding and control elements. Note that the anti-stick properties are system characteristics (see also Section 3.8). Therefore, coefficients of friction depend on the combination of friction partner, contact pressure, slide paths, slide velocity, and ambient temperature in addition to material properties. The various modified POM grades exhibit significant differences in their slide/friction behavior; besides frictional forces and wear also different stick/slip behavior and squeaking noise. The large tribological differences in application demand a large number of tribologically modified POM grades. Selection of appropriate grades for each application requires caution.

Grades with Increased Thermal Conductivity, Metal Detectability, and Anti-Static Modification

Thermal and electric conductivity can be increased by the addition of steel fibers, aluminum or bronze powders. The addition of magnetic metals, such as steel powder or steel fibers, offers grades with metal detectability, *e. g.*, for food processing. Special conductive carbon blacks or carbon nano-tubes are suitable to improve electric conductivity. In order to improve anti-static properties, special anti-static agents (*e.g.*, organic surfactants) are used.

6.8 Polyamides (PA)

Matthias Scheibitz, BASF SE, Ludwigshafen, Germany

6.8.1 Polyamides, Homopolymers (AB and AA/BB Polymers); (PA 6, 11, 12, 46, 66, 69, 610, 612 (PA 7, 8, 9, 1313, 613))

Polyamides are high-molecular materials in which the monomer units are joined by acid amide groups. In general, amide groups are created by the chemical conversion of a carboxylic acid with an amine under dehydration (condensation; Equation 6.9). This conversion is an equilibrium reaction. Water will break the amide bond under appropriate conditions.

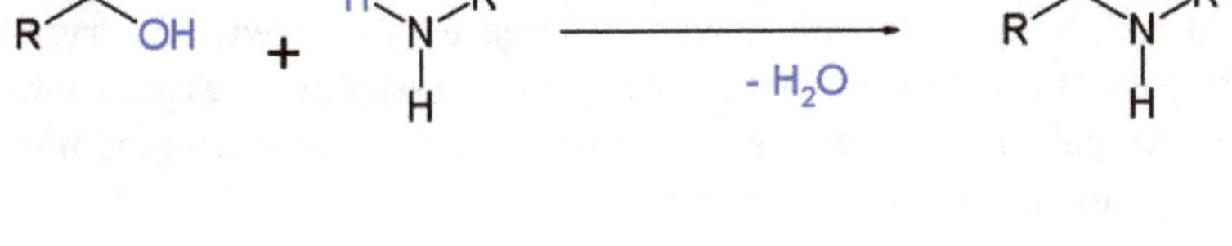

(6.9)

6.8.1.1 Chemical Composition

Polyamides can be classified following different criteria.

In some polyamides, the monomers contain both functional groups necessary for the condensation reaction depicted in Equation 6.9 (AB polymers). Typical representatives are PA6 on the basis of ε-caprolactam or ω-aminocaproic acid and PA12, which is produced from laurolactam or ω-amino lauric acid. PA11 (from ω–amino undecane acid) is part of this class.

Alternatively, the amine and carboxylic acid groups can be distributed to two monomers (AA/BB polymers). Examples include PA66, which is produced from hexamethylene diamine and adipic acid, and PA46, which is created by polycondensation of tetramethylene diamine with adipic acid. PA610 and PA612 are also of technical relevance. Other polyamides of type AB and AA/BB can be produced and have already been characterized, but they have only little commercial relevance.

Both compositions are considered homopolymers because they have only one repeat unit (either AB or AA/BB). Copolyamides contain additional monomers. An example is PA6/66, which contains three monomers: caprolactam, hexamethylene diamine, and adipic acid.

Polymeric Structure

We distinguish between aliphatic *(e. g.*, PA6, PA66), semi-aromatic (*e. g.*, 6T/6, 6T/6I/66), and aromatic polyamides (aramids). In the first case, the polymers contain only aliphatic hydrocarbon units (linear or branched) between the amide groups, while partial and purely aromatic types contain monomers with conjugated systems that are derived from benzene (see Table 6.23).

Micro-Structure

It is also possible to classify polyamides according to their melting and freezing behavior which is determined by their micro-structure.

When the material forms crystalline domains during cooling from the melt, it is called a *semi-crystalline polyamide* because in the solid state the material contains amorphous domains without long-range order in addition to the crystalline domains. Here, the structure of the monomers and their combination as well as the cooling conditions determine the ratio of crystalline and amorphous domains. Aliphatic homopolyamides without branched monomers are typically highly crystalline (PA46, PA66, PA6) and crystallize quickly, while semi-aromatic polyamides freeze more slowly and crystallize to a lesser degree. The melting points or ranges of the crystallites are characteristics of the respective polyamides. The melting temperatures increase with the number of amide groups and the degree of symmetry.

The degree of crystallinity also depends on the cooling conditions during processing. Fast cooling (quenching) results in the formation of very small structures, and the degree of crystallinity overall is lower (transparent), while slow cooling results in the formation of large superlattices and a higher degree of crystallinity (opaque or non-transparent). In practice, crystallinity degrees higher than 50% are rare; PA6 reaches between 30 and 40%, PA46 up to 70%. The degree of crystallinity not only determines the optical properties of polyamides, but also their mechanical characteristics, see Table 6.24.

Amorphous polyamides do not exhibit a melting point. They freeze glass-like from the melt, and in the solid state they do not exhibit a long-range order between the repeat units. They are also referred to as micro-crystalline materials. Here, the glass transition temperature T_g indicates beginning chain mobility above which the polymer chains are able to rotate around single bonds. Semi-aromatic copolyamides and polymers with monomer units that impede or prevent crystallization because of their branched structure typically exhibit high T_gs.

According to this classification, PA66 for example is identified as an aliphatic, semi-crystalline homopolyamide.

Table 6.23 Structure and Components of Major Polyamides

Micro-structure	Acronym, ISO 1874	Composition
Semi-crystalline, aliphatic	**AB-polymers –[-NH-$(CH_2)_x$-CO-$]_n$-**	
	PA6	X = 5
	PA11	X = 10
	PA12	X = 11
	AA/BB-polymers –[-NH-$(CH_2)_x$ – NH-CO-$(CH_2)_y$ – CO-$]_n$-	
	PA46	X = 4, Y = 4
	PA410	X = 4, Y = 8
	PA66	X = 6, Y = 4
	PA610	X = 6, Y = 8
	PA612	X = 6, Y = 10
	PA1010	X = 10, Y = 8
	PA1012	X = 10, Y = 10
	PA PACM 12	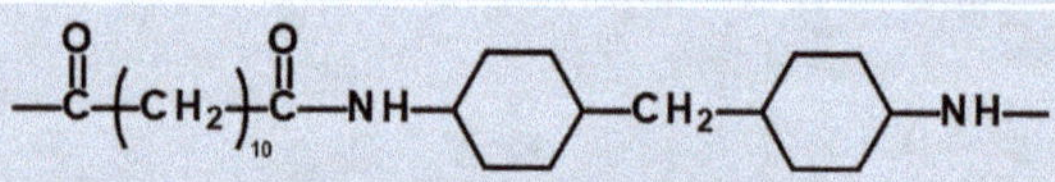
Semi-crystalline, aliphatic	**Block-copolyamides –[-NH-$(CH_2)_x$-CO-$]_n$-[O-$(CH_2)_{2\text{-}4}$-O$]_o$–[-CO-$(CH_2)_x$-NH-$]_n$-, –[-NH-$(CH_2)_x$-CO-$]_n$-[NH-{$(CH_2)_{2\text{-}4}$-O-$(CH_2)_{2\text{-}4}$-}-NH$]_o$–[-CO-$(CH_2)_x$-NH-$]_n$-**	
	TPE-A	X = 6, 11, 12
Semi-crystalline, semi-aromatic	PA MXD6	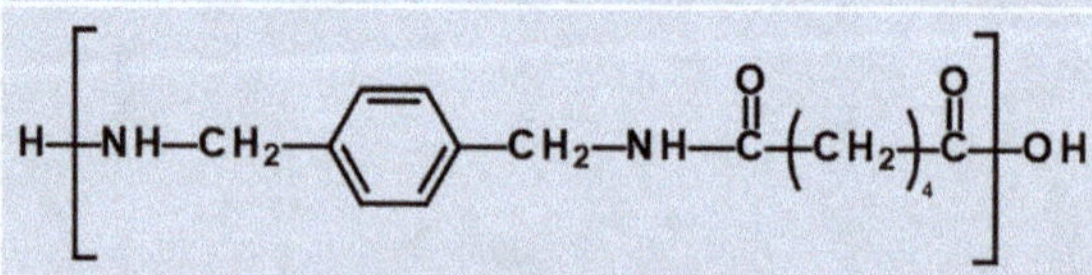
		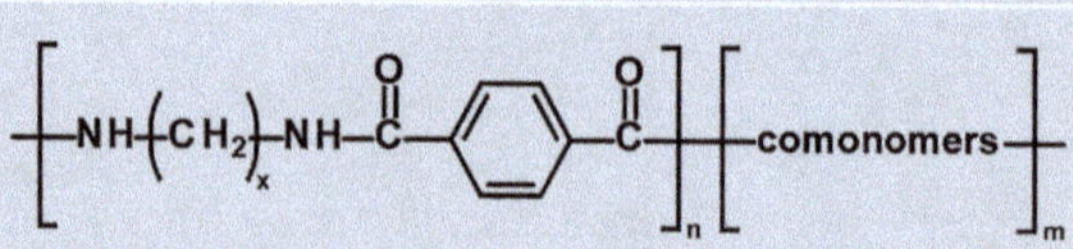

Components	Density [g/cm^3]	Water absorption [%][1)]	T_g [°C][2)]	T_m [°C]
ε-Caprolactam	1.13–1.14	2.6–3.4	45	220
ω-Aminoundecanoic acid	1.03	0.9	46	189
Lauryl lactam	1.02	0.8	37	177
Tetramethylene diamine, adipic acid	1.18–1.19	3.7	72	295
Tetramethylene diamine, sebacic acid	1.09	2.0	70	248
Hexamethylene diamine, adipic acid	1.12	2.5–3.1	47	260
Hexamethylene diamine, sebacic acid	1.07–1.08	1.4	40	222
Hexamethylene diamine, 1,12-dodecanedioic acid	1.06	1.3	65	218
Decamethylene diamine, 1,10-decanedioic acid	1.05	1.1	37	200–202
Decamethylene diamine, 1,12-dodecanedioic acid	1.03		43	191
Bis(4-aminocyclohexyl)methane, 1,12-dodecanedioic acid	1.02	1.5	140	250
Polyamide block: lauryl lactam or ω-aminoundecanoic acid or caprolactam Polyether block: polytetrahydrofuran or polypropylene glycol or polyethylene glycol	1.00–1.03	0.5–4.5 [3)]	<–40	122–205
m-Xylylene diamine, adipic acid	1.21	3.1 [3)]	75	243

Table 6.23 Structure and Components of Major Polyamides *(continued)*

Micro-structure	Acronym, ISO 1874	Composition
Semi-crystalline, semi-aromatic	PA4T	X = 4
	PA6T	X = 6
	PA9T	X = 9
	PA10T	X = 10
Amorphous, partially aromatic	PA NDT/INDT	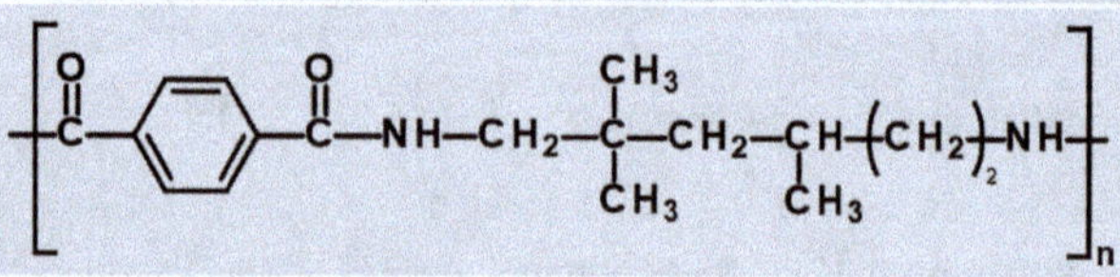
	PA6I/6T	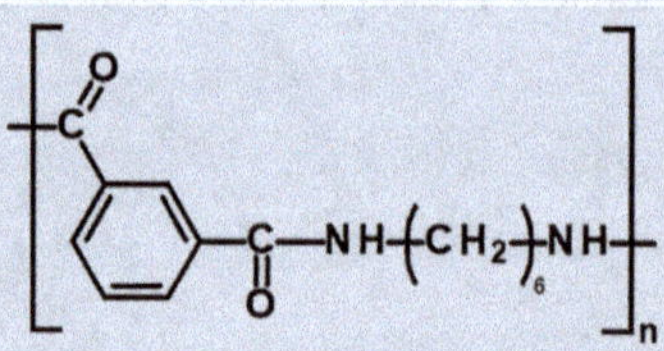
	PA12/MACM I	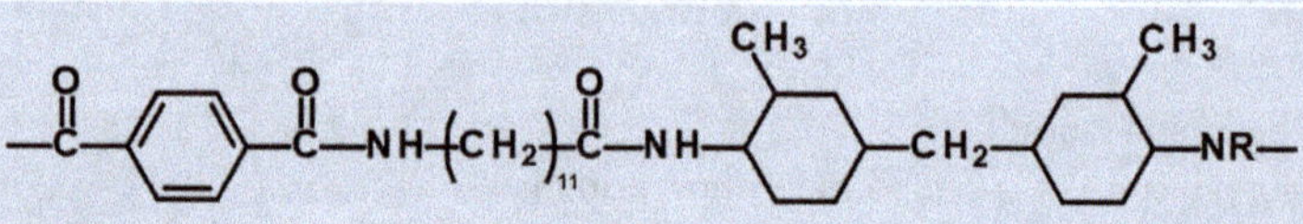
Crystalline, aromatic	PPTA	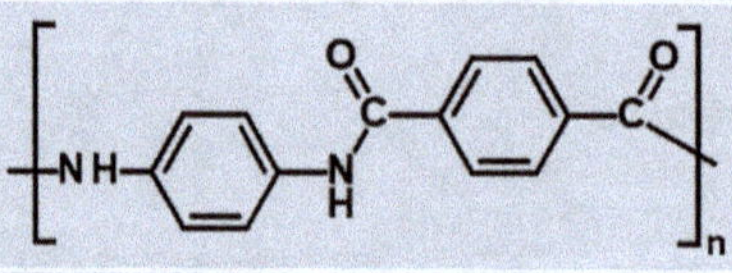
Semi-crystalline, aromatic	PMIA	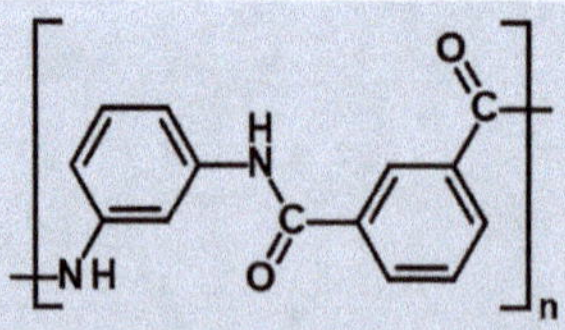

1) 23 °C, 50% rel. humidity, 2) DSC ISO 11357-1/-3, 3) 20 °C, 65% rel. humidity 4) degradation.

Components	Density [g/cm³]	Water absorption [%][1)]	T_g [°C][2)]	T_m [°C]
Tetramethylene diamine, terephthalic acid Comonomers: hexamethylene diamine, adipic acid			125	325
Hexamethylene diamine, terephthalic acid Comonomers: caprolactam or adipic acid or adipic acid and isophthalic acid or 2-methylpentamethylene diamine			95–125	295–325
Nonamethylene diamine, terephthalic acid Comonomer: 2-methyloctamethylene diamine	1.14	0.25	125	300
Decamethylene diamine, terephthalic acid Comonomer	1.10	0.15	125	285
2,2,4-/2,4,4-Trimethylhexamethylene diamine (isomer mixture), terephthalic acid	1.12	3.2	153	
Hexamethylene diamine, isophthalic acid, terephthalic acid	1.18	2	125	
Lauryl lactam, 3,3′-dimethyl-4,4′-diaminodicyclohexylmethane isophthalic acid	1.06	1.5	160	
Phenylene diamine, terephthaloyl chloride	1.44	6 [3)]		550 [4)]
Phenylene diamine, isophthaloyl chloride	1.38	5 [3)]	270	425 [4)]

Table 6.24 Properties of Aliphatic Homopolyamides, Unfilled

Property	Unit	Aliphatic homopolyamides									
		PA6		PA12		PA46		PA66		PA610	
		Dry	Cond.[1)]	Dry	Cond.[1)]	Dry	Cond.[1)]	Dry	Cond.[1)]	Dry	Cond.[1)]
Density	g/cm^3	1.12–1.14		1.01–1.03		1.18		1.13–1.15		1.06–1.09	
Tensile modulus of elasticity	MPa	2600–3200	750–1500	1300–1600	900–1200	3300	1000	2700–3300	1300–2000	2000–2500	1100–1350
Yield stress	MPa	70–90	30–60	45–60	35–40	100	55	75–100	50–70	60–65	50
Elongation at yield	%	4–5	20–30	4–5	10–15	10	20	4.5–5	15–20	4.5–5	10–20
Nominal elongation at break	%	15	>50	>50	>50	40	>50	30		>50	>50
Melting temperature	°C	220–225		175–180		295		255–260		215–225	
Heat deflection temperature HDT/A 1.8 MPa	°C	55–80		40–50		190		70–100		55–65	
Thermal coefficient of linear expansion, longitudinally (23–55 °C)	10^{-5}/K	7–10		10–12		8.5		8.5		9–13	
Thermal coefficient of linear expansion, across (23–55 °C)	10^{-5}/K	10–14		14		11		9–13		10–14	
Dielectric constant at 100 Hz		3.9–4.2	12–20	3.7–4	5–6	3.9	22	3.2–4	5–11	3.5	4

Property	Unit	Aliphatic homopolyamides									
		PA6		PA12		PA46		PA66		PA610	
		Dry	Cond.[1]	Dry	Cond.[1]	Dry	Cond.[1]	Dry	Cond.[1]	Dry	Cond.[1]
Dielectric loss factor at 100 Hz	$\cdot 10^{-4}$	60–250	2100–3500	300–700	800–1000	70	8700	50–150	1000–2400	70–150	1000–1800
Specific volume resistivity	Ohm · m	$>10^{13}$	$>10^{10}$	10^{13}	10^{11}	10^{13}	10^{10}	$>10^{12}$	$>10^{10}$	$>10^{13}$	10^{10}
Specific surface resistivity	Ohm	$>10^{12}$	$>10^{10}$	$>10^{13}$	10^{12}		10^{13}	$>10^{13}$	$>10^{10}$		$>10^{16}$
Dielectric strength	kV/mm	30	25–30	27–29	28–32	25	15	40	25–35		44
Comparative tracking index CTI/A		600		600		400		600		600	
Water absorption at 23 °C, saturation	%	9–10		1.3–1.7		13.5		8–9		3.2–3.6	
Moisture absorption 23 °C/50% rel. humidity	%	2.5–3.4		0.7–1.1		3.7		2.6–3		1.4	

[1] Specimen storage at 25 °C/50% rel. humidity until constant weight.

Nomenclature

For simplification, polyamides are identified according to their composition:

Polyamides based on ω-aminocarboxylic acids and their respective lactams are identified as PA X. Here, X denotes the number of carbon atoms in the monomer. PA6 therefore describes the polymer from ω-aminocaproic acid which is formed during polymerization by the addition of water to ε-caprolactam.

Polyamides consisting of unbranched diamines and dicarboxylic acids are identified by PA YZ. Here, Y denotes the number of carbon atoms in the diamine and Z the number of carbon atoms in the dicarboxylic acid. Therefore, the polymer from hexamethylene diamine and adipic acid is identified by PA66.

Copolyamides are identified using a similar approach. The homopolyamide combinations of the monomers contained in the polymer are noted successively. An example is PA66/6, which is formed by three monomers (hexamethylene diamine, adipic acid, and caprolactam). The potential homopolymers PA66 (hexamethylene diamine and adipic acid) and PA6 (ε-caprolactam) lead to the notation PA66/6 for the copolyamide.

There are a number of specific letter combinations for particularly important monomers, in particular those with aromatic basic structures, *e. g.*, T for terephthalic acids, I for isophthalic acid, or MXD for m-xylylene diamine, see Table 6.23.

6.8.1.2 Properties

The properties of polyamides are determined by highly polar amide groups (-CONH-) and the hydrocarbon residues between them (Figure 6.9).

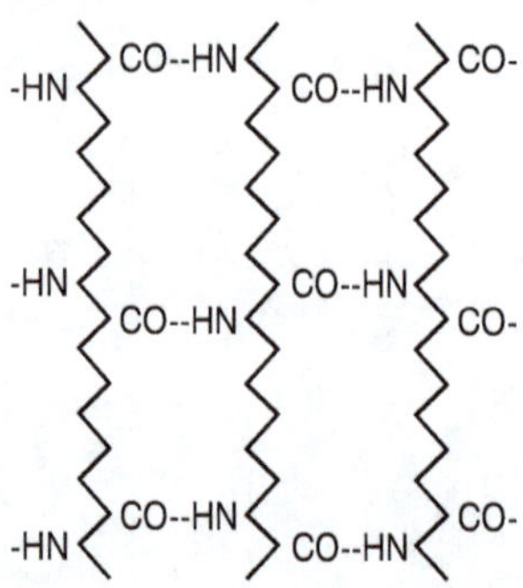

Figure 6.9
Hydrogen bridge bonds in PA 6

The amide groups form hydrogen bridge bonds in the amorphous domains and thus create the ordered structure in the crystallites. They are therefore responsible for the material's toughness, heat resistance, and high Young's modulus. The polar amide functionalities in the amorphous domains are able to interact with smaller, polar molecules, in particular with water. In polyamides, the resulting water absorption, which occurs only in the amorphous domains, is called conditioning. It causes a certain volume expansion as well as a reduction in glass transition temperature and influences mechanical properties. Similar effects are seen for the interaction with short-chain alcohols such as methanol, ethanol, and ethylene glycol.

The structure of the hydrocarbon residuals also influences a number of characteristics. Water absorption decreases with increasing length, *i. e.*, with increasing ratio of CH_2- to CONH-groups. Figure 6.10 shows the dependency of moisture equilibrium of selected polyamides on relative humidity.

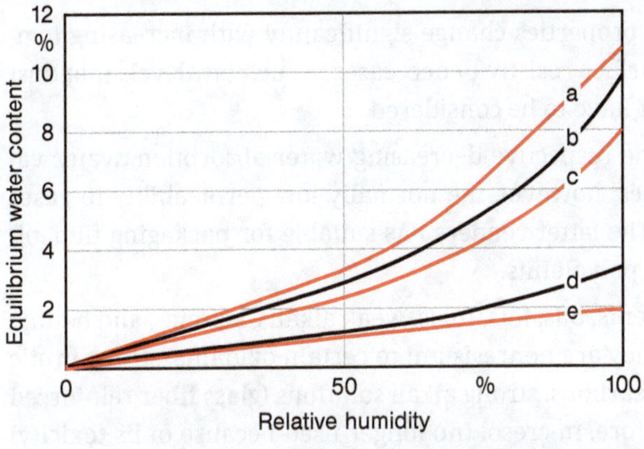

Figure 6.10 Storage conditions and water content for different PA grades

a: PA 6, b: PA 6/610/66, c: PA 66, d: PA 610, e: PA 11

With increasing number of methylene groups (-CH_2-) and thus increasing distance between amide functionalities, the intermolecular forces decrease. Therefore, PA12 is softer than PA6 and melts at lower temperatures. However, aromatic compounds and cyclo-aliphates between the amide groups in the polymer chain result in an increase in melting point and glass transition temperature.

In addition, symmetric hydrocarbon residuals promote higher degrees of crystallinity, which in turn results in higher strength, stiffness, and hardness. However, water absorption and toughness will decrease. This is seen in the comparison of properties of the higher crystalline PA66 and the less crystalline PA6 (see Table 6.24).

The toughness of polyamides is determined by their structure and molecular weight. It increases with increasing chain length.

Other properties are characteristic for semi-crystalline polyamides. When dry, immediately after thermoplastic processing, they are hard and more or less brittle. After moisture absorption from the atmosphere (conditioning at 25 °C und 50% rel. humidity) or after storage in water, they turn more ductile and abrasion resistant; stiffness and strength decrease. Technical data sheets therefore provide characteristics for dry and conditioned materials. Water absorption results in increases of both volume and dimensions, which has to be considered during component design.

Slip- and wear properties of PA are very good: they exhibit good dry run properties; they are not sensitive to contamination and chemical resistant. The glass transition temperatures of aliphatic, unbranched polyamides are around or a little above room temperature (Table 6.23). However, their crystallinity allows for application close to their melting temperature. Polyamides are often modified by additives and offered as reinforced or filled grades with tailored properties. Grades with high glass fiber filler content exhibit good load bearing properties even above T_g (low creep) at high constant loads.

Both electrical and mechanical properties change significantly with increasing temperatures and water content. Surface resistivity decreases to such low levels that dust or electrostatic attraction do not have to be considered.

With decreasing polarity and the respective decreasing water absorption, water vapor permeability of PA decreases; however, the normally low permeability to gases (O_2, flavors) slightly increases. The latter renders PAs suitable for packaging film, often as composite film, *e. g.*, with polyolefins.

Polyamides are resistant to solvents, oils, fats, fuels, weak alkalis, ketones, and boiling water (they can be sterilized); they are nonresistant to certain oxidants, strong protic solvents, and chlorinated hydrocarbons, strong alkali solutions (glass fiber reinforced grades) and strong acids. Therefore, m-cresol (no longer used because of its toxicity) and hexafluoroisopropanol (HFIP) as well as sulfuric and formic acid are good solvents for polyamides. Natural colorants (tea, coffee, fruit juices, etc.) can stain PA. PA should be stabilized for long-term use above 100 °C with, *e. g.*, phenolic, amine, or metal halide stabilizers, for outdoor applications with approx. 2% carbon black. The surface of glass fiber reinforced PA is less weather resistant and long-term exposure (several years) can lead to surface erosion (graying). PAs not modified with flame retardants will continue to burn after the ignition source is removed.

Amorphous polyamides exhibit the typical properties of amorphous materials, such as transparency, isotropy, non-warping, and little shrinkage. Their service temperature reaches up to their glass transition temperature. Therefore, T_g-increasing monomers are preferred for polycondensation (property comparison, see Table 6.24).

6.8.1.3 Processing

In general, polyamides can be processed using any thermoplastic processing technology. However, the most common processing methods are injection molding and extrusion. Injection molding allows for the economic manufacture of complex components in large numbers from polyamide. Extrusion is used to manufacture films, semi-finished products, pipes, profiles, sheets, and monofilaments, while extrusion blow molding is used for pipes with complex geometries and hollow parts (*e. g.*, tanks). Semi-finished products are mostly processed into finished components by machining.

Fibers and Monofilaments

A major part of all polyamides produced world-wide is used produce textile and carpet fibers and technical yarns via melt spinning. Here, the melt is continuously pressed through a spinneret by gear pumps, and the fibers are subsequently drawn to various extents between godets.

Film Extrusion

Flat films are manufactured using chill-roll processes while tubular films are blown.

Multi-layer tubular films are produced on regular extrusion machines from polyamides, polyolefins (in particular polyethylene), and the respective bonding agents. PA6/66 copolyamides are preferred, in particular for non-symmetric 3-layer films (PE/HV/PA), because they are well suited for the manufacture of thermoformable, flexible composite films with good transparency.

Flat films are typically extruded from PA6. They are extruded with thicknesses ranging from 20–200 µm through slot dies that can be as wide as 2000 mm. Film quality, *i. e.*, tolerances, dimensional stability, machine runability, transparency, and thermoformability, are determined by processing conditions. The temperatures at the chill rolls should range between 20–40 °C for thermoforming films/sheets with high transparency, and between 80–120 °C for films with high dimensional stability.

Injection Molding

Polyamides are typically injection molded 20–40 °C above their melting point. The material has to be sufficiently dry to avoid a significant decrease in molecular weight. Because of the high melting point, residence time in the injection molding machine should be kept short to avoid polymer degradation. Mold temperatures should be selected above T_g (dry) to ensure good crystallinity (Table 6.25).

Melt Viscosity

Polyamides exhibit melt viscosities strongly dependent on temperature and shear rates (see Table 6.25) in the processing temperature range. Typical viscosities range between 10 and 1000 Pa·s. The higher the molecular weight, the higher the melt viscosity and thus the lower the flowability. Mineral- or glass fiber reinforcement increase the viscosity as a function of reinforcement content. The melt viscosity may change over time. For example, accelerated viscosity decrease caused by degradation is observed with melts that contain too much moisture, are too hot, or are under too high mechanical shear forces.

Table 6.25 Processing Parameters for Selected Polyamides

Polyamide	Injection molding			Extrusion	Extrusion blow molding	
	Melt temperature [°C]	Mold temperature [°C]	Shrinkage [%]	Melt temperature [°C]	Melt temperature [°C]	Mold temperature [°C]
6	230–280	80–90	0.5–2.2	240–300	250–260	80
11 and 12	190–280	40–80	0.5–1.5	230–290	200–230	70
46	295–330	80–120	0.2–2.0			
66	260–320	80–90	0.5–2.5	250–300	270–290	90
610	230–280	80–90	0.5–2.8	230–290	230–250	80
612	230–290	30–100	0.5–2.2	230–290		
PACM 12	280–300	60–80	0.6–0.8	265–295		
MXD6	250–280	100–140				
6T/6	310–330	90	0.5–1.1			
6T/6I	330–350	≥140				
NDT/INDT	260–290	70–90	0.4–0.6	250–280	240–255	40

Thermal Stability of the Melt

The thermal stability of commercially available grades is excellent under appropriate processing conditions. Under common processing conditions the material is neither attacked nor changed. Only extended residence times may cause degradation of the polymer chain. Table 6.25 provides an overview of typical processing parameters for selected polyamides. No appreciable color changes occur as long as the melt is not in contact with oxygen. In contact with air, *e. g.*, open nozzles or process interruptions, the surface is likely to discolor quickly.

Pre-Treatment, Drying

Polyamides have to be dried prior to processing. If the moisture content is too high, quality may suffer (surface quality, decrease in mechanical properties, *e. g.*, by breaking the molecular chains). The recommended maximum moisture content for injection molding is 0.15% (typically recommended is 600 ppm), for extrusion 0.1%. Drying times, typically 4–8 h, depend on the moisture content. Optimal drying temperatures range from 80 to 120 °C. Light pellets and thermally sensitive colored grades should be dried at moderate temperatures up to 80 °C to avoid change in color tinge.

Casting

Caprolactam and laurolactam monomers – with the addition of catalysts and activators – can be *in situ* polymerized into high-molecular cast-PA6 or cast-PA12 by pressure-less anionic ring-opening polymerization at 100 to 200 °C in heated molds. Here, it is particularly important to exclude moisture. This procedure is used to produce components with large volumes. It is possible to cast large, thick-walled components with weights up to 2000 kg in easy to manufacture molds, such as those used for casting metals. Rotational components, such as pipes, are manufactured by centrifugal casting, hollow components (*e. g.*, tanks) by rotational casting.

Subsequent Processing

Polyamide components can be economically joined using a variety of methods. They can be easily joined with special screws designed for plastics that create their threads (self-cutting and thread forming screws). Rivet joints are readily achieved. Metal inserts have proven suitable for high-load bearing screw joints. These inserts are either molded in or subsequently inserted in suitable openings via ultrasound or inductive heating. Snap- and press-fits also result in high-load bearing joints. The excellent elasticity and strength of polyamides, even under elevated temperatures, render them particularly suitable for these designs.

All technologies developed for the welding of thermoplastics can be applied with polyamides. Adhesive solvents and lacquers based on phenol or resorcinol solutions, formic acid, solid adhesives with or without crosslinking (reaction or 2-component adhesives), polymerization adhesives, and contact adhesives are particularly suitable for bonding applications.

Components made from PA can also be joined with rubber or other elastomers, potentially after surface treatment. Semi-finished PA products can be machined with all common machining tools.

6.8.1.4 Applications

Low-viscosity polyamides, in particular PA6 and PA66, are well suited for textile fibers because of their durability and resistance. They are used for high-quality textiles, hosiery, swim and high-tech outdoor wear. In addition, medium-viscosity PA fibers are used for high-quality carpets; high-viscosity polyamide fibers are used for airbags and tire cord.

High-viscosity polyamides, in particular PA6 and PA6/66 copolymers, are well suitable for the manufacture of flat (slot die films) and tubular films (blown film) for packaging applications. They offer high strength, excellent thermoformability, high heat resistance (also against sterilizing temperatures), and excellent gas barrier properties, in particular against oxygen, and flavoring substances.

Polyamides are also used for the manufacture of stretched monofilaments that are used as technical wires, fishing lines, tennis strings, and bristles.

Because of their very good thermal and chemical resistance, dynamic strength, toughness, and good continuous-use properties, polyamides (typically with reinforcing modifications, see Section 6.8.4) are predestined for application in modern automotive engineering. Examples of these applications can be found in a variety of areas:

Motor and Transmission

Intake pipes and charge air distributors, end caps for charge air coolers, charge air lines, cylinder head covers, engine covers, air mass flow meter, oil pans, oil filter housings, oil sensors, chain guides, cam belt covers, transmission controls, sensors, ball bearing retainers, gear wheels, fasteners.

Cooling Circuits

Radiator end caps, thermostat housings, cooling water pipes, blower wheel, fan guards.

Fuel Systems

Fuel filter housings, fuel inlets, quick couplings.

Mountings

Engine mountings, torque bracket, pendulum support, transmission cross beams, body panels.

Interior

Pedals and pedal boxes, levers and controls, loudspeaker covers, door handles, seat structures.

Exterior

Structural components, exterior door handles, mirror bases, hub caps, front ends, crash-absorbers.

Electrical Systems

Cable ties, -clips, and -connectors, bulb sockets, fuse boxes, brush carriers, cable ducts, actuators and drives.

Polyamides are used in almost all areas of industrial energy technology, the electronics and household appliance industries thanks to their good electrical insulation properties, their favorable slide/friction behavior, extraordinary good mechanical strength in combination with good flammability behavior of flame retardant modified grades.

Energy Technology

Highly insulating switches and housings, terminal strips and connectors, energy distribution systems, cable ducts and fasteners, protectors and power switches, coil bobbins, circuit breakers.

Electronics

Plug-in connectors, electrical and mechanical components for computer equipment and telecommunication, capacitor cans.

Appliance Technology

Components of household appliances such as switches, magnetic valves, plug connections, programmed control devices; housings for electric hand tools; electrical equipment and housings for large household appliances such as washing machines and dish washers and small appliances such as coffee makers, kettles, and hair dryers.

Photovoltaics

Sockets and plug connectors.

High mechanical load capacity combined with good toughness, but most important the wide options for tailoring of the material, provide a wide variety of applications in the consumer and industrial product area. These applications mostly have high mechanical requirements where polyamides substitute traditional materials such as metal or wood.

Construction Technology

Wall dowels and facade anchors, fasteners for facade engineering and in solar technology, and thermal insulation web for windows.

Sanitary Technology

Handles, hardware, ventilators, continuous flow heaters, fittings, water meter housings.

Household

Seating furniture, chair rollers, cooking utensils, furniture hardware.

General Equipment Manufacturing

Ball bearing housings, gear wheels, gear boxes, seals, housings, flanges, connectors, screws, sliding components.

Materials Handling Technology

Rolls, bushings, transport containers, conveyor belts and chains.

6.8.2 Modifications

Polyamides are modified by additives, fillers, and reinforcing agents to meet specific requirements of the final application and to improve their already good characteristics, see Table 6.26 and Table 6.27.

Table 6.26 Properties of Reinforced and Filled Polyamides; Conditioned 23 °C, 50% Rel. Humidity

Polyamide/Filler	Filler load [%]	Density [g/cm^3]	Fracture stress [MPa]	Elongation at break [%]	Tensile modulus of elasticity [MPa]	Heat deflection temperature HDT/A [°C][1)]
PA6		1.13	40 [2)]	> 50 [3)]	1000	60
Short glass fibers	15	1.23	70	15	3500	190
	30	1.36	115	8	6200	210
	50	1.55	150	4.5	11,000	215
Short glass fibers/minerals	10/20	1.37	55	10	4200	195
Minerals	40	1.49	60	30	3800	90
Glass spheres	30	1.35	35	> 50	1200	65
PA66		1.13	50 [2)]	> 50 [3)]	1100	75
Short glass fibers	30	1.36	130	5	7200	250
	35	1.41	150	5	8500	250
	50	1.56	180	3.5	12,500	250
Short glass fibers/minerals	25/15	1.48	80	6	6100	225
Carbon fibers	20	1.22	160	6	12,000	245
Minerals	40	1.44	65	10	3700	110
PA610		1.08	50 [2)]	> 50 [3)]	1100	65
Short glass fibers	30	1.31	110	6	6700	200
PA11		1.03	41 [2)]	> 50 [3)]	1200	58
Short glass fibers	30	1.25	134	6	7300	175

Polyamide/Filler	Filler load [%]	Density [g/cm^3]	Fracture stress [MPa]	Elongation at break [%]	Tensile modulus of elasticity [MPa]	Heat deflection temperature HDT/A [°C][1]
PA12		1.01	40 [2]	> 50 [3]	1100	50
Short glass fibers	30	1.23	120	5	7500	170
Glass spheres	30	1.25	37 [2]	> 50 [3]	1800	55
PA46		1.18	55	> 50 [3]	1000	190
Short glass fibers	30	1.41	115	7	6000	290
	50	1.62	160	5	10,000	290
PA6T (6T/6I)		n/a				
	30	1.44	170	2	11,000	280
	50	1.65	240	2	17,500	285

[1] Dry, [2] Yield stress, [3] Nominal elongation at break.

Table 6.27 Property Comparison of Selected Aliphatic Homopolyamides, Filled and Modified

Property	Unit	Aliphatic homopolyamides, filled and modified								
		PA6 GF30		PA66 GF30		PA6-I Impact modified		PA46 GF50		PA12 Plasti-cized
		Dry	Cond.[1]	Dry	Cond.[1]	Dry	Cond.[1]	Dry	Cond.[1]	
Density	g/cm^3	1.35		1.36		1.01–1.13		1.62		1.00–1.05
Tensile modulus of elasticity	MPa	9000–10,800	5600–8200	9100–10,000	6500–7500	1100–2800	450–1200	16,000	10,000	220–750
Yield stress	MPa					25–80	20–45			15–35
Elongation at yield	%					4–5	15–30			20–45
Nominal elongation at break	%					> 50	> 50			> 50

Table 6.27 Property Comparison of Selected Aliphatic Homopolyamides, Filled and Modified *(continued)*

Property	Unit	Aliphatic homopolyamides, filled and modified								
		PA6 GF30		PA66 GF30		PA6-I Impact modified		PA46 GF50		PA12 Plasti-cized
		Dry	Cond.[1]	Dry	Cond.[1]	Dry	Cond.[1]	Dry	Cond.[1]	
Ultimate strength	MPa	170–200	100–135	175–190	115–200			250	160	
Elongation at break	%	3–3.5	4.5–6	2.5–3	3.5–5			2.7	5	
Melting temperature	°C	220–225		255–260		220		295		160–175
Heat deflection temperature HDT/A 1.8 MPa	°C	190–215		235–250		45–70		290		40–50
Thermal coefficient of linear expansion, parallel (23–55 °C)	10^{-5}/K	2–3		2–3		8.5–15		2.5		12–17
Thermal coefficient of linear expansion, perpendicular (23–55 °C)	10^{-5}/K	6–8		6–8				4		
Flammability UL94 for 1.6 mm thickness		HB		HB		HB		HB		HB
Dielectric constant at 100 Hz		3.8–4.4	7–15	4	8	3–4	5–14	4.3	16	4–24

Property	Unit	Aliphatic homopolyamides, filled and modified								
		PA6 GF30		PA66 GF30		PA6-I Impact modified		PA46 GF50		PA12 Plasticized
		Dry	Cond.[1)]	Dry	Cond.[1)]	Dry	Cond.[1)]	Dry	Cond.[1)]	
Dielectric loss factor at 100 Hz	$\cdot 10^{-4}$	100–150	2000–3000	140	1300–2300	100–140	500–3000	70	6000	900–3500
Specific volume resistivity	Ohm·m	10^{13}	10^{11}	10^{13}	10^{11}	$>10^{13}$	$>10^{10}$	10^{12}	10^{8}	10^{9}–10^{11}
Specific surface resistivity	Ohm	$>10^{13}$	$>10^{11}$	$>10^{13}$	$>10^{11}$	10^{10}–10^{12}	10^{8}–10^{10}		10^{13}	10^{11}–10^{15}
Dielectric strength	kV/mm	35–40	25–35	40	35	30–35	25–30	30	20	20–35
Comparative tracking index CTI/A		400–600		400–600		600		300		600
Water absorption at 23 °C, saturation	%	6.0–6.7		5.0–5.5		6.5–9.0		6.75		0.8–1.5
Moisture absorption 23 °C/50% rel. humidity	%	1.4–2.0		1.0–1.7		1.8–2.7		1.85		0.4–0.7

1) Specimen storage at 25 °C/50% rel. humidity until constant weight.

The broad application spectrum of technical parts only opened up to polyamides once their properties could be optimized using reinforcing glass or carbon fibers, micro glass spheres, and mineral fillers (and their combinations). Short glass fibers with diameters of 10–12 µm play by far the most important role. They are equipped with a size based on amino silanes as bonding agent and cause an increase in strength, stiffness, and dimensional stability at elevated temperatures while reducing elongation at break and creep. The anisotropic shrinkage caused by fiber orientation (aspect ratio of approx. 20) during injection molding and the resulting warpage can be reduced by the addition of glass spheres or mineral fillers. The most important minerals are – potentially surface modifying – mica, talcum, wollastonite, and kaolin. Their addition, even in exclusively mineral-reinforced grades, causes less shrinkage that is more isotropic and thus less warpage.

The injection molding of polyamides with already good flow behavior can be improved by the addition of lubricants that reduce internal friction and demolding forces. The addition of nucleating agents leads not only to a more homogeneous crystallite structure, but also to accelerated freezing of the melt and thus shorter cycle times.

Below the glass transition temperature – that is room temperature for dry PA6 and PA66, below 0 °C for conditioned grades – polyamides are less impact resistant. If high impact resistance must be achieved, impact modification is required. To reach this goal, elastomers with low glass transition temperature (*e. g.*, ethylene acrylate maleic acid anhydride or maleinated ethylene propylene or ethylene propylene diene copolymers) are compounded into the polyamide. Here it is crucial to achieve a uniform, fine distribution of the elastomer in the polyamide matrix and the chemical bonding to the polyamide, which is obtained by carboxylic acid-, carboxylic acid ester-, or anhydride groups. The fact that this generally also increases melt viscosity, decreasing the material's flowability, needs to be taken into consideration.

Service temperatures above 100 °C or excessive processing temperatures may cause thermooxidative damage to polyamides. This damage can be slowed by phenolic, amine, or copper halogenide containing stabilizers, which will allow for the mechanical properties to be retained at a high level for longer. More current developments in this area provide grades with strengths that do not decrease below 50% of the initial level even after 3000 hours at 220 °C.

Polyamides can be stabilized against UV damage by the addition of antioxidants (thermal stabilizers) and UV absorbers (*e. g.*, carbon black).

Unreinforced aliphatic polyamides can reach V-2 classification according to UL94. Because the addition of fillers and reinforcing agents lowers this classification, these grades receive flame retardant modifications. Major flame retardant additives are red phosphor, organic phosphor compounds, organic halogen compounds, melamine cyanurate, and magnesium- or aluminum hydroxide. Note that the addition of flame retardants influences the mechanical properties and typically causes a reduction in impact resistance.

Polyamides can also easily be colored, although it is important to ensure that colorants exhibit sufficient thermal stability.

The stiffness of polyamides, in particular of PA11 and PA12, but also of PA6, can be significantly reduced by the addition of plasticizers such as benzene sulfonic acid n-butyl amide (BBSA). This results in highly flexible products, in particular for applications in extruded pipes. Caprolactam is used as a plasticizer for PA6. The volatility of the plasticizer has to be taken into account in applications at elevated temperatures.

The blending of polyamides with other polymers facilitates the improvement of particular characteristics such as reversible water absorption, notched impact strength, heat resistance under load, warpage. However, this often comes at the cost of reducing other properties. Shrinkage, warpage, and water absorption can also be improved by blending polyamides with an amorphous thermoplastic such as ABS. Heat resistance under load can be improved by blending with amorphous polymers that exhibit high glass transition temperatures such as polyphenylene ether (PPE).

6.8.3 Copolyamides

Copolyamides consist of several monomers, and their selection and ratio allow for tailoring of the copolymer's properties over a wide range, see Table 6.28. With the exception of few combinations of isomorphic monomers with similar structures and comparable distance of the functional groups (*e. g.*, terephthalic acid and adipic acid), crystallinity and melting point of the homopolymers decrease with the addition of even small amounts of a comonomer.

The melting temperature reaches its minimum at approx. equimolar amounts of the components hexamethylene diamine/adipic acid, and caprolactam. At this point, the degree of crystallinity reaches its lowest value (micro-crystalline, transparent) as are chemical and heat resistance, and the solubility is increased. Therefore, these copolymers are not suitable as thermoplastic materials; however, in combination with flexible monomers, such as sebacic acid, they can be used as coatings, leather finishes, wire coatings, or melt adhesives.

Table 6.28 Property Comparison of Selected Semi-Aromatic Polyamides, Copolyamides, and Polyether Block Amides

Property	Unit	Semi-aromatic polyamides, copolyamides, and polyether block amides									
		PA 6T/6		PA 9T	PA NDT/INDT	PA66/6	PA6/66	PEBA 12		PEBA 6	
		Dry	Cond.[1)]			Dry	Dry	Shore D 40–55	Shore D 55–65	Dry	Cond.[1)]
Density	g/cm^3	1.16		1.14	1.12	1.14	1.14	0.99–1.02	1.02–1.03	1.03	
Tensile modulus of elasticity	MPa	3100	3000	2720	2800	3000	2300	70–250	270–450	90–250	60–140
Yield stress	MPa	80		85	90	80	55		20–25		
Elongation at yield	%	5		4	8	3.9	4		30–35		
Nominal elongation at break	%	8		10	> 50	35	> 50	> 50	> 50	> 50	> 50
Stress at 50% elongation	MPa							10–20		10–15	8–12
Melting temperature	°C	295		300		245	195	140–155	160–170		
Heat deflection temperature HDT/A 1.8 MPa	°C	95		125	120	60	50	> 50	50–55	34	
Thermal coefficient of linear expansion, longitudinally (23–55 °C)	10^{-5}/K	6–8			5.4			18–23 [2)]	14–18 [2)]	15–20 [2)]	

Property	Unit	Semi-aromatic polyamides, copolyamides, and polyether block amides									
		PA 6T/6		PA 9T	PA NDT/INDT	PA66/6	PA6/66	PEBA 12		PEBA 6	
		Dry	Cond.[1]			Dry	Dry	Shore D 40–55	Shore D 55–65	Dry	Cond.[1]
Thermal coefficient of linear expansion, across (23–55 °C)	10^{-5}/K	6–8			5.3						
Dielectric constant at 100 Hz					4.6			60–11		4	6
Dielectric loss factor at 100 Hz	$\cdot 10^{-4}$				250			400–1300		300–500	950–1100
Specific volume resistivity	Ohm · m	10^{13}	10^{12}		$>10^{13}$	$>10^{13}$	$>10^{13}$	10^{10}–10^{12}		10^{11}	10^{10}
Specific surface resistivity	Ohm		10^{13}		$>10^{15}$			10^{12}–10^{13}		1012	1011
Dielectric strength	kV/mm	38	35		25			30–40		35–40	30–35
Comparative tracking index CTI/A		600			600	600	600	600		550–600	
Water absorption at 23 °C, saturation	%	7		2	7.5	10.8	10.2	0.6–1.5		3.5–5.0	
Moisture absorption 23 °C/50% rel. humidity	%	1.8		0.25 [3]		2.7	3	0.3–0.7		1.0–1.5	

[1] Specimen storage at 23 °C/50% rel. humidity until constant weight, [2] tensile strength, [3] 20 °C, 65% rel. humidity.

6.8.3.1 Semi-Aromatic, Semi-Crystalline Copolyamides (Polyphthalamides, PPA)

An important subgroup of copolyamides are the so-called polyphthalamides, which have to contain at least 55 mol-% terephthalic acid or isophthalic acid or a combination of both in their polymer chain. They are semi-crystalline and high heat resistant grades. In general, they are identified using the same syntax as for polyamides; however, the acronym PPA has found wide-spread acceptance. PPAs are distinguished by their higher glass transition temperature and a higher melting point of the crystalline domains. The respective homopolyamides 4T, 6T, 9T, and 10T exhibit melting points above their degradation temperatures so that copolymerization is the only path to obtain semi-crystalline polymers than can be processed like thermoplastics. The introduction of one or more comonomers causes a decrease in crystallinity and a loss in stiffness above glass transition temperature that is more pronounced than for high-crystalline aliphatic homopolymers such as PA66 or PA46. The mechanical properties, such as strength, stiffness, and hardness, remain mostly unchanged up to glass transition, which occurs between 70 and 130 °C. Compared to aliphatic polyamides, PPAs exhibit better dimensional stability because of their reduced water absorption, less warpage, and even better chemical resistance. The most important representatives of this material class are listed in Table 6.23.

Applications are stiff, dimensionally stable technical parts in machine engineering, automotive, and electronics as well as functional elements in contact with chemicals und high performance at high service temperatures.

6.8.3.2 Semi-Aromatic, Amorphous Copolyamides

Crystallization can be completely suppressed by implementation of un-symmetric comonomers, such as branched, aliphatic or substituted cyclic diamines or isophthalic acid. Stiff amines in combination with isophthalic acid or terephthalic acid will condense and form amorphous polyamides with high glass transition temperatures. These are completely transparent, even at high layer thicknesses, and exhibit high temperature resistance. Examples include the copolyamide of terephthalic acid and a mixture of isomers from 2,2,4- and 2,4,4-trimethylhexamethylene diamine and the copolymer of isophthalic acid, bis(3-methyl-4-aminocyclohexyl)methane, and laurolactam (see Table 6.23). Stiffness and shear modulus remain more or less constant up to glass transition temperature (Figure 6.11). Water absorption has little influence on mechanical properties, similar as for semi-crystalline, semi-aromatic polymers.

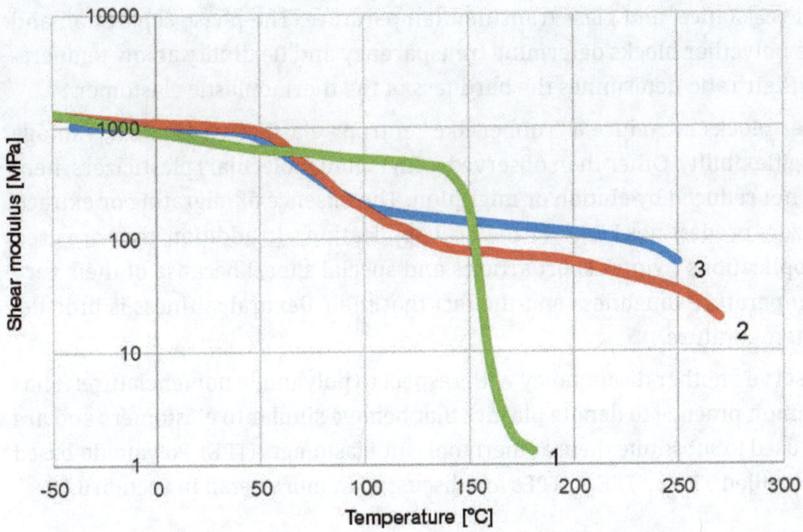

Figure 6.11 Shear modulus of PA NDT/INDT (1), PA6T/6 (2), and PA66 (3) as a function of temperature.

6.8.3.3 Elastomeric Block Copolyamides (Polyether Block Amides, PEBA)

The introduction of flexible soft segments in the polymer chains of semi-crystalline aliphatic polyamides creates 2-phase thermoplastic elastomers. Polyamide elastomers based on polyethers (polymers with oxygen bridges in the chain) in particular have gained commercial importance. These polyether block amides (PEBA) contain “rigid” thermoplastic and “soft” elastomeric units that are connected by amide and ester groups (Figure 6.12).

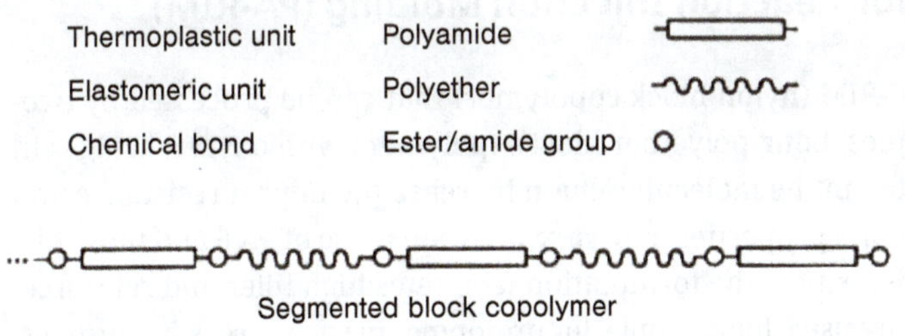

Figure 6.12 PEBA as thermoplastic elastomer

Component selection influences the physical properties. Here, the type of “rigid” polyamide block (PA12, PA11, PA6, PA66, PA612) influences the melting range, water absorption, chemical resistance, and density, while the block length influences the melting range, degree of crystallinity, and transparency. Polytetrahydrofuran (PTHF), polyethylene glycol (PEG), and polypropylene glycol (PPG) are used as polyether blocks and determine elasticity, modulus, water absorption, anti-static behavior, chemical and thermal resistance, and glass transition temperature. The phase separation and length of the polyether blocks determine transparency and flexibility at low temperatures, while their ratio determines the hardness of the thermoplastic elastomer.

The polyether blocks introduce a "rubber-like" entropy elasticity with the advantage of preserving flexibility. Other than observed with (small molecular) plasticizers, here flexibility is not reduced by elution or migration. The absence of migrating or extractable plasticizers predestines PEBA for medical applications. In addition, they are used for sports applications (winter sport articles and special shoes) because of their very good low-temperature toughness and the fact that their flexural stiffness is little dependent on temperature.

Here, we observe another discrepancy with respect to polyamide nomenclature: it has become common practice to denote plastics that behave similar to elastomers and are increasingly used to substitute them as thermoplastic elastomers (TPE). Polyamide-based TPEs are also called TPA or TPE-A. TPEs are discussed in more detail in Section 6.20.

6.8.4 Cast Polyamides (PA6-C, PA12-C)

The properties of cast polyamides compare to those of polycondensated PA 6 and PA 12. However, the polymer chains are more branched and the residual monomer content may be higher because they are not extracted. Stiffness may be higher and thermoplastic processing may be impeded or in special cases even impossible because the crystallization behavior is different and the molecular weight often higher. Volume shrinkage for cast polyamides is approx. 15% and has to be taken in consideration during mold design.

6.8.5 Polyamides for Reaction Injection Molding (PA-RIM)

PA-RIM is in fact an NBC-RIM (nylon block copolymer) that can be processed by two-component RIM techniques. Long polyether blocks (polymers with oxygen bridges in their chains) implemented in the molecular chain increase the impact resistance and decrease stiffness; therefore, properties may vary between those of PA 6 and those of a PA elastomer. The low viscosity of the formulation facilitates high filler and reinforcement loads. These resins are used, for example, for prototypes in vacuum casting systems.

6.8.6 Aromatic Polyamides, Aramids

By definition, in aramids more than 85% of amide groups are directly bonded with aromatic rings. In contrast to aliphatic and semi-aromatic polyamides they are not produced by melt polycondensation, but rather by reaction of the inert aromatic diamines with the respective dicarboxylic acid dichlorides in polar solvents.

Poly(p-phenylene terephthalamide) [PPTA, p-aramid] is used exclusively as a fiber material and exhibits the highest specific strength of all commercial fibers. Aramid fibers are discussed in more detail in Section 7.3.1.3.

6.9 Aromatic (Saturated) Polyesters

Chemical Composition

The chains of thermoplastic (saturated linear) polyesters contain ester groups in regular intervals, see Equation 6.10. In most cases, they are produced as condensation polymers of dicarboxylic acids and diols or their derivatives.

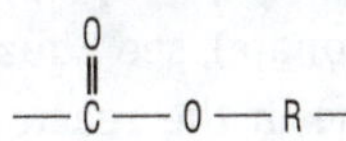

Ester group

(6.10)

Polyesters containing aromatic groups (benzene rings) are basic polymers for engineering materials. The aromatic rings stiffen the molecular chains, thereby raising heat distortion and melting temperatures. The more frequently these groups occur in the chain, the higher are these temperatures. Fully aromatic polyesters (PAR, polyarylates) have outstanding thermal stability.

Semi-aromatic polyesters are not attacked by either aliphatic hydrocarbons nor by ethanol and higher alcohols. They absorb only minimal amounts of water and are physiologically inert. Because they contain saponifiable ester groups, they are degraded by alkalis. Their resistance to oxidizing acids and to long-term exposure to water and water vapor above 70 °C is limited. The resin has to be completely dried to avoid damage by hydrolysis during thermoplastic processing.

6.9.1 Polycarbonates (PC)

Klaus Horn

6.9.1.1 Polycarbonates Based on Bisphenol A (PC)

Chemical Composition

Two processes are of technical significance for the production of bisphenol A polycarbonate: the interfacial polycondensation and the melt condensation method. During the phase boundary process, polycarbonate is produced by the reaction of bisphenol A (from phenol and acetone) with phosgene in a mixture of water and organic solvents (2 phases) and a catalyst, see Equation 6.11. PC remains in the organic phase and sodium chloride in the aqueous phase. After repeated washing of the organic phase with water, the polymer is separated from the solvent by distillation or precipitation. The molecular weight is controlled by the added amount of monophenol (chain terminator, typically phenol).

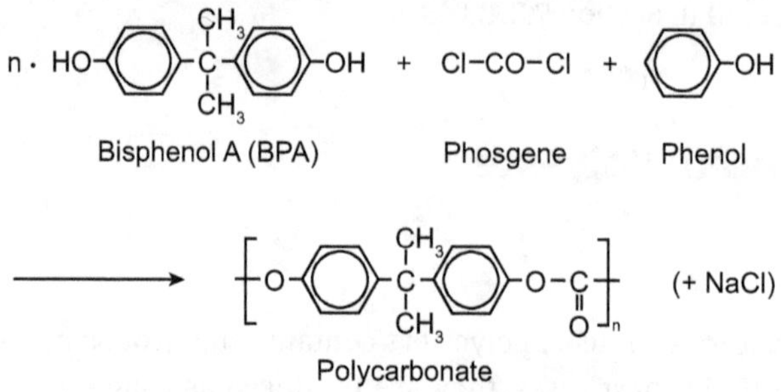

(6.11)

During the melt condensation process, bisphenol A is condensated with diphenyl carbonate in the melt under separation of phenol (from diphenyl carbonate), see Equation 6.12. The molecular weight is controlled by the residence time in the reactor. Currently, the melt condensation process is regaining significance. The molecular weight of common polycarbonates ranges from approx. 20,000 to 35,000 g/mol. Below a MW of 20,000, PC does not exhibit thermoplastic properties, whereas above 35,000 thermoplastic processing is difficult.

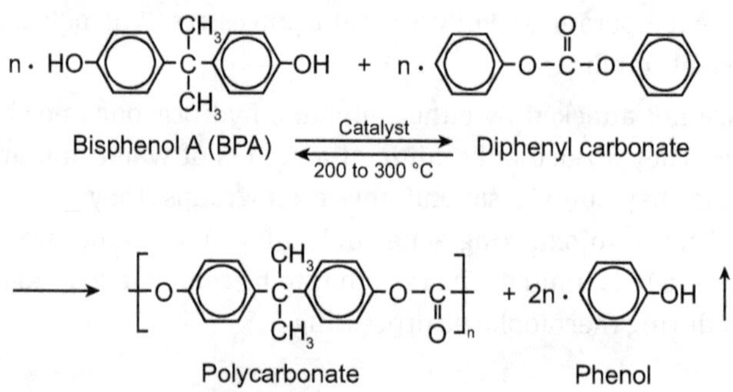

(6.12)

Processing

PC can be processed and post-treated by all standard thermoplastic processing methods. The most common method is injection molding, *e. g.*, for the manufacture of optical data storage devices or molded components for electronics or automotive applications. Extrusion of PC is typically used for the manufacture of solid, twin-walled, and corrugated sheets as well as profiles. In order to finish sheets for outdoor applications, they are coated with UV-protective layers via coextrusion. Injection and extrusion blow molding are used for packaging applications (bottle production). Common melt temperatures for injection molding range from 280–320 °C (mold temperatures from 80–120 °C), for extrusion from 240–300 °C, depending on application and molecular weight of the manufactured component.

All commercially available polycarbonates have in common that they require drying of the pellets prior to processing to avoid surface defects (flow marks) and property decreases. Residual moisture must not exceed 0.02%. For optical applications of PCs, the requirements for moisture content are even higher; here, contents of < 0.01 wt.-% are recommended. Drying time depends on the actual situation; 4 h in a convection oven at 120 °C is recommended as a guideline.

Properties

Polycarbonates distinguish themselves as amorphous thermoplastic materials by their high transparency. Theoretically, the maximum transmission is 90.2% (4 mm thickness). The combination of transparency with high heat distortion temperature and excellent toughness renders polycarbonate the ideal material for many applications.

This property combination is enhanced by PC's good compoundability with other thermoplastics such as ABS and PBT (blends).

PC is impact resistant in a temperature range from −150 °C to 135 °C and exhibits high stiffness and strength. The impact resistance values of notched unreinforced grades differ depending on specimen geometry and testing standard but reach the highest level compared to other transparent thermoplastics. With increasing molecular weight, toughness increases while the flowability of the polymer melt decreases. PC's Vicat softening temperature of 145 to 150 °C provides a level of temperature resistance that allows it to withstand most temperatures created by direct sun exposure, boiling water, or by electronic devices without damage. The long-term temperature resistance of the UL-temperature index (RTI) is 115 °C (with impact) and 125 °C (without impact). For electric/electronic applications, heat resistance as determined by ball indentation tests (IEC 60695-10-2) is crucial. The values achievable by PC at 135 °C are sufficient for many electric/electronics applications. PC can undergo superheated or ethylene oxide sterilization (120 °C), and for special grades high-energy radiation (gamma radiation) can be used (application in medical technology). To increase stiffness, PC can be reinforced using glass or carbon fibers or fillers. The addition of UV protecting agents increases weather resistance. For outdoor applications, sheets are coated with UV-protective layers via coextrusion. PC is relatively flame retardant (UL94 V-2); grades modified with halogen-free flame retardants (even transparent grades) can reach UL94 V-0 classification.

Property comparison, see Table 6.29.

Table 6.29 Property Comparison of Polycarbonate and PC Blends

			Polycarbonate and PC Blends						
Property	**Test conditions**	**Unit**	**PC Non-reinforced**	**PC-GF20 Reinforced**	**PC BPA/TMC Non-reinforced**	**PC+ABS Non-reinforced**	**PC+ABS-FR Non-reinforced**	**PC+ABS-GF 20 Reinforced**	**PC+PBT Non-reinforced**
Density	–	kg/m^3	1.20	1.34	1.12–1.18	1.08–1.17	1.18	1.29	1.20
Melt flow rate	300 °C; 1.2 kg	cm^3/10 min	3–75	3–10	5–45 [2)]	5–30 [3)]	12–40 [6)]	210 [7)]	15–30 [3)]
Tensile modulus of elasticity	1 mm/min	MPa	2350–2400	5600–6200	2400	2000–2500	2600–2800	7500	1800–2200
Yield stress	50 mm/min	MPa	60–68	–	68–78	48–63	60–65	120	40–55
Elongation at yield	50 mm/min	%	5.5–6.5	–	6.2–6.9	3.6–5.0	3.5–4.0	2.0	4.0
Nominal elongation at break	50 mm/min	%	> 50	–	> 50	> 50	> 30 to > 50	–	> 50
Failure stress	5 mm/min	MPa	–	85	–	–	–	120	–
Failure strain	5 mm/min	%	–	2.6	–	–	–	2.0	–
Charpy-impact strength	23 °C	kJ/m^2	N	35–60 (Charpy)	N	N [4)]	N [4)]	38 [4)]	N
Charpy-notched impact strength	23 °C	kJ/m^2	50–80	8–12	5–16	40–48	10–35 [4)]	8 [4)]	40–70
Vicat	50 N; 50 °C/h	°C	143–150	145–150	158–218	110–135	90–116	128	–

			Polycarbonate and PC Blends						
Property	**Test conditions**	**Unit**	**PC Non-reinforced**	**PC-GF20 Reinforced**	**PC BPA/TMC Non-reinforced**	**PC+ABS Non-reinforced**	**PC+ABS-FR Non-reinforced**	**PC+ABS-GF 20 Reinforced**	**PC+PBT Non-reinforced**
Heat deflection temperature HDT/A	1.8 MPa	°C	120–130	136–140	147–187	95–110	78–98	119	60–85
Coefficient of linear expansion, longitudinally	23 to 55 °C	10^{-4}/K	0.65–0.0	0.30	0.70	0.70–0.90	0.68–0.76	0.30	0.90–1.10
Coefficient of linear expansion, across	23 to 55 °C	10^{-4}/K	0.65–0.70	0.70	0.70	0.70–0.90	0.68–0.80	0.70	0.90–1.10
Flammability UL 94	1.5 mm	Class	HB/V-2/ V-0 [1)]	V-2 to V-0	HB to V-0	HB	V-0	HB	HB
Glowing wire test (GWFI)	1.5 mm	°C	800–960	960	800–960	700–850 [5)]	960	800 [5)]	700 [5)]
Glowing wire test (GWIT)	1.5 mm	°C	850–960	–	–	–	775–850	–	–
Rel. dielectric constant	100 Hz	–	3.1–3.2	3.3	3.0	3.1	3.2	3.4	3.0–3.5
Dielectric loss factor	100 Hz	10^{-4}	5–15	10	7	20–35	50	25	10–15
Specific volume resistivity	–	Ohm · m	$>10^{14}$	$>10^{14}$	$>10^{14}$	$>10^{14}$	$>10^{14}$	$>10^{14}$	$>10^{18}$

Table 6.29 Property Comparison of Polycarbonate and PC Blends *(continued)*

			Polycarbonate and PC Blends						
Property	**Test conditions**	**Unit**	**PC Non-reinforced**	**PC-GF20 Reinforced**	**PC BPA/TMC Non-reinforced**	**PC+ABS Non-reinforced**	**PC+ABS-FR Non-reinforced**	**PC+ABS-GF 20 Reinforced**	**PC+PBT Non-reinforced**
Specific surface resistivity	–	Ohm	$>10^{16}$	$>10^{16}$	$>10^{16}$	$>10^{16}$	$>10^{16}$	$>10^{16}$	$>10^{16}$
Dielectric strength	1 mm	kV/mm	34	36	30–35	35–45	30–35	35	30–35
Comparative tracking index CTI/A	Test solution A	Level	225–250	175	225–600	175–275	300–350	150	100–125
Water absorption	23 °C, saturation	%	0.30	0.24	0.30	0.70	0.50	0.60	0.50
Water absorption (equilibrium)	23 °C; 50% *r. h.*	%	0.12	0.10	0.12	0.20	0.20	0.20	0.20

N = no failure
1) depending on viscosity, with or w/o flame retardants
2) MVR: 330 °C; 2.16 kg
3) MVR: 260 °C; 5 kg
4) Izod measurement
5) in 2.0 mm
6) MVR: 240 °C; 5 kg
7) Melt viscosity in Pas, 260 °C; 1000 s^{-1}

Applications

Molded components and housings for electric/electronic applications (covers and panels for switching and measuring devices, distribution points), optical data storage (CD, DVD, Blu-ray), glazing for stadium roofs and green houses, ophthalmic and other lenses, fresh water containers up to 20 l, molded components coated with scratch resistant siloxane for headlight covers and automotive glazing (side and rear, roofs, black screens, light housings).

6.9.1.2 Polycarbonate Copolymers

Polycarbonates Based on Trimethylcyclohexane Bisphenol (PC-TMC)

Besides bisphenol A, TMC bisphenol has also reached technical significance, see Equation 6.13. This material class is an example for chemically modified polycarbonates.

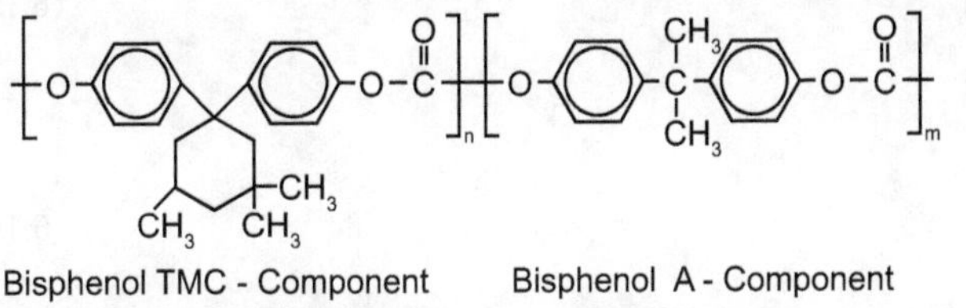

(6.13)

The incorporation of TMC bisphenol significantly reduces the mobility of the polymer chain which results in an increased glass temperature. Pure TMC bisphenol polycarbonate has a glass transition temperature of 239 °C.

Statistical co-polycarbonates from bisphenol A and TMC bisphenol can be easily prepared in practically any mixing ratio. This allows the tailoring of a continuous heat distortion range between 150 and 239 °C. Grades of practical significance exhibit Vicat temperatures between 158 °and 218 °C. Bisphenol A/TMC bisphenol co-polycarbonates exhibit property profiles that are similar to those of bisphenol A polycarbonate, including similar optical properties in addition to the wider range and higher maximum of their Vicat heat distortion temperature. With increasing TMS bisphenol content, the toughness decreases.

Property comparison: see Table 6.29.

Processing: These polycarbonates are injection molded at melt temperatures of 320–340 °C and mold temperatures of 100–160 °C, depending on heat distortion temperature. They also have to be dried at 130 °C for 4 to 12 hours.

Applications: These copolycarbonates are used in applications for which the heat distortion temperature of pure bisphenol A polycarbonate is not sufficient. For example, the latest designs in car headlights require extremely heat resistant headlight masks. In medical technology, these grades are used because they allow for superheated sterilization at 135 °C and 143 °C.

Other Chemically Modified Polycarbonates

In general, any bifunctional unit (bisphenols, dicarboxylic acid) can be used in combination with bisphenol A to produce polycarbonates. Some of these combinations have been commercially available in the past, including co-polycarbonates with bisphenol A and 4,4'-dihydroxydiphenyl sulfide (higher notched impact strength), see Equation 6.14, 4,4'-dihydroxybiphenyl (DOD) (increased low temperature notched impact strength up to – 50 °C), see Equation 6.15, and dodecane dicarboxylic acid (improved flowability, Vicat approx. 130 °C), see Equation 6.16.

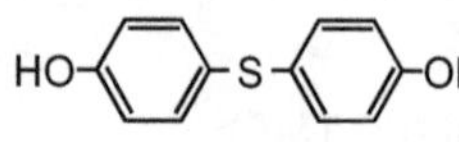

(6.14)

4,4'-Dihydroxyphenyl sulfide (Bisphenol S)

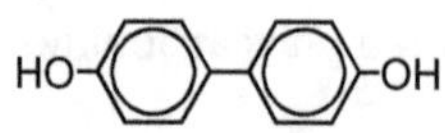

(6.15)

4,4'-Dihydroxybiphenyl

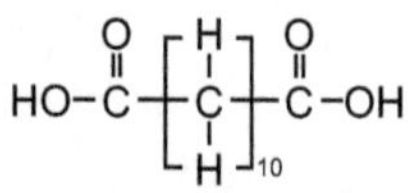

(6.16)

Dodecane dicarboxylic acid

Block Co-Polycarbonates

Siloxane co-polycarbonates based on bisphenol A can be produced with phenolic-OH terminated polysiloxane blocks using the phase boundary process. Compared to pure bisphenol A PCs, these polycarbonates exhibit higher notched impact strength at low temperatures, some down to −50 °C.

6.9.1.3 Blends

Polycarbonate can be easily blended with other polymers. These blends are produced using twin screw extruders or kneaders. The polycarbonate and the polymeric blend partner are melted at approx. 280 to 360 °C in the extruder. The polymeric melts are mixed by the twin screws, resulting in two- or multi-phase grades. Blends of polycarbonate and ABS, ASA, PET, and PBT have gained technical significance. Blending complements the properties of each blend partner.

PC+ABS Blends

In the case of PC+ABS blends, the high heat distortion temperature of PC is complemented by the better flowability and good impact resistance at low temperatures of ABS, see Figure 6.13. The Vicat heat distortion resistance of the PC+ABS blend depends on the ratio of both polymer components and ranges between that of pure ABS at approx. 100 °C and 150 °C of pure polycarbonate. Of technical significance are blends with Vicat softening temperatures of 110 to 135 °C. Flame retarded PC+ABS blends exhibit Vicat softening temperatures between 95 and 135 °C. Typically, polycarbonate represents the major phase in commercially available blends.

Property comparison: see Table 6.29.

Processing: The majority of PC+ABS blends is injection molded, in some cases they are extruded. Processing temperatures for PC+ABS blends typically range from 220 to 260 °C, but optimum conditions can be very grade-specific, depending on the PC+ABS ratio of the respective grade. The pellets should be dried at a temperature that is approx. 10 °C below the Vicat softening temperature of the specific grade (rule of thumb: 4–8 h, residual moisture content in the pellets < 0.02%).

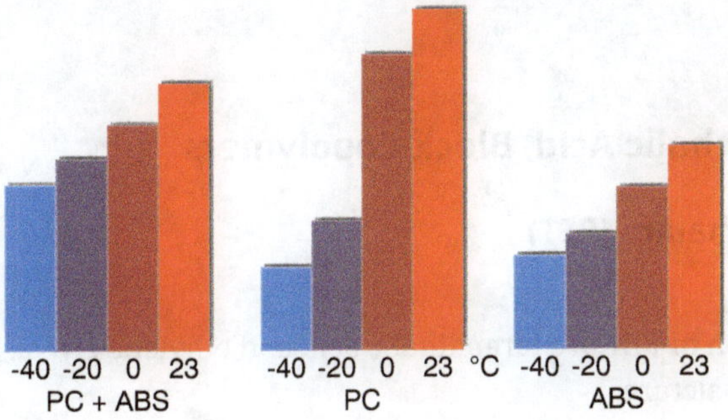

Figure 6.13 Comparison of Izod notch impact strength of ABS and PC and one of their blends

Applications: Standard-PC+ABS-Blends are used for interior and exterior components in the automotive industry. For interior applications, particularly easy flowing PC+ABS blends are used to facilitate the manufacture of linings and covers with extremely small wall thicknesses. Uncoated interior covers made from PC+ABS blends meet the higher requirements for weather and aging resistance. These blends can be coated and metallized.

A major field of applications is the electric and information technology industry, in particular for flame retarded PC+ABS grades in which the flame retardant compounds meet the requirements of the major eco labels. Easy flowing, flame retardant PC+ABS blends are used for TV housings, providing significant contributions to lowering fire damage.

PC + Polyester

PC + polyester (PBT, PET) blends combine the high heat distortion resistance and good toughness of PC with the better chemical resistance of the semi-crystalline polyester. Many blends are also modified with elastomers and/or reinforced with glass fibers or other fillers.

Property comparison: see Table 6.29.

Processing: PC+PBT blends are typically injection molded between 260 and 270 °C; PC+PET blends at 280 °C. These blends also require drying prior to processing. PC+PBT blends are dried at approx. 100 °C and PC+PET blends at approx. 110 °C (rule of thumb: 4 h). When determining drying and processing conditions, the material manufacturer's instructions need to be followed because the individual grade composition determines these conditions.

Properties: PC + polyester is distinguished by high toughness, even at low temperatures, good chemical resistance, and thus a reduced susceptibility to stress cracking. These material grades can be easily coated.

Applications: Molded components for the automotive industry (coated spoiler, radiator grill), electric/electronic applications (housings), lighting technology, sport and leisure (helmets).

6.9.2 Polyesters of Terephthalic Acid, Block Copolymers

6.9.2.1 Polyethylene Terephthalate (PET)

Chemical Composition

The raw materials for condensation of PET are terephthalic acid and ethylene glycol. Equation 6.17 shows the general structure.

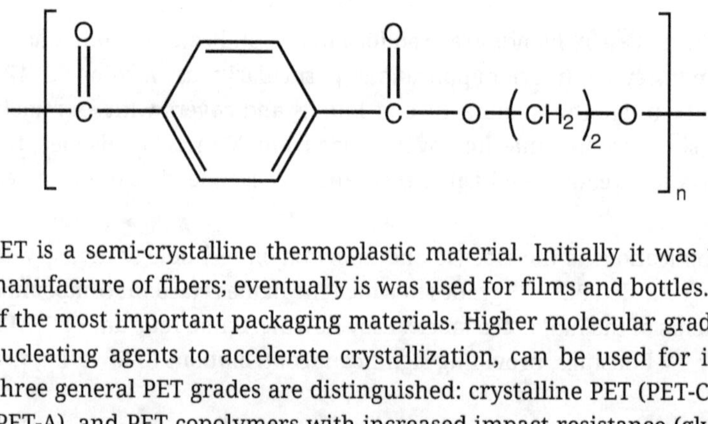

(6.17)

PET is a semi-crystalline thermoplastic material. Initially it was used only for the manufacture of fibers; eventually is was used for films and bottles. Today, PET is one of the most important packaging materials. Higher molecular grades, modified with nucleating agents to accelerate crystallization, can be used for injection molding. Three general PET grades are distinguished: crystalline PET (PET-C), amorphous PET (PET-A), and PET copolymers with increased impact resistance (glycol-modified PET, PET-G). Incorporation of voluminous comonomers, such as isophthalic acid or 1,4-cyclohexane dimethanol (CHDM) (see Equation 6.18) is used to lower crystallinity to facilitate the manufacture of transparent parts (*e. g.*, bottles).

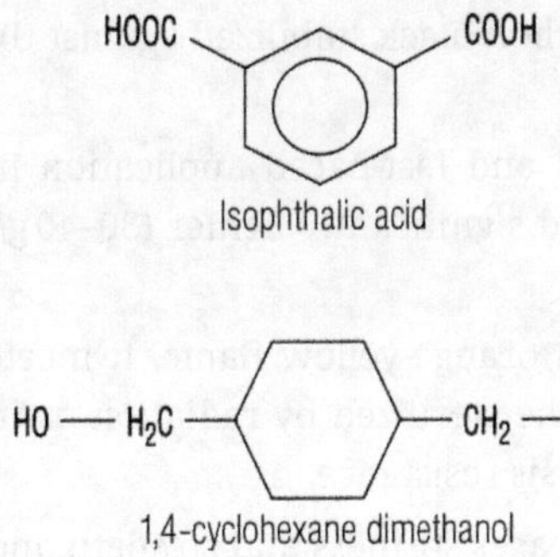

(6.18)

Processing

The major processing methods for PET are injection molding and one- and two-stage stretch blow molding for PET bottles and containers. Films, sheet, and solid profiles are extruded. Prior to thermoplastic processing, moist pellets must be dried for ≈ 10 h at 130 °C. For injection molding of amorphous parts, the melt temperature should range between 260 and 290 °C; the mold temperature should be above 60 °C for amorphous and ≈ 140 °C for semi-crystalline parts (wall thickness > 4 mm). Despite its high shrinkage of 1.2–2.5% for semi-crystalline parts, PET is well suited for overmolding of metal inserts as long as the wall thickness is adequately high. Joints can be established by the following methods: ultrasonic, friction, hot plate, and hot gas welding, and adhesive bonding with cyanoacrylate-, EP-, or PUR adhesives.

Properties

Mechanical properties depend on the degree of crystallinity, which in turn depends on the processing conditions during injection molding. Mold temperatures of 140 °C, long residence times, and annealing result in a degree of crystallinity of 30–40%. These parts exhibit high stiffness and strength below 80 °C and low creep under constant static load. However, their impact resistance is low, while slip- and wear properties are good.

Amorphous parts are desired when high transparency, high toughness, excellent slip- and wear properties, low shrinkage, and high dimensional stability are required.

At ≈ 80 °C amorphous PET components exhibit a glass transition region in which the modulus of elasticity, particularly of non-reinforced grades, decreases sharply. Highly reinforced grades exhibit structural integrity up to ≈ 250 °C. Long-term service temperature ranges from 100–120 °C. The good electrical properties are little frequency- and temperature-dependent. Because PET has low permeability for O_2 and CO_2, it is well suited for bottling alcohol-containing or carbonated beverages. It is resistant to weak acids and alkaline solutions, oils, fats, aliphatic and aromatic hydrocarbons, and carbon tetrachloride. It is not resistant to strong acids and alkaline solutions, phenol, and long-term use in hot water above 70 °C. PET is not prone to stress cracking. It exhibits good weather stability, particularly when carbon black stabilized against UV radiation.

PET's general property spectrum explains its global and fast-paced application in packaging. Compared to glass, it is non-breakable and significantly lighter (30–40 g/l for PET compared to 500–900 g/l for glass).

Without added flame retardants, PET burns with an orange-yellow flame. It meets requirements for food contact; however, it can only be sterilized by radiation or in ethylene oxide atmosphere because of its low hydrolysis resistance.

Addition of carbon- or glass fiber reinforcement increases stiffness and strength and reduces processing shrinkage to 0.4–0.8%. At the same time, anisotropy caused by shrinkage increases so that for applications in which strength is required, *e. g.*, electrical or electronic applications, micro-glass spheres are generally used.
Comparison of properties, see Table 6.30.

Applications

Staple fibers and filaments (72% of total consumption), PET bottles (disposable, reusable, hot filling), wear resistant parts (also glass fiber-reinforced PET) such as bearings, gear wheels, shafts, guides, couplings, locks, knobs. Insulating, magnetic, and anti-stick films for cast resin processing, ink ribbons for printers, carrier films for photographic film, shrink film, fibers, PET-A films for blister packaging, foamed PET-C deep drawing film for fast-food trays.

Table 6.30 Property Comparison for Polyalkylene Terephthalates and Polyester Elastomers

Property	Unit	Polyalkylene terephthalates and polyester elastomers									
		PET			PBT			(PBT + ASA)	PTT	TPC-EE	
		Amorphous PET-A	Semicrystalline PET-C	-GF 30	Nonreinforced	Elastomer modified	-GF 30		Nonreinforced	Shore D 35–50	Shore D 55–75
Density	g/cm³	1.33–1.35	1.38–1.40	1.56–1.59	1.30–1.32	1.2–1.28	1.52–1.55	1.21–1.22	1.33	1.11–1.20	1.22–1.28
Tensile modulus of elasticity	MPa	2100–2400	2800–3100	9000–11,000	2500–2800	1100–2000	9500–11,000	2500	2550	30–150	200–1100
Yield stress	MPa	55	60–80	–	50–60	30–45	–	53	61	–	–
Elongation at yield	%	4	5–7	–	3.57	6–20	–	3.6	>10	–	–
Nominal elongation at break	%	> 50	> 50	–	20– > 50	> 50	–	> 50		> 50	> 50
Stress at 50% strain	MPa	–	–	–	–	–	–	–		10–30*⁾	30–50*⁾
Failure stress	MPa	–	–	160–175	–	–	130–150	–		–	–
Failure strain	%	–	–	2–3	–	–	2.53	–		–	–
Melting temperature	°C	–	250–260	250–260	220–225	200–225	220–225	225		155–210	215–225

Table 6.30 Property Comparison for Polyalkylene Terephthalates and Polyester Elastomers *(continued)*

Property	Unit	Polyalkylene terephthalates and polyester elastomers									
		PET			PBT			(PBT + ASA)	PTT	TPC-EE	
		Amorphous PET-A	Semi-crystalline PET-C	-GF 30	Non-reinforced	Elastomer modified	-GF 30		Non-reinforced	Shore D 35–50	Shore D 55–75
Heat deflection temperature HDT/A 1.8 MPa	°C	60–65	65–75	220–230	50–65	50–60	200–210	80		–	50–55
Thermal coefficient of linear expansion, longitudinally (23–55 °C)	10^{-5}/K	8	7	2-3	8–10	10–15	3–4.5	10		15–22	10–18
Thermal coefficient of linear expansion, across (23–55 °C)	10^{-5}/K	–	–	7-9	–	–	7-9	–		–	–
Flammability UL94 for 1.6 mm thickness	Class	HB[1)]	HB[1)]	HB[1)]	HB[1)]	HB[1)]	HB[1)]	HB[1)]	HB[2) 3)]	HB[1)]	HB[1)]
Dielectric constant at 100 Hz		3.4–3.6	3.4–3.6	3.8–4.8	3.3–4.0	3.2–4.4	3.5–4.0	3.3		4.4–5	
Dielectric loss factor at 100 Hz	$\cdot 10^{-4}$	20	20	30–60	15–20	20–130	20–30	10		100–200	

Property	Unit	Polyalkylene terephthalates and polyester elastomers									
		PET			PBT			(PBT + ASA)	PTT	TPC-EE	
		Amorphous PET-A	Semi-crystalline PET-C	-GF 30	Non-reinforced	Elastomer modified	-GF 30		Non-reinforced	Shore D 35–50	Shore D 55–75
Specific volume resistivity	Ohm · m	$>10^{13}$	$>10^{13}$	$>10^{13}$	$>10^{13}$	$>10^{13}$	$>10^{13}$	$>10^{14}$		ca. 10^{10}	
Specific surface resistivity	Ohm	$>10^{14}$	$>10^{14}$	$>10^{14}$	$>10^{14}$	$>10^{14}$	$>10^{14}$	$>10^{15}$		$>10^{13}$	
Dielectric strength	KV/mm	250	30	30–35	25–30	25	30–35	30		20–25	
Comparative tracking index CTI/A		300–400	300–400	250–275	600	600	350–525	600		600	
Water absorption at 23 °C, saturation	%	0.6–0.7	0.4–0.5	0.4–0.5	0.5	0.4–0.7	0.35–0.4	0.5	< 0.4	0.6–1.2	0.4–0.8
Moisture absorption 23 °C/50% rel. humidity, saturation	%	0.3–0.35	0.2–0.3	0.2	0.25	0.15–0.2	0.1–0.15	0.2		0.3–0.6	0.2–0.4

*) = Tensile strength [1] also available as V-0 [2] with 30% GF also available as V-0 [3] at 1.5 mm thickness.

6.9.2.2 Polybutylene Terephthalate (PBT)

Chemical Composition

The chemical constitutions and properties of PBT and PET are very similar. During condensation, 1,4-butanediol is used instead of ethylene glycol, see Equation 6.19.

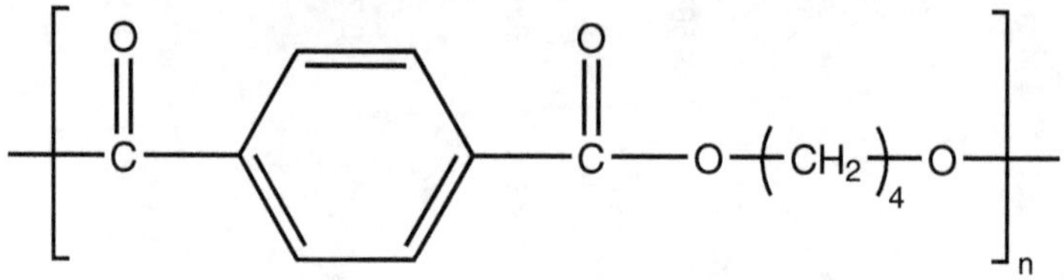

(6.19)

PBT is also semi-crystalline; however, it crystallizes faster and is therefore better suited for injection molding operations than PET.

Processing

PBT is mostly injection molded at melt temperatures of 230–270 °C. Mold temperatures below 60 °C are common; however, optimum surface quality is only reached at 110 °C. As with PET, resin drying is required. PBT can be joined by ultrasonic-, friction, hot plate-, and hot gas welding as well as by two-component adhesives.

Properties

Stiffness and strength are slightly lower than those of PET, toughness at low temperatures is slightly better. Friction and wear properties are excellent. Glass transition of the amorphous phase is at 60 °C. Maximum dimensional stability is achieved at temperatures between 180–200 °C, the long-term heat distortion temperature is between 100–120 °C. PBT is a good electrical insulator and its properties are little influenced by water absorption, temperature, or frequency. Permeability for CO_2 is significantly higher than that of PET, chemical and weather resistance and burn rates are comparable. Resistance to hot water is better. PBT grades meet the requirements for food contact.

PBT grades with the following modifications are available: free-flowing, flame retarded, higher impact resistance by elastomer modification, see Figure 6.14, reinforced or filled to increase stiffness, strength, wear resistance, or to lower the coefficient of dynamic friction.

Comparison of properties, see Table 6.30.

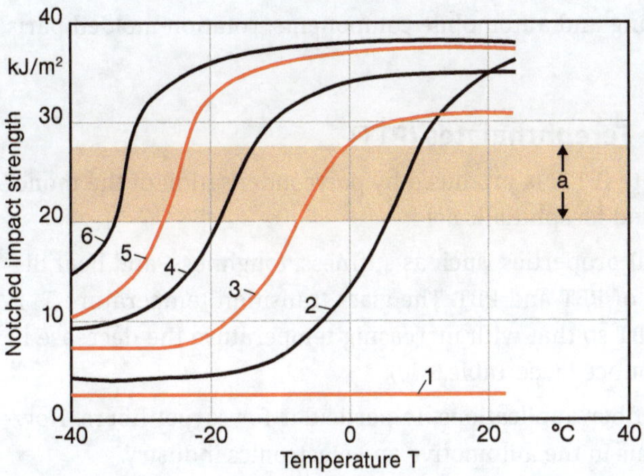

Figure 6.14 Notched impact toughness of unreinforced modified polybutylene terephthalate as a function of temperature

1: Unmodified

2: With polyacrylate rubber

3: With polyolefin graft polymer

4: With PC and polybutadiene graft polymer

5: With polybutadiene graft polymer

6: With modified polybutadiene graft polymer

a: Transition region between ductile and brittle failure

Applications

Friction bearings, roller bearings, valve parts, screws, multi-point connectors, pump housings, parts for small appliances such as coffee makers, egg cookers, toasters, blow dryers, vacuums, cooking utensils.

6.9.2.3 Cyclic Polybutylene Terephthalates (CBT)

CBT is a thermoplastic oligomer which melts at 160 °C. When heated, the material exhibits a water-like viscosity. The cyclic oligomers are opened using a catalyst (based on Zn or Ti) and polymerized to form polybutylene terephthalate (PBT). CBT can be processed between 170 and 240 °C, and is available both as single- and two-component systems. The reaction rate of the conversion of CBT to PBT depends on the concentration of the catalyst and can be adjusted so that polymerization occurs in a time window between 20 seconds and 10 minutes. The special properties of the material facilitate processing like a thermoset while resulting in thermoplastic end products. CBT blends are also produced.

Applications: CBT plastics are used for injection and rotational molding applications and primarily in composites. Examples include wings for wind turbines and composite applications in boat-building and automobile components, rotation-molded parts and fuel tanks.

6.9.2.4 Polytrimethylene Terephthalates (PTT)

Polytrimethylene terephthalate (PTT) is produced by polycondensation of the monomers 1,3-propanediol (PDO) and terephthalic acid.

PTT's mechanical and thermal properties, such as stiffness, toughness, and heat distortion, range between those of PET and PBT. The glass transition temperature T_g is slightly higher than that of PBT so that with increasing temperature the decrease in Young's modulus is less pronounced (see Table 6.30).

Applications: PTT is used for fiber applications, in particular for carpet fibers. However, there are also applications in the automotive and electronics industry.

6.9.2.5 Thermoplastic Polyester Elastomers (TPC)

See Section 6.20.2.2.

6.9.2.6 Polyterephthalate Blends (PET+: PBT, MBS, PMMA, PSU, Elastomer)

Blends with PBT, MBS, PMMA, and PSU exhibit improved processability. Blends of MBS with any of the three PET grades are used to improve impact resistance while retaining transparency. MBS copolymers (methacrylate butadiene styrene copolymers, see Section 6.6.2) are manufactured by polymerizing a rigid shell on a soft rubber core. Applications include packaging suitable for both freezer and microwave. Blends with PEN exhibit a higher heat distortion point.

PBT + PC is tougher and is increasingly used for large exterior parts in the automotive industry. PBT + PET-GF50 is extremely rigid, UV resistant, with a smooth surface. It is used for automotive windscreen wiper arms, exterior mirror fixtures. PBT + ASA-(GF) exhibits less shrinkage at a higher melt flow rate and improved surface quality. Automotive applications: headlight frames, electronics housings. Blends with EPDM, LCP, and SMA are available.

6.9.3 Polyesters of Aromatic Diols and Carboxylic Acids (PAR, PBN, PEN)

6.9.3.1 Polyarylates (PAR)

Chemical Composition, Properties

PAR are thermoplastic condensation products of pure aromatic polyesters and polyester carbonate (PEC). Equation 6.20 exemplifies a structure.

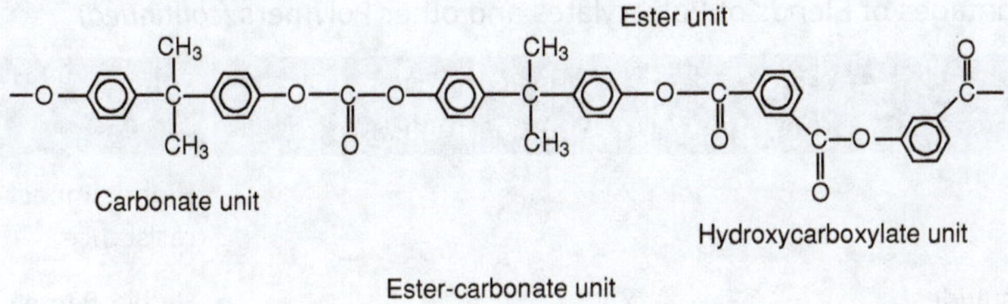

(6.20)

PARs are transparent and in their mechanical and electrical properties and chemical resistance comparable to PC. Grades for food contact are available. PARs can be sterilized. The glass transition temperature ranges from above 180 to 325 °C, the long-term service temperature is at 150 °C. By their nature, PAR are difficult to ignite, very UV resistant, and suitable for outdoor use even without stabilization. However, without UV absorbers they tend to yellow.

Processing

PAR has to be dried to a moisture content of less than 0.02% at a temperature between 120 and 130 °C before it can be processed (primarily injection molding and extrusion at melt temperatures of 340–400 °C). The mold temperature may reach up to 150 °C, shrinkage in flow direction is ≈ 0.2%, perpendicular to flow it is ≈ 0.7–0.9%. All other processing and post-processing operations resemble those of PC.

Modifications

Addition of fibers and fillers, in particular glass and potassium titanate fibers, increases strength. Talc is also used as a reinforcing filler. Flame retardant additives improve the already good burn rate (V-2 according to UL 94).

Blends with other polymers yield certain advantages, in particular with regard to processability, resistance to fuels, risk of hydrolysis, and price reduction. The lower impact resistance at low temperatures is compensated by grafted silicone rubbers, see Table 6.31.

Table 6.31 Advantages of Blends of Polyarylates and other Polymers

Blend partner	Fuel resistance	Hydrolysis resistance	Processability	Price	Other
Polyalkylene terephthalates	x		x	x	
Aliphatic polyamides	x		x	x	
Aromatic polycarbonates			x		
Polyolefins			x	x	Higher impact resistance

Table 6.31 Advantages of Blends of Polyarylates and other Polymers *(continued)*

Blend partner	Fuel resistance	Hydrolysis resistance	Process-ability	Price	Other
ABS polymers		x	x		Higher impact resistance
Polyphenylene sulfides		x			Higher flame resistance
Polyether imides	x				
Polycarbonate-siloxane block copolymers	x		x		Higher impact resistance
Grafted silicone rubber	x				Higher impact resistance
x principal improvement					

Applications

PARs are used when the heat distortion temperature of PC is not sufficient and PSU is too expensive; for example, for control panels of kitchen appliances, parts for blow dryers and microwaves, lamp housings and reflectors, tools, elements for office- and other machines. Sintered semi-finished products such as sheet, pipes, tubes, rods; transparent cast film for insulation or multi-layer laminates in electrical or electronic devices, liquid crystal monitors, adhesive tapes, UV filters; sintered or injection molded semi-finished products: functional elements in all technical fields that are exposed to elevated temperatures.

6.9.3.2 Polybutylene Naphthalates (PBN)

PBN is a polymer based on dimethyl-2,6-naphthalene dicarboxylate (NDC) and butanediol. Compared to PBT it exhibits a higher heat distortion point (solder resistant), and less permeability to methane and methanol than PA 11 or PE-HD.

Applications: Electronic devices, fuel carrying parts (copolymers (TPE) for hoses and cables).

6.9.3.3 Polyethylene Naphthalates (PEN)

PEN is a polymer based on dimethyl-2,6-naphthalene dicarboxylate (NDC) and ethylene glycol. It is amorphous and transparent and is used as an alternative to or blend partner with PET for bottles (hot cleaning for multiple use), films, and fibers. For the production of polyester bottles the use of PEN is still cost prohibitive. Only for special applications PEN was able to substitute PET because of its better barrier properties and heat stability.

Processing: injection molding, extrusion, thermoforming, blow molding.

Properties: low oxygen permeability, good UV resistance and resistance to chemicals and hydrolysis.

Applications: In the medical field it is implemented in blood sampling devices, packaging, containers.

6.10 Aromatic Polysulfides and Polysulfones (PPS, PSU, PES, PPSU)

Christian Maletzko, BASF SE, Ludwigshafen, Germany

6.10.1 Polyphenylene Sulfides (PPS)

Chemical Composition

PPS are semi-crystalline polymers in which aromatic monomer units are connected by sulfur atoms. They are distinguished by their extremely high heat stability (melt temperature up to 445 °C), high chemical resistance, and strength, see Equation 6.21. They can be produced as either cross-linkable thermosets or as thermoplastics, the latter being by far the more relevant.

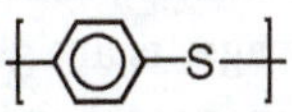

Polyphenylene sulfide (PPS) (6.21)

Processing

Injection molding: Injection molding is the most important processing method; melt temperatures ranging from 315–370 °C, mold temperatures from 25–200 °C. Above a mold temperature of 120 °C, part surfaces turn smooth and glossy; at 40 °C the highest degree of toughness is achieved. Because glass fiber reinforced resins contain less than 0.05% water, pre-drying is only necessary in special cases, *e. g.*, when hydrophilic fillers are used. Storage at 150 °C for 6 h has proven advisable. As PPS has a very low melt viscosity, even filled grades are easy flowing, facilitating molding of thin-walled parts.

Compression molding, sintering: PPS can be processed by compression molding or sintering (*e. g.*, for surface coating). Oxidative cross-linking prior to processing is used to achieve the necessary melt viscosity of the powders. The processing steps for molded parts are: compressing in the cold mold at 70 MPa, heating of the mold to 360 °C within 1 h, recompressing at 70 MPa (compressing time 2 min per 2.5 mm wall thickness), cooling at a rate of less than 10 K/h, demolding at ≈ 150 °C.

Post processing operations: Carbide tools are recommended for machining of highly filled resins. Ultrasonic and hot plate welding are possible, while high-frequency welding is not suitable. Adhesion with solvent adhesives is impossible because of PPS's high chemical resistance; however, two-component adhesives based on epoxy or polyurethane resins yield good bonds. Before lacquering, the surfaces have to be pre-treated either by flame or plasma treatment or a primer based on PUR has to be used.

Properties

Thermoplastic PPS is only little branched and therefore highly crystalline. The degree of crystallinity depends on the temperature-time history during processing. The long-term service temperature ranges from 200–240 °C, the maximum service temperature is at 300 °C. There are no known solvents below 200 °C; however, at elevated temperatures PPS can be swelled. PPS is resistant to alkalis and non-oxidizing acids, with the exception of hydrochloric acid. It is attacked by oxidizing agents such as nitric acid. Light attacks the surface of non-stabilized and non-pigmented grades. Gas permeability is higher than for other semi-crystalline thermoplastics. PPS is flame retardant by nature and with a wall thickness of 0.4 mm it is rated V-0 by UL 94.

PPS is a brittle thermoplastic and therefore rarely used as an unfilled resin for injection molding. Applications include films produced by flat sheet die extrusion and fibers. For injection molding, compounds with fibrous reinforcements, mineral fillers, or combinations thereof are preferred. The main reinforcing agents are glass- but also carbon- and aramid fibers; common fillers are calcium carbonate, calcium sulfate, kaolin, mica, talc, and silica. Filler loads up to 70 wt.-% are possible. The mechanical properties are strongly influenced by the type and load of filler/reinforcement, see Table 6.32. Blends with fluoropolymers improve the tribological properties. Conductive compounds are available.

Comparison of properties, see Table 6.32.

Applications

Micro-precision injection molding, encapsulation of chips and other electronic devices, lamp- and headlight sockets, pump housings and other parts, structural foam parts, films, CF-, AF-, GF-prepregs.

Their high heat distortion and chemical resistance makes PPS suitable for high-performance composite materials, with continuous reinforcing fibers (mainly glass, carbon, and aramid). Various coupling agents and impregnation techniques have been developed.

Table 6.32 Property Comparison of Polyaryl Sulfides and -Sulfones

Property	Unit	Polyaryl Sulfides and -Sulfones					
		PPS -GF 40	PES		PSU		(PSU + ABS)
			Unreinforced	-GF 30	Unreinforced	-GF 30	
Density	g/cm^3	1.60–1.67	1.36–1.37	1.58–1.6	1.24–1.25	1.44–1.45	1.13
Tensile modulus of elasticity	MPa	13,000–19,000	2600–2800	9000–11,000	2500–2700	7500–9500	2100
Yield stress	MPa	–	75–80	–	90	–	50
Elongation at yield	%	–	5–6	–	6–7	–	4
Nominal elongation at break	%	–	20–50	–	20– > 50	–	> 50
Stress at 50% strain	MPa	–	–	–	–	–	–
Failure stress	MPa	165–200	–	125–150	–	110–125	–
Failure strain	%	0.9–1.8	–	1.9–3	–	2–3	–
Melting temperature	°C	275–290	–	–	–	–	–
Heat deflection temperature HDT/A 1.8 MPa	°C	ca. 260	200–205	210–225	170–175	185	150
Thermal coefficient of linear expansion, longitudinally (23–55 °C)	10^{-5}/K	1.5–2.5	5–5.5	2–3	5.5–6	2	6.5
Thermal coefficient of linear expansion, across (23–55 °C)	10^{-5}/K	3.5–5	–	4–4.5	–	–	–
Flammability UL 94 with 1.6 mm thickness	Class	V-0	V-0	V-0	V-2/HB[1)]	V-0/V-1	HB[1)]
Dielectric constant at 100 Hz	–	3.9–4.8	3.5–3.7	3.9–4.2	3.2	3.5–3.7	3.1–3.3

Table 6.32 Property Comparison of Polyaryl Sulfides and -Sulfones *(continued)*

Property	Unit	Polyaryl Sulfides and -Sulfones					
		PPS -GF 40	PES		PSU		(PSU + ABS)
			Unreinforced	-GF 30	Unreinforced	-GF 30	
Dielectric loss factor at 100 Hz	$\cdot 10^{-4}$	10–20	10–20	20–30	8–10	10–20	40–50
Specific volume resistivity	Ohm · m	$>10^{13}$	$>10^{13}$	$>10^{13}$	$>10^{13}$	$>10^{13}$	$>10^{13}$
Specific surface resistivity	Ohm	$>10^{14}$	$>10^{13}$	$>10^{13}$	$>10^{15}$	$>10^{15}$	$>10^{14}$
Dielectric strength	kV/mm	20–30	20–30	20–30	20–30	30–35	20–30
Comparative tracking index CTI/A		125–150	100–150	125–175	125–150	150–175	175
Water absorption at 23 °C, saturation	%	< 0.1	1.9–2.3	1.5	0.6–0.8	0.4–0.5	0.3
Moisture absorption 23 °C/50% rel. humidity, saturation	%	< 0.05	0.6–0.8	0.6	0.25–0.3	0.15–0.2	0.1

[1)] also available as V-0

6.10.2 Polyaryl Ether Sulfones (PAES, PSU, PSU+ABS, PES, PPSU)

Chemical Composition

Polysulfones are polycondensates containing diaryl sulfone groups as their characteristic molecule groups in their chains, see Equation 6.22.

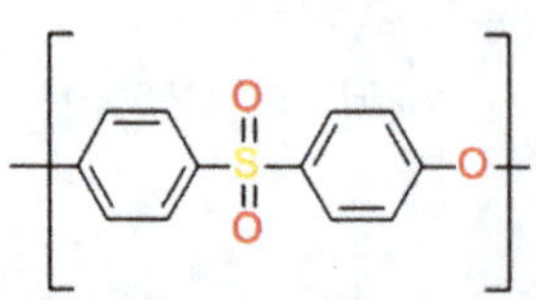

(6.22)

They are formed by the reaction of 4,4'-dihalodiphenyl sulfone with a bifunctional phenol. Varying the bisphenol component results in material grades with differing materials properties.

The main commercial products are obtained by reaction of 4,4'-dichlorodiphenyl sulfone (DCDPS) with bisphenol A, forming polysulfone (PSU), of 4,4'-dihydroxydiphenyl sulfone, forming polyether sulfone (PESU), and of 4,4'-dihydroxybiphenyl, forming polyphenyl sulfone (PPSU), see Equation 6.23. The use of hydroquinone creates polyether ether sulfone (PEESU). Copolymers can also be created by using phenolic units in the mixture.

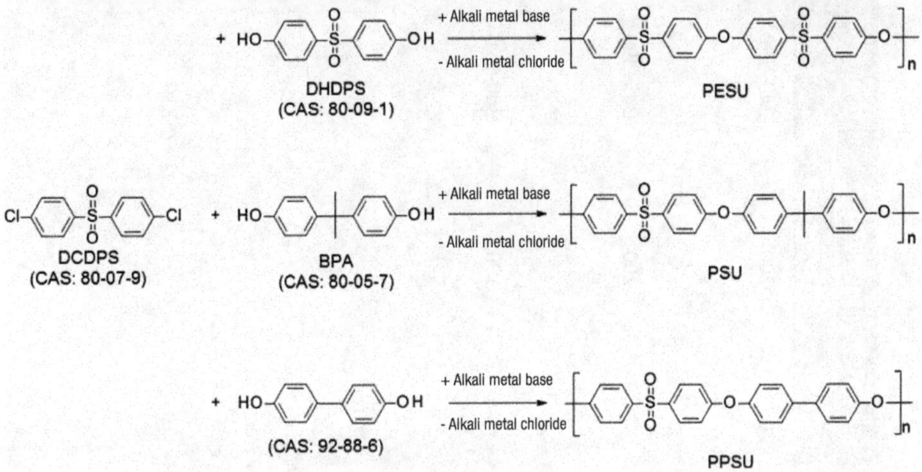

(6.23)

Chemical notations and acronyms, basic structures, service and processing temperatures are provided in Table 6.33.

Processing

PAES can be processed like any other thermoplastic material. *Injection molding* is the most common processing method. Pre-drying of the resin at 150–260 °C for 3–6 h is necessary. Because the melt is highly viscous, melt and mold temperatures have to be high during processing. This reduces molecular orientation and internal stresses, which may lead to media-induced stress crack formation. Heated tool and ultrasonic welding are used to *join* PSU parts; 2-component adhesives *(e. g.*, epoxy resins, polyurethanes, and phenolic resins) are also used. RF-welding is not suitable because of the high dielectric constant of the materials.

Properties

PAES are amorphous thermoplastics exhibiting a high degree of transparency when unfilled. Service temperatures (see Table 6.33) are higher than for common engineering plastics. PSU exhibits high elongation at break and high stiffness and strength; however, notch sensitivity is also high. Wear and slip properties are good and can be even improved by the addition of PTFE or graphite. PAES exhibit high chemical resistance *(e. g.*, hydrolysis, acidic or alkaline media); however, the material is susceptible

Table 6.33 Polysulfones and Polysulfone Modifications

Acronym	Structure	Service temperature °C		Melt temperature °C	Mold temperature °C
		Short-term	Long-term		
PES	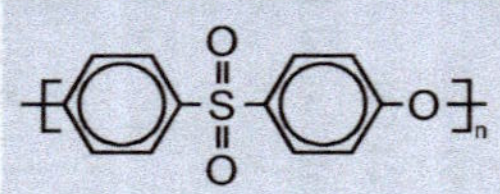	220	200	340–390	Up to 170
PES	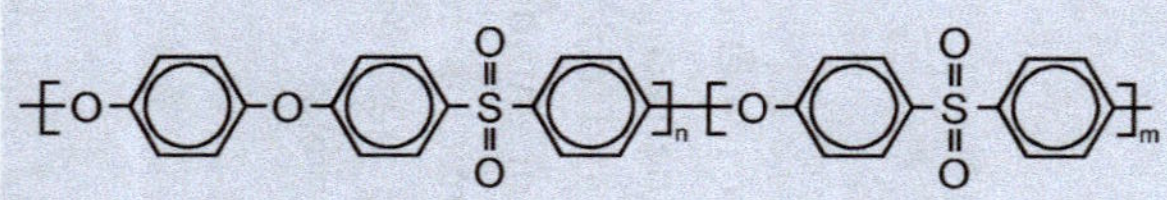	–	200	345–390	140–165
PPSU	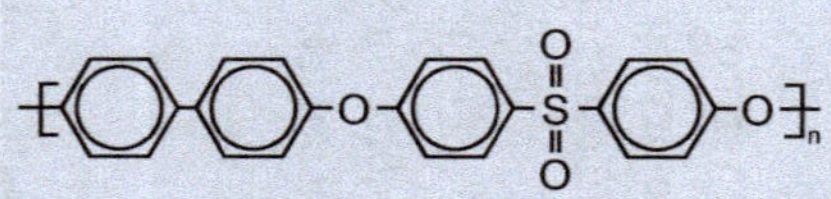	–	190	360–390	140–165
PSU	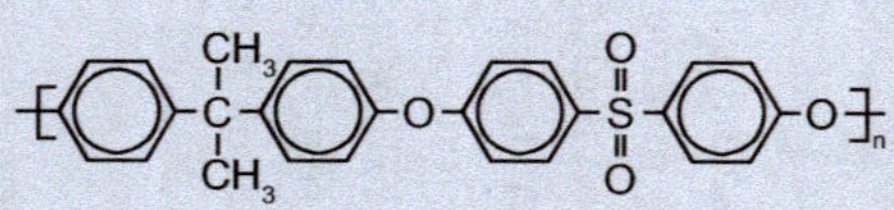	180	160	330–380	120–160
PSU + ABS		HDT-A = 149		280–310	70–120
PES + PA–GF 30				310–330	80–140

to stress cracking. For outdoor use, surfaces should be protected against UV. It is resistant to high-energy radiation. Depending on grade, PSU is classified as HB according to UL 94, while PESU and PPSU reach V-0 classification with 1.6 mm thickness (without flame retardant additives).

Typical variations of the available grades include: varying molecular weights (*e.g.*, increased melt viscosity for extrusion or membrane manufacture), reinforced grades with glass or carbon fiber contents of 10–30% to increase stiffness and strength, and grades filled with minerals to increase heat distortion temperature and to decrease stress cracking.

Comparison of properties, see Table 6.32 and Table 6.33.

Applications

Connectors, coil cores, insulators, capacitors, brush holders, alkaline batteries, printed circuit boards, temperature-stressed parts for household machines and devices, medical devices and dairy equipment, filtration membrane (micro and ultrafiltration, dialysis), lenses, spot lights, reflectors, lamp holders, protective coatings for metals.

6.11 Aromatic Polyethers, Polyphenylene Ethers, and Blends (PPE)

Chemical Composition

Modified polyphenylene ethers (or oxides, PPE, PPO), see Equation 6.24, have gained significant technical and economic importance because of their unlimited compoundability with thermoplastics such as PS, PA, or PBT. The resulting resins are manufactured below 320 °C, are competitively priced, with good impact resistance at low temperatures, and service temperatures > 100 °C. Pure PPEs have no practical industrial applications.

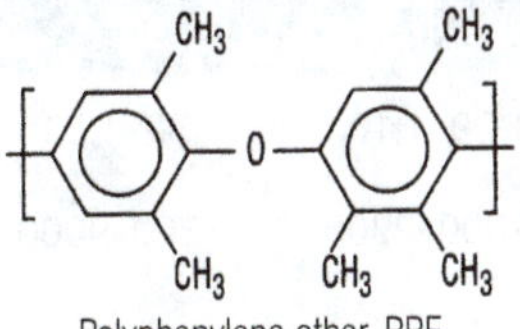

(6.24)

Polyphenylene ether, PPE

Blends of *polyphenylene ether* and impact resistant *PS* at a ratio of 1 : 1 have gained economic importance. These blends exhibit better oxidation resistance than pure PPE, which tends to rapid oxidative degradation at temperatures above 100 °C. Processability is improved; the melt temperature ranges between 260 and 300 °C. Service temperatures, however, range much lower: short-term from 120–130 °C; long-term from 100–110 °C. The styrene content causes higher stress cracking susceptibility.

Laminating or coextrusion of highly viscous (also reinforced) PPE grades with sulfur cross-linkable rubber blends and subsequent hot vulcanization creates plastic-rubber composites with a composite strength exceeding the tear strength of the vulcanized rubber.

Blends or alloys of amorphous PPE with semi-crystalline polycondensates offer optimized property profiles with better solvent- and stress cracking resistance than the amorphous and lower shrinkage and warpage than the semi-crystalline part.

Heat distortion stability up to 210 °C with excellent impact resistance at low temperatures can be reached with both reinforced and unreinforced blends based on PPE + PA 66. These grades are used for in-line paintable automotive parts, as well as for oil- and fuel resistant parts under the hood. PPE + PBT alloys absorb less water but are also less temperature resistant.

The Radlite process is used to manufacture laminar glass-mat reinforced non-wovens from PPE/PA powder in aqueous foam emulsion for large surface parts.

Foamable blends of PS-E (expandable PS) with PPE can be processed into foams with an overall density of 25–250 kg/m^3 and respective flexural strengths of 0.5–8.8 MPa (particle foams). Their heat distortion point ranges from 104–118 °C (PS-E: 95 °C).

Low-molecular PPE is easy to process and has good adhesivity and compatibility with SB copolymers and aromatic and cycloaliphatic EP resins. It is used as an additive for SB and EP in adhesives and sealant compounds as well as for coatings and surface clays with increased heat distortion point; it also improves the melt flow rate of amorphous thermoplastics such as PEI at low loads of 5 wt.-%.

Comparison of properties, see Table 6.34.

Table 6.34 Comparison of Properties for Modified Polyphenylene Ethers

Property	Unit	Modified polyphenylene ether			
		(PPE + SB)		(PPE + PA 66)	
		Unreinforced	GF 30	Unreinforced	GF 30
Density	g/cm^3		1.26–1.29	1.09–1.10	1.32–1.33
Tensile modulus of elasticity	MPa	1900–2700	8000–9000	2000–2200	8300–9000
Yield stress	MPa	45–65	–	50–60	–
Elongation at yield	%	3–7	–	5	–
Nominal elongation at break	%	20– > 50	–	> 50	–

Property	Unit	Modified polyphenylene ether			
		(PPE + SB)		(PPE + PA 66)	
		Unreinforced	GF 30	Unreinforced	GF 30
Stress at 50% strain	MPa	–	–	–	–
Failure stress	MPa	–	100–120	–	135–160
Failure strain	%	–	2–3	–	2–3
Melting temperature	°C	–	–		
Heat deflection temperature HDT/A 1.8 MPa	°C	100–130	135–140	100–110	200–220
Thermal coefficient of linear expansion, longitudinally (23–55 °C)	10^{-5}/K	6.0–7.5	3	8–11	2–3
Thermal coefficient of linear expansion, across (23–55 °C)	10^{-5}/K	–		–	
Flammability UL 94 for 1.6 mm thickness	Class	HB[1)]	HB[1)]	HB[1)]	HB[1)]
Dielectric constant at 100 Hz	–	2.6–2.8	2.8–3.2	3.1–3.4	3.6
Dielectric loss factor at 100 Hz	10^{-4}	5–15	10–20	450	420
Specific volume resistivity	Ohm · m	$>10^{14}$	$>10^{14}$	$>10^{11}$	$>10^{11}$
Specific surface resistivity	Ohm	$>10^{14}$	$>10^{14}$	$>10^{12}$	$>10^{12}$
Dielectric strength	kV/mm	35–40	45	95	65
Comparative tracking index CTI/A		200–400	175–250	600	300
Water absorption at 23 °C, saturation	%	0.15–0.3	0.5	3.4–3.5	2.7–3.6
Moisture absorption 23 °C/50% rel. humidity, saturation	%	< 0.1	< 0.1	1.1–1.2	0.8–1.2

[1)] also available as V-1 to V-0

Applications

PPE+PS-HI: Automotive instrument panels and interior lining, hub caps, radiator grill, housings, large parts for office machines, TV- and electric devices in structural foam, appliances parts, as masterbatch to improve thermal and mechanical properties of PS, halogen-free flame retardants.

PPE + PA: Instrument panels and vehicle front ends, automotive body parts.

PPE + PS-E: Energy absorbers in vehicles.

6.12 Aliphatic Polyesters (Polyglycols) (PEOX, PPOX, PTHF)

Depending on molecular weight (PEOX: 200–20,000 g/mol; PPOX: 400–2000 g/mol), polyethylene(propylene) oxides are liquid, waxy to rigid polymers. Basic chemical structure, see Equation 6.25.

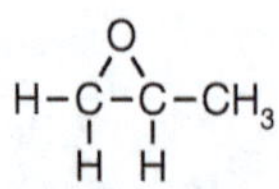

Propylene oxide

(6.25)

PEOX is soluble in water and many other solvents, colorless, odorless, and non-toxic; it is used in pharmaceutical and cosmetic products, such as skin and hair tonics, lipsticks, ointments, lubricants, anti-blocking agents.

PPOX is liquid even at higher molecular weights, in contrast to PEOX; it is used as lubricant additive and as brake- and hydraulic fluid.

PTHF (polytetrahydrofuran) is produced by cationic polymerization of tetrahydrofuran. PTHF are strictly linear polyether diols with molecular weights from 650–3000 g/mol.

6.13 Poly(aryl)ether Ketones and Aliphatic Polyketones

6.13.1 Poly(aryl)ether Ketones (Aromatic Polyether Ketones) (PAEK; PEK; PEEK; PEKEEK; PEKK)

Björn Fink

Chemical Composition

Poly(aryl)ether ketones (PAEK) contain ether(-oxygen) bridges and keto-groups (C=O) and 1,4-phenylene units. The simplest polyether ketone is depicted in Equation 6.26. PAEKs are semi-crystalline polymers whose melt temperature depends on the ratio of keto-groups, see Table 6.35. Their acronyms reflect the number and configuration of the various groups, *e. g.*, two ether- and one keto-group: polyether ether ketone, PEEK.

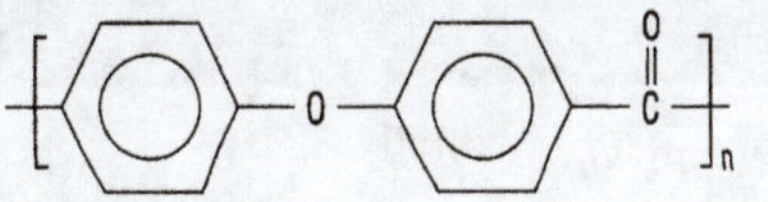

(6.26)

Polyether ketone: PEK

Processing

PAEKs are thermoplastics and can be processed by all thermoplastic processing methods. The high processing temperatures require pre-drying at 150–200 °C for at least 3 h to eliminate residual moisture. Melt temperatures for injection molding or extrusion typically range between 350–420 °C. Injection mold wall temperatures should range between 150–190 °C and thus significantly above glass transition. The mold temperature significantly influences the degree of crystallinity and thus all properties that are determined by it, including mechanical properties and chemical resistance. Special easy-flowing grades are available for thin-walled injection molded parts. Laser sintering is used to manufacture very complex components.

PAEK semi-finished products can be further processed using a variety of methods. For example, PAEK films are thermoformed and extruded products are machined. Welding and adhesive bonding are common joining methods.

Special fine powders are used to coat metal substrates. Electrostatic spray and dispersion coatings are established technologies. Post-coating, the substrate has to be heated to 400 °C. At this temperature, PAEK will melt and form a uniform coating. The substrate has to be able to withstand these temperatures.

Properties

PAEKs retain their good properties over a wide temperature range. Table 6.35 shows the crystallite melting point and glass transition temperature of selected PAEKs. The relative thermal index for a non-reinforced PEEK is at 260 °C.

Table 6.35 Thermal Properties of Selected Poly(aryl)ether Ketones

Material	Ketone group content %	Melting temperature °C	Glass temperature °C
PEEEK	25	324	129
PEEK	33	335	141
$P(E)_{0.625}(K)_{0.375}$	37.5	337	144
PEEKEK	40	345	148
PEK	50	365	152

Table 6.35 Thermal Properties of Selected Poly(aryl)ether Ketones *(continued)*

Material	Ketone group content %	Melting temperature °C	Glass temperature °C
PEEKK	50	365	150
$P(E)_{0.43}(K)_{0.57}$	57	374	157
PEKEKK	60	384	160
PEKK	67	391	165

The heat deflection temperature for non-reinforced PAEK grades is just above the glass transition. Reinforcing fibers considerably contribute to increasing the mechanical properties and thus increasing the heat deflection temperature – close to the melting point (Table 6.36).

Table 6.36 Thermal Properties

Property	Unit	PEEK	PEK	PEKEKK	PEEK – GF30	PEEK – CF30
Spec. heat capacity	kJ/kg · K	2.2	2.2	–	1.7	1.8
Average coefficient of thermal expansion below T_g	10^{-6}/K	55	55	55	45	40
Average coefficient of thermal expansion above	10^{-6}/K	140	130	125	110	100
Thermal conductivity	W/m · K	0.29	–	–	0.30	0.95
Heat distortion temperature HDT/1.8 MPa	°C	152	163	172	328	336
Continuous operating temperature (electric)	°C	260	–	–	240	–
Continuous operating temperature (mechanic static)	°C	240	–	–	240	240

PAEKs exhibit very good mechanical properties even at high temperatures. Unreinforced PAEKs exhibit ductile behavior below T_g, which facilitates applications even at very low temperatures. They also exhibit low creep and high fatigue strength. Table 6.37 and Table 6.38 show mechanical properties of selected PAEKs.

Table 6.37 Comparison of Mechanical Properties for Selected PAEKs

Property	Temperature	Unit	PEEK	PEK	PEKEKK	PEEK – GF30	PEEK – CF30
Tensile strength	23 °C	MPa	100	115	115	180	260
Tensile modulus of elasticity	23 °C	GPa	3.7	3.7	4.3	11.8	25
Strain at failure	23 °C	%	45	20	20	2.7	1.7
Flexural strength	23 °C	MPa	165	185	180	270	380
	125 °C	MPa	85	110	110	190	250
	175 °C	MPa	18	32	36	81	117
	275 °C	MPa	13	16	21	50	60
Flexural modulus	23 °C	GPa	4.1	4.2	4.1	11.3	23
Impact strength Charpy – unnotched	23 °C	kJ m^{-2}	No failure	No failure	No failure	55	45
Impact strength Charpy – notched	23 °C	kJ m^{-2}	7.0	3.8	4.0	8.0	7.0

Table 6.38 Comparison of Mechanical Properties of Selected Reinforced PAEKs

Property	Unit	Range
Uni-directional		
Fiber volume content	%	40–65
Density	g/cm^3	1.51–1.63
Tensile strength	MPa	1570–2600
Young's modulus	GPa	78–155
Strain at failure	%	1.8–2.0
Quasi-isotropic, compression molded fabric 0°/90°/±45°, fiber volume content: 60%		
Tensile strength at 23–200 °C	MPa	1100–700

PAEKs exhibit good electrical properties over a wide temperature and frequency range. The high-purity PAEKs exhibit low outgassing characteristics and are suitable for food contact. Special grades are rated bio-compatible according to USP class VI.

Because of their excellent resistance to hydrolysis, medical instruments from PAEK can be autoclave sterilized repeatedly without appreciable loss of properties.

Chemically, PAEKs are largely resistant to non-oxidizing acids, alkalis, oils, fats, brake fluids, and other media encountered in automotive and aviation applications.

Without flame retardant additives, PAEKs achieve V-0 classification according to UL 94 and have an extremely low smoke density in comparison to other thermoplastics.

Good tribological properties, such as low wear and an almost constant coefficient of friction over a wide temperature range, can be further improved by the addition of sliding additives such as PTFE and graphite.

Studies regarding the permeability of different gases showed excellent barrier properties of PAEK films (Table 6.39).

As for most polymers with aromatic structures, PAEK's UV resistance is limited.

Table 6.39 Permeability of Semi-Crystalline PEEK Films; 100 μm Thickness at 1 bar Pressure

Gas	Unit	Permeation rate
Carbon dioxide	cm^3 m^{-2} per day	424
Helium	cm^3 m^{-2} per day	1572
Hydrogen	cm^3 m^{-2} per day	1431
Methane	cm^3 m^{-2} per day	8
Nitrogen	cm^3 m^{-2} per day	15
Oxygen	cm^3 m^{-2} per day	76
Water vapor	cm^3 m^{-2} per day	4

Applications

Applications of PAEK are defined mainly by the good property spectrum over a wide temperature range. Demanding applications in almost all sectors of the industry have been established because of the balanced property profile. The following components are examples of parts developed from PAEK for various industries:

Mechanical engineering: Valves and valve seats, pipes, thrust washers, gaskets, gears, pumps, guides for oven grates, filling pipes, conveyor elements, bearings, coatings for tools (for the production of golf ball cores, shoe soles), coatings for cook ware (rice cookers, frying pans), coatings for conveyor belts, tapes.

Electrical: Cable insulation, sensor housings, applications in semiconductor manufacturing (such as wafer carriers, CMP rings and test sockets), deep-drawn films for loudspeaker cones.

Medical technology: Instruments for (minimally invasive) surgery (*e.g.* bipolar forceps, instrument handles) and dental equipment, compressor seals, compressor plate, spacers, gears for gear pumps, impellers, dental fixation devices. Special grades used with respective certification as materials for permanent implants or screws for fixation.

Transportation: Wheel covers for aircraft, cable ties, cable holders, tank lids, tank vents, door handles, impellers, brackets for wires and cables, pipes, insulation films, plungers, pistons, valve seats, bushings, wing caps, wings, rotors, thrust washers, cable drums, fuel lines, bushings, lamp bases, worm gears, bearings, gears in air conditioners, seat adjustments and sealing and sliding elements, window lifters, actuators.

6.13.2 Aliphatic Polyketones (PK)

6.13.2.1 Chemical Constitution

PK is synthesized using a special catalyst system consisting of carbon monoxide, ethylene, and propylene; see Equation 6.27. Ethylene and propylene molecules are statistically distributed in the chain. PK is semicrystalline with a degree of crystallization in the range of approximately 30–40%.

$$(-CH_2-\underset{\displaystyle R}{\underset{|}{CH}}-\overset{\displaystyle O}{\overset{\|}{C}}-)_n \qquad (6.27)$$

$R = H \text{ or } CH_3$

6.13.2.2 Processing

PK absorbs approximately 0.5% of moisture in air and 2.3% in water and, therefore, should be pre-dried at 60 °C for 4 h to avoid surface defects. Its primary processing method is injection molding; melt temperatures: 240–270 °C, mold temperatures: 60–80 °C (20–120 °C). PK crystallizes quickly, allowing for short cycle times; post-tempering is not necessary; processing shrinkage: 1.8–2.2%, and little or no post-shrinkage. Film, sheet, and profile extrusion, blow molding, rotational casting, and coatings are possible processing methods.

6.13.2.3 Properties and Applications

PK exhibits high impact resistance and elongation at break (25%), short-term service temperatures of 180 °C, and is resistant to aliphatic hydrocarbons, saline solutions, weak acids and bases as well as to media used in automotive applications. Its wear resistance is higher than that of PA or POM, in particular as friction partner for other plastics. Fire-retarded grades reach V-0 at thicknesses of 3.2 mm and 1.6 mm. For a comparison of properties, see Table 6.40.

Table 6.40 Properties Comparison of Aliphatic Polyketones

Property	Unit	PK	PK-GF 30
Density	g/cm^3	1.24	1.46
Tensile modulus of elasticity	GPa	1.4	7.3
Yield stress	MPa	60	–
Failure stress	MPa	–	1.20
Elongation at break	%	–	3
Melting temperature	°C	220	220
Heat deflection temperature, HDT/A 1.8 MPa	°C	100	215
Flammability according to UL 94 at 1.6 mm thickness	Class	HB	HB
Dielectric loss factor at 100 Hz	10^{-4}	250	500
Specific volume resistivity	Ohm · m	10^{11}	10^{11}
Specific surface resistivity	Ohm	10^{14}	10^{14}
Dielectric strength	kV/mm	18 (d = 1.6 mm)	24 (d = 1.6 mm)
Comparative tracking index, CTI/A	–	600	600
Water absorption at 23 °C, saturation	%	2.2	–
Moisture absorption at 23 °C/50% rel. humidity, saturation	%	0.5	0.3

Applications: fuel systems in automotives, pipe systems, and electrical and electronic applications (flame-retarded, halogen- and phosphorus-free grades).

6.14 Aromatic Polyimides (PI)

Chemical Composition

Polyimides exhibit the highest heat distortion point. They contain a characteristic imide-group, see Equation 6.28.

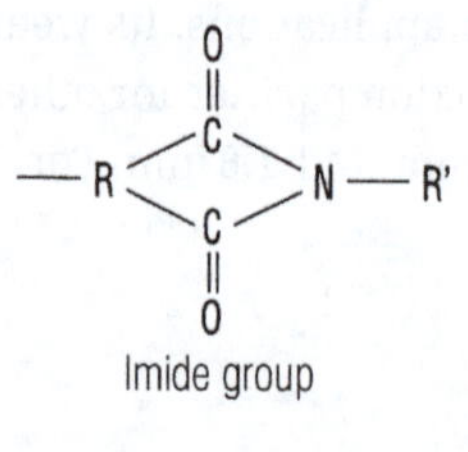

(6.28)

The polycondensation of aromatic diamines with aromatic dianhydrides creates fusible products in an intermediate stage that are transformed into an insoluble and no longer fusible state by heating. The polycondensation products are available both as thermosets and as thermoplastics. The latter cannot be processed like thermoplastics, and special techniques have to be employed: compression molding or sintering of a powdered intermediary product. Some grades can be injection molded or extruded at high temperatures (≈ 350 °C). Polyaddition products are created from pre-polymers with unsaturated aliphatic end groups, which are saturated by thermal polymerizing groups. These grades exhibit slightly lower heat distortion points than the polycondensates. They are used to manufacture composites in heated presses or autoclaves or are processed by winding. During curing, volatile materials may form. Grades for spraying or coating, *e. g.*, of sheet metal, as well as for adhesive bonding and welding are available.

Table 6.41 and Table 6.42 provide basic structures and properties of selected thermosetting and thermoplastic polyimides.

6.14.1 Thermosetting Polyimides (PI, PBMI, PBI, PBO, and Others)

6.14.1.1 Polyimides (PI)

The production of pure aromatic PI in the form of insoluble and non-melting resins by polycondensation requires special processes because of the need to remove the volatile by-products of the reaction. Resin manufacturers offer solid semi-finished products, precision sintered parts, as well as 7.5–125 mm thick films for electric applications and cable insulation for aeronautical applications that are fatigue resistant at temperatures from −240 to 260 °C in air and up to 315 °C in vacuum or inert atmosphere. The same applies accordingly for very high-molecular polyimides based on carboxylic acid dianhydrides and diisocyanates and polybenzimidazoles (PBI) that are produced under elimination of CO_2. PI shrink film, electrical and/or heat conductive and films with adhesive PFA layers are also available. Similar films made from urea derivatives (parabanic acid) are heat resistant up to 155 °C.

Comparison of properties, see Table 6.42.

Table 6.41 Polyimide and Selected PI Modifications

Basic grade	Acronym	Basic structure	Service temperature °C		Notes
			Short-term	Long-term	
Thermosets					
Polyimide	PI	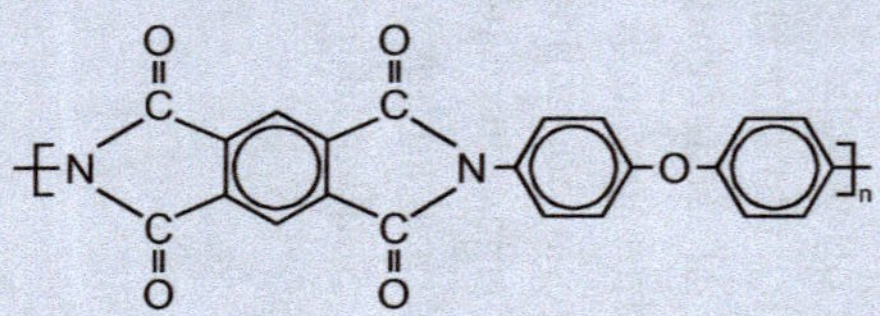	< 400	260	
Poly(bis-maleinimide), (polyamino-bis-maleinim-ide)	PBMI (PAMI)	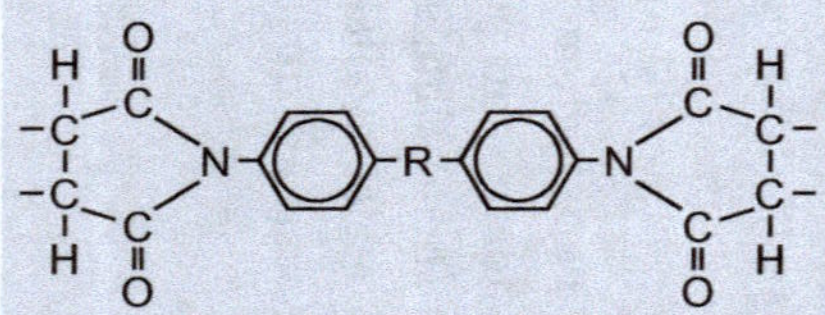	250	190	
Polybenzim-idazole	PBI	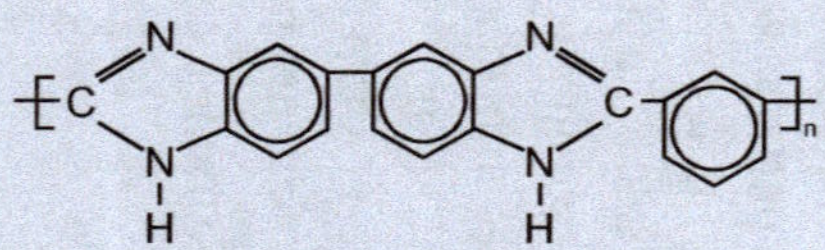	< 500 w/o air < 300 w. air		Semi-conductor polymer
Polyoxadia-zobenzimid-azole	PBO	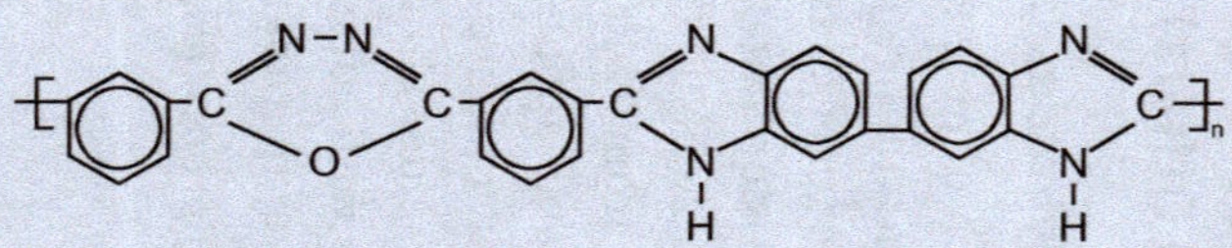	< 500		

Basic grade	Acronym	Basic structure	Service temperature °C		Notes
			Short-term	Long-term	
Thermoplastics					
Polyamide imide	PAI	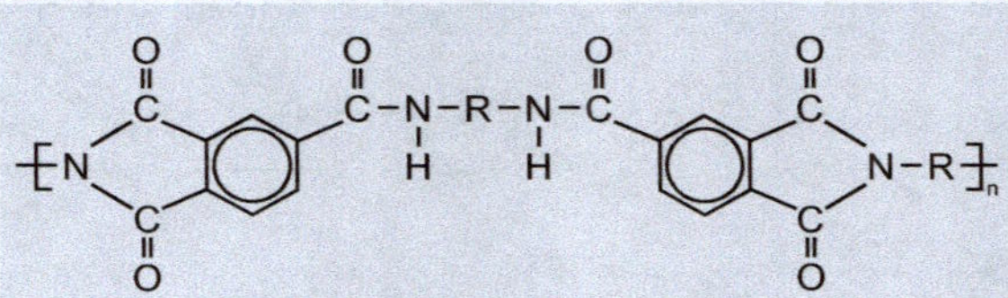	300	260	
Polyether imide	PEI	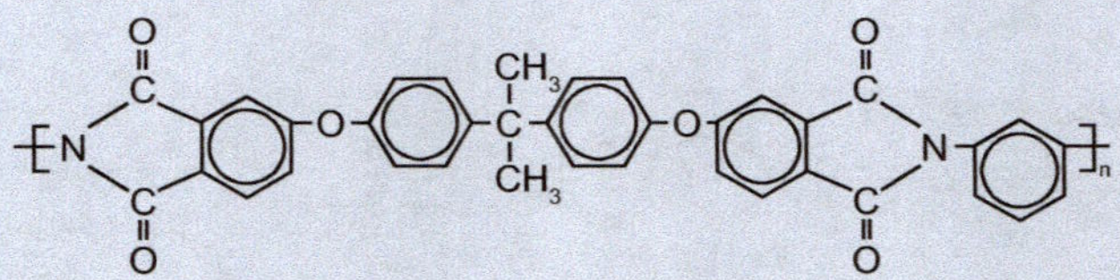	> 200	170	
Polyimide sulfone	PISO	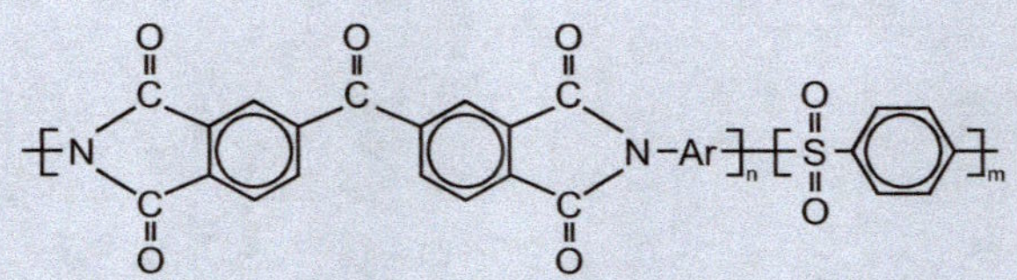 Ar: 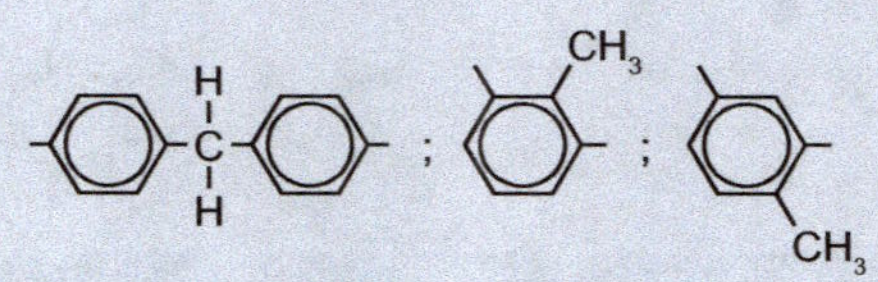	> 250	210	

Table 6.41 Polyimide and Selected PI Modifications *(continued)*

Basic grade	Acronym	Basic structure	Service temperature °C		Notes
			Short-term	Long-term	
Polymethacrylimide	PMI	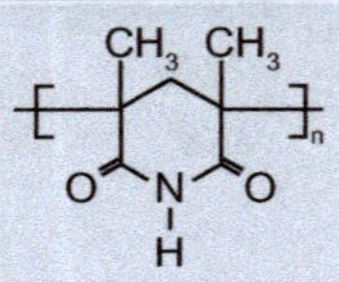		180	Rigid foam
PMI modified	PMMI	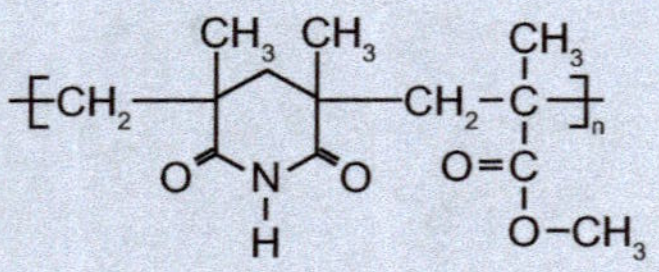		120–150	
Polyester imide	PESI	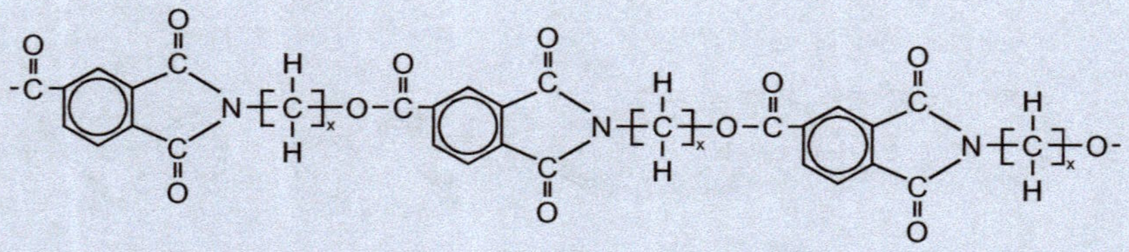		200	Semiconductor polymer
Ladder polymers					
Polyimidazopyrrolone (pyrrone)		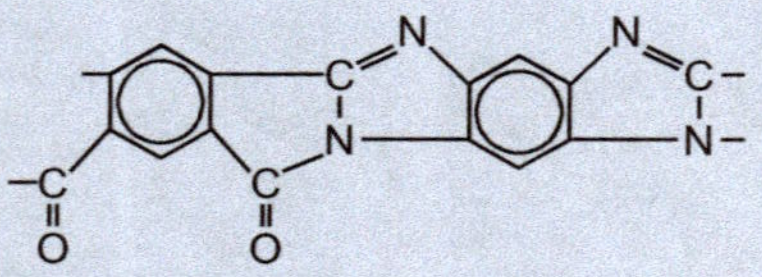			Ladder polymer
Polycyclone		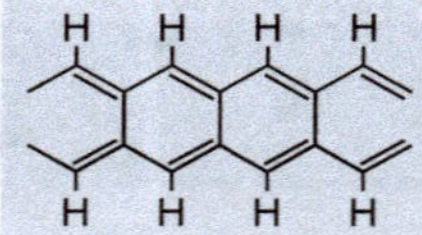	1000 (Fiber)		Ladder polymer

Table 6.42 Comparison of Properties for Thermosetting and Thermoplastic Polyimides

Property	Unit	Thermoset	Polyaryl sulfides and -sulfones				
		PI	PAI			PEI	
			Unreinforced	-GF 30	-CF 30	Unreinforced	-GF 30
Density	g/cm^3	1.43	1.38–1.40	1.59–1.16	1.45–1.50	1.27	1.49–1.51
Tensile modulus of elasticity	MPa	2300	4500–4700	12,000–14,000	24,500	2900–3000	9000–11,000
Yield stress	MPa	210	–	–	–	85–100	–
Elongation at yield	%		–	–	–	6–7	–
Nominal elongation at break	%		–	–	–	30– > 50	–
Stress at 50% strain	MPa		–	–	–	–	–
Failure stress	MPa		150–160	205–220	250	–	150–165
Failure strain	%	88	7–8	2–3	1.2	–	ca. 2
Melting temperature	°C		–	–	–	–	–
Heat deflection temperature HDT/A 1.8 MPa	°C	> 400	275	280	280	190	205–210
Thermal coefficient of linear expansion, longitudinally (23–55 °C)	10^{-5}/K	20	3.0–3.5	1.6	0.9	5.5–6.0	2
Thermal coefficient of linear expansion, across (23–55 °C)	10^{-5}/K		–			–	5

Table 6.42 Comparison of Properties for Thermosetting and Thermoplastic Polyimides *(continued)*

Property	Unit	Thermoset	Polyaryl sulfides and -sulfones				
		PI	PAI			PEI	
			Unreinforced	-GF 30	-CF 30	Unreinforced	-GF 30
Flammability UL 94 for 1.6 mm thickness	Class		V-0	V-0	V-0	V-0	V-0
Dielectric constant at 100 Hz	–		3.5–4.2	4.4	–	3.2–3.5	3.6
Dielectric loss factor at 100 Hz	$\cdot 10^{-4}$		10		–	10–15	15
Specific volume resistivity	Ohm · m		$>10^{15}$	$>10^{15}$		$>10^{15}$	$>10^{14}$
Specific surface resistivity	Ohm	$>10^{15}$	$>10^{16}$	$>10^{16}$		$>10^{16}$	$>10^{15}$
Dielectric strength	kV/mm	200	25	25–35	–	25	25
Comparative tracking index CTI/A					–	150	150
Water absorption at 23 °C, saturation	%	3.5				1.25	0.9
Moisture absorption 23 °C/50% rel. humidity, saturation	%	2.0				ca. 0.5	ca. 0.4

6.14.1.2 Polybismaleinimides (PBMI)

PBMIs are created by condensation with p-diamines and are either self-curing through terminal double bonds or they cure with reaction partners while gradually cross-linking. The formation of volatile products during polycondensation necessitates special measures during processing. Outstanding properties are high heat deflection temperature, radiation resistance, self-extinguishing after removal of the ignition source, chemical resistance, and efficient processability. Transfer molding resins contain 20–50% glass-, graphite-, carbon-, or synthetic fibers or PTFE to improve slip and wear behavior.

Compression molding requires mold temperatures from 190–260 °C, transfer molding requires approx. 190 °C. Injection molding grades can be processed on injection molding machines designed for thermosets with mold temperatures ranging from 220–240 °C and cycle times from 10–20 s/mm wall thickness. Sintering powders are processed in stepwise processes. Annealing for 24 h at 200–250 °C is always required.

Applications: Substitution of metals, *e.g.*, for vacuum pumps, integrated circuits, printed circuits, electric cable connectors, missile nosecones.

6.14.1.3 Polybenzimidazoles (PBI)

PBIs belong to a group of polymers that do not adhere to the ladder principle of the double chain but rather contain additional single chains. They are formed by melt condensation. In vacuum they exhibit a small decrease in weight only above 500 °C, while in regular atmosphere fast degradation starts above 200 °C. Heat resistance increases when hydrogen in the labile NH group is substituted by a phenyl group.

During condensation, PBI adhere very well to metal surfaces and are therefore used as heat resistant metal adhesives. Here, oxidation of PBI at elevated temperatures can be neglected because of the limited exposure of the adhesive layer to environmental influences. PBIs are resistant to all solvents, oils, acids, and bases; they are therefore used as coatings for surface protection in special applications. PBI are used in aerospace and military applications.

6.14.1.4 Polytriazines

Equation 6.29 (R = carbon group) shows the triazine ring, which is resonance-stabilized similar to the benzene ring and therefore high temperature resistant. In industrial practice, materials are often characterized as polytriazine although (with R = O) they are in fact polycyanurates. Successively cross-linkable polymers are formed with a R = bisphenol A remnant and typically modified with bismaleinimide.

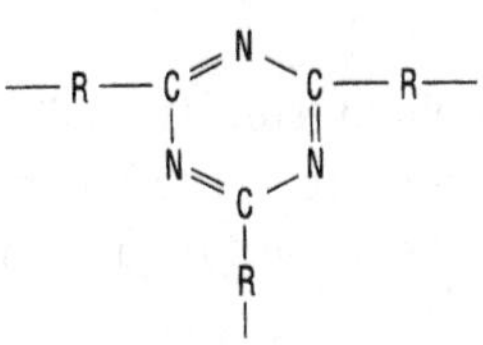

(6.29)

Triazine ring

As pre-polymers they can be melted at 30–130 °C with low viscosity; they are easily processable and, once cured, reach a glass transition temperature of 200–300 °C. Their excellent mechanical, electrical, and dielectric properties lead to applications as copper laminated composites (carrier for printed circuits), other applications in the microelectronic field, insulation for large engines, insulation- and bearing material in the automotive and aeronautical industries.

6.14.2 Thermoplastic Polyimides (PAI, PEI, PISO, PMI, PMMI, PESI, PARI)

6.14.2.1 Polyamide Imides (PAI)

PAI is a special thermoplastic material used primarily in the cooling industry and for aeronautical and aerospace applications. Its characteristics include 220 N/mm^2 tensile strength and 6% elongation at break at −96 °C, 66 N/mm^2 (reinforced up to 137 N/mm^2) tensile strength at 232 °C, and high creep rupture strength. It is resistant to aliphatic, aromatic, chlorinated and fluorinated hydrocarbons, ketones, esters, ethers, high-energy radiation, weak acids and bases. PAI is attacked by water vapor and bases at elevated temperatures. Resins modified with PTFE or graphite are tried and tested for unlubricated bearings with minimal friction coefficients up to 250 °C. For optimum properties in service, parts must be polymerized by curing for several days at 250 °C.

In contrast to polyimides, which are difficult to process, PAI can be injection molded into parts with intricate geometries; PAI can also be extruded. Melt temperatures range from 330–370 °C, mold temperatures from 200–230 °C. Pre-drying for 8–16 h at 150–180 °C is required. To increase wear resistance, it is recommended to cure parts at 245–260 °C for 24 h to 5 days.

Modification with graphite and PTFE are used as bearing materials; glass- and carbon fibers and minerals are used to increase stiffness and strength and to reduce cost.

Processability can be improved by mixing with low-viscous engineering thermoplastics such as PSU, PEI, PA, PPS, and PC.

Comparison of properties, see Table 6.42.

Applications: Components under mechanical or electrical loads up to 260 °C such as impellers for hydraulic or pneumatic pumps, bearings and housings for fuel gages, bearings, solutions in polar solvents for wire coating and as glues, nano-filtration membranes with TiO_2 filler.

6.14.2.2 Polyetherimides (PEI)

Because of its excellent melt flow rate, PEI can be injection molded (runner-less), injection blow molded, extruded, and foamed. With good creep rupture strength under load and at elevated temperatures, good electrical properties, the highest flame resistance of all thermoplastics other than fluoropolymers measured by oxygen index, and reasonably priced, it is suitable for a wide range of applications. PEI is amorphous, amber transparent when unpigmented, and soluble in methylene chloride and trichloroethylene. It is resistant to alcohols, automotive and aeronautical fuels, lubricants and cleaners, even when under load. It is also resistant to acids and weak bases (pH < 9), to hydrolysis by hot water and steam, UV and high-energy radiation.

PEI is typically processed by injection molding at melt temperatures from 375–425 °C and mold temperatures from 100–150 °C. Pre-drying for 7 h at 120 °C or for 4 h at 150 °C is required. For compounds with up to 40% glass fiber content with high melt viscosity the maximum injection pressure of the molding machine should range between 1500 and 2000 bar. PEI grades with release agents exhibit an increase in flow distance of 10%. Grades modified with mineral fillers and/or with PTFE exhibit improved slip and wear properties. Grades reinforced with up to 40% glass or carbon fibers have increased stiffness and strength.

Comparison of properties, see Table 6.42.
Applications: High voltage breaker housings, multi-way connectors, parts for microwaves, solder bath resistant parts, parts for pistons and brake cylinders, carburetor housings, bearings, gear wheels.

Blends of PEI and PEC (polyester carbonate, see Section 6.9.3) are used because of their thermal properties, hydrolytic stability, and stain resistance in applications such as microwave dishes and automotive headlight reflectors.

Blends with free-flowing PPE (≤ 5 wt.-%) improve processability (melt temperature for 30% glass fiber reinforced grades from 360–380 °C to 310–340 °C) without having major impact on mechanical properties.

6.14.2.3 Polyimide Sulfones (PISO)

PISO is transparent, with a glass transition temperature from 250–350 °C, and a service temperature of 208 °C; it is resistant to all common solvents and easier processable than the PI system it is based on. PTFE- and graphite filled injection molding grades based on PISO are available (processing temperature < 370 °C). Unmodified linear PISOs with the structure depicted in Table 6.41 (T_g > 310 °C) no longer flow. Film is cast from dimethylformamide solution.

6.14.2.4 Polymethacrylimides (PMI, Rigid Foams)

In a first processing step, methacrylonitrile and methacrylic acid monomers together with chemical (formamide) or physical blowing agents (isopropanol) are used to cast a polymer. In a second processing step, sheets with densities ranging from 30–300 kg/m^3 are foamed during a thermal process at temperatures between 170 and 220 °C. The imide reaction takes place during the foaming process.

Delivery forms: Sheet of max. 2500 mm × 1250 mm, 1–65 mm thickness, with densities from 30–300 kg/m^3 (not available for all grades).

Properties: Closed-cell, vibration resistant rigid foam with high stiffness and high heat distortion point.

Properties depend on grade and density:

- Modulus of elasticity: 20–380 N/mm^2
- Compressive strength: 0.2–15.7 N/mm^2
- Shear strength: 0.4–7.8 N/mm^2
- Heat distortion point: up to 215 °C
- Long-term service temperature: up to 180 °C

All grades are resistant to most solvents and fuel components; they are not resistant to bases.

Excellent creep behavior, even at elevated temperatures, and therefore suitable for autoclave manufacture of sandwich parts at 180 °C and 0.7 N/mm^2. Also suitable for co-curing processes, which means curing of the complete sandwich structure (prepreg layers and core) in one step, and as thermo-elastically pre-formable sandwich core for in-mold pressing. The latter technique is based on the generation of pressure by the foam core in the closed mold. With standard PMI grades, this pressure is generated by post-foaming; with pre-pressed grades the pressure is created by the expansion of the core. Because the core expansion is constricted in the closed mold, pressure builds and presses the cover layer of the sandwich against the mold surface.

PMI rigid foam can be easily adhesively bonded (both with itself and with common cover layer laminates) by two-component adhesive systems.

Applications: Core material for structural parts built in series for aircrafts (flaps, tabs, stringer), helicopters (rotor blades), in shipbuilding (cores for hulks), automotive (cores for body parts); sandwich cores for high performance bicycle frames and rims; core for cross-country and downhill skis; cores for tennis racks; parts for antennae and satellites, X-ray transmitting stretchers for X-ray labs.

6.14.2.5 Polymethacrylate Methylimides (PMMI)

The insertion of methyl methacrylate in the PMI chain creates a transparent plastic material with certain advantages compared to PMMA: increased heat distortion point (120–150 °C), higher stiffness, strength, chemical resistance, and index of refraction (1.53). The linear coefficient of expansion is lower, light transmission is at 90%. PMMI can be injection molded or extruded at 270–300 °C and exhibits good compatibility when co-processed with PET, PA, PVC, S/AN, ABS, even without the use of compatibilizers.

Applications: Head- and rear lights, instrument covers, optical fibers, sun roofs and side windows in automotives, packaging for food, cosmetics, and medical supplies.

6.14.2.6 Polyesterimides (PESI)

Polyesterimides can be injection molded. The resins are soluble in organic solvents and can be crosslinked to make them suitable for wire coating in the electronic industry. These wires are long-term temperature resistant up to 200 °C.

6.15 Self-Reinforcing Liquid Crystalline Polymers (LCP)

6.15.1 Chemical Constitution

Chemically, LCPs are largely based on the following polymers:

- Polyterephthalates and -isophthalates; LCP-PET, LCP-PBT (see Section 6.9.2).
- Poly(m-phenylene isophthalamides); LCP-PMPI (see Section 6.8).
- Poly(p-phenylene phthalamides); LCP-PPTA (see Section 6.8).
- Polyarylates; LCP-PAR (see Section 6.9.3.1).
- Polyester carbonates; LCP-PEC (see Section 6.9.3).

Research is also being conducted on base materials such as polyazomethines, polyesteramides, and polyesterimides.

Certain materials, polymers among them, form rigid crystalline regions in the liquid state or when in solution. This highly ordered state in solution or in the melt is in stark contrast to the disordered macromolecules of conventional polymers. LCPs form a liquid crystalline state between the liquid and the solid state called "mesomorphous". The rigid crystalline areas in a molecular chain are called "mesogenous". Only after LCPs are heated above the so-called isotropic melting temperature, an isotropic liquid is obtained. An LCP consisting of mesogenous areas only in the main chain cannot be processed like a thermoplastic, because it melts at 400–600 °C, which is above degradation temperature. Selective introduction of disturbances between the mesogenous areas of these molecules results in "thermotropic" main chain LCPs which melt at 220–400 °C. Such disturbances may take the form of flexible $-CH_2$ sequences, built-in angles (*e. g.*, by isophthalic acid), parallel shifts (2,6-hydroxynaphtholic acid, HNA, among others, "crankshaft polymers") or voluminous substitutions. Monomer units creating rigid linear chain segments are suitable for main chain LCPs, see example in Equation 6.30.

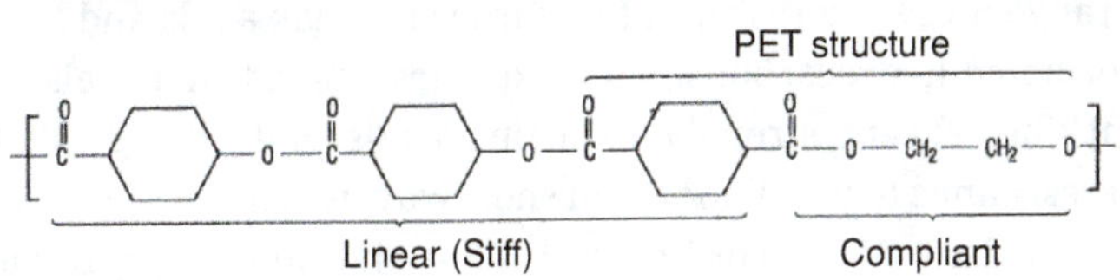

(6.30)

In side chain polymers, the mesogenous areas are located in side chains using flexible spacers to graft them to meltable macromolecules. In the melt, mesogenes can be oriented in the electric field (birefringence), which makes side chain LCPs suitable as "functional polymers" in electro-optical information storage systems.

Polyester-based LCPs are almost exclusively used for injection molding and extrusion and for applications with fillers and reinforcements, such as glass and carbon fibers, mineral, graphite, or PTFE modification. In addition, conductive and metal-coatable grades are available. Liquid crystalline polyamides (so called aramids) are difficult to melt and are spun into fibers directly out of solution.

6.15.2 Processing

LCPs are mostly injection molded. Extrusion, extrusion blow molding, and coextrusion are also used. For injection molding, melt temperatures range from 280–330 °C, mold temperatures range from 70–130 °C. Perfectly closing molds are important to avoid flash-related problems during injection molding. Gating is critical because of the highly oriented state of the melt; weld lines are weak spots. LCP parts (both injection molded and extruded) can be machined by all standard methods; they can be joined by adhesion with adhesives suitable for polyester and by ultrasonics. LCPs suitability for metal coating makes them material of choice for MID (molded interconnect devices) parts and three-dimensional electronic circuit boards.

6.15.3 Properties

LCPs for thermoplastic processing exhibit low viscosity melts so that delicate parts with thin walls can be manufactured. LCPs have a good melt flow rate, flame resistance, dimensional stability at elevated temperatures, chemical resistance, low thermal expansion, and good mechanical properties. During processing, the molecules are oriented in strain and shear flows, resulting in highly anisotropic properties and a so-called "self-reinforcing" effect in the direction of orientation. Most data on toughness and strength (up to 240 MPa), the high modulus of elasticity (40 GPa), and the low coefficient of thermal expansion are only available in the direction of molecular orientation. The long-term heat distortion temperature without mechanical load ranges from 185–250 °C, depending on grade.

In a wide temperature range, LCPs are resistant to hydrolysis, weak acids and bases, alcohols, aromates, chlorinated hydrocarbons, esters, ketones, and all chemicals usually causing stress cracking, except strongly oxidizing acids and strong alkalis. Weather resistance and resistance to γ-radiation and short wavelengths are good. LCP are intrinsically flame retardant (V-0 according to UL 94) and other than for low tracking resistance they exhibit very good electrical properties.

Glass- and carbon fiber reinforced resins exhibit slightly higher strength and stiffness. Mineral fillers reduce the anisotropy of properties in injection molded parts. LCPs are added to other thermoplastics to enhance the melt flow properties (5–30%). On the other hand, addition of low amounts of other thermoplastics is said to decrease the dependency on orientation for LCP properties.

Newer resins (with fewer ions) are offered for electronic applications. They are used to minimize corrosion or short circuits caused by electrically charged particles migrated to the surface in metal coated conductor paths with narrow distances.

LCPs have good barrier properties against gases and water vapor, which is used in coextruded films for packaging applications.

Comparison of properties, see Table 6.43.

6.15.4 Applications

Applications in electronics and electrical industries, connectors, electro-mechanical components, automotive applications, glow plugs, chip card readers, sunlight sensors, couplers for optical fibers, bearings, gaskets, functional elements for pumps, measuring devices, parts in contact with fuels (under the hood), precision parts for aggressive environments, MID, multi-layer films with good barrier properties. Application of LCPs in PEM (polymer-electrolyte-membrane) fuel cells is currently under research.

6.16 Ladder Polymers: Two-Dimensional Polyaromates and -Heterocyclenes

Chemical Constitution

Ladder polymers consist of linear chains, which are cross-linked at constant intervals (see examples in Table 6.41, pyrrone and polycyclone). This composition makes for their extremely high temperature resistance and stiffness.

In principle, the synthesis consists of the connection of monomers with ring structures, which are predominantly aromatic (benzene-) rings or nitrogenous heterocyclenes. During synthesis, bi-functional or multi-functional polymer blocks are created, which will form ladder-like or three-dimensional network structures under further crosslinking. Both multi-stage synthesis and processing are difficult and expensive. Ladder polymers can neither be processed like thermoplastics nor in solution. Shape forming has to take place from the soluble or fusible starting material prior to molecular or ladder crosslinking, respectively.

Table 6.43 Comparison of Properties for Liquid Crystalline Copolyesters

Property	Unit	LCP-PC	LCP-PET		
		Unreinforced	Unreinforced	-GF 30	-CF 30
Density	g/cm³		1.40	1.6	1.5
Tensile modulus of elasticity	MPa	2500–4300	10,400	16,100	23,000
Yield stress	MPa		–	–	–
Elongation at yield	%		–	–	–
Nominal elongation at break	%		–	1.8	–
Stress at 50% strain	MPa		–	–	–
Failure stress	MPa	62–94	156	188	167
Failure strain	%	5.6–3.7	2.6	2.1	1.6
Melting temperature	°C		280	280	280
Heat deflection temperature HDT/A 1.8 MPa	°C	125–135	168	232	240
Thermal coefficient of linear expansion, longitudinally (23–55 °C)	10^{-5}/K		-0.3	-0.1	-0.1
Thermal coefficient of linear expansion, across (23–55 °C)	10^{-5}/K		+6.6	+4.7	+5.2
Flammability UL 94 for 1.6 mm thickness	Class		V-0	V-0	V-0
Dielectric constant at 100 Hz	–		3.2	3.4	–
Dielectric loss factor at 100 Hz	$\cdot 10^{-4}$		160	134	–
Specific volume resistivity	Ohm · m		$>10^{14}$	$>10^{14}$	50–100
Specific surface resistivity	Ohm		$>10^{13}$	$>10^{13}$	10^{4}
Dielectric strength	kV/mm		39	43	–
Comparative tracking index CTI/A			150	125	–
Water absorption at 23 °C, saturation	%		< 0.1	< 0.1	0.1
Moisture absorption 23 °C/50% rel. humidity, saturation	%		< 0.05	< 0.05	0.06

Linear Polyarylenes

Linear or branched polyphenylenes in their final stage are insoluble and non-melting. Their long-term service temperatures – except against molten alkali metals – range from 200–300 °C, under oxygen exclusion up to 400 °C.

Poly-p-Xylylenes (Parylenes)

These are dielectrics for long-term use at temperatures from −160 to 275 °C in form of thin films or coatings. The polymers are polymerized at 600 °C in vacuum from gasified xylene (Equation 6.31) dimer radicals or from chlorinated xylene and precipitated on cold surfaces. Because of their $-CH_2$-group contents they are oxidation resistant only up to approx. 80 °C and resistant to all organic solvents up to 150 °C. Polymonochloro-para-xylylene is a chlorinated variant of poly-p-xylylene; it can be dissolved in chloronaphthalene and can therefore be used as part coating.

Poly-p-hydroxybenzoates

Linear polyaryl esters (see Section 6.9.3) are polymerized from p-hydroxybenzoic acid (Equation 6.31) monomers, which contain only aromatic rings besides the -O-CO- ester groups. The polymer melts at 550 °C and can be used by flame-spraying or, rendered malleable by additives, processed as a bearing material. Copolymers with approx. 40% diphenol terephthalate are self-reinforcing injection molding materials. The long, relatively rigid macromolecules order themselves in the melt to form "thermotropic" liquid crystalline phases. Parts injection molded from this anisotropic melt are by an order of magnitude more rigid and stiffer in flow direction than perpendicular to flow. Lyotropic liquid crystalline solutions of non-melting aramids are used to produce fibers.

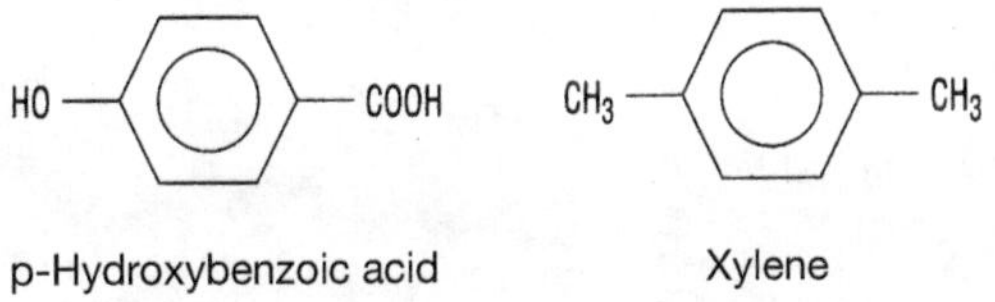

(6.31)

Polyimidazopyrrolone, Pyrrones

Polypyrrones are true ladder polymers and exhibit good resistance to high-energy radiation. They were developed for applications in space.

Polycyclones

Polycyclobutadiene and polycycloacrylonitrile both belong to this group of ladder polymers. They can neither be processed like thermoplastics nor in solution and are used in particular for fibers that exhibit little strength, but are short-term temperature resistant up to 1000 °C. In air, beginning at 200 °C, fast oxidation results in a rapid decrease in strength; therefore, the high heat distortion point can only be utilized if the fibers are either imbedded in a resin matrix or metallized.

Other Ladder Polymers

The synthesis of long-chain ladder polymers with heterocyclic cores, similar to PI, is still an area of increased research. Their configurations and nomenclature is shown in Table 6.44. They are used as impregnation or adhesive resins for printed circuit boards, in microelectronics, and as radiation-crosslinking photoresist coatings for printed circuits on semiconductor chips in the computer industry, which can resist temperatures up to 400 °C and other stresses in subsequent operations. The syntheses also utilize the versatile reactivity of multi-functional isocyanates. The goal is to create "open-chain" prepolymers that are liquid crystalline in concentrated solution or in the melt. They are applied to the board in a pre-ordered fashion, and then can be induced to form insoluble and non-melting end products, or undergo cyclization.

Poly-2,6-diphenyl-phenyleneoxide belongs to the PPE group (see Section 6.11), although it contains only aromatic rings. It has a T_g = 235 °C and a T_m = 480 °C and is spun out of organic solution to form fibers for filter fabrics and for high-frequency cable insulation papers.

Table 6.44 Heterocycles for the Synthesis of Ladder Polymers

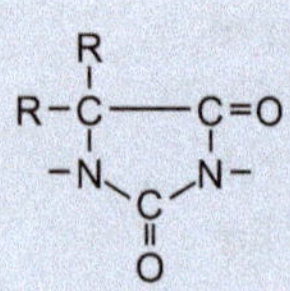	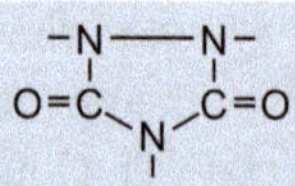	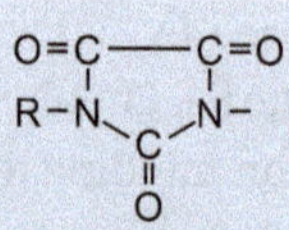	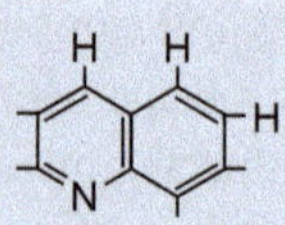	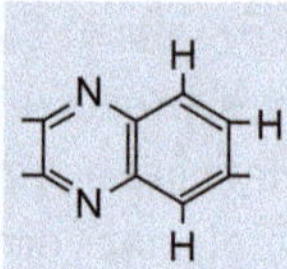	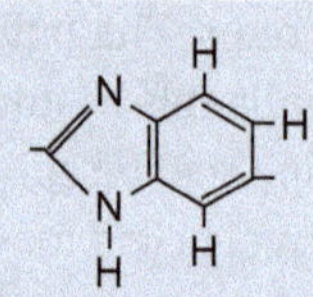
Hydantoin	Urazole	Parabanic acid	Quinoline	Quinoxaline	Benzimidazole

6.17 Polyurethanes (PUR)

6.17.1 Fundamentals

By their chemical constitution, polyurethanes are never "pure" polymers, such as PVC from vinyl chloride or polyethylene from ethylene, but rather chemically-structurally mixed polymers. The urethane group, lending its name for this polymer class, is often only a minor constituent of the macromolecule. Even "PUR" products without any urethane group at all are known. However, all variations have the same "polyisocyanate-chemistry" in common. The property profile of PUR resins is mostly determined by characteristics other than the urethane structural characteristics. This fact constitutes their unique variety among polymers: polyurethanes are primarily known as foams, elastomers, and solid parts. In addition, they play a significant role in technical applications as PUR-coatings and paints, adhesives, plasticizers for other polymers, and as fibers.

6.17.1.1 Chemical Constitution

Isocyanates are characterized by the energy-rich, highly reactive isocyanate group -N=C=O. This group reacts *exothermally* not only with hydrogen-active compounds, but under suitable conditions even with itself. How fast an NCO-group reacts with its partner depends on the partner as well as on the structure of the residual molecule, to which it is bonded. The most important hydrogen-active compounds exhibit OH- or NH_2-functions, they are alcohols or amines. Water plays an important role as an OH-component in the production of foams. During the metathesis with isocyanates, gaseous CO_2 is created which acts as a foaming agent. CO_2 is also created when NCO-groups react with organic acids, which are characterized by a carboxyl group -COOH. In addition, unused or excessive isocyanate in the reaction mixture can lead to *secondary subsequent reactions* with already formed primary reaction products. This will create new chemical structures, which may strongly influence the properties of the manufactured polyurethane.

Isocyanates also react with each other, creating yet additional structural elements in the polymer with the respective effects on the polymer properties: isocyanurate structures (trimerisates of isocyanate) cause a drastic decrease in flammability in rigid foams for the construction industry, which is utilized in particular in rigid polyisocyanurate foams manufactured with pentane blowing agents. The -N=C=O group can also react with *compounds lacking active hydrogen* such as epoxides. This results in products that have gained importance as cast resins. Table 6.45 summarizes the principles of polyisocyanate chemistry.

Table 6.45 Principles of Polyisocyanate Chemistry

R−N=C=O	Isocyanate group
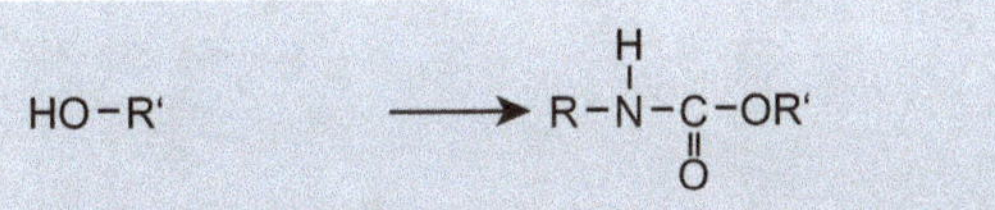	Urethane
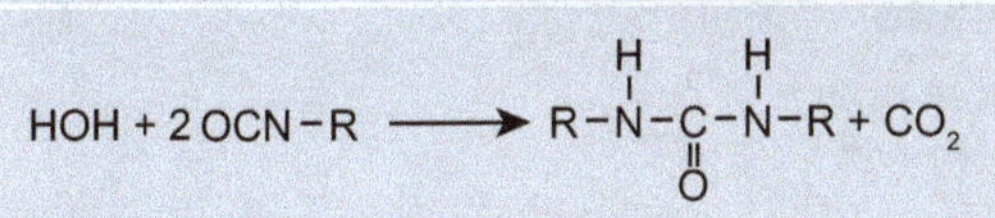	Symmetric urea
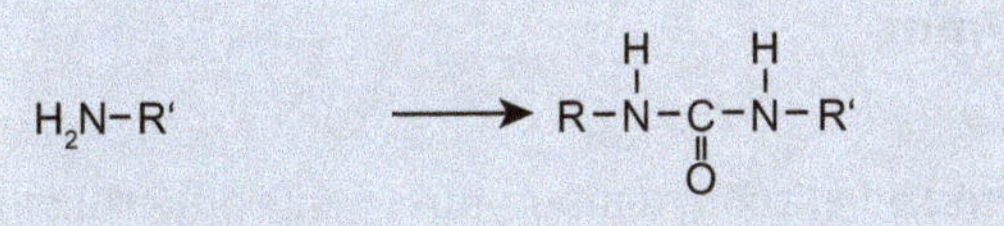	Asymmetric urea
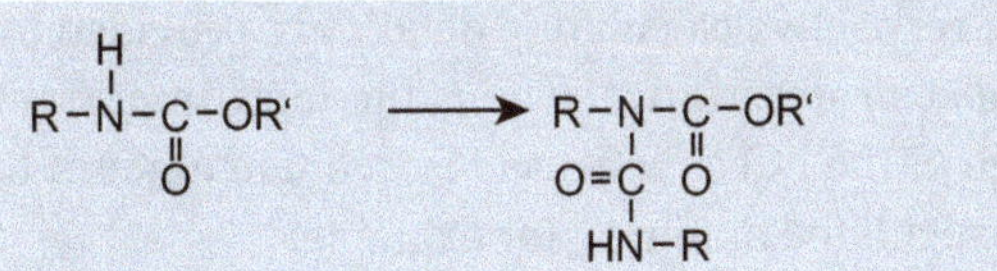	Allophanate

Reaction	Product
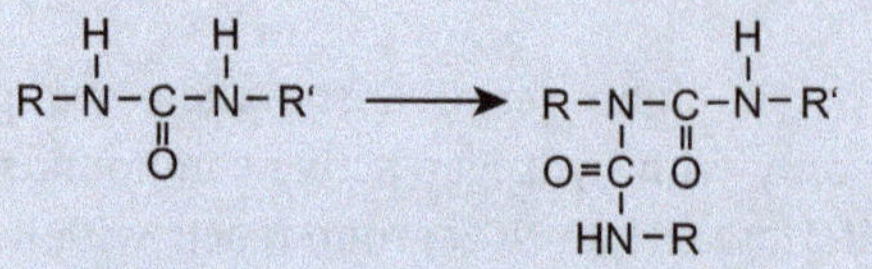	Biuret
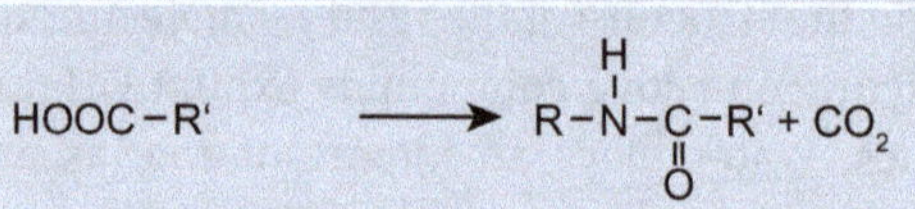	Amide
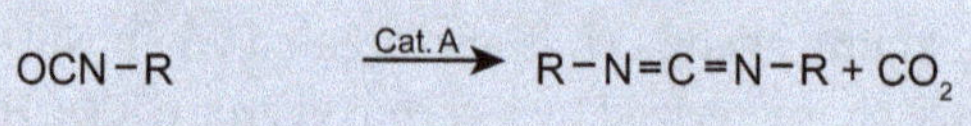	Carbodiimide
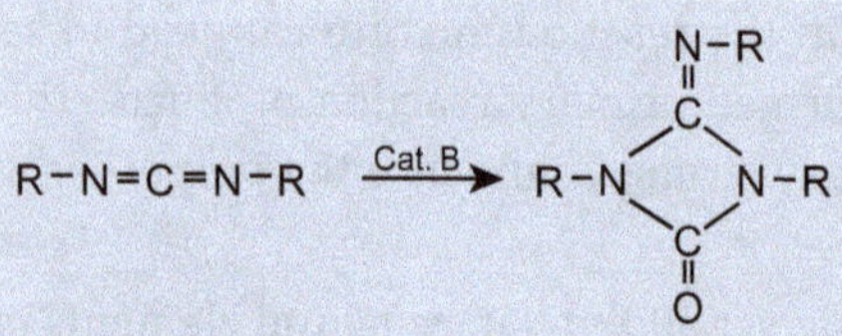	Uretonimine
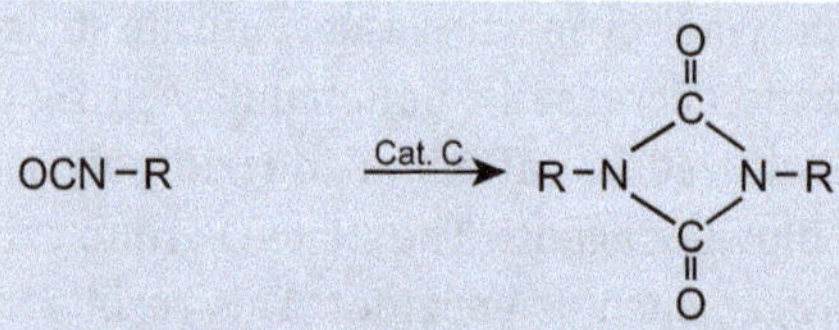	Uretdione
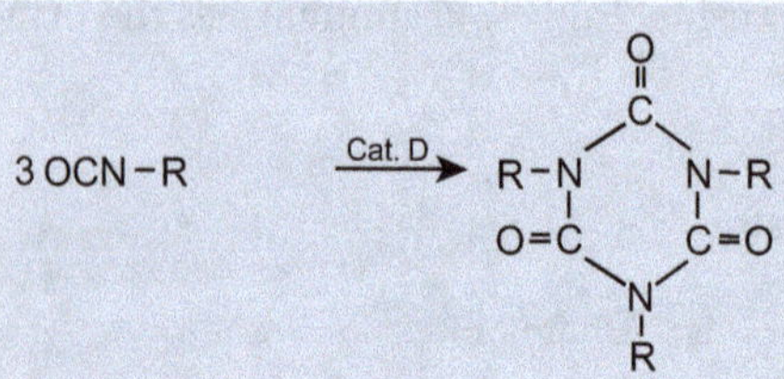	Isocyanurate
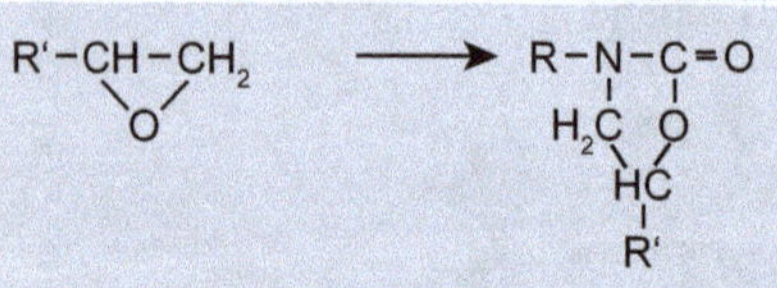	Oxazolidone

6.17.1.2 Manufacture of the Polymer

(See also Section 4.7.)

The generally liquid raw materials are mixed at room temperature, while adhering to exact stoichiometric volume ratios. The respective calculation details are provided by the resin suppliers. The *reference number* or descriptive code is the most important parameter: it references the percentage of the effectively used isocyanate amount to the stoichiometric amount (that is the calculated amount); for example:

- Reference number = 100: used amount of isocyanate equals calculated amount
- Reference number = 110: used amount of isocyanate is 10% higher than calculated amount
- Reference number = 90: used amount of isocyanate is 10% lower than calculated amount

Varying the reference number makes it possible to influence the processability of the reaction mixture and the properties of the PUR grade – within certain limits. The raw material suppliers often recommend an optimal reference number range. The PUR manufacturer is free to develop an individual recipe. Two-component systems, containing a polyisocyanate on the one hand and a polyol component on the other hand, are economically (or for logistic or ecological reasons) preferred. The polyol component typically also contains all further additives necessary for processing. For some foam processes, the blowing agents must be added separately. PUR raw materials are delivered in hobbocks, barrels, reusable containers, or tankers.

There are two general processing techniques for PUR, based on the specifically adjustable difference in reactivity, both for isocyanates and polyols:

- *One-shot-process.* The complete amount of polyisocyanate and the complete amount of polyol and additives are mixed in one shot and reacted.
- *Pre-polymer process.* The complete amount of polyisocyanate is combined with a fraction of polyol in a first step. In a second step, this product, which still contains reactive NCO-groups, is reacted with the remaining polyol and the necessary additives.

Although the one-shot process is faster, the slower pre-polymer process has the advantage of creating better ordered PUR polymers. Which process to use depends on the type of PUR to be produced and on the infrastructure of the manufacturer; in particular, whether it is capable of producing the NCO pre-polymers.

In general, processable pre-polymers or so-called *modified isocyanates* are offered by the resin suppliers together with the remnant polyol and additives as formulations and as 2-component systems. The PUR manufacturer only completes the second step of the pre-polymer process and can thus utilize the advantage of the one-shot process.

In addition to OH- and NH_2-groups, there are also special -CH_2-compounds with an active hydrogen group that can react with isocyanates to form so-called *"disguised isocyanates"*. At room temperature, these products are inert towards polyols and decompose only at elevated temperatures into the (-CH_2-active) disguise medium and isocyanate, which can now react with the polyol. Disguised isocyanates are used for one-component systems (varnishes), among others.

Because of the flexibility in reactivity of the polyisocyanate and polyol components together with an extensive range of additives, various processing techniques are available for the production of polyurethanes, such as: casting, spraying, foaming, injection molding, extrusion, rolling (rubber industry), coating, and spinning to produce: blocks, sheet, molded parts, foaming of hollow parts, composites, coatings, impregnations, films, fibers, and others.

6.17.1.3 Flammability

Polyurethanes are flammable like all organic materials; both manufacturing and use therefore come with a certain fire hazard.

During storage and handling of the *raw materials* the fire hazard is minimal because of the relatively high flashpoints of these products. However, during, for example, the *production* of flexible foam blocks, the high risk of self-ignition has to be taken into account. Incorrect metering can push the already exothermic PUR reaction through discoloration/burning in the core to self-ignition. To comply with insurance regulations, newly foamed blocks have to be stored in fire-proof rooms until they are completely cooled.

Depending on their *application*, polyurethanes can be equipped with flame-retardant properties. They are subject to standard flame testing and classification.

In addition, other fire-related hazards are also classified: melting behavior (dripping, burning), smoke density and -toxicity, etc.). Not all fire testing methods are internationally standardized; in general, they depend on the country as well as on the application (furniture, flooring, construction, automotive, etc.).

The interpretation of flammability test results requires particular caution. The results do not allow conclusions with regard to all potential applications and their associated flammability risk.

6.17.1.4 Raw Material Handling, Safety

Depending on the amount of material produced, the equipment for the *production of PUR* may be subject to local or federal regulations. The pertinent regulations and Hazmat declaration standards apply for transport, loading and unloading operations and storage of the raw materials to minimize environmental impact.

In addition, local and federal legislation regulates the workplace safety issues for employees handling these raw materials (approx. 500,000 worldwide). The raw material suppliers offer extensive information, including information on inertization of unused raw material and up-to date safety data for each product.

6.17.1.5 Environmental Protection, Safety, and Recycling

PUR end products do not pose any hazardous risks for environment or health as long as they are *used as intended*. In general, PUR has a longer life span than its raw materials: once the product is at the end of its service life, it will enter the waste stream as does PUR processing waste, such as cuttings, sprues, rejects, etc. All this PUR waste can be reused.

Table 6.46 summarizes the most important means of PUR waste recycling. While *mechanical recycling* is generally performed directly by the PUR producer, *feedstock recycling* requires detailed chemical knowledge, typically only available on an industrial scale. Depending on local conditions, *thermal recycling* (incineration) can be performed by PUR producers or through modern incineration plants.

Table 6.46 Methods for the Recycling of PUR Waste

Mechanical recycling	Feedstock recycling	Thermal recycling (energy recovery)
Press-bonding	Glycolysis	Consumer waste incineration
Particle board	Hydrolysis	Fluidized bed furnace
Powdering	Pyrolysis	Revolving cylindrical furnace
Injection molding	Hydrogenation	Pyrolysis
Extrusion-forging	Glass production	Metallurgical recycling

An example of *mechanical recycling* are flake composites from PUR flexible foam production waste: among other applications they are used for gym mats, special upholstery, and sound absorbing elements. Also marketed are particle boards from shredded rigid foam waste. RIM-PUR pellets from automotive waste are used in certain building applications.

Glycol production is a typical example of *feedstock recycling*.

6.17.2 Raw Materials and Additives

The raw material system of polyurethanes consists of three components: polyisocyanates, polyols, and additives. The PUR manufacturer using ready-formulated 2-component systems will not encounter the latter.

6.17.2.1 Di- and Polyisocyanates

Polyurethane chemistry relies on only a few different grades of basic isocyanate (see Table 6.48), typically diisocyanates. Higher functional triisocyanates are special products, *e. g.*, used for coatings and adhesives. Polymeric diphenylene methane-4,4'-diisocyanate (PMDI) is another higher than difunctional isocyanate. Quantitatively it is the most important polyisocyanate, even surpassing TDI. The following special products find increasing applications: para-phenylene diisocyanate (PPDI) for PUR-TPE (TPU) and bis-(isocyanatomethyl)-norbornane (NBDI) for coatings.

The extensive range of polyisocyanates offered by raw material suppliers is based almost exclusively on the modifications of the base isocyanates listed in Table 6.47.

Table 6.47 Di- and Polyisocyanates of Technical Importance in Polyurethane Chemistry

TDI		NDI
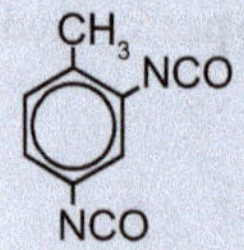	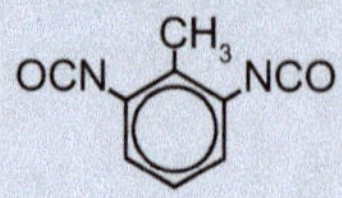	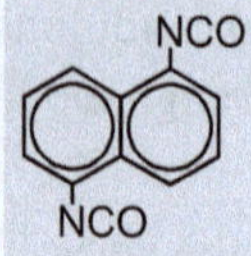
Toluylene-2,4-diisocyanate + toluylene-2,6-diisocyanate		Naphthalene-1,5-diisocyanate
80% + 20% = TDI 80/20		
65% + 35% = TDI 65/35		
MDI		
MDI-monomer		MDI-polymer (PMDI)
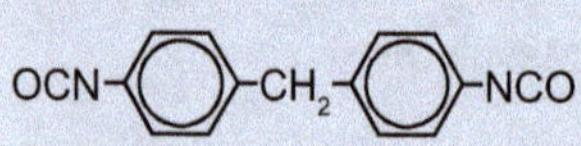		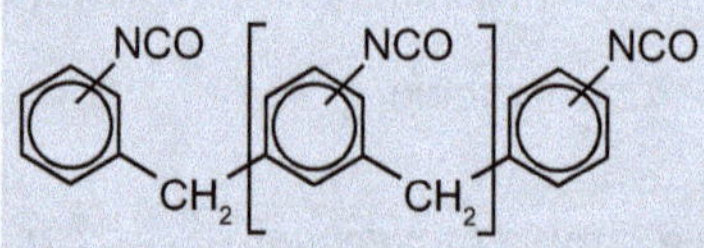
Diphenylmethane-4,4′-diisocyanate		
IPDI		HDI
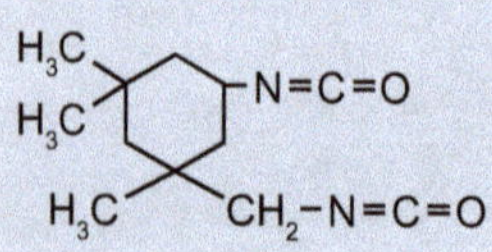		$O{=}C{=}N{-}(CH_2)_6{-}N{=}C{=}O$
Isophorone diisocyanate		Hexane-1,6-diisocyanate

6.17.2.2 Polyols, Polyamines

These products are the most important reaction partners of isocyanates. The variety of grades available exceeds the number of isocyanate products by orders of magnitude. The appropriate combination of different polyols, or polyamines and/or crosslinking agents where appropriate, requires knowledge and experience and facilitates the wide variety of available polyurethanes. In rigid foam formulations, polyols based on rape seed oil, castor oil or other renewable sources find increasing application.

Polyether polyols are the most commonly used backbones for polyurethanes. A wide palette of long- and short chained polyether polyols with 2–8 OH-groups per molecule (functionality) are available facilitated by the technically easy access to alcohols (di- and higher functionality) and epoxides (propylene- and/or ethylene oxide). In polyurethanes, this covers a property spectrum reaching from little crosslinked/linear-flexible to highly crosslinked/rigid. Polyether polyols are largely hydrolysis resistant; they require stabilizing additives to protect against photooxidation.

Polyester polyols are used to a much lesser extent compared to polyether polyols. Their production is more expensive, and at comparable chain lengths they exhibit higher viscosity. On the other hand, they are less susceptible to photooxidation, but more hydrolysis prone. Their undisputed advantage is their contribution to higher strength in the final PUR products.

Numerous polyether polyester polyols based on long-chain fatty acids are used on an industrial scale.

Polyols with organic fillers are special grades that find a surprisingly broad field of applications. They are known both as graft-, SAN-, or polymer polyols and as PHD-polyols. They are milky-white, stable dispersions of styrene-acrylonitrile polymers in the first and of polyureas in the second case. They provide flexible foams with relatively low densities, higher indentation hardness and elasticity.

Poly(amines) (*e.g.*, polyoxypropylene diamine) are highly reactive reaction partners for isocyanates. They are either higher or lower molecular compounds with two or more amino- (NH_2-) groups. An example of the former is polyether amine, the latter lead to crosslinking agents.

6.17.2.3 Crosslinking Agents and Chain Extenders

The NH_2-groups help incorporating urea segments into the PUR structure (see Table 6.45), which provide strength and improved temperature resistance to the polymer. In addition, amino compounds can be used to increase the polyaddition reaction rate. Lower molecular bi- and higher functional alcohols, such as butanediol or glycerin, can also be used as so-called OH-crosslinking agents to tailor the crosslinking density of the polymer and influence its properties, *e.g.*, its density.

6.17.2.4 Additives

In addition to the base raw materials polyisocyanate and polyol, the manufacture of PUR also requires certain additives. Polyurethane manufacturers typically do not come into contact with these additives unless they develop and produce their own in-house formulations. Almost all required materials are included in the 2-component systems ready formulated by the raw material suppliers:

Catalysts are used to accelerate the rate of reaction; for delayed acceleration (imidazo derivatives) so-called latent catalysts are used. For this purpose, typically tertiary amines and/or organo-zinc compounds are used.

Surfactants, such as emulsifiers, effect better miscibility of the normally incompatible reaction partners polyisocyanate/polyol/water and together with catalysts facilitate uniform PUR reaction rates. Special organic silicon compounds are used as *foam stabilizers* and/or for *cell size control* in foam manufacturing. They stabilize the growing cells until the foam is cured. They also control open- or closed-cell development and cell size.

Blowing agents are used to produce foam from liquid, viscous resins. There are chemical and physical blowing processes. The former are based on isocyanate-water reaction (Equation 6.32) and create gaseous CO_2 blowing agents. During the physical blowing process, the exothermically reacting resins are foamed by the evaporation of low-boiling liquids that are added to the mixture. Environmental considerations have led to the use of hydro(chloro)fluorocarbons (HCFCs) and/or hydrocarbons (pentane, cyclopentane), which have less ozone depletion potential than the CFCs (chlorofluorocarbons) used in the past.

$$R-N=C=O + H_2O \longrightarrow \left[R-NH-\overset{\overset{O}{\|}}{C}-OH\right] \xrightarrow{R-NCO} R-NH-\overset{\overset{O}{\|}}{C}-NH-R + CO_2 \qquad (6.32)$$

The use of i-hexane results in pronounced skin formation in flexible integral foams and thus in good end use properties, see Table 6.48.

Table 6.48 Properties of Flexible Integral Foam Produced with i-Hexane

Property	Unit	Density (kg/m³)			
		200	300	400	600
Skin density	kg/m³	500	–	950	1000
Core density	kg/m³	120	–	290	500
Hardness	Shore A	39	59	70	80
Shrinkage	%	2.78	1.52	1.01	0.25
Compression hardness at 10% deformation	kPa	20	66	115	306
Skin tensile strength	MPa	3.3	5.1	6.1	8.1
Core tensile strength	MPa	0.37	0.83	1.1	2.6
Skin elongation at break	%	1	120	125	150
Core elongation at break	%	125	135	120	130
Core compression hardness	kPa	17	60	0	270

Most processes use conventional multi-component metering devices to add 4 wt.-% liquid CO_2 in addition to 3–5 wt.-% water to the polyol mixture to produce extremely flexible block foams with uniform cell structures.

Flame retardants are added to decrease the flammability of PUR. Typical candidates are inorganic (aluminum hydroxide, magnesium hydroxide, aluminum oxide hydrates, ammonium polyphosphates), or organic chlorine, bromine and/or phosphorus (sometimes also nitrogen) containing compounds: tetra bromide benzoic acid esters or aliphatic bromine compounds.

Fillers (*e.g.*, carbon black, chalk, silicates, barite, wollastonite) can be used to lower material cost of polyurethanes. *Glass fibers* are most often used as reinforcing fillers to improve physical properties. Lately, hemp, jute, etc. have also gained importance.

In many PUR applications, *antiaging agents* are indispensable to protect against photooxidation and hydrolysis. They are available in a wide variety of products, *e.g.*, lactones and substituted phenols.

Coloring agents are available as pastes or doughs. They are compounds of inorganic or organic colorants or pigments in polyols.

Antistatic agents decrease the electrostatic charge; biocides protect PUR against microorganisms (bacteria, fungi).

Release agents for fast and easy ejection of PUR molded parts from the mold are included in the polyol formulation as so-called internal release agents. Typically, materials used are Ca-, Mg-, Al-, and Zn-salts of fatty acids such as stearates, laureates, and others. In general, surface treatment of the mold is also required.

6.17.3 PUR Polymers

Figure 6.15 provides an overview of the *different forms* of PUR products. It also exemplifies that there are no general *"properties of polyurethanes"*. Rather, the question can be answered for *one particular polyurethane*, preferably in connection with its actual *application*. The information provided in Table 6.49 therefore also only provides a first indication of the properties of different foam types.

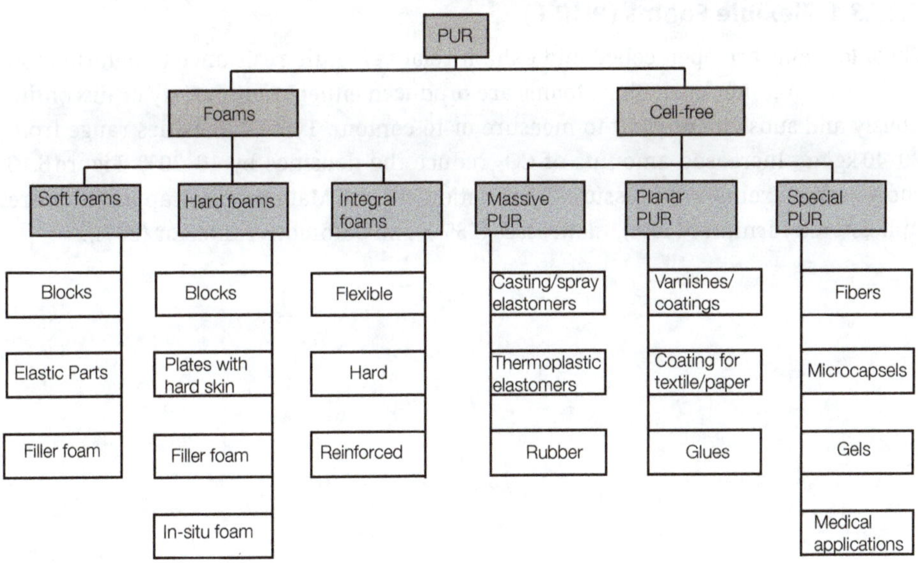

Figure 6.15 Different forms of polyurethane products

Table 6.49 Comparison of Properties for PUR Foams

Foam type			Density	Tensile strength	Elongation at break	Compression hardness[1]	Thermal conductivity
			kg/m^3	kPa	%	kPa	W/mK
Flexible block foam	Polyester		35	160–120	200–450	6.5	
	Polyether	High load	36			6.0	
		Normal	36	100–180	100–400	4.5	
		High elastic	34	50–120	80–200	2.7	
Molded foam		Cold foam	38–44	110–152	125–150	3.3–5.6	
		Hot foam	33	90–112	190	4–4.7	
Semi-rigid foam		Energy absorber	80–100	350–450	10–25	150–320	
Rigid block foam			32	200	4	200	0.021
			50	270	5.2	350	0.021
			90	900	5	700	0.027

[1] Compression stress at 40% compression

6.17.3.1 Flexible Foams (PUR-F)

Flexible foams are open-celled and exhibit relatively little resistance to deformation under compression load. Block foams are produced either continuously or discontinuously and subsequently cut to measure or to contour. Typical densities range from 20–40 kg/m^3. Increased amounts of CO_2 reduce the densities by 10–20%. Figure 6.16 shows some exemplary pressure-deformation curves. Main areas of application are upholstered furniture (40%), mattresses (25%), and automotive interior (20%).

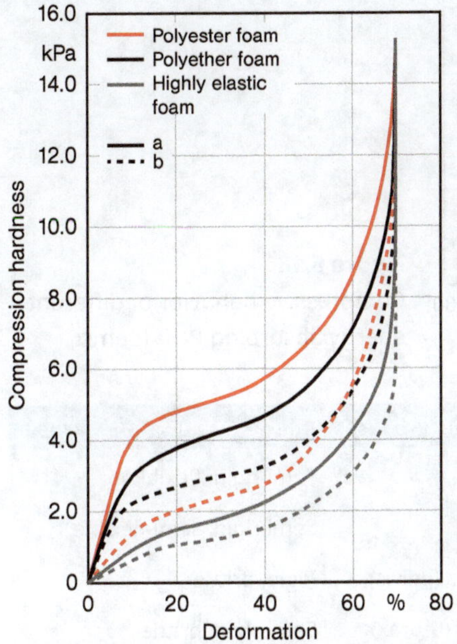

Figure 6.16
Hysteresis loops
a: Load line
b: Relief line

Molded foams are produced as either *cold foam* or *hot foam* (*i. e.*, without or with external heat input until final curing), depending on the resin system. The densities range from 30–300 kg/m^3 depending on application.

Flexible molded PUR foams are mainly used for seats in automotive and aircraft construction; they are also used for upholstered furniture and for technical products (*e. g.*, sound-absorbing material). Flexible *filling foams* with densities ranging from 150–200 kg/m^3 are used as automotive interior and dashboard padding material. They are manufactured by back foaming of PVC- or ABS skins in closed molds.

The compressive stress behavior of PUR foams can be influenced physically by the density (see Figure 6.17) and chemically by the foam system (see Figure 6.18 (and table below), exemplified by energy-absorbing foams).

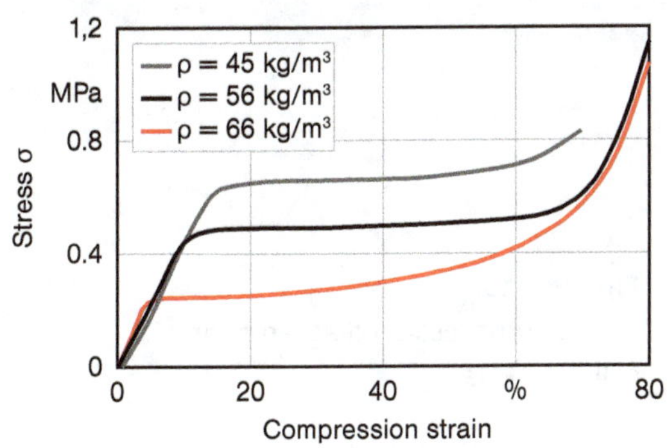

Figure 6.17
Influence of density on material properties of flexible PUR

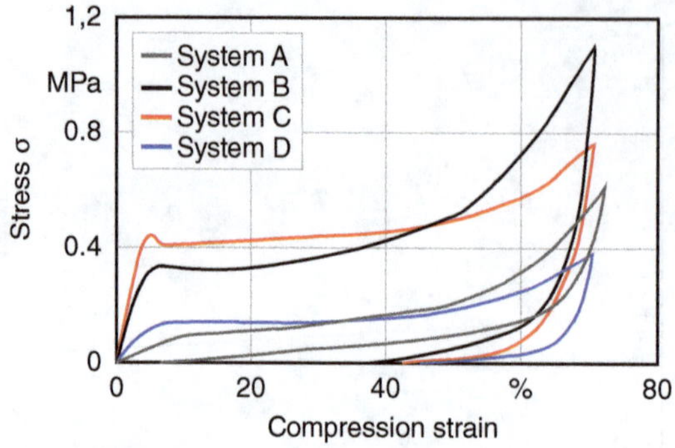

Figure 6.18
Compressive behavior of different energy absorbing PUR foams

PUR foam	Typical application	Characteristics
Bayfill, System A	Dash boards	Semi-rigid, flexible
Bayfill, System B	Bumpers	Semi-rigid, semi-flexible
Bayfill, System C	Side impact panels (pelvis)	Rigid, brittle
Bayfill, System D	Side impact panel (thorax)	Semi-rigid, brittle

Cellular PUR elastomers (TPU, see Section 6.20.2.5) represent an intermediate state compared to solid PUR. A water-containing reaction mixture is filled into pressureless steel molds and then foamed under separation of CO_2. These molded parts are particularly suitable for automotive cushioning because of their extremely progressive characteristics (see Figure 6.19).

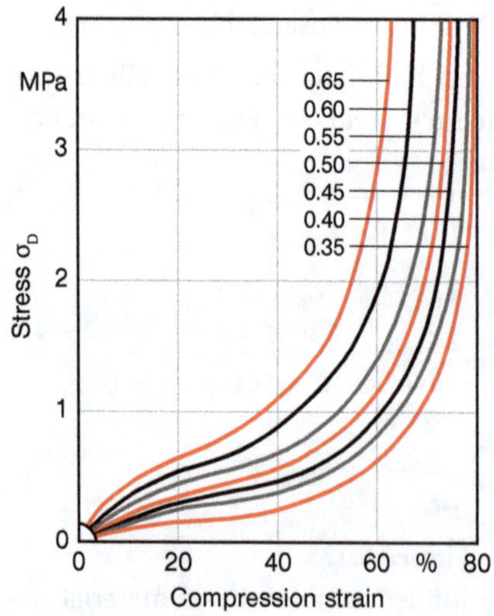

Figure 6.19
Stress-compression diagrams for cellular TPUs

6.17.3.2 Rigid Foams (PUR-R)

Rigid foams exhibit relatively high resistance to deformation under compression load. Their most important properties are their very low thermal conductivity caused by the insulating gases in the *closed cells* and their ability to form solid composites with almost any flexible or rigid skin layer. Rigid *foam blocks* are manufactured either continuously or in batch processes as semi-finished products and subsequently cut to measure depending on application. They range in density from 30–90 kg/m^3 and are used for various insulation applications, automotive interiors, and others.

Comparison of properties, see Table 6.49.

Foamed panels with protective outer skins are produced either continuously on double belt lines or discontinuously. The outer skins can be either flexible or rigid. Examples of carrier materials include Kraft paper, aluminum, particle or plaster boards, coiled steel or beaded sheet metal, and coated glass fiber mats. Areas of application are: building/construction as roofing insulation, sheet rock and plaster boards, sandwich elements for industrial buildings and cooling facilities. The core density of the foams ranges from 30–40 kg/m^3. Foams are introduced into hollow parts or hollow spaces as liquid reaction mixtures and fill these spaces after foaming. With densities from 30–60 kg/m^3 they serve as thermal insulators in cooling appliances of all sorts, for hot water boilers and district heating pipes. *In situ* foams for roofing and wall insulation, window- and door installation, and sealing is produced directly on site by mobile foaming machines or by pressure vessels from single- or two-component foam systems. Densities are ≈ 30 kg/m^3; spray foams for roof sealing purposes have densities up to 55 kg/m^3.

6.17.3.3 Integral Foams (PUR-I)

While flexible and rigid foams are fundamentally different in their chemical structure (highly crosslinked/little crosslinked), *integral foams* represent a *variation in production technology:* they are available both as rigid and as flexible *molded foams* only. The liquid, highly reactive mixture is injected into a closed mold, in which foam generation and mold temperature control are used to produce parts with cellular cores and solid skins. The transition from core to skin is not abrupt but rather continuous (see Figure 6.20). This technique of foaming highly reactive PUR systems with particularly short holding times is generally called *reaction injection molding (RIM)*; it is used to produce both large parts as well as parts with minimum dimensions and with flow length-to-wall thickness ratios of up to 1 : 1000.

Flexible integral foams with densities from 200 to 1100 kg/m^3 are used for exterior body parts, bicycle seats, shoe soles, and others.

Rigid integral foams with comparable densities are used for technical parts, in construction, furniture, and sport- and leisure products of various designs and purposes.

Glass fiber reinforced integral foams (R-RIM) with densities from 1.0–1.4 g/cm^3 are used for automotive body parts and housings.

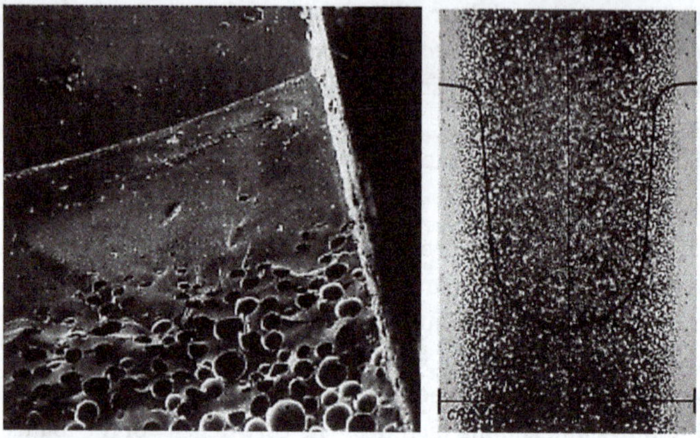

Figure 6.20 Integral foams

Left: Foam structure

Right: Density transition across specimen section

PUR-R-RIM is used to produce stiff, laminar parts, *e.g.*, automotive interior side panels. Long fibers, fabrics, non-woven material, or fiber mats, for which bio-renewable materials such as flax, sisal, or coir (coconut fibers) are increasingly used, are sprayed or impregnated with the PUR mixture and laid into the mold where they are compression molded.

Table 6.50 provides an overview of the properties of various integral foam systems, while Table 6.51 shows the dependency of certain properties on density and provides an example of flexible integral foam properties in automotive applications.

R-RIM parts (with 20–25% wollastonite reinforcement) can be coated in-line, *e.g.*, for exterior automotive panels. The plastic parts are connected to the car body prior to electro-immersion coating. Shrinkage is 0.8 or 0.57%, respectively, for temperature loads of 2 times 45 min at 200 °C. These systems are also suitable for thin-wall technology: automotive door sills with 2 mm thickness and a length of 2 m.

Table 6.50 Property Comparison for PUR Flexible Integral Foam, R-RIM

PUR-System	Filler	Filler content	Density	Shore-hardness	Flexural modulus of elasticity	Tensile strength	Elongation at break	Application, see below
		wt.-%	g/cm³		MPa	MPa	%	
Flexible		0	1.05–1.10	D 66	500–760	26–33	135–150	A
	Glass flakes	20	1.20	D 75	1550	31	30	A
	Mica	22	1.25	D 75	1720	30	25	A
	Glass fibers, average fiber length 180 µm	15	1.20	D 68	950	31	140	A
		20	1.24	D 70	1400	28	130	A
		25	1.30	D 71	1700	27	120	A
Shoe soles		0	0.55–0.60	A 55–60		8–10	550–460	B
		0	0.60	A 60		25	600	C
Rigid	Glass mat 225 g/cm²	30	0.25		250	5.8 [1)]	2.3	D
	300 g/cm²	20	0.48		900	25 [1)]	2.2	E
	600 g/cm²	20	1,0		3500	110 [1)]	2.2	F
	1800 g/cm²	32	1.4		650	170 [1)]	2.8	G
S-RIM[2)]		25	1,05		4200	70	1.9	H
LFI[3)]		25–30	0.7–1.0		4500–6000	54–68	2.2–1.8	H

[1)] Flexural strength; [2)] Structural RIM; [3)] Long fiber injection.

A Car panels
B Integral shoe soles
C Shoe soles
D Automotive roof linings
E Automotive roof frames, -columns/door-/seat-/trunk liners, rear window shelf
F Car dashboard frame, consoles, seat frame, seat-back, spare tire and engine cover
G Automotive seat shells and frames, bumper beams
H Sun roof

Table 6.51 Property Comparison of Flexible PUR Integral Foam in Automotive Applications (RIM)

Property	Unit	Application (see below)					
		A		B	C	D	
		Core	Skin				GF 22
Density	g/cm^3	0.120–0.175		0.7	0.95–1.08	1.0–1.1	1.22
Tensile strength	MPa	0.27–0.50	1.7–3.5	5.7	14–31	17–28	23
Flexural modulus	MPa			17	75–720	40–350	1300
Elongation at break	%	220–125	220–120	230	220–140	> 300–230	130
Shore hardness				A 75	D 39–69	D 33–57	D 67

Application:
Type A: Motorcycle seats, interior automotive safety equipment
Type B: Bumper skins, door protection strips
Type C: Bumpers, truck splash boards, external automotive body parts
Type D: Car body parts

6.17.3.4 Solid PUR Polymers, PUR-S

Solid PUR materials are available as molded parts, sheet, films, and as "thick" coatings (> 1 mm). For *cast and spray elastomers*, the liquid (or molten) raw materials are mixed and cast in open molds. Depending on raw material type, the part will cure at room temperature (cold cast systems) or at elevated temperatures (60–130 °C: hot cast systems). PUR elastomers are available in Shore hardnesses ranging from 10A to 70D. They are used for solid tires, rotor bearings, sieve bottoms, sealings, and casings, and as coatings in shipbuilding, automotive, and in building/construction.

Table 6.52 provides an overview of properties for selected systems.

Table 6.52 Property Comparison of Cast PUR Resins; A: Polyether-Urethane, B: Polyester-Urethane

Property	Unit	A, rigid	B, rigid	A/B, flexible	A, flexible	B, flexible
Shore D		86	85	61	20	21
Shore A		100	99	96	70	71
Tensile strength	MPa	86	60	20	4.5	
Elongation at break	%	6.5	5.9	87	73	200
Glass transition temperature	°C	138	90	47	15	-21

Thermoplastic PUR elastomers (TPU) are discussed in Section 6.20.2.5.

PUR rubber is produced by rubber processing technologies, *e. g.*, on rollers (also cross-linked with sulfur), and molded, pressed or extruded to form parts with Shore hardnesses from 50A to 60D. Examples are shocks, grindings disks, and pump linings.

Non-cellular, laminar PUR grades produced in thin layers (< 1 mm) are typically coatings (DD-coatings) – solvent or water based, also UV curable – for textiles, paper, leather, and other substrates, and PUR adhesives and special polyisocyanates for particle board production. A special class are *PUR fibers*, used for textiles and swim wear.

Last but not least are special areas of applications such as PUR micro-capsules (*e. g.*, for copy paper), PUR gels, medical applications in dialysis equipment, artificial blood vessels, and bandages replacing casts.

6.18 Biopolymers and Derivatives

Hans-Josef Endres

Bioplastics

In this book, plastics that are produced from renewable raw materials are mentioned according to their chemical structure in the appropriate sections. Because bioplastics have attracted a lot of attention lately, we will briefly address the overall topic here.

Unfortunately, a number of different terms are used for bioplastics, such as biopolymers, green plastics, ecoplastics, biodegradable plastics, bio-based polymers, etc., and at the same time, there are varied and confusing discussions about the correct use of these terms. The best general definition for the term "bioplastic" is currently a polymer material that possesses at least one of the following two properties:

I. Made from bio-based (renewable) raw materials

II. Biodegradable

Following this definition, there are the following three basic bioplastic/biopolymer groups:

1. Degradable petrochemical bioplastics
2. Degradable (primarily) bio-based bioplastics
3. Non-degradable bio-based bioplastics

Biodegradable plastics can be based on both petrochemical and renewable raw materials. The degradability of biopolymer materials is determined exclusively by their chemical and physical microstructure and not by the origin of the raw materials or the polymer manufacturing process. This means that biopolymers do not necessarily have to consist exclusively of renewable raw materials. Biodegradable biopolymers can also be produced from petrochemical raw materials such as polyvinyl alcohols, polycaprolactone, various polyesters, polyesteramides, etc. (Figure 6.21, bottom right). Conversely, not all biopolymers based on renewable raw materials are necessarily biodegradable; these include, for example, highly substituted cellulose acetates, vulcanized rubber, casein plastics, bio-based polyamides, or linoleum (Figure 6.21, top left). Typical examples of the last group of bio-based and simultaneously biodegradable bioplastics are starch-based plastic blends, polyhydroxyalkanoates (PHA), or PLA (polylactic acid).

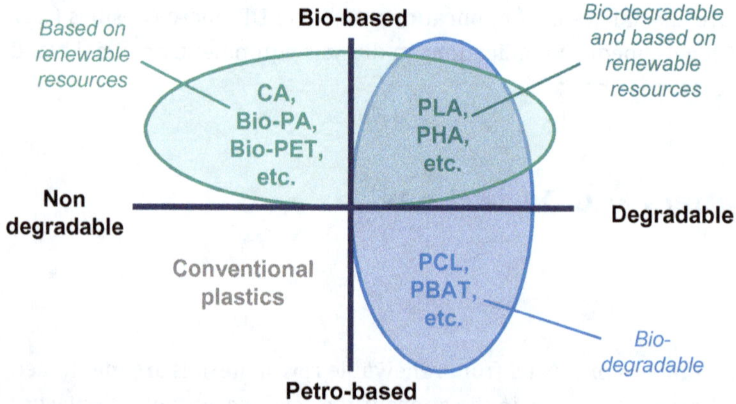

Figure 6.21 Bioplastics and the three fundamentally-different biopolymer groups [1, modified]

To avoid misunderstandings when using the terms "biopolymer" or "bioplastic," a precise nomenclature is essential. It is therefore recommended to specifically refer to biodegradable or bio-based bioplastics. These two different approaches are still being pursued today and form the technical basis for various bioplastics. Degradability refers to a functional property or disposal option at the end of the material's life cycle, regardless of the origin of the raw materials. Bio-based, on the other hand, describes exclusively the origin of the polymer raw materials at the beginning and makes no statement about degradability at the end. This means that during the production of renewable raw materials as polymer input, CO_2 is extracted from the atmosphere and stored in the biopolymer for its lifetime. At the end of its life, the captured carbon is converted back into the same amount of CO_2, for example, through composting or incineration.

The first technical, industrially used polymer materials, whose development began over 100 years ago, were all bio-based, as petrochemical raw materials were not yet available at that time. These so-called "old-economy" bioplastics were based on the renewable plant-based raw materials cellulose and natural latex, or on animal proteins. With increasing industrialization at the end of the 19th century, the availability of raw materials for the production of these materials became an important issue.

The natural raw materials available at that time were modified during the material production process to create the first truly durable polymer materials with a property profile that was completely new for the time, without them being explicitly referred to as bioplastics. This production of durable polymer materials from degradable organic raw materials was the beginning of the success story of plastics. These old-economy bioplastics therefore belong to the group of bio-based, non-degradable bioplastics. Of the bioplastics of the old economy, only natural rubber, regenerated cellulose and cellulose derivatives (cellophane, viscose, celluloid, cellulose acetate), and, in smaller quantities, linoleum are economically important on the plastics market today (Figure 6.22). Today, the plastics industry is largely dependent on oil- or coal-based raw materials.

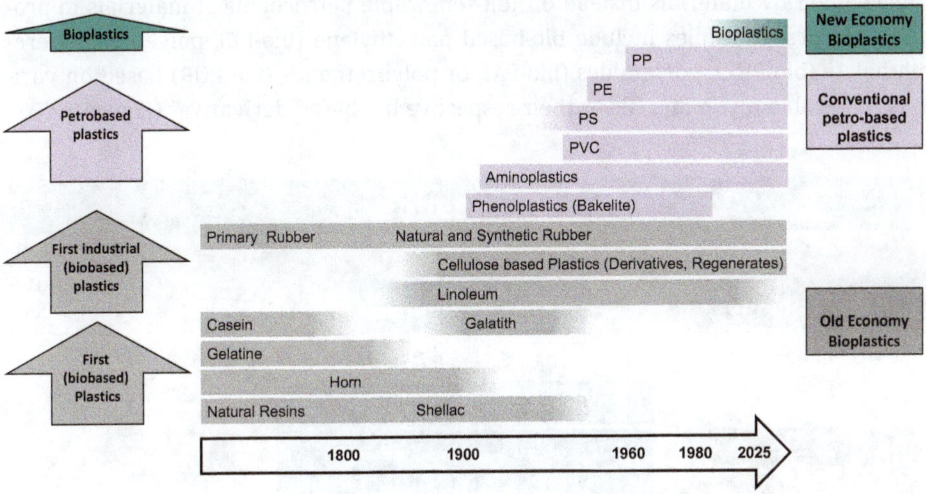

Figure 6.22 Timeline of bioplastics with eras [2, modified]

The renewed development of novel bioplastics (so called "new economy" bioplastics), which began intensively in the mid-1980s, was initially driven by the then-existing waste problem and the unsatisfactory disposal situation of conventional plastics. The goal of these developments was therefore to develop degradable bioplastics as a solution to the waste problem. Currently, there are increasing applications in medicine, landscaping and horticulture, wastewater treatment, etc., where degradability under the respective environmental conditions enables additional functions, such as plowable films for agriculture, in aquatic environment degradable nets and fishing gears, bioresorbable implants or suture materials for surgery, or the targeted release of active ingredients (fertilizers, medicinal substances).

With current developments in the field of bioplastics, the availability of raw materials, *i.e.*, independence from petrochemicals and often politically critical countries of origin, is once again becoming an important issue. Within this group of non-degradable, bio-based bioplastics, one development has emerged strongly and technically very successfully over the past 10 years: the so-called drop-in solutions. Simply put, this is the approach of maintaining established synthesis methods based on petrochemical raw materials and replacing petrochemical starting materials as completely as possible with biogenic raw materials or intermediates. The goal is to produce bio-based plastics that are similar to known conventional plastics in terms of their processing and usage properties. Due to their identical chemical structure, the drop-ins exhibit the same property profiles as their petrochemical counterparts with the same additives. This means that when replacing conventional plastics with the corresponding drop-ins, no changes in processing, application, and especially recovery or recycling are to be expected. This is less about acute or current raw material availability than about strategically secured, long-term availability, *i.e.*, the use of bio-based and renewable raw materials instead of non-renewable petrochemical materials to produce plastics. Examples include bio-based polyethylene (bio-PE), polyethylene terephthalate (bio-PET), polyamides (bio-PA), or polyurethanes (bio-PUR) based on various renewable raw materials or their respective bio-based derivatives (Figure 6.23).

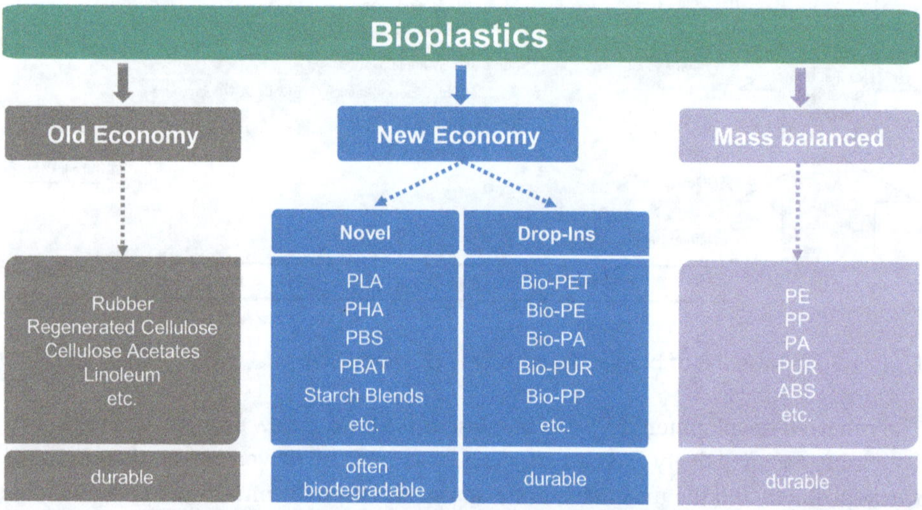

Figure 6.23 Traditional (old economy), novel (new economy) and mass balanced bioplastics [2, modified]

This means that the bioplastics of the new economy consist of two basic groups: (1) chemically novel biopolymers, *i. e.* (1) polymer types that were largely unknown in the plastics sector until a few years ago (*e. g.*, novel bio-based polyesters such as PLA (polylactic acid) or PBAT (polybutylene adipate terephthalate)) and (2) so-called "drop-ins," which are chemically identical to their petroleum-based counterparts but are partially or entirely bio-based (Figure 6.23). The best-known examples are currently bio-PET and bio-PE, measured by commercial market share. In addition, work is currently underway on other drop-ins, including in the PP sector, thermosets (*e.g.*, bio-based EP resins), and elastomer polymer materials (*e.g.*, bio-based EPDM or bio-based polyurethanes).

Since drop-ins are indistinguishable from their petrochemical counterparts in terms of their technical properties, the question of the proportion of renewable or biogenic raw materials is becoming increasingly important, particularly for drop-ins but also for other bioplastics. This question can currently best be answered using radiocarbon dating (also called the C14 method or radiocarbon method) according to ASTM D6866. ^{14}C dating is based on the radioactive decay of the carbon isotope ^{14}C and has been used to determine the historical age of carbon-containing organic materials up to approximately 50,000 years old. In nature, carbon occurs in three isotopes: ^{12}C, ^{13}C, and ^{14}C. In contrast to ^{12}C and ^{13}C, which occur in higher concentrations in inorganic compounds, ^{14}C is not stable and is therefore also referred to as radiocarbon. According to Libby, its half-life is approximately 5568 ± 30 years [3]. It is reformed in the upper atmosphere at the same rate as the decay rate and incorporated into biomass during the metabolic processes of photosynthesis. However, in dead and mineralized biomass, the amount of ^{14}C decreases over time due to radioactive decay, since no new ^{14}C is absorbed. This means that petrochemical raw materials or petroleum-based plastics no longer contain any "young" ^{14}C due to the relatively short half-life of the carbon isotope ^{14}C. The ^{14}C content or the ^{14}C:^{12}C ratio is therefore a measure of bio-based carbon and thus an indicator of the proportion of renewable raw materials in the material.

Mass balancing is another, not entirely uncontroversial, approach to bio-based plastics (see Figure 6.23), which has been increasingly promoted by the chemical industry in particular over the past decade. The so-called mass balance method, which is a purely certification-based method, allows the combination of renewable feedstocks with existing production infrastructure without requiring a complete restructuring of existing plastics production systems and value chains. Instead of producing bioplastics entirely separately, manufacturers introduce renewable feedstocks such as bio-naphtha or biogas into the same supply chain as fossil feedstocks. Physical blending of bio-based and fossil feedstocks is acceptable, with certification systems ensuring that the allocated proportions correspond to the actual inputs. The proportion of bio-based ingredients is then recorded and allocated to the final product through a balancing and certification process. This process does not reflect the actual physical mass flows, but merely allocates the amount of biogenic inputs at the beginning to the final product.

In the end, mass-balanced bioplastics are plastics consisting of a mixture of renewable and fossil feedstocks, with the proportion of renewable feedstocks allocated through a certified balancing approach with specific certification rules by a third-party provider. This method ensures that a certain proportion of bio-based raw materials is integrated into the production system, even if the final product itself is not entirely bio-based.

The key steps of this hybrid mass-balanced approach are:

1. Procurement of renewable raw materials
   - Bio-based naphtha (from vegetable oils, waste fats, or plant residues)
   - Bio-based gas (*e.g.*, biogas from organic waste)
   - Second-generation biomass (non-food sources such as wood residues)
2. Co-processing with fossil raw materials
   - The bio-based raw materials are fed into existing petrochemical plants together with fossil materials.
   - The raw materials are mixed in large-scale crackers to produce monomers (*e.g.*, ethylene, propylene).
3. Mass balance allocation
   - Since bio-based and fossil-based molecules can no longer be physically distinguished and traced when processed together, the mass balance approach is used, which records and reports the proportional amount of biogenic carbon.
   - Certified systems ensure that in proportion to the input share a portion of the renewable sources is allocated to the finished plastic.
4. Polymerization and plastics production
   - The certified monomers (ethylene, propylene, etc.) are polymerized into established plastics, meaning the finished materials exhibit the well-known properties of conventional plastics.
5. Certification and labeling
   - The certified plastics are then supplied to industries such as packaging, automotive, and electronics. A surcharge is charged for the certificates confirming the same amount of bio-based input.

Mass-balance bioplastics offer a practical and scalable way to reduce the environmental impact of plastics production without compromising performance. While this method has gained traction as an enabler of sustainable materials, it has significant limitations regarding transparency, dependence on fossil fuels, and long-term sustainability. This is the reason why this approach is controversial in politics and science. More robust solutions – such as fully bio-based, biodegradable plastics – may be necessary to drive real change.

On the other hand, mass-balance bioplastics offer a viable solution for the gradual defossilization of the plastics industry. Crucially, these mass-balanced bioplastics exhibit the same processing, usage, and even recycling properties as their conventional counterparts, while simultaneously contributing to reducing CO_2 emissions and the consumption of fossil resources. Therefore, the use of bioplastics can be effectively promoted without technical changes throughout the entire value chain, including existing production routes, technical infrastructure, and distribution structures. Despite challenges in raw material sourcing and certification, mass-balance could therefore serve as a bridging technology to facilitate the transition to a fully bio-based circular economy.

Although the market share of bioplastics is continuously growing, it currently amounts to a maximum of 2% for new economy types and approximately 10% of the total plastics market for old and new economy types as well as mass-balanced bioplastics.

6.18.1 Cellulose and Starch Derivatives: CA, CTA, CP, CAP, CAB, CN, EC, MC, CMC, CH, VF, PSAC

6.18.1.1 Chemical Constitution

The term "cellulose derivatives" includes all plastics that are derived from cellulose, which is a carbohydrate that forms a framework for all organic matter. Cellulose is extracted from wood by extracting the lignin and also, typically, the pectin. The seed hairs of cotton provide cellulose in almost pure form. Cotton linters (short-haired fiber waste) are usually employed as raw material in the plastics industry. Cellulose derivatives are generated by etherification (with alcohols) or esterification (with acids) of cellulose. Here, up to three OH-groups of the ring system (glucose remnant) can react; see Table 6.53. The properties are commonly modified by the type of reaction, the degree of substitution, and the addition of plasticizers.

Table 6.53 General Structure of Cellulose Derivatives

Plastic	Chemical description	Basic structure 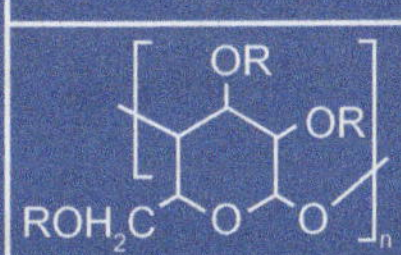 R
CA, [CTA]	Cellulose (tri)acetate	$-COCH_3$
CP (CAP)	Cellulose propionate	$-COCH_2CH_3$
CAB	Cellulose acetobutyrate	$-COCH_3/-CO(CH_2)_2CH_3$
CN	Cellulose nitrate	$-NO_2$
MC	Methyl cellulose	$-CH_3$
EC	Ethyl cellulose	$-CH_2CH_3$
CMC	Carboxymethyl cellulose	$-CH_2COOH$
	Hydroxyethyl cellulose	$-(CH_2)_2OH$

CA is created by esterification with glacial acetic acid, acetic acid anhydride, and sulfuric acid or zinc chloride as catalysts and usually methylene chloride as solvent. In further steps, acetic acid contents of approximately 44–61% can be achieved, resulting in low to highly viscous grades.

For *CTA*, each of the three OH-groups of the glucose remnant is esterified with acetic acid. Due to the use of petro-based acetic acid and petro-based plasticizers, the proportion of biogenic carbon for CA is usually max. 50%.

CAP is created by esterification of cellulose with propionic acid. It contains 55–62% propionic acid in addition to 3–8% acetic acid (hence the A in CAP).

Other than in CA, for *CAB* the cellulose is also esterified with butyric acid. The acetic acid content ranges from 19% to 23%; the butyric acid content, from 43% to 47%.

CN, often falsely called nitrocellulose, is created by conversion of cellulose with a mixture of sulfuric and nitric acid, which allows the setting of the nitrogen content between 10.6% and a maximum of 16%.

MC (EC) is created by a conversion of 1.3–1.5 (2.0–2.6) of the possible three OH-groups of the glucose remnant with methyl chloride (ethyl chloride). The starting material here is alkali cellulose, which is created by dipping of cellulose in 18% caustic soda.

For *CMC*, alkali cellulose is converted with the sodium salt of monochloroacetic acid, with up to 6–10 carboxymethyl groups for each 10 glucose units.

Oxyethyl cellulose is created by conversion of cellulose with ethylene oxide.

CH (cellulose hydrate) is generated by intramicellar swelling of cellulose in caustic soda. This only extends the distance of the chain molecules; no chemical conversion takes place.

6.18.1.2 Processing

The common cellulose esters (CA, CAB, CAP) can be processed like any other thermoplastic material. Their flow behavior is good and allows for pin-point gating. Tools must be well vented to avoid plasticizer deposits. Pre-drying for approximately 3 h at 60–90 °C or for approximately 1.5 h in a flash dryer is required. For injection molding, the water content should be below 0.15%; for extrusion, it should be below 0.05%. Plasticizer vapors must be extracted. Melt temperatures for *injection molding* range from 180 °C to 230 °C; mold temperatures, from 40 °C to 80 °C, where the higher temperatures apply for more rigid products with less plasticizer content.

Films and sheets can be *extruded* with flat-sheet dies. The *chill-roll process* is used to cast films thinner than 0.8 mm. Automotive trim is produced by *extrusion sheathing* of 50–100 µm aluminum films with CAB (using a coupling agent).

CAB and CAP powders are used for *rotational molding* and metal *fluidized bed coating*.

Machining, polishing, welding (ultrasonic is less suitable), adhesion with glues (CA and CAP: 50% methyl glycol acetate and 50% ethyl acetate) or with two-component adhesives based on PUR or EP resins, coating, printing, and metallizing are possible.

6.18.1.3 Properties and Applications

Cellulose (Tri)Acetate, CA (CTA)

With increasing acetate contents, the viscosity of *CA* increases:

- Approx. 44–48% – low viscosity, application: textile production, printing inks
- Approx. 52–56% – medium viscosity, application: molding grades, coatings, adhesives
- Approx. 56–61% – high viscosity, application: safety and electro-insulation films

To facilitate thermoplastic processing, 15–20 (38) wt % plasticizers are added to the resin: dimethyl, diethyl, dimethyl glycol phthalate; trichloroethyl phosphate in combination with dimethyl or diethyl phthalate; and only in combination with these, dibutyl, diisopropyl, and di-2-ethylhexyl phthalate.

The properties of CA can be tailored in a wide range, depending on the type and amount of plasticizer used. With increasing plasticizer content, the heat distortion point decreases and the melt flow rate as well as the creep tendencies under mechanical load increase. These grades are very suitable for tool handles, because they relieve the stresses created by shrinking the coating on a metal surface. CA is crystal clear and transparent and can be dyed in brilliant and deep colors. It is tracking-resistant and, because of its water content, it is antistatic (not attracting dust). CA is resistant to fats, oils, mineral oils, and aliphatic hydrocarbons; the resistance to fuels, benzene, chlorinated hydrocarbons, and ethers depends on the composition. Its alkali and acid resistance is low. CA cannot be made permanently weather-resistant. It is little susceptible to stress cracking because stresses are relieved. For a comparison of properties, see Table 6.54.

Table 6.54 Properties Comparison of Cellulose Derivatives

Property	Unit	Cellulose derivative			
		CA	CAB	CP	EC
Density	g/cm³	1.26–1.32	1.16–1.22	1.17–1.24	1.12–1.15
Tensile modulus of elasticity	MPa	1000–3000	800–2300	1000–2400	1200–1300
Yield stress	MPa	25–55	20–55	20–50	35–40
Elongation at yield	%	2.5–4	3.5–5	3.5–4.5	-
Heat deflection temperature, HDT/A 1.8 MPa	°C	45–80	45–95	45–95	ca. 50
Thermal coefficient of linear expansion, parallel (23–55 °C)	10^{-5}/K	10–12	10–15	11–15	10

Table 6.54 Properties Comparison of Cellulose Derivatives *(continued)*

Property	Unit	Cellulose derivative			
		CA	CAB	CP	EC
Flammability UL94 for 1.6 mm thickness	Class	HB	HB	HB	HB
Dielectric constant at 100 Hz	–	5–6	3.7–4.2	4.0–4.2	ca. 4
Dielectric loss factor at 100 Hz	10^{-4}	70–100	50–70	50	100
Specific volume resistivity	Ohm · m	10^{10}–10^{14}*	10^{10}–10^{14}*	10^{10}–10^{14}*	10^{11}–10^{13}*
Specific surface resistivity	Ohm	10^{10}–10^{14}*	10^{12}–10^{14}*	10^{12}–10^{14}*	10^{11}–10^{13}*
Dielectric strength	kV/mm	25–35	32–35	30–35	ca. 30
Comparative tracking index, CTI/A	–	600	600	600	600
Water absorption at 23 °C, saturation	%	3.5–4.5	2.0–2.3	1.9–2.8	–
Moisture absorption at 23 °C/50% rel. humidity, saturation	%	ca. 1.2	0.6–0.8	0.6–0.9	–

* Values for conditioned/dried material

CTA exhibits higher water and heat resistance than CA; however, it is barely processable like a thermoplastic. Some of its applications are electrical insulation film and acetate rayon.

Applications for molding resins: insulating tool handles, writing utensils, combs, clasps, buttons, eyewear frames (cut from sheet), and hollow fibers for dialysis.

General applications (also for CAP and CAB): fibers, films, coatings, adhesives, and low-profile additives for sheet molding compounds (SMC) and bulk molding compounds (BMC).

Cellulose Propionate, CP (CAP)

CAP resins are also compounded with 2–5 wt % plasticizer: dibutyl and di-2-ethylhexyl phthalate or adipate, dibutyl and dioctyl sebacate, dibutyl and dioctyl acetate. For applications with food contact, citric acid and palmitic acid esters are used.

The general properties resemble those of CA. Weather resistance of stabilized CAP is better than that of stabilized CA. Monomeric plasticizers tend to migrate at elevated temperatures. Copolymers of CAP with EVA graft polymers exhibit better stability, stiffness, creep resistance, heat distortion point, and migration resistance. Polymer modification content may reach 5–30 wt %. The good solubility of UV and IR absorbers is used for applications as optical fillers. For a comparison of properties, see Table 6.54.

Applications: similar to those of CAB – model railway components, powder boxes, high-quality frames for sunglasses and other glasses, stencils, and toothbrush handles.

Cellulose Acetobutyrate (CAB)

To make CAB suitable for thermoplastic processing, the types and amounts of plasticizers used are comparable to those for CAP. The general properties compare to those of CA and CAP; however, stabilized CAB exhibits better weather resistance. Copolymerization with EVAC (see Section 6.2.4) enhances the property levels as in the case of CAP. For a comparison of properties, see Table 6.54.

Applications: automotive steering-wheel sheathing, control knobs, handles, exterior lights, dome lights (thermoformed), advertising signs, ski goggles, trim, and earmarks for livestock.

Cellulose Nitrate (CN), Celluloid

CNs with nitrogen contents of 10.6–10.8% are alcohol-soluble thermoplastics; with 12–12.2% nitrogen, they are ester-soluble. Nitrogen contents of more than 13% up to a maximum of 16% result in guncotton. CN with a nitrogen content of approximately 11% is compounded with camphor to produce celluloid.

Celluloid is the oldest thermoplastic material manufactured in large amounts. Today, it is only used for the production of table tennis balls, toys, toothbrushes, and combs, because of its high flammability. Other areas of application for CN include bonding agents for coatings and adhesives and coating material for artificial leather.

Methyl Cellulose (MC)

MC with 1.3–1.5 OH-groups per glucose remnant is soluble in cold water; it coagulates at elevated temperatures depending on the molecular weight, degree of etherification, and concentration. MC with more than 2 OH-groups is not soluble in water.

Applications: impregnation resins, surface films on paper (fat- and oil-resistant), thickener, adhesives, and paint binders.

Ethyl Cellulose (EC)

EC is available in various viscosity grades. High- and medium-viscosity grades are used for the manufacture of films and molded parts; medium- and low-viscosity grades are used for coatings and adhesives. For a comparison of properties, see Table 6.54.

Carboxymethyl Cellulose (CMC)

In contrast to MC, all viscosity grades of CMC are soluble in warm and cold water.

Applications: adhesives, paint binders, additives for detergents and drilling fluids, aid in the textile industry, emulsifying agent, and thickener.

Hydroxyethyl Cellulose

Hydroxyethyl cellulose exhibits the same properties in water as CMC and is used in the same areas of application.

The various methyl, ethyl, carboxymethyl, and hydroxyethyl cellulose ethers are not used as plastics materials but mostly as functional polymers in a variety of applications in the textile, building materials, cosmetics, laundry and food industries.

Cellulose Hydrate (CH), Cellulose Film; Vulcanized Fiber (VF)

CH is also called regenerated cellulose.

Properties: Cellulose film exhibits low permeability to water vapor, gases, oils, etc.

To produce vulcanized fibers, absorptive papers or cellulose are swelled in a highly concentrated warm zinc chloride solution, stacked to form panels, washed out, and subsequently pressed.

Applications: crystal clear films (cellulose films), which can be weather-stabilized by double-sided coating, and highly loaded parts such as gear wheels, grinding wheels, and suitcase reinforcements.

Polysaccharides (PSACs); Thermoplastic Starches

Today, almost 50% of the starch produced is used in non-food applications. In the beginning, starch was employed only as a filler in thermoplastic materials to obtain biodegradable grades: The starch content decayed, causing the decomposition of the matrix resin into small particles (PE + starch for films used for plastic bags). Lately, plastics are produced from starch without using synthetic compounds. Only biodegradable plasticizers (water), oils, and fats are added.

PSACs are biodegradable and compostable resins. Due to their poor processing and usage properties, they are often blended with other materials or as copolymers. Pellets are the common delivery form; they can be injection-molded and blown into films.

Properties: density: 1.1–1.2 g/cm^3, Young's modulus: 1000–2600 MPa, tensile strength: 30 MPa, elongation at break: 600–850% (films), modulus of elasticity: 1000–2600 MPa, and Vicat A/50 73 °C up to 103 °C.

Applications: compostable packaging and perishable consumer goods (films and molded products).

6.18.2 Polyhydroxy Fatty Acids / Polyhydroxyalkanoates (PHAs)

This group of materials includes polyhydroxybutyrate (PHB) and polyhydroxyvalerate (PHV), which are created by a variety of bacteria as storage substances. This fact is used today for the industrial production of PHB. Both materials are considered biodegradable.

Polyhydroxybutyrate is a linear polymer; it is water-resistant, mostly resistant to fats and oils, and impermeable to oxygen, carbon dioxide, and water vapor. It is highly crystalline, exhibits properties similar to those of polypropylene (it is slightly more brittle and stiffer than PP), and can be injection-molded or extruded.

Properties of compounds based on starch: density: 1.17–1.25 g/cm^3, tensile strength: 15–27 MPa, and Young's modulus: 900–2000 MPa.

Applications: laboratory analyses, medical applications, and injection-molded parts.

6.18.3 Casein Polymers, Casein Formaldehyde (CS, CSF), Artificial Horn

Casein is an important protein in milk. Caseins are used for glues and as binders in paints. Curing with formaldehyde turns casein extracted from skim milk into a polymer that is chemically related to natural horn and resembles it in terms of appearance and machinability. CSF is mainly used for jewelry and for buttons and buckles.

6.18.4 Polylactides, Polylactic Acids (PLAs)

Polylactic acid (PLA) is a polyester that does not occur naturally. It is manufactured from sugar via a multistep synthesis, by fermenting sugar into lactic acid and then polymerizing it to PLA. PLA is transparent, crystalline, stiff (Young's modulus: up to 5000 MPa), and has a high mechanical strength; it can be processed using conventional thermoplastic processes (such as thermoforming and injection molding). It exhibits relatively high moisture absorption, which usually requires preliminary drying.

There are two stereoisomeric variations of lactic acid, L-lactide and D-lactide, and, consequently, two variations of PLA (PDLA and PLLA). Depending on how these building blocks are incorporated in the PLA chain, the morphology of the material varies from amorphous to crystalline, with its corresponding influence on the thermal and mechanical properties.

PLA is used in the packaging industry – for example, in yoghurt containers – because the material is biodegradable. It is also used for absorbable implants and sutures.

More recently, PLA has also been used as a blend partner for conventional plastics such as PMMA.

References for Section 6.18

[1] Endres, H.-J., Siebert-Raths, A., *Engineering biopolymers*, Hanser, Munich (2011)

[2] Endres, H.-J. (2017). Bioplastics. In: *Biorefineries. Advances in Biochemical Engineering/Biotechnology*, Wagemann, K., Tippkötter, N. (eds), Vol. 166, Springer, Cham. *https://doi.org/10.1007/10_2016_75*

[3] Godwin, H., Half-life of radiocarbon, *Nature*, 195 (1962), p. 984

6.19 Electrically Conductive/Luminescent Polymers

For many electro-technical applications, the intrinsic good insulation properties of polymeric materials are a prerequisite. However, polymeric materials with tailored electric conductivity properties are required in certain applications; for example, to avoid electrostatic charging, to provide electromagnetic shielding for plastics housings, and for the production of electrodes, light-emitting diodes, field-effect transistors, and others.

Conductive thermoplastic compounds are created by the incorporation of conductive fillers or reinforcements. Here, we will cover conductive polymers.

6.19.1 Chemical Constitution

Intrinsically conductive polymers (ICPs) have been a research topic for many years and find increasing applications. These organic materials consist of conjugated double bonds (alternating single and double bonds), as perfectly displayed in polyacetylene (see Table 6.55). The addition of electron donors (Na, K, Cs) or acceptors (I_2, $SbCl_5$, $FeCl_3$, among others), atoms or molecules that release (reduction) or accept (oxidation) electrons, results in increased electron mobility and conductivities up to 10^5 S/cm. As in metals, single free electrons remain and they are no longer bound to the atom cores but rather slide along the molecules and transport electric charges. This process is called "doping" in semiconductor technology. Equation 6.33 exemplifies this with polyacetylene, PAC (polyenes).

$$
\begin{array}{lll}
\text{(Acceptor } I_2\text{):} & \nearrow\ 3\ I_2 \ldots CH=CH-CH=CH-CH=CH^{(+)} - 2\ I_3^- & \text{(Cation)} \\
\ldots CH=CH-CH=CH-CH=CH & & \\
\text{(Donor Na):} & \searrow\ Na \ldots CH=CH-CH=CH-CH=CH^{(-)} - Na^+ & \text{(Anion)}
\end{array}
\qquad (6.33)
$$

Doping of Polyacetylene (PAC)

Table 6.55 General Structures of Intrinsically Conductive Polymers (ICPs)

Polymer	Acronym	Chemical structure	Spec. volume resistivity [Ω · m]	Oxidation or reduction agents, counter ions
Polyacetylene	PAC	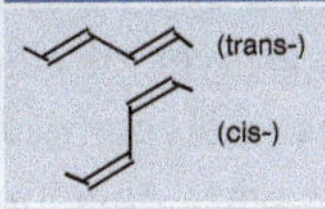	10^{-3}–$6 \cdot 10^{-6}$	I_2, AsF_5, Li, Na
Polypara-phenylene	PPP	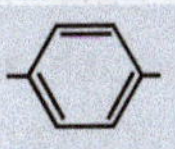	10^{-2}–10^{-3}	AsF_5, Li, K, Na
Polyphenylene sulfide	PPS	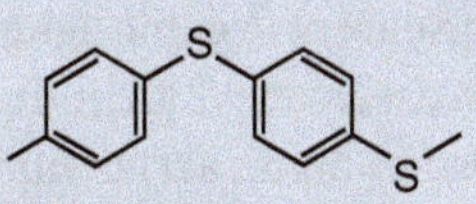	$1 \cdot 10^{-2}$–$2 \cdot 10^{-3}$	AsF_5
Polyaniline	PANI	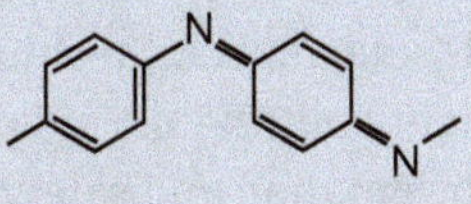	$3 \cdot 10^{-2}$–$5 \cdot 10^{-3}$	HCl, BF_4^-
Polyparaphenyl vinylene	PPV	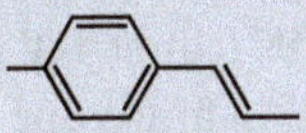	$3 \cdot 10^{-1}$–$2 \cdot 10^{-4}$	AsF_5
Polyphenylene butadiene	PPB	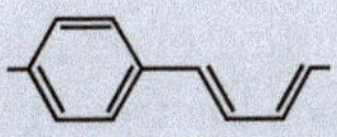		
Polyparapyridine	PPYR	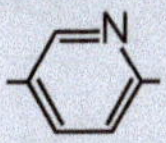		
Polyparapyridine vinylene	PPYV	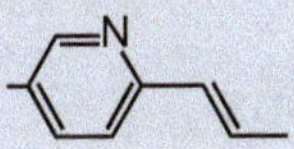		
Polypyrrole	PPY	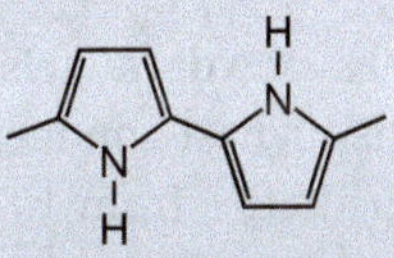	10^{-2}–$5 \cdot 10^{-3}$	BF_4^-, ClO_4^-, Tos^-, $FeCl_4^-$
Polythiophene	PT	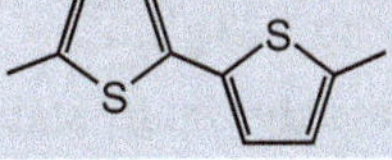	10^{-1}–10^{-3}	BF_4^-, ClO_4^-, Tos^-, $FeCl_4^-$
Polyfuran	PFU	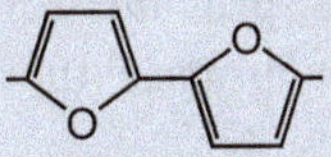		$CF_3SO_3^-$
Polyethylene dioxythiophene	PEDT	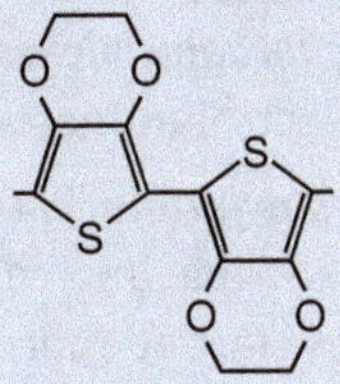	ca. $3 \cdot 10^{-2}$	
Polyacene		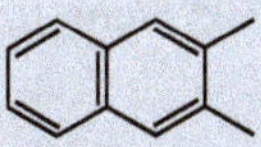		

Electroluminescence is defined as the ability of certain materials to emit light when voltage is applied. This has led to the development of OLEDs (organic light-emitting diodes). They consist of a quartz or glass wafer that, in the simplest case, carries three layers: an indium-tin oxide layer as anode, a thin film of luminescent polymer (*poly-3,4-ethylene dioxythiophene*), and, finally, a layer of calcium, magnesium, or aluminum as cathode. First-generation conductive polymers were completely insoluble and non-melting. Soluble conductive polymers were developed to cast thin layers of conductive polymers, second-generation polymers, such as those based on polyphenylene vinylene. Today, flexible, three-dimensionally formable, and transparent electroluminescence systems (TOLED: transparent organic light-emitting device) are available, consisting of a multi-layer film. An electrically conductive polymer is spin-coated, dipped, sprayed, or vacuum-deposited on a carrier film with a thickness of typically 0.15–0.8 mm. When AC is connected to the electrode, the intermediate layer will glow. It is characteristic for this so-called "smart surface technology" that it hardly consumes energy and does not emit heat. Light-emitting plastic parts are suitable for automotive interior and console lighting, orientation and safety signage, and for folding monitors. OLEDs can be applied at temperatures from −4 °C to +70 °C and, because of their chemical variability, they can be produced in any color, including white.

Photoluminescence is created by the splitting of double bonds under the absorption of light. When the double bonds reconnect, the freed bond energy is emitted as light. Masterbatches based on PE, PP, PA, PS, and other engineering plastics are also available.

6.19.2 Properties and Applications

Polyacetylenes (PACs) are available as powders, gels, or films (also transparent, stretchable up to 600%, resulting in its highest conductivity of more than 10^5 S/cm). They are insoluble and exhibit conductivities from 10^{-5} S/cm to 10^5 S/cm, depending on the polymerization process. Prior to doping, the densities range from 0.4 g/cm^3 to 0.9 g/cm^3; after doping, from 1.12 g/cm^3 to 1.23 g/cm^3.

Polyphenylenes: Non-doped polymers are thermally stable up to approximately 450 °C. Positive doping with, for example, $FeCl_3$ provides conductivities of approximately 10^2 S/cm. The products are sensitive to hydrolysis; those with negative K-doping are also sensitive to oxygen.

Polyparaphenylene vinylenes (PPVs), polyparaphenylenes (PPPs): Their applications include planar light-emitting diodes and electroluminescence.

Polyheteroaromates, polypyrroles (PPYs), polyfurans (PFUs), polythiophenes (PTs): These materials are electro-chemically polymerizable and can be stripped from the anode as films. Self-supporting film thicknesses start at approximately 30 µm. Coatings are as thin as 0.01 µm; conductivities, from 10^{-3} S/cm to 10^2 S/cm; the tear strength,

20–80 N/mm^2; and the elongation at break, 10–20%. Their applications include antistatic equipment (PSU films), heating bands, fuses, sensors, and batteries.

Polyanilines (PANIs) can achieve a conductivity of approximately 10 S/cm by doping with various acids in an aqueous solution.

Polyphenylene amines are insoluble and non-melting polymers. As finely dispersed compounds in coatings (nanopowders), they shift the corrosion potential of iron, steel, aluminum, zinc, stainless steel, and copper towards the precious metals, thus slowing the corrosion process, in particular when the metal surface suffers minor damages. Other applications are solar cells, membranes for gas separation, fuel cells, batteries, and sensors.

Polyethylene dioxythiophenes (PEDTs) are polymerized in solution as an insoluble, non-melting powder with powder conductivities of up to 30 S/cm. The surface resistivity σ_e of a layer with a thickness of approximately 0.1 µm on PC or glass is approximately 1000 Ω. Their applications include solid electrolyte condensers and conductive coatings on plastics or glass by oxidative polymerization with, for example, iron(III) toluenesulfonate solution in n-butanol.

Polyethylene dioxythiophene polystyrene sulfonates (PEDT/PSSs): Oxidative polymerization of the monomeric PEDT in the presence of aqueous polysulfonic acid creates colloidal solutions of PEDT that can be processed. Potassium peroxide disulfate is used as an oxidizing agent. The minimal surface resistivity σ_e is approximately 150 Ω. Their applications include transparent, antistatic coatings of photographic films, glass, light-emitting diodes, and OLEDs.

6.20 Thermoplastic Elastomers (TPE)

Acronyms, see Table 6.56.

Table 6.56 Overview of Thermoplastic Elastomers

Acronym	Chemical description
TPE	**Thermoplastic elastomers**
TPA	**Polyamide TPE**
TPA-EE	TPA, soft segments with ether- and ester bonds
TPA-ES	TPA with polyester soft segments
TPA-ET	TPA with polyether soft segments
TPC	**Copolyester TPE**

Table 6.56 Overview of Thermoplastic Elastomers *(continued)*

Acronym	Chemical description
TPC-EE	TPC, soft segments with ether- and ester bonds
TPC-ES	TPC with polyester soft segments
TPC-ET	TPC with polyether soft segments
TPO	**Olefin TPE**
TPO-(EPDM+PP)	Ethylene/propylene/diene + polypropylene
TPO-(EVAC+PVDC)	Ethylene/vinyl acetate + polyvinylidene chloride
TPS	**Styrene TPE**
TPS-SBS	Styrene/butadiene block copolymer
TPS-SIS	Styrene/isoprene block copolymer
TPS-SEBS	Styrene/ethene/butene/styrene block copolymer
TPS-SEPS	Styrene/ethene/propene/styrene block copolymer
TPU	**Urethane TPE**
TPU-ARES	Aromatic rigid segments, polyester soft segments
TPU-ARET	Aromatic rigid segments, polyether soft segments
TPU-AREE	Aromatic rigid segments, soft segments with ether and ester bonds
TPU-ARCE	Aromatic rigid segments, polycarbonate soft segments
TPU-ARCL	Aromatic rigid segments, polycaprolactone soft segments
TPU-ALES	Aliphatic rigid segments, polyester soft segments
TPU-ALET	Aliphatic rigid segments, polyether soft segments
TPV	**TPE with crosslinked rubber**
TPV-(EPDM-X+PP)	Highly crosslinked EPDM + PP
TPV-(NBR-X+PP)	Highly crosslinked acrylonitrile/butadiene
TPV-(NR-X+PP)	Highly crosslinked natural rubber + PP
TPV-(ENR-X+PP)	Highly crosslinked epoxidized natural rubber + PP
TPV-(PBA-X+PP)	Crosslinked polybutylacrylate + PP
TPZ	**Other TPEs**
TPZ-(NBR+PVC)	Acrylonitrile/butadiene rubber + polyvinyl chloride

6.20.1 Physical Constitution

TPEs combine the elastomeric properties of crosslinked elastomers with those of rubbers (see Section 6.23) with the advantage of thermoplastic processability.

Their composition allows the classification in two groups:

Polymer blends consist of a "rigid" thermoplastic polymer matrix, into which either non-crosslinked or crosslinked elastomer particles are incorporated as a "flexible phase". Examples are thermoplastic polyolefin elastomers (TPO or TPV) that consist of PP with up to 65% of incorporated ethylene-propylene-[diene] rubber (EP[D]M).

Graft- or copolymer molecules contain thermoplastic sequences A and elastomeric sequences B. Both components A and B are incompatible and demulsify locally so that the rigid A sequences form physical crosslinking points in the continuous matrix of flexible B sequences. An example is styrene block copolymer TPS, in which blocks of polystyrene (S) and butadiene (B) alternate: SSSSSSSS-BBBBBBBB-SSSSSSSS.

At service temperature, the flexible B sequences are above their glass transition temperature (freezing point); however, the rigid A sequences are below their glass transition temperature (for amorphous polymers) or their melt temperature (for semi-crystalline polymers). Above their transformation temperature, the A sequences soften so that TPEs can be processed like thermoplastics.

6.20.2 Chemical Constitution, Properties, Applications

Compared to rubbers, TPEs offer the following advantages: being suitable for thermoplastic processing and therefore also for recycling, weldability, transparency for some grades, colorability. An important area of application is therefore rigid-flexible combinations in 2-component injection molding (see Section 4.3.5.4) and in coextrusion (see Section 4.2.3.9). Table 6.56 provides an overview of acronyms and chemical designations, while Table 6.57 provides a comparison of properties.

Table 6.57 Property Comparison of Thermoplastic Elastomers (TPE)

TPE	Type	Density	Shore hardness		Service temperature °C			Glass transition temperature T_g	Resistance against[1)]				
TPA	PA6	1.0–1.2	> 65	< 75		85	−60	80	2	1	2	2	2
TPA	PA12	1.1–1.3	> 85	< 85	150	120/150	−65	160/220	2	1	3	4	2
TPC[2)]		0.9–1.0	> 50	< 70		115	−50	160	4	4	2	1	2
TPO	EVAC-PVDC		> 55	< 80	100		−40						
TPO	PP-EPDM-X	0.94–1.0	> 35	< 75	145	125	−50	160	3	3	1	1	1
TPV	PP-NBR-X		> 45	< 70	110		−40	160					
TPS	SBS	0.9–1.1	> 27	< 50	90	80	−40	95	2	4	2	2	2
TPS	SEBS		> 10	< 75	150	130	−50						
TPU	AREE	1.1–1.3	> 65	< 80	120	100/80	−50	130–200	1	1	3	4	1

1) 1 = excellent, 2 = good, 3 = adequate, 4 = poor
2) see also Table 6.30

Table 6.58 provides guidelines on which plastic and rubber grades may be substituted by TPEs in certain applications.

Table 6.58 Examples of Substitutions of Plastics and Rubbers by TPEs

Plastic	Rubber	TPE
PE, PTFE	CR	TPA
PE, PTFE	CR, NBR, EPDM, ECO	TPC
PVC-P, PC+PBT	NR, SBR	TPO
PVC-P, PA+PPE, PC-Blends, PUR	NR, CR, SBR, NBR, EPDM, ECO	TPV
PVC-P, PUR	NR, CR, SBR, EPDM	TPS
PE, PP, PTFE, PVC-P, ABS, PUR	NR, CR, SBR, EPDM, NBR	TPU

In the following sections, both TPE groups will be described in detail (see also the sections covering the base polymers).

6.20.2.1 Copolyamides (TPA)

Polyamide elastomers are block copolymers with rigid polyamide blocks and flexible polyether- (TPA-ET) or polyester- (TPA-ES) blocks or with both flexible blocks (TPA-EE). Components for the production of TPA-ET are lactam, dihydroxypolytetrahydrofuran, and dicarboxylic acid; for the production of TPA-EE: laurolactam, adipic acid, and bis(hydroxymethyl)cyclohexane.

TPEs are resistant to oils and fuels, sensitive to organic solvents and thermooxidative attack.

Applications: seals and gaskets, tubes in automotive and medical applications, sports goods, processing aids for TPU, antistatic agents for thermoplastics.

6.20.2.2 Copolyesters (TPC)

TPCs are block copolymers consisting of flexible polyalkylene ether diols and/or long-chain aliphatic dicarboxylic acid ester segments and semi-crystalline PBT segments. The properties range from rubber-like to highly flexible engineering plastics. Service temperatures range from −4 to 100 °C; heat stabilization is required at higher temperatures. TPCs are resistant to fuels, lubricants, and hydrolysis, and can be stabilized against UV and weather influences.

Prior to *processing* the pellets should be dried; TPC can be injection molded or extruded at temperatures of ≈ 220 °C with minimum residence times to avoid degradation.

Applications: membranes, hoses for compressed air and hydraulics, cable sheathing, bellows, caps, coupling and drive elements, seals, soles for ski boots and soccer shoes, rolls, bearings, and fasteners in automotive applications.

6.20.2.3 Polyolefin Elastomers (TPO)

TPOs are blends typically based on isotactic PP and ethylene-propylene rubbers (EPDM), see also Section 6.2.6.

Applications: bumpers, mud flaps, cables, shoes, soles.

TPO-(EVAC-PVDC) is a chlorine-containing, thermoplastic olefin-elastomer based on alloys of PVDC-rigid domains with a partially crosslinked flexible EVAC-copolymer matrix. It exhibits Shore hardnesses in a range from A 60 to 70 and is used to substitute oil-resistant vulcanized nitrile (NBR)- or chloroprene (NCR)-rubbers in a wide range of service temperatures (−40 to 120 °C) in static applications. These TPOs are age-, weather-, and ozone resistant to a higher degree than vulcanized rubbers. They are not suitable for applications under high dynamic loads, such as tires, because of their high mechanical damping properties. TPO grades are available as pellets with 10% carbon black filler, even in pastel tones, and can be processed at 170 °C on corrosion-resistant PVC processing machines as well as on rubber processing machines.

Applications: hoses, seals, conveyors, wire and cable sheathing, automotive engineering.

6.20.2.4 Polystyrene Elastomers (TPS)

Anionic polymerization with, *e.g.*, lithium butyl, allows the successive creation of blocks of styrene and butadiene and then again styrene. This results in a three-block copolymer TPS-SBS. Polymers of the SIS type (I = isoprene) can be produced by a similar method, TPS-SIS. The elastic properties of these polymers are caused by the fact that the polystyrene chain segments aggregate to rigid domains, while the polybutadiene segments aggregate to flexible rubber domains (2-phase plastics). The rigid domains act as physical crosslinking points, which will dissolve in the temperature range of the PS melting point to form a melt that can be processed like any thermoplastic.

Because of the susceptibility of the BR- and IR-chains to oxidation, these polymers are usually hydrogenated. This transforms SBS-three-block copolymers into styrene-ethene butene-styrene-three block copolymers (TPS-SEBS) and SIS into styrene-ethene propene-styrene-three block copolymers (TPS-SEPS). However, these products no longer exhibit the high elasticity of SBS and SIS grades, respectively, but are more rigid.

TPS are substitutes for vulcanized rubber with the advantage of effective thermoplastic processability: tubes and profiles, medical devices, injection molded soles in the shoe industry, cable insulation and sheathing grades, sound absorbing elements under the hood, bellows.

TPS are also used as flexile components in rigid-flexible-multi component injection molding. Good to excellent bonding is achieved with the following plastics: PE, PP, PS, ABS, PA, PPO, PBT.

Incorporation of expandable hollow spheres into the TPS-SEBS pellets allows for the production of fine pored foams with densities of 0.5 g/cm³ that can be injection molded or extruded for handles, wheels, or profiles.

6.20.2.5 Polyurethane Elastomers (TPU)

TPUs are block copolymers that are created by polyaddition of long-chain diols (chain extenders: 1,4-butanediol or 1,6-hexanediol; long-chain diols: polyether diols, polyester diols, polycarbonate diols) with aliphatic diisocyanates (IPDI, HDI). They are considered "bio-stable". Figure 6.24 and the table below show the influence of the formulation on the shear modulus.

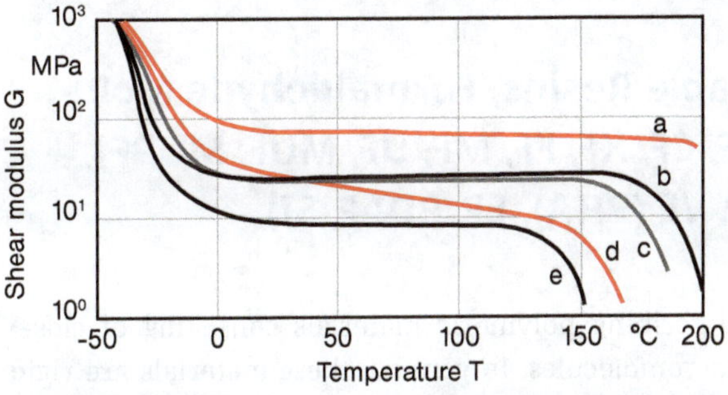

Figure 6.24 Influence of formulation on the shear modulus of TPU

Formulation	a	b	c	d	e
	Mole				
Poly(ethylene adipate)	1	1	1	1	1
NDI	4	2			
TDI			2		
MDI				4	2
1,4-Butanediol	2.6	0.85		2.6	
Spec. isobutyl ester			0.85		
1,4-bis(β-hydroxyethoxy)benzene					0.85

TPUs exhibit high wear resistance and flexibility, even at low temperatures, damping, transparency, resistance to fats, oils, and solvents, and high long-term service temperatures. Applications: tubes, hinge seals, attenuators, shoe industry, impact modification of polar thermoplastics, migration-resistant plasticizers for PVC.

6.20.2.6 Polyolefin Blends with Crosslinked Rubber (TPV)

These TPEs are crosslinked with flexible rubber segments. They exhibit increased elasticity and resistance. Oil-resistant and foamable grades are available. Applications: door- and window seals, air intake manifolds in automotive applications.

6.20.2.7 Other TPEs, TPZs

This category includes TPEs based on PVC with NBR [TPZ-(NBR + PVC)] or PBA [TPZ-(PBA + PVC)]. They are used for extruded seals. TPEs with crosslinked NBR are considered TPVs.

6.21 Thermosets, Curable Resins, Formaldehyde Molding Resins (PF, RF, CF, XF, FF, MF, UF, MUF, MUPF), Other Resins (UP, VE (PHA), EP, PDAP, SI)

Thermosets is a general term describing polymeric materials consisting of close-meshed, spatially crosslinked macromolecules. In general, these materials are rigid and behave as elastics up to their degradation temperature so that that they cannot be processed like thermoplastics. Molding and forming occurs either at the same time or prior to chemical crosslinking (curing).

Raw materials for thermosets are reaction resins (curable resins), which will react either at room temperature after the addition of curing agents or at elevated temperatures without curing agents. Curing can also be initiated by electron- or UV radiation.

Curable molding compounds are resins that can be processed by compression, transfer molding, extrusion, or (for special resins) by injection molding under simultaneous formation of macromolecules (crosslinking). They are filled and reinforced (some to a high degree). Delivery forms: dust-free grounds, pellets, free-flowing rods, and granules.

Prepregs (pre-impregnated) are laminar or tape-like materials, into which reinforcement in the form of mats, fabrics, or rovings can be incorporated (possible orientation from anisotropic to uniaxial).

6.21.1 Chemical Constitution

6.21.1.1 Formaldehyde Molding Resins (PF, RF, CF, XF, FF, MF, UF, MUF, MUPF)

The polycondensation of formaldehyde resins will be exemplified by phenol formaldehyde (PF). It begins with fundamental reactions (see Equation 6.34 to Equation 6.36), which involve the formation of various cross-linkages during curing:

Addition to phenol alcohols:

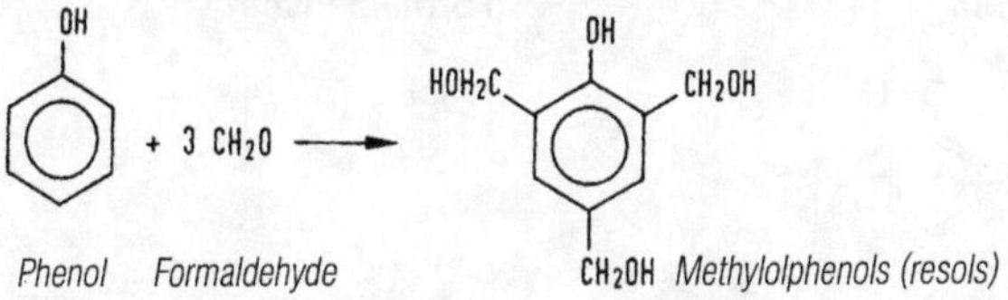

(6.34)

Condensation under elimination of water:

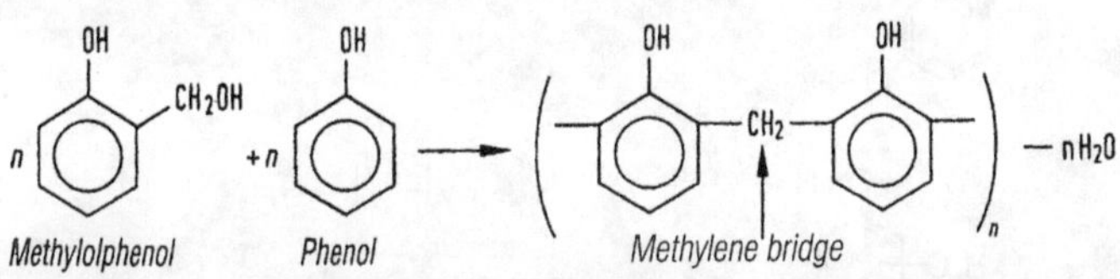

(6.35)

The condensation is performed step-wise to facilitate the escape of volatile reaction products. In stage "A" (resol), the reaction product is still soluble and can be melted. In stage "B" (resistol), the product can only be swelled and softens only at elevated temperatures, while in stage "C" (resite) complete crosslinking has occurred (Equation 6.36). The product is insoluble and cannot be melted. The manufacturing of molded parts can be based on stage A resins, though mostly it is based on stage B resins.

Condensation with further elimination of water:

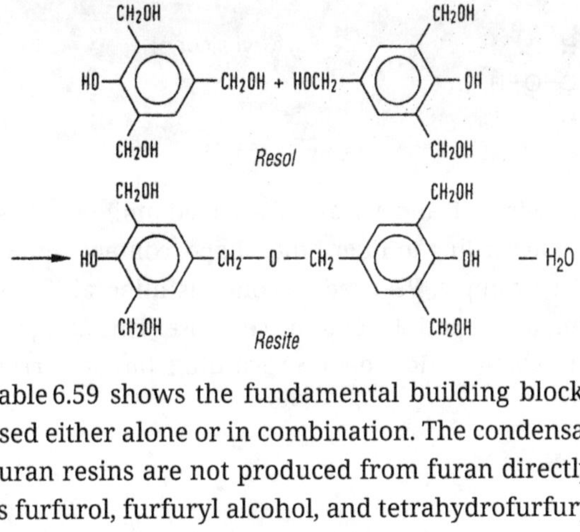

(6.36)

Table 6.59 shows the fundamental building blocks of formaldehyde resins that are used either alone or in combination. The condensation process is similar to that of PF. Furan resins are not produced from furan directly, but from furan derivatives, such as furfurol, furfuryl alcohol, and tetrahydrofurfuryl alcohol.

Table 6.59 Fundamental Building Blocks of Formaldehyde Resins

Resin	Acronym	Basic structure	
(Formaldehyde)			
Phenol formaldehyde	PF	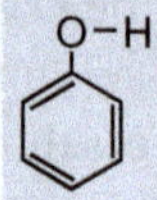 +	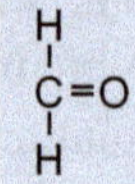
		Phenol +	Formaldehyde
Cresol formaldehyde	CF	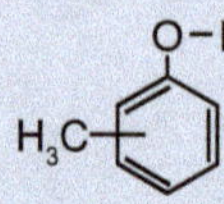	Cresol
Resorcinol formaldehyde	RF	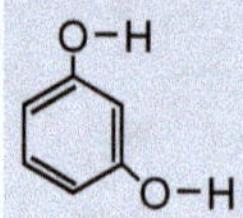	Resorcinol
Xylenol formaldehyde	XF	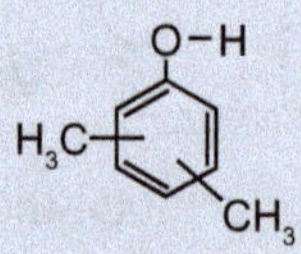	Xylenol
Urea formaldehyde	UF	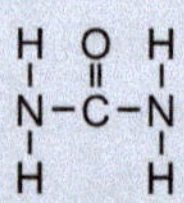	Urea
Melamine formaldehyde	MF	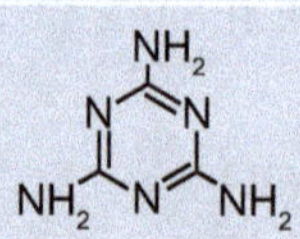	Melamine
Furfuryl alcohol formaldehyde, furan resin	FF	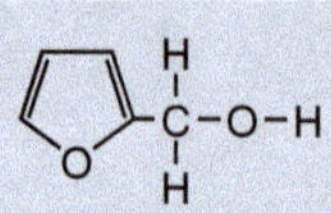	Furfuryl alcohol

PF, CF, RF, and XF are also called phenolics; UF and MF are also called aminoplastics. They are categorized in ISO 14526-30 according to filler type, shape, content, and in certain cases minimal properties. For example, PF Type 13 contains mineral fibers, whereas Type 31 contains wood flour. MF Type 130 contains cellulose. This categorization will be used in the tables in this chapter. More details regarding the categorization is provided in Table 6.63.

6.21.1.2 Unsaturated Polyester Resins (UP)

Unsaturated polyesters are soluble polyesters that can be melted (the building blocks of the polyester chains are connected by oxygen bridges) and that contain at least one unsaturated component. They are typically copolymerized in a mixture with polymerizable compounds, such as styrene or other monomeric vinyl-, allyl-, or acrylic compounds and peroxide compounds to form rigid, insoluble polymeric materials that can no longer be melted.

The first step in the manufacture of a linear polyester (pre-condensate) is the pre-condensation of an acid, *e.g.*, maleic acid anhydride, and an alcohol, *e.g.*, ethylene glycol, see Equation 6.37.

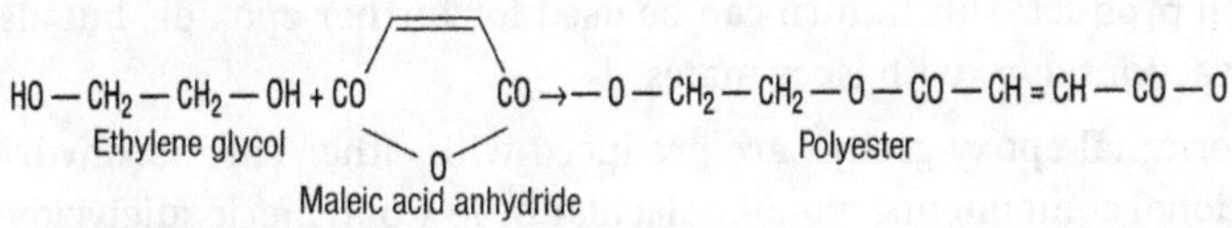

(6.37)

The use of high-order glycols, such as propylene glycol or butanediol, leads to more flexible and water resistant plastics. Linear polyesters can be crosslinked with the help of the double bond present in the maleic acid group, see Equation 6.38:

HOOC – CH = CH – COOH (6.38)

Maleic acid

Crosslinking typically occurs in a solution of monomeric compounds and continues until there is no solvent left. The transition (polymerization) from the unsaturated stage to the saturated, spatially crosslinked stage occurs only when the mixture is catalyzed, *e.g.*, by peroxides, see Equation 6.39.

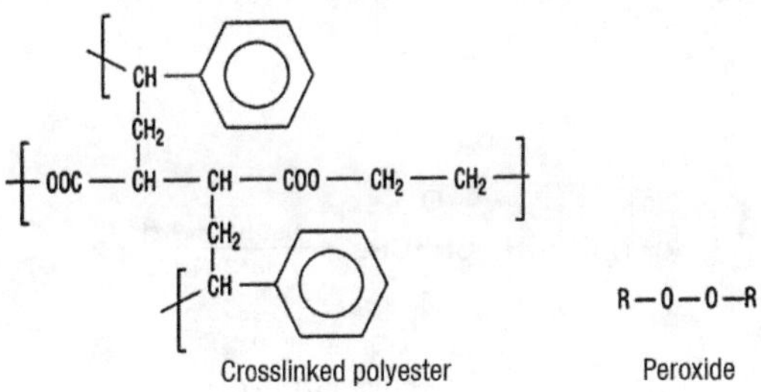

(6.39)

UP resins based on dicyclopentadiene (DCPD) exhibit a higher heat distortion point, lower styrene content, less fiber visibility and surface tack than standard UP.

6.21.1.3 Vinyl Ester Resins (VE); Phenyl Acrylate Resins; Vinyl Ester Urethanes (VU)

VE and VU are reaction resins from, *e.g.*, bisphenol A-glycidyl ethers, see Table 6.60, or epoxidized novolacs, with terminal esterified acrylic acid and/or methacrylic acid, see Table 6.4, No. 11 and 16. Dissolved in styrene (or a similar solvent), they cure during processing with special peroxide-cobalt-amine systems by crosslinking copolymerization with the solvent (similar to UP resins). The terminal crosslinking creates vibration-proof tough-rigid products.

VE and VU exhibit better mechanical properties and resistance compared to UP.

6.21.1.4 Epoxy Resins (EP)

The richness of epoxide chemistry (Table 6.61) relates to the capacity of the epoxy (or oxirane) group to combine with "active" hydrogen (alcohols, acids, amides, amines) under suitable reaction conditions and with suitable catalysts. The hydrogen is displaced relative to the epoxy oxygen in such a way that it gives rise to an active HO group in the addition product. This in turn can be used for further epoxide, but also for other addition reactions, *e. g.*, with isocyanates.

Epoxy resins with terminal epoxy groups are produced with either epichlorohydrin or bi- or multi-functional components, mono-molecular or low polymeric "diglycidyl" compounds (G in Eq. 1–5 in Table 6.60). Using different reactions, these products can also be directly incorporated in pre-products, *e. g.*, as seen in Eq. 6 in Table 6.60.

Table 6.60 EP Chemistry

Basic addition reaction of EP chemistry		
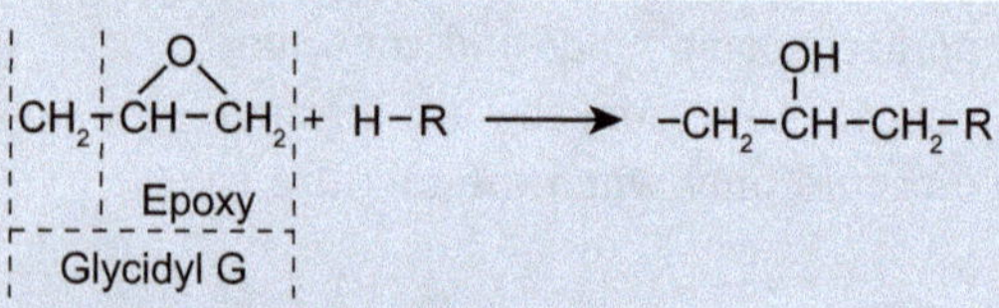		
Examples of di-epoxy prepolymers		
(1) Bisphenol glycidyl ether types		
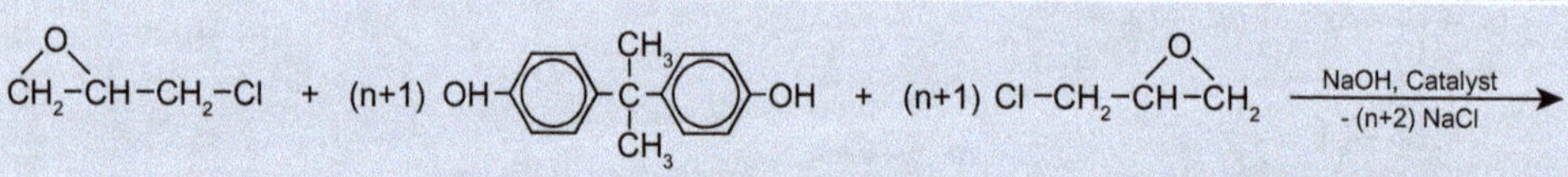		
Epichlorohydrin	Bisphenol A	Epichlorohydrin
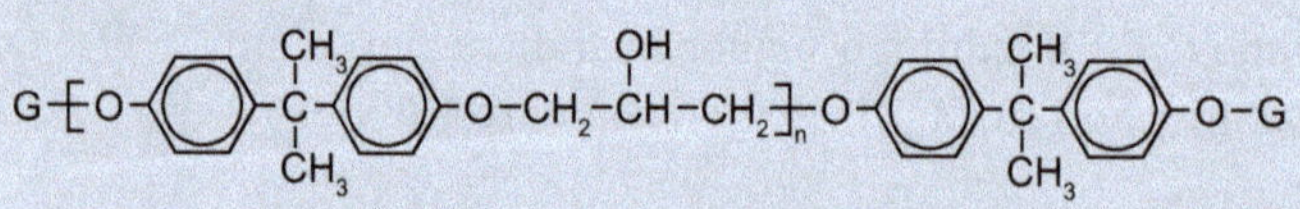		
"Dian"-resin: n = 0: Bisphenol A-diglycidyl ether (BADGE), 0 < n < 10: liquid to solid resins		
Variations: Flame retarded by halogenated dian or other bisphenols		

(2) Epoxy (cresol) novolac types	
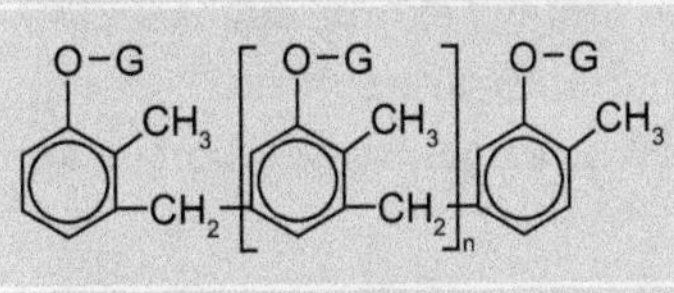	
(3) Aliphatic diglycidyl ether types	
Flexible resins with, *e. g.*, n = 4	
(4) Diglycidyl amine derivatives	
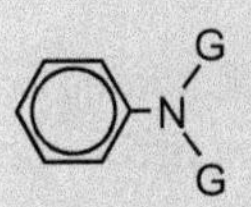	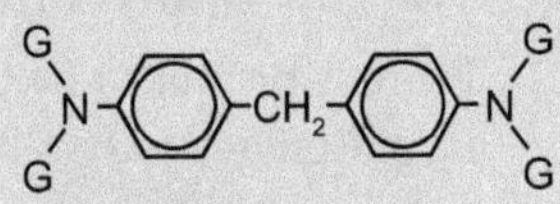
Diglycidyl aniline (bi-functional diluent)	*Multi-functional* for HT-coating resins
(5) Cycloaliphatic diglycidyl ester types	
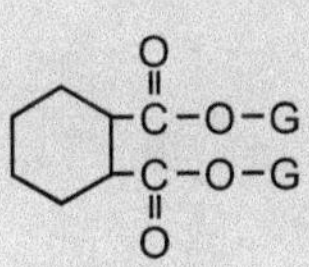	Hexahydrophthalic acid diglycidyl ester for arc- and tracking resistant HT-resins
(6) Cycloaliphatic types with directly linked EP groups	
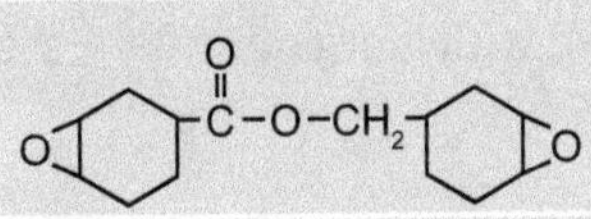	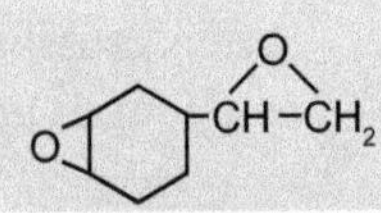
3,4-Epoxycyclohexylmethyl 3,4-epoxycyclohexanecarboxylate Materials for electro-technical encapsulating and casting resins	Vinyl cyclohexene dioxide

Reactants for the synthesis of highly crosslinked EP resin products include bi- or multi-functional low molecular weight products containing active hydrogen. They do not simply act as catalysts by initiating crosslinking of pre-polymers, as is the case of curing agents and accelerators for UP resins. Rather they are chemically incorporated into the macromolecule by addition to the epoxy group. It is therefore necessary to mix exactly measured quantities of the crosslinking components during epoxy resin processing. An additional possible variation is the use of "reactive diluents" with only one epoxy group to reduce the viscosity of the liquid resins and to increase the flexibility of the molding materials.

Liquid aliphatic polyamines and polyamidoamines are predominantly used to *cold-cure* liquid epoxy resins; tertiary amines are used as catalytic curing accelerators. For *warm curing* > 80 °C, either aromatic amides or their derivatives, or anhydride of phthalic acids or related acids (for flame retarded laminates) are used, sometimes together with accelerators. Systems cured with aliphatic amines exhibit highest resistance against chemicals; those cured with aromatic amines are solvent resistant, while those anhydritically cured show the highest resistance against weathering and acids.

Many reactants, particularly amines, are corrosive and otherwise hazardous chemicals; their use is regulated and both manufacturers and the chemical industry offer specific guidelines for their use.

6.21.1.5 Diallyl Phthalate Resins, Allyl Esters (PDAP)

PDAP is produced by the conversion of allyl groups ($CH_2{=}CH{-}CH_2$) with phthalates or isophthalates to two different grades of molding compounds, identified as ortho- or meta-grades. Because of their low viscosity, they are pre-polymerized during the production of molding compounds. Crosslinking occurs primarily peroxidically, sometimes with the addition of accelerators such as co-naphthenate or amines. They cure under pressure at 140–180 °C.

6.21.1.6 Silicone Resins (Si)

Silicones, of general structure shown in Equation 6.40, contain oxygen and silicon in their chains.

Alkyl-, aryl-, or chlorine alkyl groups are used as organic remnants "R". When trifunctional organo(trichloro)silanes or silicon tetrachloride are used during polycondensation, crosslinked or branched polymers are created. Filled with inorganic fillers, the compounds are compression molded at 175 °C.

$$-\overset{\displaystyle R}{\underset{\displaystyle R}{\overset{|}{\underset{|}{SiO}}}}- \qquad (6.40)$$

6.21.2 Processing, Forms of Delivery

Curable molding compounds are generally used in the form of compounds consisting of hot-curing resins with organic or inorganic fillers, reinforcements, colorants, or other additives. Special cold-curing molding compounds are compression molded (cold) and subsequently cured in an oven. Curable molding compounds are available as: dust-free grounds, pellets, free-flowing rods, or granules. Aminoplastics are considered fast molding resins.

The following acronyms and notations are established:

- *GMC* (granulated molding compounds) and PMC (pelletized molding compounds), respectively: dry, free-flowing granulated or pelletized (rods) molding compounds.

- *BMC* (bulk molding compounds): wet, doughy-fibrous compounds, chemically thickened, ISO 8606.
- *DMC* (dough molding compounds): wet, doughy-fibrous compounds with increased filler content, ISO 8606.
- *SMC* (sheet molding compounds, resin mats) and SMC-R (R = random) are impregnated (prepreg) molding compounds with 2-dimensionally, non-oriented reinforcing fibers (typically 25–50 mm long, glass fiber rovings 25–65 wt.-%), flowable both perpendicular and parallel to fiber orientation. Production, see Section 4.8.2.1.
- *SMC-D* (D = direct) containing 75–200 mm long fibers, oriented longitudinally; barely flowing longitudinally.
- *SMC-C* (C = continuous) containing continuous longitudinal fibers; longitudinally not able to flow.
- *Fabric prepregs* are impregnated woven or nonwoven fabrics; glass fiber content 35–60 wt.-%.
- *Prepreg tapes*: made from continuous fibers; unidirectional reinforced tapes.

UP molding compounds contain ≈ 40% fine mineral fillers together with 12–25 (65) wt.-% glass fibers. For special compression molding compounds, organic fillers and textile fibers are also added. GMC are available styrene-free, usually with diallyl phthalate as crosslinking agent. Shelf life until processing is ≈ 1 year. BMC and DMC contain resins dissolved in styrene. Though they are more difficult to meter, they do not reduce the length of the reinforcing fibers during processing; shelf life several months.

EP molding compounds are produced from the melt as granular, flaked, or rod-shaped compounds, sometimes colored with (chemically linked) inorganic fillers. Depending on curing agent, they have a limited shelf life above 20 °C. Because of their excellent flowing properties, these compounds are particularly suited for injection compression molding at low pressures, see Table 6.61. Because the curing reaction is accelerated above T_g (≈ 45 °C), the compounds must be quickly transferred to the mold at temperatures only slightly above the melting temperature of the resins (70 °C).

EP-prepregs are produced by impregnating glass, carbon, or synthetic fibers with resin solutions and subsequent evaporation of the solvent.

The molding compounds are plasticized under pressure and at elevated temperatures in the plasticizing unit of the processing machine or in the mold, respectively; they are then molded and cured under pressure. Pre-heating of the compounds improves and accelerates processing.

Table 6.61 provides data about processing conditions for the most common curable molding compounds. Condensation resins, such as PF, UF, MF, require mold venting and higher processing pressures and temperatures compared to easier flowing UP- and EP resins. Injection molding cycle times of these molding compounds are shorter than those of thermoplastics, because the molded parts can be ejected from the mold while still hot and because the exothermic reaction facilitates rapid heating with wall thicknesses > 4 mm.

Table 6.61 General Guidelines for Processing of Curable Molding Compounds

Process	Compression molding		Injection molding					
Molding compound	Compression temperature	Pressure	Barrel temperatures[1]		Mold temperatures	Back pressure	Injection pressure	Holding pressure
			Conveying zone	Nozzle				
	°C	bar	°C	°C	°C	bar	bar	bar
PF, Type 11–13	150–165	150–400	60–80	85–95	170–190	Up to 250	600–1400	600–1000
Type 31	155–170	150–350	70–80	90–100	170–190	300–400	600–1400	800–1200
Type 51, 83, 85	155–170	250–400	70–80	95–110	170–190	Up to 250	600–1700	800–1200
Type 15, 16, 57, 74, 77	155–170	300–600						
MF, Type 131	135–160	250–500	70–80	95–120	150–165	300–400	1500–2500	1000–1400
Type 150–152	145–170	250–500	70–80	95–105	160–180	Up to 250	1500–2500	800–1200
Type 156, 157	145–170	300–600	65–75	90–100	160–180	Up to 150	1500–2500	800–1200
MF/PF, Type 180, 182	160–165	250–400	60–80	90–110	160–180		1200–2000	
UP, Type 802	130–170	50–250	40–60	60–80	150–170	Without	200–1000	600–800
EP	160–170	100–200	ca. 70	ca. 70	160–170	Without	Up to 1200	600–800
PE, crosslinking	120 °C melting-, up to 200 °C crosslinking temperature		135–140	135–140	180–230			

[1] The melt temperatures cannot be determined accurately because of the contribution by friction and exothermic reaction.

6.21.3 Properties

6.21.3.1 General Properties

Curable molding compounds contain not only resins but also combinations of fillers and reinforcements: wood flour, cellulose, cotton fibers, cotton fabric cuttings, mineral fibers, stone powder, mica, short and long fibers, glass mats, and others.

Cured thermoset parts exhibit combinations of properties that make them suitable for engineering applications (see Table 6.62):

- Appropriate selection of fillers (content 40–60 wt.-%) make it possible to meet a wide range of end use requirements and enhances properties.
- Suitable for highly flame retarded electrical insulating applications.
- Stiffness and strength up to high temperatures barely decreasing, at low temperatures unchanged.
- Long-term service temperatures ranging from 100 to > 200 °C, combined with short-term overload capacity of several 100 K result in good dry-run characteristics and high safety reserves at extremely high temperatures.
- Dimensional accuracy, also very little shrinkage of parts – depending on resin grade and processing method – of < 0.1%, particularly if molding shrinkage is mitigated by tempering.

The brittle fracture behavior of highly crosslinked molding compounds can be compensated by appropriate part design. In addition, new PF molding compounds with up to 2% elongation at break and good thermal stability are now available.

Molding compounds with inorganic fillers have a higher long-term service temperature and better moisture resistance than compounds with organic fillers. The molding compounds' notch sensitivity decreases with increasing length (or volume) of the fillers; however, at the same time, processing becomes increasingly more difficult, particularly with respect to the free-flowing properties of the compound, the flow properties of the melt, and the surface quality of the finished part.

The properties of the molded compounds (cured molding compounds) are controlled by the type of fillers and reinforcements.

6.21.3.2 Phenoplastic Molding Compounds (PF, CF, RF, XF)

Cured PF compounds darken after exposure to sunlight or extended exposure to elevated temperatures, turning yellowish-brown. Therefore, they are offered only in dark colors.

All PF molding resins are resistant to organic solvents, fuels, fats and oils, even at elevated temperatures. They have only limited resistance to more concentrated acids and alkalis, and PF compounds with organic fillers are more susceptible to attack af-

Table 6.62 Basic Characteristics of Reaction Resin Compounds

Properties	Resin	UP		
	Reinforcement	GF - short fibers	GF < 20 mm	GF
	Delivery form	Pellets	Rods, BMC	Pellets
	Identification	Types[1] 802/804	Types[1] 801/803	Ceramic-like
Density	g/cm³	1.9–2.1	1.8–2.0	ca. 2.1
Mechanical				
Tensile strength	MPa	> 30	> 25	ca. 35
Flexural modulus	GPa	10–15	12–15	5–9
Impact resistance (Charpy)	kJ/m²	4.5–6	22	5–6
Notched impact resistance (Charpy)	kJ/m²	2.5–4	22	3–4
Thermal				
Heat deflection temperature HDT/A 1.8 MPa	°C	> 200	200–260	ca. 265
Service temperature, long-term	°C	> 160	150	> 200
Thermal coefficient of expansion	$10^5 \cdot K^{-1}$	2–4	2–5	1.5–3.0
Thermal conductivity	W/K · m	0.8	0.4	0.9
Electrical				
Specific volume resistivity	Ohm · cm	10^{12}	10^{12}	10^{14}
Surface resistivity	Ohm	10^{12}	10^{10}–10^{12}	10^{13}
Dielectric strength	kV/cm	120–180	130–150	–
Dielectric constant	50 Hz – 1 MHz	4–6	4–6	4.5–7
Dielectric loss factor	50 Hz – 1 MHz	0.04–0.01	0.06–0.02	0.02
Tracking resistance		> 600	> 600	> 600
Water absorption 4 d, 23 °C	mg	45	100–60	30–40
Water absorption 24h/23 °C	%	0.1–0.5	< 0.5	0.2

1) Types 803 and 804 flame resistant.
2) el. only for 5 types.
3) SMC-R to SMC-C, 25–65 GF.

	EP		DAP		SI
GF- chopped strand mats	Mineral	GF	Mineral	GF	Quartz
Chopped strand mats	Ground pellets or rods				
SMC, Type[2] 830–834	Not standardized				
1.7–2.4	1.6–2.0	1.6–2.1	1.65–1.85	1.7–2.0	1.9
50–230[3]	30–85	35–140	35–56	42–77	ca. 40
1–7	11–13	10–25	7–10	8.5–11	10–18
50–70	6–7	9–100	–	–	4.3–4.6
40–60	2	3–100	–	–	–
180– >200	107–260	107–260	160–288	166–288	>250
150	150–>200	150–>200	150–180	150–180	300
1–4	2–4	1–3	2–7	2–5	3.1–3.5
0.5	0.6	0.6	0.3–1.0	0.2–0.6	0.4–0.6
10^{12}–10^{15}	10^{14}–10^{16}	10^{14}–10^{16}	10^{11}–10^{15}	10^{13}–10^{16}	10^{16}
10^{10}–10^{11}	$>10^{12}$	$>10^{12}$	10^{10}–10^{14}	10^{10}–10^{14}	
130–150	130–180	140–150	–	–	250
4–6	3.5–5	3.5–5	4–5	4–4.5	3.4–3.6
<0.1– <0.01	0.07–0.01	0.04–0.01	0.007–0.015	0.004–0.009	0.001–0.002
>600	>600	>600	>600	>600	>600
<100	30	30	–	–	–
0.1–0.3	0.03–0.2	0.04–0.2	0.2–0.5	0.1–0.3	0.1

ter longer exposure than those filled with inorganic materials because of their increased water absorption. The rate of water absorption is so small for all PF compounds that they are not damaged by short-term exposure to water. Adequately cured moldings should show no signs of damage after boiling in water or dye solutions for half an hour. Inorganic compounds should be used for insulating parts exposed to weather or damp conditions because the slow water absorption of organic filled compounds causes a decrease in electrical insulating properties.

6.21.3.3 Aminoplastic Molding Compounds (UF, MF)

A resin content of approx. 60% is necessary to obtain a sufficient melt flow rate in *UF molding compounds*. The light colored compounds, filled with bleached cellulose, make for translucent molding compounds with good coloring properties and light resistance. The parts are taste- and odorless, resistant to food and oils, but less resistant to acids and alkalis.

Parts made from *MF molding compounds* filled with 60% α-cellulose are white and in all other colors resistant to light-, hot water-, and dish-washing detergents. Their scratch resistance is higher than that of MPF (melamine phenolic resin compounds) or UF. Grades 152, 153, and 154 are insulation materials that are tracking resistant, moisture resistant, and resistant to high mechanical loads. The inorganic filled grades 155 and 156 exhibit arc and incandescence resistance and are almost nonflammable. MF is attacked by acids and strong alkalis; it is largely resistant to fuels, oils, solvents, and alcohols. Only special MF compounds can be injection molded or compression injection molded. Because of their high shrinkage, MF parts are susceptible to stress cracking.

6.21.3.4 Melamine Phenolic Molding Compounds (MPF)

MPFs are mixed resin molding compounds that do not match MF with regard to tracking, distortion, temperature, and light resistance, but that are produced in light colors and meet the requirements for color fastness of many applications. Advantageous compared to MF is the significantly reduced shrinkage.

6.21.3.5 Melamine Polyester Resin Molding Compounds (MF+UP)

Parts molded from MF + UP (filled with either cellulose or with a mixture of organic/inorganic materials) combine the brilliant color, tracking and arc resistance of MF with the minimal shrinkage, crack resistance, and increased dimensional stability under high temperature loads of UP.

6.21.3.6 Polyester Resin Molding Compounds (UP)

Under appropriate processing temperatures and low pressures, UP resins for molding compounds have good melt flow properties. They quickly and completely cure by polymerization without separation of volatile by-products. The cured parts exhibit very little shrinkage and reliably maintain their dimensions; they are also not susceptible to stress cracking. They are light resistant in any light colored pigmentation, resistant to alcohols, ethers, gasoline, lubricants, and fats; conditionally resistant to benzene, esters, weak acids, and boiling water; nonresistant to alkalis and strong acids (more corrosion resistant binders, also phenacrylate or vinyl ester resins, see Section 6.22.5). They exhibit high glass transition temperatures and good electrical properties with high tracking resistance and low dielectric losses. Molding compounds with inorganic fillers may be arc and incandescent resistant (see Table 6.63).

While GMC, PMC, BMC, DMC, and SMC-R exhibit largely isotropic mechanical properties in planar direction, SMC-C and SMC-D can be used to produce parts with optimized stability in the direction of load. Long-fiber strands inserted lengthwise into the mold result in molded parts with extremely high oriented resistance to loads.

Processing shrinkages of 0.2–0.4% result in unacceptable rough surfaces, particularly in large automotive parts. Addition of 5–25% PVAC, PS, PMMA, or CAB dissolved in styrene results in low profile resins (LP resins). During curing of the UP resin, the styrene dissolved in the thermoplastic evaporates and its vapor pressure compensates for the shrinkage of the UP resin, thus creating smooth surfaces; characteristic values, see Table 6.63.

6.21.3.7 Vinyl Ester Molding Compounds (VE)

VEs exhibit higher elongation at break than UP (3.5–6%), and long-term service temperatures of 100–150 °C. They are resistant to 37% HCl and 50% NaOH solution, liquid chlorine, chlorine dioxide, hypochlorite in all concentrations, hydrocarbons, and oxygen-containing organic media. The resins are temperature resistant to a higher degree and also resistant to chlorinated and aromatic solvents.

6.21.3.8 Epoxy Resin Molding Compounds (EP)

EP molding compounds are distinguished by little processing shrinkage and basically no post-shrinkage even at elevated temperatures, together with high dimensional accuracy. Because of EP's good melt flow properties, it is possible to embed very fine metal particles that will not undergo deformation during processing. Also, dense and well adhering coating of large metal parts can be achieved.

EP is resistant to fuels and hydraulic oils. Characteristic values, see Table 6.63.

Table 6.63 General Characteristics for Standardized Curing Molding Compounds

Type #	Filler group	Density	Max. service temperature without additional load	
		g/cm^3	Hours to days °C	Months °C
Phenoplastic molding compounds according to ISO 14526				
11.5–16	Mineral, granular or fibrous	1.8–2.1	160–170	130–150
30.5–32 85	Wood flour	1.4	140	110
51–84	Organic fibers, strands, chips	1.35	140–150	110–120
Aminoplastic molding compounds according to ISO 14527/8/91)				
UF: 131, 131.5				
MF:	Cellulose			
152				
150	Wood flour	1.5	110	70
152.7	Cellulose			
153, 154	Cotton fibers, chips		100	80
155	Rock flour,	2.0	140	110
156–158	mineral fibers	1.7–1.8		
214/215	Rock flour, mineral fibers		Strength stability under heat (Martens temperature test)	
Polyester resin molding compounds ISO 14530				
801, 803	Glass fibers	1.8		
			200	160
802, 804	Short glass fibers	~2.0		
Polyester sheet molding compounds				
830–834	Mainly glass mats	1.8–2.1	200	160

1) Aminoplastic-phenoplastic molding compounds (Types 180–183, acronym MP) range between PF- and MF-compounds with regard to their characteristics.

2) These values refer to specimens with dimensions 120 x 15 x 10

Applications	Mechanical characteristics[2)]		Thermal characteristics	
	Tensile strength GPa	Young's modulus GPa	Coefficient of thermal expansion $10^{-6} \cdot K^{-1}$	Thermal conductivity W/m · K
Parts unsusceptible to moisture and heat, reinforced with mineral fibers suitable for high loads, suitable for many applications, including housings and electrical insulation	15–25	7–15	20–30	0.7
	~30	6–8	30–40	0.3
Technical parts; compared to Type 31 higher notch resistance and (orientation dependent) strength	~25	4–9	30–40	0.3
Light colored parts, electrical and insulation material	~30	6–10	50–60	~0.35
Table ware				
Tracking resistant electrical parts, type selection depending on other loads	15–30	8–13	20–40	~0.7
Impact resistant housings; covers; insulation under high loads	25	12–15	35	0.6
	30	9–11	20–30	~0.7
Large parts for high mechanical loads	60–200	5–8	15–30	~0.2

6.21.3.9 Diallyl Phthalate Molding Compounds (PDAP)

PDAPs exhibit particularly good electrical properties, even at elevated temperatures, as well as good heat stability (to more than 200 °C) and weather resistance. Most molding compounds are self-extinguishing. Characteristic values, see Table 6.63.

6.21.3.10 Silicone Resin Molding Compounds (SI)

They belong to the group of high temperature resistant plastics and exhibit a heat deflection temperature of 250–300 °C after curing. Characteristic values, see Table 6.63.

6.21.4 Applications

6.21.4.1 Phenoplastic Molding Compounds (PF, RF, CF, XF, FF)

Special compounds filled with wood flour and rubber with increased impact resistance (but significantly reduced heat distortion temperature) are used for parts in electrical applications, such as wire plugs, housings and sockets. Resins filled with short glass fibers (heat distortion point up to 180 °C) are used in telecommunication and electronics. Compounds filled with minerals and increasingly also filled with glass fibers that have short-term service temperatures up to 280 °C, long-term up to 180 °C, are used for temperature- and dishwasher resistant handles and fittings of small appliances, pots and pans, and in automotive applications (fuel resistant), *e.g.*, for ignition electronics, carburetor heads, valve covers, complete cooling fluid pumps, multiple air intake manifolds (manufactured, *e.g.*, by lost core techniques), brake- and clutch reinforcing pistons. Brake and clutch pads are short-term resistant to temperatures up to 600 °C; rocket parts are exposed to similar high loads. High quality insulation resins (short-term 400 °C) are also available as copper adhesive grades for collectors with high rotational speeds. Other special application areas include slip ring assemblies for ship generators, water pump impellers, (liquid-) gear drives.

Molding compounds filled with copper powder are good heat conductors; they can be magnetized with iron or barium ferrite; suitable for shielding of X-rays with lead powder; compounds with relatively high graphite filler content are used for semi-conductors, pump parts, or lubricating pins in automotive axles.

Glass fiber reinforced PF-prepregs und SMC have gained importance in transportation, in particular in airplane equipment, tunnel-, and mine facing, because of their temperature resistance (T_g = 300 °C), low smoke gas density and relative nonflammability. The SMC webs are saturated with highly concentrated liquid resol and special curing agents; they require 6–8 days of curing time and then have a shelf life of approx. two months.

6.21.4.2 Aminoplastic Molding Compounds (MF, UF, MPF)

MF: Compounds filled with inorganic materials (types 155 and 156, also filled with glass fibers) are arc and incandescent resistant. They do not burn, and are therefore suitable for rooms with high fire hazards and for ship building. MF molded parts are attacked by acids and strong alkalis; they are largely chemically resistant to fuels, oils, solvents, and alcohols.

Only special grades of MF molding compounds are suitable for injection and compression injection molding. Because of their high post-shrinkage, they are susceptible to stress cracking.

UF: Preferably in white for closures in cosmetic applications, sanitary objects, small appliances, and for electrical insulation. UF resins compounds do not meet requirements for food contact (dishes, etc.) because of their post-processing release of small amounts of formaldehyde.

MPF: Light colored parts for electrical applications, small appliances, screw connections.

MF + UP: Electrical-, appliance-, control units applications; light sockets.

Main area of application are parts under high loads (with complicated geometries) in electrical and electronics applications. Automotive electrics, headlight reflectors, and parts of appliances. Other than grades 801 and 802, grades 803 and 804 are flame retarded (UL 94 V-0, non-burning according to ASTM D 635), but otherwise similar. Molding compounds similar to type 804 can be compression molded, injection compression molded, and injection molded into parts with dimensional stability up to 200–240 °C (Martens). Their dimensional stability and mechanical and electrical properties are hardly affected by elevated temperatures and humid/hot climates, and they meet the same requirements as ceramics or steatite.

6.21.4.3 Unsaturated Polyester Resin Molding Compounds (UP)

UP-SMC (resin mats): The flow behavior of the resin mats allows the production of large double-curved parts, even with bosses and ribs, such as cabs, moon roofs, engine coverings, tailgates, bumpers, seat shells for trucks and cars, large parts for the interior of ships and aircraft, and others; type 834 fulfills the respective fire protection requirements. To avoid bubble formation during subsequent stove enameling of such parts, SMC/IMC (in-mold coating) is used to apply an ≈ 0.1 mm thick skin of liquid PUR glue (under a pressure of 400 bar) to the part's surface. For automotive parts in the so-called crush zone, modified flexible LP-SMCs with low Young's modulus are available. Low-density, high impact resistant structural foam parts ($d \geq 1\ g/cm^3$) are produced by the addition of ≈ 1% micro-encapsulated fluoroalkanes to the prepreg. Continuous mat prepregs (Unipreg) are particularly suitable for thin-walled complicated parts. With special UP- or phenacrylate resins, also carbon fiber reinforced, they almost reach the property profile of the much more expensive high-tech EP prepreg composites (see Table 6.64). Automotive bumpers and door frames are examples of applications for these HMC (= high modulus continuous) advanced SMCs.

Table 6.64 Characteristics of Selected Fiber Reinforced Composite Parts

Products		Pultruded pipes and profiles			
Resin/fiber group		UP/GF	EP/GF	EP/AF	EP/CF
Reinforced by	Rovings				
Resin content	%		ca 30		
Spec. weight	g/cm^3	1.9	2.1	1.4	1.6
Water absorption	%	ca. 1	0.2–0.3	0.5	0.2
Tensile strength	MPa	700	700	1300	1400
Elongation at break	%	2	2	1.8	0.6
Young's tensile modulus	GPa	35	35	75	130
Service temperature range	°C	<150	<180	–	–
Linear thermal expansion	$10^{-6} \cdot K^{-1}$	10	10	0	0.2
Thermal conductivity	$\frac{W}{m \cdot K}$	0.20	0.24	–	–

Glass reinforced surface mats are used as top coats for wood materials under high loads. Adhesive prepregs for the electronics industry are UP resin laminates on flexible carriers that will melt during coil winding at 100–120 °C and then cure during annealing.

The most highly mechanically loaded glass mat prepregs are useful only for flat or uni-directional curved moldings. Similar high mechanical values (750 N/mm^2 flexural strength, 200 kJ/m^2 notched impact strength) are achieved by directional forming of cross-wound, filler-free XMC mats with 65–72% glass fiber content.

Prepregs consisting of light-curing UP resins – filled with aluminum hydroxide for opaque products – and chopped mats or cut rovings with up to 35% glass fiber content are produced continuously (0.5–6 mm thick) between PVAL cover films in one process step. The leather-like soft material can be stored at room temperature for several weeks, as long as it is protected from light and UV radiation.

These prepregs are used to thermoform (at 80–90 °C) parts with limited degrees of deformation. Standard thermoforming machines can be used; however, larger parts (boat hulls, spoiler, covers) can be formed in simple box molds with edge clamps and vacuum connection. The molded parts are generally stable enough to be cured in an

Fiber reinforced plastic sheet						
EP/GF		PI/GF	EP/GF	EP/AF	PI/CF	
Bisphenol type	TGDA type					
Fabric			Unidirectional fabric	Fabric		Unidirectional fabric
			32	–	50	40
	1.9–2.0		2.1	–	1.5	1.6
0.2	0.2	0.2	0.2	–	0.8	0.8
300	350	350	900	350–500	550/450	1400
2	2	2	< 2	–	–	–
23	22	21	35	29	–	–
≤ 130	155	200	130	generally −55 to +80 °C		
–	–			0	–	

adjacent lamp area with 40–50 s exposure time for 4 mm thickness. After that, the cover films can be easily removed; lamination is also possible during the thermoforming process.

6.21.4.4 Epoxide Resin Molding Compounds (EP)

EP, dry compounds: Generally used for high-quality precision parts, down to very small dimensions. They are used particularly with metal inserts; in the electronics industry for collectors and coating of wrapped capacitors. Coating of electronic semi-conductor components (chips) by compression injection molding in multi-cavity molds with 240–360 cavities constitutes the largest consumption of this material. The grades used for this application have to be highly pure, silica-filled, and with high T_g (*e. g.*, by benzophenonetetracarboxylic acid anhydride) curing molding compounds.

EP-woven prepregs: Laminates and composites with honeycombs or other core materials, particularly in the aerospace industry; copper laminated for printed circuits.

EP-unidirectional prepregs, tapes: Applications as construction material for laminated (oriented) autoclave-cured tail units in airplanes, compression elements in space applications, high-performance sports gear.

6.21.4.5 Diallyl Phthalate Molding Compounds (PDAP)

PDAP molding compounds: for electronic components in military and space equipment; in these applications in particular, the high dimensional stability up the 200 °C and the temperature- and weather stability as well as the excellent electrical properties under extreme conditions and climate variations are utilized.

6.21.4.6 Silicone Molding Compounds (SI)

Electronic components for high-temperature applications.

6.22 Curable Casting and Laminating Resins

While the production of parts made from curable molding compounds and prepregs essentially consists of only a physical molding and a chemical curing process using delivered (sometimes also made in-house) molding compounds, the reactive resin compounds for casting or laminating have to be mixed in-house. In general, the resin types used here are the same as for molding compounds and prepregs.

6.22.1 Phenoplastics (PF, CF, RF, XF)

6.22.1.1 Phenolic- (PF), Cresol- (CF), Resorcinol- (RF), Xylenol-Formaldehyde Resins (XF)

Chemical Constitution

The chemical constitution of these materials has already been covered in Section 6.21.1. Condensation of the phenoplastics PF, CF, RF, and XF can be catalyzed by:

- *Acids with formaldehyde deficiency:* these resins are also called novolacs. They are soluble in and dilutable with organic solvents. Curing (cross-linking) primarily occurs with hexamethylene tetramine (hexa curing agent), see Equation 6.41.
- *Alkalis with formaldehyde excess:* here, resol, resitol, und resite states are passed through successively, see Eqs. 5.34–5.36.

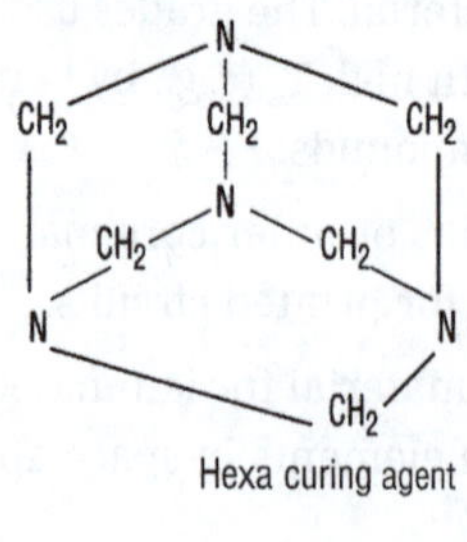

Hexa curing agent

(6.41)

Processing, Applications

Phenolic resins are available in the form of solids, liquids, dissolved, but also as free-flowing powders and as stable aqueous solutions. A large area of application are thermosetting molding compounds, see Section 6.21. After curing at elevated temperatures, unfilled resols create translucent parts, *e.g.*, cutlery handles; addition of bicarbonate or other gas-releasing additives creates foams. Resitols are used with fabric inserts to produce sheets, gear wheels, or grinding wheels. Phenolic resins are used to produce a wide variety of coatings; etherification (n-butyl alcohol) or esterification (fatty acids) result in plasticized/elastic resins with increased elasticity and improved solubility and compatibility.

When bisphenol A or analogs are used as phenolics components, the resulting resins do not yellow. The reaction of novolacs with epichlorohydrin results in poly-functional epoxies (see 1 in Table 6.60). Cross-linking of phenolic resins with epoxies results in high-quality internal coatings for cans and containers for food and beverage contact.

Aqueous coatings of phenolic resins are based on resols with carboxyl and amine functionality; they are primarily used in automotive applications and as repair coatings.

Additional applications for phenolic resins include: wood composites, particle board, thermosetting molding compounds, see Section 6.21, core and molding sand binder, insulating materials, abrasives, putties, adhesives, flame resistant products, tanning agents, ion-exchanger (after condensation of acidic or basic functional groups), foams, fibers, and coatings, high-quality brake pads (up to 600 °C resistant), binders for iron or barium ferrite (magnetic) or PB (radiation shielding) and glass fiber reinforced prepregs and free-flowing SMC for the automotive, aircraft, and mining industry.

Phenolic foams result from the exothermic, acid-catalyzed curing of resols, usually in the presence of low-boiling (halogen) hydrocarbons as blowing agents. Block and conveyor production are common. Both open-cell and closed-cell foams are produced.

6.22.2 Aminoplastics (UF, MF)

6.22.2.1 Urea (UF), Melamine Formaldehyde Resins (MF)

Chemical constitution: To tailor certain properties, resins are modified with polyalcohols, sugar, lactams, acrylates, alcohols, sulfites, and others. Combinations with other resin types such as PF, EP, and UP are common.

Melamine foams with densities of approx. 7 kg/m^3 are the curing result (with an acid curing agent) of MF pre-condensates with an emulsified, low-boiling blowing agent.

Properties: UF and MF exhibit high scratch, temperature (UF: 90 °C, MF: 150 °C), and chemical resistance as well as good light-fastness.

Applications: Thermosetting molding compounds, impregnating resins, glues, binders, crosslinking agents, chipboard finishing, insulation materials, coating resins, parquet sealing, enamel, paper and leather additives, nonwoven binders, foam as insulation, nutrient and active ingredient carriers.

6.22.3 Furan Resins (FF)

Chemical constitution: Reaction of strong acids (HCl, H_2SO_4) with furfuryl alcohol results in polycondensation to a solid resin. There are a number of variation options: the etherification of the methylol group of furfurol alcohol and the co-condensation with PF, UF, MF resins, as well as with ketones and aldehydes.

Properties: excellent solvent as well as acid and alkali resistance, heat and dimensional stability, tendency to yellowing.

Applications: thermosetting molding compounds, laminating resins, putties, foundry and building material binders, foams, adhesives, grinding and polishing materials.

6.22.4 Unsaturated Polyester Resins (UP)

Chemical Constitution, Processing

By varying the acids, glycols, and vinyl monomers, the characteristics of the resins can be tailored to the desired application: *e. g.*, addition of styrene or vinyl acetate results in denser cross-linking networks and thus harder resins; methyl methacrylate, on the other hand, creates long, flexible blocks between the polyester chain and correspondingly soft resins. Particularly hard and densely crosslinked specialty resins contain, *e. g.*, α-methyl styrene or allyl derivatives, see Table 6.4.

UP resins are stabilized so that they are durable for several months in cool and light-protected storage. When using a peroxide catalyst, the crosslinking temperature ranges between 80 and 100 °C and can be lowered to room temperature by using activators. Depending on the formulation, the curing time can vary from a few minutes up to hours. The pot life is about 60% of the curing time. Curing agent and activator must never be mixed together (explosion hazard!).

Liquid resins are flammable (in a low hazard class). Peroxide pastes and amine accelerators are corrosive. Peroxides decompose when used improperly. Protective occupational safety and health regulations must be followed, including limiting styrene emissions and appropriate workplace venting and exhaust air cleaning systems.

After solidification of the resin, the cold curing process with peroxide and accelerator is not yet complete. Molded UP parts that have to meet good aging, chemical and warm water resistance or food contact requirements have to go through a post-curing process for 5 h at 80 °C up to > 100 °C in hot air. Subsequently, molded parts have to be stored in dry conditions for a minimum of one to two weeks at > 20 °C before use. UP resin floor screeds and coatings take several days before they are fully mechanically and chemically resilient. Water, fillers, and pigments may influence the reaction process.

Alkyd resins are polyester resins modified with fatty acids or their (fatty) oils (triglycerides) and represent a special group within the polyester resins. They are used as binders for paints.

Properties, Applications

The processes described in Section 4.8 for the manufacture of long fiber reinforced products are used for parts and semi-finished products made from UP resins; for properties see Table 6.64 and Table 6.65. Other areas of application are thermosetting molding compounds, casting compounds, lightweight concrete, and foams. The properties of the resins can be largely tailored by chemical modifications to meet application requirements.

Unfilled resins shrink by 6 to 8 vol.-% during curing. For hot compression processes, there are low-shrinkage (low profile) two-phase systems available, see Section 6.21.3.5. Unfilled cured resins are transparent, glass-fiber filled resins are translucent (alignment of refractive indices caused by MMA addition). Above 140 °C de-polymerization may be initiated, and above 400 °C gases ignite, compounding materials without flame retardants will continue to burn. Highly mineral filled compounds are practically non-flammable.

Cross-linking via visible light of GF-UP laminates up to 20 mm thickness, *e.g.*, using standard fluorescent lamps, is feasible with "VLC"- (visible light curing-) UP resins with light sensitizer additives that are stable when stored under the exclusion of light. The procedure is applicable using light tunnels for the continuous production of laminate sheets between cover films, using vacuum injection processes with exposure to light through transparent upper molds, using winding processes with lamp panels. It can also be applied for roof coatings and repairs utilizing sunlight. The operating parameters can be set so that the parts do not require post-curing, facilitating significant energy savings compared to peroxide annealing. Opaque, highly filled, or carbon or aramid fiber reinforced products cannot be light cured.

Table 6.65 Standard Characteristics for GFP Laminates

Property	**UP-GF resin**				
Glass content wt.-%	**25**	**45**	**50**	**65**	**65**
Textile glass composition	**Mat, quasi-isotropic**		**Fabric, roving longitudinal = across < 450 g/m²**		**Fabric, roving, 90% longitudinal**
Density g/cm³	1.35	1.45	1.60	1.80	1.80
Tensile strength MPa	70	140	200	3000	500
Young's modulus MPa	5000	9000	10,000	19,000	28,000
Flexural strength MPa	120	180	220	350	550
Coefficient of linear expansion $10^{6} \cdot K^{-1}$	35	25	18	15	12
Thermal conductivity W/K · m	0.15	0.23	0.24	0.26	0.26
Property	**EP-GF resin**				**For comparison: Cr-Ni-steel plate**
Glass content wt.-%	**50**	**65**	**65**	**67–78**	
Textile glass composition	**Fabric; rovings longitudinal = across**		**Rovings, almost 100% longitudinal**	**S-glass 92% longitudinal, special resin**	
Density g/cm³	1.60	1.80	1.80	1.8–2.0	8.0
Tensile strength MPa	220	350	700	1300–1700	ca. 500
Young's modulus MPa	10,000	18,000	30,000	ca. 60,000	195,000
Flexural strength MPa	280	400	800	1200–1600	220
Coefficient of linear expansion $10^{6} \cdot K^{-1}$	18	15	12		
Thermal conductivity W/K · m	0.24	0.26	0.26		

At room temperature, standard resins based on orthophthalic acid and simple diols are resistant to water, saline solutions, diluted acids, and hydrocarbons. They are not resistant to chlorinated hydrocarbons, many solvents, alkalis, and concentrated and oxidizing acids; heat distortion temperature is approx. 70 °C.

Flame retardant modified resins contain brominated or highly chlorinated acid components (HET acid) and antimony trioxide within low-smoke and low-viscosity resins. They are filled with aluminum hydroxide $Al(OH)_3$ at a ratio of 1 : 1.8.

Resins with increased corrosion resistance based on iso- or terephthalic acid and neopentyl glycol (Equation 6.42) are suitable for thin films under continuous warm water exposure (swimming pools) and exhibit good mechanical properties.

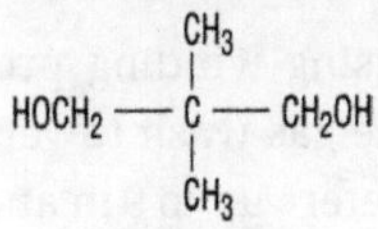

Neopentyl glycol

(6.42)

Resins with high hydrolysis and saponification resistance that are based on bisphenol A are long-term resistant to water and 20% hydrochloric acid up to 100 °C, to 70% sulfuric acid, 20% alkali solution up to 80 °C; HDT 110 to 125 °C.

Flexible resins with aliphatic dicarboxylic acid units $HOOC(CH_2)_xCOOH$ (x = 4: adipic acid; x = 7: azelaic acid) for compounding, or modified with isocyanate, are suitable, for example, for impact-resistant floor coverings.

Highly transparent weather resistant resins contain methyl methacrylate (MMA) as comonomers.

When curing, "hybrid" resins from relatively low molecular weight UP resins with hydroxyl end groups and diisocyanate dissolved in styrene will form products that can be tailored from long chain-crosslinked soft elastomers to rigid and impact resistant materials with property profiles that cannot be achieved by UP or PUR alone (*e. g.*, σ_B = 80 to 95 MPa, ε_B = 7 to 12%, $E \geq 3$ GPa, σ_{bB} = 142 to 160 MPa). The low-viscosity, highly fillable 2-component resins, where each component contains the crosslinking catalyst for the other, are processed using resin transfer molding (RTM) or resin injection molding (RIM) processes, or by pultrusion to manufacture gelcoats and foams. Structural vehicle parts are applications of increasing importance.

Water-emulsifiable UP resins are available as casting resins with 40 to 70% content for wood-like furniture parts.

Addition of 3 to 4% PVDC microspheres in boat building laminating resins results in lightweight composites similar to syntactic foams. Because PVDC dissolves in styrene, these resins have to be used immediately after mixing of the components.

Structural foam mat laminates, d_R 0.4 to 0.8 g/cm^3, are produced using low-pressure processes and nitrogen-splitting blowing agents together with special inhibitor/accelerator systems. The system is used for the manufacture of PMMA bathtubs; here, resin fibers are sprayed with additional blowing agents to create microcellular support layers. The binder in UP resin lightweight concrete is foamed to d_R 0.05 to 0.2 g/cm^3 using a CO_2-splitting blowing agent.

6.22.5 Vinyl Ester Resins (VE); Phenacrylate Resins

Chemical Constitution, Processing, Properties: see Section 6.21.1.3

Applications: Large GRP pipes and equipment parts produced using winding processes, *e. g.*, for continuous operation at 110 °C, self-supporting flue gas wash towers exposed to H_2SO_4-, HCl-, and HF-containing condensates with diameters up to 9 m and 35 m height with PHA-matrix for chemical protective layers and load-bearing laminate. Also available are fast-curing specialty resins used to process with preformed glass fiber reinforcements in reaction transfer molding for body parts and other vehicle components. They take advantage of the high fatigue strength of PHA resins.

6.22.6 Epoxy Resins (EP)

Chemical Constitution: see Section 6.21.1.4

Properties and Applications

Epoxy resins developed for specific applications exhibit high adhesive strength with almost all other materials, low shrinkage during cross-linking, good corrosion and temperature resistance, and good electrical properties. Approximately 50% of EP products are used as solvent-free liquid or (powder) enamel coatings for surface protection (also aqueous as EP-dispersions) and as structurally resilient adhesively bonded joints, approximately 20% each for reactive resin concrete and adhesive mortar and in the electrical/electronics industry. Melting matrix resins are increasingly used to impregnate carbon and aramid structural composites in the aerospace industry (see Section 6.21). Combinations of EP resins with high-temperature resistant thermoplastics are also used for this purpose, *e. g.*, as interpenetrating networks (IPN); they are multiple times tougher than EP.

Standard resins are liquid to solid diglycidyl ether with chemical compositions according to (1) or (2) in Table 6.60; grades with compositions shown in (3) are flexibilizing resins; cyclo aliphates – (5) and (6) – and multifunctional amine derivatives (4) are used for high-temperature electronics, adhesives, coatings, and specialty composites. Fast UV-curing cycloaliphatic resins are highly resistant products that adhere to almost any substrate.

Depending on *application*, we distinguish between multi-purpose resins, casting resins, impregnation resins for electronical applications, laminating resins for highly stressed fiber reinforced structural materials (see Table 6.64 and Table 6.65), tooling resins, and resins for protective coatings. Heat cured epoxy resin compounds with high-strength and high-temperature resistant fiber reinforcement can withstand temperatures up to 240 °C long term. While the temperature resistance of normal, in particular cold cured, epoxy resins is lower, it still compares well to that of other comparable materials. Filled epoxy resins stand out for their negligible volume shrinkage, dimensional stability, hardness, and abrasion resistance, which renders them particularly suitable for applications such as measuring and testing gauges, masters, and as mold materials for metals and plastics. Because of EP's good adhesive bonding to any substrate, the use of release agents is required. Adhesive strength and curing of liquid EP resins with aliphatic polyamines are compromised by wet surfaces. Specialty curing agents, in particular polyaminoamide adducts, are water resistant so that resins containing them will adhere to wet substrates, and some may even be processed under water or in aqueous solution.

For construction applications, special curing agents designed for low temperatures and "masked", water-activated (ketimides), curing agents are available. The reactive epoxy groups are able to crosslink with many other materials, including tar products or liquid polysulfide rubbers (see Section 6.23.7). The generally good chemical resistance of EP resins can be tailored by the use of crosslinking agents; their alkali resistance is particularly high.

EP-foams (with 0.03 to 0.3 g/cm^3) are created from powder resins using chemical blowing agents, or from liquid resins using physical blowing agents under constant stirring. "Syntactic" EP resin foams are used to embed electronic components; these resins exhibit creep strength up to 200 °C, are pressure resistant up to 60 bar for deep sea applications, and can also be used as elastic mold resins.

6.22.7 Dicyclopentadiene Resins (DCPD)

DCPD is a low-viscosity liquid monomer and can be polymerized to form an olefinic thermoset by splitting and recombination of internal C=C ring double bonds. The Metton-RIM process adds a metathesis catalyst on the one hand, and an activator and an inhibitor on the other hand to the monomer. After mixing within the tailored latency period, the components polymerize and can be demolded within a few seconds. The mechanical characteristics (E_b = 1.9 GPa, σ_B = 34 MPa) and creep resistance at elevated temperatures (HDT = 85 °C, GT = 95 °C) can be improved by glass fibers or other reinforcing additives. The material is hydrophobic, can be easily coated, and is impact resistant to approx. −30 °C; applications: large parts (≥ 2.3 kg) for special vehicles, sports equipment.

6.22.8 Diallyl Phthalate Resins (PDAP)

Cast resins with high transparency, impact, scratch, and abrasion resistance, typically used for lightweight eye glass lenses.

6.22.9 Hydrocarbon Resins

Soluble resins, typically produced by cationic, sometimes also by thermal, polymerization of unsaturated hydrocarbons. Highly unsaturated resins, such as copolymers of mostly butadiene and styrene, can be mixed with monomers and crosslinked via peroxide catalysts (both heat activated and cold, Buton resins). These casting resins are used in microwave technology because of their favorable dielectric properties (radar, 10^9 Hz).

Bisdiene resins are the result of the chemical reaction of cyclopentadiene sodium with an aliphatic dihalogen compound through a pre-polymer stage in the presence of polybutadiene at elevated temperatures. These hydrocarbon resins are used for printed circuit boards because of their extraordinary good dielectric characteristics up to approx. 200 °C.

6.23 Elastomers

For the plastic materials identified by acronyms only, Table 1.1 in Section 1.1 provides the chemical names and the sections in Chapter 6 in which they are covered.

6.23.1 General Description

Elastomers, colloquially often referred to as rubbers, are largely amorphous at room temperature and exhibit a low glass transition temperature. Uncrosslinked they exhibit thermoplastic properties, *i. e.*, with increasing temperature they become softer and the rubbery elasticity caused by the tangled structure decreases gradually. Using vulcanization with sulfur or a comparable chemical or physical process, the macromolecules are crosslinked – wide-meshed to elastomers (soft rubber) or close-meshed to hard rubber (ebonite). The former exhibit high, reversible elongation properties. Elastomers take an intermediate position between the thermoplastics that are still able to flow and the rigid thermosets. They no longer soften at elevated temperatures and are therefore not processable like thermoplastics.

The basis for classification of rubbers in ASTM D1418 and ISO 1629 is different than the one for plastics and therefore leads to different acronyms. In the following short descriptions, the acronyms are matched with their related chemical descriptions and their basic chemical structure.

Natural rubber and analogous synthetic elastomers, polymers of conjugated dienes (*e. g.*, isoprene, butadiene, chlorobutadiene) as well as copolymers of conjugated dienes and vinyl derivatives (*e. g.*, styrene, acrylonitrile) are the most important elastomer grades with more than 80% market share. They contain numerous unsaturated double bonds in their molecular chain, of which only a fraction is saturated when conventionally vulcanized with sulfur and organic plasticizers to form soft elastomers. The more double bonds are left in the vulcanizate, the lower is its oxidation- and weather resistance, although it can be significantly improved by the addition of antioxidants. Products with low diene content, *e. g.*, those produced by polymerization of isobutene with small amounts of isoprene, exhibit respective improved aging resistance; however, because of the low number of double bonds, their vulcanizability is rather low. Products with low double bond contents can also be created by ring-opening polymerization, *e. g.*, trans-1,5-polypenteneamer (from cyclopentene by ring-opening polymerization) and polyocteneamer (from cyclooctene). The latter plays a certain role as a processing aid rather than as a rubber. Copolymerization of ethylene and propylene creates ethylene-propylene elastomers that are completely saturated and therefore cannot be vulcanized with sulfur (EPM). Terpolymerization of ethylene and propylene with non-conjugated dienes (in particular ethylidene norbornene) produces saturated polymer chains with double bonds in the side chain, which can be vulcanized with sulfur (EPDM), or more advantageously with peroxides (R-O-O-R'). EP elastomers exhibit excellent aging resistance and play a significant commercial role. Other synthetic elastomers are: propylene oxide elastomer, polyphosphazene, and polynorbornene. Polymers without double bonds require different vulcanization methods: some require alkaline crosslinking agents, others require oxidative crosslinking with peroxides. Both saturated and unsaturated polymers are crosslinked with electron beams. Most grades of synthetic rubber can be blended; however, a combined polymerization by sulfur or peroxides is not possible

6.23.2 General Properties

A relevant criterion for the selection of elastomers for a specific application is their temperature resistance, in particular in automotive applications. Figure 6.25 shows the increased demands on temperature and oil resistance for high-performance products in the automotive industry according to internationally recommended ASTM standards. This trend is caused largely by increasing temperatures under the hood.

The mastication of bales, the customary form of raw material delivery for rubber processing, requires much work. The effort to make synthetic elastomer available as more easily processable crumbs or immediately extrudable or injection moldable powdered pellets, with additives necessary for vulcanization already added has so far proven unsuccessful. Although relatively low molecular "liquid rubbers" with reactive end groups offer easier processing methods, such as those of liquid plastic pre-products, the rubber industry has yet to embrace these opportunities.

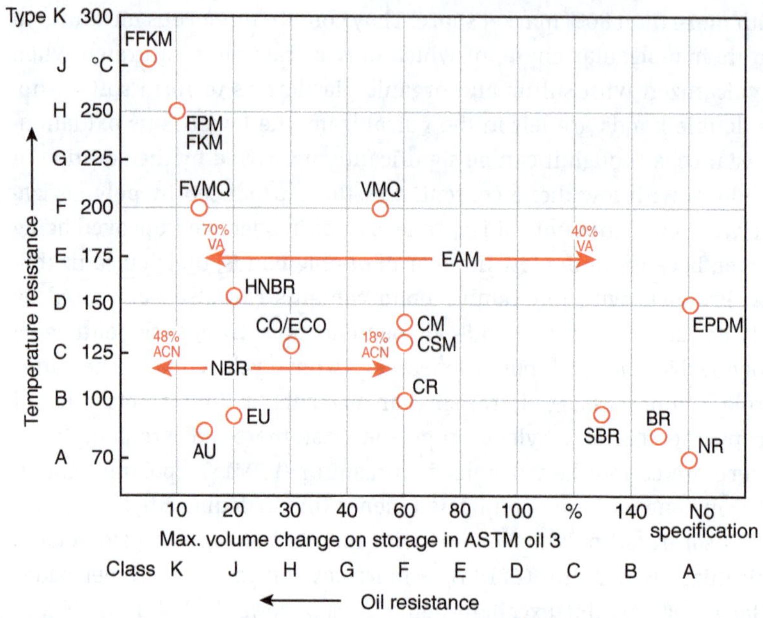

Figure 6.25 Classification of elastomers depending on temperature and oil resistance

The "reinforcement" of the formulation by active fillers is a very important issue in rubber technology. These fillers are typically carbon blacks for dark and highly dispersed silica for light-colored products. Sometimes the fillers are already added to the latex. Conventional higher molecular synthetic rubber extended with oil results in easily processable formulations (oil extended rubber). Many rubber formulations also contain plasticizers.

In the following, the most important elastomers are briefly described. Table 6.66 and Table 6.67 provide an overview of selected properties and chemical resistance for the most important rubber grades.

Table 6.66 Simplified Schematic Property Comparison for Selected Vulcanized Elastomer Grades (Scale: 1 = excellent, 2 = very good, 3 = good, 4 = sufficient, 5 = poor)

	NR	BR	CR	SBR	IIR	NBR	EPDM	CSM	EAM	ACM	ECO	AU	VMQ	FKM	FVMQ
Tensile strength	1	5	2	2	4	2	3	3	3	3	3	1	4	3	4
Elasticity	1	1	2	4	5	4	3	5	5	5	3	2	2	5	3
Wear resistance	3	1	2	2	4	3	4	3	4	4	4	1	5	4	5
Weather/ ozone resistance	5	5	4	5	3	5	1	1	1	2	2	2	1	1	1

	NR	BR	CR	SBR	IIR	NBR	EPDM	CSM	EAM	ACM	ECO	AU	VMQ	FKM	FVMQ
Thermal resistance	5	5	4	4	3	4	2	3	2	2	3	3	1	1	2
Low *T* flexibility	1	1	3	2	3	4	2	4	4	5	3	4	1	5	2
Gas permeability	4	4	2	4	1	2	3	2	2	3	2	2	5	2	5
Service *T* (°C)*	100	100	120	110	130	120	140	130	170	160	130	130	200	210	180

* for optimum compound ≈ 1000 h

Table 6.67 Chemical Resistance of Selected Vulcanized Elastomer Grades (up to 100 °C)

	NR	BR	CR	SBR	IIR	NBR	EPDM	CSM	EAM	ACM	ECO	AU	VMQ	FKM	FVMQ
Paraffin hydrocarbons	C	C	B	C	C	A	C	C	A	A	B	A	C	A	A
Fuels	C	C	C	C	C	B	C	C	C	B	B	B	C	A	B
Aromatics	C	C	C	C	C	C	C	C	C	C	C	C	C	A	B
Chlorinated hydrocarbons	C	C	C	C	C	C	C	C	C	C	C	C	C	A	C
Motor oils	C	C	B	C	C	A	C	C	B	A	B	B	B	A	A
Hypoid oils	C	C	C	C	C	B	C	C	C	A	B	B	C	A	A
Mineral lubricants	C	C	C	C	C	B	C	C	B	A	B	B	B	A	A
Alcohols	A	A	A	A	A	A	A	A	B	B	B	B	B	C	C
Ketones	A	A	B	A	A	C	A	B	C	C	C	C	C	C	C
Esters	B	B	C	B	C	C	C	C	C	C	C	C	C	C	C
Water	A	A	A	A	A	A	A	A	A	B	B	C	B	A	C
Acids (dilute)	A	A	A	A	A	A	A	A	B	C	B	C	B	A	C
Alkalis (dilute)	A	A	A	A	A	A	A	A	B	C	B	C	B	A	C
Brake fluids	A	A	B	A	A	C	A	C	C	C	C	C	A	C	A

A: little or no attack (max. +10% change in volume)
B: weak to medium attack (max. +25% change in volume)
C: strong attack (more than 25% change in volume)

6.23.3 R-Elastomers (NR, IR, BR, CR, SBR, NBR, NCR, IIR, PNR, SIR, TOR, HNBR)

Unsaturated chains, partially or completely made from diolefins.

Natural Rubbers (NR)

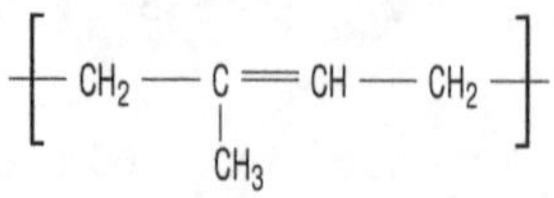
(6.43)

Even when unfilled, NR (Equation 6.43) exhibits high strength and elasticity. Thermal range: long-term −50 to +70 °C, special grades up to +90 °C; short-term up to 120 °C. NR exhibits little mechanical damping, is not susceptible to creep up to 50 °C, is not resistant to oil, and must be ozone-stabilized.

Applications: rubber springs, truck tires.

Isoprene Elastomers (IR)

IR (Equation 6.43) is a synthetic version of NR and their properties are comparable. IR exhibits slightly higher elasticity.

Applications: same as NR.

Butadiene Elastomers (BR)

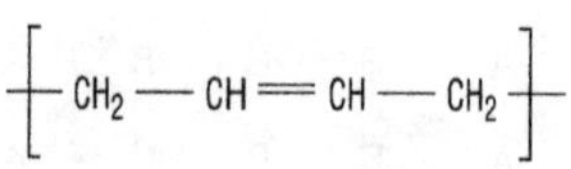
(6.44)

BR (Equation 6.44) is typically blended with other rubbers. It is available in wear- and low temperature resistant grades.

XBR is a *carboxylic group containing BR (-COOH).*

Applications: automotive running treads.

Chloroprene Elastomers (CR)

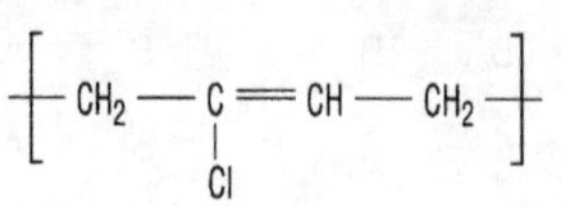
(6.45)

CR (Equation 6.45) exhibits improved aging resistance compared to NR, IR, and SBR; it is flame resistant and to some degree resistant to oils and fats. Thermal range: long-term −40 to +110 °C, short-term up to +130 °C. Long-term storage below 0 °C results in reversible hardening by crystallization. Low gas permeability. Styrene chloroprene rubber contains an additional CH–C_6H_5 group; SCR and XCR contain a carboxylic group, see XBR.

Applications: bellows, grommets, cooling water hoses, sealing profiles in construction, roofing materials, conveyer belts, cable sheathing, protective gear.

Styrene Butadiene Elastomers (SBR)

$$\left[\underset{\displaystyle C_6H_5}{\underset{|}{CH}} - CH_2 - CH_2 - CH = CH - CH_2 \right] \tag{6.46}$$

SBR (Equation 6.46) is produced by cold polymerization of 75% butadiene and 25% styrene. It is available in numerous grades with different stabilizers for both light-colored and dark products. It often substitutes NR as a general purpose rubber and represents the largest rubber class. Thermal range: long-term −40 to +100 °C, short-term up to +120 °C. SBR is not resistant to mineral oils.

XSBR is a carboxylic group containing SBR.

Applications: tire formulations, cable sheathing, technical rubber parts, hoses and profiles, foam rubber.

SBR lattices are available in special formulations for paints (Litex) and carboxylized as binders for needle felt and back coatings for carpets.

Nitrile-Butadiene Elastomers (NBR)

$$\left[\underset{\displaystyle CN}{\underset{|}{CH}} - CH_2 - CH_2 - CH = CH - CH_2 \right] \tag{6.47}$$

NBR (20–50% acrylonitrile content, Equation 6.47) gains resistance to oils, fats, fuels with increasing ACN content. At the same time, it becomes less flexible at low temperatures. Thermal range: long-term −30 to +100 °C, short-term up to +130 °C, special grades −40 to +150 °C. Low gas permeability. NBR + PVC is ozone resistant. NBR + phenolic resin is tough-elastic and hot water resistant.

XNBR is an NBR containing carboxylic groups.

Applications: most important sealing material in automotive and mechanical engineering; hoses, brake pads.

Nitrile-Chloroprene Elastomers (NCR)

$$\left[\underset{\displaystyle CN}{\underset{|}{CH}} - CH_2 - CH_2 - \underset{\displaystyle Cl}{\underset{|}{C}} = CH - CH_2 \right] \tag{6.48}$$

NCR (Equation 6.48) exhibits slightly better oil resistance than CR.

Butyl Elastomers (IIR, CIIR, BIIR)

$$\left[-CH_2-\underset{CH_3}{\overset{CH_3}{\overset{|}{\underset{|}{C}}}}-CH_2-\underset{CH_3}{\underset{|}{C}}=CH-CH_2- \right] \tag{6.49}$$

IIR (Equation 6.49) is a copolymer. A very high isobutene content is responsible for good chemical and aging resistance and low gas permeability (air); low isoprene content for vulcanizability. Thermal range: long-term −40 to +130 °C. Chlorobutyl(CIIR)- and bromiumbutyl (BIIR) rubbers show good processability and aging resistance as well as low air permeability.

Applications: inner tubes, inner liners of tires, gas-proof membranes, pharmaceutical plugs, roofing material.

Isoprene Styrene Elastomers (SIR)

$$\left[-\underset{C_6H_5}{\underset{|}{CH}}-CH_2-CH_2-\underset{CH_3}{\underset{|}{C}}=CH-CH_2- \right] \tag{6.50}$$

Special rubber (Equation 6.50) of little commercial interest.

Polynorbornene Elastomers (PNR)

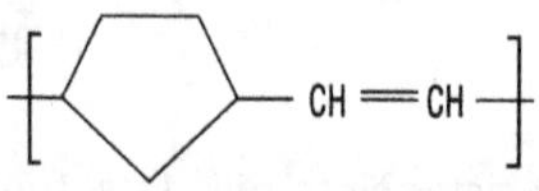

(6.51)

Plasticizers and fillers can easily be added to PNR (Equation 6.51), which results in very soft elastomers. Temperature resistance and susceptibility to creep are unfavorable compared to NR; ozone resistance is similar to NR.

Applications: soft roller coatings, foam rubber.

Trans-Polyoctenamer Elastomers (TOR)

$$\left[-CH_2-(CH_2)_5-CH=CH- \right] \tag{6.52}$$

TOR (Equation 6.52) is primarily used as a blending component with other rubbers with loads ranging from 5–30%. These blends provide better filler absorption and distribution, a significant reduction in viscosity of the blends at processing temperatures, and a considerable increase in melt flow rate.

Hydrated NBR Elastomers (HNBR)

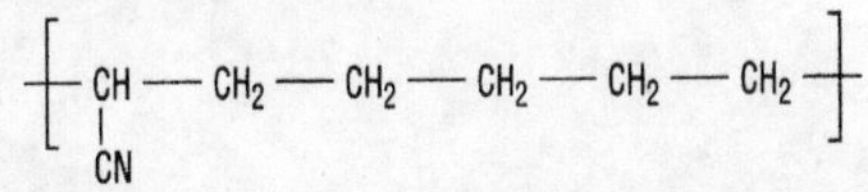

(6.53)

HNBR (Equation 6.53) is peroxidically crosslinked, up to 150 °C significantly more oil-, oxidation-, and wear resistant than NBR.

Applications: automotive, oil production equipment.

6.23.4 M-Elastomers (EPM, EPDM, AECM, EAM, CSM, CM, ACM, ABR, ANM, FKM, FPM, FFKM)

Saturated chains of polymethylene type.

Ethylene Propylene (Diene) Elastomers (EPM, EPDM)

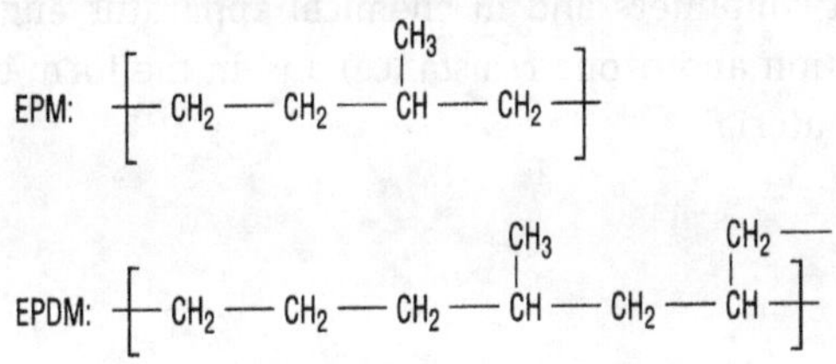

(6.54)

EPDMs (Equation 6.54) are atactic copolymers of ethylene and propylene, see Section 6.2.6.3. EPM can only be crosslinked with peroxides; EPDM is also vulcanizable with sulfur. Because the unsaturated sites are not located on the main chain in terpolymers, they exhibit the chemical resistance and – with appropriate stabilization – the good weather-, ozone-, and aging resistance of saturated polyolefins. Thermal range: long-term −40 to +130 °C, special grades up to +150 °C. Highly pure grades are obtained by polymerization with metallocene catalysts.

Applications: automotive exterior parts, solid and foam sealing profiles, O-rings, hoses, cable sheathing.

Ethylene Acrylic Ester Elastomers (AECM)

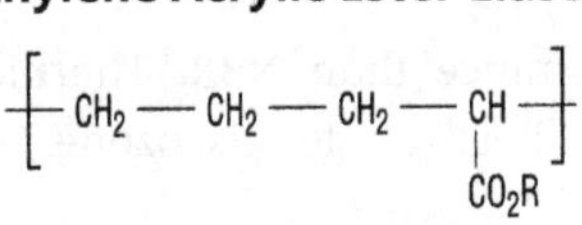

(6.55)

AECM (Equation 6.55) is an elastomer with excellent temperature resistance und medium oil resistance. Thermal range in air: long-term −25 to +170 °C, short-term up to +200 °C.

Applications: special seals in automotive applications, cooling water hoses, cables.

Ethylene Vinyl Acetate Elastomers (EAM)

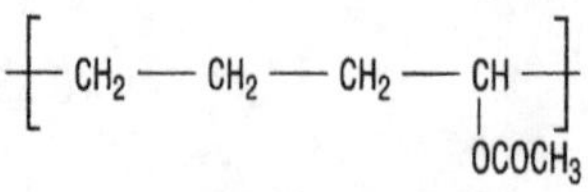

(6.56)

EAM (Equation 6.56) combines high temperature resistance with limited oil resistance. Toughness at low temperature is low. Thermal range: long-term −10 to +150 °C.

Chlorosulfonated PE Elastomers (CSM)

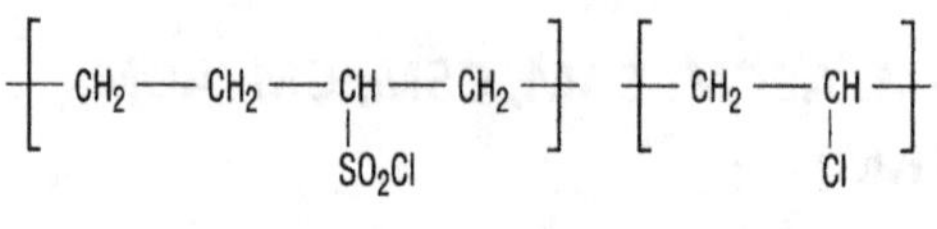

(6.57)

CSM (Equation 6.57) is a light-colored special rubber with good weather- and chemical resistance. Thermal range: long-term −20 to +120 °C. Self-vulcanizing, mineral filled compounds are used as spraying mixtures or weldable webs for roofing and linings.

Other applications: lining of transport containers and in chemical apparatus engineering, cable insulations (good oxidation and ozone resistance) and in the form of blends with other rubbers as coating material.

Chlorinated PE Elastomers (CM)

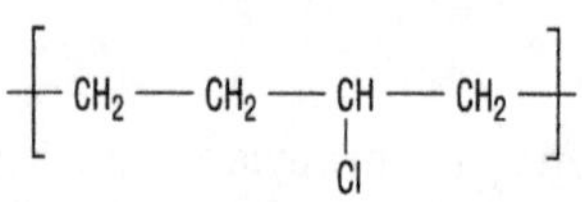

(6.58)

CM (Equation 6.58) exhibits slightly better hot air and hot oil resistance as well as a lower brittleness temperature than CSM; it is also less expensive.

Applications: technical rubber parts, cable sheathing, hoses for engines, impact modifier for PVC.

Acrylate Elastomers (ACM, AEM, ANM)

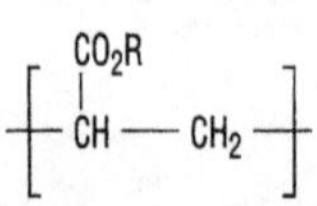

(6.59)

ACM (Equation 6.59) exhibits higher temperature resistance than NBR. Thermal range: long-term −25 to +150 °C, short-term up to +170 °C. Resistant to oils, ozone, UV radiation.

Applications: engine seals in automotive applications, self-vulcanizing acrylate latex for adhesion of non-woven materials.

Fluoro Elastomers (FKM)

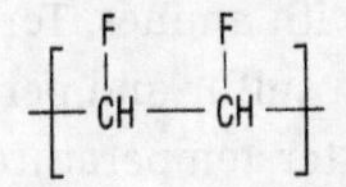

(6.60)

FKM (Equation 6.60) is resistant to most liquids. Thermal range: long-term −20 to +200 °C, short-term up to +250 °C.

Applications: special seals.

Propylene Tetrafluoroethylene Elastomers (FPM)

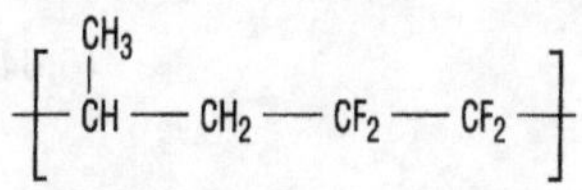

(6.61)

FPM is a copolymer of polypropylene and tetrafluoroethylene and exhibits extremely high chemical resistance, Equation 6.61.

Perfluoro Elastomers (FFKM)

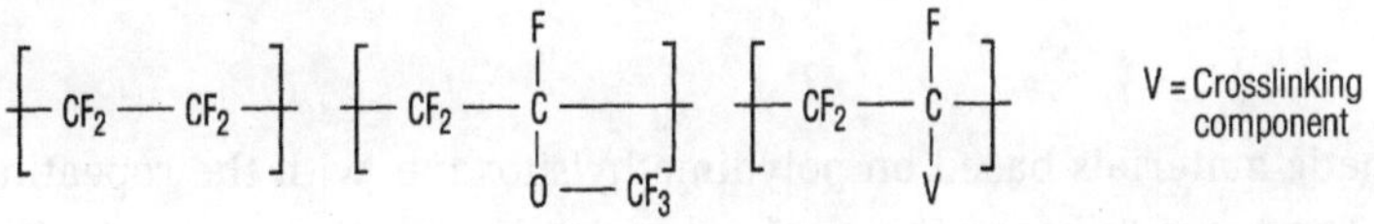

(6.62)

Because of its very high price, FFKM (Equation 6.62) is only used for special seals in the chemical industry or in oil drilling equipment.

6.23.5 O-Elastomers (CO, ECO, ETER, PO)

Chains with oxygen.

Epichlorohydrin Homopolymer-, Copolymer- and Terpolymer Elastomers (CO, ECO, ETER)

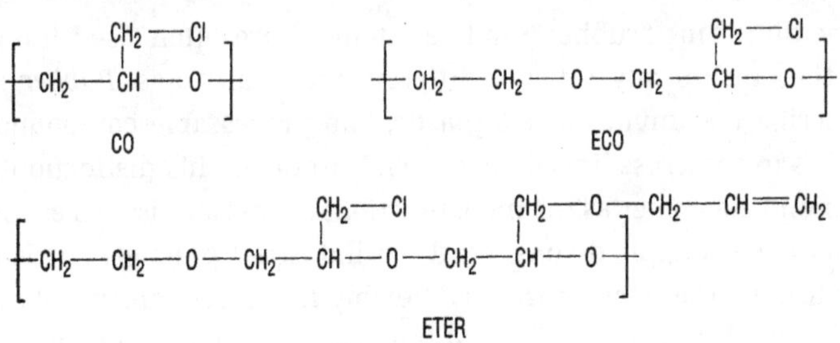

(6.63)

Epichlorohydrin elastomers are homopolymeric (CO) or with ethylene oxide copolymerized (ECO) polyethers, Equation 6.63. They can be crosslinked with amines. Terpolymers containing vinyl groups (ETER) are also vulcanizable with sulfur and peroxide. Compared to NBR, they exhibit similar oil- and slightly better temperature resistance. Low temperature impact strength and elasticity are significantly higher. Thermal range: long-term −40 to +120 °C.

Applications: hoses and special seals in automotive applications.

Propylene Oxide Elastomers (PO)

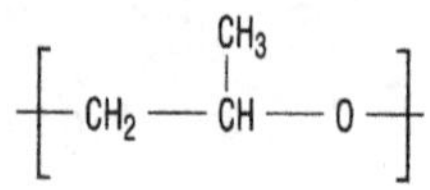

(6.64)

PO (Equation 6.64) exhibits good heat and cold temperature resistance, but only low oil resistance.

6.23.6 Silicone Elastomers

Bernd Pachaly

Silicones are synthetic materials based on polydimethylsiloxane, with the repeating unit $[(CH_3)_2Si\text{-}O\text{-}]_n$. The name "silicone" was coined by analogy to "ketone" in hydrocarbon chemistry, even though silicones do not contain true carbonyl groups. The term reflects a historical naming convention rather than an actual chemical structure. Although silicones have been known for over 100 years, their industrial use only began around 1950 with the development of suitable monomer synthesis routes.

Silicones are available in the form of fluids (oils), resins, and elastomers. In technical applications, silicone elastomers are the most commonly used; therefore, this section focuses on this material class. The term "silicone elastomers" also formally includes elastic sealants and adhesives used in the construction industry, such as "sanitary silicones". While their chemistry is the same, these specific applications are not covered in this section. For a comprehensive review of the silicone material class, see [1].

In the rubber industry, the terms "rubber" and "elastomer" are often used interchangeably, although they technically refer to different material states. *Rubber* is commonly used to describe the unvulcanized, plastic, and processable compound, whereas *elastomer* refers to the cross-linked, elastic end product. This distinction – rubber as the precursor and elastomer as the final, functional material – is well established in theory but not always applied consistently in industrial practice. Particularly in the context of silicones, these terms are used flexibly. In this section, we follow the technical distinction but also acknowledge that the broader and more widely accepted term "silicone elastomers" includes very soft, gel-like materials that would not typically be classified as "rubber" in the traditional sense.

Silicone elastomers are classified into three main types, distinguished by their consistency and cross-linking systems:

- High-consistency rubber (HCR)
- Liquid silicone rubber (LSR)
- Room-temperature vulcanizing (RTV) silicone rubber, available as one-component (RTV-1) or two-component (RTV-2) systems

When networks are formed from pure polydimethylsiloxanes, their mechanical strength is significantly lower than that of organic elastomers. Even when using high-molecular-weight polymers, the resulting elastomers exhibit low tear strength – less than 1 MPa. This is surprising at first, considering that the Si–O bond energy is 443 kJ/mol, much higher than the C–C bond energy of 346 kJ/mol. The reason lies in the weak intermolecular interactions between the polymer chains, which also explains the low glass transition temperature of around −120 °C.

The outstanding properties of silicone elastomers are achieved by incorporating reinforcing fillers, predominantly based on precipitated or fumed silica with specific surface areas of 100 m^2/g to 400 m^2/g. Initially hydrophilic, the surface of silica can be partially made hydrophobic to produce fillers that provide excellent reinforcement while maintaining low viscosity, allowing for liquid or pasty rubber compounds. For highly transparent and low-viscosity silicone elastomers, MQ resins (*silsesquioxanes*) are also used as fillers.

In automotive and mechanical engineering applications, the term “VMQ” has become established for silicone rubbers, as defined by DIN ISO 1629, referring to the polymer base. In this section, the terms “HCR”, “LSR”, and “RTV-1/2” will be used consistently.

In 2019, the global market size for silicone elastomers, from the rubber manufacturers’ perspective, was approximately €5.2 billion or 920,000 tons. The market had been growing at an average annual rate of 4.1%. Asia accounted for the largest regional share (41%), followed by North and South America (31%), and Europe and other countries (28%). In terms of value, the market was divided as follows: 38% HCR, 17% LSR, and 45% RTV. In volume terms, the shares were 67% HCR, 13% LSR, and 20% RTV.

In terms of market value, the four largest end-use industries for silicones were automotive (including rail and aerospace), with 25%; followed by power generation and distribution, with 17%; electrical and electronics, with 15%; and medical technology, with 12%. Other notable sectors included lighting and LED applications (8%), kitchenware (5%), and a range of industries such as renewable energy and rapid prototyping. A diverse group of additional applications made up the remaining 14%, encompassing areas like mechanical engineering, equipment manufacturing, sporting goods, toys, sanitary applications, building profiles, food, oil and gas, and defense. When looking at volume rather than value, the leading industries were slightly different: the automotive sector accounted for 28%, while power generation and distribution followed closely, with 25%.

Chemical Structure and Curing Systems

The curing step converts the plastic, uncross-linked silicone rubber into an elastic, permanently cross-linked material. While the older distinction between "hot curing" and "cold curing" rubber is no longer used, the specific chemical reaction that initiates cross-linking remains essential and varies widely depending on the system.

Radical Curing

In radical cross-linking, organic peroxides are added as initiators. When heated, these form free radicals that initiate cross-linking reactions. If the silicone polymer contains only methyl groups on the silicon atoms, nonspecific peroxides such as dibenzoyl peroxide are used. When a small number of vinyl groups are deliberately incorporated into the polymer chain, vinyl-specific peroxides like dicumyl peroxide can be applied. The decomposition temperature of the peroxide determines the processing temperature, typically between 90 °C and 175 °C.

The advantages of radical cross-linking include a long shelf life of the ready-to-use rubber compounds and a robust and reliable curing process. Disadvantages include a limited control over the cross-link density and the formation of cleavage products, which can cause odors or surface residues – these by-products must be removed via post-curing (thermal aftertreatment). Despite this, radical curing remains the standard method for HCR.

Addition Curing

Addition curing is based on a platinum-catalyzed hydrosilylation reaction between silicon-vinyl groups and silicon-hydride groups (Si–H), both introduced during polymer synthesis in small amounts. Since this reaction can occur spontaneously at room temperature, silicone rubbers are either formulated as two-component systems (LSR, RTV-2), mixed shortly before processing, or formulated with temperature-activated platinum catalysts (*e. g.*, in HCR).

In two-component systems, inhibitors such as vinylcyclohexanol are added to control the pot life (*i. e.*, processing window). A special variant of addition curing is UV-initiated cross-linking, where the platinum catalyst is activated by UV light. This enables one-component liquid silicones (UV silicones) that cure at room temperature but must be processed in darkness until exposure.

The advantages of addition curing are a fast, stoichiometric, and by-product-free reaction and the fact that there is no need for post-curing, even for high-purity medical applications, provided low-volatility polymers are used. They are suitable for short cycle times, particularly at elevated temperatures. Addition curing is the preferred system for LSR and RTV-2.

Condensation Curing

Condensation curing is classified into two types: RTV-1 (one-component) and RTV-2 (two-component).

RTV-1 silicones are based on terminally OH-functionalized polydimethylsiloxanes, with cross-linkers of the type $RSiX_3$, where X can be acetoxy, oxime, alkoxy, or amine. During compounding, $RSiX_3$ reacts with the Si–OH groups, forming pre-cross-linking bonds without initiating full cross-linking. A catalyst – usually an organotin compound – is added to accelerate both the initial reaction and the final cross-linking, which is triggered by moisture from the air. The by-products of this reaction are HX-type compounds, such as acetic acid, oximes, or alcohols.

RTV-1 systems are widely used in construction applications (*e.g.*, sealants and adhesives) but are less common in technical applications due to their long curing times and odorous by-products. However, they are difficult to replace for elastic adhesive applications.

RTV-2 silicones use the same base chemistry but do not require ambient moisture. Cross-linking is initiated by mixing the two components. Typical by-products are water (H_2O) or alcohols (HX). A unique aspect of RTV-2 cured rubber is its pure Si–O–Si network, free from organic bridging structures, which can act as weak points in other systems.

Other Curing Systems

Sulfur-based curing, standard in organic rubbers, is no longer relevant for silicones. Although it can be applied to vinyl-rich silicones, the resulting products are inferior to those cured with peroxides.

Dehydrocondensation (reaction of Si–OH and Si–H with hydrogen release) is usually a side reaction in addition curing. However, it is deliberately used for the production of closed-cell silicone foams, where hydrogen acts as an internal blowing agent.

Radiation cross-linking, for example, using gamma (γ) radiation, can induce cross-linking at doses of 20–50 J/g by generating radicals and breaking bonds. However, this method is not industrially relevant, as it leads to an uncontrolled formation of polar groups at the surface and it produces gaseous by-products such as hydrogen, methane, and ethane. Still, these reactions are important in evaluating the radiation resistance of silicone elastomers.

Properties

Silicone elastomers possess a unique combination of properties that clearly distinguish them from organic elastomers (see Table 6.68). Although reinforcing fillers can enhance their tensile strength to as much as 12 MPa and provide tear resistance values of up to 55 N/mm, the inherently low intermolecular forces between the silicone polymer chains define much of their behavior.

Table 6.68 Typical Property Ranges for Dense Silicone Rubbers (Excluding Foams and Gels) with No Special Additives to Modify Their Functional Performance

Property	HCR	LSR	RTV-2
Shore A hardness [-]	10–85	5–90	10–60
Tensile strength [MPa]	5–12	7–11	2–8
Elongation at break [%]	200–900	200–1000	200–700
Tear resistance [N/mm]	15–45	15–55	5–15
Service temperature range [°C]	−60 to 300	−60 to 250	−60 to 200
Density [g/cm³]	1.1–1.4	1.1–1.2	1.1–1.4
Compression set at 180 °C [%]	5–35	10–40	20–50
Dielectric strength [kV]	20–28		
Resilience [%]	30–70		
Thermal conductivity [W/mK]	0.2–0.3	0.2–0.3	0.1–0.2
Relative permittivity er [-]	2.5–3.1		

Due to their low surface energy, silicone elastomers are hydrophobic, which means that they repel water and easily separate from other materials. That gives them excellent release properties. They retain their elasticity even at extremely low temperatures and show minimal permanent deformation under pressure. These materials are also highly permeable to gases and water vapor, and their resistance to ionizing radiation exceeds that of all other polymer classes. Being naturally transparent, silicone elastomers are easy to color in translucent tones.

In terms of chemical resistance, silicones perform well when exposed to acids and polar solvents. However, they show lower resistance to strong bases and tend to swell significantly in nonpolar solvents or fuels. Electrically, silicone elastomers are excellent insulators with a high dielectric strength, but they can be made electrically conductive through the incorporation of appropriate fillers such as carbon black. Their flame resistance is notably high, making them suitable for safety-critical applications. The thermal conductivity can be adjusted depending on the formulation, enabling their use in thermal management systems.

Silicone elastomers also stand out due to their mechanical durability. They can be formulated for specific damping characteristics and are exceptionally stable under long-term dynamic loads. For instance, pressure points on silicone keypads can withstand millions of actuations without noticeable changes in the mechanical properties. Their physiological inertness makes them ideal materials for applications involving contact with food, drinking water, or the human body, including medical devices.

Production Technologies

Monomers

The wide variety of silicone products is based on a common foundational technology. The initial step, known as direct synthesis, involves the reaction of silicon powder with methyl chloride in a fluidized-bed reactor at temperatures between 260 °C and 360 °C, using copper as a catalyst. This process yields a mixture of methylchlorosilanes, with dimethyldichlorosilane [$(CH_3)_2SiCl_2$] as the desired product.

Through hydrolysis, this compound is converted into a mixture of cyclic and linear siloxanes, which serve as the base for polymer production. At this stage, the two main industrial approaches diverge: one separates the cyclic siloxanes, particularly octamethylcyclotetrasiloxane (D4), via distillation and uses them as the monomer for further polymerization. The other route removes the cyclic siloxanes and proceeds with a mixture of short-chain linear oligomers ($HO–[(CH_3)_2Si–O]_n–OH$). Neither route is fundamentally superior. In the D4 route, the cyclic siloxanes must be highly purified to remove trifunctional components. In the linear route, the dimethyldichlorosilane must be purified at an earlier stage for the same reason.

Polymers

Silicone polymers for HCR typically have a degree of polymerization between 5000 and 10,000, corresponding to viscosities between 10,000 Pa·s and 50,000 Pa·s. A defined amount of vinyl-methyl-functional siloxane is added to the base polymer so that the final material contains 0.02 mol % to 2 mol % vinyl groups. If D4 is the base material, the polymer is synthesized through an alkaline ring-opening polymerization. If the base material is the linear oligomer mixture, an acid-catalyzed polycondensation is used. In both cases, the resulting terminal OH groups are typically removed through endcapping, which stabilizes the viscosity and prevents unwanted cross-linking during rubber compounding and storage. By contrast, polymers used for LSR and RTV do not require such high molecular weights and therefore have a lower viscosity and are easier to process.

Elastomers

Once the polymers are made, they are converted into rubber compounds in a process known as compounding, which differs significantly depending on whether HCR, LSR, or RTV is being produced.

HCR, as a solid rubber, is highly viscous and requires intensive shear forces to be processed. Initially, the polymer and the reinforcing filler (often, up to 45% fumed or precipitated silica) are mixed. To facilitate mixing, low-molecular-weight siloxanes are added; they act as temporary plasticizers and also partially make the filler surface hydrophobic. The resulting compound is then supplemented with curing agents: either peroxide initiators for radical curing or platinum catalysts and Si–H cross-linkers for addition curing. Further additives may include color pigments, adhesion promoters, thermal stabilizers, flame retardants (*e. g.*, aluminum trihydrate), and conductive fillers. Compounding is carried out in batch kneaders, on two-roll mills, or in specialized continuous mixers. The finished HCR is delivered as blocks, granules, or endless profiles, cords, or tapes.

LSR is a pasty, liquid rubber. Because of its lower viscosity, the required shear forces are lower than for HCR. Around 30% hydrophobic silica is used as a filler, either pretreated or made hydrophobic in situ during mixing. Making it hydrophobic is critical for maintaining the viscosity stability and ensuring a reproducible processing window for injection molding. The polymers have a lower molecular weight and are typically richer in vinyl groups; mixtures of various chain lengths are often used. The final compound consists of two components (A and B) containing the platinum catalyst and the Si–H cross-linker, which are mixed only immediately before being processed. As in the case of HCR, LSR compounds also include additives to modify functional properties. They are manufactured in batch planetary mixers or specialized continuous systems and typically supplied in drums, although bag-in-box systems are also available.

RTV-2 is produced similarly to LSR, often using the same mixing equipment. Depending on the application, it can be adjusted to be flowable or thixotropic. RTV-1, on the other hand, must be processed under exclusion of moisture and is filled immediately after compounding into cartridges, foil packs, drums, or special tankers (*e.g.*, piston tankers) for delivery.

Modified Fluorosilicones and Phenylsilicones with Special Properties

For applications where the properties of standard polydimethylsiloxanes are not sufficient, modified base polymers are available. One such modification includes fluorosilicones (also called FVMQ according to DIN ISO 1629), in which some or all dimethyl units are replaced by trifluoropropyl-methyl units. These are produced by adding trifluoropropene to methylhydrogensiloxanes. The resulting materials offer significantly improved resistance to nonpolar media such as fuels and oils and can be used from −70 °C to +170 °C. However, above 150 °C and under oxidative conditions, fluorosilicones may release toxic trifluoropropionaldehyde, which must be considered during processing.

For applications requiring elasticity at very low temperatures, phenyl-methyl silicones (PVMQ, DIN ISO 1629) are used. Here, methyl groups are partially replaced by phenyl groups, allowing operation between −100 °C and +250 °C. These materials are available in all standard curing systems, including HCR, LSR, and RTV.

Processing

Gravity Casting

The simplest method for processing silicone elastomers is gravity casting, using flowable RTV silicones. This method is employed, for instance, in the production of casting molds for rapid prototyping or for overmolding electronic components.

Dispensing and Jetting

Dispensing involves the precise dosing of pasty silicones – usually RTV rubber – through small nozzles onto a defined area of a component. The silicone is applied in continuous lines, with the nozzles moved in programmed x–y patterns. This enables the precise application of sealing, bonding, or potting materials on electronic assemblies. When the component is not physically contacted during application, the process is called jetting – small droplets are ejected from the nozzle towards the surface. If the application system is also capable of movement along the z axis, then three-dimensional structures can be built layer by layer using stable silicones. This capability is used in additive manufacturing (see Section 4.10), particularly in material jetting processes for making molded parts. Both dispensing and jetting are widely used in the production of electronic devices.

Compression Molding

Compression molding is a processing method used primarily for HCR and particularly suitable for producing larger, less detailed parts in medium production volumes. The tooling costs are significantly lower than in injection molding. The process begins with manual placement of the HCR compound into the mold cavity, which is preheated. The cavity is then closed using a pressure plunger, which compresses the material into the desired shape. After curing, the part is removed from the mold – usually manually – and the process cycle repeats.

Injection Molding and Two-Component Molding

Injection molding of LSR offers high automation and short cycle times (see Section 4.3.5.7). The two components of the addition-curing LSR system are mixed in a 1:1 ratio under pressure in a mixing block. Color pastes and functional additives can also be added at this stage. To prevent premature curing in the injection unit, the mixed material must be kept cold until it enters the mold – this is referred to as cold-runner technology. Once inside the mold, the material cures in seconds at temperatures between 170 °C and 220 °C. Because of the low viscosity and injection pressures up to 2500 bar, precision molds with very tight tolerances are essential. LSRs are also suitable for multi-component injection molding, where specially formulated self-bonding types can be combined with other materials. Components weighing less than 0.1 g can be manufactured with this process. HCR can also be injection-molded, but the high viscosity requires specialized screw-fed systems for material handling.

Extrusion

Silicone extrusion is a continuous production process used mainly for HCR (see Section 4.2). The rubber is transported through a cooled extruder barrel via a rotating screw and pushed through a die that gives the material its final shape. The extrudate is then passed directly into a heating zone or salt bath, where it undergoes vulcanization. The shape is fixed only after this thermal treatment, resulting in products such as cords, tubes, or profiles. Wall thicknesses of less than 0.1 mm are achievable. By adding chemical blowing agents, silicone foams with densities below 0.5 g/cm^3 can also be produced using this process.

Calendering

Calendering involves feeding HCR silicone through a system of heated rollers arranged in sequence. The material is processed between the roller gaps, producing flat silicone sheets that are used as intermediate products for further processing steps like stamping gaskets. Multi-layer laminates can also be produced by reinforcing the silicone sheets with textile fabrics, such as those used in conveyor belts. In addition, materials like textiles or paper can be coated with silicone in the calendering process. However, for very thin films (below 0.1 mm), calendering is not suitable. In these cases, doctor blade or slot die techniques using LSR or RTV silicones are preferred.

Applications

Automotive Industry

In modern passenger vehicles, silicone elastomers are used in over 40 different components, and a single car can contain several kilograms of silicone. They are found in hoses for cooling systems and turbochargers, in seals and cable grommets as well as in damping elements for suspension systems and engine mounts. In the field of electric vehicles, silicones are particularly valuable – used in high-voltage cables and for thermal management of lithium-ion batteries, where they serve as thermally conductive encapsulants.

Generation and Distribution

Due to their outstanding electrical insulating properties, flame resistance, and dielectric strength, silicone elastomers play a major role in power systems. They are used in high-voltage insulators, cable joints, terminations for underground cables, and fire-resistant cable sheathing. Silicones are also employed in photovoltaic modules, both for sealing and potting purposes, ensuring long-term durability even in demanding environments.

Electronics

The electronics sector represents one of the largest application areas by volume. Silicone elastomers are widely used in keypads for keyboards, remote controls, and handheld devices. They also serve as potting materials in electronic assemblies, such as control units, power modules, and charging equipment. In high-contrast displays, silicones are applied for optical bonding, and in high-performance LEDs, they are used for lenses and encapsulation. Their transparency and UV stability make them ideal for secondary optics and outdoor electronics.

Medical Technology

Because of their inherent biocompatibility, silicones are broadly employed in medical applications, including uses as implants and in direct contact with blood. They are used to manufacture tubing, catheters, balloons, and valves as well as respiratory masks. In baby care, silicones are found in bottle nipples and pacifiers. In wound care, they are used as skin adhesives and in wound dressings, providing gentle adhesion and breathability.

Other Industries

The versatility of silicone elastomers is evident in countless everyday products. Examples include scale-free showerheads, kitchenware such as baking molds and microwave-safe dishes, irrigation components, and milking liners used in agriculture. They are also used in oven door seals and many other household or industrial applications where their unique properties – rather than price – are the decisive factor.

Outlook

Current innovation efforts in silicone elastomer development are not only focused on enhancing application-specific properties – such as thermal conductivity, heat resistance, adhesion, or optical clarity – but, increasingly, on sustainability-related topics. These include improving the recyclability of silicones, reducing their environmental persistence, minimizing volatile emissions, and enabling the use of renewable raw materials and energy sources.

One particularly promising area of future development is the use of electroactive silicones in smart applications. These materials are being explored for use as generators, sensors, and actuators, where their elasticity and electrical responsiveness can be harnessed in advanced technologies.

Given the continuous emergence of new applications and the growing demand from developing regions, the global volume growth for silicones is expected to continue at an average annual rate of 4% to 5% in the coming years.

References for Section 6.23.6

[1] Winnacker/Küchler, Chemische Technik: Prozesse und Produkte, Vol. 5: Organische Zwischenverbindungen, Polymere, Chapter 6: Silicone, 1095 ff., WILEY-VCH, Weinheim (2005)

6.23.7 T-Elastomers (TM, ET, TCF)

Chains with sulfur.

Polysulfide Elastomers (TM, ET)

$$\left[S - CH_2 - CH_2 - S \right] \tag{6.65}$$

TM/ET (Equation 6.65) exhibits excellent solvent resistance (esters, ketones, aromates) and aging resistance. Odor and low strength and heat resistance are disadvantages.

Applications: container linings, sealants compounds (liquid rubber).

Thiocarbonyl Difluoride Copolymer Elastomer (TCF)

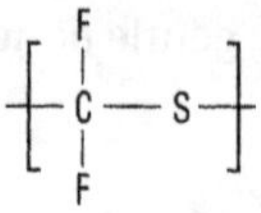

(6.66)

6.23.8 U-Elastomers (AFMU, EU, AU)

Chains with oxygen, nitrogen, and carbon.

Polyester (AU)-/Polyether (EU) Urethane Elastomers

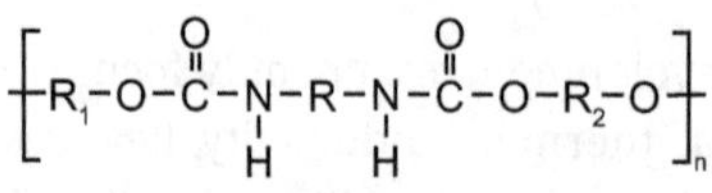

(6.67)

R_1 = Polyester chain (AU), R_2 = Polyether chain (EU)

Other than the rest of the elastomers, polyurethanes are produced by polyaddition of low molecular and thus low viscous pre-products, see Section 6.17. Processing takes the form of either thermoplastic molding of special pellets (thermoplastic PUR elastomers), casting of reactive mixtures, or the classic methods of elastomer processing: production of a pre-polymer, addition of fillers and crosslinking on rolls with subsequent vulcanization in heated molds.

AU and EU (Equation 6.67) exhibit exceptional high strength, flexibility, and elasticity; their properties range between those of the other elastomers and thermoplastics and thermosets, respectively. They are wear resistant, oil-, fuel-, and ozone resistant; they exhibit relatively high mechanical damping. AU exhibits better water resistance than EU.

Applications: hydraulic seals, chain wheels, rollers, solid wheels, bellows, drive belts, fenders.

6.23.9 Polyphosphazenes (PNF, FZ, PZ)

Chains with alternating phosphorus and nitrogen.

Fluoro-Phosphazene Elastomers (PNF)

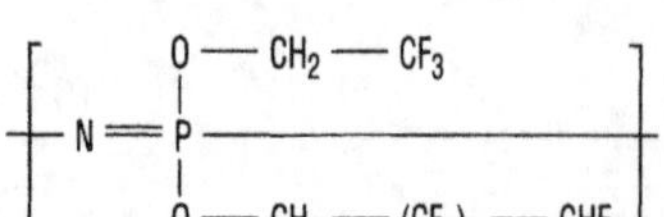

(6.68)

PNFs are produced by ring-opening polymerization of trimer chlorinated phosphazene with subsequent exchange of the chlorine atom by alkoxy side groups, Equation 6.68.

PNF containing substituents with highly fluorinated unsaturated sites can be crosslinked with sulfur as well as with peroxides. The main chain containing only sulfur and nitrogen makes PNF absolutely resistant to oxygen and ozone. PNF is rubber-elastic from −50 to +150 °C; in the case of fire, it does not melt, drip, or develop smoke; it is resistant to fuels, oils, and hydraulic liquids.

Applications: seals for high-altitude airplanes and arctic oil- and fuel lines.

Halogen-free grades with similar good fire behavior are used as open- or closed cell foams for ship insulation and aircraft upholstery.

Other polyphosphazenes, FZ, PZ.

FZ: phosphazene elastomer with fluoroalkyl- or fluoroxyalkyl groups.

PZ: phosphazene elastomer with phenoxy groups.

6.23.10 Other Rubbers

Polyperfluorotrimethyltriazine Elastomer (PFMT)

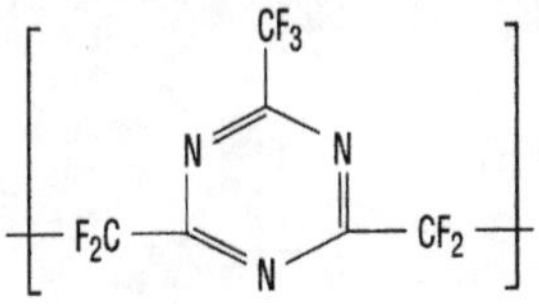

(6.69)

7 Additives, Fillers, and Fibers

7.1 Additives

Norbert Grünewald

Polymers are rarely processed as pure raw materials. More often, a polymer contains several additives to aid during processing, add color, or influence the structure, properties, or the chemical/aging resistance. These include lubricants, anti-blocking agents, release agents, processing aids, stabilizers, antistatic agents, electrically conductive additives, flame retardants, colorants, impact modifiers, plasticizers, bonding agents, fillers, and reinforcing and blowing agents. Residuals of these additives as well as of synthesis additives, such as catalysts, emulsifiers, precipitants, curing agents, and accelerators, may remain in the molded component (intentionally or not) and affect its serviceability. Currently, the annual global consumption of additives (excluding pigments and fillers) is approximately 35,000 kt, with an annual growth rate of 5.6% until 2032. The effectiveness of any additive depends on its solubility in the respective polymer, which is determined by the chemical structure and the crystallinity of the polymer as well as by processing temperatures. When assessing the use of additives, a wide range of factors have to be considered: synergistic effects between certain additives, undesired reactions with each other and with the polymeric material, and safety, hygiene, and toxicological aspects. Additives may also affect post-processing steps such as printing, labeling, or coating. The choice of additives always depends on the intended use of the product.

7.1.1 Slip and Anti-blocking Agents, Release Agents

Slip agents prevent plastics (*e. g.*, films) from sticking together in the final product. Additionally, some slip agents reduce the melt viscosity of molding compounds (internal lubricants) and/or act as lubricants (external lubricants) between the melt and metal surfaces, even at low concentrations of less than 0.5%. A prerequisite for slip effects is that the molecules must contain linear or branched hydrocarbon chains of more than 12 carbon atoms. The introduction of polar groups specifically influences solubility and compatibility. The majority of the slip agents is used for polyolefins, PS, ABS, and PVC. For quality control, the coefficient of friction (COF) is typically measured. Since slip agents migrate slowly, the COF should only be measured 7–10 days after production (*e. g.*, extrusion). Table 7.1 provides an overview of common slip agents.

Anti-blocking agents prevent the sticking of films to one another; typically, they tend to migrate to the surface. An example is polydimethylsiloxane (PDMS), available as 50% masterbatch with PP, PS, POM, PE, TPE, ABS, and PA. Both chalk and talcum may serve as anti-blocking agents by maintaining dry surfaces, in particular in compounds containing plasticizers. Cross-linked silicone resins in powder form (spherical) are added to plastic compounds to provide uniformly rough surfaces, thus preventing films from sticking. A list can be found in Table 7.2.

Release agents that are sprayed into molds rely on similar mechanisms but are used less frequently. For the production of components that require post-processing/finishing operations, the release agent should be silicone-free. PTFE – as powder or dry spray – is also used as internal slip or release agent. Wax, metal soaps, and silicones are also used as release agents.

In general, all liquids improve the abrasion and slip behavior of plastics. However, the slip agents have to be chemically neutral and must not cause stress cracking in the plastic material. Fats and oils based on silicone are the most commonly used, for example, with PP, PS, PVC, ABS, POM, PMMA, PC, PBT, PPO, PA 6, and PA 66.

Table 7.1 Slip Agents for Plastic Melts

	Slip agent	Characteristics
Internal lubrication ↓	Fatty acid ester ■ Butyl stearate ■ Glycerin mono-oleate ■ Pentaerythrityl tetrastearate (PETS)	High volatility Mainly used in PVC PETS does not cause clouding
	Fatty alcohol	High volatility, used in PVC
	Metal soaps, such as calcium stearate	Used in PVC, polyolefins, also as release agent and nucleating agent (PA, polyester, PUR) Plate-out tendency Calcium stearate requires short plasticizing times
Internal/ external lubrication	Fatty acid amides, fatty acid diamides ■ Amide wax ■ Oleic acid amide ■ Erucic acid amide	Amide wax: very versatile Oleic acid amide: slip agent in PE-LD Erucic acid amide: in PP
↑	Fatty acid complex ester (oligomeric esters from fatty acids, dicarboxylic acids and polyols)	Used in PVC
	Montan wax (long-chained, with lengths of C22 to C34)	Can be used with many plastics (PS, PP, PET, PBT, POM, PC, PA, TPU), compatible, high temperature resistance, improved surface smoothness and gloss, release agent
External lubrication	Polyethylene wax, oxidized	Used in extrusion, blow molding, PVC as release agent
	Polypropylene wax, oxidized	Predominantly used as compatibilizer
	High-molecular polyethylene wax (PE-LD, PE-HD)	Used in PVC, for transparent bottles, masterbatch medium
	Fluoroelastomers	Rarely used, increased abrasion resistance
External lubrication (non-migrating)	Siloxane	Permanent effect

Table 7.2 Anti-blocking Agents for Plastic Melts

	Anti-blocking agent	Chemical composition
Inorganic	Natural silica	Silicon dioxide (SiO_2) – mined
	Synthetic silica	Silicon dioxide (SiO_2) – manufactured
	Talc	Magnesium silicate – mined
	Calcium carbonate	Calcium carbonate ($CaCO_3$) – mined
	Ceramic beads	Aluminum silicate ceramic – manufactured
	Kaolin/clay	Aluminum silicate – mined
	Mica	Aluminum potassium silicate – mined
Organic	Bisamide	Ethylene bisstearamide (EBS)
	Secondary amide	Stearyl erucamide
	Primary amide	Stearamide, erucamide
	Organic stearate	Gylcerol monostearate (GMS)
	Metallic stearate	Zinc stearate
	Miscellaneous	Silicone, PTFE

7.1.2 Stabilizers

Plastics are subjected to a wide range of influences during processing and subsequent use. Depending on their molecular structure and morphology, they undergo degradation reactions that are induced by thermo-mechanical, thermo-oxidative, and photo-oxidative processes.

7.1.2.1 Antioxidants

The largest amount of stabilizers is used to prevent oxidative degradation in polymers. They are most important for polyolefins. To a lesser extent, they are also relevant for polystyrenes, polycarbonates, polyesters, polyamides, polyvinyl chlorides, and polyacetals as well as polyurethanes and elastomers. With the increasing demand for a circular economy in plastics, their significance is growing for all polymers in the context of re-additivation. Exposure to heat, irradiation, and mechanical loads, such as shearing, creates radicals that react with oxygen to form peroxide radicals, which, in turn, will split off a hydrogen atom and transform into hydroperoxides before further degradation and formation of radicals. With sufficient oxygen availability, hydroperoxides and alkyl radicals as well as peroxy-, alkoxy-, and hydroxyl radicals will

be created in non-stabilized plastics. These chain reactions result in the degradation of plastic materials. The thermo-oxidative process is mainly determined by the solubility and diffusion rate of oxygen in the plastic. The need for thermo-stabilization is also driven by higher service temperatures and the increasing implementation of recycled material streams that often require antioxidants for extended stability.
We distinguish between primary and secondary antioxidants, depending on their respective mode of action. Primary antioxidants (radical scavengers) contain reactive H atoms that react with free radicals. This group includes, for polyolefins, phenols and amines as well as aromatic amines. Secondary antioxidants degrade hydroperoxides, terminating the chain reaction and thus stabilizing the oxidation reaction. Organic phosphites, phosphonites, and thioester represent this group. Antioxidants based on lactones (*e.g.*, vitamin E) are used to improve the processability of polyolefins.

Required antioxidant loads range from 0.05 wt % to 1 wt %. Note that antioxidants may cause (dis)coloration and may interact with other additives or fillers. Although the aging process may be significantly delayed by the incorporation of antioxidants, it cannot be offset or neutralized. Comprehensive information on a wide range of antioxidants and their applications can be found in the *Plastics Additives Handbook*, 6th edition, by H. Zweifel, R. D. Maier, and M. Schiller (Hanser, 2009).

7.1.2.2 UV Stabilizers

Exposure to UV light creates extremely reactive radicals in plastics, which, in turn, initiate degradation reactions and molecular rearrangements in the polymer chains. In combination with oxygen, water, or other environmental impacts, the damaging mechanism of photo-oxidation is further intensified. The results include yellowing, loss of transparency or gloss, chalking, change in color, cracks, or decrease in mechanical properties. UV absorbers at 0.1 wt % to 2.0 (10) wt % are incorporated as UV stabilizers. They absorb the ultraviolet spectrum of light and dissipate it as low-level heat. Typical representatives of this group are substituted benzophenones, salicylic acid esters, hydroxyphenyl benztriazoles, triazines, benzylidene malonate, and carbon black. Other currently widely used UV stabilizers include sterically hindered amines (HALS, hindered amine light stabilizers) incorporated at 0.1 wt % to 1.0 wt % as polymeric and oligomeric types. Halogen-containing flame retardants and acidic media interfere with HALS. Here, NOR-HALS (alkoxyamine-HALS), which combines UV stabilization and flame retardancy, is better suited.

Metal complexes (Ni, Zn) and sulfur-containing compounds, such as dialkyldithiocarbamate or dialkyldithiophosphate, have proven to be effective peroxide decomposers. A range of active components is combined in UV stabilizer masterbatches. Nickel chelates are now rarely added to polymers as quenchers, where they deactivate radicals and dissipate energy in the form of radiation and heat. Various active ingredients are combined in UV-protection masterbatches, requiring the addition of heat stabilizers.

7.1.2.3 Heat Stabilizers

Heat stabilizers are used for all applications exposed to thermal stress. Phenols serve a dual function, acting both as primary antioxidants and as heat stabilizers, making them widely used in many types of plastics.

In the past, PVC was exclusively stabilized with lead and cadmium. Lead and cadmium stabilizations (barium-cadmium (BaCd) and lead-barium-cadmium (PbBaCd) stabilizations) have now been replaced in Europe and in products imported into Europe by lead-free stabilizers: for highly transparent bottles by tin, for PVC-P by barium-zinc (BaZn) instead of BaCd, and for food applications by calcium-zinc (CaZn) with organic co-stabilizers. The type of stabilizer used influences the plate-out effect during PVC processing (formation of a hard deposit, in particular at the screw tip).

7.1.2.4 Foam Stabilizers

Typically, PUR foam is rather unstable during formation and would collapse with the advancing reaction unless foam stabilizers are added. Foam stabilizers are generally organo-silicon compounds (polyether-polysiloxane) which exhibit surface-active effects but can also act as emulsifiers. Foam stabilizers regulate the foam structure, open and closed cell formation, and cell size.

7.1.3 Static Inhibitors, Antistatic Agents

Most plastics exhibit surface resistivities greater than 10^{15} Ohm and are therefore able to build up electrostatic charges. As a result, they may attract dust or promote spark formation. Antistatic agents are used to lower the surface resistivity to less than 10^{11}–10^{10} Ohm.

Internal antistatic agents (fatty acid esters and amides, ethoxylated amines, diethanol amides, ethoxylated alcohols, and alkyl sulfonates and alkyl phosphates) are incorporated into the plastic directly or via masterbatches. They are not compatible with the plastic and migrate to the surface of the manufactured component, where they form a hydrophilic layer that absorbs ambient water and thus becomes conductive. Antistatic masterbatches containing the active agents listed above (in concentrations ranging from 3% to 50%), tailored for PE, PP, PS, and PA, are available. Glycerin monostearates, also with ethoxylated amine as synergist, are used to provide polyolefins with antistatic properties.

External antistatic agents (e. g., solutions of quaternary ammonium salts) are applied to the surface by dipping or spraying; there, they form a thin, electrically conductive film after the solvent volatilizes. However, this film is typically not very durable because it is susceptible to abrasion and does not rely on a “depot effect” as seen in internal antistatic agents.

Permanent antistatic properties are supplied by blending hydrophilic polymers with the matrix (*e. g.*, polyamide copolymers), creating both surface and volume conductivity.

Adding high-conductivity carbon black (approx. 15%, for active carbon black 5–10 wt %) to thermoplastic materials, in particular to polyolefin compounds, results in specific volume resistivities ranging from 10^3 Ohm · cm to 10^6 Ohm · cm. Applications include sheathing materials and electric panel heaters. To achieve better conductivity, the carbon black content must be increased accordingly.

Conductive additives that are able to reduce the electrical resistivity by up to 1 Ohm · cm even in low concentrations are used to shield electronics from high-frequency emissions. Aluminum flakes coated with bonding agents, micro steel fibers, silver-coated glass fibers and spheres, super-fine, highly effective nickel-coated graphite fibers (diameter: 8 µm + 0.5 µm Ni layer), gold-graphite powder, special carbon blacks, and carbon fibers are used to meet regulatory requirements regarding electromagnetic interference and radio-frequency shielding (10 kHz to 140 GHz range) with medium to good shielding results (> 60 db). Interior housing surfaces are coated with conductive paints, 5 µm Al-vacuum deposition (Elamet process), metal sprays, or chemo-galvanizing processes (see Section 4.15) to provide shielding. See also Section 6.19.

7.1.4 Flame Retardants

When plastic materials burn, pyrolysis creates free radicals that initiate radical chain reactions with oxygen and thus induce degradation. Flame retardants effectively interrupt this mechanism using the following modes of action:

a) Physically by diluting the substrate with nonflammable substances

b) Release of nonflammable gases

c) Cooling of substrate and/or fire effluents

d) Chemical interruption of the combustion process in the gas phase

e) Protective layer created by charring, intumescence

f) Cross-linking of the substrate

Typically, several mechanisms are employed simultaneously. Table 7.3 provides a list of important flame retardants sorted by active elements and grouped according to their primary mode of action. Flame retardants can be additive (physically blended into the polymer) or reactive (chemically bound to the polymer backbone during synthesis). Reactive flame retardants – such as brominated phenols or tetrabromophthalic acid anhydride – offer improved permanence and thermal stability. Additive systems, on the other hand, are widely used and may act as plasticizers or fillers, depending on the compatibility.

Table 7.3 Basic Flame Retardants

Group	Chemical	Application	Mechanism – see also text
Halogens (chlorine and bromine), restricted in the EU	▪ Bromine compounds, often with antimony trioxide as synergist[1)] ▪ Polybrominated diphenyl ethers (PBDEs), such as decabromodiphenyl ether (BDE) ▪ Tetrabromobisphenol A (TBBPA) and derivates ▪ Hexabromocyclododecane (HBCD) ▪ Brominated polystyrene ▪ Brominated phenols ▪ Tetrabromophthalic anhydride	Electronics, cable, wires Thermoplastics and thermosets, elastomers, PUR foams	a, b, d Cause a disruption of the radical mechanism; brominated hydrocarbons offer higher efficiency than chlorinated hydrocarbons
Phosphorus	▪ Triphenyl phosphate (TPP), Resorcinol bis (diphenyl phosphate) (RDP), Bisphenol A bis(diphenyl phosphate) (BDP) ▪ Red phosphorus ▪ Ammonium polyphosphate (APP) ▪ DOPO and derivatives ▪ Aluminum diethyl phosphinate	Housings for electronic devices, PVC, PPO, HIPS PC/ABS, EP/UP gelcoats, PUR flexible and rigid foams, PA, polyester fibers, PE, EVA	b, e, f
Nitrogen	Melamine and derivates, such as melamine cyanurate and melamine polyphosphate	PUR foam, PA, HIPS	a, e also alternative for halogenated hydrocarbons
Metal hydroxides and salts (Al, Mg, Zn)	▪ Aluminum hydroxide (ATH) stable up to approx. 200 °C ▪ Magnesium hydroxide (MDH) stable up to approx. 300 °C	PVC, polyolefins for cable applications Thermosets for electronics, seats, and paneling	a, c ATH and MDH often receive an organic coating for better compatibility

Group	Chemical	Application	Mechanism – see also text
Others	Boron compounds with Al, Mg, Zn Zinc borate with antimony trioxide or aluminum hydroxide or red phosphorus	PVC-P, polyolefins, elastomers, PA, epoxy resins	c, b glass-like coating
	Expandable graphite, 250-fold	PUR foams, elastomers	e
	Zinc compounds such as zinc sulfide and zinc chloride with ammonium polyphosphate or zinc borate	PVC, PP, PA	e
	Nanocomposites based on aluminum silicate clay minerals (montmorillonite)	PMMA, PP, PS, PA	Prevent dripping and promote charring, special processing
	Silicon compounds	PP, PE, PA, EP, PES, textiles, cable insulations, PUR foam	c, e. f
	Talcum, chalk		Not flame retardant as such, "thins" polymers and thus reduces fire load

[1] Antimony trioxide alone is not a flame retardant; it synergistically increases the temperature range of a flame retardant.

In recent years, oligomeric and polymeric flame retardants have gained importance due to their low volatility and reduced tendency to migrate. Moreover, synergistic combinations, such as antimony trioxide with brominated compounds or zinc borate with phosphorus systems, are frequently used to enhance the effectiveness while minimizing the dosage.

In 2013, the global consumption of flame retardants exceeded 2 million metric tons. The largest markets are in building and construction, followed by transportation and electronics. While brominated flame retardants dominate the Asian market, aluminum hydroxide (ATH) holds the largest share in Europe and North America, with approximately 50%. Going forward, the greatest demand is expected from the electronics industry (*e. g.*, smartphones, tablets).

Due to growing environmental and health concerns, many halogenated flame retardants have been restricted or banned, particularly in the EU. Under the European Waste Electrical and Electronic Equipment Directive (WEEE), substances such as Pb,

Hg, Cd, Cr (VI), PBB, and PBDE must not exceed 0.1% by weight per homogeneous material. This also impacts commonly used additives like TBBPA, for which further regulation is expected. Melamine cyanurate and other nitrogen-based compounds may also face restrictions. In the US, regulatory actions have been taken at the state level (*e.g.*, California, Maine). This regulatory pressure drives the development of alternative products. Fastest growing is the demand for phosphoric acid esters and a variety of inorganic flame retardants. Although the proportion of halogen-containing products is declining, the absolute consumption is still growing because of increasing use in Asia.

The international standard ISO 1043-4 classifies flame retardants by group codes; see Table 7.4. This is especially important for the identification and traceability of recycled plastics, for which halogenated substances must often be avoided.

Table 7.4 ISO 1043-4 Coding of Flame Retardants

Code No.	Flame retardant
10–29	Halogen compounds (chlorine, bromine, fluorine)
30–39	Nitrogen compounds
40–49	Organic phosphorus compounds
50–59	Inorganic phosphorus compounds
60–69	Metal oxides, metal hydroxides, metal salts
70–74	Boron and zinc compounds
75–79	Silicon compounds
80–89	Graphite, among others
90–99	Not assigned

For further reading, refer to the *Plastics Flammability Handbook*, 4th edition, by J. Troitzsch and E. Antonatus (Eds.) (Hanser, 2021).

7.1.5 Colorants

Colorants for plastics are comprised of specially compounded *insoluble pigments* and of plastic-*soluble dyes*. Figure 7.1 demonstrates the significance of pigments for the coloring of plastics in comparison to dyes. Pigments are classified as organic or inorganic depending on their chemical constitution. In contrast to pigments for special effects (see below), absorbing pigments reflect only the colors that they do not absorb (subtractive color) so that the color impression is created by the complementary color.

Because of their irregular form and light absorption, absorbing pigments exhibit only one color (independent of the sight angle) and no gloss.

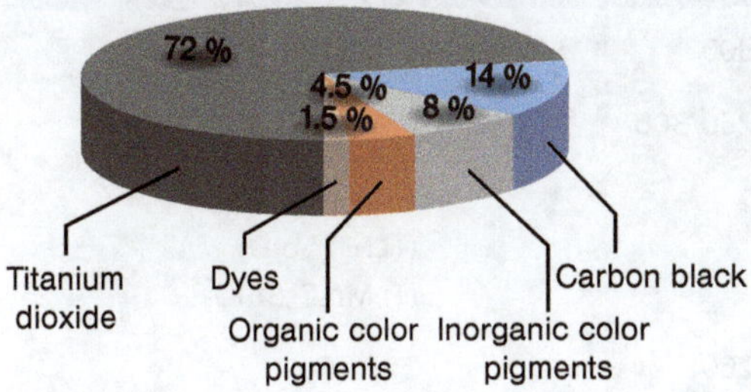

Figure 7.1 Proportions of different colorants in current applications

Inorganic pigments (Table 7.5) include metal oxides, sulfides, chromates (except CrVI), and carbon blacks (the last are categorized as inorganic despite their chemical constitution).

Table 7.5 Inorganic Pigments

Pigment	Chemical family	Temperature resistance in PE-HD [°C]	Chemical formula
White	Lithopone (zinc sulfide + barium sulfate)	300	$ZnS \cdot BaSO_4$
	Titanium dioxide (rutile, anatase modification)	300	TiO_2
	Zinc oxide		ZnO
Black	Carbon blacks		
	Iron oxide black	240	Fe_3O_4, $(Fe,Mn)_2O_3$
	Spinel black	300	$Cu(Cr,Fe)_2O_4$, $Cu(Cr,Mn)_2O_4$, $(Fe,Co)Fe_2O_4$
Yellow/orange	Chrome rutile yellow	300	$(Ti,Sb,Cr)O_2$, $(Ti,Nb,Cr)O_2$, $(Ti,W,Cr)O_2$
	Iron oxide yellow	220/260	α-$FeO(OH)$, γ-$FeO(OH)$
	Nickel rutile yellow	300	$(Ti,Sb,Ni)O_2$, $(Ti,Nb,Ni)O_2$
	Bismuth vanadate/ molybdate	280	$BiVO_4Bi_2MoO_6$
	Zinc ferrite		$ZnFe_2O_4$

Table 7.5 Inorganic Pigments *(continued)*

Pigment	Chemical family	Temperature resistance in PE-HD [°C]	Chemical formula
Brown	Chrome iron brown	300	$(Fe,Cr)_2O_3$
	Iron oxide(manganese-) brown	260/300	$(Fe,Mn)_2O_3$
	Rutile brown		$(Ti,Mn,Sb)O_2$, $(Ti,Mn,Cr,Sb)O_2$
	Zinc ferrite brown	260	$ZnFe_2O_4$
Red	Iron oxide red	300 (1200)	α-Fe_2O_3
	Molybdate red	260/300	$Pb(Cr,Mo,S)O_4$
Green/blue	Chrome oxide green	300	Cr_2O_3
	Cobalt(spinel-)green	300	$(Co,Ni,Zn)_2(Ti,Al)O_4$
	Cobalt blue	300	$CoAl_2O_4$, $Co(Al,Cr)O_4$
	Ultramarine blue	300	$Na_8(Al_6,Si_6,O_{24})S_x$
Metallic	Aluminum	300	Al
	Copper	260	Cu/Zn alloys

Azo and polycyclic pigments, among others, represent the most significant families of *organic pigments*. All azo pigments are characterized by their common azo group (-N=N-), which is responsible for the selective absorption of light. Because the entire molecular structure determines the specific color of an azo pigment, they cover a wide color spectrum, ranging from yellow, orange, brown, and red to violet. Some azo pigments contain carcinogenic amine components and may therefore not be used for plastic coloration applications. Polycyclic pigments also cover a wide color spectrum, ranging from orange, brown, red, violet, and blue to green. Important representatives of polycyclic pigments include quinacridones (red) and phthalocyanines (blue/green).

Polyol dyes are used to color PUR foams. Here, chromophores (constituents of a dye with excitable electrons such as conjugated double bonds) are chemically bound to polymer chains with terminal hydroxyl groups that react with isocyanate. The results are homogeneously colored polyurethanes (foams, rigid and rubber-elastic components – the latter can also be manufactured with transparent coloration).

The following properties determine the application and processing conditions of pigments: light and weather fastness and heat and chemical resistance. Other undesirable issues to be considered are warpage caused by pigments and migration of pigments to the surface of manufactured components. Pigments may also considerably affect flow, curing, and electrical properties. Compatibility with the plastic is another important criterion for the selection of suitable pigments. The coloration created by pigments is determined by the particle size and distribution in the plastic. The smaller the particles, the more vivid the color. Inorganic pigments generally exhibit higher light fastness and better temperature resistance with a low migration tendency. Organic pigments offer more luminous colors.

Several plastics yellow under UV exposure and are therefore not light-/colorfast as such. This results in a darkening, in particular of lighter colors.

Pigments are available in powder form; however, they are difficult to disperse in plastic materials. Therefore, plastics are often colored using pigment pastes with binders and/or plasticizers. These are available either as pelletized masterbatches or in liquid plastics.

Over the past decades, the colorant market has seen a continuous increase in products that provide special optical effects in addition to coloring plastics. These products include metal and pearlescent pigments, fluorescent colorants, and, last but not least, optical brighteners.

7.1.5.1 Pigments for Special Effects

Pigments that provide metal and pearlescent effects are "special effect" pigments. Metal pigments consist of small aluminum, bronze, or gold platelets that reflect light like mirrors. Pearlescent pigments are based on silicon or aluminum oxide platelets with thin (few-nanometers thick) titanium or iron oxide layers. TiO_2-pearlescent pigments are often used. Pearlescent pigments partially reflect light directly at the smooth surface of the pigment, while another part of the incident light is reflected multiple times at the phase boundaries of the platelet-like particles. This creates the same interferences that are generally observed with thin layers (where the layer thickness matches the wavelength of visible light).

Major influencing factors for special effect pigments include the transparency of the plastic matrix, the complete dispersion of the pigment platelets, and the parallel orientation of the platelets in the plastic. When aiming for highly vivid colors, fillers and semicrystalline plastics should be avoided. PS and PMMA are particularly well suited for special effect pigments. Depending on the particle size, effects ranging from silk gloss to glimmer can be achieved. Unbalanced flow paths and flow lines disturb surface effects. Pearlescent pigments are added to thermoplastics in concentrations of approximately 0.5 wt% to 2 wt%; concentrations in films or thin layers may be considerably higher.

Color-variable pigments (CVP) are also used to vary the color of a plastic surface with changing view angles. These pigments consist of a reflecting Al layer, a SiO_2 layer responsible for the effect, and a Fe_2O_3 layer (approx. 25 nm thick) that increases reflection. The thickness of the SiO_2 layer determines the achievable color range (see Table 7.6).

Table 7.6 Color Changes Caused by Different SiO_2 Layer Thickness, CVP Pigments

SiO_2 thickness [nm]	Color, top view	Color, slant view
310 to 340	Green/gold	Reddish gray
360 to 400	Red	Gold
410 to 420	Violet	Green
430 to 440	Copper	Red

7.1.5.2 Fluorescent Colorants

Fluorescence can be defined as a shift in the energy of incident illumination from shorter to longer wavelengths (such as blue to yellow) which can make the fluorescent color appear brighter (more saturated) than it could possibly be by reflection alone. With very few exceptions – for example, zinc sulfide – there are no fluorescent pigments. Therefore, soluble fluorescent dyes, such as xanthenes (*e.g.*, fluorescein, rhodamine), are used. To ease the dispersion of the dyes, they are compounded into meltable polyester/polyamide systems and subsequently added like insoluble pigments to plastic materials (masterbatches, temperature resistance: 260–280 °C). They are compatible with polyolefins and polystyrene; the compatibility with ABS, PA, PC, PMMA, and PVC-U needs to be confirmed. For PVC-P and PUR, non-melting thermoset pigments are typically used (heat resistance up to 210 °C). Fluorescent dyes – also in combination with titanium dioxide – are used as optical brighteners to improve the degree of whiteness of plastics.

A detailed description of colorants and their processing can be found in *Coloring of Plastics*, by A. Müller (Hanser, 2012).

7.1.6 Additives for Laser Labeling

Depending on the absorption of laser light, plastics vary in terms of their suitability for laser labeling. The addition of suitable pigments turns plastics that originally cannot – or can only barely – be laser-labeled, such as PE, PP, POM, PA, and PMMA, into materials that can be uniformly labeled. These pigments enable color-intensive reactions with the laser beam in many plastics, which are induced by reactions in the polymer but partially also by reactions with the pigment. The color scale of components that can be labeled ranges from black to colored to transparent components.

7.1.7 Impact Modifiers and Plasticizers

Impact modifiers are added to brittle plastics to increase their elasticity, which is particularly important at temperatures below 0 °C. A typical example is the addition of elastomers such as EPDM or SBS to PP or PS. However, combinations of rigid plastics also exist, such as LDPE in PA6.

In the field of blends, compatibilizers are used to create a compatible copolymer from two predominantly incompatible plastics. This is mainly achieved using anhydride-modified compatibilizers, as shown in Table 7.7.

Plasticizers and flexibilizers (see Section 6.4.2) are used for rubber-elastic PVC compounds, to control the toughness of rigid plastics, and for coating resins. The current discussion of health risks posed by phthalic-acid-containing plasticizers has caused a decline in demand in Western Europe and driven the development of alternative plasticizers. However, the demand in Brazil, Russia, India, and China is still expected to increase.

7.1.8 Bonding Agents

Bonding agents form molecular bridges at the interfaces between an inorganic additive and an organic polymer matrix. They contain hydrolyzable groups to bond with inorganic materials and organo-functional groups in the same molecule. Semi-organic silanes and titanates are required for glass-fiber-reinforced thermosetting resins, while specific organo-functional groups are necessary for thermoplastic composite materials. Inorganic additives are pretreated (coated, also with stearates) in a separate processing step. Bonding agents are used to improve adhesion between plastics or paints and metallic or other surfaces and for the manufacture of composite films in which the individual layers do not exhibit adequate adhesion (Table 7.7).

Table 7.7 Bonding Agents (in Parentheses: Percentage by Weight of Components) Nomenclature: See Table 6.4

Bonding agent	Bonding between
Anhydride-modified EMMA, EAA	PE, EVAL, EMI, PET, PC
Anhydride-modified EVAC	PE, EVAL, PS, EMI, PET, PC
Acid-/acrylic-modified EVAC	PE, EVAL, EMI, PET
Anhydride-modified PE	PE, EVAL, EMI
Anhydride-modified PP	PP
EBA (89/7/4)	PE, PP, PBT
EBA (80/20)	PE, PVDC

Table 7.7 Bonding Agents (in Parentheses: Percentage by Weight of Components) Nomenclature: See Table 6.4 *(continued)*

Bonding agent	Bonding between
EVAC (95/5 to 70/30)	PE, PP, EVAL, PBT, PA
SB/PE block copolymer[1)]	PE, PS, SB
SB/PP block copolymer[1)]	PP, SB
SB blends	SB, PO, PC, PMMA, PA
Vinyltris(methoxyethoxy)silane	PE, steel

[1)] For coextrusion processes, these bonding agents are compounded with polyolefins and PS.

Compatibilizers, in contrast, are designed to improve the miscibility of two otherwise immiscible polymers. They are particularly important in the production of plastic blends made from noncompatible or poorly compatible components, such as in multilayer films.

In plastic recycling, difficult-to-separate mixed plastics often arise. By adding suitable compatibilizers, these mixtures can be converted into compatible blends with good properties.

7.1.9 Blowing Agents and Kickers

We generally distinguish between physical and chemical blowing agents. They create a tailored "filling" of plastic components with gas bubbles as a disperse phase. Blowing agents are employed for foamed film extrusion, thermoplastic foam injection molding, and the MuCell® process. They are also required for the manufacture of expanded polystyrene and polyurethane foams. Table 7.8 lists the characteristics and respective areas of application of chemical blowing agents.

Table 7.8 Chemical Blowing Agents for Plastics and Rubbers

Blowing agent	Abbreviation	Decomposition temperature in air [°C]	Generated gas [ml/g]	Used in	Blowing gas
Azodicarboxylic acid diamide	ADC	205–215	220	PVC, PE, PP, ABS, PS, PA, PPE, TPE	N_2, CO, (NH_3, CO_2)
Azodicarboxylic acid diamide, activated	ADC activated	155–215	150–220	PVC, PE, PP, EVAC, PS, ABS, EPDM, NBR, SBR, CR, IR, NBR/PVC	N_2, CO, (NH_3, CO_2)

Blowing agent	Abbreviation	Decomposition temperature in air [°C]	Generated gas [ml/g]	Used in	Blowing gas
p-Toluenesulfonyl hydrazide	TSH	155–165	115	EVAC, EPDM, SBR, IR, CR, NBR, NR	N_2, H_2O
4,4′-Oxybis(benzenesulfonyl hydrazide)	OBSH	155–165	120–125	PE-LD, EVAC, NBR/PVC, EPDM/CR, SBR, CR, NBR, NR, PUR	N_2, H_2O
2,4,6-Trihydrazino-1,3,5-triazine	THT	245–285	185	ABS, PE, PP, PA, PET, PPE	N_2, NH_3
p-Toluenesulfonyl semicarbazide	TSSC	226–235	120–140	PP, ABS, PE-HD, PVC-U, PPO, PS-HI, PA	N_2, CO_2, NH_3
5-Phenyltetrazole	5-PT	240–250	190–210	ABS, PPE, PC, PA, PBT, LCP	N_2
N,N′-Dinitrosopentamethylene tetramine	DNPT	195	190–200	NR, SBR, PS-HI, CR, BR, PVC-P	N_2, (NH_3) HCHO
Sodium hydrogen carbonate	$NaHCO_3$ Citric acid	150–230	185–240	PS, ABS, PA, PE, PP, PVC-U	CO_2, H_2O

Table 7.9 uses azodicarboxylic acid diamide (ADC), the most known chemical blowing agent, to illustrate different areas of application. Some blowing agents react exothermically (ADC); others, endothermically (sulfohydrazides). Peroxide batches, possibly in combination with ADC, are also used. Blowing agent batches contain between 10% and 40 (70)% blowing agent. Carrier materials for these batches are PE-LD (predominantly for applications in polyolefins such as EVA) and universal carriers (EVA or EVA wax), which enable use in polystyrene and engineering thermoplastics. In order to reduce degradation temperatures and control gas generation and the porous structure, chemical blowing agents are combined with kickers (accelerators, starters) and cell regulators. Typical kickers are metal oxides such as zinc oxide or zinc oxide/stearic acid. Processing of chemical blowing agents may create nonvolatile degradation products that could lead to changes in taste or odor or corrosion.

Table 7.9 Examples of Azodicarboxylic Acid Diamide (ADC) Applications

Application	wt %
Prevention of sink marks when injection molding components with wall thickness differentials	0.05–0.15
Foaming of thermoplastics for weight reduction, TSG, TSE	0.3–1
Foaming of PVC plastisols	0.7–2
Expanded rubber, sponge rubber	2–6
Foam rubber, expansion process	3–15
Extrusion with subsequent expansion of cross-linked PO	3–20
Foam blocks PO, PVC-P, and elastomers	3–20

Physical blowing agents are inert gases dissolved in the plastic melt. They include nitrogen and low-boiling liquids such as *n*-butane, *i*-pentane (requires explosion prevention measures), cyclopentane (in the past, also chlorofluorocarbons, CFC), and N_2, CO_2, H_2O. Physical blowing agents require higher technical efforts (metering unit), their solubility decreases with increasing temperatures, and they typically achieve lower densities (PS sheets – physical: 30 kg/m^3 to 250 kg/m^3, chemical: 250 kg/m^3 to 600 kg/m^3).

7.1.10 Nucleating Agents

The properties of semicrystalline plastics are fundamentally determined by the type and degree of their crystalline content. Therefore, the addition of nucleating agents is used to influence crystallinity and crystal structure in plastics (Table 7.10 and Table 7.11): They increase the temperature at which melt crystallization begins (faster melt solidification), the rate of spherulite growth (10–5000 mm/min), and the crystalline content in the material. They also decrease the spherulite size. This allows for a shorter cycle time. To achieve a nucleation effect, the particles must be around 1 µm in size; larger particles do not contribute to nucleation. Concentrations of 0.5% are sufficient. Nucleating agents typically are insoluble, inorganic fillers such as metal oxides, metal salts, silicates, and boron nitrides. Fillers and reinforcing agents (talcum, synthetic silica, kaolin) as well as pigments and dyes may also have nucleating effects. "Clarifiers" are nucleating agents that are soluble in the melt (PP), where on cooling they form a three-dimensional "fiber network" whose surface has a nucleating effect. The created nucleating density is magnitudes higher than those created by insoluble agents, so that the resulting products are highly transparent (PP films).

Table 7.10 Nucleating Agents[1)]

Nucleating agent	Trade name	Manufacturer
Organic sodium phosphates	Mark NA / Mark	Adeka Argus / Witco Corp.
Ionomer	Aclyn	Allied Signal
Ionomer	Surlyn	DuPont
Aluminum tert-butylbenzoate	Sandostab/AL-PTBBA	Sandoz Chemicals / Shell
Talcum	Secupur Nucleating	Secupur
Sodium benzoate (NB)	[2)]	Gabriel-Chemie

1) *Ullmann's Polymers and Plastics: Products and Processes*, Volume 1, Wiley-VCH (2016)
2) For masterbatches by Gabriel-Chemie

Table 7.11 Nucleating Agents: Sorbitol

Nucleating agent	Abbreviation	R_1[1)]	R_2[1)]	T_m [°C]	Trade name	Manufacturer
Dibenzylidene sorbitol	DBS	H	H	225	Millad 3905	Milliken Chemical
Bis(*p*-ethylbenzylidene)sorbitol	EDBS	C_2H_5	H	235	NC-4	Mitsui Chemicals
Bis(*p*-methylbenzylidene)sorbitol	MDBS	CH_3	H	242	Millad 3940	Milliken Chemical
Bis(3,4-dimethylbenzylidene)sorbitol	DMDBS	CH_3	CH_3	278	Millad 3988	Milliken Chemical

1) Chemical structure sorbitol

7.1.11 Anti-bacterial Agents, Fungicides

Anti-bacterial substances are employed to prevent damage caused by bacteria, fungi, and other microorganisms. The global increase in the standard of living and stricter health and safety regulations are continuously driving demand. However, there is no universal solution. The use of masterbatches that eliminate or delay the growth of bacteria (such as E. coli, salmonella, and listeria) is recommended for plastic products, particularly polyolefins. However, these are always specialized for specific strains. Most damage in PVC-P products is caused by fungi and their excretion products. Suitable corrective measures include the compounding of fungicides into the PVC-P that will migrate to the surface of the molded component.

7.2 Fillers

Norbert Grünewald

Fillers not only serve as extenders to reduce plastic material consumption and thus costs, but they also improve processability (shorter cycle times, by increasing thermal conductivity) and other characteristics, such as stiffness, impact resistance, heat resistance, electrical conductivity, and dimensional stability. They may also reduce thermal expansion. Elongated and fibrous fillers also increase strength, whose extent depends on the mechanical properties of the filler and its length/thickness ratio (aspect ratio – see Table 7.12) as well as size and distribution. Some fillers also act as additives (for example, aluminum trihydrate is also a flame retardant), making an unambiguous differentiation between fillers and additives difficult. In general, when concentrations exceed 5%, the substance is considered a filler rather than an additive. However, this limit is fluid and not always meaningful, in particular with respect to nano-fillers that are added in significantly lower concentrations.

Table 7.12 Aspect Ratios of Selected Additives

Form	Aspect ratio	Example	Note
Sphere	1	Glass spheres Silicate spheres	Filler
Cube	1	Calcium carbonate Feldspar	
Rectangle	2–4	Silicates Barium sulfate	Transition from filler to reinforcing agent
Platelet	5–10 (100)	Mica Talcum Kaolin Graphite Aluminum hydroxide	Reinforcing agent
Fiber	> 10	Asbestos Wollastonite Whiskers Glass fibers Carbon fibers Others	

In 2021, the global consumption of fillers was approximately 50 million t. The most important fillers globally are calcium carbonate (elastomers, PVC processing, UP resins), GCC (ground calcium carbonate), PCC (precipitated calcium carbonate), and carbon black (tire industry), followed by kaolin and talcum (PP production). Other fillers, such as mica, silica, and wollastonite, together represent only approximately 9% of the global market. Currently, an increasing trend towards highly filled plastics (up to 85 vol %) and nano-scaled fillers can be observed. The highest demand for GCC is expected in Asia, while the highest demand for PCC is expected from Europe. The recovery of the tire industry in the US will drive a higher demand for carbon black there.

Fillers typically are inorganic or organic solid particles. Organic fillers include wood flour (high content of connected fibers) and cellulose powder (pulverized single fibers) that are incorporated into phenolic, amino, and UP resins but also in PP and PVC, with filler contents of up to 65%.

Table 7.13 lists a selection of inorganic fillers and Table 7.14 compares the effect of fillers and reinforcing agents on the property profile of plastics (abbreviations: see also ISO 1043).

Table 7.13 Characteristics of Inorganic Fillers

Filler	Form/ Abbreviation	Diameter [µm]	Density [g/cm^3]	Ratio *l/d*	Spec. surface [m^2/g]	Surface tension [mN/m]	Young's modulus [GPa]
Silica (pyrogenic) $SiO_2 \cdot n\ H_2O$	Spheres	0.01	2.2	1	380	1200	
Aluminum hydroxide (aluminum trihydrate) $Al(OH)_3$		1–32	2.4		0.9–3.0		
Aluminum nitride AlN	Powder	20–150					
Aluminum silicate AlSi	Spheres/sph	< 300	0.63	1			
Calcium carbonate, chalk $CaCO_3$	Cubes/cub	3–7, (0.7) < 30	2.7	ca. 1	1–15 (PCC: 32–40)	200	35
Calcium sulfate $CaSO_4$	Cubes	4	2.96				
Glass (see also Table 7.14)	Spheres/sph	5	2.5	1	1.3	1200	70
Glass	Hollow spheres / sph	5–500	0.2–0.8				

Table 7.13 Characteristics of Inorganic Fillers *(continued)*

Filler	Form/ Abbreviation	Diameter [μm]	Density [g/cm^3]	Ratio *l/d*	Spec. surface [m^2/g]	Surface tension [mN/m]	Young's modulus [GPa]
Mica $(K,Na,Ca)(Al,Mg,Fe,Li)_{2\text{-}3}(OH)_2(Si,Al)_{4\text{-}5}O_{10}$	Platelets/fl	< 20	2.85	20–100			
Kaolin $Al_2(Si_2O_5)(OH)_4$	Platelets	1–10		< 10	1040		
Ceramics	Hollow spheres	10–300	0.4–0.7				
Nano-whisker	Whiskers	0.35	3.3	50		650	280
Talc (talcum) $Mg_3Si_4O_{10}(OH)_2$	Platelets/fl	< 30	2.8	2–20	3.6	120	20
Wollastonite $Ca_3[Si_3O_9]$	Whiskers	< 20	2.85	10–50			

Table 7.14 Effect of Fillers and Reinforcing Materials on the Characteristics of Plastic Materials

	Textile glass	Asbestos	Wollastonite	C-fibers	Whiskers	Synthetic fibers	Cellulose	Mica	Talcum	Graphite	Sand/quartz powder	Silica	Kaolin	Glass spheres	Calcium carbonate	Metal oxides	Carbon black
Tensile strength	+ +	+		+	+ –			+	O					+			
Compressive strength	+								+		+			+	+		
Young's modulus	+ +	+ +	+ +	+ +	+			+ +	+		+	+		+	+	+	+
Impact strength	– +	–	–	–	–	+ +	+	– +	–		–	–	–	–	– +	–	+
Reduced thermal expansion	+	+			+			+	+		+	+	+			+	

	Textile glass	Asbestos	Wollastonite	C-fibers	Whiskers	Synthetic fibers	Cellulose	Mica	Talcum	Graphite	Sand/quartz powder	Silica	Kaolin	Glass spheres	Calcium carbonate	Metal oxides	Carbon black
Reduced shrinkage	+	+	+	+				+	+	+	+	+	+	+	+	+	+
Improved thermal conductivity		+	+	+					+	+	+	+			+		+
Improved resistance to thermal distortion	+ +	+	+	+ +				+	+				+		+	+	
Electrical conductivity				+						+							+
Electrical resistivity			+					+ +	+			+	+ +			+	
Heat resistance			+					+	+		+	+	+			+	+
Chemical resistance		+	+					+	O	+			+	+			
Improved abrasion resistance				+				+	+	+			+				
Extrusion rate	– +	+						+					+		+		
Machine abrasion	–	O			O	O	O		O	O	–			O	O		O
Cost savings	+	+	+				+	+	+	+	+ +	+	+	+	+ +		
	Fibrous fillers and reinforcing materials							Platelet-like types			Spherical fillers						

+ + strong effect, + weaker effect, O no effect, – negative effect

7.2.1 Inorganic Fillers

7.2.1.1 Carbon Black

Carbon black is produced on an industrial scale by pyrolysis of hydrocarbons using furnace, channel, or gas black processes. Depending on how the carbon black is produced and by varying process parameters, the size of the primary particles (spherical, 10–300 nm) as well as the aggregate formation, and thus the property profile of the carbon black, can be controlled. For example, carbon blacks with small primary particles and highly branched aggregates provide good electrical conductivity. For many applications, carbon black undergoes specific post-treatments: the compatibility with resins and surfactants is improved by subsequent oxidation of the base carbon black for high color depth coatings. Standard carbon blacks are internationally classified (abrasion-resistant, electrically conductive, etc.) by the ASTM standard D 2516.

Applications: Using different types of carbon black in tire applications allows for the optimization of three characteristics: roll and skid resistance and abrasion. Carbon black is also employed as a filler in conveyor belts and as black pigment in printing inks, coatings, and for plastics coloring, where it also increases UV stability. It is also used as material for electrodes and as a conductive black in electric applications.

7.2.1.2 Calcium Carbonate ($CaCO_3$)

Natural, ground calcium carbonate (GCC) from chalk, limestone, or marble and synthetic, precipitated calcium carbonate (PCC) are used in either uncoated or coated form. They serve as cost-effective fillers for increasing heat stability and surface quality, for example, in PVC, PP, polyesters, thermoplastic elastomers, and BMC and SMC (Table 7.15). Calcium carbonate ($CaCO_3$, Mohs hardness = 3, ρ = 2.7 g/cm^3) provides a high degree of whiteness, low abrasiveness in processing equipment, and simple dispersion. It is nontoxic, taste- and odorless, and approved for packaging applications with food contact. Coated versions positively affect mechanical properties and offer low plate-out and high heat resistance (600 °C).

Table 7.15 Applications of Calcium Carbonate

Material	Application	Average particle diameter (D_{50} values)[1)] [µm]
Natural, ground calcium carbonate (GCC)		
PVC, flexible	Cables, floor coverings, profiles, films	Coated and non-coated, different particle sizes
PVCplastisols	Wallpaper, synthetic leather, tarps	1.5 to 40
	Underbody coating systems	0.4

Material	Application	Average particle diameter (D_{50} values)[1)] [µm]
PVC, rigid	Pipes, window and shutter profiles	Surface coated < 1.5
PP	Outdoor furniture, (microporous) films, sewer pipes	Typically surface coated 1.4 to 3.5
PE	Anti-blocking agent for film manufacturing (content: up to 10%) Grinding balls in pigment masterbatches (TiO_2) to prevent pigment agglomeration PE-HD hollow parts (5–15% for higher stiffness, better printability)	Typically surface coated < 3
PS	Films (content: 5–10%) Improving drawability	
EBA base	Fireproof cables	
UP resin	SMC and BMC applications (content: 30–40%)	1.5–10
Synthetic, precipitated calcium carbonate (PCC)		
PVC plastisols	Underbody coating Sealing compounds	Stearic acid coated, 0.07 to 2
PVC, rigid	Improvement of gloss	

1) According to Omya International AG

7.2.1.3 Talc/Talcum

Talc is a natural, hydrated magnesium silicate that is available in a variety of modifications (fibers, platelets, whiskers, and steatite). As a filler it is used only in platelet structure. Pure talcum exhibits good slip properties and is very soft (Mohs hardness = 1, ρ = 2.9 g/cm^3) because of its layered structure (between two silica layers there is an intermediate layer of magnesium oxide and hydroxide). Talc is commercially available in finely ground form with particle diameters from 1 µm to 15 µm. It is incorporated as a filler in polyolefins with contents up to approximately 75%. Even small amounts of fine talc (1 wt % to 5 wt %) have a nucleating effect in PP and increase the Young's modulus and flexural strength while reducing the impact strength. Special talcum products are HAR types (high aspect ratio talc), submicron-talcum (D_{50}: approx. 0.3 µm to 0.6 µm), and nano-talcum. The use of more expensive HAR types results in a more pronounced improvement in the mechanical properties and heat resistance than with standard types. PP filled with submicron-talcum exhibits a more balanced mechanical property profile (stiffness, heat resistance, impact resistance).

Applications: Talcum is the most important filler for PP and PP copolymers (approx. 200,000 t/a) and is used in particular because of its reinforcing effect. It is also used as an extender for thermosetting compounds and reaction resins. The addition of special talcum types facilitates the machining of some plastics. In films, talcum is also employed as an anti-blocking agent. Talcum is also used in combination with glass fibers in plastics.

7.2.1.4 Silica

Pyrogenic silica is hydrolyzed at 1000 °C from volatile chlorosilane that is exposed to a hydrogen-oxygen flame. Initially, highly viscous SiO_2 primary particles are created in the flame (size: 5 nm to 30 nm); they subsequently fuse to form larger aggregates (100 nm to 1000 nm) and, finally, form tertiary structures ranging from 1 µm to 250 µm in size. The shear-thinning (or thixotropic) properties of silica allow for the selective control of the flow properties of liquid systems.

Applications: Silica is used in unsaturated polyesters, PVC plastisols, and coatings as a thickener and to prevent sedimentation. In natural and synthetic rubbers, it is used as reinforcement; in silicone elastomers, to increase the dielectric properties; and in insulation gels, to achieve hydrophobicity.

7.2.1.5 Aluminum Trihydrate

Aluminum hydroxide [$Al(OH)_3$], also known as aluminum trihydrate (ATH), is a filler and flame retardant. Its platelet-like structure leads to high-quality surfaces and improves the dielectric strength and creep resistance.

7.2.1.6 Kaolin

Kaolin (Mohs hardness = 2; calcinated kaolin: Mohs hardness = 6 to 8) is a hydrated aluminum silicate. For elastomers, it is the second most important filler, after carbon black. Depending on the particle size and the specific surface, different degrees of "elastomer hardness" can be achieved.

Applications: The use of kaolin in UP resins reduces shrinkage and improves the surface quality. In PVC plastisols and sealing compounds, kaolin is employed in combination with calcium carbonate to tailor the rheological properties. Calcinated kaolin is used in the cable industry to improve insulation resistance.

7.2.1.7 Mica

Mica refers to a group of crystallizing silicate minerals with the general chemical formula $(K,Na,Ca)(Al,Mg,Fe,Li)_{2-3}(OH)_2(Si,Al)_{4-5}O_{10}$ (the atoms in parentheses can replace each other in any combination; however, they always remain in the same proportion with respect to the other atom groups). In commercial terms, the most important mica types include muscovite, a potassium-aluminum silicate; phlogopite, a magnesium-iron-aluminum silicate; and biotite, which has the highest iron content of these micas. Mica can be easily split between its layers and therefore exhibits a relatively low Mohs hardness in the direction parallel to its layers (2 parallel, 4 in all other directions). Depending on its aspect ratio – which, in turn, is determined by the production process; for example, spiral jet milling, counter jet milling, and agitator ball milling – mica takes on a functional or a filler role in plastics. Surface silanization of mica facilitates the compounding into plastics.

Applications: Mica is used in epoxy and phenolic resins to improve mechanical, thermal, and electrical properties. Mica-modified PP is employed in automotive applications. Since the mid 1980s, coated mica is used in automotive coatings and cosmetics to achieve the so-called "pearlescent" effect (see also Section 7.1.5).

7.2.1.8 Wollastonite

Wollastonite (chemical formula: $Ca_3[Si_3O_9]$) is a silicate and offers a variety of technical applications because of its fiber- to needle-like crystals and its high melting point of 1540 °C. Its high aspect ratio (10–20) renders wollastonite not only a filler but also a reinforcing agent (see Section 7.3.1). Increasingly, HAR-wollastonite types with low grain-size upper limit are being developed. These new types decrease the abrasive effects in PP compounds while maintaining good mechanical properties and a high scratch resistance.

Applications: The plastics industry accounts for approximately 3% of the global wollastonite consumption, in which it is mainly used as a filler for PP, PA, and PU but is also employed to improve the stiffness and flexural strength. One of the main application areas of wollastonite is in the ceramics, where it is used to improve the mechanical properties of white ceramics.

7.2.1.9 Solid and Hollow Micro-Spheres

Spherical additives improve the flow characteristics of molding compounds during processing, and the shrinkage and dimensional stability in highly filled products. The surfaces of components molded from compounds with micro-spheres appear smoother than those with added fibers. Micro-spheres are isotropic so that properties are uniformly distributed in the molded component. When added to plastic materials, they cause uniform stresses in the material.

Spheres are classified into the solid, hollow, and porous categories. Spheres with diameters below 300 µm to 500 µm are called micro-spheres; spheres with diameters larger than 0.5 mm (so-called macro-spheres) can also be used. Solid glass micro-spheres, also called by the trade name "Ballotini", improve the Young's modulus, compression strength, hardness, and surface smoothness. "Cenospheres" are a mixture of solid and hollow spheres from flue ash of coal-burning power plants. "Micro-balloons" are boron silicate or silica hollow spheres. Low weight percentages (2% to 4%) of these voluminous light fillers in thermoplastic or thermoset resin compounds may result in weight savings of 20% to 30%. At the same time, these fillers – with a wall thickness of only 0.5 µm to 2 µm – are pressure-resistant and increase the impact resistance. Table 7.16 provides an overview of the characteristics of spheres used as fillers in plastics.

Table 7.16 Characteristics of Spheres

Characteristic	Solid spheres	Hollow spheres	Porous spheres
Size [µm]	0.5–850	0.5–500 (5000)	25–16,000
Melt temperature [°C]	740	1200–1500 (2100)	approx. 700
Density [g/cm^3]	2.45–2.55	0.14–0.8	0.25–0.74
Apparent density [g/cm^3]	–	0.35–1.1	0.14–0.43
Thermal conductivity [W/m·K]	1.05	0.09–0.11	0.08
Compressive strength [N/mm^2]	–	1.7–28	0.4–1.9

7.2.2 Nano-fillers

Nano-fillers are defined by primary particle sizes that do not exceed several hundred nanometers; they have a long-standing application history in the plastics industry. For many decades, established nanostructured materials, such as carbon black, silica, whiskers, and pigments (*e. g.*, TiO_2, generally processed in the form of agglomerates), have represented billion-dollar markets globally – even though the growth rates have been stagnant. Nanoparticles exhibit an extremely high aspect ratio and are therefore suitable not only as fillers but also as reinforcing agents, because the effectiveness of these agents increases with increasing aspect ratio. Another important factor determining the effectiveness of nano-fillers is their uniform, fine dispersion in the plastic material. Lately, enormous progress has been made in terms of dispersion technology and nano-filler modification, thus achieving optimized properties at low filler levels:

- Improved mechanical properties
- Improved abrasion and scratch resistance
- Reduced flammability
- Antistatic properties
- Anti-microbial properties

- Improved barrier properties
- Higher temperature resistance
- Reduced thermal contraction and expansion

Increased development efforts have focused on nanocomposites with PA 6, PA 12, PP, PMMA, PUR, and EP. Here, nano-fillers on the base of layered silica, carbon particles, oxides (*e.g.*, Al_2O_3, SiO_2), and metal particles (*e.g.*, Ag, Au) were studied. A dynamic market growth is expected for these types of nano-fillers, provided that manufacturing processes and processing technologies can be adapted accordingly and the effect of these materials on living organisms proves to be nonhazardous (in terms of toxicity).

7.2.2.1 Layered Silica

Layered silica such as montmorillonite (an aluminum silicate) is a nano-filler of particular interest because of its high aspect ratios (100–500). One of the particular properties of montmorillonite is its excellent swelling in water: Water molecules permeate between the montmorillonite layers and separates them. When plastic nanocomposites are manufactured using layered silicates, the organic-modified silicate layers are penetrated by the polymer instead of water. As a result, nanometer-sized, thin platelets (size: $1\,nm \cdot 200\,nm$) are uniformly distributed in the plastic matrix. The high aspect ratio of the silicate platelets reinforces the plastic and facilitates a barrier effect against gases and liquids. In addition, nano-fillers improve the flammability properties of the composite (reduced flammability, increased ignition point). Layered silicates act as UV blockers, increase the chemical resistance, and stabilize films against water exposure. The addition of 2 wt% to 5 wt% causes a distinct increase in the tensile strength and heat resistance in PA. To reach the same level of strength, the addition of 15% short glass fibers or 30–40 wt% conventional filler would be required.

7.2.2.2 Carbon Particles

Depending on their nature, carbon particles are classified as carbon blacks (see Section 7.2.1), carbon nanotubes (CNTs), expanded graphite (EG), spherical particles, and carbon nano-fibers (CNFs). Expanded graphite exhibits a layered structure and is mostly used as a flame retardant (expanded layer thickness: 100–400 nm). Graphite spheres are used as fillers in reflectors because they improve the thermal conductivity, optimize shrinkage properties, and cause less interference with the light path.

7.2.2.3 Nanotubes

Nanotubes exhibit diameters smaller than 100 nm; however, typical diameters reach only a few nanometers (they can be as small as 0.3 nm). With a length of several micrometers, they are more than 10,000 times thinner than a human hair. Currently, carbon nanotubes (CNTs or "buckytubes", in reference to fullerenes) have been the focus of development efforts. They are cylindrical carbon structures consisting of one-atom thick layers of graphene that are rolled at specific, discrete angles (*e. g.*, with an ideal hexagonal structure in linear tubes). Nanotubes may be single- or multiple-walled and open or closed at the ends (the latter allowing them to be filled, *e. g.*, with lead, silver, or inert gases). Depending on their manufacturing conditions, nanotubes may form larger, intertwined fiber bundles, buckypapers, or yarns.

Their structure determines whether nanotubes exhibit metallic or semi-conductive properties. They provide current-carrying capacities approximately 1000 times higher than copper wires. CNTs are used to improve the electrical conductivity as well as mechanical characteristics. With a density of 1.3 g/cm^3 to 1.4 g/cm^3 (and under ideal theoretical conditions), CNTs exhibit an approximately 130 times better ratio of tensile strength to density than steel.

Research on carbon nanotubes (CNTs) continues to focus on electronic applications such as transistors, circuits, and displays. In plastics, CNTs enhance the mechanical properties and the electrical conductivity.

Some industrial applications already exist, such as in fuel lines to prevent sparking or in plastic components for electrostatic dissipation and more efficient painting processes. CNTs are also used in high-performance composites, including tennis rackets and bicycle frames. Plastic films with CNT layers for electrical conductivity and heat generation are still in development. While some applications have reached large-scale use, many potential uses remain in the research phase. High production costs and challenges in mass manufacturing continue to be limiting factors.

Because the elongated structure and the proportions of nanotubes resemble those of asbestos, the toxic effect of nanotubes is currently under investigation.

7.2.2.4 Fillers for Magnetic, Thermal, and Electrical Properties

The property profile of plastics can be tailored by adding magnetizable or electrically or thermally conductive fillers. Here, we encounter significant differences in the dependence on the filler content to achieve certain properties. While magnetic and thermal properties disproportionally depend on the filler volume content, electrical conductivity exhibits a distinct threshold with respect to filler volume content. As soon as this so-called percolation threshold is exceeded, electrically conductive compounds reach their maximum, stable conductivity. This effect is caused by the formation of a continuous network of conductive pathways, created by filler particles in contact with or very close to each other (approx. 10 nm); see Figure 3.78.

Key factors for the processing properties of these compounds and the magnetic, electrical, and mechanical properties of the injection-molded components are the type, amount, and distribution of the filler in the plastic matrix.

Increasing filler content results in an almost linear improvement of the magnetic properties; however, it comes at the expense of a reduction in the processing properties and the mechanical properties in service. Plastic-based injection-molded permanent magnets are therefore typically manufactured with filler contents of 50 vol % to 65 vol %. Multi-component injection molding processes offer novel approaches to the efficient manufacture of integrated permanent magnets with locally adhered or movable elements by combining magnetizable compounds with other plastics.

Magnetizable compounds of economic significance are mainly hard magnetic fillers such as ferrites (strontium/barium ferrite) and rare earths (NdFeB, SmCo). Depending on whether the magnetic properties rely on the orientation of the filler particles, we distinguish between iso- and anisotropic fillers. Anisotropic magnetic fillers, in particular ferrites, undergo orientation in the melt, which is achieved by a magnetic field inside the cavity during injection molding. Preferred plastic matrices are low-viscosity, semicrystalline thermoplastics such as PA, PBT, and, in some applications, also higher-temperature-resistant thermosets such as PA 6T and PPS in combination with viscosity-lowering and adhesion-promoting additives.

Electrically conductive fillers are compounded into plastics in order to shield mobile phones, computers, medical equipment, and aeronautical electronics against electromagnetic fields; see Section 3.5.5. Electrically conductive antistatic additives include graphite and carbon black (active carbon black is added at 5 wt % to 10 wt %). Metals (Al, Cu, Zn, Ag, Ni, and brass), which provide better protection than powders, fibers, or fabrics, also play an important role in electromagnetic shielding. In addition, nickel-graphite and gold-graphite powders as well as silver-coated spheres are incorporated into synthetic rubbers (silicone rubber) at contents of up to 65 wt % (see Section 7.1.3). There are also efforts underway to manufacture electrically conductive, transparent plastics from compounds with indium-tin-oxide (ITO) nano-powders.

Many materials may be considered as *thermally conductive fillers* that also provide electrical conductivity, such as carbon black and metals. However, if only thermal conductivity is desired, fillers such as boron nitride and aluminum oxide are used. Compared to carbon black and graphite, but also metals, these fillers are very expensive. While non-filled plastics are insulators and exhibit thermal conductivities of only 0.2–0.6 W/m·K, compounds with thermal conductivities of up to 15 W/m·K are commercially available; see Sections 3.4.3 and 3.10.8. Here, the volume content reaches a maximum of 20%, although compounds with up to 60 vol % filler can be manufactured. The thermal conductivity of these compounds depends on the filler content, the filler geometry and particle size, and the thermal conductivity and orientation of the filler; see Section 3.2.2.5. While filled cast resins have been used in the electronics industry for decades, heat-conductive thermoplastics are gaining ground. Major applications include injection-molded cooling elements, for example, for LED lighting and housings for electronics components.

7.3 Fibers

Fibers are categorized in terms of their origin, length, composition, or their use. We distinguish between staple fibers, fibers of limited length, and continuous fibers (filaments). The term "synthetic or man-made fibers" includes all fibers manufactured from natural or synthetic polymers. Figure 7.2 provides a classification of synthetic fibers. Natural fibers, such as flax, wood, and jute have found increasing applications as fillers and reinforcing materials in the recent past (see Section 7.3.1.2). They are used as reinforcing fibers (see Section 7.3.1) and for the manufacture of textile products such as clothing, home textiles, and carpets (see Section 7.3.2).

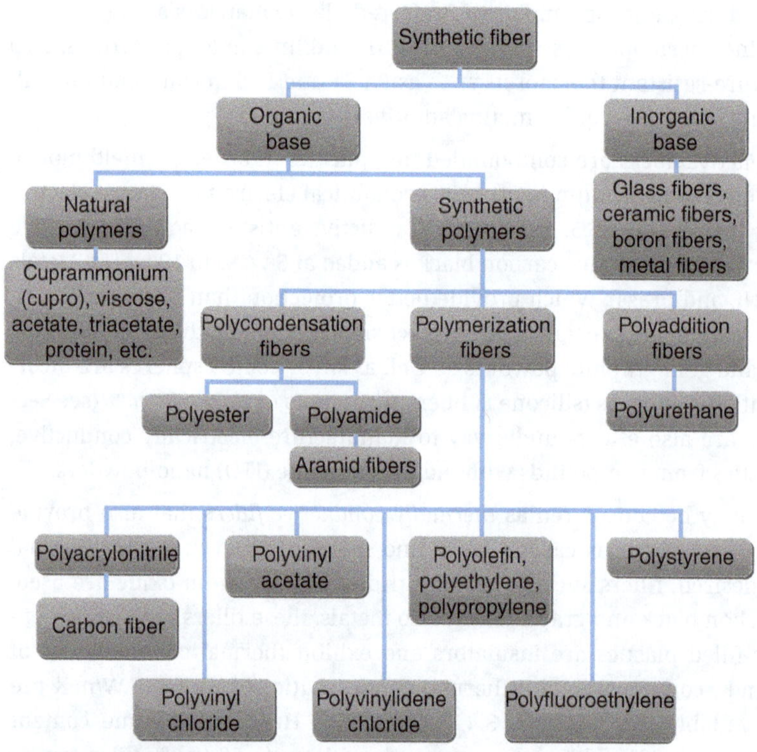

Figure 7.2 Classification of synthetic fibers

7.3.1 Reinforcing Fibers

Plastics exhibit lower strength and stiffness compared to metals. This disadvantage can be significantly improved by the addition of reinforcing fibers. Reinforcing fibers are predominantly compounded with thermoplastics and thermosets where they have specific effects on the mechanical characteristics: in thermoplastics they increase the Young's modulus and typically lower impact resistance. In brittle plastics, such as thermosets, they may increase impact resistance. The use of electrically conductive fibers allows for conductivity and electromagnetic shielding properties to be adjusted within a wide range (see also). Rather than compounding only one specific fiber type, the implementation of fiber mixtures and hybrid reinforcements, *e.g.*, aramid fibers with glass/carbon fibers or carbon fibers with PBO (Section 7.3.1.3), has become increasingly popular. In addition, natural fibers are also used for reinforcement. Numerous other fibers, such as polycrystalline boron, silicon carbide, aluminum- and zirconium oxide, steel, aluminum, and tungsten fibers are used as reinforcement. Table 7.17 and Figure 7.3 provide an overview of the most important synthetic fibers based on plastics.

Table 7.17 Characteristics of Reinforcing Fibers

Group	Fiber	Density g/cm^3	Tensile strength GPa	Young's modulus GPa		Elongation at break %	Coefficient of expansion $10^{-6} \cdot K^{-1}$	
			Parallel	Parallel	Perpendicular		Parallel	Perpendicular
Glass, GF	E-Glass	2.6	2–3.5	ca. 70	ca. 70	3.0–4.8	5	5
	S-Glass	2.48	4.3	87	87	5	2.9	2.9
Carbon, CF	HM1	1.96	1.75	500	5.7	0.35	−1.5	15
	HM2	1.8	3.0	300		1.0	−1.2	12
	HT	1.78	3.6	240		1.5	−1	10
	HST	1.75	5.0	240	15	2.1	−1	10
	IM	1.77	4.7	295		1.6	−1.2	12
	HMS		3–4	350–600		> 2		
Aramid, AF	p-Aramid, LM	1.44	2.8–3.7	59–90		4.0–4.3	−2.3	40
	p-Aramid, HM	1.45	2.9	127–170	5.4	1.9–2.1	−4.1	52
	m-Aramid	1.38	0.65	17		22		

Table 7.17 Characteristics of Reinforcing Fibers *(continued)*

Group	Fiber	Density g/cm³	Tensile strength GPa	Young's modulus GPa		Elongation at break %	Coefficient of expansion $10^{-6} \cdot K^{-1}$	
			Parallel	Parallel	Perpendicular		Parallel	Perpendicular
PBO	AS	1.54	5.8	180		3.5		
	HS	1.56	5.8	270		2.5	−6	
PE		0.97	2.7–3.6	89–116			−12.1	

GF: E-glass = electrical, S-glass = strength; CF: HM = high modulus, HMS = high modulus/strength, HT = high tenacity, HST = high-tensile, IM = medium modulus, LM = low rigidity; AF: LM = low modulus fiber, HM = high modulus fiber; PBO: AS = standard spinning fiber, HS = high modulus fiber

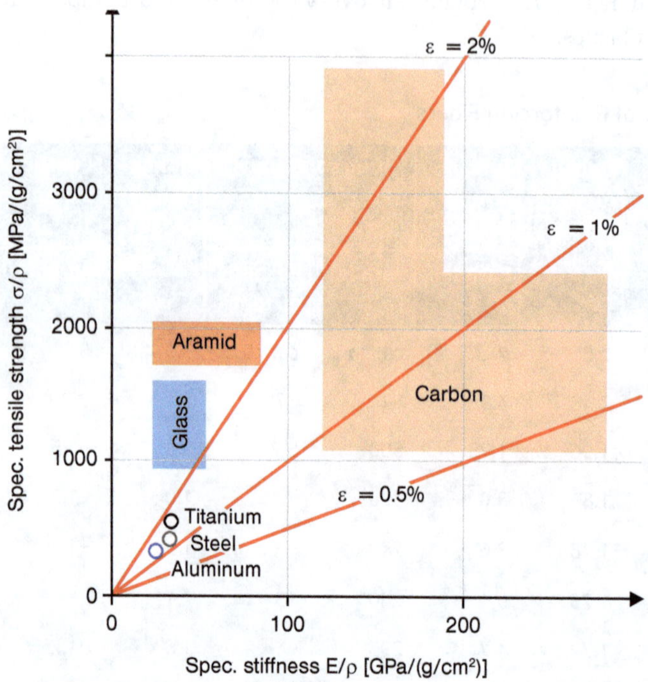

Figure 7.3 Comparison of selected fiber and weight specific characteristics for aramid and carbon fibers, E-glass, steel, titanium, and aluminum

Reinforcing fibers are processed in form of short-, long, and continuous fibers as well as in fiber products such as mats, fabrics, non-wovens, and tapes.

The only naturally occurring inorganic fibers are mono- and polycrystalline asbestos and wollastonite (see Section 7.2.1.8). Asbestos (ancient Greek *asbestos* for: indestructible, everlasting, indelible) is a general term for different, naturally occurring, fibrous silicate minerals that can be found in many locations imbedded in the earth's crust. Asbestos exhibits high strength (500–1000 N/mm^2), is heat and acid resistant, and can be woven. These properties made it suitable for shipping, insulation, construction, and automotive tire applications. Today, with scientific evidence of the health hazards posed by asbestos, many countries are prohibiting its use (among them the European Union and the United States). Currently, asbestos primarily poses a disposal problem.

7.3.1.1 Synthetic, Inorganic Fibers, Glass Fibers (GF)

With a share of 90%, glass fibers are dominant within the group of synthetic, inorganic fibers. Also included in this group are boron, beryllium, silicon carbide, and metal fibers, which will not be discussed in more detail.

Glass Fibers (GF)

Glass fibers consist of silicon-, boron-, or aluminum oxides; the vast majority (90%) is produced via a nozzle drawing process. Here, the required components, such as silica sand, limestone, kaolin, dolomite, boric acid, and hydrofluoric acid are melted in a furnace at approx. 1300 °C, before the liquid glass is passed through 400–4800 spinnerets, resulting in fibers with a thickness of 1–2 mm. Then the viscous fibers are continuously drawn at haul-off rates from 1200 to 3600 m/min until they reach the desired diameter of 3.5–24 µm. Size is applied to protect the fibers from mechanical and chemical loads as well as from exposure to moisture. This size (*e. g.*, a silane size) also serves as a bonding agent and may account for as much as 2 wt.-%, although 0.3–0.8% are more typical. Rovings or strands are created by collating 100 to 250 of these glass fibers. Continuous rovings may be used in subsequent processing steps such as weaving or pultrusion. They may also be processed into staple fibers, short fibers, or mats with weights of 250–900 g/cm^2. Short staple fibers are manufactured by exposing the glass fibers exiting the spinneret to blasts of air. They are mainly used for light (30–150 g/cm^2) resin-bound mats or non-wovens for resin-rich, smooth surfaces. Another manufacturing process for glass fibers is the so-called rod-drawing process.

Glass types: The general term textile glass fibers includes all glass fibers for textile applications. Figure 7.4 proves an overview of the various glass fiber products. Low-alkaline E(lectrical)-glass has a market share of 90%. E-glass is an excellent electrical insulator with high transparency for radio waves. Increasing the SiO_2-, MgO-, or Al_2O_3 portion to 40–70% results in improved mechanical characteristics. These more expensive glass types for high-performance composites are commercially available with the labels R- (resistance), S- (strength), and M- (modulus) glass. Also available are C- (corrosion) glass with higher acid resistance, and E-CR (corrosion resistant) glass. Hollow glass fibers are drawn from capillaries.

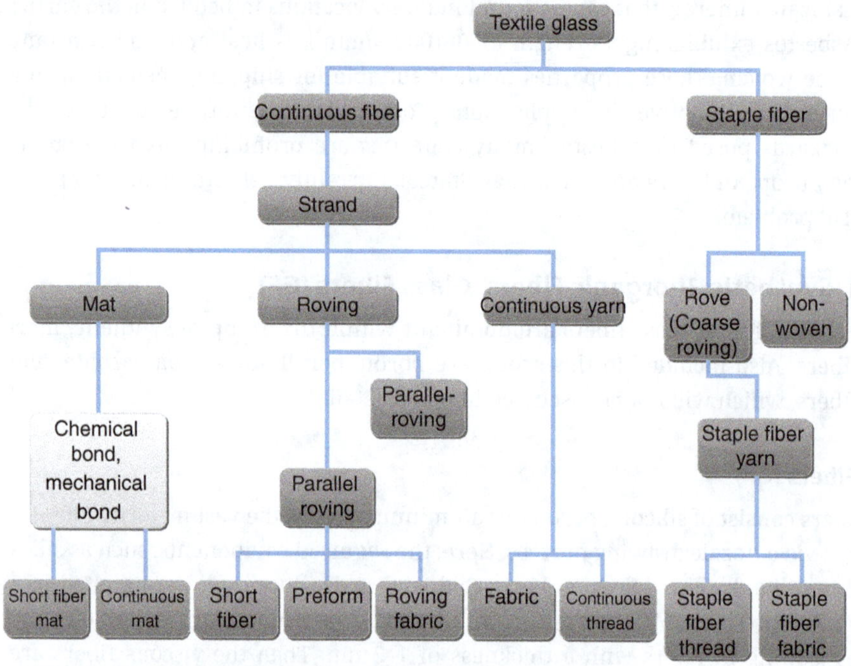

Figure 7.4 Overview of glass fiber products

Forms of delivery: Glass fibers are available in the form of staple and continuous fibers and are sold as yarns, threads, rovings, mats, and fabrics. Cut fibers are processed as short glass fibers (approx. 140–210 µm) and long glass fibers (approx. 8–25 mm). Prepregs, resin pre-impregnated rovings, have a large market share.

Glass fibers play an important role in *reinforcing of thermoplastics and thermosets* (cast resins). Short- or long fiber *reinforced thermoplastics* are predominantly injection molded; however, they are also extruded, blow molded, and thermoformed. Today, high-value applications are unimaginable without short fiber reinforced PA, PP, PBT, PC, and many other materials with fiber contents ranging 15–50 wt.-% (and more). Fiber contents below 15 wt.-% result in negligible reinforcing effects; above 50 wt.-% fiber content, uniform wetting of the fibers (necessary to create smooth surfaces) becomes increasingly difficult.

In order to process long glass fibers, they have to undergo suitable pre-treatment. We distinguish between two approaches: G-LFRT (employing rod-shaped granulates) and a direct process, D-LFRT. Because of the differences in mechanical load during processing, G-LFRT fibers have a length of approx. 2–6 mm while D-LFRP fibers are 8–10 mm long. The most common long fiber reinforced material is PP with 30–60 wt.-% glass fiber content. Other matrix materials include PE-HD, PA, and special types such as PC/ABS, PPS, PBT, and POM. Thermoplastics are often reinforced using glass mats (GMT processes). Here, the majority of applications use PP; however, there is a limited number of PET und PA 6 applications, reinforced with 30–40 wt.-% glass fiber mats.

Thermosetting matrix materials for glass fiber composites (GFC) are predominantly unsaturated polyester, epoxy- and vinyl ester resins (Table 7.18). Still in the development stage are aerosols as extremely light matrix materials (ρ = 0.4 g/cm^3 compared to common matrix materials with ρ = 1.17–1.25 g/cm^3) that are organically degradable.

Table 7.18 Characteristics of Various Resins

Resin	Density g/cm^3	Tensile strength GPa	Young's modulus GPa	Elongation at break %	Form of delivery	Processing temperature °C
Epoxy resins (EP)	1.1–1.25	70–90	2.8–6.0	6–8	Liquid	RT to 230
Unsaturated polyester resins (UP)	1.12–1.25	50–70	3.5–5.0	≤ 3	Liquid	RT to 160
Vinyl ester resins (VE)	1.07	75–85	3.4–3.8	3.5–7	Liquid	RT to 175

The common *processing methods* used for the manufacture of fiber reinforced components include standard processes such as manual lay-up, spray lay-up, fiber winding, compression molding, and pultrusion, see Section 4.8. Resin injection methods, also called liquid composite molding, such as RTM–processes (resin transfer molding) or LFI (liquid fiber injection), are experiencing solid growth rates. They allow for effective wetting of glass fiber reinforcement with matrix material during the forming process. Glass fibers are cut either mechanically (ultrasound cutters) or thermally (laser cutters). Dry semi-finished fiber products for complex or thick-walled components are produced using either sewing technologies for multi-layered fabrics, adhesion or welding, or braiding/knitting technologies. Preforms, for example, are manufactured using yarns that will subsequently melt at elevated temperatures and thus create an adhesive bond between layers. Glass fiber reinforced resins allow for the following maximum glass content: mats (hand lay-up: 35 wt.-%, compression molding: 50 wt.-%), fabrics: 66 wt.-%, rovings (uniaxial): 75 wt.-%.

Properties: In contrast to plastic or aramid fibers, glass fibers exhibit amorphous structures and isotropic mechanical characteristics. Up to the point of failure (break), glass fibers exhibit ideal linear elastic behavior. Elongation at break for glass fibers ranges between 1.5 and 3%. Glass fibers exhibit high compressive strength, high tensile strength, and a comparatively low Young's modulus (Table 7.17). The Young's modulus of pure glass fibers is in the range of that of aluminum and approx. a third of that of steel. Glass fibers exhibit high heat resistance. Continuous exposure to up to 250 °C does not reduce mechanical properties. The softening temperature of E-glass is above 625 °C. Its heat conductivity is higher than that of other fibers but significantly lower than that of metals. Glass fibers have good dielectric properties.

Applications: The share of glass fiber reinforced plastic components constitutes 90% of the global market of fiber reinforced materials. *Short glass fibers* are used for components with high strength requirements. This advantage, in combination with low weight, easy handling during processing, and a number of other tailored characteristics make short glass fiber reinforced plastics the material of choice for applications in the automotive, appliance, and electrical equipment industry. Examples for components with *long glass fiber reinforcement* include operating devices in automotive applications (battery mountings, clutch pedals, gear shift levers, air intake manifolds, cable trays, instrument panels/dashboards), in electronic applications (housings with electromagnetic shielding), in machine construction (gear wheels), and components for the sport and leisure industry (such as commodity surf boards, snow boards, and skis). Floor coverings, for example, are manufactured from fiber reinforced thermoplastics using GMT processes.

Glass fiber reinforced thermosets are used in automotive applications such as spoilers, hoods and hatches, fenders, roof modules for convertible hard tops, oil pans for utility vehicles, aquariums, panels, refrigerated train cars, rotor blades for helicopters and wind turbines, carriers for high voltage insulators, sail and motor boats, caravans, gliders, pipe systems and containers in nuclear plants and the chemical industry.

7.3.1.2 Natural, Organic Reinforcing Fibers

Renewable materials are used as fillers and reinforcing agents in plastics. Lately, these materials have come to the fore in light of environmental considerations (CO_2-neutrality, promotion of sustainable, ecological agriculture practices). However, often overlooked is the fact that natural fibers have been used in industrial applications for a long time – for sound technical and economic reasons. They are primarily used with processing technologies at the periphery of the plastics processing industry, *e. g.*, compression molding. Natural fibers are used in both thermoplastic and thermosetting matrices. They also find increasing applications in injection molding pellets, SMC, and others. In general, natural fibers exhibit low density, *i. e.*, good specific characteristics (light weight construction), good damping behavior (acoustic components in automotive applications), low shrinkage, safety for food contact applications, and low wear. A disadvantage is the low thermal resistance, which may cause problems during processing and prohibits applications with high-temperature melting plastics. Analogous to synthetic fiber reinforcements, fiber/matrix adhesion and the effect of respective bonding agents play an important role.

Another disadvantage of natural fibers frequently mentioned are the season-related changes in quality of these agricultural products. With increasing quality awareness and the availability of objective and comparable testing protocols, this is no longer an obstacle preventing the use of natural fibers in technically demanding applications. Table 7.19 provides a quick overview of the most important, technically relevant natural fibers. Figure 7.5 systematically explains the biological relationships.

In terms of volume, wood is the most significant renewable filler. In so-called wood plastics compounds (WPC), wood flour is incorporated at approx. 70 vol.-%, predominantly in thermoplastic matrices such as PE, PP, but also PVC. Most all applications are exterior wood substitutions (*e.g.*, decking) that are extruded. Although the processing cost are higher than for traditional wood products, WPCs are advantageous in terms of design options and improved weatherability. Compared to all-plastic products, WPCs are far more cost effective because of the high level of inexpensive filler material.

Table 7.19 Characteristics of Natural Fibers

Fiber	Plant base	Density	Length	Width	Tensile strength	Young's modulus	Elongation at break
		in g/cm³	in µm	in µm	in N/mm²	in N/mm²	in %
Jute**	*Corchorus olitorius*	1.42	42.3	34.8	570	17,000	3.2
Sisal**	*Agave sisalana*	1.30	35.9	85.3	430	4600	11.4
Flax A** (short steep)	*Linum usitatissimum*	1.48	34.1	16.9	790	19,000	4.5
Flax B** (steep)	*Linum usitatissimum*	1.46	–	19.0	870	14,600	6.6
Hemp**	*Cannabis sativa*	1.50	49.5	26.8	830	13,000	8.0
Ramie*	*Boehmeria nivea*	1.54	129.2	18.5	1250	36,000	3.8
Cotton*	*Gossypium barbadense*	1.51	27.3	15.4	620	12,000	7.3
(Lyocell 6.7 dtex)	Industrially produced fiber based on cellulose	1.54	56.8	23.3	820	8600	16.7

Average values for length and width of single fiber (*) and fiber bundle (**)

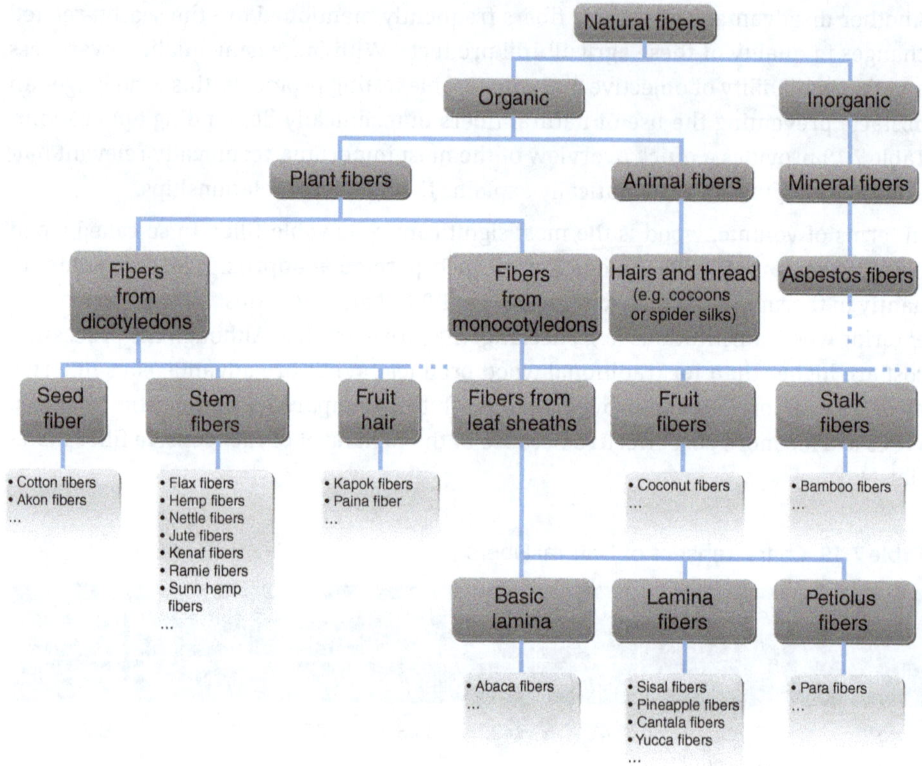

Figure 7.5 Systematic overview of natural fibers[1]

7.3.1.3 Synthetic, Organic Reinforcing Fibers, Carbon Fibers (CF), Aramid Fibers (AF), and Others

Carbon Fibers (CF)

Carbon fibers are synthetic products that are manufactured in several processing steps (heating, oxidation, and carbonization) from polymeric precursor fibers. During production, the graphitic structures are perfected above 1800 °C. However, the distance between layers always remains larger than in pure graphite structures. Precursor for high-quality carbon fibers are polyacrylonitrile (PAN) fibers, which account for 90–95% of the market. Figure 7.6 schematically illustrates the manufacturing steps. Other precursors are petroleum pitch and cellulose fibers such as rayon. Petroleum pitch forms mesomorphic, nematic phases at temperatures > 350 °C while eliminating hydrogen. Spinning of theses melts results in highly longitudinally oriented fibers. Additional elimination of hydrogen, carbonization, and subsequent heating to

[1] Müssig, J., Baur, E., From the Plant to the Natural Fibre-reinforced Part, Kunststoffe plast europe 7/2004, pp. 99–102

3000 °C creates highly graphitized fibers from rather inexpensive raw materials. Isotropic pitch precursors are used for "general-purpose" short fibers with low mechanical characteristics to substitute asbestos and reinforce concrete. Since 2010, the number of fiber manufacturers has steadily increased, meeting the global demand of 46,500 t in 2013. The global demand for carbon fibers is predicted to reach 89,000 t by 2020.

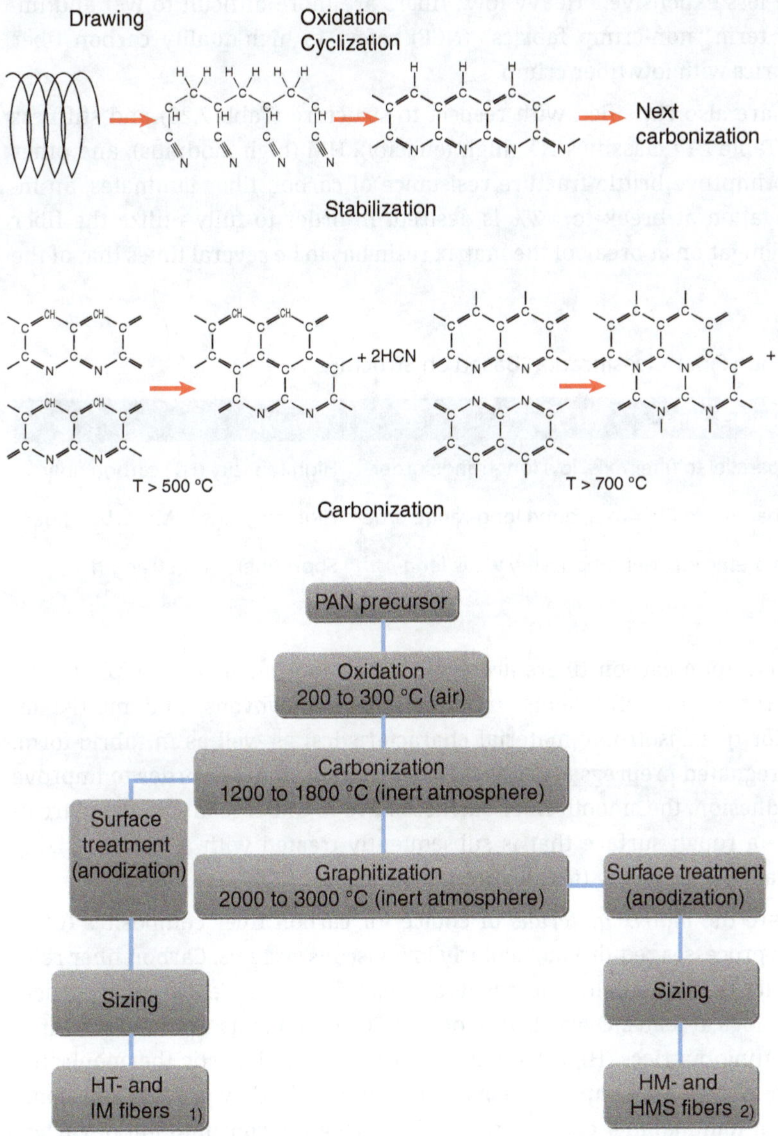

Figure 7.6 Chemical reactions and flow for C-fibers from PAN precursors

1) HT: high tenacity, IM: intermediate modulus

2) HM: high modulus, HMS: high modulus/strength, type/characteristics see Table 7.17

Fiber types: The specific manufacturing process and the degree of carbonization determine the type of carbon fiber. A carbon fiber has a diameter of approx. 5–8 µm. In general, 1000 to 48,000 single fibers (filaments) are combined to form a roving before they are wound on spools. Depending on the number of filaments in a tow and their fineness we distinguish between low-filament and multi-filament (heavy tow) fibers. Fibers with more than 24 k (= 24,000 filaments) are referred to as "heavy tow" fibers. They do not reach the high quality of fibers made from "small"-, "low"- or "flat tow" fibers, but are less expensive. "Heavy tow" fibers are more difficult to wet and impregnate. The term "non-crimp fabrics" (NCF) refers to high-quality carbon fiber multi-axial fabrics with low fiber crimp.

Carbon fibers are also classified with respect to structure (Table 7.20) and stiffness and strength. Table 7.17 classifies HT (high tenacity), HM (high modulus), and other fiber types. To improve brittle fracture resistance of carbon fiber laminates, an increase in elongation at break to > 2% is desired. In order to fully utilize the fiber strength, the elongation at break of the matrix resin has to be several times that of the fiber.

Table 7.20 Carbon Fiber Classification Based on Structure

Structural characteristics	Fiber type
Layers mostly parallel to fiber axis, low long-range order	High tenacity (HT) carbon fiber
Layers mostly parallel to fiber axis, good long-range order	High modulus (HM) carbon fiber
No discernable preferred orientation, very weak long-range order	Short fibers, low strength

Form of delivery: Spun carbon fibers are available as continuous and staple fibers. They are offered as monofilaments, unidirectional non-wovens, and multi-axial non-wovens (for quasi-isotropic material characteristics) as well as in fabric form. Resin pre-impregnated prepregs also have a large market share. In order to improve fiber-matrix adhesion, the smooth fiber surfaces have to undergo an oxidizing treatment to create a rough surface that is subsequently treated with a suitable sizing agent to facilitate the following (textile) processing steps.

Epoxy resins are the *matrix materials* of choice for carbon fiber composites (CFC). Resin injection processes require particularly low-viscous systems. Carbon fiber reinforced vinyl ester resins and phenolic resins are used for components under particularly high tribological loads. Overall, 75% of all CFC components are manufactured with thermosetting matrices. High temperature resistant and elastic thermoplastics (*e. g.*, PEEK) are also gaining importance as matrix materials; they provide additional improvement in flammability characteristics. Following the consumption of carbon fibers, approx. 72,000 t CFC were manufactured in 2013 globally.

The common *processing methods* used for the manufacture of carbon fiber reinforced components include standard processes such as manual lay-up, fiber winding, and pultrusion. Resin injection methods, also called liquid composite molding, are experiencing solid growth rates. They facilitate the effective wetting of the carbon fiber reinforcement with the thermosetting matrix material during the forming process. Dry semi-finished fiber reinforced products for complex or thick-walled components are produced using either sewing technologies with carbon yarns for multi-layered fabrics, or braiding/knitting technologies. Preforms, for example, are manufactured using yarns that will subsequently melt at elevated temperatures. Carbon staple fibers are used for injection molded components.

Properties: The weight specific strength and stiffness characteristics of carbon fibers are higher than those of metal-, E-glass-, and aramid fibers (Table 7.17). Therefore, carbon fibers are used when both improved mechanical properties and low weight are required. Carbon fibers also provide good electrical (approx. $6.7 \cdot 10^4$ S/m) and thermal conductivity, excellent abrasion and corrosion resistance. Carbon fibers offer high fatigue strength the dynamic properties of carbon fiber laminates are better than those of any other material (*e. g.*, aluminum, steel). Other than in glass fibers, the characteristics of carbon fibers are highly anisotropic. This applies also to the coefficient of thermal expansion – a fact that can be used deliberately during the design of components. The many varied application opportunities for carbon fibers come at the expense of high material costs, the hygroscopic behavior, and high wear of tools and cutting devices.

Applications: The achievable weight reductions represent the crucial economic advantage of carbon fibers and explain the increasing application of carbon fiber composites for structural and finishing components in the aerospace industry. In 2013, aeronautical applications made for 30% of the carbon fiber consumption: fighter jets and commercial airplanes as well as helicopters. While wind turbines typically are still manufactured with glass fiber reinforced plastics, rotor blade lengths of 40–50 m require carbon fibers reinforcement. Globally, wind turbines accounted for 14% of carbon fiber consumption in 2013. Equipment for high performance/professional sports represent another important area of application for carbon fibers (in 2013 globally 14%). That includes golf, tennis, racing bicycles, skiing (poles, high performance skis, and snowboards), aquatic sports (paddle, masts, and plastic sails), and fishing poles. Carbon fiber reinforced composites are still in their infancy in automotive applications. To meet reduction requirements in terms of CO_2 emissions, the manufacture of transportation vehicles consumed 11% of global carbon fibers in 2013 (cars: 46%, trucks: 18%, racing cars: 15%, passenger trains: 13%, others: 8%). They were used in auto body segments, interior trim, car roof segments, and roof and roof rack structures.

Glass fiber composites with UP and acrylic resins still dominate in machine and plant construction applications – areas in which extreme weight reduction is not of comparable importance. However, carbon fiber composites are making steady inroads here. One example is corrosion-free pressure vessels. Adhesively bonded carbon fiber laminates are used in construction applications to increase the flexural strength of concrete; they are used for bridge and building reinforcement. Carbon-reinforced concrete has also gained increasing prominence for new, filigree architectural structures. Carbon fiber reinforced composites are used in medical technology for implants and for components that require increased X-ray transparency. Carbon fiber non-wovens are used for electromagnetic shielding and resistance heaters.

Aramid Fibers (AF)

Aramid fibers are golden/yellow organic fibers that consist of linear, aromatic polyamides. The characteristics of these **ar**(omatic poly)**amides** are determined by the regular arrangement of phenylene rings, amid groups, and the hydrogen bridges between chains. The term aramid does not apply to every aromatic polyamide (polyaramide) with aromatic groups in the main chain. The U.S. Federal Trade Commission defined aramids as long-chain synthetic polyamides, in which at least 85% of the amide groups are directly linked to two aromatic rings. Polyphenylene terephthalamide (PPTA), the initial polymer, cannot be melt spun because its melt temperature is higher than its degradation temperature. Therefore, viscous 20% solutions of PPTA in concentrated sulfuric acid are spun into highly crystalline fibers with strong molecular orientation. The fiber properties vary depending on the presence of a liquid crystalline phase. The individual fiber filaments are collected, washed, neutralized, and sized to improve their handling properties. Globally, the production volume of aramid fibers reached approx. 50,000 t/a.

Fiber types: We distinguish between meta-aramids [poly(m-phenylene isophthalamide); Teijinconex™ and Nomex™] and para-aramids [poly(p-phenylene isophthalamide); Kevlar™]. Commercially available fibers include high modulus (HM) fibers and low modulus fibers (HE, high energy). High modulus fibers are manufactured by a drawing process immediately after melting spinning.

Properties: Aramid fibers stand out for their very high strength, high impact resistance, high elongation at break, and good vibration damping properties. However, these characteristics are highly anisotropic. The compression strength in fiber direction is significantly lower than tensile strength. Therefore, aramid fiber composite components are suitable for light weight construction applications under high tensile loads, but not recommended for components predominantly under flexural or compression loads. Aramid fibers exhibit low densities (ρ = 1.45 g/cm^3) and thus high weight specific strength. Similar to carbon fibers, aramid fibers exhibit a negative coefficient of thermal expansion in fiber direction because of their high molecular orientation. The fibers are heat- and fire resistant. The glass transition temperature of aramid fibers is at approx. 300 °C, and at temperatures above 400 °C the fibers begin to char. Aramid fibers are hygroscopic (up to 7%) and UV-sensitive; they are difficult to process because of their high ductility. Fabrics can be cut only with special, micro-teethed scissors, and aramid reinforced laminates require water jet cutters for rework or finishing operations with acceptable results. Organometallic coupling agents, based on titanium and zirconium, are used to enhance the adhesion of aramid fibers to epoxy, vinyl ester and polyurethane matrices.

Applications: Typical applications for aramid fibers are ballistic protection, such as shrapnel protection and bullet-proof vests, protective helmets, fire-proof protective clothing, armored vehicles, but also in high-performance sport (competition kayaks). Aramid fibers are also used as reinforcement in high-performance tires. Other applications include high-duty track and tension ropes. Aramid fibers are often used in optimized hybrid reinforcements in combination with carbon and/or glass fibers.

When incorporated in materials, such as PA, POM, PBT, and PC, aramid fibers improve tribological properties. Aramid fibers in PTFE-thermoplastic combinations significantly reduce wear factor and friction coefficient.

Honeycomb cores made from aramid fibers are used for hulls, racing car bodies, panels for boats, antennae, airplane containers and stabilizers, missile wings, fuel tanks, and helicopter rotor blades.

PAI-, PBI Fibers

Other high-temperature textile fibers, such as PAI (polyamide imide) and PBI (polybenzimidazole) are dry spun from specific solvents (*e.g.*, NMP, *N*-methylpyrrolidone). They exhibit 10–50% elongation at break, are temperature resistant up to > 200 °C, infusible, char above 500 °C with little gas formation, are inflammable, and resistant to chemicals, common solvents, and fuels and lubricants

Applications: fire proofing (PBI with 15% water absorption), bullet and shrapnel protection, airplane interiors, filter materials, packaging, seals and gaskets, friction linings (substituting asbestos), technical papers for honeycomb cores, and electrical and heat insulation.

PBO Fibers

Poly(p-phenylene-2,6-benzobisoxazole) (PBO) is another high-performance fiber. It has been commercially available since 1998. The fibers were developed in Japan under the tradename Zylon® and are offered in two *fiber types*, Zylon AS (spin fiber) and Zylon HM (high modulus fiber). The HM fibers undergo an additional drawing process.

Properties: PBO exhibits good mechanical properties. Both Young's modulus and tensile strength exceed those of aramid fibers (see Table 7.17). The heat resistance is extremely high (PBO 650 °C; p-aramid 550 °C, m-Aramid 400 °C). PBO does not burn because it has an extremely high limiting oxygen index (LOI) (Table 7.21). Another advantageous characteristic is its good chemical resistance. Disadvantages of PBO include low UV-resistance and difficult processability because PBO is difficult to cut. PBO is also very expensive and is therefore often processed mixed with other fibers, *e. g.*, high-strength PE fibers.

Table 7.21 Comparison of Limiting Oxygen Index (LOI) for Different Fibers

	Polyester	Aramid	PBI	PBO
LOI	17	29	41	68

Applications: PBO is used in protective gear, such as bullet-proof vests, fire fighter outfits, and others. In sporting goods, PBO is used in sails, tennis rackets, hot air balloons, and ski poles. Additional applications are found in tire reinforcement and the aerospace industry.

Polyethylene Fibers

High-molecular PE (UHMW-PE) is used to manufacture high-performance PR fibers in a special spinning/drawing process (Dyneema®, Spectra®). At the same weight, these fibers offer up to 15 times more tensile strength than steel (2500 to 3000 N/mm^2) and are 40% stronger then aramid fibers. High fiber-parallel orientation of the molecular chains (95%) and degree of crystallinity (85%) facilitate this outstanding level of stiffness and strength. The high degree of orientation renders the fibers anisotropic, and thus the coefficient of thermal expansion extremely negative. High-strength PE fibers have a density of 0.95 to 0.97 g/cm^3 and are thus slightly lighter than water. The fibers are very durable and show high resistance to abrasion, moisture, UV radiation, and chemicals. Similar to aramid fibers, PR fibers exhibit high ductility and are therefore used as hybrid fabrics in high-impact and ballistic applications. A disadvantage of PE fibers is their low compression strength parallel to fiber direction. The fibers should be exposed to tensile loads only; transverse loads will quickly split the fiber in longitudinal direction. The upper thermal service limit is at 70 °C (short-term, *e. g.*, curing temperature of laminates, up to 125 °C). PE fibers exhibit medium UV resistance; however, it is superior to that of aramid fibers.

Applications: PE fibers are used for the manufacture of fishing lines, fishing nets, climbing harnesses, bullet-proof vests, motorcycle helmets, and other devices used in rock climbing. High-strength PE fibers are also particularly suitable for running rigging (no stretch, non-decomposing), for paragliders and kites, and for winch launching of gliders. Ropes come with a length of 400 m and low mass and have therefore substituted steel ropes in mountain and other rescue operations. In composite materials, PE fibers are typically used in mixed fabrics with carbon fibers.

PET Fibers

PET fibers can be used to reinforce PP compounds. These compounds are known under the name Rialene and exhibit very good notched impact strength at temperatures as low as −30 °C. Adding PET fibers simultaneously improves both Young's modulus and impact strength. The same effect is seen with the addition of wollastonite. Energy absorption to break (shatter proof) is twice as high as in glass fiber reinforced compounds, and despite increase in impact resistance modification, heat resistance is increased by 30%. High elongation at break and a significantly lower coefficient of expansion facilitate high assembly reliability. The high flexural modulus of elasticity renders the material suitable for hinge-like applications. As an alternative to PET fibers, PEN fibers are also used as reinforcements.

7.3.2 Fibers, Yarns, Bristles, Tapes

The fibers commonly used in the clothing and carpet industries exhibit Young's moduli too low (PE approx. 14 GPa, PA approx. 6 GPa) to be suitable for fiber composite applications (see Section 7.3.1). These fibers are also referred to as synthetic fibers because they are manufactured from synthetic polymers.

Monofilaments are used for the manufacture of sewing threads, twines, or ropes. Multifilaments (smooth or textured), yarns, and twines from spun fibers are used to manufacture fabrics, knits, or other goods for clothing or home textiles. Besides standard manufacturing methods for carpets, tufting methods are also used. Here, the loops are secured on the backside by a coat of rubber or plastic. Fibers or continuous filaments are also used to manufacture non-woven fabrics, which are solidified by either mechanical or thermal treatment. Synthetic paper and water-repellent textile fabrics can be manufactured from filaments.

Monofilaments with 360 to 1100 dtex are also called *bristles*. They are used for brushes, brooms, and paint brushes. Monofilaments with diameters exceeding 0.08 mm are used for fishing lines, tennis rackets, ropes, and similar products. Extruded films are cut in longitudinal direction to form stretched *film tapes* or *flat yarns*, drawn at high ratios (1 : 12) split into longitudinal fibrils, and wound to tear resistant, coarse binding twine. These and flat yarns are also used to manufacture textile fabrics – sack cloth, carpet backing fabric, wall coverings.

The raw materials used here are the same polymers used for molding processes. Depending on their respective melting point or thermal stability, synthetic fibers are processed using the following technologies:

Melt Spinning Processes

- Here, raw materials such as PA, PE, and PP are melted in an extruder in the absence of oxygen, pushed through spinnerets, solidified by cooling in air, and subsequently drawn. The haul-off rates vary between 500 and 4000 m/min.

Dry Spinning Processes

- This technology pushes a spinning solution through spinnerets into a hot-air duct. The filaments solidify when the solvent evaporates. The haul-off rates reach up to 1000 m/min. Examples are acetate and a variety of acrylic fibers.

Wet Spinning Processes

- Here, the polymeric spinning solution is pressed through spinnerets into a precipitation bath, where the fibers solidify (precipitate). After drawing, the wet fibers are wound. This technology is used for viscose- and triacetate fibers as well as for certain acrylic fibers, including modacrylic fibers. It is gaining increasing popularity with the development of special fiber types with high temperature and flame resistance. Haul-off rates are up to 150 m/min.

Depending on fiber type, all three technologies spin through spinnerets with varying numbers of holes and hole diameters. Spinnerets/dies with one hole produce monofilaments; spinnerets with up to several hundred holes are used to manufacture continuous filaments, while multi-hole spinnerets are used to manufacture tows that are typically turned into staple fibers. Subsequently, monofilaments, continuous fibers, and tows are drawn and exposed to an additional heat treatment to achieve desired properties. Staple fibers are converted to yarns and twines using a number of different technologies.

The by far largest manufacturer of synthetic fibers is China, followed by Taiwan and the U.S. In 2013, the global production of synthetic fibers amounted to 55,800,000 t, whereas that of natural fibers was 29,700,000 t. In terms of quantity, the most important fibers are polyester, polyamide, polyacrylonitrile, and polypropylene fibers. Today, it is common to mix fibers for textile applications to achieve improved service, cleaning, and physiological properties. To meet specific requirements, variations in the fineness and fiber length of the filaments, din number of filaments, as well in the form of spun fibers can be made. For example, gloss and feel of fibers can be determined by the geometry of the cross section (round, angular, star-shaped, etc.) of the filaments. It is also possible to spin bi-component filaments from two different thermoplastics. Here, the two materials may be arranged side by side or in a so-called sheath/core, concentric structure. In the following, we will provide a quick overview of the applications for different fibers.

Polyester Fibers

The most common polyester fiber is made of polyethylene terephthalate (PET), but PBT and PTT are also spun into fibers. Melt-spun, hot drawn PET fibers are highly crystalline and exhibit low shrinkage. They are particularly suitable for shape-retaining and crease-, chemical-, and UV resistant textiles. Dyeing these textiles requires special processes. In 2013, polyester quantitatively represented the largest share of synthetic fibers with a production volume of 29,000,000 t. China and Southeast Asia are the predominant producers of synthetic fibers. Besides virgin polyester, recycled polyester is gaining importance in fiber manufacturing. In 2009, reclaimed PET bottles (7,500,000 t) were converted into 5,900,000 t pellets. Of these, 3,400,000 t were used for the manufacture of synthetic fibers.

Applications: All areas of textile application, tire cords, conveyor belts, rot-proof nonwoven fabrics for road construction and roof sheeting, coated fabrics for textile construction applications. PBT- and PTT fibers for home textiles (Kodel®-type) because they allow for easier dyeing.

Polyamide Fibers

Polyamide fibers are predominantly manufactured using melt spinning processes. The most important PA fibers are based on PA 66 (also known as nylon) and PA 6. They exhibit high abrasion and tensile strength, weather and alkali resistance, as well as good elasticity. When polyamides are used in high-moisture environments, PA types with low water absorption characteristics should be selected.

Applications: PA fibers are used in all areas of textile applications – carpets, tire cords, lines, strings, bristles, belts, tapes, and ropes. Fully aromatic polyamides (polyaramides, PMPI, PPTA) are used for high-strength, temperature resistant fibers for fire protection gear and plastic reinforcement.

Polyacrylonitrile Fibers

Spun PAN fibers, consisting of copolymers in dimethylformamide solution with a minimum 85% acrylonitrile content, wet or dry spun, also belong to the group of important synthetic fibers for general applications. They provide good dyeing and crimping properties, high UV and weather resistance, good acid and alkali resistance, are abrasion and tear resistant while providing elasticity. Although PAN fibers see increasing demand as carbon fiber precursors, their overall demand is declining because the low prices of polypropylene and polyester fibers have taken over the market for other applications.

Applications: Textiles for clothing (knit and woven fabrics), home textiles (upholstery fabrics, blankets, tapestry), awnings and fabrics for outdoor furniture, canopies, tents, acid-proof textiles, carpets, and special types for asbestos replacement. PAN fibers are also precursors for the manufacture of carbon fibers. Modacrylic fibers and threads, copolymers consisting of 40 to 60% acrylonitrile and 60 to 40% vinyl chloride and/or vinylidene chloride are also used for non-flammable safety gear, window treatments, carpets, and upholstery fabrics.

Polyolefin Fibers

Polyolefin fibers are produced from isotactic PP (approx. 90%) and PE-HD (10%). Section 7.3.1.3 provides more detail regarding the specific tensile strength of high-performance polyethylenes (Dyneema®, Spectra®) compared to other materials.

Applications: Ropes, nets (buoyant) for the shipping and fishing industries and for technical components. PP-fibrillated fibers (> 97% isotactic content) are used as agricultural twines and asbestos substitutes. PP non-woven fabrics are used for erosion control in road and hydraulic construction. Flat yarns, cut from film and drawn for increased strength, are laminated as reinforcing elements in composite films. They are also used to manufacture fabrics for sacks and bags that are subsequently extrusion coated with PE.

Polyvinyl Alcohol Fibers

Drawn polyvinyl alcohol fibers (PVA), wet spun from aqueous solution into saline precipitation baths, exhibit excellent mechanical properties; however, they are sensitive to water and moisture. PVA fibers are insolubilized with formaldehyde and can be added to acrylonitrile-(co)-polymer or polyamide yarns for water absorption control. Filaments with increased mechanical properties are manufactured using dry spinning technologies.

Applications: fishing nets, conveyor belts, additives for papers and textile composites.

Polytetrafluoroethylene Fibers

PTFE fibers are incorporated into other thermoplastics such as POM or PA. These materials are then used for the development of friction bearings, guide rails, gear wheels, seals, strippers/deflectors, insulator bushings, chain guides, and rollers. PTFE fibers are also combined with stainless steel monofilaments for filter applications in the pharmaceutical industry (Ytek®).

Applications: PTFE fibers are used exclusively or in mixed fabrics as membranes, cable sheathing, textiles, filter and architectural fabrics, and protective clothing. In biomedical applications, PTFE-coated yarns are used as dental floss, PTFE fabrics are used as OR-textiles and tubes. [Note: PTFE is the base material for Gore-Tex® clothing, in which it is manufactured in extremely fine membrane layers with very fine pores. The pore size allows for water vapor to pass while liquid water does not penetrate. The resulting fabrics are used for water- and wind-proof yet breathable clothing.]

Polyurethane Fibers

A variety of applications requires elastic fibers. Besides synthetic and natural rubber fibers, synthetic elastane (spandex) fibers (*e.g.*, Lycra®, Dorlastan®) are manufactured in melt and solution spinning processes. The material used here is a block-copolymer with urethane- and ether segments with high elasticity (between 500 and 800%).

Applications: clothing, including sports and swim wear.

Micro-Fibers

Increasingly, micro-fibers as well even finer ultra-fine and ultra-super-fine polyester, polyamide, and other polymeric fibers are used for specific applications. Micro-fibers distinguish themselves with unique characteristics, including softness, pleasant feel, good drapability, gloss, volume, and excellent surface characteristics.

Finer micro-fibers are manufactured on an industrial scale using indirect spinning processes. The 2-step process first produces a bi-component fiber in order to subsequently dismantle the bi-component structure and extract one component (dependent on process, different approaches are used).

Fiber fineness is measured using the international standard "tex" (1 tex = 1 g/1000 m; 1 dtex = 1 g/10,000 m). Fibers with fineness ranging from 0.1 to 1.0 dtex are generally labeled micro-fibers, whereas those with fineness < 0.3 dtex are called ultra-micro-fibers (Table 7.22). In general, any fiber thinner than silk may be described as a micro-fiber. However, the specific characteristics of micro-fibers become evident only at fineness levels of less than 0.5 dtex. Minimum fineness currently achievable is approx. 0.0001 dtex. These fibers are called ultra-superfine micro-fibers (for comparison: fineness of the most common fine natural fibers: natural silk 1.2 dtex, new wool 3 to 6 dtex, cotton 1.5 to 2.5 dtex).

Table 7.22 Classification of Fibers with Respect to Linear Mass Density

	Coarse fibers	Medium fine fibers	Fine fibers	Micro-fibers	Super-micro-fibers and ultra-superfine fibers
dtex	> 7.0	7.0 to 2.4	2.4 to1	1.0 to 0.3 (0.1)	< 0.3

8 Material Properties

This chapter provides comparisons of the most important properties of plastic materials in table format. When possible, tables and charts referring to specific classes of plastic materials were placed in the respective chapter.

Because material names and properties have already been discussed in the previous chapters, this chapter will not offer additional explanations.

8.1 Processing Characteristics and Tolerances

Table 8.1 Processing-Relevant Characteristics for Selected Plastic Materials (Injection Molding)

Plastic material	Processing temperature °C	Pre-drying °C/h	Mold temperature °C	Shrinkage %	Length of flow path[1] at 2 mm wall thickness
PE-LD	160–220	–	20–60	1.5–5.0	550–600
PE-HD	180–250	–	10–60	1.5–3.0	200–600
EVAC	130–240	–	10–50	0.8–2.2	320
PP	200–270	–	20–90	1.3–2.5	250–700
PB	220–290	–	10–60	1.5–2.6 –	300–800
PIB	150–250	–	50–80	1.5–3.0	
PMP	280–310	–	ca. 70		
PS	170–280	–	10–60	0.4–0.7	200–500 –
SAN	200–260	85/2–4 –	50–80	0.4–0.7	200–500
SB	190–280	70–80/2[2]	10–80	0.4–0.7	320 –
ABS	200–275	70–80/2–4	50–90	0.4–0.7	
ASA	220–260		50–85	0.4–0.7	

Table 8.1 Processing-Relevant Characteristics for Selected Plastic Materials (Injection Molding) *(continued)*

Plastic material	Processing temperature °C	Pre-drying °C/h	Mold temperature °C	Shrinkage %	Length of flow path[1] at 2 mm wall thickness
PVC-U	170–210	–	20–60	0.4–0.8	160–250
PVC-P	160–190	–	20–60	0.7–3.0	150–500
PMMA	190–270	70–100/2–6	40–90	0.3–0.8	200–500
POM	180–230	110/2[2]	60–120	1.5–2.5	500
PA 6	240–290	80/8–15[2]	40–120	0.8–2.5	400–600
PA 66	260–300	80/8–15[2]	40–120	0.8–2.5	810
PA 610	220–260	80/8–15	40–120	0.8–2.0	–
PA 11	200–250	70–80/4–6	40–80	1.0–2.0	–
PA 12	190–270	100/4[2]	20–100	1.0–2.0	200–500
PA 6-3-T	250–320	80–90/10	70–90	0.5–0.6	–
PC	270–320	110–120/4	80–120	0.6–0.8	150–220
PET	260–300	120/4[2]	130–150 20[3]	1.2–2.0 0.2[3]	200–500
PBT	230–280	120/[2]	40–80	1.0–2.2	250–600
PPE + PS	240–300	100/2	40–110	0.5–0.8	260
PSU	340–390	120/5 –	100–160	0.6–0.8	–
PPS	320–380	160/5	20–200	ca. 0.2	–
PES	320–390		100–190	0.2–0.5	–
PVDF	250–270	–	90–100	3–6	–
PTFE	320–360	–	200–230	3.5–6.0	–
PFA	380–400	–	95–230	3.5–5.5	80–120
PEEK	350–390	150/3	120–150	ca. 1 –	–
PAI	330–380	180/8	ca. 230	0.4–0.7	–
PEI	340–425	150/4	65–175	ca. 1	–
PEK	360–420	150/3	120–160		–
CA	180–220	80/2–4	40–80	0.4–0.7	350
CP	190–230	80/2–4	40–80	0.4–0.7	500
CAB	180–220	80/2–4	40–80	0.4–0.7	500
PF Type 31	60–80	–	170–190	1.2	–
MF Type 131	70–80	–	150–165	1.2–2	–
MF/PF	60–80	–	160–180	0.8–1.8	–
Type 180/82		–			–
UP Type 802	40–60	–	150–170	0.5–0.8	–
EP Type 891	ca. 70		160–170	0.2	

Plastic material	Processing temperature °C	Pre-drying °C/h	Mold temperature °C	Shrinkage %	Length of flow path[1] at 2 mm wall thickness
TPO	180–200	75/2 or 65/3	10–80	1.5–2.0	–
SBS	175–250	–	10–90	0.3–2.2	–
TPA	170–230	110/2–4 or	15–80	1.0–2.0	–
TPC	160–220	100/3–6			–
TPS	180–220				–
TPU	180–220		20–40	0.8–1.5	–
		110/0.5 or			
		100/2			

[1] At average melt temperature, melt pressure, and injection mold wall temperature.
[2] Not necessary when delivered in pre-dried form.
[3] For amorphous grades

Table 8.2 Additional Shrinkage Information

Plastic material	%	Plastic material	%
PP-GF 20	1.2–2.0	PBT-GF-30	0.5–1.5
PS-HI	0.4–0.7	PPE+PS-GF-30	0.2
SAN-GF 30	0.2–0.3	PSU-GF40	0.2–0.4
ABS-GF 30	0.1–0.3	PPS-GF-40	0.2
POM-GF 30	0.5–1.0	PES-GF-40	0.15
PA 6-GF 30	0.2–1.2	PEK	0.7–0.9
PA 66	0.2–1.2	PEK-GF-30	0.3–0.8
PET-A	0.2	PC-GF-30	0.2–0.4
PET-GF-30	0.2–2.0		

Table 8.3 Compatibility of Various Plastic Materials During Melt Processing

	PE-HD	PE-LD	PP-H	PP-HI	PVC	PS	S-B	ABS	ASA	SAN	PA	PBT	PET	POM	PC	PMMA	PSU	PES	PAEK	TPU	(PPE+S-B)
PE-HD	M																				
PE-LD	M	M																			
PP-H	I	I	M																		
PP-HI	M	M	M	M																	
PVC	I	I	I	I	M																
PS	C	C	C	C	I	M															
S/B	C	C	C	C	I	M	M														
ABS	I	I	I	I	M	I	I	M													
ASA	I	I	I	I	M	I	I	M	M												
SAN	I	I	I	I	M	I	I	M	M	M											
PA	C	C	C	C	D	C	C	C	C	C	M										
PBT	C	C	C	C	D	I	I	I	I	I	I	M									
PET	C	C	C	C	D	I	I	I	I	I	I	M	M								
POM	I	I	I	I	D	I	I	I	I	I	I	I	I	M							
PC	I	I	I	I	D	I	I	M	M	M	C	M	M	I	M						
PMMA	I	I	I	I	M	I	I	M	M	M	I	I	I	I	M	M					
PSU	D	D	D	D	D	I	D	M	M	M	C	D	I	D	M	D	M				
PES	D	D	D	D	D	D	D	D	D	D	D	D	D	D	M	D	M	M			
PAEK	D	D	D	D	D	D	D	D	D	D	D	D	D	D	D	D	M	M	M		
TPU	I	I	I	I	C	I	I	M	M	M	D	I	I	M	M	M	D	D	D	M	
(PPE+S/B)	C	C	C	C	D	M	M	I	I	I	C	I	I	I	C	I	D	D	D	I	M

Compatibility

- M Miscible or compatible
- I Incompatible
- C Addition of commercially available coupling agent necessary
- D Decomposition or crosslinking

Table 8.4 Suitability of Plastic Material Combinations for Two-Component Injection Molding

	Plastic	PE-HD	PE-LD	EVAC	PP	PPE + PS	PS, PS-HI	SAN	ASA	ABS	PVC	PVAC	PMMA	POM	PA 6	PA 66	PC	PC + ABS	PC + PBT	PET	PBT	PPSU	PPO + PS	CA	TPA	TPC	TPO	TPS	TPE-Spec.	TPS-SEBS	TPU
1	PE-HD	+																													
2	PE-LD	+	+																												
3	EVAC	+	+	+																											
4	PP	–	+	+	+																										
5	PPE + PS				○	+																									
6	PS, PS-HI	–	–		–	+	+																								
7	SAN	–	–		–	–	–	+																							
8	ASA	–	–	+	–	–	–	+	+																						
9	ABS	–	–	○	–	–	–	+	+	+																					
10	PVC	○		–	○	–	○	+	+	+	+																				
11	PVAC							○	○	○		+																			
12	PMMA	○	○		○		–	+	+	+	+		+																		
13	POM	○	○		○		–							+																	
14	PA 6	○	○		○		–								+																
15	PA 66	○	○		○		–								+	+															
16	PC	–	–		–		–	+	+	+	+		○		○	○	+														
17	PC + ABS	–	–		–	–	○	+	+	+	+						+	+													
18	PC + PBT	–	–		–	–	○	+	+	+							+		+												
19	PET							+	+	+							+	+	+	+											
20	PBT	–	–		–	–	–	+	+	+	+						+			+	+										
21	PPSU																+					+									
22	PPO + PS																						+								
23	CA	–	–	○	–	–	–	+	+	+	+	–									+			+							
24	TPA				○		–			–	–			–	+	+	–			–	–				+						
25	TPC				–		–			–	–		–	–	○	○	–			+	+					+					
26	TPO				+		–			–	–		–	–	○	○	–			–	–						+				
27	TPS	+	+		+		○			+	○		○	–	○	○	+			○	○	–						+			
28	TPE-Spec.	+	+			+	+			+					+	+	+	+	+	+	+			+					+		
29	TPS-SEBS	+	+		+	+	+	+	+	+			+	+	+	+	+	+	+		+		+							+	
30	TPU	–	–	–	–	–	–	+	+	○	+		–	+	+	+	+	+	+	+	+			+							+

+: Good bond, ○: Possible bond, –: No bond

Table 8.5 Post-Processing and Finishing Operations for Selected Plastic Materials

Plastic material	Post-processing, finishing[1,2]								
	Electro-plating	Spray-painting and printing	Hot stamping	Vapor deposition	Solvent adhesion	Ultrasonic welding	Hot plate welding	Vibration welding	High-frequency welding
PE-LD	–	+	++	+	–	–	+	+	–
PE-HD	–	+	++	+	–	+[3]	+	+	–
EVAC	–	+	++	+	–	o[3]	+	+	o
PP	++[4]	+	++	+	–	+	+	+	–
PB	–	+	++	++	+	+[3]	+	+	–
PS	–	++	++	+	++	+	+	+	–
SAN	–	++	++	+	++	+	+	+	o
SB	+[4]	++	++	+	++	+	+	+	o
ABS	++[4]	++	++	+	++	+	+	+	+
ASA	–	++	++	+	++	+	+	+	o
PVC-U	+[4]	++	++	+	++	+	+	+	+
PVC-U-E	–	+	++	+	++	o	+	o	+
PVC-P	–	+	++	–	++	–	+	–	+
PVC-P-E	–	+	++	–	++	–	o	–	+
PMMA	–	++	++	+	++	++	++	++	
POM	–	+	++	+	++[5]	++	++	++[3]	

Plastic material	Post-processing, finishing[1,2]								
	Electro-plating	Spray-painting and printing	Hot stamping	Vapor deposition	Solvent adhesion	Ultrasonic welding	Hot plate welding	Vibration welding	High-frequency welding
PA 6	–	++	++	+	++[5]	+	+	+	+
PA 66	–	++	++	+	++[5]	+	+	+	+
PA 610	–	++	+	+	++[5]	+	+	+	+
PA 11			++	++	++[5]	+	+	+	+
PA 12	–	+	++	++	++	+	+	+	+
PA 6-3-T	–	++	++	+	++	+	+	+	+
PC	+	++	++	+	++	+	+	+	–
PET	–	++	++	+	–	+	+	+	–
PBT	–	+	++	–	–	+	+	+	–
PPE+PS	+	++	++	+	++	+	+	+	–
PSU	+	++	++	+	++	+	+	+	–
PPS	–	+	++	+	–	+	+	+	–
PES	–	++	++	+	++	+	+	+	–
PVDF	–	++	++	+	++	o[3]	+	o	–
PTFE	–	+	+		–	–	–	–	–
PFA	+	+	+	+	–	–	+	o	–
PEEK		++	++	+		+	+	+	–

Table 8.5 Post-Processing and Finishing Operations for Selected Plastic Materials *(continued)*

Plastic material	Post-processing, finishing[1,2]								
	Electro-plating	Spray-painting and printing	Hot stamping	Vapor deposition	Solvent adhesion	Ultrasonic welding	Hot plate welding	Vibration welding	High-frequency welding
PAI		++	++	+		+	+	+	–
PEI		++	++	+		+	+	+	–
PEK		++	++	+		+	+	+	–
CA	–	++	++	+	++	o	+		+
CP	–	++	++	+	++	o	+		+
CAB	–	++	++	+	++	o	+	–	o
PF Type 31		++	–	+	–	–	–	–	–
MF Type 131		++	–	+	–	–	–	–	–
MF/PF Type 180/182		++	–	+	–	–	–	–	–
UP Type 802		++	–	+	–	–	–	–	–
EP Type 891		++	–	+	–	–	–	–	–
TPO	–	+	++	+	–	o[3]	+	+	
TPA	–	++	++	+	++	o[3]	+	+	
TPE	–	++	++	+	–	o[3]	+	+	

Plastic material	Post-processing, finishing[1,2]								
	Electro-plating	Spray-painting and printing	Hot stamping	Vapor deposition	Solvent adhesion	Ultrasonic welding	Hot plate welding	Vibration welding	High-frequency welding
TPS	–	++	++	+	+	+	+	+	
TPU	–	++	++	+	++	o[3]	+	+	

1) Adhesion ++ = possible without pre-treatment, + = possible with pre-treatment, – = not possible
2) Welding + = possible, o = possible under certain conditions, – = not possible
3) Ultrasonic welding only possible for near-field applications
4) For certain grades
5) Not recommended

Table 8.6 Weldability of Different Material Combinations

Hot plate welding \ Ultrasonic welding	ABS	ASA	CA	EVAC	PA6	PA66	PC	PE-HD	PE-LD	PMMA	POM	PP	PPE + PS	PS	PS-HI	PBT	TPU	PVC-P	SAN	PC + PBTB	PC + ABS	PPS	LCP
ABS	+																						
ASA		+																					
CA			+																				
EVAC																							
PA6	–				(+)\+																		
PA66	–				(+)	(+)\+																	
PC	+				–	–	+																
PE-HD	–	–	–				–	+															
PE-LD	–	–	–				–		+														
PMMA	+				–	–	(+)			+													
POM											+												
PP	–	–					–					+											
PPE+PS	–												+										
PS	–				–	–	–							+									
PS-HI	–				–	–	–								+								
PBT	○				–	–				–						+							
TPU																	+						
PVC-P	○									○								+					
SAN	(+)									○				○	○				+				
PC+PBTP	○				–	–	(+)			(○)		–	–			(+)				+			
PC+ABS	+				–	–	+			(+)		–	–								+		
PPS	–				–	–	(○)									–						(+)\+	
LCP																							○

+ : Good bond, ○ : Possible bond, – : No bond, ( ) = Under certain conditions

Table 8.7 Adhesive Bonding of Selected Plastic Materials

		Polarity	Solubility	Possibility of	
Adhesive bond	**Plastic material**	**+ polar**	**+ soluble**	**Diffusion bonding**	**Adhesion bonding**
		– non-polar	**– insoluble/ hardly soluble**		
Good	PS	+/–	+	+[1]	+
	PVC-U	+	+	+	+
	PVC-P	+	+	+[2]	+
	PMMA	+	+	+/–	+
	PC	+	+	+	+
	ABS	+	+	+	+
	CA	+	+	+	+
	PUR	+	–	–	+
	UP	+	–	–	+
	EP	+	–	–	+
	PF	+	–	–	
	UF-MF	+	–	–	+
Under certain conditions	PA	+	–	+/–	+
	POM	+	–	–	+
	PET	+	–	–	+[3]
	Rubber	+	–	+/–	+
Difficult	PE	–	–	–	+/–[4]
	PP	–	–	–	+/–[4]
	PTFE	–	–	–	+/–[4]
	SI	+/–	–	–	+

[1] Not possible for foamed PS;
[2] follow instructions of PVC-P supplier;
[3] after pre-treatment with caustic soda (80 °C, 5 min);
[4] only after pre-treatment, PP + PE reach parent material strength, low peel strength

8.2 Overview of Mechanical, Thermal, and General Electrical Properties

Table 8.8 Mechanical and Thermal Characteristics of Plastic Materials

Plastic material	Density g/cm³ ISO 1183	Mechanical characteristics			Thermal characteristics		
		Tensile strength N/mm² ISO 527	Strain at break % ISO 527	Tensile modulus MPa ISO 527	Linear coefficient of expansion $K^{-1} \cdot 10^6$	Thermal conductivity W/mK	Specific heat kJ/kg K
PE-LD	0.914–0.928	8-23	300–1000	200–500	230–250	0.32–0.40	2.1-2.5
PE-HD	0.94–0.96	18–35	100–1000	700–1400	120–200	0.38–0.51	2.1-2.7
EVAC	0.92–0.95	10–20	600–900	7–120	160–200	0.35	2.3
IM	0.94	21–35	250–500	180–210	120	0.24	2.20
PVK	1.19	20–30	–	3500	–	0.29	–
PP	0.90–0.907	21–37	20–800	1100–1300	110–170	0.17–0.22	2.0
PB	0.905–0.920	30–38	250–280	250–350	150	0.20	1.8
PIB	0.91–0.93	2–6	>1000	–	120	0.12–0.20	–
PMP	0.83	25–28	13–22	1100–1500	117	0.17	2.18
PS	1.05	45–65	3–4	3200–3250	60–80	0.18	1.3
SAN	1.08	75	5	3600	80	0.18	1.3
SB	1.05	26–38	25–60	1800–2500	70	0.18	1.3
ABS	1.04–1.06	32–45	15–30	1300–2700	60–110	0.18	1.3
ASA	1.04	32	40	1100–2600	80–110	0.18	1.3
PVC-U	1.38–1.55	50–75	10–50	1000–3500	70–80	0.14–0.17	0.85–0.9
PVC-P	1.16–1.35	10–25	170–400	–	150–210	0.15	0.9–1.8
PTFE	2.15–2.20	25–36	350–550	410	120–250	0.25	1.0
PFEP	2.12–2.17	22–28	250–330	350	80	0.25	1.12
PCTFE	2.10–2.12	32–40	120–175	1050–2100	60	0.22	0.9
PETFE	1.7	35–54	400–500	1100	40	0.23	0.9
PMMA	1.17–1.20	50–77	2–10	1600–3600	70–90	0.18	1.47
POM	1.41–1.42	62–70	25–70	1600–3200	90–110	0.25–0.30	1.46
PA 6	1.13	70–85	200–300	1400	60–100	0.29	1.7
PA 66	1.14	77–84	150–300	2000	70–100	0.23	1.7
PA 11	1.04	56	500	1000	130	0.23	1.26
PA 12	1.02	56–65	300	1600	110	0.23	1.26
PA 6-3-T	1.12	70–84	70–150	2000	80	0.23	1.6
PA 612	1.06–1.07	–	–	450–2500	–	–	–
PA 46	1.10	–	–	1850–3300	–	–	–

Plastic material	Density g/cm^3 ISO 1183	Mechanical characteristics			Thermal characteristics		
		Tensile strength N/mm^2 ISO 527	Strain at break % ISO 527	Tensile modulus MPa ISO 527	Linear coefficient of expansion $K^{-1} \cdot 10^6$	Thermal conductivity W/mK	Specific heat kJ/kg K
PC	1.2	56–67	100–130	2100–2400	60–70	0.21	1.17
PET	1.37	47	50–300	3100	40–60	0.24	1.05
PBT	1.31	40	15	2000	60	0.21	1.30
PPE+PS	1.06	55–68	50–60	2500	60–70	0.23	1.40
PSU	1.24	50–100	25–30	2600–2750	54	0.28	1.30
PPS	1.34	75	3	3400	55	0.25	–
PES	1.37	85	30–80	2450	55	0.18	1.10
PAI	–	100	12	4600	36	0.26	
PEI	1.27	105	60	3000	62	0.22	
PI	1.43	75–100		3000–3200	50–60	0.29–0.35	
PEEK	1.32	90	50	3600	47	0.25	
CA	1.30	38	3	2200	120	0.22	1.6
CP	1.19–1.23	14–55	30–100	420–1500	110–130	0.21	1.7
CAB	1.18	26	4	1600	120	0.21	1.6
VF	1.1–1.45	85–100	–	–	–	–	–
PUR	1.05	70–80	3–6	4000	10–20	0.58	1.76
TPU	1.20	30–40	400–450	700	110–210	1.7	0.5
PF	1.4	25	0.4–0.8	5600–12,000	10–50	0.35	1.30
UF	1.5	30	0.5–1.0	7000–10,500	50–60	0.40	1.20
MF	1.5	30	0.6–0.9	4900–9100	50–60	0.50	1.20
UP	2.0	30	0.6–1.2	14,000–20,000	20–40	0.70	1.20
DAP	1.51–1.78	40–75	–	9800–15,500	10–35	0.60	–
SI	1.8–1.9	28–46	–	6000–12,000	20–50	0.3–0.4	0.8–0.9
PI	1.43	75–100	4–9	23,000–28,000	50–63	0.6–0.65	0.8
EP	1.9	30–40	4	21,500	11–35	0.88	

Table 8.9 Electrical Properties of Plastic Materials

Plastic material	Specific volume resistance Ω cm IEC 60093	Electric surface resistance Ω IEC 60093	Dielectric constant IEC 60250		Dielectric loss factor tan δ IEC 60250		Dielectric strength kV/cm IEC 60243-2	Comparative Tracking Index IEC 60112		
			50 Hz	10^6 Hz	50 Hz	10^6 Hz		KA	KB	KC
PE-LD	> 10^{17}	10^{14}	2.29	2.28	$1.5 \cdot 10^{-4}$	$0.8 \cdot 10^{-4}$	–	3b	> 600	> 600
PE-HD	> 10^{17}	10^{14}	2.35	2.34	$2.4 \cdot 10^{-4}$	$2.0 \cdot 10^{-4}$	–	3c	> 600	> 600
EVAC	< 10^{15}	10^{13}	2.5–3.2	2.6–3.2	0.003–0.02	0.03–0.05	620–780	–	–	–
IM	> 10^{16}	10^{13}			$6–10 \cdot 10^{-4}$	$6/10 \cdot 10^{-4}$				
PVK	> 10^{16}	10^{14}	–	3	$< 4 \cdot 10^{-4}$	$< 5 \cdot 10^{-4}$	–	3b	> 600	> 600
PP	> 10^{17}	10^{13}	2.27	2.25	$7 \cdot 10^{-4}$	$6 \cdot 10^{-4}$	500–650	3c	> 600	> 600
PB	> 10^{17}	10^{13}	2.5	2.2	$4 \cdot 10^{-4}$	–	–	3c	> 600	> 600
PIB	> 10^{15}	10^{13}	2.3	–	$7 \cdot 10^{-5}$	$3 \cdot 10^{-5}$	–	3c	> 600	> 600
PMP	> 10^{16}	10^{13}	2.12	2.12			700	3c	> 600	> 600
PS	> 10^{16}	> 10^{13}	2.5	2.5	$1–4 \cdot 10^{-4}$	$0.54 \cdot 10^{-4}$	300–700	1–2	140	150–250
SAN	> 10^{16}	> 10^{13}	2.6–3.4	2.6–3.1	$6–8 \cdot 10^{-3}$	$7–10 \cdot 10^{-3}$	400–500	1–2	160	150–260
SB	> 10^{16}	> 10^{13}	2.4–4.7	2.4–3.8	$4–20 \cdot 10^{-4}$	$4–20 \cdot 10^{-4}$	300–600	2	> 600	> 600
ABS	> 10^{15}	> 10^{13}	2.4–5	2.4–3.8	$3–8 \cdot 10^{-3}$	$2–15 \cdot 10^{-3}$	350–500	3a	> 600	> 600
ASA	> 10^{15}	> 10^{13}	3–4	3–3.5	0.02–0.05	0.02–0.03	360–400	3a	> 600	> 600
PVC-U	> 10^{15}	10^{13}	3.5	3.0	0.011	0.015	350–500	2–3b	600	600
PVC-P	> 10^{11}	10^{11}	4.8	4–4.5	0.08	0.12	300–400	–	–	–
PTFE	> 10^{18}	10^{17}	< 2.1	< 2.1	$< 2 \cdot 10^{-4}$	$< 2 \cdot 10^{-4}$	480	3c	> 600	> 600
PFEP	> 10^{18}	10^{17}	2.1	2.1	$< 2 \cdot 10^{-4}$	$< 7 \cdot 10^{-4}$	550	3c	> 600	> 600
PCTFE	> 10^{18}	10^{16}	2.3–2.8	2.3–2.5	$1 \cdot 10^{-3}$	$2 \cdot 10^{-2}$	550	3c	> 600	> 600
PETFE	> 10^{16}	10^{13}	2.6	2.6	$8 \cdot 10^{-4}$	$5 \cdot 10^{-3}$	400	3c	> 600	> 600

Plastic material	Specific volume resistance Ω cm IEC 60093	Electric surface resistance Ω IEC 60093	Dielectric constant IEC 60250		Dielectric loss factor tan δ IEC 60250		Dielectric strength kV/cm IEC 60243-2	Comparative Tracking Index IEC 60112		
			50 Hz	10^6 Hz	50 Hz	10^6 Hz		KA	KB	KC
PMMA	$> 10^{15}$	$> 10^{15}$	3.3–3.9	2.2–3.2	0.004–0.04	0.004–0.04	400–500	3c	> 600	> 600
POM	$> 10^{15}$	$> 10^{13}$	3.7	3.7	0.005	0.005	380–500	3b	> 600	> 600
PA 6	10^{12}	10^{10}	3.8	3.4	0.01	0.03	400	3b	> 600	> 600
PA 66	10^{12}	10^{10}	8.0	4.0	0.14	0.08	600	3b	> 600	> 600
PA 11	10^{13}	10^{11}	3.7	3.5	0.06	0.04	425	3b	> 600	> 600
PA 12	10^{13}	10^{11}	4.2	3.1	0.04	0.03	450	3b	> 600	> 600
PA 6-3-T	10^{11}	10^{10}	4.0	3.0	0.03	0.04	350	3b	> 600	> 600
PA 612	$> 10^{15}$	10^{13}	4.0	3.6	–	–	–	–	–	–
PA 46	$> 10^{15}$	10^{13}	4.0	3.8	–	–	–	–	–	–
PC	$> 10^{17}$	$> 10^{15}$	3.0	2.9	$7 \cdot 10^{-4}$	$1 \cdot 10^{-2}$	380	1	120–	260–
PET	10^{16}	10^{16}	4.0	4.0	$2 \cdot 10^{-3}$	$2 \cdot 10^{-2}$	420	2	160 –	300 –
PBT	10^{16}	10^{13}	3.0	3.0	$2 \cdot 10^{-3}$	$2 \cdot 10^{-2}$	420	3b	420	380
PPE+PS	10^{16}	10^{14}	2.6	2.6	$4 \cdot 10^{-4}$	$9 \cdot 10^{-4}$	450	1	300	300
PSU	$> 10^{16}$	–	3.1	3.0	$8 \cdot 10^{-4}$	$3 \cdot 10^{-3}$	425	1	175	175
PPS	$> 10^{16}$	–	3.1	3.2	$4 \cdot 10^{-4}$	$7 \cdot 10^{-4}$	595	–	–	–
PES	10^{17}	–	3.5	3.5	$1 \cdot 10^{-3}$	$6 \cdot 10^{-3}$	400	–	–	–
PAI	10^{17}	–	–	–	–	–	–	–	–	–
PEI	10^{18}	–	–	–	–	–	–	–	–	–
PI	$> 10^{16}$	$> 10^{15}$	–	–	–	–	–	1	> 300	> 380
PEEK	$5 \cdot 10^{16}$	–	–	–	$3 \cdot 10^{-3}$	–	–	–	–	–

Table 8.9 Electrical Properties of Plastic Materials *(continued)*

Plastic material	Specific volume resistance Ω cm IEC 60093	Electric surface resistance Ω IEC 60093	Dielectric constant IEC 60250		Dielectric loss factor tan δ IEC 60250		Dielectric strength kV/cm IEC 60243-2	Comparative Tracking Index IEC 60112		
			50 Hz	10^6 Hz	50 Hz	10^6 Hz		KA	KB	KC
CA	10^{13}	10^{12}	5.8	4.6	0.02	0.03	400	3a	> 600	> 600
CP	10^{16}	10^{14}	4.2	3.7	0.01	0.03	400	3a	> 600	> 600
CAB	10^{16}	10^{14}	3.7	3.5	0.006	0.021	400	3a	> 600	> 600
VF	10^{10}	10^{8}	–	–	0.08	–	–	–	–	–
PUR	10^{16}	10^{14}	3.6	3.4	0.05	0.05	240	3c	–	–
TPU	10^{12}	10^{11}	6.5	5.6	0.03	0.06	300–600	3a	> 600	> 600
PF	10^{11}	$> 10^{8}$	6	4.5	0.1	0.03	300–400	1	140–180	125–175
UF	10^{11}	$> 10^{10}$	8	7	0.04	0.3	300–400	3a	> 400	> 600
MF	10^{11}	$> 10^{8}$	9	8	0.06	0.03	290–300	3b	> 500	> 600
UP	$> 10^{12}$	$> 10^{10}$	6	5	0.04	0.02	250–530	3c	> 600	> 600
DAP	10^{13}–10^{16}	10^{13}	5.2	4	0.04	0.03	400	3c	> 600	> 600
SI	10^{14}	10^{12}	4	3.5	0.03	0.02	200–400	3c	> 600	> 600
PI	$> 10^{16}$	$> 10^{15}$	3.5	3.4	$2 \cdot 10^{-3}$	$5 \cdot 10^{-3}$	560	1	> 300	> 380
EP	$> 10^{14}$	$> 10^{12}$	3.5–5	3.5–5	0.001	0.01	300–400	3c	> 300	200–600

Table 8.10 Temperature Characteristics of Plastic Materials

Plastic material	Glass fiber content %	HDT/A °C	Service temperature °C			T_S/T_M °C	T_g °C
			max. short-term	max. continuous	min. continuous		
PE-LD	0	ca. 35	80–90	60–75	−50	110	−30
PE-HD	0	ca. 50	90–120	70–80	−50		
PE-UHMW	0	ca. 50	150	100	−260		
PE-X	0	40–60	200	120			
EVAC	0		65	55	−60		66
COC	0						60–180
EIM	0		120	100	−50		
PP	0	55–70	140	100	0 to −30	160–170	0 to −10
PP	30	120	155	100	0 to −30	160–170	0 to −10
PB	0	55–60	130	90	0		
PIB	0		80	65	−40		−70
PMP	0	40	180	120	0	245	
PDCPD	0	90–115					
PS	0	65–85	75–90	60–80	−10		95–100
PS-(M)	0	95				270	
SAN	0	95–100	95	85	−20		110
SB	0	72–87	60–80	50–70	−20		
ABS	0	95–105	85–100	75–85	−40		80–110
ASA	0	95–105	85–90	70–75	−40		100
PVC-U	0	65–75	75–90	65–70	−5		85
PVC-C	0		100	85			85
PVC-P	0		55–65	50–55	0 to −20		ca. 80
PVK	0	150–170	170	150	−100		173
PTFE	0	50–60	300	260	−270	327	127

Table 8.10 Temperature Characteristics of Plastic Materials *(continued)*

Plastic material	Glass fiber content %	HDT/A °C	Service temperature °C			T_S/T_M °C	T_g °C
			max. short-term	max. continuous	min. continuous		
PCTFE	0	65–75	180	150	−40		
PVDF	0	95–110		150	−60	140	40
PVF	0			120	−60	198	−20
ECTFE	0	75	160	140	−75	190	45
ETFE	0	75	200	155	−190	270	
ETFE	25	210	220	200		270	
FEP	0		250	205	−200	290	
PFA	0	45–50	250	200	−200		
THV	0			130	−50	160–180	
PPE mod.	0	135	120–130	100–110			
PPE mod.	30	160					
PPE+PA 66	0		210				
PMMA	0	75–105	85–100	65–90	−40		105–115
AMMA	0	73	80	70			80
POM-H	0	100–115	150	110	−40	175	25
POM-H	30	160	150	110	−60	175	25
POM – copolymer	0	110–125	110–140	90–110		165	
POM – copolymer	30	160	110–150	90–110		165	
PA 6	0	55–85				220	55
PA 6	30	190–215	140–180	80–110	−30	220	55
PA 11	0	55	140–150	70–80	−70	185	50
PA 610	0	90	140–180	80–110		215	55–60

Plastic material	Glass fiber content %	HDT/A °C	Service temperature °C			T_S/T_M °C	T_g °C
			max. short-term	max. continuous	min. continuous		
PA 612	0		130–150	80–100			55–60
PA MXD6	30	228	190–230	110–140		240	85–100
PA 6-3-T	0	120	130–140	80–100	−70	240	150
PA 6T	0		120–130	70–90		(500)	
PA 6/6T	30	250		155		295	115
PA PACM 12	0	105		ca. 100		250	140
PA 6T/6I	0					330	130
PA PDA-T	0					500	
PPA	0	120					
PPA	33	270		> 160			
PPA	45	290					
PMPI	0		260				
PPTA	0		> 250	> 200			
PC	0	125–135	115–150	115–130	−150	220–260	150
PC	30	135–150	115–150	115–130	−150		
PC-TMC	0			150			150–235
PC+ABS	0	105					
PC+ASA	0	109					
PET	0	80	200	100–120	−20	255	98
PET	30	200–230	220	150		255	98
PBT	0	65	165	100		255	60
PBT	30	200–210	220	150		225	60
PET+PS	0		200	100	−20		

Table 8.10 Temperature Characteristics of Plastic Materials *(continued)*

Plastic material	Glass fiber content %	HDT/A °C	Service temperature °C			T_S/T_M °C	T_g °C
			max. short-term	max. continuous	min. continuous		
PBT+PS	0		165	100	−30		
PAR	0	155–175	170	150		420	
PPS	0	135	300	200–240		285	85
PPS	30	255	300	200–240		285	85
PES	0	200–215	180–260	160–200			225
PES	30	210–225	180–260	160–200			225
PSU	0	170–175	170	160	−100		190
PSU	30	185	180	160	−100		190
PPSU	0						221
PPE mod.	0	135	120–130	100–110			
PPE mod.	30	160					
PPE+PA 66	0		210				
PPE+PS	0	115–130	120	100	−30		140
PPE+PS	30	137–144	130	110	−30		140
PAEK	0	200				380	170
PAEK	30	320		240–250		380	140–170
PEK. PEKEKK	0	170	300	260		365–380	175
PEEK	0	140	300	250		335–345	145
PEEK	30	315	300	250		335–345	145
PEEEK	0					324	110
PEEKEK	0					345	148

Plastic material	Glass fiber content %	HDT/A °C	Service temperature °C			T_S/T_M °C	T_g °C
			max. short-term	max. continuous	min. continuous		
PEEKK	0	103	260	220		365	167
PEEKK	30	165	300	250		365	167
PEKEK	0					384	160
PEKK	0		350	260		391	165
PI	0	280–360	400	260			250–270
PI	30	360	400	260			250–270
PI – molding compound	various		400	260	−240		
PBMI	40	> 300	250	190			
PBO	0			< 500		525	
PAI	0	280	300	260	−260		240–275
PAI	30		300	260	−260		240–275
PEI	0	190–200		170	−170		215
PEI	30	195–215	180	170	−170		215
PISO	0		> 250	210		250–350	273
PMI (foam)	0			180			
PMMI	0	130–160		120–150			
PESI	0			200			
LCP	0	170		220		285	
LCP	30	230		220		285	
LCP	0	180–240		185–250		275–330	160–190
LCP-A	50	235					
LCP-C	50	250					

Table 8.10 Temperature Characteristics of Plastic Materials *(continued)*

Plastic material	Glass fiber content %	HDT/A °C	Service temperature °C			T_S/T_M °C	T_g °C
			max. short-term	max. continuous	min. continuous		
PUR – cast resin	0		70–100	50–80			15–90
PF Type 31, 51. 74. 84	various	160–170	140	110–130			
PF Type 13	various	170	150	120			
PF Type 4111	various	240		170			
UF	various		100	70			
MF Type 150/52	various	155	120	80			
MF Type 156	various	180					
MPF 1206	various	190		160			
MPF Type 4165	20–30	165					
UP Type 802/4	10–20	250	200	150			
UP Type 3620	various	110					
UP Type 3410	various	270		200			
EP Type 891	ca. 20		180	130			
EP Type 8414	25–35	150	180	130			
PDAP	various	160–280	190–250	150–180	–50		
SI	30	480	250	170–180	–50		

8.3 Electrical Characteristics

Table 8.11 Electrical Characteristics of Conductive Plastic Compounds

Plastic material	Filler	Filler content %	Specific volume resistance ρ_e Ω · cm	Specific surface resistance σ_e Ω	Shield damping 30–1000 MHz dB
PE	Carbon black	7	7		18–20
PP	Carbon black	20	10–20		
PS	Carbon black	25	10^1		
ABS	Al chips	30	ca. 10		
PVC	Carbon black	7	2		35–40
PVC	Polyaniline	30	1		> 40
POM	Carbon black		20	$3 \cdot 10^3$	
PA 66	C-fibers, nickel plated	40	10^{-2}–10^{-1}		64–78
PA 66	Steel fibers	5	1–10		30–40
PA 66	Steel fibers	15	10^{-2}–10^{-1}		43–68
PA 66	Polyaniline	30	4		
PC	C-fibers	10	10^1–10^2	10^5–10^6	
PC	C-fibers	20	10^1–10^2	10^1–10^4	
PC	C-fibers	30	10^0–10^1	10^1–10^4	
PC	C-fibers	40	10^0–10^1	10^0–10^1	
PC	Steel fibers	5	1–10		28–36
PC	Steel fibers	15	10^{-2}–10^{-1}		39–58
PET	C-fibers	30	10^0–10^1	10^1	
PBT	C-fibers	30	10^0–10^1	10^1	
PPS	C-fibers	40	10^0–10^1	10^1	
PPS	Carbon minerals		400	$5 \cdot 10^2$	
PPS	C-fibers, nickel plated		10^{-2}–10^{-1}		45–70
PEI	C-fibers	30	10^1	10^1–10^4	

Table 8.12 Conductivity of Intrinsically Conductive Plastic Materials, ICPs

Plastic material	Acronym	Doping	Specific volume resistance ρ_e Ω·cm
Polypyrrole	PPY	BF_4	0.01
Polypyrrole	PPY	SO_3	0.006
Polyacetylene	PAC	AsF_5	0.001
Polyacetylene, stretched	PAC	AsF_5	0.003
Polyaniline	PANI	BF_4	0.1
Poly(p-phenylene)	PPP	AsF_5	0.002
Poly(p-phenylene)	PPP	Na	0.00033
Polyfuran	PFU	CF_3SO_3	0.02
Poly(p-phenylvinylidene) stretched	PPV	AsF_5	0.002
Polythiophene	PT	CF_3SO_3	0.05

8.4 Optical Characteristics

Table 8.13 Properties of Transparent Plastic Materials

Material	Light transmission	Index of refraction n_D at 20 °C	Young's modulus MPa	Yield stress MPa	Density g/cm^3
Crown glass	Crystal clear	1.4–1.6			
Flint glass	Crystal clear	1.53–1.59			
Water	Crystal clear	1.33			
PE	Transparent to opaque	1.51	200–1400	8–30	0.915–0.96
EIM	Transparent	1.51	150–200	7–8	0.94–0.95
COC	Crystal clear	1.53	3100		1.02
PP	Transparent to opaque	1.5	800–1100		0.9
PMP	Crystal clear to opaque	1.46	1200–2000	10–15	0.83–0.84

Material	Light transmission	Index of refraction n_D at 20 °C	Young's modulus MPa	Yield stress MPa	Density g/cm³
PS	Crystal clear	1.58–1.59	3100–3300	42–65	1.05
SB	Opaque		2000–2800	25–45	1.03–1.05
ABS	Transparent to opaque	1.52	220–3000	45–65	1.03–1.07
SAN	Crystal clear to opaque	1.57	3600–3900	70–85	1.08
SMMA	Transparent		3400	70–83	1.08–1.13
SMSA	Transparent		3500	60	1.07–1.17
SBS	Crystal clear		1100–1900		1.0–1.2
PVC-U	Crystal clear to opaque	1.52–1.54	2900–3000		1.37
PVC-HI	Crystal clear to opaque		2300–3000	40–55	1.36
PTFE	Opaque	1.35	400–700		2.13–2.33
PVDF	Transparent to opaque	1.42	2000–2900	50–60	1.76–1.78
PCTFE	Opaque	1.43	1300–1500		2.07–2.12
PBA	Transparent	1.467			
PMMA	Crystal clear	1.49	3100–3300	62–75	1.18
MBS	Transparent		2000–2600		1.11
MABS	Transparent		2000–2100		1.08
PMMA-HI	Crystal clear		600–2400	20–60	1.12–1.17
POM	Opaque	1.49	2800–3200	60–75	1.39–1.42
PA 6/11/12/66	Transparent to opaque	1.52–1.53			
PA 6-3-T cond.	Transparent	1.57	2800–3000	80–90	1.12
PC	Crystal clear	1.58–1.59	2400	55–65	1.2
PBT	Opaque	1.55	2500–2800		1.30–1.32
PET-A	Crystal clear to transparent	1.57	2100–2400	55	1.34

Table 8.13 Properties of Transparent Plastic Materials *(continued)*

Material	Light transmission	Index of refraction n_D at 20 °C	Young's modulus MPa	Yield stress MPa	Density g/cm³
PET	Transparent to opaque		2800–3000	60–80	
PET-G	Opaque		1900–2100		1.23–1.26
APE (PEC)	Transparent	1.57–1.58	2300	65	1.15–1.18
PSU	Transparent to opaque	1.63	2500–2700	70–80	1.24–1.25
PES	Transparent to opaque	1.65	2600–2800	80–90	1.36–1.37
PPE+PS	Opaque		1900–2700	45–65	1.04–1.06
PF	Transparent	1.63			
UP	Crystal clear	1.54–1.58			
EP	Crystal clear to opaque	1.47			
CA	Crystal clear	1.47–1.50	1000–3000	25–55	1.26–1.32
CAB	Crystal clear	1.48	800–2300	20–55	1.16–1.22
CAP	Crystal clear	1.47	1000–2400	20–50	1.17–1.24
CAP	Crystal clear to transparent		1000–2100		1.19–1.22

8.5 Resistance to Environmental Influences

8.5.1 Water, Humidity

Table 8.14 Water Absorption (to Saturation) and Coefficients of Diffusion for Water

Plastic material	Water absorption		Coefficient of diffusion, D
	Standard climate 23 °C/50% r. h.	Water 23 °C	10^{-6} mm²/s
PE-LD		0.002–0.2	ca. 0.14
PE-HD		0.002–0.2	ca. 0.74
PP		> 0.02	ca. 0.24
PMP		ca. 0.05	
PS		0.2–0.3	
SAN		ca. 0.2	
ABS		ca. 0.7	
PVC-(E)	ca. 0.18	0.5–3.5 (60 °C)	
PVC-(S)		0.3 (60 °C)	
PMMA		1.6–2	
POM	ca. 0.3	ca. 0.6	
PPE + PS		ca. 0.15	
PPE + PS-GF	ca. 0.03	ca. 0.15	
PC	ca. 0.2	ca. 0.4	
PET	ca. 0.35	0.5–0.7	
PET-GF 33	ca. 0.2	0.25	
PBT	ca. 0.45	ca. 0.45	
PBT-GF 33	0.1–0.2	0.1–0.2	
PSU	ca. 0.25	ca. 0.6	
PA 6	2.8–3.6	9–10	ca. 0.4
PA 9		ca. 2.5	

Table 8.14 Water Absorption (to Saturation) and Coefficients of Diffusion for Water *(continued)*

Plastic material	Water absorption		Coefficient of diffusion, D
	Standard climate 23 °C/50% r. h.	Water 23 °C	10^{-6} mm²/s
PA 11	0.8–1.2	ca. 1.8	
PA 12	0.7–1.1	1.3–1.9	
PA 66	2.5–3.5	7.5–9	ca. 0.2
PA 68	ca. 3	4–4.5	
PA 610	1.5–2	3–4	
PA 612	1.3–2	2.5–2.8	
PA 6-3-T	2.6–3	6.2–7	
PA cast	2.3–2.7	7–8	ca. 0.32
PA 6-GF 30	1.5–2	ca. 6	ca. 0.4
PA 66-GF 30	1–1.5	ca. 5.5	ca. 0.2
PA 11-GF 30	ca. 0.54	ca. 1.2	
PA 12-GF 30	ca. 0.45	ca. 1.1	
CA		3.8–5	
CAB		2–2.5	
CP		2.3–2.7	
UP		ca. 0.4	
UP-GF		0.5–2.5	
EP	0.5–0.8	0.7–1.5	0.2–0.3
EP-GF 55	0.3–0.5	ca. 0.8	
PAI		0.22–0.28	
PVDF		ca. 0.25	
PES	ca. 0.15	ca. 2.1	
PI	ca. 1.2	ca. 3	

8.5.2 Chemical Resistance

Table 8.15 Chemical Resistance

Plastic material acronym	Water	Acids, bases					Inorganic salts	Halogens	Oxid. compounds	Solvents										Fuels	Mineral oils	Fats, oils
		Weak	Strong	Hydrofluoric acid	Weak	Strong				Paraff. hydrocarbons	Halogen-alkanes	Alcohols	Ethers	Esters	Ketones	Aldehydes	Amines	Org. acids	Aromat. compounds			
PE-LLD	+	+	+	+	+	+	+	−	−	⊗	−	○	−	○	○			+	−	⊗	○	⊕
PE-LD	+	+	+	+	+	+	+	−	−	⊗	−	+	○	⊕	⊕	⊕	+	+	⊕	⊕	⊕	⊕
PE-HD	+	+	+	+	+	+	+	−	−		⊗	+	○	+	+			+	⊕	⊕	⊕	+
PE-C	+	+	+		+	+	+	○	○	⊕	⊗	+	○	−	○	⊗	+	⊕	⊗	○		○
EVA		+	⊗		+	+	+	−	⊗	⊗	−	⊕			−	+	⊕	⊕		○	○	
PIB	+	+	+	+	+	+	+	○	○	−	−	+	−	−	⊗	⊕	+	+	−	−	−	
PP	+	+	⊕	⊕	+	+	+	⊗	−	⊕	⊗	+	○	⊕	⊕	+	+	⊕		⊕	+	+
PMP	+	+	+		+	+	+	−	⊗	−	⊗	⊕	−	⊗	○	+		⊗	⊕	○	⊕	+
PS	+	+	⊕	⊕	⊕	+	+	−	○	⊗	−	+	⊗	−	−	⊗	+	⊕	⊗	⊗	○	+
SB	+	+	○	○	+	+		−	−	⊕	−		−	−	−				−	⊗	○	+
SAN	+	+	⊕	+	+	+	+	−	−	⊕	−	⊕	−	−	−	⊗	+	⊕	−	⊕	+	+
ABS	+	+	⊕	+	+	+	+	−	⊗	○	−	⊕	−	−	−	⊗	+	+	−	+	+	+

Table 8.15 Chemical Resistance *(continued)*

Plastic material acronym	Water	Acids, bases					Inorganic salts	Halogens	Oxid. compounds	Solvents										Fuels	Mineral oils	Fats, oils
		Weak	Strong	Hydrofluoric acid	Weak	Strong				Paraff. hydrocarbons	Halogen-alkanes	Alcohols	Ethers	Esters	Ketones	Aldehydes	Amines	Org. acids	Aromat. compounds			
PVC-U	+	+	+	+	⊕	+	+	○	⊕	⊕	⊗	+	⊗	–	–	⊗	⊕	⊕	⊗	⊗	+	+
PVCVAC		+	○		+		+		○		–	+	⊕	–	–	+	+		–	+		+
PVC-P	+	+	⊕		+	○	+	○	⊕		–	⊗	⊗	–	–	+	+	⊕	–	–	○	○
PTFE	+	+	+	+	+	+	+	+	+	+	+	+	+	+	+	+	+	+	+	+	+	+
PCTFE	+	+	+	+	+	+	+	⊕	⊕		⊗	+	–	⊗	+	+		⊗	⊗	+	+	+
PMMA	+	⊕	⊗	○	+	+	+	○	○	⊕	–	○	⊕	–	○	+	+	⊗	⊗	⊕x	+	+
AMMA	+	+	+	○	⊕	⊕		○	○	⊗		+	+	+	–				⊕	+	+	+
POM	+	⊕	–		+	+	+	–	⊗	+	⊗	+	+	⊕	+	⊕		⊕	⊕	+	+	+
PPE + PS	+	+	+		+	+	+											+				
CA	+	+	–	–	○	–	+	–	–	+	⊗	○	+	–	–	○	⊗	⊗	⊕	⊕	+	+
CTA	+	⊕	–	–	○	–	+	–	–	+	⊗	○	+	–	–	+	⊕	⊗	⊕	⊕	+	+
CAB	+	+	–	–	⊕	○	+	–	⊗	+	–	⊗		–	–	+	⊕	⊕	–	+	+	+

Plastic material acronym	Water	Acids, bases					Inorganic salts	Halogens	Oxid. compounds	Solvents										Fuels	Mineral oils	Fats, oils
		Weak	Strong	Hydrofluoric acid	Weak	Strong				Paraff. hydrocarbons	Halogen-alkanes	Alcohols	Ethers	Esters	Ketones	Aldehydes	Amines	Org. acids	Aromat. compounds			
CP	+	○	–	–	–	–	+	–	–	+	–	–	–	–	–	+	⊕	⊕	–	+	+	+
PC	+	+		⊕	–	–	⊕	+	○	⊕	–	⊕	–	○	○	–	–	○	–	⊕	+	+
PET		+	⊕		⊕		+		+	+	⊗	⊕	+	+	+			+	+	+	+	+
PBT		○	⊗		+	+	+			+	⊗	+	+	○				⊕	⊕	+	+	+
PA 6	+	–	–	–	⊕	+	+	–	–	⊕	⊕	⊕	+	+	+	⊕	+	⊕	⊕	+	+	+
PA 12	+	–	–	–	⊕	⊕	+	–	–	+	⊕	⊕	+	+	+	○	+	⊕	⊕	+	+	+
PA 66	+	–	–	–	⊕	+	+	–	–	⊕	⊕	⊕	+	+	+	⊕	+	⊕	⊕	+	+	+
PA 610	+	–	–	–	⊕	+	+	–	–	⊕	⊗	⊕	+	+	+	⊕	+	⊕	⊕	+	+	+
PA arom.	+	⊗	⊗	○	⊕	+	+		+	+	⊕	⊗	+	+	⊕	○		⊗	⊕	+	+	+
PSU	⊕	+	+		+	+	+		+	⊕	–	+			–				–			
PF	+	○	–		○	–					⊕	⊕	⊕	⊕	⊕				⊕	⊕	+	+
UF	+	⊗	–		○	–					⊕	⊕	+	⊕	⊕				⊕	⊕	+	+
MF	+	○	–		⊕	–					⊕	+	⊕	⊕	⊕				⊕	⊕	+	+

Table 8.15 Chemical Resistance *(continued)*

Plastic material acronym	Water	Acids, bases					Inorganic salts	Halogens	Oxid. compounds	Solvents										Fuels	Mineral oils	Fats, oils
		Weak	Strong	Hydrofluoric acid	Weak	Strong				Paraff. hydrocarbons	Halogen-alkanes	Alcohols	Ethers	Esters	Ketones	Aldehydes	Amines	Org. acids	Aromat. compounds			
UP	+	⊕	○	–	⊕	–	+			+										+	+	+
EP	+	⊕		+	⊕	⊕	+	+	–		○	+	+		⊕		⊕	+		+	+	+
TPU	+	⊕	○	–	⊕	⊕	+	–		+	⊗	⊕	⊕	⊕	+	+	+	○	⊕	+	+	+

+ resistant, ⊕ resistant to limited resistance, ○ limited resistance, ⊗ limited resistance to not resistant, – not resistant

8.5.3 Stress Cracking Resistance

Table 8.16 Recommended Media for Stress Cracking Tests for Selected Plastic Materials

Plastic material – acronym	Media causing stress cracking	Exposure time
PE	Surfactant solution (2%), 50 °C	> 50 h
	Surfactant solution (2%), 70 °C	48 h
	Surfactant solution (5%), 80 °C	4 h
PP	Chromic acid, 50 °C	
PS	n-Heptane	
	Petroleum/gasoline, boiling range 50–70 °C	
	n-Heptane : n-propanol (1:1)	
SB	n-Heptane	
	Petroleum/gasoline, boiling range 50–70 °C	
	n-Heptane : n-propanol (1:1)	
	Oleic acid	
SAN	Toluene : n-propanol (1:5)	15 min
	n-Heptane	
	Carbon tetrachloride	
ABS	Dioctyl phthalate	
	Toluene : n-propanol (1:5)	
	Methanol	15 min
	Acetic acid (80%)	20 min
	Toluene	1 h
PMMA	Toluene : n-heptane (2:3)	15 min
	Ethanol	
	N-Methylformamide	
PVC	Methanol	
	Methylene chloride	30 min
	Acetone	3 h

Table 8.16 Recommended Media for Stress Cracking Tests for Selected Plastic Materials *(continued)*

Plastic material – acronym	Media causing stress cracking	Exposure time
POM	Sulfuric acid (50%), local wetting	Up to 20 min
PC	Toluene : n-propanol (1:3 to 1:10)	3–15 min
	Carbon tetrachloride	1 min
	Caustic soda (5%)	1 h
PC + ABS	Methanol : ethyl acetate (1:3)	
	Methanol : acetic acid (1:3)	
	Toluene : n-propanol (1:3)	
PPE + PS	Tributyl phosphate	10 min
PBT	Caustic soda	
PA 6	Zinc chloride solution (35%)	20 min
PA 66	Zinc chloride solution (50%)	1 h
PA 6-3-T	Methanol	
	Acetone	1 min
PSU	Ethylene glycol monoethyl ether	1 min
	Ethyl acetate	
	1,1,1-Trichloroethane : n-heptane (7:3)	
	Methyl glycol acetate	
	Carbon tetrachloride	
	1,1,2-Trichloroethane	1 min
	Acetone	1 min
PES	Toluene	1 min
	Ethyl acetate	1 min
PEEK	Acetone	
PAR	Caustic soda (5%)	1 h
	Toluene	1 h
PEI	Propylene carbonate	

8.5.4 Environmental Influences

Table 8.17 Resistance of Plastic Materials

Plastic material	Weathering		Resistance to	
	Non-stabilized	Stabilized	Microorganisms	Macroorganisms
PE-LD	3	1–2	1–2	2–3
PE-HD	3	1–2	1–2	2–3
PP	3	2	1–2	2–3
PB	3	2		
PMP	3			
PS	3	2	1	2–3
SB	3			
SAN	3			
ABS	3	2	1	2–3
ASA	3	2		
PVC-U	2	1–2	1–2	1
PVC-P	3	2	2–3	3
PVDC	2–3			
PTFE	1		1	1–2
PCTFE	1		1	1–2
PVF			1	1–2
PVDF	1			
PMMA	1		1–2	1–2
POM	3			
PPO	3			
CA	2–3		2–3	3
CAB	2–3		2–3	3
PC	2			
PET	2			

Table 8.17 Resistance of Plastic Materials *(continued)*

Plastic material	Weathering		Resistance to	
	Non-stabilized	Stabilized	Microorganisms	Macroorganisms
PA	3		1–2	1–2
PSU	3			
PI	3			
PUR	3		3	1
GF-UP	1–2		1–2	1
EP	2		1	1–2
PF	1–2		1–3	1
MF	1–2		2–3	1
UF	3		1–3	3
Resistance: 1: excellent, 2: average, 3: little resistance				

8.5.5 Migration and Permeation

Table 8.18 Water Vapor and Gas Permeability

Plastic material	Temperature	Film thickness	Water vapor	N_2	Air	O_2	CO_2	H_2	Ar	He	CH_4
	°C	µm	$g/m^2 \cdot d$	$cm^3/m^2 \cdot d \cdot bar$							
PE-LD	23	100	1	700	1100	2000	10,000	8000			
PE-LD	20	25	5			5400 (23 °C)					
PE-HD	25	40	0.9	525	754	1890	7150	6000			
PE-HD	23	100	0.6–1	700	1100	1600–2000	10,000	8000			
EVAC. 20% VAC	23	100	455	1400		4000	17,000				
EVAL	20	20				0.2–1.8					
EIM	25	25	25			9300					
PP	23	100	0.7–0.8			600					
PP	20	25	ca. 3.5	2300 (23 °C)							
PP-O	20	25	ca. 1.3								
PP	25	40	2.1	430	700	1900	6100	17,700	1480	1920	
PS	23	100	12	2500		1000	5200				
PS-O	25	50	14	27	80	235	800	1260			

Table 8.18 Water Vapor and Gas Permeability *(continued)*

Plastic material	Temperature	Film thickness	Water vapor	N_2	Air	O_2	CO_2	H_2	Ar	He	CH_4
ABS	23	100	27–33	100–200		400–900					
ASA	20	100	30–35	60–70		150–180	6000–8000	50			100–110
PS-HI	23	100	13	4000		1600	10,000				
SBS	23	250	6	230		830	4300				
PVC-U	23	100	2.5	2.7–3.8		33–45	120–160				
PVC-U	20	40	7.6	12	28	87	200				
PVC-P	20	40	20	350	550	1500	8500				
PVDC	25	25	0.1–0.2	1.8–2.3	5–10	1.7–11	60–700	630–1400			
PVDF	23	60	2.4 (90 µm)	28		94		345		975	
PVF	23	25	50	3.8		4.7	170	900			
PCTFE	40	25	0.4–0.9	39		110–230	250–620	3400–5200			
ECTFE	23	25	9	150		39	1700				
ETFE	23	25	0.6	470		1560	3800				
PTFE	23	300	0.03	60–80	80–100	160–250	450–700			1700–2100	
PAN	20	20	ca. 80			15					

Plastic material	Tem-perature	Film thickness	Water vapor	N_2	Air	O_2	CO_2	H_2	Ar	He	CH_4
POM	23	80	12	5	8	24	470	210			12
PA 6 and 66	23	100	10–20	1–2		2–8	80–120				
PA 6	20	20	30			30 (23 °C)					
PA 11 and 12	23	100	2.4–4	0.5–0.7		2–3.5	6–13				
PC	23	25	4	680		4000	14,500	22,000			
PET	23	40	4.5–5.5 (50 °C)	6.6	12	30	140	850	16	1170	8
PET	23	100	5	4		25	90				
PET-0	23	25	0.6	9–15		80–110	200–340	1500			
PSU	23	25	6	630		3600	15,000	28,000			
PI	23	25	25	94		390	700	3800			
PBI	25	100	6	500		1800	6000	7100			
PF	20	40	45								
PF Type 31			300–560								
MF Type 152	20	40	400								
TPU	23	25	13–25	550–1600		1000–4500	6000–22,000				
CH	20	25	500			250 (23 °C)					

Plastic material	Tem-perature	Film thickness	Water vapor	N_2	Air	O_2	CO_2	H_2	Ar	He	CH_4
CH coated	20	25	5.5			150 (23 °C)					
CA	25	25	150–600	470–630	1800–2300	13,000–15,000	14,000				
CAB	25	25	460–600	3800		15,000	94,000				
SBS+PS 50/50	23	250	4.4	110		530	2300				
PET-G/BS+PS/PET-G 10/230/10	23	250	4		30,190	700					

Table 8.19 Oxygen Permeability of Barrier Films Depending on Temperature and Humidity

Film	Temperature	O_2 permeability in $cm^3/m^2 \cdot d \cdot bar$ at rel. humidity in %					
	°C	0	50	75	85	95	100
EVOH	23	0.1	0.24		1.5		15
PVDC	23	2.0	2.2	2.2	2.2	2.2	2.2
PAN	23	9	12	12	12	12	12
Composite film PE-LD/EVAL/PE-LD 20/10/30 µm	23	0.3	0.45	1.5		15	
	30	0.8	1.4	5		38	
	40	1.7	4	16		80	

Table 8.20 Permeation Values for Blends of Amorphous PA/EVOH and PA/PA 6; Film Thickness 25 µm

PA content: 0 → 45%	Mol-% ethylene in EVOH					
O_2 ($cm^3/m^2 \cdot d \cdot bar$)	32		38		44	
dry: 23 °C/35% *r. h.*	< 0.1 → 0.2		0.2 → 0.4		0.4 → 1.8	
humid: 23 °C/80% *r. h.*	1.1 → 2.3		2.9 → 4.0		5.3 → 5.7	
	% PA blended with PA 6					
O_2 ($cm^3/m^2 \cdot d \cdot bar$)	0	20	30	50	80	100
0 °C/0–5% *r. h.*	15	15	15	15	15	13
0 °C/95–100% *r. h.*	55	30	20	8	6	4
30 °C/0–5% *r. h.*	60	59	59	59	59	57
30 °C/95–100% *r. h.*	225	210	180	136	84	
H_2O ($g/m^2 \cdot d$)						
23 °C/95% *r. h.*	186	92	77	44	30	28

8.5.6 Flammability of Plastic Materials

Table 8.21 Flammability Classification According to UL 94

Plastic material acronym	With flame retardant	Classification UL 94	Specimen thickness mm
PE-LD		HB	
PE-LD	X	V2	
PE-HD		HB	
PE-HD	X	V2	
PP		HB	
PP	X	V2	
PS		HB	
SAN		HB	
SB		HB	
SB	X	V2	
SB	X	V0	
ABS		HB	0.8–3.2
ABS	X	V0	0.8–3.2
ASA		HB	
PVC		V0	
POM		HB	
PA 6		V2	
PA 6	X	V0	
PA 6-GF		HB	0.8–3.2
PA 66		V2	≥ 0.8
PA 66	X	V0	
PA 61		V2	
PC		V2	1.6–3.2
PC	X	V2/V0	1.6 / ≥1.6

Plastic material acronym	With flame retardant	Classification UL 94	Specimen thickness mm
PC-TMC		HB	1.6–3.2
PC-TMC	X	V2/V0	1.6–3.2
PC + ABS		HB	1.6
PC + ABS	X	V0	1.6
PBT		HB	0.8–3.2
PBT	X	V2/V0	≥ 0.8/0.8
PBT-GF		HB	0.8–3.2
PBT-GF	X	V0	
PES		V0	
PES-GF		V0	
PAI		V0	≥ 0.2
PAR		V2	1.6–3.2
PAR		V0	≥ 0.4
PPS	X	V0	≥0.4
PPS-GF40		V0/5V	≥ 0.4/≥ 1.7
PPS-M65 (metal)		V0/5V	≥ 0.8/≥ 6.1
PEEK		V0	≥ 2
PEK		V0	≥ 0.8
PEI		V0	≥ 0.4
PEI-GF		V0	≥ 0.25
PSU		HB/V0	≥ 1.5/≥ 4.47
PSU	X	V0	≥ 1.5
PPE+PS		HB	≥ 1.6
PPE+PS	X	V1/V0	≥ 1.6

8.6 Friction and Wear Behavior

All characteristics provided in the following tables were determined by pin-on-disk tests without lubricants (technically dry) at a sliding speed of v = 0.5 m/s and a contact pressure of p = 4 N/mm².[1] Unless noted otherwise, the ambient temperature was T_a = 23 °C. When mold temperatures are not explicitly stated, the injection molding temperature T_{mold} was selected according to CAMPUS (see Chapter 3). Roughness Ra always refers to the steel surface, as the polymer surface does not factor here. For more detailed explanations, see Section 3.8.

Table 8.22 Typical Wear and Friction Coefficients for Plastic vs. Plastic

	Typical wear coefficient k [10^{-6} mm³/Nm]					Coefficient of friction f [-]			
Material combination (pin on disk)	*< 1*	*1–10*	*10–50*	*50–100*	*> 100*	*< 0.25*	*0.25–0.5*	*0.5–0.75*	*0.75–1.0*
PA 66 vs. POM	x						x		
PBT vs. POM	x						x		
POM vs. PA 66	x						x		
POM vs. PBT		x					x		
PBT vs. PA		x							x
POM vs. POM					x		x		
PA 66 vs. PA 66					x				x
Source: see Section 3.8									

[1] The characteristic values can be found in [12, 14, 16–18] in Section 3.8.

Table 8.23 Typical Wear and Friction Coefficients for Plastic vs. Steel (100 Cr6. HRC > 60, Ra = 1.5–1.7 µm)

	Typical wear coefficient k [10^{-6} mm^3/Nm]					Coefficient of friction f [–]			
Material combination (pin on disk)	*< 1*	*1–10*	*10–50*	*50–100*	*> 100*	*< 0.25*	*0.25–0.5*	*0.5–0.75*	*0.75–1.0*
POM TF	x					x			
POM mod.	x					x			
PPS CF 30	x					x			
PA 66 TF Si	x					x			
PEEK GF 30	x						x		
PEEK CF 10 TF 10 Gr 10	x						x		
PA 66 CF 20	x						x		
PA 66 CF 30	x						x		
PA 66 Ar 15 TF	x						x		
PA 66 CF 15 TF	x						x		
PA 66 CF 30 TF Si	x						x		
PE-UHMW		x				x			
POM		x					x		
POM PE-UHMW		x					x		
PBT		x					x		
PET		x					x		
PA 66 PE		x					x		
PA 66 GF 30		x					x		
PA 66 GF 30 TF		x					x		

Table 8.23 Typical Wear and Friction Coefficients for Plastic vs. Steel (100 Cr6. HRC > 60, Ra = 1.5–1.7 µm) *(continued)*

	Typical wear coefficient k [10^{-6} mm^3/Nm]					Coefficient of friction f [-]			
PPA GF 50		x					x		
PPA GF 30 TF		x					x		
PEEK CF 30		x					x		
PEEK		x						x	
POM (T_{mold} = 50 °C)			x				x		
PPS			x				x		
POM GF 10			x				x		
POM GF 20			x				x		
POM GF 26 PE-UHMW			x				x		
PE-HD			x				x		
POM GK 10			x					x	
POM GK 20			x					x	
POM GF 20 TF			x					x	
PA 66			x					x	
POM GF 26				x				x	
POM GK 30				x				x	
ABS					x		x		
PC					x			x	

Source: see Section 3.8

Table 8.24 Typical Wear and Friction Coefficients as a Function of Steel Roughness for Plastic vs. Steel (100 Cr6. HRC > 60)

	Typical wear coefficient k [10^{-6} mm^3/Nm]					**Coefficient of friction f [–]**			
Material combination (pin on disk)	*< 1*	*1–10*	*10–50*	*50–100*	*> 100*	*< 0.25*	*0.25–0.5*	*0.5–0.75*	*0.75–1.0*
PA 66-X (Ra = 0.5–3.5 µm) DC = 74%		x						x	
PA 66 (Ra = 3.5 µm)		x						x	
PA 46 (Ra = 3.2 µm)		x						x	
PPS (Ra = 0.5–2.0 µm)			x				x		
PA 12 (Ra = 3.2 µm)			x				x		
PA 66 (Ra = 1.7–2.5 µm)			x					x	
PA 6 (Ra = 3.2 µm)			x					x	
POM (Ra = 2.5 µm)			x						
PPS (Ra = 2.2 µm)				x					
PA 66 (Ra = 0.5 µm)					x				

"-X" = irradiation crosslinked; degree of crosslinking (DC) determined by gel content, see Section 3.8

Table 8.25 Typical Wear and Friction Coefficients as a Function of Temperature During Processing and Testing for Plastic vs. Steel (100 Cr6, HRC > 60, Ra = 2.5 µm)

	Typical wear coefficient k [10^{-6} mm^3/Nm]					Coefficient of friction f [–]			
Material combination (pin on disk)	*< 1*	*1–10*	*10–50*	*50–100*	*> 100*	*< 0.25*	*0.25–0.5*	*0.5–0.75*	*0.75–1.0*
PA 66 (T_{mold} = 40 °C)			x					x	
PA 66 (T_{mold} = 80–100 °C)		x						x	
PA 66-X (T_{mold} = 40–100 °C) DC = 74–54%		x						x	
PA 66 (T_a = 23 °C)		x						x	
PA 66 (T_a = 50–100 °C)			x					x	
PA 66 (T_a = > 120 °C)					x				
PA 66-X (T_a = 23–210 °C) DC = 74%		x					x		

"-X" = irradiation crosslinked; degree of crosslinking (DC) determined by gel content.
T_a = ambient temperature during pin on disk test;
T_{mold} = mold temperature during injection molding
Source: see Section 3.8.

Index

Symbols

A

B

C

D

E

F

G

J

K

L

M

N

O

P

Q

R

S

www.ingramcontent.com/pod-product-compliance
Lightning Source LLC
LaVergne TN
LVHW081930200926
848484LV00009B/2398
* 9 7 8 1 5 6 9 9 0 1 7 6 2 *